797,885 Books

are available to read at

www.ForgottenBooks.com

Forgotten Books' App
Available for mobile, tablet & eReader

ISBN 978-1-334-53223-8
PIBN 10769910

This book is a reproduction of an important historical work. Forgotten Books uses state-of-the-art technology to digitally reconstruct the work, preserving the original format whilst repairing imperfections present in the aged copy. In rare cases, an imperfection in the original, such as a blemish or missing page, may be replicated in our edition. We do, however, repair the vast majority of imperfections successfully; any imperfections that remain are intentionally left to preserve the state of such historical works.

Forgotten Books is a registered trademark of FB &c Ltd.
Copyright © 2017 FB &c Ltd.
FB &c Ltd, Dalton House, 60 Windsor Avenue, London, SW19 2RR.
Company number 08720141. Registered in England and Wales.

For support please visit www.forgottenbooks.com

1 MONTH OF FREE READING

at

www.ForgottenBooks.com

By purchasing this book you are eligible for one month membership to ForgottenBooks.com, giving you unlimited access to our entire collection of over 700,000 titles via our web site and mobile apps.

To claim your free month visit:

www.forgottenbooks.com/free769910

* Offer is valid for 45 days from date of purchase. Terms and conditions apply.

English
Français
Deutsche
Italiano
Español
Português

www.forgottenbooks.com

Mythology Photography **Fiction** Fishing Christianity **Art** Cooking Essays Buddhism Freemasonry Medicine **Biology** Music **Ancient Egypt** Evolution Carpentry Physics Dance Geology **Mathematics** Fitness Shakespeare **Folklore** Yoga Marketing **Confidence** Immortality Biographies Poetry **Psychology** Witchcraft Electronics Chemistry History **Law** Accounting **Philosophy** Anthropology Alchemy Drama Quantum Mechanics Atheism Sexual Health **Ancient History** **Entrepreneurship** Languages Sport Paleontology Needlework Islam **Metaphysics** Investment Archaeology Parenting Statistics Criminology **Motivational**

34th Congress, 1st Session. | HOUSE OF REPRESENTATIVES. | Ex. Doc. No. 12.

REPORT

OF THE

COMMISSIONER OF PATENTS

FOR THE YEAR 1855.

ARTS AND MANUFACTURES.

ILLUSTRATIONS BY M. C. GRITZNER.

VOLUME II.

WASHINGTON:
CORNELIUS WENDELL, PRINTER.
1856.

In the House of Representatives, *February* 18, 1856.

Resolved, That there be printed, for the use of the members of the House of Representatives, fifty thousand copies of the Mechanical part of the Patent Office Report, and ten thousand copies of the same for the use of the Patent Office.

Attest: WM. CULLOM, *Clerk*.

347339

CONTENTS.

IV.—DESCRIPTIONS AND CLAIMS—Continued.

V.—ILLUSTRATIONS.

XIV.—LUMBER.

No. 12,318.—Russell Jennings.—*Improvement in Augers.*—Patented January 30, 1855. (Plates, p. 210.)

This instrument has a double-twisted pod A and two floor-lips B B; the latter project beyond the positions in which they have heretofore been placed, and the spur *a*, instead of being situated at the outer front corner of the cutting-edge of the floor-lip, where the latter from its necessary thinness is weakest, is projected from its hinder part or heel, where it is strongest. Thus is effected a new relation between the cutting-edge of the floor-lip and the spur, so that the one does not interfere with the operation of the other.

Claim.—So constructing the cutting-edges of a double-twist auger-bit that the vertical scores shall follow the chisel—*i. e.*, so that the cutting-edges of scores and chisel shall never intersect the worm or helix of the shaft at the same point.

No. 12,551.—Isaac W. Hoagland.—*Ship-Augers, etc.*—Patented March 21, 1855. (Plates, p. 210.)

The inventor says: I do not claim making the cutting portion of the auger detached from the screw portion, irrespective of the precise mode of attachment shown.

I *claim* attaching the cutting portion B of the auger to the screw portion, as shown and described, viz: by means of the dovetail notch formed by the shoulder *b* and inclined end *d*, dowel *f*, and screw *e*. (See engraving.)

No. 12,575.—Charles W. Cotton.—*Improvement in Attaching Augers to Handles.*—Patented March 21, 1855. (Plates, p. 210.)

Claim.—Attaching or securing augers to handles by having a metallic tube B placed around the centre of the handle, and having a transverse rectangular taper hole *a* made through the handle and tube, and a metallic band C placed around the tube B, and turning loosely thereon, said band having slots *c d* made through it, a part of the slot *d* being of taper form; the shank of the auger being placed in the hole *a*, and through the slots *c d* in the band, and secured in the handle by turning said band and causing the edges of the taper portion of the slot *d* to pass in the notches or recesses *f f* in the shank, as shown and described.

No. 12,583.—RANSOM COOK.—*Improvement in Machines for Turning the Lips of Augers.*—Patented March 27, 1855. (Plates, p. 210.)

In using this machine, the cam-lever *c* is first raised to nearly a vertical position, thus allowing the clamps *b* to open. An auger or bit, with the lips shaped as shown in figure 5, being red-hot, is placed in one of the crimping-dies *a*, with the lips projecting beyond the die and towards the wrench, figures 3 and 4. The upper end of the cam-lever is then brought quickly down, thus forcing the other crimping-die against the auger, and firmly holding it between the two dies. A quick turn is given to the screw-shaft *f*, which brings the wrench in the hub of the wrench-wheel *d* into an embrace with the end of the auger, the centre of which enters the hole in the wrench, while the lips pass into the slots on its side. The workman then seizes one of the handles to the wrench-wheels and turns it towards himself, when the wrench, keeping the auger straight by means of its hold of the centre, turns or bends the lips into the desired position, both being turned at the same time and angle, while the shoulders of both are left in the same line.

The inventor says: I do not claim any of the several and separate parts of the machine described, except the wrench; but I *claim* the combination of the screw-shaft, or its equivalent, with the wrench and crimping or clamping dies, as substantially set forth, and for the purpose mentioned. I also claim the shape of the wrench, substantially as described, for the purpose of turning the lips of boring implements.

No. 12,868.—EBENEZER W. NICHOLS.—*Mode of Securing Brace-Bits in their Sockets.*—Patented May 15, 1855. (Plates, p. 210.)

The figure represents the bit R as being fastened in the stock S. To loosen the bit, the nut *a* is turned (thereby screwing down the screw-shank of the spring) until the projection B of the spring springs back into the recess *c* in the stock, when the bit can be drawn out. The pin *g*, entering a groove in the outside of nut *a*, serves to confine said nut to the end of the stock.

The inventor says: I do not claim the projection B *d* upon the spring *f*; but I do *claim* the burr-nut *a* (or its equivalent), in combination with the spring *f*, operating upon the wedge principle (by the use of the screw) the projections B *d*, for the purpose and in the manner herein described.

No. 13,261.—L. H. GIBBS.—*Expanding Auger or Bit.*—Patented July 17, 1855. (Plates, p. 210.)

The plate B is received into the slot D in the auger A; and by inserting the lower pin *g* into one of the series of holes in the plate B, the cutting-lip *j* on plate B can be set further out or in, so as to bore larger or smaller holes.

Claim.—1st. The adjustable plate B, with the rib *d*, the lip *j*, the index-holes *c c c c* in it, combined with the auger A, with slot D, the tapering-pins *g g*, and set-screw *h*, as described and for the purposes set forth.

No. 13,925.—Guillaume Henri Talbot.—*Improvement in Auger-Handles.*—Patented December 11, 1855; patented in England August 25, 1855. (Plates, p. 210.)

The nature of this improvement will be understood from the claim and engravings.

The inventor says: I do not claim the manner described of giving a revolving action in either direction to the auger-bit or boring-tool, by reversible pawls and ratchets, operating in connexion with a vibrating handle, apart from the relative arrangement and form of handle specified, as such is common to drill-stocks; but I *claim* in gimlet or auger-handles the arrangement substantially as specified, within the body of the said handle, which crosses the bit of the ratchets a b, and pawls a^1 b^1, with their reversing gear, for operation of the bit or bit-socket in either direction, either by a revolving or vibratory action of the gimlet-handle, on pressure of the hand applied on both sides of the axial line of the bit, and under the usual clutch of the hand on the handle, over the centre line of the bit, and whereby the actuating pawls, ratchets, and accompanying devices form no obstruction, and are protected from injury or derangement, substantially as set forth.

No. 13,943.—John Gourlay.—*Adjustable Crank-Brace for Augers.*—Patented December 18, 1855. (Plates, p. 211.)

The full lines represent the instrument when used as a single-crank brace, and the dotted lines as arranged when used as a double-crank brace.

The inventor says: I am aware of W. P. Barnes's invention, and therefore only *claim* the particular method of varying the length of leverage in handles, as set forth.

No. 13,998.—John P. Rollins.—*Improved Extension-Bit.*—Patented December 25, 1855. (Plates, p. 211.)

The lower end of lip B and that of the cutting-spur C are pivoted at F to the centre-stock A. D is a hollow fine-threaded screw, screwing into stock A and turning in the cutting-spur C. E is a coarse-threaded screw fast in lip B, and screwing into the hollow screw D. The threads on E are twice as coarse as those on D. By turning screw D, the spur C and lip D will move the same distance, either towards or from the centre spur A, and thus the bit can be expanded or contracted.

The inventor says: I do not claim the invention of movable cutters; but I *claim* the manner in which the lip and cutter are set or secured for operation, when being adjusted, without the use of separate screws for that purpose, and in the manner described.

No. 12,452.—William L. Young.—*Machine for Cutting Barrel-Heads.*—Patented February 27, 1855. (Plates, p. 211.)

The different pieces of the barrel-head are confined between the face-plate 3 and the disc 4. The action of hand-wheel 7 brings the projecting points (with which the circular row of elliptic springs 10 is provided, which springs are arranged in the disc 4) in contact with the different pieces of the barrel-head, pressing them firmly against the plane surface of the driving face-plate 3. 13 and 14 are the tools for cutting the barrel-head.

Claim.—The centrally pressing toothed-springs, as arranged; and als their combination, with a disc free to vibrate on its axle, for the purposes set forth.

No. 12,543.—Archibald H. Crozier.—*Machine for Cutting Barrel-Heads.*—Patented March 21, 1855. (Plates, p. 211.)

The disc H, which forms the lower part of the rotating clamp H K, has a shaft I, which is stepped in box *e*, the latter serving as a fulcrum when it (the clamp) is traversed towards or from the cutters $c\ c^1$, which is done by means of a lever J. Disc K is provided with spurs, which penetrate the heading when the disc is forced down by screw L, turned by hand-wheel M, so as to clamp heading *g*. Screw L is connected with disc K so as to allow it to vibrate upon the end of the screw, and the nut in which the screw turns is pivoted in bar P. Shaft I and screw L form the arms or links of a toggle-joint; that is, the clamp is carried from the cutters by lever J, and the pieces of heading are placed upon disc H, and screw L is turned so as to clamp them fast; they are then moved with the clamp towards the cutters by lever J until screw L is in line with shaft I, which movement forces the discs against the heading, the bar P having sufficient spring to prevent the breaking of any of the parts.

Claim.—The described machine for cutting out and forming the heads of barrels and other similar articles; first, in arranging and operating two rotating cutters $c\ c^1$ so as to cut scores in the opposite sides of the rotated heading *g* at the same time—one cutter being arranged and operated so far in advance of the other that the latter cutter may cut so far into the heading and into the score made by the former without nterfering with it (the first cutter,) so as to sever the surperfluous portions of the heading from the head, at the same time that they cut it circular and bevel or form the edge to fit the croze in the cask, substantially as described. 2d. Traversing or vibrating the clamp edgewise after the heading is placed in it, to bring the heading in contact with the cutters, and to remove the head from the cutters after it is formed, so as to take it out of the clamp and insert material to form another head and bring it into contact with the cutters without stopping them (the cutters) during the operation or time occupied in making the change. 3d. The revolving-clamp in combination with the rotating-cutters, arranged and operated substantially as described and for the purposes set forth.

No. 13,016.—Jos. D. Spiller.—*Bench-Rest.*—Patented June 5, 1855. (Plates, p. 211.)

The bench-rest consists of a bar A with a flat head B, and slides within case D. When the upper surface of one of the teeth of rack F is bearing against the tooth on latch G, not only should the curved front part of escapement-pallet N rest against the angular corner of one of the rack-teeth, but the lower edge of the said curved part of the pallet should be a short distance below the corner of the next tooth below. By depressing thumb-slide I, the latch G will be turned on fulcrum H so as to move its tooth out of the rack, and the bench-rest will slide upwards by the action of spring E; the pallet N will then catch against the next tooth below. By pressing down the thumb-slide so as to remove the latch from the rack, and at the same time pressing down upon the bench-rest, this rest can be moved downward. The top B of the rest is provided with wedge-teeth *f* for the purpose of seizing and holding the end of a board when borne against it. When it is desirable to be used as a rest without teeth, the plate S can be moved out sufficiently to prevent the teeth on B from coming into contact with the board. (See figure 3.)

Claim.—Combining with a bench-rest mechanism, substantially as described, for not only elevating said rest with an intermittent rotary motion during successive pressure on a spring thumb-slide, applied to said bench-rest, as set forth, but for enabling said bench-rest to be moved downward whenever necessary, in manner and for the purpose as specified.

Also, combining with the serrated top plate of the bench-rest a plain slide-plate, combined and made to operate therewith substantially as specified.

No. 13,678.—A. HOTCHKIN.—*Bench-Hook.*—Patented October 16, 1855. (Plates, p. 211.)

Claim.—The construction of the bench-hook, as shown and described, viz: having the catch or stop C attached by a joint *b* to a plate B, said catch or stop being provided with a shank *d*, against which a spiral spring *g* acts, and also provided with a segment-bar D, having holes *h* in one side, in which a spring pawl E catches and retains the catch or stop in the desired position.

No. 12,437.—WILLIAM B. EMERY.—*Method of Adjusting Cylinders in Boring-Machines.*—Patented February 27, 1855. (Plates, p. 211.)

The hub *a* of dial-plate A projects both ways, and is fastened to the axle Q of the cylinder N by means of set-screw D; dial-plate B fits over the inner end of hub *a*.

To bore, for instance, four parallel spiral lines of holes around the cyliner N, the boring instrument is brought down, making a hole; the cylinder is then turned one-fourth of a revolution by the notches upon A, while B remains stationary, being held by catch G, and the second hole is bored; then another one-fourth of a revolution, and so forth, making four holes, each of which is the first of the spiral line of holes. The carriage P is then moved a sufficient number of notches of ratchet L to space the holes longitudinally; then by the notches on B the cylinder is revolved any required part of a revolution, when the cylinder is again ready for the boring of four more holes, as at first. This process is repeated until the whole length of the cylinder has been successively under the boring instrument.

Claim.—1st. The dividing dial and its catch, in combination with the ratchet and pawl, or their equivalents, substantially and for the purpose as described.

2d. The compound dial-plate and catches, substantially and for the purpose as described.

No. 12,598.—Urias Kimble.—*Tool for Boring Hubs to Receive Boxes.* —Patented March 27, 1855. (Plates, p. 212.)

This invention consists in constructing a shaft with about one-third of the upper end square, the remaining part round, and provided with a screw-thread, on which is run a nut *c* with spurs on the bottom of the nut, which rest on an oval-shaped box 5, which has a hole through it to receive the shaft. An adjustable gauge 3 is attached to the shaft, which is kept in its place by a set-screw. An adjustable knife 4 is also attached to the shaft, and held in its place by a set-screw. The shaft is turned by handle 2.

The object of this improvement is to cut a gane in the hubs of wagon-wheels of any desired size and depth. The gauge determines the depth, and the knife the size of the gane. The box 5, which is of the right size to fit in the hub when reamed out ready to set the box, is calculated to steady the shaft, and may be adjusted, as it is oval, so that the knife will start in the centre, whether the hub is reamed out true or not. Previous to cutting the gane, it is to be driven into the hub sufficiently tight to steady the shaft when in motion. The nut prevents the knife from going in too fast; the spurs on the bottom of the nut prevent its turning when the shaft is in motion.

The inventor says: I do not claim the shaft, the adjustable knife, or the adjustable gauge, as they have been known before.

I *claim* the oval-shaped box, with the nut with spurs on the under side resting on the oval-shaped box, in combination with the shaft, the knife, and the gauge, for the purpose set forth.

No. 12,677.—H. C. Garvin and J. H. King.—*Tool for Boring Hubs.* —Patented April 10, 1855. (Plates, p. 212.)

A, knife for boring out hubs; B, slide in which the knife is fastened; C, mortise for the reception of the knife; D, bolt that fastens the knife while in operation, to be firmly secured; E E, cross-heads, through which slide B passes; F, friction rollers fastened upon cross-heads E, with flanges to form a groove around the hub, before the apparatus is put to work, so that the knife, in performing the work, cannot throw off the apparatus; G G, rods and taps to regulate and fasten the apparatus to the hub, and keep it secure to its bearing while at work; H stay to keep the knife firm to its bearing during the boring; I, nut that feeds the knife while in operation.

Claim.—The apparatus for boring wagon and carriage hubs for reception of boxes, (narrow and through boxes,) as herein described, using for that purpose the aforesaid apparatus, or any other substantially the same.

No. 12,776.—CHAUNCEY COWDRY, ORRIN TOLLS, and CHAUNCEY C. TOLLS.—*Wheelwrights' Boring and Tenoning Machine.*—Patented May 1, 1855. (Plates, p. 212.)

The inventor says in his specification: For boring a hub, place frame R into position fig. 3, fasten the bub upon the index G^7, adjust the graduated scale of sliding-tubes T, until the proper size is at the required height, by means of screw U; also graduate the height of frame C accordingly, place the index point upon the required number of spaces on the index plate, regulate the position of the movable stop L, so that the lever G will stop the forward movement of the auger; when the hole is deep enough, press lightly with one hand upon the lever G, turning with the other hand crank O, until lever G strikes the stop L; then draw back the lever G, raise the index point, turn the index and hub one space, and repeat the operation till the hub is bored.

After the spokes are driven in, place the hub into position fig. 2, fastening it upon the index as above; raise support B^2 so that the spoke will rest firmly in the notch (see fig. 6); operate with the hollow auger as in boring, placing the movable stop so that the shoulder of the tenon will be at the required distance from the hub. When the tenon is made, turn back screw D^4, and bring the support B^2 into position, shown by dotted lines in fig. 2, so that the spokes may pass over it when the index and hub are turned; turn to the next spoke, raise the support B^2, &c., repeating the operation as above.

Claim.—1st. The combination and arrangement of the frame R, with the scale of graduated and sliding-tubes T and screw U, as described.

2d. The combination of the hinged support B^2 with the sliding screw-clamp C^3, substantially as described.

3d. The combined arrangement of the several parts, substantially as described and set forth.

No. 12,808.—JAMES TEMPLE, assignor to ISRAEL WARD and JAMES TEMPLE.—*Machine for Boring Fence-Posts.*—Patented May 1, 1855. (Plates, p. 213.)

The bolsters F F^1 F^2 can be raised or lowered by bringing nearer together or by spreading out their bases, which are made adjustable by means of slots *c* and set-screws G. As the bolsters are raised or lowered, the spur-gears L L roll around the periphery of the pinion H, which lowers with the bolsters, but still remain in gear with said pinion, the distance between their respective journals and the journal of the long pinion not changing, all of which have their bearings in the bolsters. Lever M and cross-head N, working against the collars *e* upon the auger-shanks K, serve to force up the augers into the wood; the spur-wheels slide on the long pinion, but at the same time they are rotated by it.

Claim.—The supporting of the long pinion and auger-shanks in the adjustable-hinged pillar-blocks, such as described, so that the augers may be set to bore holes at variable distances apart, while the spur-wheels on their shanks shall still keep in gear with the long pinion, as described.

No. 13,158.—Adolph Brown and Felix Brown.—*Machine for Boring and Turning Wood.*—Patented July 3, 1855. (Plates, p. 213.)

Square pieces of wood are put through the square holes of wheels L and upon the centre-bits *g*, and the machine is set in motion. Wheel K, acting upon the wheels L, turns them and the wood round. Screw R, worked by wheels Q and P, moves support T, and consequently all the pieces of wood, towards the supports C and E, while they are revolving. The wood comes first in contact with the knives *h* attached to support C, by which the outside of the wood is turned off. The wood passes then through holes in support C, and is acted upon by boring-tools *b* fast on support E, whereby the wood is bored out the required depth for a box; then the wood comes in contact with the knives *n* fast on support E, which cut the neck for the reception of the cover. When the wood has been brought in thus far, the toothed portion of wheel P has left the pinion Q, and the screw R, with support T, remains stationary, and the wood will only turn without moving along. During this time the cam *x* acts upon projection *p* of ring F in such manner as to press the knives *m* fast to said ring against the wood, cutting thereby the wood through.

Claim.—1st. The manner of guiding and turning round the wood by passing said wood through suitable holes made through the wheels L, thereby allowing the wood to be fed up to the tools at the same time the same is turned round.

2d. The ring F, with the tools for cutting off the wood attached, arranged, and worked as specified.

No. 13,243.—John Young, of C.—*Machine for Boring Posts and Pointing Rails.*—Patented July 10, 1855. (Plates, p. 213.)

The driving-belt is placed on pulley W when the machine is to be used for boring; and when the augur is to be withdrawn, the belt is shifted to the loose pulley V, and the tightening pulley M is drawn on the belt on pulleys T G, which will enable the auger and chips to be withdrawn (without stopping the machine), as the belt will revolve the pulley X and its screw-shaft P, which will withdraw the frame R which supports the auger. When it is desired to bore posts, the four screws *c c c c* are withdrawn, and the saw-table O is removed. Preparatory to pointing, the screw *a* is withdrawn, when the boring part can be removed. Then the table O, with the circular saw C, is returned, and the four screws *c* are tightened. The driving-belt is placed on pulley J, the middle of the rail in the clamp H, the dog Q is drawn down the teeth of clamp H entering the rail, the carriage L is drawn out, and the rail and carriage run to the saw, which will then cut the straight side of the rail. Now the carriage is returned, the table with the clamp and rail is pushed from over against pin 5, and then run up to the saw, at the same time depressing the treadle I, which throws up the gauge on the table marked *b*, which gauges the end of the rail.

The inventor says: I do not claim any of the separate devices set forth.

But I *claim* the reversible clamp and feed-carriage, in combination with the boring apparatus, substantially as set forth for the purposes mentioned.

No. 13,606.—A. Wyckoff and E. R. Morrison.—*Improved Boring-Machine.*—Patented September 25, 1855. (Plates, p. 213.)

Motion being given to the toothed wheel F, the auger D will rotate, and also the worm J, the augur turning in a reverse direction to the worm, which latter receives its motion by means of gearing E G and band and pulleys L K. As the bits *b* cut the wood, the chips pass within the auger, and the worm passes the chips out at the back end of the auger. B^1 represents the log fastened to carriage B.

Claim.—1st. The employment or use of the tubular or hollow auger, constructed as shown, for the purpose specified.

2d. The combination of the tubular or hollow auger D and worm or screw J, arranged substantially as shown for the purpose specified.

No. 12,149.—Louis Koch, assignor to Theo. Pincus.—*Machine for Manufacturing Wooden Boxes.*—Patented January 2, 1855. (Plates, pp. 214 and 215.)

In the illustration of this machine (which is designed to feed a number of pieces of wood simultaneously to the working-tools), there is represented only one pair of pieces of wood A^3 A^3, with the corresponding boring and cutting-tools i i^1, g g^1, h h^1, and f f^1. In the position represented, the wood A^3 on the left-hand side has been sufficiently far bored out, and the recess for the cover turned on; while on the right-hand side the wood is clear of the tools g^1 and h^1, and ready to be cut off, in which position the wood is now held stationary. The machine being then set in motion by means of driving-shaft I, the tool-holders and shaft O will be set revolving by proper gearing (indicated in the drawings). While O turns from 1 to 2, lever H^1 is acted on by projection o^2 t on cam D^1, and presses frame N^1, with cones M^1 attached, forward, thereby compressing the spring r^1. By this motion of the frame, arms n^1 are forced up the inclined surface of cones M^1, turning the spindles l^1, and consequently the knives f^1, so as to bring the latter towards the centre of the wood, cutting off the box. While O turns from 2 to 3, lever H^1 falls back to its former position; the springs r^1 force back frame N^1, which relieves arms n^1, the springs o^1 throwing back the knives f^1 to the outside of the wood and out of action. The lower end of lever X^1 has come in contact with pin p^1, and thereby has been thrown out of one of the grooves on pulleys U^1, setting thereby the latter free to be turned. Spring v^1 of pulley W^1 fits also into one of the teeth on pulleys U^1, thereby locking the two pulleys together. While O turns from 3 to 4, projection d^1 w^1 on cam R^1 has acted on lever E^1, pressing thereby pin s^1 fast to pulley W^1 upwards;

and U^1 being locked to W^1, the shaft V^1 is partly turned, thereby advancing the wood towards and into tool-holder K^1 as much as the depth of the hole the wood has been bored out. While O has turned from 1 to 4, levers F and F^1, having rested on the concentric part of the surface of cam S^1, have remained stationary. While O turns from 4 to 1 lever F^1 is acted on by cam S^1, and moves, by its connexion with projection T^1, fast on frames R (said frames connecting the supports Q Q^1); then supports Q Q^1 from the right hand to the left, forcing thereby on the right hand the wood into tool-holder K^1 and against boring-tool g^1, and afterwards against tool h^1, whereby the wood is bored out and the groove for the cover turned on; and on the left hand the wood is brought out of the tool-holder K clear of *g*. While O turns from 4 to 1, cam R^1 acts still on lever E^1, giving to the same and therefore to pulleys U^1 W^1 (as those two pulleys are still locked together), and consequently to the wood between the ratchet-wheels e^1, a second motion, in combination with the one just described, and which said motion is equal to the thickness of the bottom of the box. These two motions which the wood receives from supports Q Q^1 and from upper shaft V^1, are, together, equal to the whole length of the box. During this movement of the wood, the knives i^1 turn the outside of it. The motion of Q Q^1 has brought the lower end of lever X^1 clear of pin p^1, and it is now through spring m^1 pressed against the periphery of pulley U^1, and falls, at the moment the motion of Q Q^1 is finished, into the next groove of U^1, holding thereby the same, and consequently the shaft V^1, stationary, until X^1 touches again pin p^1 in the backward motion of the supports. While O has turned from 3 to 1, lever E has moved down the incline on cam R^1. Pulley W is then acted on by weight Z, pulling it around so as to keep pin *s* always against lever E. During this interval, spring *v* slides over the back of a tooth on pulley U until it falls into the next tooth. The upper end of X is held by spring *m* in one of the grooves on pulley *l*, preventing thereby the same from turning, while pulley W is turning backwards, as just described. During the time O has made one half a revolution, as just described, the wood on the left-hand side has been brought clear of borer *g*, and the boxes on this side are ready to be cut off; while from the wood on the right-hand side one set of boxes have been cut off, the wood been again sufficiently advanced, and has been bored out, and the recess for the cover turned on. By the second half revolution of O the same operation as above will be repeated on opposite sides.

To make different sized boxes, cams R^1 S^1 D^1 must be changed for others differently divided as regards the space each projection occupies, and pulleys U U^1 must be changed for others having a greater number of grooves and teeth.

Claim.—1st. The mode of making different-sized boxes on the same machine by the mere change of the cams R^1 S^1 D^1 and the pulleys U U^1, corresponding to the size of the boxes, as described.

2d. Cutting off the boxes, when finished, by tools *f* or f^1, fastened to spindles *l* or l^1; said spindles being attached to the tool-holders, and worked by an arm *n* or n^1 fast on the end of the spindles, as described.

3d. The construction and application of the frames N or N^1, with cones M M^1 respectively attached, actuating through the arms n or n^1 the spindles l or l^1, and consequently the tool f or f^1; said frames being worked by cam D^1 and levers H or H^1, as described.

4th. The arrangement and connexion of the supports Q Q^1, provided with shafts and ratchet-wheels, between which latter the wood, out of which the boxes are to be made, is held; said supports being worked by cams S^1 and levers F or F^1 for feeding the wood to the tools, and releasing the same, as specified.

5th. The construction of the pulleys U and W or U^1 and W^1, worked by cams R^1 and levers E and E^1, as well as by weights Z or Z^1, as described; said pulleys, when connected, acting upon the upper shafts V or V^1, running in the supports Q or Q^1, for approaching the wood up to the tool-holders after the completion of each set of boxes.

No. 13,921.—Isaac M. Singer.—*Improved Machine for Carving Wood, &c.*—Patented December 11, 1855. (Plates, p. 215.)

The pattern B is secured to bench *a*. Block *d* is fitted to slide on rail *c* for adjustment, and can be held in place by screw *e*. To the upper part of this block is jointed a short spindle *f*, turning on pin *g*, and this spindle forms the axis on which the two systems of pentagraph-levers can turn horizontally, whilst the connexion of the spindle with the block *d* admits of the vibration in a vertical plane. *h i* is the horizontal, and a^1 b^1 the vertical pentagraph. *k* is the tracer connected to one angle of the pentagraphs. The table *m*, with the block to be carved upon it, is jointed to the opposite angle of the pentagraphs at *n*. The table slides in ways *p p*, on the top of a carriage *q*, which in turn slides in ways *r r*, at right angles to ways *p*, so that the block can be moved horizontally in any direction. The bed *s*, with the ways *r r* on it, is secured to a standard *t*, which slides in vertical ways *u u*, the whole being suspended by a link *v* and balanced by weight *x*, so that the bed, with the table *m* and block upon it, can be moved up and down freely by the pentagraph. The tracer *k* is directed over the surface of the pattern by means of a hand-lever jointed to the end c^4 of the pentagraphs, and not shown in the engravings. The revolving cutter i^1 is mounted in a stationary frame.

The inventor says: I do not limit myself to the special construction or arrangement of parts specified, as these may be varied without changing the mode of operation of my invention.

I do not claim the combination of the tracer with the bracer which carries the block of wood to be carved, by means of one system of pentagraph-levers, as this is described in a patent granted me on the 10th day of April, 1849.

But I *claim* combining the tracer with the table which carries the block of wood to be carved, by means of two systems of pentagraph-levers operating at right angles with each other, substantially as described, whereby the block to be carved will be directed and presented to the action of the cutter in such a manner as to determine the configuration, as well in a vertical as in a horizontal direction, as set forth.

No. 13,101.—LOVELL T. RICHARDSON.—*Socket-Handles for Chisels.*—Patented June 19, 1855. (Plates, p. 215.)

The nature of this improvement will be understood from the claim and engravings.

The inventor says: I do not claim a die of any particular form; neither do I claim simply forming or constructing sockets by means of a die, for dies are used for analogous purposes.

But I *claim* constructing the sockets A with a die, so formed that a transverse partition or ledge *a* is left within the socket, said partition or ledge dividing the recess *b* which receives the handle of the implement from the recess *c* which receives the shank, so as to obtain a sufficient body or weight of metal at that part of the socket which is welded to the shank *d* of the tool or implement, for the purpose of forming a strong and durable connexion of the socket and shank, as set forth.

No. 13,734.—ARETUS A. WILDER.—*Improved Lath-Machine.*—Patented October 30, 1855. (Plates, p. 215.)

The saw *k* is arranged with its plane of motion parallel to the vertical side *m* of the frame of the machine, and at a distance from said side of the frame equal to about one-half the usual thickness of planks to be operated upon. Close to the side *m*, and extending in rear of the saw, cutters, guide and feed rollers, there is an adjustable back-rest. It consists of a board *b*, which can be set at various angles with the vertical side *m*, by means of set-screws *c*. The plank *a*, during the time it is operated upon by the saw and cutters, is slid along the back-rest by means of feed-rollers *f f*, and it is pressed close to the back-rest by means of the flanges of the guide-rollers *d* at the lower edge of the plank, and by one guide-roller *i* at the upper edge of the plank. The roller *i* is arranged on a shaft *h*, between the cutter *g* and saw *k*. The rollers *d* (in vertical lines with the feed-rollers *f*) are loose on their shafts, and are pressed against back-rest *b* by means of springs *e*. The plank thus confined to the back-rest passes first under the cutter *g*, which planes its upper edge, and then it is operated upon by the saw. As the saw cuts vertically, whereas the plank has a certain inclination given by means of back-rest *b*, the outer sides of the plank will be oblique to the cut performed by the saw.

The inventor says: What I *claim*, in re-sawing and bringing plank to an equal width at the same time, is, the flanged head-rollers *d d*, with their springs, or equivalents, in combination with the adjustable back-rest, for the purposes before described.

No. 13,846.—JOEL P. HEACOCK.—*Improvement in Cooper's Tool.*—Patented November 27, 1855. (Plates, p. 215.)

In using the tool the barrel is held in the ordinary way, applying the throat-side *e* of the blade e e^1 to the stave as near the upper hoop as possible, and working from you, shaving down to the bilge. To finish the dressing, the tool is to be turned so that the drawing-knife edge e^1 will be in contact with the barrel, drawing towards the hoop that is upward, and so continuing around the barrel.

Claim.—Constructing a cooper's tool in the manner substantially as described, for the purpose of over-shaving, as set forth.

No. 13,200.—JOHN POWER.—*Cork-Machine.*—Patented July 3, 1855. (Plates, p. 215.)

The nature of this invention will be understood from the claims and engravings.

Claim.—1st. Supporting the head of the knife-carriage near its centre upon a bearing-screw *h*, and applying adjustable-screws *i i* near its end, for the purpose of making it adjustable at different heights, and either level or at different inclinations, substantially as set forth.

2d. Giving rotary-motion to the mandrel D, or its equivalent, which revolves the cork by means of a band *l*, connected with the knife-carriage, and carrying a weight *o*, the weight serving to keep tight the bands and transmit motion from the knife-carriage to the cork during the cutting operation, and also to draw back the carriage after the cutting operation, substantially as described.

No. 13,714.—WM. R. CROCKER.—*Machine for Manufacturing Corks.*—Patented October 30, 1855. (Plates, p. 216.)

C is a rapidly revolving hollow shaft, with the cutters *t* at its front end. The shaft *d* slides in stands *b*, and a frame *h* is fastened to it by a set-screw *s;* so that the frame and the rod *m* (which latter is fastened to the frame *h* and fits loosely in the cavity of the shaft C) slide together with the shaft *d*. The piece of cork is held against plate *e*, and the shaft *d* slid forward against the cutter, until the hub *r* comes into contact with stand *b*, when the cutter will have cut through the cork. The shaft *d* is then drawn back, when the rod *m* will push the cork just cut out of the cavity of the cutter. The cutter-tube is slotted for cutting conical corks. The concavity being entirely cylindrical, the outside is slightly bevelled off to produce a knife-edge, which has a tendency while cutting to crowd the lips of the cutter inward towards the centre ; and as these are made very thin and have a spring-temper, they are capable of contraction as they are forced into the block, cutting it gradually smaller or conical.

The inventor says : I *claim* the application of the revolving cylindrical cutters, to cut corks from a block or slab, as described, whether the cutters are slit to cut tapering or conical, or unslit to cut cylindrical corks.

I do not claim a cylindrical cutter; but this mode of construction, use, and application, allowing myself the privilege of arranging the same in detail, while the principle and distinguishing characteristics are retained.

No. 13,096.—Jonah Newton.—*Method of Securing Cutters to Rotary Discs.*—Patented June 19, 1855. (Plates, p. 216.)

Claim.—Securing the cutters *a* to the disc or plate A, as shown and described, viz: having the cutters with semi-circular form, with ledges or projections *d* on their back or convex sides, the ledges or projections being fitted in grooves *c* in the semi-circular edges of the projections *b* of the plate A; the front or concave sides of the cutters having grooves *e* in them, to secure the nuts or segment heads *f*.

The projections *b*, cutters *a*, and nuts or heads *f*, having screws *h* passing through them for the purpose of allowing the cutters to be adjusted properly, and also securing firmly the cutters to the disc or plate.

No. 13,169.—Isaac B. Hartwell.—*Machine for Cutting Cavities Spherical, Ellipsoidal, &c.*—Patented July 3, 1855. (Plates, p. 216.)

To the edge of the spherical shell *o* are attached the cutters *b c*. P is the driving-pulley firmly connected with gear-wheel Q, both P and Q revolving loosely on the axis of the rocking-frame J K, and (by means of intermediate gearing R T U) giving motion to the cutter-shell *o*. Wheel V, with hand-lever W, is fast on the axis of the frame J K. Lever W serves to set the frame J K. By means of shaft X, pulleys Y Z, and a small pinion on the axis of Z, a slow feed-motion is given to gear-wheel V, by which J K is brought forward and outward. The pinion and wheel V can be disconnected by hand-lever *i*. Dy these means a spherical groove *g* can be cut in block s^1, which is fast on sliding-table k^1. When the block is brought under the cutter-shell the machine is set in motion, when the shell (moving on two axes) cuts its way in the wood until axis *a* is brought to a horizontal position. The pinion is then disconnected from V, and J K brought back to a vertical position; the shell is raised by means of screw and crank H I, and the block is advanced, and the next groove is cut, and so forth. Finally, the block is brought under the circular saws K K, on shaft H, in the oblique frame D E, sliding in oblique grooves. The frame and saws are pressed down until the latter have made a longitudinal cut to the depth of *c* and *d*. The wood remaining between *c* and *d* is then split out, and the ellipsoidal cavity is to be smoothed and finished.

1st. I *claim* the spherical shell or cutter O turning at the same time on the axis of the sphere of which the shell is a part, so as to cut a spherical groove, or a convex and concave surface of less extent than a quadrant of the superfices of a sphere, yet corresponding in shape to the convex surface of a spherical section formed by two planes passing through the sphere at right angles.

2d. I claim the method of giving a compressed motion to the spherical shell or cutter, by means of the tight-gear wheel V, figure 1, and the loose-gear wheel Q revolving on the axes of the rocking-frame J K, so as to be in connexion with the pinion R in all necessary positions of the rocking-frame J K.

3d. I claim the use of circular-saws set in an oblique sliding-frame, in connexion with the spherical cutter, for the purpose of cutting straight grooves to connect with the spherical grooves at each end of the block of wood.

No. 12,122.—Thomas H. Burley.—*Dovetailing-Machine.*—Patented January 2, 1855. (Plates p. 216.)

Eigure 1 represents a top view, and figure 3 an end elevation of the machine. There is a guage-table with two inclined fronting-guides $c\ c^1$ corresponding to two sets of saws $a\ a^1$; so that when the piece to be mortised is laid on the table, and with it brought up to the edges of the saws a, they will cut obliquely into the piece, (illustrated in figure 3,) and also cut a straight edge e on the bottom of the mortise. The piece being then brought to the saws a^1, the other oblique side and the bottom edge of the mortise are cut. In tenoning, the piece is slit to the proper depth by the common saw g, then carried to the inclined table h, passed under the chisels k, and held fast by bar m. The chisels have two cutting edges at right angles to each other, and these are inclined so as to be parallel to the inclination of the table h. The chisels k having cut one side of the tenon, the piece is removed to the side h^1 of the table, where chisels k^1 complete the other side.

Claim.—1st. The inclined fronting-guide, in combination with the oblique cutting-edges of the saw-teeth, in the manner and for the purposes set forth.

2d. The double-inclined tables, in combination with the series of vertical chisels, in the manner and for the purposes set forth.

No. 13,522.—John J. Haley. — *Dovetailing-Machine.* — Patented September 4, 1855. (Plates, p. 217.)

Upon the sliding-frame D the guides E E are raised, for giving obliquity to the cut of the chisels, requisite in forming the dovetail. The front guide for the set of chisels is for directing them in cutting the right side of the mortise, while the guide immediately behind directs the chisels at an opposite angle and side. F F are the chisel-stocks, sliding the guides E E; $a\ a$ the chisels, secured to said stocks. Power being applied to pulley G, it moves pinion d, spur e, shaft e^1, and pinion f, in whose face the crank-pin g is affixed, which moves the pitman h in elevating the chisel-stock; this pinion drives another by its side on shaft H, in whose face the other crank-pin i is secured, which raises the other chisel-stock. The cam I upon shaft H works between jaws $l\ l$ projecting from sliding-bar m. The horizontal chisels o are secured to a cross-piece n at the end of bar m, and thus receive a reciprocating motion for the purpose of removing every chip or cut.

I *claim* the forming of a dovetail, either as a mortise or a tenon, at a single operation, by angularly placed reciprocating chisels $a\ a$, in combination with horizontally placed chisels $o\ o$, arranged substantially as set forth.

I claim giving a reciprocating motion to the chisels $o\ o$, by the snail cam I on shaft B, in combination with chisels $a\ a$, gear $d\ e\ f$, and pitman-rods *h h*, for the purpose of actuating the chisels in unison with each other, in the manner described.

I claim the arrangement of the angular guides E E, in combination with the guides E^1, for the purpose of effecting the under-cut or sides of the dovetail.

I claim the arrangement and combination of the angular guides E E, and chisels *a a*, on stocks F F, with the horizontal chisels *o o*, and guides *m*, and snail I on shaft H, for producing the dovetail and completing the mortise, in the manner set forth.

No. 13,574.—Amos P. Hughes.—*Dovetail Key-Cutter.*—Patented September 18, 1855. (Plates, p. 217.)

The wood is thrust through the tube B against and past the edges *a* of the dovetail-cutters C C. The size of the tube is to be determined by the size of the dovetail keys which are to be cut.

b represents one of the dovetail keys.

I *claim* the combination of the two angular V-shaped and adjustable cutters with the guiding-tube, or its equivalent, substantially in the manner and for the purpose specified.

No. 13,696.—John Bell.—*Improved Dovetailing-Machine.*—Patented October 23, 1855. (Plates, p. 217.)

A number of boards *r* are set in box E, and clamped by means of screws *d*. The box is then set on the inclined plane D in such a manner that the tongues *h* will fit into grooves *g*. When the cutting-chisels A are set in motion, they strike the edges of the boards in a direction which is oblique to their sides, and thus the grooves and tongues cut into said boards are oblique in two different directions; and though all the corresponding sides of the tongues are parallel to each other, and thus form a parallelopiped, still none of its angles are right-angles, but all oblique ones. The object of this is, that when two boards *n n* are fitted together by dovetails, as described, it will be impossible to separate them in any horizontal or vertical direction; they can only be separated in a direction parallel to the lines *p*, and this separation is prevented by nailing a bottom-board *o* to the sides *m n*. The distance between each two chisels A is equal to the width of the chisels, so as to cut the tongues and grooves to fit into each other. It is obvious that the depth of each groove and the height of each tongue must be equal to the thickness of the boards, to form a finished surface when fitted together. For that purpose it is necessary that the distance of the inclined planes D from the centre of shaft B can be regulated.

Claim.—The combination of the box, clamp, or frame E, or its equivalent, for holding the pieces to be dovetailed or tenoned, with the series of rotating cutters, substantially as described.

Also, in combination with the double inclined tables, the double set of rotating cutters, having the planes of the edges of the cutters working parallel with said tables, substantially in the manner and for the purpose set forth.

No. 13,180.—Paul Peckham.—*Machine for Dressing Conical Tapering Surfaces.*—Patented July 3, 1855. (Plates, p. 217.)

The object of this improvement is to dress a tapering stick of wood so that, while it has one flat surface, the rest of it, in transverse section, may be in the form of a segment of a circle. The flat side of the stick rests on support C, its smaller end foremost. The support is then elevated, to raise the stick up to the cutters *a* attached to revolving tubular cutter-stock B.

I *claim* the arrangement and application of the depressing rest C, with respect to the hollow tubular cutter-stock, substantially as described, so as to enable a person to dress or round a tapering stick of wood, in manner as specified.

No. 13,568.—Thomas Durden.—*Machine for Felling Trees.*—Patented September 18, 1855. (Plates, p. 218.)

The machine having been placed in position, the jaws of the dog (turning on fulcra *c c*) are forced into auger-holes formed in the tree, and confined by means of projections *b* at the front ends of the jaws, and a wedge *x*. Motion being communicated to shaft J, the cutters will be revolved by means of gearing L, M, H, G, and fed forward by means of the screw-thread K on shaft J, working through a female screw in carriage D. The cutters C act first, and cutters C^1 follow after and prepare the way for each new cut.

I *claim* the employment of cutters C C C^1 C^1, of the peculiar form shown, in combination with the feeding arrangement K L M, substantially as and for the purpose set forth.

I likewise claim providing each of the jaws of the dog with a projection *b*, and arranging and operating them as shown, for the purpose set forth.

No. 12,174.—Warren Wadleigh.—*Improved Machine for Cutting Irregular Forms.*—Patented January 2, 1855. (Plates, p. 218.)

Arbors H H^1 H are passed through the front rail G, and corresponding dead-centres J through the rear rail G^1 of a carriage F G F G^1, which carriage slides on rails C C^1. The two outside arbors and dead-centres support the patterns I I, and the middle arbor (or middle arbors, as there may be any desirable number of them) and dead-centre support the material to be operated upon. The worm-gear K L serves to revolve the arbors, and with them the patterns and material. The cutter h^1 is mounted on a cutter-bar *e*. The ends of this bar are square, and slide through boxes *i*, and one of these ends beyond its box *i* is connected to a revolving crank by means of a connecting-rod *f*. Thus the cutter-bar and cutter receive a reciprocating motion. The boxes *i* are fitted to pieces *j*, which slide vertically in brackets *h h*, which brackets are fastened to the rails C C^1 of the main frame of the machine. The top parts of the boxes *i* rest against the patterns I I, and are kept in continuous and close contact with them by spiral springs, the lower ends of which are attached to the sliding supports *j*, and the upper ends to the stationary brackets *h*. Thus the boxes *i*, the cutter-bar, and the cutter will be vibrated vertically, corresponding to the shape of the patterns I. The wedge-shaped ends *n n* of a bent rod m^1 pass between the boxes *i* and the square ends of the cutter-bar *e*. The screw *o* is so attached to rod m^1 by means of a washer and pin *p*, that it can be freely turned; its threaded part passes through a female screw *q*, which is fitted to move up and down in the stand *k*, which is fastened to the rail C. By means of turning this screw *o*, the operator causes the wedge-shaped ends *n n* of rod m^1 to traverse between the cutter-bar ends and the boxes, thereby elevating or depressing the cutter-bar and cutter, (while the machine is in operation,) so as to cut the article larger, the same size, or smaller than the patterns.

Claim.—1st. A reciprocating cutter with one or two edges guided or governed by one or more patterns, so as to cut the rough blocks or pieces of wood, or other material, into the form required, substantially as described.

2d. The wedges *n n*, or their equivalents, so constructed and arranged as to enable the operator to vary the distance between the pattern and cutter-bar while the machine is in motion, for the purposes set forth, substantially as described.

No. 12,192.—Wm. J. Casselman.—*Improved Machine for Turning Irregular Forms.*—Patented January 9, 1855. (Plates, p. 219.)

When the work-tables D (two or more) and the pattern-table D^1 revolve, the tracer *d* and the tools *e e* are caused, by the movement of the carriage G on the slide F, to move slowly across the face of the pattern and work, and the tracer, as it ascends and descends in tracing over the undulating surface of the pattern, gives by means of the levers J J K K a corresponding movement to the cutters, and causes them to cut the work to the form of the pattern.

The inventor says: I do not claim the suspension of a tool from a lever which transmits to it a movement given to the tracer by passing over the undulating surface of the pattern; neither do I claim the employment of a tool thus suspended above a revolving work-table; but I do *claim* the particular mode, herein described, of arranging and combining a pattern-table, two or more work-tables, a tracer, and a number of cutting-tools to correspond with the number of work-tables; that is to say, the work-tables and pattern-table being arranged with their axes in the same plane, and the tracer, cutting-tools, and the levers connecting them being all attached in such a way to a carriage, which has a movement in a direction perpendicular to the axes of the revolving tables, but parallel to the plane of the said axes, that the points of the cutters and tracer stand in the same plane, or in a plane near to and parallel with the plane of the axes of the tables, and will all bear at all times the same relation to each other and to the pattern and work.

No. 12,884.—J. S. BARBER, assignor to ROBERT J. MARCHER.—*Machine for Cutting Irregular Forms.*—Patented May 15, 1855. (Plates, p. 219.)

To insert a blank, the table C is lowered by releasing lever E; the blank of the general form of pattern F is secured to this pattern, so that as the rollers *q* bear upon the periphery of the pattern, the cutters *r* shall operate upon the blank. The table is then raised, the driving pinion *t* thrown into gear with the cogged circumference of the table, and the blank slowly rotated, while the cutters are operating upon it, (one upon the inside of the blank, the other upon its outside,) completing the oval during a single revolution of the table.

The inventor says: I do not claim the revolving cutters with rollers upon their shafts, when the latter revolve in fixed bearings; neither do I claim the forms or patterns to which the blanks are secured. But I do *claim* the within described machine for turning ovals, consisting essentially of the sliding cutters *r*, in combination with the table C and pattern F, connected together, and operating in the manner substantially as herein set forth.

No. 13,076.—AVERY BABBETT.—*Machine for Cutting Irregular Forms.*—Patented June 19, 1855. (Plates, p. 219.)

The bearings B of the cutter-shafts *c* are vertically adjustable. Crank H, shaft I, and pinion J serve to move carriage G along on ways F. Shaft L (to which index W is fastened) extends upwards and terminates in two horizontal pivots, on which the frame M vibrates. M can be set horizontal or at any desired angle by means of index N. The material is secured between the points S and R. Index U is keyed fast to shaft Q, and extends through the bearing towards R; and from the lower side of said shaft is projected plate V, on which the material is rested to prevent it from turning or changing its position relatively to shaft Q, except by altering the position of index U. The cutter-shafts *c* carry knife-discs D, of which, however, the engraving represents only one on shaft c^1. By means of using on the various knife-discs knives of proper shapes, and properly altering the positions of the indices, forms similar to the one represented in fig. 2 can be produced by the successive action of the knife-discs.

Claim.—The machine specified, for the purpose of producing angular irregular forms, substantially as set forth.

No. 13,386.—DANIEL DUNLAP.—*Improved Cutter-Head for Irregular Forms.*—Patented August 7, 1855. (Plates, p. 219.)

The lower end *a* of each cutter E is received in an angular recess *b* in the head B, while the conical ends of screws *c* press against the bevelled upper edge *a* of the cutter.

The inventor says: I do not claim merely applying to a plane-iron a continuance to gauge its depth of cut; nor do I claim the combination of knives in any manner with a rotary cutter-head, so that said head shall serve as a guide or directrix to the form or pattern carrying the material to be dressed.

But I *claim* combining with or arranging in connexion with the rotary guide B, and each of its knives, in manner as described, the cylinder crescent-gauge D, whereby, while the pattern or former is borne against the guide-head, the material will not only be reduced by successive cuts, until brought down to its proper depth, but the danger of accident diminished, as specified.

I also claim the described improved mode of applying and securing each of the cutters to its stock or supports, whereby, by a force acting longitudinally on them, they are not only held in such direction, but at the same time are pressed laterally against the curved inner faces of the gauges D D, in manner and for the purpose as specified.

No. 13,511.—P. H. Wait.—*Machine for Cutting Irregular Forms.*—Patented August 28, 1855. (Plates, p. 220.)

The pieces of stuff S S are secured at the upper part of each frame *a*. The pinion *m* gives rotary motion to the two pieces S S, screw-shaft G, and pattern H; and the screw-shaft G (when clamped by nut F) slides the rollers C and cutter-disc D along on their shafts. As the rollers C bear against the pattern in consequence of spring I, the pattern, as it rotates, will move the lower ends of the frames out and in, according to its shape, and the pieces at the upper ends of the frames will be moved in a corresponding manner towards and from the cutter-disc, so as to be cut to the shape of the pattern.

The dotted lines in fig. 1 represent the circles of the various gear-wheels.

The inventor says: I do not claim the pattern H, nor the means of turning irregular-formed articles by means of a pattern; for this has been done in various ways.

But I *claim* the employment or use of two vibrating or oscillating frames placed upon a rod or shaft B, and operated by means of the pattern H bearing against the sliding rollers or discs C—said pattern, as it rotates, moving the stuff at the upper parts of the frames towards and from the cutter-disc L, the cutter-disc and rollers, or discs, being moved by means of the screw-rod or shaft G and nut F as shown.

No. 13,897.—Chester C. Tolman, assignor to James Sargent and Daniel P. Foster.—*Gimlet.*—Patented December 4, 1855. (Plates, p. 220.)

The nature of this invention will be readily understood by reference to the claim and engraving.

I *claim* constructing the lower or outer of the two screw-threads or flanches B B of the gimlet in rounded or curved parabolic form, and having the sides of said portions of the threads or flanches brought to a sharp or cutting-edge, the screw or worm *c* being used or not as desired.

No. 12,429.—Jacob Pierson.—*Machines for Manufacturing Hoops.*—Patented February 20, 1855. (Plates, p. 220.)

The log is placed between the centres c^1 and E^1, the former being adjusted so as to make the top of the log nearly level; the carriage D being traversed so as to bring the head-stock E under the cutter g^3, the carriage L is lowered by the rack *d* so as to let the rest *x* bear upon the log, and the machine set in motion, when the log will traverse under the cutter g^3, which will form and plane a hoop which is severed from the log by the saw *o*. The crooked rest X is fastened to carriage L, and is bent so that each end may rest and traverse on the log operated upon, and support the carriage L; and it is intended that the carriage shall have full liberty to rise and fall, and accommodate itself to the curves in the log, so as to cut the hoops as near parallel with the grain of the wood as possible. Besides, the saws and cutter pass entirely beyond the end of the log in each direction, so that the weight of the carriage is supported alternately by the opposite ends of the rest X.

Claim.—1st. A vibrating or traversing frame, carrying a rotary-cutter, so constructed and arranged that the cutter may be made or allowed to plane or cut its full depth, or a proper depth, in crooked as well as straight logs, so as to make the hoop or other article formed by the cutter parallel, or nearly parallel, with the grain of the wood, substantially as described.

2d. In combination with the frame and cutter, mentioned in the first claim, the circular-saw O, so arranged and operated as to separate the hoop or article formed by the above-mentioned cutter from the log, substantially as described.

3d. In making the rests or guides *x* which govern the position of the traversing-frame, rotating-cutter, and saw O, to traverse on the log, substantially as described, so as to cut the hoop or other article parallel, or nearly parallel, with the grain of the wood.

No. 13,097.—Royal Parce.—*Machine for Cutting Locks and Tapering ends of Wooden Hoops.*—Patented June 19, 1855. (Plates, p. 220.)

The bed-piece A (on which the hoop rests) has an opening E, the front of which has the shape of the lock to be cut in the hoop; B is the cutter-jaw, which works between guides D, and is attached thereto by rod G passing through the guides, said cutter forming a hinge; C C are the cutters, with which the front and side of B are faced, corresponding with the shape of E, the knife at the side cutting the hoop transversely, and the other longitudinally.

The handle F of the jaw passes down through the opening in the bed-piece, so as to be out of the way of the hoop.

The inventor says: I do not claim the use of gauges for obtaining the length of hoops between the locks, but only as used in connexion with my machine; but not being aware that there is in use any device or machine for cutting locks, other than the edge, saw, or knife—

I *claim* the whole combination as described, and especially the principle of cutting locks in wooden hoops by means of knives, or other cutting apparatus, having substantially the form of the lock required to be made, and cutting both transversely and longitudinally by the same movement, no matter in what other combination found.

No. 13,746.—ANDREW BLAIKIE and WALTER CLARK.—*Lath-Machine.*—Patented November 6, 1855. (Plates, p. 220.)

The block from which the laths are sawed is placed upon the feed-table I, and its end is placed or pressed between the lower pair of feed-rollers E E^1 by the attendant. The lower feed-rollers feed the block to the saw D, which cuts a lath from the block, and the block passes between the upper feed-rollers, the plate J entering the saw-kerf. When the lath is sawed from the block, it is directed or guided (by means of inclined-planes *j*) into the receptacle L, and the block falls upon the inclined-table B and descends by its own gravity to the bottom of said table, where the attendant stands, and is again placed upon the feed-table I, and another lath sawed from it, and so on till the block is entirely sawed into laths.

The stuff is represented in strong broken lines.

The inventors say: We do not claim separately the feed-rollers, for they are in common use; but we *claim* the arrangement of the saw D, feed-table I, return-table B, separating-plate J, deflecting or guide-plates *j*, and feed-rollers E E^1, for the purpose specified.

No. 12,224.—LUTHER WENTWORTH.—*Lathe for Turning Fancy Handles, etc.*—Patented January 9, 1855. (Plates, p. 221.)

The dotted lines represent the collar E and cutter *c* (the latter ready for cutting) after they have been moved by means of the connecting-rods F and cam I.

The inventor says: I do not claim the revolving mandrel, carrying cutters to revolve around the work while the cutter is stationary; but I do *claim*, 1st. The within-described mode of arranging and operating the cutters *c* and *d*; that is to say, attaching them to arms D and D^1 which revolve with the mandrel, and are attached to collars E and E^1, which are allowed to slide upon the mandrel C, but not permitted to turn with it, and so guiding the said arms by the inclined slots *i* and studs *j*, or their equivalents, that the sliding-movement of the collars upon the mandrel, produced by cams I I^1, or pattern-wheels, will move the cutters to and from the centre of the work, for the purpose of turning mouldings or grooves at intervals, or giving an irregular profile to the article being turned, as herein fully set forth.

I do not claim hanging a rotary-saw in a swinging-gate, nor allowing the saw-spindle a longitudinal movement, under the control of a spring; but I do claim, 2d. A saw S, arranged as described upon the lathe in a swinging-gate U, which is weighted at W opposite the saw, to throw the saw to an inoperative position, but which is tilted to throw the saw into operation at the proper time to cut off the finished articles from the stick by means of a lever X, actuated by a wiper *v* on a wheel Y, which is attached to one of the feed-rolls, or otherwise so driven as to make one revolution while the stick moves the length of one of the articles to be turned, as herein set forth.

No. 12,423.—William Stephens.—*Improved Slide-Rest for Lathes.*—Patented February 20, 1855. (Plates, p. 222.)

Claim.—Attaching the puppet-head J to the lathe, as shown and described, viz: by having a sector-frame I attached to the socket or collar H, and having an arm K at the lower part of the puppet-head, the lower end of the arm being secured to the lower end of the sector-frame, and the puppet-head fitting or working on the arc *b* of the sector-frame. The puppet-head being operated or moved by a screw-rod N, or its equivalent, and secured at any desired point on the arc by a set-screw L, by which the puppet-head may be so adjusted as to allow articles to be turned between the centres of the spindle and mandrel as in ordinary lathes; or the puppet-head be used as a slide-rest for facing or cutting plates on a chuck, as described. (See engraving.)

No. 12,662.—Warren Aldrich.—*Improved Lathe.*—Patented April 10, 1855. (Plates, p. 222.)

The splined-shaft A, which extends the length of the table, is driven from the head-stock and gives motion through pinion *a*, attached to the carriage, to the gears B B^1. The gear B^1 turns loosely on the end of the hollow screw *b*, and carries the same by friction, being clamped at will by the nut *d*, worked by a screw on the end of the rod *c* within the hollow screw *b*. (The left end of screw *b* is represented in section in order to show said rod *c*.) The apparatus for supporting the tool-post consists of three portions. A block E traverses the carriage in the direction of the screw *b*; a block F revolves upon the upper surface of the block E and around the central spindle D, being secured in any position by two bolts held in a circular dovetail groove in the block E. Another block G, holding the tool-post, traverses the block F in any direction determined by the amount of circular motion of the latter around the spindle D, which, if equal to half a revolution, will reverse the motion of the cutting-tool. The gibs, inserted where these blocks bear against their respective sides, may be tightened by the screws *i* and *h*, so as to prevent any traversing motion.

Suppose the gear B^1 to be clamped by the nut *d*, and the screw *b* to be in revolution; the screw tightened at *i*, in the lower block, to prevent motion, the screw *h* being loose. The spindle D, having worm-gears at each end, will revolve by the action of the screw *b*, and, by means of the upper worm-gear working in the screw *n*, will traverse the block G in any direction allowed by the adjustment of the block F upon the block E. If the screw *h* be tightened, and the screw *i* be loosened, the spindle D cannot revolve, and the block E, carrying with it blocks F and G and the tool, will traverse the main carriage in the direction of the screw *b*.

Figure 1 represents a side view; figure 3 a cross-section of carriage; and figure 2 a horizontal section of the same on line A B.

The inventor says: I do not claim a combination of a tool-rest or carriage, a rotary-carriage, and a sliding-carriage together, and with a mechanism of such character or combination as will impart to such tool-post carriage an automatic traverse motion in whatever position its supporting or turning carriage may be disposed and fixed on the carriage by which it is sustained; nor do I herein claim the peculiar mechanism connected with the three carriages, and described in the specification of my letters patent, dated March 15, A. D 1853, such combination being composed of what are therein exhibited as the splined-shaft U, the movable bevel-gear V, the vertical shaft and its bevel-gears W Y Y, the horizontal shaft X, gears *c g*, and screws I and its female screws, affixed to the tool-rest carriage H.

Having invented a simpler combination of mechanical parts for such purpose, and which, although an equivalent to an element or device in the combination wherein it is employed, is by no means analogous to such device or element, and is much superior in many respects,

I herein *claim* such combination, when used in connexion with the three carriages, as described; the same consisting of a long screw *b*, a vertical shaft D, two worm-gears H^2 I^2, and a screw *n*, arranged, applied, and made to operate together, substantially as specified.

No. 12,742.—C. A. Noyes.—*Improved Slide-Rest for Lathes.*—Patented April 17, 1855. (Plates, p. 222.)

No description required.

Claim.—Constructing the slide-rest as herein shown, viz: having the top H of the sliding-box C rest upon a shaft I, and inclining or tilting said top by means of the screw E, toothed-wheel F, pinions L L G, screw-rods K K, and nuts J J, substantially as herein shown, whereby the edge of the cutting-tool, which is secured on the upper surface of the top H, may be raised or lowered as desired, and presented in a proper position to the article to be turned.

No. 12,747.—Chester Van Horn.—*Improved Slide-Rest for Lathes.*—Patented April 17, 1855. (Plates, p. 222.)

No description required.

The inventor says: I do not claim the carriage B, nor any mode of operating the same; neither do I claim the transverse movement of the tool-block C on the carriage B, for these are common to most slide-rests; but I do *claim* forming the tool-block C of two parts *c d*, and connecting said parts together by a dovetail or its equivalent, so that the upper part *c* may slide or work on the lower part *d*, the faces of the two parts *c d* that are connected being oblique or inclined, as herein shown, and the part *c* being moved or operated by a screw E, or its equivalent, for the purpose of elevating or depressing the tool G, as herein described. (See illustration.)

No. 12,874.—E. K. Root.—*Improved Slide-Lathe.*—Patented May 15, 1855. (Plates, p. 223.)

The tool-carriage *a* receives its longitudinal motion from shaft *b* and worm *c*. The arbor of wheel *g* passes through the front of the carriage and there carries pinion *i*, which engages wheel *j*, the cogs of which engage the cogs of an auxiliary rack *k*, fitted to slide longitudinally in a recess in the carriage. Wrist-pin *l* projects from the top of said rack, and is fitted to block *m*, which block slides in groove *n* in the under face of circular-plate *o*, fitted to turn accurately in a recess in the bottom of the tool-post slide *p*, which latter slides between ways *q q* on top of the carriage, and at right angles to the motion of the carriage. This plate *o* can be turned, and consequently groove *n* set at any desired angle to the line of motion of the carriage, by means of worm-gear *r*. It is secured in the desired position by a set-screw *s*. The bed-plate *u* of the tool-post *t* is connected with the slide *p* by a wrist-pin *w*, so that in turning said bed the proper angle can be given to the tool.

If the groove *n* has an inclination to the line of motion of the carriage, a motion will be given to the tool-post slide *p* towards or from the line of the axis of the mandrel, and by a proper inclination of said groove any desired taper can be turned. When the groove is set parallel with the line of motion of the carriage, the lathe will operate like the ordinary slide-lathe to turn cylinders.

The inventor says: What I *claim*, as my invention for turning tapers on slide-lathes, is giving to the tool-post slide a motion towards or from the line of the axis of the mandrel, by means substantially as herein described, or any equivalent therefor, in combination with the longitudinal feed-motion of the carriage, and derived therefrom, or bearing a certain relation thereto, substantially as described.

No. 13,787.—Eli Horton.—*Lathe-Chuck.*—Patented November 13, 1855. (Plates, p. 223.)

The nature of this improvement will be understood from the claims and engravings.

Claim.—In combination with the opening H on the front plate for the introduction of the solid jaws, the hub L on the back plate for closing said opening and retaining the jaws in their respective slots, substantially as described.

Also, the locating of the circular rack G in the deep recess or groove formed between the flanges F Z, which not only form a tight casing to protect it from chips, filings, &c., but also support it, as well as the shanks of the screw-bolts, substantially as set forth.

No. 12,580.—CHAS. F. BAUERSFELD.—*Clamp and Mouth-piece for Lumber-jointing Machines.*—Patented March 27, 1855. (Plates, p. 223.)

a a^1 are a pair of clamps connected to the same handle *b* by means of levers *c* c^1 and links *d* d^1. *f* is a gauge, forming at its front edge a mouth-piece for the rotating-cutter *g*. This gauge is, by means of a pair of parallel arms, held in a longitudinal or other desired direction, whether moved towards or from the stuff; its position being adjusted to suit the particular cutter, and worked by means of bridle *i* and hand-screw *j*.

Claim.—1st. Two or more clamps so arranged and connected, as described, as to be simultaneously and equally applied to or withdrawn from the different parts of a portion of furniture to be jointed by the means of a single handle.

2d. The parallel motion fixed in any desired position by means of the bridle and screw, as described.

No. 12,861.—FRANCIS P. HART.—*Gauge for Slitting Lumber.*—Patented May 15, 1855. (Plates, p. 223.)

In gauging obliquely for articles of tapering form, one of the wheels *b* is clamped by nuts *c* c^1 tight up to the shoulder *i;* and the binding-screw C in the stock A is loosened, to allow the screw-shaft B to turn in the stock. The shaft is adjusted in the stock by a scale *d*, so as to bring the teeth of the wheel *b* at a distance from face *e* of the stock equal to the width of one end of the article to be gauged. The gauge is then run along the stuff, and the wheel, receiving rotary motion through the bite of its teeth on the stuff, turns the screw, and, according to the direction of its revolution, either increases or diminishes the distance between wheel *b* and face *e*, thus scribing obliquely to the grinding edge. By changing the size of wheel *b*, the taper will be greater or less. In gauging parallel, the wheel *b* is left free by the nuts *c* c^1, and stock and shaft are secured together by binding-screw C.

The slider D and scriber *f* are for the purpose of gauging mortises. Fig. 3 represents the scriber *f* (when not required for use) as being folded down within the groove in the shaft B. When thrown outwards, the scriber is kept at right angles by the square shoulder *g*, and when cutting is kept up by reason of its edge being bevelled on the opposite side to the shoulder. In gauging curved work, the guide-piece F is screwed out by means of screw *j*, and its rounded end forms the bearing, and allows the gauge to follow any curved line.

Claim.—1st. The employment of a rotary cutter secured to the shaft of the gauge, when the said shaft screws into the stock, and is made capable of turning freely therein, as described, for the purpose of gauging taper-work.

2d. Attaching the adjustable scriber *f* by a hinge-joint constructed with a shoulder *g*, substantially as described, to the slider which carries it, so that it may be rigid when extended for mortising, but may fold into the recess in which the said slider works when the gauge is used for other purposes than mortising.

3d. The employment of a round-faced guide-piece F fitted to slide within the stock of the gauge, so as to be withdrawn into it when the gauge is to be used for straight work, but to be protruded from it when required to serve as a guide for gauging curved, circular, or irregular work, as herein fully set forth.

No. 13,342.—Albert Walcott.—*Machine for Dressing Lumber from the Log.*—Patented July 24, 1855. (Plates, p. 223.)

The nature of this invention consists in planing, tonguing and grooving, matching, or otherwise dressing in the solid timber before sawing, or in connexion with sawing after the logs have been sawed into cants or thick plank of the desired thickness to make the width of the desired dressed stuff; or, in other words, the dressing and sawing at the same time, the dressing only preceding the sawing by the distance between the dressing-machinery and the saw as the cants or plank are moved forward past the dressing-machinery first, and then the saw. In the engraving, C represents the plank on carriage A travelling on ways B; D, dressing-cylinder; E and F, cutters for tonguing and grooving; G, saw for sawing off the dressed stuff.

The inventor says: I do not claim sawing lumber into cants or planks, the carriage, or any particular construction of carriages; the cylindrical cutters, or any particular construction of cutters or planes, for surfacing, or for tonguing, grooving, matching, rabbetting, beading, or moulding; neither do I claim the circular saw: as they have all been used before in other various modifications.

I do not limit myself to any precise form or arrangement of parts, nor to any particular device for moving or operating them; for these may be varied to an almost unlimited extent without changing the principles of my invention, as set forth.

I do not limit myself to the rotary cylindrical cutters for surfacing; as cutters in the face of a wheel may be used, or even stationary knife-stocks used for surfacing.

Neither do I limit myself to the use of the circular saw for sawing off stuff; as a reciprocating saw may be used, and many other similar variations may be made by any competent or skilful machinist, without essential or substantial variation from the character of my invention, as set forth.

I *claim* the particular arrangement and combination of mechanism for manufacturing and dressing out lumber from the log, cant, or plank, as the case may be, by successive operations, and in manner substantially the same as set forth and described; not confining myself, however, to any particular arrangement of mere mechanical details or devices to effect the desired result—that is a piece of lumber finished or partially so, as the case may be, for building, and other purposes.

No. 13,142.—Joseph Sykes.—*Wheelwrights' Guide-Mandrel.*—Patented June 26, 1855. (Plates, p. 223.)

The journals are allowed to turn freely on the bearings on frame F, and the spokes S are driven truly in the hub, because it is truly centred on the mandrel, in consequence of the fianches *d* on the plates H being fitted in the V-shaped recesses *c*.

In case of iron hubs, cones I may be used instead of plates H, the cones being fitted on the mandrel and forced into the ends of the hole in the hub by turning nut D.

Claim.—The combination of the mandrel A with its permanent and loose journals B E, and the circular plates H H, or cones I I, either plates or cones being used as circumstances require; the above parts being arranged as shown, and for the purpose as set forth.

No. 12,276.—Louis Francis Groebl.—*Improved Marquetry.*—Patented January 23, 1855. (Plates, p. 224.)

This invention consists of composing floors of thin and small pieces of wood of variegated shape, so as to form various patterns. The pieces are all fitted together by tongues and grooves around their whole outline. This is an extremely durable, cheap, and highly ornamental floor, as the forms, colors, and grain of the pieces admit of infinite diversification to represent the various ornamental devices. As these floors are from time to time varnished or waxed, they always look as if new, and they are much easier cleaned than carpets. The engraving clearly represents the manner of joining the pieces together—*a* being the tongue on the piece A, and *b* being the groove on piece B.

The inventor says: I *claim* the marquetry described, in which the different pieces of which it is composed are firmly united at their adjoining edges, so as to secure the advantages described.

But I make no claim to the invention of tonguing and grooving, nor to forming an ornamental design or style of decoration, by making combinations of wood of various forms or colors.

No. 12,297.—Leopold and Joseph Thomas.—*Match-Machine.*—Patented January 23, 1855. (Plates, p. 224.)

The fluted rollers *d d* which are mounted in a sliding carriage *b b*, gradually draw in the block of wood from which the matches are to be cut, and which is placed between them. The carriage moves the block against a row of cutters *j* (which have the form of hollow cylindrical punches). These cutters cut the matches out, and those matches which are being cut push the matches previously cut through the holes in the stationary brass block *l*, (which holes are opposite and correspond with the holes of the cutters,) and through corresponding holes in the wheel *m*. This wheel *m* receives (by means of pawl and ratchet-wheel *t*) an intermittent motion, so as to present a new row of holes to every new row of matches. The carrier-wheel *m*, with its holes, revolves in lose contact with a roller *n*, the surface of which is supplied with phosphoric preparation, so that the roller will deposite small portions of the preparation in the holes as they pass by the roller. The matches, in being pushed through these holes, carry with them the phosphoric substance deposited there, and the ends of the matches, when passed through the wheel, are covered with the said substance. The matches fall then on a sliding shelf o, which carries them laterally towards a concave semi-circular stop v^1, until the corresponding concave semi-circular lip v on the shelf closes against the stop, thus holding the matches in a round bunch, ready to have the box and lid placed over them. A number of boxes are placed in the vertical guide w, and a number of lids in the vertical guide w^1, opposite to guide w. As soon as the bunch of matches is ready to be boxed, the plungers x x^1 move towards each other, press a box and a lid through cylindrical openings at the lower end of the guides, and push them over the bunch of matches and close the box.

The shoving-head c^1 which works in a groove in the table a of the machine, is connected with the end of the sliding shelf by a bolt d^1, which, having a spiral spring e^1 around it, ordinarily keeps the end of c^1 in contact with the extremity of shelf o with so much force that the motion of the lever drawing the shoving-head back also withdraws the shelf without separating its contact with the shoving-head. When, however, the sliding shelf is arrested in its progress by the closing of v v^1, the stroke of the moving-cam not being exhausted, the shoving-head is drawn still further back, the spiral spring e^1 being compressed, and the shoving-head being drawn away from the end of the shelf. The effect of this independent motion of c^1 is to press between and against the short arms of the levers y y^1 which causes the ends of the long arms of these levers to press inwards, thus forcing the plungers x x^1 through the guides. As soon as cam z causes lever a^1 to return, the shoving-head is drawn back by spring e^1, and again comes in contact with the end of the shelf; and the springs s^1 s^1 cause the plungers to fly back, and the lever a^1 returns the shelf to its original position before wheel m.

Claim.—1st. The use of the sliding carriage, with the feed-rollers, for the purposes described.

2d. The combination of sliding shelf, shoving-head, levers, and plungers, for the purpose of packing the finished matches in boxes.

3d. The carrier-wheel and roller, for applying the phosphoric composition to the matches by machinery.

No. 12,482.—F. A. GLEASON.—*Machine for Cutting Mitre and other Joints.*—Patented March 6, 1855. (Plates, p. 224.)

The operation of this machine is as follows: The tonguing stock (fig. 2) is taken off, the carriage m brought forward, and the bed s moved so near the mitre-saw F, that the dovetail groover G shall cut through the desired portion of the work. The machine is then set in motion and the work carried past the saw, which cuts the mitre; the clearing-knife b cuts away the part running back of the saw, which would otherwise run against the chuck, and the groover cuts a dovetail groove through. The carriage is then placed opposite the saw, and the bed is brought back (by means of adjusting screw *o*) so far that the piece just grooved, laid loosely upon it, shall drop down to the position shown in fig. 3, which determines the position of the tongue.

The groover is fixed in the centre of the chuck, and is provided with two cutting-blades, on the ends of which are lips for cuting off the chip. The saw F is fastened by turning a shoulder angular, or under cut, (see fig. 3 at *a*,) and then cutting away sections of it (see fig. 5). The saw is turned out and cut to correspond with it, then placed on and turned a little backwards, which fastens it. The tonguing stock is attached to the head-stock B, by shank H being placed in stirrup I, (see fig. 4,) and fastened by set-screw *f*. The bed *s* may be set at any angle by means of a set-screw and slotted arm *t*.

Claim.—1st. The rotary dovetail groover, as described, or its equivalent.

2d. The mitre-saw F, with the clearing-knife *b*, fixed upon the same chuck, and concentric with the groover; also the manner of fastening the saws, as described.

3d. The tonguing-stock, with its saw and bevel-cutter, or their equivalents; also the manner of attaching it to the head-stock.

4th. The carriage, with its movable bed, which may be adjusted to any angle required.

No. 12,796.—Mathew Spear.—*Mitre-Box.*—Patented May 1, 1855. (Plates, p. 225.)

The two supporters *a a* and a^1 a^1 turn about a common pivot, in combination with a saw-guide *d* attached to the same pivot. The stuff to be sawed is laid upon these supporters, so as to extend through the saw-guide, and made to rest against one or both of the edge-supporters H H, as the case may be. The lower surfaces of these edge-supporters, which are placed one on each of the supporters *a a*, a^1 a^1, are provided each with a lip I, which lip is to be placed in either one of the grooves a^2 b^2 c^2 e^2 f^2. When so placed, the edge-supporter is confined in place by a clamp-wedge g^2, which may be passed in and through either of the holes h^2, so that the wedge will operate against the side of the groove.

To saw stuff at a right-angle, it should be borne against the inner edges of the edge-supporters when their lips are confined in the grooves c^2, the machine being entirely opened, as represented in the figures. In order to mitre stuff to a right angle, the machine is to be closed entirely, while the lips are in the grooves c^2. A piece may also be sawed at a right-angle, with the machine being so closed, by placing the lips in the grooves f^2, these grooves being in line with one another, and at right-angles to the saw-guide when the machine is closed. For mitreing at a greater angle, the lips may be placed in grooves b^2, the angle being determined by opening the machine and setting it at the angle required, and then clamping it to the arch beam A by means of wedge-clamps D D^1. The arc A (clamped to the saw-guide *d* by a wedge E) slides in recesses B C^1 in the two supporters. The supporters being clamped to the arc, the saw-guide is then moved so as to bisect the angle formed between the two clamp D D^1.

Each of the grooves e^2 is at right-angles with the saw-guide, when this is placed in contact with the supporter a or a^1, of the opposite groove, and the two outer edges of the movable edge-pieces K K are set at a right-angle to each other. These edge-pieces K K, arranged on the outer edges of the supporters a a^1, can be slid towards or away from one another, and fixed in position by clamp-screws i^2.

In order to cut stuff at any angle to which the outer edges of the machine may be placed, the edge-supporter is fixed in one of the grooves e^2, and the saw-guide moved up against the lumber-supporters a or a^1; then if the stuff be placed with one edge against the inner edge of the edge-supporter, and so as to project through the saw-guide, it will be sawed to the angle required, (corresponding with the angle of the outer faces of pieces K K,) by the saw running in the saw-guide.

The commencement of the inner scale N is placed where the inner edge of the clamp D^1 intersects the arc A, when the machine is closed. The termination of this scale is where the said edge of the clamp intersects the arc, when the two outer faces of K K are arranged at an angle to each other, which is the supplement to that which they make with one another when the machine is closed. The middle of this scale is marked zero.

In order to mitre to the supplement of any angle which the outer faces of K K make with one another, first observe the distance on scale N from the zero to the inner side of D^1, then move the lumber-supporter a^1 until said side of D^1 is at the same distance from the zero on the opposite half of scale N, the saw-guide being moved into bisecting the angle formed between the two edge-supporters placed in grooves b^2.

a^4 are flanges on the under-side of the pieces K K. The dovetailed feet c^4 are made to slide into the lower parts of the lumber-supporters, and to project from the lower surfaces. On recovering these feet from their dovetailed sockets, the flanges a^4 may be used for obtaining the angle of the edges of any piece of board that may be placed between them, the same being to adapt the machine for cutting at such angle or the mitre of it.

Each edge-supporter has an adjustable gauge L, with a bent or right-angular head k^2, which, when the machine is used for mitreing to the supplement of an angle not contained in the scale N, may be turned down into position fig. 5, so as to serve to increase the bearing surface against which the stuff is to be placed that is to be mitred, it being understood that such stuff is to be supported against the inner end of the edge-supporter, and to project through the saw-guide. The support-block m^2 has a projection l^2, which may be inserted in either of the holes h^2, and then the gauge may be used to regulate the length of the piece to be cut, such piece being made to abut at one end against the head of said gauge, while the other is made to overlap the saw-guide. The rod of gauge L slides in support-block m^2, and is fastened in position by clamp-screw n^2.

Claim.—The additional improvement, viz: The sliding index arc

A, as combined with the lumber-bearers, or supporters a a^1, and the saw-guide, and made to operate therewith, essentially as specified.

Also, the combination of the extra grooves f, with the lumber-supporters, the same being for the purpose as above set forth.

Also, the combination of the grooves e^2 and said lumber-supporters, such being for the purpose as set forth.

Also, the combination of the adjustable gauge with the edge-supporter, the same being to determine the length of the stuff to be operated upon.

Also, the above described mode of constructing the head k^2 of the adjustable gauge, so that it may serve to increase the bearing for the stuff during the operation of mitreing the supplement of an angle, as described.

Also, the movable edge-pieces K K, in combination with the lumber-bearers.

Also, combining with the curved arc A the inner index scale, for the purpose of enabling a person to adjust the machine for the purpose of mitreing to the supplement of any angle required, as specified.

Also, combining with the movable pieces K K the projecting lips a^4 a^4, the same being for the purpose above specified.

No. 12,862.—Lorton Holliday.—*Mitre and Bevelling Machine.*—Patented May 15, 1855. (Plates, p. 226.)

The two discs D D^1, turning on a common pivot E, can be set at any angle in a horizontal plane by means of pointer E^1, scale C, slot a, and set-screw b. These discs carry an upright frame G, which revolves together with the discs. The saw-guide I is hinged to the lower end of said frame, and can be secured in any position, forming any angle with a vertical line, (for instance, in the position indicated in dotted lines in fig. 2,) by means of set-screw H.

Claim.—The manner herein described and shown of arranging and combining the several parts constituting the mitre-box herein described. This arrangement and combination rendering the saw-guide capable of being adjusted in the path of a horizontal circle, as well as in the path of a vertical circle, to any angle desired, and enables the saw to cut a bevel lap on the strip or board simultaneous with the sawing of the mitre or angle, and also indicates the angle cut, substantially as set forth.

No. 12,956.—Geo. W. La Baw.—*Mitre-Machine.*—Patented May 29, 1855. (Plates, p. 226.)

The material Z or z is secured to bench A by gauges B and lever D, or by gauges K. A depression of the treadle I will force the heads F down, and the knives X H H G will cut four different parts of joints, as represented in fig. 2.

Claim—The combination and arrangement in the manner described, or in any other manner equivalent thereto, of the several specific parts, or their equivalents, of the described mitre and cutting machine, without limiting myself to any particular arrangement of parts, for the purpose set forth.

No. 13,127.—Jonas S. Halsted and Cornelius J. Ackerman.—*Carpenter's Mitre and Bevel Square.*—Patented June 26, 1855. (Plates, p. 226.)

In order to lay out or mark the joint, the ends of the pieces are first squared. The side-pieces or stiles D are marked or laid out by getting the point *e* on the moulding *d* by actual measurement; the flanch or ledge *b*, on the inner edge of the handle A, is then fitted in the groove or kerf formed by the bead or moulding *d;* and the handle is moved till the point *e* touches the upper edge of the handle. The two lines *f g* are then drawn, *f* being a mitre or an angular line of 45°, and the line *g* a horizontal one. The vertical line *h* is obtained without marking, as it is on the shoulder *i* of a recess in the face of the stile. Thus by a single application of the implement, the side-pieces or stiles are marked. The head-pieces E are marked or laid out as shown in figures; the point *e* being obtained, the inner edge of the handle is placed against the shoulder *i* of the recess, and the blade C is adjusted in line with the bevel at the upper end of the handle A, and the implement is moved till the point *e* touches the upper edge of the blade C. The mitre-line *j* across the bead or moulding, and a point *k* on the shoulder *i*, are thereby obtained; and by reversing the implement, the line *l* is obtained at right-angles with the shoulder *i*.

Claim.—The construction of the implement as shown and described, viz: having a ledge or flanch *d* project on each side of the inner edge of the handle A, the upper and lower ends of said handle being cut or bevelled at an angle of 45°; the handle being provided with a blade B, which is attached at right-angles to it, and also provided with an adjustable blade C, as shown, and for the purpose as set forth.

No. 12,287.—R. P. Benton.—*Improvement in Feeding Mortising-Machines.*—Patented January 23, 1855. (Plates, p. 226.)

The stuff *q* is secured in slide X by set-screw *p;* and as the slide R, together with the slide X and with the stuff *q*, is fed towards the revolving-cutter *b*, the slide X receives a reciprocating motion by means of crank Q, and the stuff is consequently moved back and forth in a direction lengthwise of the mortise; and the length of this vibration can be so adjusted by adjusting crank Q, by means of set-screw *e*, as to cut the mortises the required length. A small plate *k*, arranged on the under side of slide R, fits between the threads of screw S, and thus the slide receives its motion. When the stuff has been fed towards the cutter the required distance, the bent-lever W, (against which slide R then strikes,) lever N, spring *r*, and bar V, serve to raise the plate *k* from screw S, and to draw back the slide R to its original position.

Claim.—Feeding the stuff to be mortised to the cutter *b*, in the manner substantially as shown, viz: by means of a rotating screw-rod *s* operating upon a slide R and an adjustable crank Q, which gives a reciprocating motion to the slide X; the above parts operating conjointly as shown, and for the purpose as set forth.

No. 12,563.—Elihu Street.—*Mortising and Tenoning Machine.*—Patented March 21, 1855. (Plates, p. 227.)

l is a plane-iron for cutting tenons; *h* another plane for smoothing the ends of boards; S a saw; C a chisel; and all these tools are attached to the sliding-frame *b*.

Claim.—The improvement on a machine for mortising, tenoning, sawing, and smoothing, by combining certain tools together, used by carpenters in the manufacture of doors, sash, and blinds, as described.

No. 12,691.—Benjamin T. Norris.—*Machine for Mortising Blinds.*—Patented April 10, 1855. (Plates, p. 227.)

A frame B (carrying the racks h h^1 and detents *d e*) slides in proper ways on the stationary frame A. Projections R S, extending from frame B, serve to attach one end of the stock C (which is to be mortised) firmly to frame B. Friction-rollers *a a*, supported by frame A, serve to guide the stock in its way. (Figure 1 exhibits only one stock and one pair of friction-rollers, the other being left out to show the cutting instrument in full view). Rack *h* is firmly attached to frame B; rack h^1 is allowed (by means of slots and screws) sufficient longitudinal motion to either depress or raise the detents *d e*, with which it is connected by levers, as apparent from the illustration. Lever D, which has its fulcrum in frame A, carries two pawls *g* and *c*. Lever D being moved backwards from its position in figure 1 towards its upright vertical position, the pawl *c* will push the rack *h*, and with it frame B and stock C, backward. At the same time pawl *g* follows lever D, and rack h^1 is allowed to follow the impulse of a spring which draws rack h^1 backward sufficiently far to depress detents *d e* till they stop the backward motion of frame B, by coming in contact with the upper end of the next one of the index-rods E. Now moving lever D forward to its original position, pawl *g* pushes forward rack h^1 the length of its longitudinal play, thereby elevating the detents *d e*, and draws the frame B and stock C forward the length of the distance between each two mortises. The stocks of the cutters or drills *n n*, which are rapidly revolved, by means of pulley J and band O, are separate from each other, and are made to slide within pulley J with which they revolve. The detent being stopped by the index (as above described), the cutter-frame is drawn up the inclined ways H H to the highest point of the mortise to be cut, the drills *n n* are thrown out on both sides into the stocks C C, the depth of a mortise (which is done by turning screw L, whereby levers P P are depressed, the lower arms of said levers throwing out the cutters), then the cutters are allowed to descend (being drawn down the ways H H by the weight of the cutter-frame), and when the mortise is cut, the cutters are thrown back by unscrewing L.

Claim.—The manner of constructing and operating the parts carrying the machine, viz: the hollow pulley and spindle within the same, arranged as described, together with the levers attached, operated by the means described, or other suitable device. I also claim, as new, the manner of operating and guiding the stock by the combination of the lever with the racks and the movable detents, as herein described; but do not claim the index constructed of bars, as described, that having been in use before.

No. 13,184.—EDWARD Q. SMITH.—*Method of Cutting Straight or Curved Mortises.*—Patented July 3, 1855. (Plates, p. 228.)

The clamp-apparatus 15 on table B holds the piece to be mortised, and the table slides on ways 9. The pitman *i* gives lateral movement to the frame J and *n* that carries the spindle *k* and chisel 18. The crank-arm *m* is adjustable, so as to produce different lengths of mortises. The two ends of pieces 6 6 (which give the curve to the mortise) are attached to frame *n* by thumb-screw 21 in slot 22; the outside ends of said pieces 6 6 work in guides 12 12, which guides are stationary with the exception of revolving around their centres.

In order to make a curved mortise (see 6, fig. 5), the segment-pieces are to be placed at an angle (see figs. 2 and 3); and when a reverse curve (see 8, fig. 8) is required, the segment-pieces are placed in a reverse position (see 8, fig. 3). When a straight mortise is required (see 7, fig. 5), the segments are placed in a right line (see 7, fig. 3). The mandrel and chisel receive their rotary motion by means of belt 13.

Claim.—The two segment-pieces 6 6 placed at an angle to each other, or straight, so as to direct the lateral movement of the mortising-chisel in a curved or right line, and thereby form a curved or straight mortise, as mentioned, all substantially for the purposes set forth.

No. 13,271.—JOSEPH A. PEABODY.—*Machine for Mortising Window-Blinds.*—Patented July 17, 1855. (Plates, p. 228.)

The nature of this invention consists in making all the angular mortises in stiles to window-blinds to receive the slats at one operation, by providing the machine with a swinging bar, to which the stile is properly secured, and a horizontal sliding-carriage B, to which a series of revolving mortising-chisels E F, sufficient to mortise the whole length of the stile, are adjusted.

Claim.—The bar or carriage N which carries the blind-stile, and which is moved by lever or otherwise, and the changeable and adjustable arms *o o*, or their mechanical equivalents, one end of each of them being connected to the bar N, while their opposite ends are so connected by pins, or otherwise, to the machine, that these arms are changeable and adjustable so as to impart any desired angle to the mortises, essentially in the manner and for the purposes set forth.

Also, the carriage B, or its equivalent, which may be vibrated or moved by lever, or otherwise, for carrying a series of revolving mortising-chisels; this carriage, and the chisels attached to it, being so moved that the chisels will form or cut all the angular mortises in one window-blind stile at one operation, essentially in the manner and for the purposes set forth.

No. 13,594.—Ezra Gould.—*Improved Method of Regulating Length of Stroke in Mortising-Machines.*—Patented September 25, 1855. (Plates, p. 228.)

The chisel is attached to arbor D. If the lever F be so operated that the part of the lever below its fulcrum bears upon the collar *w* of the plate *t*, said plate will clamp the pinion *s* between itself and the pinion *r*, and pinion *n* on shaft *a* will give motion to shaft *o*, which will be slower than shaft *a*; the pinion *q* turns pinion *k*, and arm E will consequently be turned, and draw plate *e* from the centre of the pulley to the outer end of the slots *j* and *c*, and arbor D will then have its greatest throw or length of vibration. By moving lever F in the opposite direction, pinion *l* will be clamped between pinion *n* and plate *m*, and pinion *l* will then turn pinion *r*; and as the pinion *l* is larger than *r*, the shaft *o* will rotate faster than shaft *a*, and arm E will consequently be moved in the opposite direction, and plate *e* will be moved to the inner end of the slots *j* and *c*, and arbor D will be stationary while A revolves. By this variable throw the stuff may be adjusted upon and taken from the bed-piece without stopping the driving-pulley A, and the chisel may also be turned without stopping the pulley.

Claim.—Attaching the connecting-rod C to a curved slotted arm E by means of the plate *e* and pins *f i*; the plate *e* working in a slot *c* in the pulley A, and the arm E operated by means of the gearing *l n p s* and *q k*, arranged as shown, or in an equivalent way, for the purpose specified.

No. 13,663.—Hezekiah B. Smith.—*Mortising Machine.*—Patented October 9, 1855. (Plates, p. 299.)

The timber is placed upon bed F^2; then pressing lever N^2 down by the foot, the clutch K^2 is moved so as to clutch with and revolve pulley W, shaft J^2, and (by means of bevel-wheels I^2 H^2) screw-shaft G^2, which will move the carriage B and chisel Y^2 down until the chisel penetrates to the proper depth into the timber, when the pulley Q (previously set at the proper height in slot *i*) will come in contact with the bent part *j* of lever O^2, which will push off the lever and unship clutch K, so as to stop the downward motion of the chisel-carriage. The upward motion of the carriage is stopped by pulley P^2 coming in contact with the part *j* of the lever O^2.

The inventor says: I *claim*, 1st. Moving the chisel-carriage B to and from the wood to be mortised by power, essentially in the manner and for the purposes set forth.

2d. I claim, in combination with bent lever O^2, clutches K^2, *a*, *b*, *c*, and *d*, pulley-stops P^2 and Q^2, or their mechanical equivalent, by which the said chisel-carriage B will stop its own motion at or near any desired point, substantially in the manner and for the purposes set forth.

No. 13,759.—LOOMIS E. PAYNE and ORRIS PIER.—*Mortising-Machine.*—Patented November 6, 1855. (Plates, p. 229.)

The carriage *c* which supports the cutter *g* receives a reciprocating motion by means of pitman *v* and the double-plated eccentric (with the adjusting and tightening-screw *u* at its centre) at the end of shaft *x x*. The table d^1 which carries the timber is moved forward on ways by means of rack h^1 attached to the under side of the table and pinion *i* at the head of shaft j^1; l^1 is a ratchet at the bottom of shaft j^1. Just above the ratchet, a flat arm m^1 passes from the shaft on to a supporting-rod n^1. About midway of shaft j^1, a lever t^1 passes out; this lever encircles the shaft, and when pushed to the left engages the pinion with the rack, and is held in this position by a bent rod u^1, which is also drawn to the left at the same time, and catches against the lef end of the piece v^1. As soon as this rod is pushed out, the spring *w* draws the arm t^1 and disengages the pinion and rack. A slotted piece x^1 is secured to the under side of the table d^1. This has, at its inmost end, a projection extending downwards; and when the rod u^1 is forced into the engaging position, its upper end is brought in a line with this projection. The object of this arrangement is to act as a gauge to the mortise.

Claim.—A double semicircular mortise-bit or gauge, arranged so as to clear itself thoroughly in its action, and this in combination with the double-eccentric plate, to regulate the motion to and fro of said mortise-bit; the whole being combined and operating substantially as set forth.

No. 12,243.—GEO. M. RAMSAY.—*Moulding-Machine.*—Patented January 16, 1855. (Plates, p. 230.)

The piece *x* to be operated upon is placed against the side-rest *b*, and is carried past the cutter-head *f* by means of the feed-rollers *h i*; the front end of the piece comes in contact with the spring *n* and brings it (together with the feed-roller *m* attached to its end) into the position represented in fig. 1 in dotted lines. As soon as the rear end of the piece has passed the guide-roller *l*, the spring *n* throws the piece over against the side-rest *c* into position *y*, when the roller *m* (whose cog-wheel m^1 is now thrown into gear with cog-wheel m^2) commences to revolve, and feeds the piece back to and past the cutter-head. Thus the piece is operated upon on both its sides by once passing through the machine. Above the cutter-head there is placed a hollow cone *z* containing within it a fan *g* attached to and revolving with the shaft of the cutter-head. The wings of the fan are placed at such angles as to produce a current of air downwards, for the purpose of blowing the shavings underneath the machine.

The inventor says: I do not claim the mere fact of working the two opposite sides of the same piece of stuff by once passing to and through the machine. But I *claim* the automatic reversible feed, or its equivalent, whereby the two opposite sides of the same piece of stuff are worked, as described, by the double action of one cutter-head, by once passing the stuff to and through the machine; also the arrangement of the cone and fan, all operated and operating substantially in the manner and for the purpose set forth.

No. 12,248.—C. B. Morse.—*Moulding-Machine.*—Patented January 16, 1855. (Plates, p. 230.)

The nature of this invention is plainly stated in the claims, and fully illustrated in the engravings.

The inventor says: I disclaim constructing a cylindrical rotating cutter-head with a separating joint athwart its middle, as such is not new. I *claim* constructing the cutter-head of two flanged discs, with slots or openings *g* through one of the discs to admit of cutters *e* being attached to the other part, and partially marked by the flange of the perforated disc, as described, in combination with cutters *f* and *l* in openings through the rim or flanges, and secured respectively to each disc, so as to present a cutting edge over the whole space caused by the opening or closing of said discs, by means of nut E and set-screws *j*, said combination favoring a current inward from the edges of the cutters to fill the partial vacuum formed in the interior of the head by the rotation of the same, thereby causing a speedy inward removal of the shavings from the cutters, and admitting of the double action of the same, as set forth. Also, the adjustable shields *s* in combination with the feed-rollers F, for preventing the said rollers from lifting the piece operated upon against the cutters, when the feed is not continuous and the extremity of the piece reaches the roller, as set forth.

No. 12,916.—Robert J. Marcher.—*Tool for Grooving Mouldings.*—Patented May 22, 1855. (Plates, p. 230.)

The nature of this improvement will be understood from the claim and engravings.

The dotted lines in fig. 1 represent the position when the tool-stock D stands vertical, and the cutter *b* is raised from the moulding out of the groove *g* just made, when the instrument is ready to be moved along by hand or otherwise the proper distance to commence the next groove.

Claim.—Forming or cutting transverse and parallel grooves *g* in concave portions *f* of mouldings B, by means of a tool-stock D attached to a plate C by a pivot E, which is at the centre of a circle of which the concave forms a part. The cutter *b* being attached to the lower end of a slide L^1, which is operated or pressed down when the cutter acts upon the moulding by a spring H, and elevated upon the return motion of the cutter by raising the lever G, the cutter *b* having a stop or guard *c* adjoining it for the purpose of regulating the depth of the cut, as herein shown and described.

No. 12,917.—Robert J. Marcher.—*Tool for Grooving Mouldings.*—Patented May 22, 1855. (Plates, p. 231.)

When the cutter *e* has cut the groove, the lever L is raised by the hand, and the cutter is raised out of the groove, and the frame H is then drawn back by the hand, the slot *d*, in consequence of the pin *c* fitting in it, causing the lower end of the tool-stock to clear the face of the moulding. The slot *d* gives an arbitrary motion to the tool-stock, and causes it to rise and fall so as to conform to the face of the moulding; but the spring K forces the slide J down independently of the motion of the tool-stock, and compensates for any variation in the regularity of the slot or moulding.

The broken lines represent the position of the cutter and other parts when raised out of the groove.

Claim.—The reciprocating tool-stock I, slide J, with cutter *e* (one or more) and stop or guard e^1 attached to its lower end, and having a spring K acting upon its upper end, and the slot *d* in the plate F, the above parts being arranged substantially as herein shown, and for the purpose as herein set forth.

No. 13,602.—H. Schevenell and Richard S. Schevenell.—*Improved Machine for Cutting Ornamental Mouldings.*—Patented September 25, 1855. (Plates, p. 231.)

N are the strips to be operated upon, and are fastened to carriage B; spring I presses the cutters K upon the strips, and the ends of plate J (which is attached to the gate F) upon the peripheries of the patterns L L. Thus, as the carriage moves along, the gate and cutters will be moved up and down by the inequalities in the periphery of the pattern-wheels, and the strips will be cut in a waved manner. When the ends of the strips have passed the cutters, the projection *g*, on the carriage back of the gate, will strike the wedges M, and the gate will be forced upward beyond the reach of the patterns L. The finished strips can then be removed, others inserted, and the carriage moved back to its original position. During the latter movement the stop *g*, at the opposite end of the carriage, will strike the wedges and again force them out from underneath plate J.

The inventors say: We do not claim the reciprocating gate or slide F, with cutters attached, for they have been previously used; but we *claim* the combination of the reciprocating gate or slide F, rotary patterns L L, and the inclined planes or wedges M M, the above parts being arranged substantially as shown, for the purpose specified.

No. 12,234.—William C. Hopper.—*Bench-Plane.*—Patented July 16, 1855. (Plates, p. 231.)

The chisel *c* is placed in front of its wedge *e*, and rests against the front shoulders *d* in the cavity of the plane, in combination with the use of a mouth-piece *a* on the face of the plane, in front of the edge of the chisel, to serve as a rest for the chisel, and to confine the throat of the plane.

By this means the chisel is to be held firmly in its place throughout its whole length, which prevents its having any spring and causes it to work smoothly.

Claim.—The constructing of planes with the chisel or bit set in front of its wedge, in combination with the use of a mouth-piece, substantially in the manner and for the purpose described.

No. 12,787.—George E. Davis.—*Bench-Plane Stock.*—Patented May 1, 1855. (Plates, p. 231.)

The body of the plane is made of metal, which, being very thin, presents little or no impediment to the shavings passing out, as they are cut from the wood. The lip I fills the recess caused by the bevel on the edge of the plane-iron in the ordinary plane; thereby preventing its catching and filling with portions of wood as it is used, and presenting a smooth surface to the wood.

Claim.—The metal plane-stock, having a formation of a lip I in the back part of its throat, so as to fill the recess which would otherwise be below the level of the cutting-irons, so as to present a continuous smooth surface to the plane, excepting the edge of the cutting-irons, and throat forward of them for the outward passage of the shavings, essentially in the manner and for the purposes set forth.

No. 13,174.—M. G. Hubbard.—*Method of Hanging Plane-Stocks and their Mouth-Pieces.*—Patented July 3, 1855. (Plates, p. 231.)

The nature of this improvement will be easily understood from the claims and engravings.

Claim.—The construction and attachment of the plane-stock and pressure-bar (*b*) to the bed-frame of the planing machine, by which the pressure of the bar is regulated, and the bar and plane-stock can be turned back to sharpen the plane, (see figure 3); also, jointing the floating or front plane to springs or bars turning on a pivot to rise or fall, as above set forth, over the inequalities of the boards, as above specified.

No. 13,381.—Leonard Bailey.—*Plane-Scraper.*—Patented August 7, 1855. (Plates, p. 231.)

The nature of this improvement will be understood from the claim and engravings.

Claim.—Combining the scraper, or plane-cutter, with the stock, by means of the movable holder and its adjusting mechanism, substantially as specified.

No. 13,575.—Horace Harris.—*Improved Plane-Bit.*—Patented September 18, 1855. (Plates, p. 232.)

A is the plane-iron, and B the bit.

I *claim* the adjustment of the cap and bit, with the grooves at each side, and of the thumb-screw at the top of the cap and bit for the regulation of the cut of the bit, while the iron is held fast in the stock by the wedge-fastening.

No. 13,626.—Hiram Taylor and John C. Taylor.—*Cooper's Crozing-Plane.*—Patented October 2, 1855. (Plates, p. 232.)

The stock 2 (cast in one piece) serves as a guide for the bit 3, and is provided with a groove for the reception of said bit and the wedge 4; the object of which arrangement is to simplify and cheapen the construction.

The inventors say: We do not claim adjusting a bit by a wedge;

But we *claim* casting the stock in one piece, as described, and combining therewith a wedge, for the purposes set forth.

No. 13,745.—I. Henry A. Bleckmann.—*Bench-Plane Iron.*—Patented November 6, 1855. (Plates, p. 232.)

The cutting-iron D is placed between two iron plates A B, and can be regulated by means of set-screws J and slots M. To keep the plates more firmly together they are provided with screws C. By this means well-tempered steel plates can be used, which have not been injured by welding them to other plates.

Claim.—The placing of a piece or a plate of steel between plates of iron, forming a plane-iron, for the purpose and in the manner above described.

No. 13,757.—William Nixon.—*Improved Cutter-Head for Rotary Planers.*—Patented November 6, 1855. (Plates, p. 232.)

The cutters a a^1 are attached to the cutter-stock g, and are provided at their cutting-edge with a double bevel a c and a b. In front of the cutting-edge, and at the termination of the bevel at b, the cutter-stock rises abruptly, thereby throwing the shavings back and preventing the cutter from entering deeper into the material to be planed, so that it supplies the place of a cap-iron and a firm support for the cutter.

Claim.—The double bevel of the cutter in combination with the bevel on that part of the stock or cylinder which is in front of the cutter, so that the stock may act as a cap-iron to the cutter, and to clear the shavings, as set forth.

No. 13,957.—John P. Robinson.—*Plane for Finishing Grooves in Patterns, &c.*—Patented December 18, 1855. (Plates, p. 232.)

By having the plane-stock of a triangular form, and the cutting-iron of corresponding shape, rounded grooves D can be cut, as will be understood from the engravings.

Claim.—Constructing the plane-stock A of triangular or three-sided prismatic form, the two lower sides forming a greater or less angle with each other, and the plane-iron B fitted in the stocks as shown, for the purpose set forth.

No. 12,162.—M. G. HUBBARD.—*Improved Mode of Hanging the Knife in Planing-Machines.*—Patented January 2, 1855. (Plates, p. 232.)

This improvement consists in hanging the front knife *b*, which is to remove the surface of the material to be planed, (that contains the grit, &c., which would dull the succeeding knives,) on the ends of two arms *a* that are atached to the permanent frame a considerable distance in front of the knife, so as to allow the knife a play up and down, without much deviation of the inclination of the edge of the knife.

Claim.—Hanging the first knife on arms projecting from the stock, horizontally, or nearly so, by which it is attached to the frame, substantially in the manner and for the purpose set forth.

No. 12,211.—CYRUS B. MORSE.—*Rotary Planing and Matching Machine.*—Patented January 9, 1855. (Plates, p. 233.)

The board (in the condition in which it leaves the saw-mill) is placed on its edge by the side of the belt *m*. The divisions on the scales *k*, to which the upper edge of the board reaches, are then noted. Should the ends be unequal in width, the lever *o* is moved until it points to the same number on arc P, which is shown by the narrowest end of the board on one of the scales *k*, and secured there by letting the pin p^1 drop into the corresponding hole of the arc. The other end of lever *o* will then have moved the carriage M the proper extent to permit the cutters J to operate on said width of the board. The machine being in motion, the cleat *n* will carry the board forward beyond the facing-wheel A, and between the two sets of cutters J, which will cut the grooves.

As the board leaves the tonguing cutters, the cleat *n* on the opposite portion of belt *m* comes in contact with the small stud m^2 on rock-shaft T, slightly turning said shaft and thereby expanding spring O^2; which spring, by its contraction when the stud slips from the cleat, causes the hammer n^2, on an arm of shaft T, to strike the ball U, and thus announce the finish of the board.

The inventor says: I do not claim any particular form, size, or number of the mechanical devices; neither do I limit myself to any exact combination or arrangement of the same, so long as the objects are obtained without changing the principle of operation. But I do *claim* the combination and arrangement of the following mechanical elements for the purpose of preparing, or reducing and tonguing plank or boards, whether in combination with planing or grooving the same or not; that is, the adjustable cutter-carriage H, carrying the reducing and tonguing cutters J, graduating lever O, segmental scale P, and scales *k k*, with the indicating apparatus T, U, n^2, or their equivalents, when arranged and combined for the objects herein set forth.

No. 12,438.—Wm. B. Emery.—*Method of Adjusting Stuff in Planing-Machines.*—Patented February 27, 1855. (Plates, p. 233.)

B are the teeth projecting from the bed-plate A, and the pieces D to be planed are fastened between the said teeth by wedges *c*. The whole is then passed under the cutters of any planing machine which will reduce the surfaces of the pieces D to one common plane, and all the pieces to a common width and a common angle. By turning the pieces the several sides may be successively planed.

Claim.—1st. The bed-plate A, of iron or other suitable material, provided with teeth projecting from it, and adjusted at suitable angles, together with the wedges, or their equivalents, constructed substantially as described.

2d. The bed-plate and wedges, substantially as described, in combination with any suitable planing-machine.

No. 12,880.—Leonard Tilton.—*Device for Adjusting Planing-Machinery.*—Patented May 15, 1855. (Plates, p. 233.)

The lower feed-rollers D run in fixed bearings. The bearings *a*, of the upper feed-rollers D^1, are movable. The upper ends of the vertical sliding-rods *b* have short horizontal arms *c* attached to them, which arms pass through the bearings *a* above the journals of the rollers D^1, so that the bearings can turn on said arms. Thus, by raising or lowering the rods *b*, the distance between the feed-rollers can be regulated so as to suit different thicknesses of board; and one bearing can be raised higher than the bearing on the other end of the feed-roller, so as to give it an inclination (as shown in dotted lines in figure 2) to correspond with boards which are thinner at one side than at the other.

The cutting-cylinder B is raised or lowered by turning shaft M, both ends of the cylinder being operated upon at the same time; and by turning the rod O, the clamps N are made to bind against the rods *i i* and secure the cylinder at the desired height, all which is apparent from the figures.

The cutters R and R^1 cut the tongue and groove. R works in stationary bearings. The bearing *p r*, of the axis of cutter R^1, is sufficiently elongated to allow a guide-rod T in front of the cutter, and a screw-rod S in rear of it, to pass through said bearing *p r*. By turning the screw-rod, the cutter R^1 can be moved nearer or farther off the cutter R, so as to suit different widths of boards.

The inventor says: I do not claim any of the parts of the within described machine, irrespective of the devices herein shown, for admitting of the adjustment of the feed-roller and cutters, with the exception of these devices. The parts described are substantially the same as the Woodworth and other planing-machines. I do not claim the feed-rollers and cutting-cylinders for planing planks or boards, for they have been previously used.

But I do *claim*, 1st. Hanging the axes or journals of the upper feed-rollers D^1 in suspended bearings *a*, attached to the rods *b*, in the manner and for the purpose as herein shown.

2d. Adjusting the cutting-cylinder B, by having its bearings *h h* attached to vertical slide-rods *i*, operated by the bevel-wheels *m l* and screws *j*, and securing the cylinder at the desired height by means of the clamps N N and rod O, as herein shown and described.

3d. The employment or use of the guide-rod T and guide *r*, arranged as shown, for the purpose of keeping the adjustable cutter R^1 firm and steady while operating, and causing it to be adjusted with facility.

No. 13,345.—Nelson Barlow.—*Improved Method of Feeding Planks to Planing-Machines.*—Patented July 31, 1855. (Plates, p. 233.)

E and F are the feed-rollers; M the plank; D the cutter-cylinder. The under feed-roller E is attached to the main frame of the machine in an unyielding position; the upper feed-roller F has its bearings in the frame B, its under side being in true line with bearing-plate *b*; frame B is attached to stands C, which latter can be swung forward or backward, the cutter-cylinder shaft serving as a fulcrum; adjusting-screws and guides allow the frame B to be depressed or raised to suit planks of different thicknesses. The plate *b*, at the lower part of frame B, extends from side to side between the stands, and bears upon the upper surface of the plank. After the plank passes the cutter cylinder, it rests upon, and is supported by, roller G. As this roller is connected with the stands, and they are fulcrumed on the cutter-cylinder shaft, it follows that the roller occupies a fixed relative position to the under side of the plank and to the cylinder, no adjustment of this being necessary when changes are made for different thicknesses.

Claim.—The self-adjusting frame B, connected by axles, or any equivalent means, to the main frame, when combined with the cylinder and fixed rollers, as specified.

No. 13,536.—Solomon S. Gray, assignor to Himself and S. A. Woods. —*Improved Universal Dog for Planing-Machines.*—Patented September 4, 1855. (Plates, p. 234.)

This clamp can at any instant be converted into a stationary clamp without in the least limiting its capabilities as a swivelling or pivoted clamp. Figure 2 represents the clamp when arranged for planks of different lengths.

By arranging the screw F, which forces the dogs against the ends of the planks, above the bed I of the machine, and of the dogs which hold the planks, the dogs are forced down upon the bed, and the plank is prevented from rising.

I *claim*, 1st. The arms $f\ f^1$, in combination with a pivoted clamp, whereby it is rendered rigid when desired, as described.

2d. I claim placing the screw, which forces up the clamp, above the level of the dogs, for the purpose set forth.

No. 13,618.—SETH C. HURLBUT and WESTEL W. HURLBUT.—*Improved Feed-Motion for Planing-Machines, &c.*—Patented October 2, 1855. (Plates, p. 234.)

The spur-wheels 10 are fast to the shafts of the vertical feed-rollers A^1 A^1, B^1 B^1, and are moved by worm-wheels L. Thus it will be seen the feed-rollers can move sideways so as to accommodate various thicknesses of stuff, without getting out of gear.

The inventors say: We *claim* the application of the worm-wheels in connection with the spur-wheels attached to the shafts of the feed-rollers, to effect their proper revolution, and to admit of their opening apart to receive various thicknesses of lumber, as above described.

This application we claim as novel and as our invention, in connexion with the feed-rollers of a planing-machine.

No. 13,808.—JAMES A. WOODBURY.—*Improvement in Planing-Machines.*—Patented November 13, 1855. (Plates, p. 234.)

b c d are the pressure and guide-rollers; e e^1 are the edge-cutters, placed upon the shafts *f*; *g h* the bearings of said shafts. The edge-cutters can be moved either towards or away from the board, or from one side of the machine to the other, by means of the horizontal screw-shafts *i k* which pass through the bearings g g^1, the upper screw-shafts *i* working in a female screw formed in bearing *g*, and the lower *k* in a female screw formed in bearing g^1. *l l*, *m m*, are the guides against which the board bears as it passes through the machine, one guide being provided on each side of the machine. Thus the board can be fed in upon either side of the machine, the edge-cutters being changed from one side to the other, as may be desired, by applying a winch to the ends of the screw-shafts by which they are actuated.

Claim.—Making both of the edge-cutters adjustable, in such manner that the board can be fed in from either side of the machine, double guides being provided for the purpose, as specified.

No. 13,947.—WILLIAM W. JOHNSON.—*Machine for Planing Felloes.*—Patented December 18, 1855. (Plates, p. 234.)

The arms E and F are hinged at O, and graduated and constructed as seen in the engraving; by which means the centre upon which the plane C performs its arc is kept concentric with the radius of the required felloe.

Claim.—The combination of the lever E, sliding in the arms A, graduated as shown, with the graduated lever F and hollow cylinder or barrel G, E and F being hinged for the purposes set forth, or any device which is substantially the same.

No. 13,237.—CHARLES M. SWANY.—*Gauge for Stair-Rails.*—Patented July 10, 1855. (Plates, p. 234.)

The two tubes B^1 B^1 are set the required distance apart by means of set-screws *b*, and the front of the guide F is placed against the side of the stair-rail. By moving the tool along, two lines are described at the same time at equal distances apart. In consequence of the guide F yielding, it will conform to the variable inclination of the side of the rail, it being understood that at the curves or "wreaths" the position of the side varies in a somewhat spiral manner. Thus the workman can square or dress the rail without making the guide-lines from the centre as heretofore.

The inventor says: I do not claim the adjustable spring-pencils separately; for they, or their equivalents, have been previously used.

But I *claim* the combination of the yielding or movable gauge F and adjustable spring-pencils D D, arranged as shown and for the purpose set forth.

No. 12,891.—George B. Ambler.—*Improvement in Wooden Saddle-Trees.*—Patented May 22, 1855. (Plates, p. 234.)

The object of placing the gullet-hook D, with its flat part *c*, within the slot B, and securing it by a screw E and screw-thread b^1, (as shown in the figures,) and having its flat part form, as it were, part of the tree-head itself, and its top and bottom stand even with the *upper* and *under* surfaces of the tree-head, instead of placing the gullet-hook *under* the head and confining it by a nut, is to make the appearance of the front part of the saddle more symmetrical, and to lessen the liability of the gullet-hook to work loose and turn.

Claim.—Providing the slot B in the centre of the head of a wood saddle-tree, and the screw-thread in the top strap C^1; said slot serving to receive the end *c* of the gullet-hook, and the screw-thread serving to firmly hold the screw, substantially as and for the purposes set forth.

No. 12,797.—Solomon P. Smith.—*Machine for Clamping Sash, etc.*—Patented May 1, 1855. (Plates, p. 235.)

The sash J to be clamped is laid horizontally upon the upper part of the frame A, two adjoining sides of the sash being against the heads or bars B B. The outer end of the lever H is then depressed by the foot, and the head F, underneath the frame A, is raised, and the levers e^1 e^2 force outward the lower ends of the clamps D, and cause their upper ends to press against two sides of the sash, which is consequently firmly secured between clamps and heads D and B. The heads B, by being secured to studs C, may be adjusted nearer or farther from frame A, so that different sized articles may be clamped with the same apparatus. The studs C rest upon cross-pieces *a* of the frame, and are adjustably secured to them by bolts *b* passing through plates *c*, attached to the slides C, and through holes a^1 in the cross-pieces *a*.

The inventor says: I do not claim separately the adjustable heads B B, neither do I claim operating the clamps D by a toggle-joint, irrespective of their connexion with the adjustable heads and the arrangement herein shown; but I do *claim* the employment or use of the clamps D D D D, operated by levers e^1 e^2, connected to a head F, and forming what is commonly termed a toggle-joint, in combination with the adjustable heads B B, the above parts being arranged and operating in the manner and for the purpose as herein shown and described.

No. 12,126.—THOMAS J. FLANDERS.—*Construction and Mode of Driving Circular Saws.*—Patented January 2, 1855. (Plates, p. 255.)

Driving-shaft H revolves face-gear D, which drives spur-gear B C, the teeth of which gear between the saw-teeth or into holes N in the saw-plate A^2. Thereby the saw-plate is revolved without being provided with a central shaft, and it is steadied by means of guide-rollers *c c*. The saw-plate A^2 is provided with projections to which the teeth are riveted, which teeth are fitted to recesses made alternately in each side of the plate, to one-half of the thickness of the latter, the projections and teeth being only one-half the thickness of the plate also.

Claim.—1st. Supporting a circular saw edgewise, and operating it at the same time by means of two spur-gears, or their equivalents, arranged and operated at right-angles to the saw, so that the said gears act upon the plate of the saw between the teeth or through holes in the plate to propel and support it edgewise at the same time, as described, thereby dispensing with the shaft, or its equivalent, heretofore used to propel and support the saw (see figures 2 and 3).

2d. Crimping or corrugating the plates of saws, substantially as described, (see figure 1,) so that they will require little or no setting, and to make them stiffer; also, that the bands may run in contact with the sides of the score-cut, and support and steady themselves so as to be less liable to be swerved by knotty, cross-grained, or hard places in the wood or material sawed, and at the same time run with less friction and power.

3d. Making the teeth of crimped or corrugated saws separate from the plate, and fastening them into the recesses formed by the crimping or corrugating, or between the bends in the plate of the saw, so that the saw may be supported and steadied by roller-guides, or otherwise, substantially as described.

4th. Making the teeth of saws one-half the thickness of the plate, or less, by taking off the side of the tooth and the opposite side of the next tooth, as described, so that they will cut one-half, or less than one-half, the thickness of the saw-kerf, so that the saw will run at a higher speed without heating, and execute a given quantity of work with two-thirds or less of the power heretofore required, substantially as described.

No. 12,170.—M. A. TAPLIN.—*Improved Method of Flanging a Path-finding Saw.*—Patented January 2, 1855. (Plates, p. 235.)

The upper end of the saw D traverses a vertical slit *m* in a pendulum block U, secured by a pivot *n*, immediately over the saw, and in the same plane. This block extends downwards as far as it can, and clear the top of the largest logs; and it has also a vertical slot *s* at right-angles to the slit *m*, in which a cross-head *w*, secured to the upper end of the saw, traverses, to keep the saw vertical in one direction, while the slit keeps it so in the opposite direction. If the saw should at any time tend to run slightly to one side, as all saws do occasionally, the block U will yield by turning on its pivot *n* to allow the saw to move aside, or to assume a curve in the log. By this means the binding and heating of the saw is to be avoided.

Claim.—The pendulum block to support and guide the upper end of the saw, substantially as described.

No. 12,176.—LYSANDER WRIGHT.—*Sawing-Machine.*—Patented January 2, 1855. (Plates, p. 236.)

The saw Q is confined between two cross-heads P and T. The lower cross-head T is vertically reciprocated by the usual means of pitman, crank, etc. The upper cross-head, which slides in ways L L, is constantly drawn upwards, in consequence of its being connected with a pulley U by means of a strap I[1], and said pulley (by means of a smaller pulley H, on the same axle, and straps I I) being connected to a spring C. The spring is attached to, and the axle of the pulleys is supported by, the stationary frame F G. The ways L L are fastened to a bed K, which is pivoted to a shoe 10, which latter is adjustably attached to the frame F. The hold-fast M, together with the guide-block O, is supported by rod M[1], the upper screw-threaded end of which passes through the threaded hub of pinion N, which can be revolved by means of gear-wheel N[1] and nut E, thereby raising or lowering the hold-fast. The set-screw J, which works through a vertical slot in the frame F, serves to place the shoe, bed, and slides higher or lower, so as to suit different lengths of saws. The rake is given to the saw by throwing out the bed, (see figure 3), which is pivoted to the shoe, and fixing it in this position by means of set-screw A[1].

The considerable difference between the diameters of the pulleys U and H, to which are respectively connected the spring and the upper cross-head, is the reason that, even during a long way described by the cross-head and saw, the way described by the front end of the spring will be comparatively a short one, and thus the spring will never be over-strained.

Claim.—The two pulleys, varying in size, for the purpose set forth; also, the arrangement and combination of the guides, cross-head, hold-fast, and guide-block, shoe, and screw, for the purpose of raising and lowering, and to give the rake, all substantially as set forth.

No. 12,197.—WM. B. EMERY.—*Mode of Arranging and Driving Circular Saws.*—Patented January 9, 1855. (Plates, p. 236.)

The axis F is guided by the arm N, so as to keep axis F at its proper distance from the main driving axis J, while vibrating to accommodate the motions of the saw-mandrel B, which mandrel B is sustained by a carriage D, which is allowed to slide on rods E. The dotted lines represent the two extreme positions of the mandrel and other parts.

Claim.—The combination of the three axes B F and J with the frame

M, and the guide N, arranged substantially in the manner and for the purpose described.

No. 12,337.—Pinney Youngs.—*Improvement in Sawing-Machines.*—Patented January 30, 1855. (Plates, pp. 236 and 237.)

The log to be sawed is secured between the dogs Y Y on the bar X, and motion is given to saw-shaft C; the saw B and shaft E are made to rotate, and the carriage W moves on (see arrow), the belt *o* being over the fast or working-pulley J on the shaft I. When the log has passed the saw B, the lower end of pawl w^1 strikes against the stop y^1, and the upper end of said pawl clutches the projection v^1 on the face of the toothed wheel A^1, which is consequently turned, as is also the opposite wheel A^1, in consequence of connexion d^1 e^1 f^1 h^1. The screw-shafts *z z* are turned, and the bar X and log are moved or set to the saw at a distance corresponding to the distance the wheels A^1 were moved. A pin a^2 in the carriage now acts against the slide *x*, and the belt-shipper L is moved; the belt *o* is thrown off the working-pulley J and on one of the loose pulleys K, the cross-belt *p* being at the same time thrown on the working-pulley J; and the carriage moves back. The slide *x*, when moved, acts against the button O; and the belt-shipper, being connected to the button, is operated accordingly. The position of the rods *g g* is also changed at the same moment, as the sliding-plate *j* is connected to the button by arm z^1, and the levers D D are shifted or changed, and the guides which were previously in contact with the saw are thrown out from it, and the opposite pair brought in contact with it at its opposite end; one pair of guides being at all times against the cutting edge of the saw.

On the outer end of each shaft B^1, and between the outer surface of the side-piece of the carriage and the toothed wheel A^1, there is secured a lever l^1. The upper ends of these arms bear against pins m^1 m^1, which are attached to slides n^1 n^1, which work on segments o^1 o^1 attached to the outer side-piece of carriage W. The spiral springs j^1 j^1 keep the levers l^1 l^1 against the pins m^1 m^1. On the extreme end of each shaft B^1 there is secured a collar p^1, having an arm q^1 provided with a slot r^1 at its lower end, in which slot r^1 a pin s^1, attached to the outer surface of a small plate t^1, fits; this plate t^1 has a projection u^1 on its inner face, which fits over the circular projection v^1; on the outer surface of wheel A^1, and directly below projection u^1, there is a pawl w^1, the upper end of which bears against the outer edge of the projection v^1 when said pawl w^1 is in the proper direction.

Claim.—1st. The employment or use of two pairs of guides *e e* secured to the ends of levers D D, and arranged as shown, or in an equivalent way, so that said levers will be operated by the movement of the carriage, and each pair of guides brought alternately in contact with the saw near its cutting edge; the levers D D being operated simultaneously with the reversing movement of the carriage, for the purpose of allowing the saw to be properly guided or stayed while cutting in either direction, as set forth.

2d. The combination of the toothed wheels A^1 A^1, arms or levers

l^1 l^1, g^1 g^1, and pawls w^2 w^1 attached to plates t^1 t^1; the arms q^1, plates t^1, and pawls w^1 forming a clutch, and so arranged as to operate the wheels A^1 and rotate the screw-shafts z, as shown and described, for the purpose of properly setting the log or timber to the saw; the movement of the wheels A^1 being regulated by adjusting the pins n^1 on the segments o^1, or in an equivalent way, so as to give the required set to the log or timber.

No. 12,412.—GEORGE P. KETCHAM.—*Method of driving Pairs of Reciprocating Saws.*—Patented February 20, 1855. (Plates, p. 237.

The two positions represented in fig. 1 (one in full and the other in dotted lines) are the two extreme positions of lever E.

Claim.—Operating the saw-sashes B B by means of the inclined wheel or cam D and lever E, with its pendents or projections *b b*; the parts to be operated being connected to the ends of the lever E by rods or pitman G; the above parts being constructed and arranged substantially as shown. (See engraving.)

No. 12,493.—LINUS STEWART.—*Improvement in Mode of Constructing Saw-Plates and Setting Teeth therein.*—Patented March 6, 1855. (Plates, p. 237.)

a a are two thin steel plates, between which are fastened plates *b b* at equal distances apart; in the space or mortise between the plates *b b* is put the bit *c*, which is secured by means of key D; the bit *c* is made with flanges on the sides E E, which project slightly over the thickness of the plate to give clearance to the saw.

Claim.—The improved mode of constructing saw-plate and fastening of the bits therein, as described; that is, the bits shall be so made and arranged with projections on each side equal to the set of the saw, and fastened therein with a key or other known modes of securing the same.

No. 12,664.—NELSON BARLOW.—*Saw-Teeth.*—Patented April 10, 1855. (Plates, p. 237.)

A recess D is formed on the front angle of the tooth, extending inwards as far as the roots of the teeth, if desirable, and parallel with the side of the saw; the shape of which being concave or angular, gives acute edges upon the sides of the teeth; to give the best effect to which improvement, the back angles of the teeth are rounded in a form corresponding with the other, so as to enlarge the cutting-surface of the points, and to render each tooth so formed a perfect cutter on the different angles presented to the wood.

The inventor says: The advantage of this improvement is, that the portion of wood removed is retained between said cutting-edges, and carried forward; the tendency being to draw inward from the cutters, the wood at the same time being cut smoothly upon its sides.

Claim.—The within described improvements, consisting of the recessed space D and combined cutters upon the sides of saw-teeth, and also the rounded form given to the outer points of the teeth, when arranged, formed, and operating substantially as herein described.

No. 12,679.—CHARLES B. HUTCHINSON.—*Mode of Guiding Reciprocating Saws.*—Patented April 10, 1855. (Plates, p. 237.)

G is a guide-plate of the thickness of the saw-plate, and firmly secured at each end to the stationary cross-timbers H. The saw C slides through guide-straps E projecting from plate G. Straps D, attached to the saw-plate, pass around the guide-plate, to hold the saw-plate snugly to its place in front of the guide-plate. B is the pitman.

Claim —The use of a thin guide-plate for holding and guiding the saw, placed immediately behind and in the same plane with it, and following it through the log, whether the same be used by itself or in connexion with any other means of holding and guiding the saw.

No. 12,705.—HIRAM WELLS.—*Device allowing Circular Saw-Spindles to yield.*—Patented April 10, 1855. (Plates, p. 237.

This improvement consists in providing the saw-spindle A in its boxes B with an angular frustro-conical groove *a*. A spring-guide *b*, sliding freely in hole *c*, bears with its upper part *f* against the central part of groove *a*. This arrangement allows the spindle a slight lateral movement, by one of the sides of the grooves slightly depressing the guide. The guide, however, having a tendency to again assume its central position in the groove, will draw the spindle back to its proper position. The shoulders *k* prevent an excess of lateral motion by eventually abutting against the sides of the projection *f* of the guide.

Claim.—Arranging within the box or bearing, and combining with it and the saw-spindle, substantially as specified, the guide, its spring, and the compound frustro-conical grooves provided with shoulders, as described ; the whole constituting a device of great simplicity of construction, and of much advantage in not only allowing a circular saw while in operation to move laterally, but to limit such movement of it, and subsequently restore it to its normal or original position.

No. 12,809.—ELIAS A. TUBBS, assignor to E. A. TUBBS and H. T. CROXON.—*Machine for Sawing Fire-Wood, &c.*—Patented May 1, 1855. (Plates, p. 238.)

Starting with the saw elevated, as seen in fig. 4, and the shaft O and lever T in the relative position seen in fig. 3, the log is thrown on and secured in the movable clamp C, and power is applied to the main shaft E. The finger Q, upon the shaft O, having already closed the clutch L, the drum K feeds the log into the machine; and as the crank g^1 is in its highest position, the stationary clamp C^1 is open, the lever D being raised clear of the bent ends of the jaws a^1 b^1. The log is thus permitted to feed through this clamp until it strikes against the stop Z.

by which means the shaft X is revolved; and the arm B^1, being thereby depressed, presses the lever T out of the notch i, and into one of the teeth of the wheel R, which clutches the shaft O, with the cog-wheel N, and this shaft is caused to make a semi-revolution, by which it is brought into the position shown in fig. 2, the lever T being disengaged from the wheel R, and entering the notch h of the plate U. While this is taking place, the shipping finger Q has opened the clutch L, by which means the feed is stopped, and the pin q^1 has depressed the rod r^1, permitting the crank Q^1, which supports the saw, to escape from the catch S^1, and the saw descends upon the log and commences to operate; the crank g^1, also having descended to its lowest position, depresses the lever D^1, and closes the clamp C^1 upon the log, by which it is held firmly until the cut is made. While the saw is operating upon the log, the parts remain in the position seen in fig. 2, the crank Q^1 not being raised sufficiently high by the descent of the saw to bring the teeth of the wheel R^1 into gear with those upon the wheel P. So soon, however, as the cut is finished, the weight of the saw causes it to descend so as to raise the crank Q^1 to its highest position, whereby the wheels R^1 and P engage, the former being driven by the latter a semi-revolution, and the saw and its frame are raised into the position seen in fig. 4, the crank Q^1 resting against the stop S^1, as before. While this is taking place the shipper U^1, in revolving, presses the lever t^1 against the lever T, by which means the shaft O is again set in motion; the shipping finger, striking against the prong g, closes the clutch for the purpose of operating the feed, and the clamp C is opened as already explained; when the whole log is thus sawed up, and the rod m striking against the apparatus opens the clutch L, and stops the feed.

Claim.—1st. The method, substantially herein described, of bringing the saw into operation by the pressure of the log upon the stop Z, as set forth.

2d. The method, substantially herein described, of causing the weight of the saw, after it has passed through the log, to bring into operation the mechanism which raises it out of the way preparatory to making another feed.

3d. The method, herein described, of operating the clamp C^1 by means of the spring-bar D^1, whereby the clamp is rendered capable of holding logs of varying thicknesses without constant re-adjustment, as set forth.

4th. The device, herein described, for the purpose of stopping and starting the feed at the required moment, consisting essentially of the combination of the shaft O, the lever T, the wheel N, and the shipping-finger Q, constructed and operating in the manner substantially as herein set forth.

No. 12,821.—Simon Ingersoll.—*Machine for Sawing or Felling Trees.*—Patented May 8, 1855. (Plates, p. 239.)

Motion is transmitted from shaft and crank I to lever H J, and from it, by the link K, to the saw-frame C a C, which turns on pivot E, secured in slide F. At the same time that the saw has this circular motion, the weight M, through the cord and slide, gives the saw a parallel forward motion, and thus feeds it up to the tree as fast as the sawing is performed.

Claim.—The manner, herein described, of giving to the segmental saw its reciprocating action during the forward feed of the same by means of the pitman H, connected by link-rod K to the saw-frame, and by joint or link-rod J to the side carrying the saw or saw-frame, and giving forward feed thereto, substantially as and for the purpose set forth.

No. 12,843.—Francis A. Wolff.—*Method of Sawing a Log by its own Weight.*—Patented May 8, 1855. (Plates, p. 239.)

The back of the framing A, between the bluff and the saws, is planked up to form a bearing for the logs to slide against in their descent. The log, slightly slabbed on one side to prevent its rolling on the ways, is lowered down till its top end is about level with the upper chain-wheels *r r*. It is then secured to the endless chains *p p* by the dog *k* near the top, and its weight puts the saws in motion in a manner apparent from the figures. When the dog comes nearly down to the saws, another dog is attached below, and the weight taken off the upper one, so that it may be removed. As soon as the log is clear of the upper saws *h*, the mill is stopped by means of brake *d*, the log is turned on its side, and secured at its upper end to the chains, as before, and another log attached above as at first, and the saws again put in motion. The lower log is then operated upon by the lower saws *f*, whilst the upper log is acted upon by the upper saws.

The inventor says: I do not claim the principle of sawing the timber by machinery driven by the weight of the log in itself; but I do *claim* the method herein described of making the weight of a log or logs of timber propel the saws which saw them, by suspending them on endless chains, working around chain-wheels, which drive the saws, substantially as set forth.

No. 13,028.—Charles M. Day.—*Feed-Motion for Saw-Mills, &c.*—Patented June 12, 1855. (Plates, p. 239.)

An eccentric communicates a rocking motion to shaft G and rack H, and this motion is communicated to lever E by arm F; and as the lever E vibrates, the clamps *e* bind against the rim *d*, during its forward movement, and rotate the pulley D; and as the pinion *c* gears into rack *b*, the carriage B is moved a certain distance. Upon the backward movement of lever E, the clamps *e* do not bind against the rim *d*, but slip upon it. The carriage is gigged back by releasing the clamps from the pulley by operating levers K.

The inventor says: I do not claim the clamps *e e*, separately; but I *claim* the clamps *e e* attached to a lever E, and working upon a pulley D, in combination with the rack H and pinion *h*, attached to the arm F, all arranged, constructed, and operated as shown and described.

I also claim the levers K attached to the pivots *f f* of the clamps, and connected by the pin *j* and slot *k*, at their inner ends, for the purpose of relieving the clamps from the pulley D when the carriage is gigged back.

No. 13,040.—Isaac M. Newcomb.—*Sawing-Machine.*—Patented June 12, 1855. (Plates, p. 240.)

Reciprocating motion being given to frame D E F, the saw being raised above the log *c* and retained by the arm W and catch-rod X, the tongue K is in the notch of the slide N, which, following the oblique tongue, receives a transverse motion, which is communicated through lever L to pawl *o*, which operates ratchet-wheel P, and thereby gives a rotary motion to the pinions S S (one not shown in the figure) which gear into racks T, thereby causing the carriage to move along the ways until finger Z comes in contact with a pin *a* and is brought into a line with catch-rod X, when it unlocks X from arm I, and allows X and arm W to fall into the position shown in dotted lines in figure 2, and the saw to rest upon the log. When the cut is made, the hook J catches the orfward end of X, and, by its backward motion, lifts the saw from the log, when the hook at the back end of X locks upon arm I, and retains it until finger Z comes in contact with the next pin, when X is again unlocked, and the saw let again to operate upon the log.

Claim.—The bar F, secured to the arms E of the vibrating frame by a hinge-joint, and its connection with the saw-shaft G, combined and operating substantially as described.

Also, the guide-arms of the carriage H, combined and operating as described; and also the lifting-arm W and catch-rod X, combined with the hook J of the saw-shaft G; also the oblique tongue K and notched slide N, combined and operating substantially as described.

No. 13,043.—Andrew S. Rice and Guy Tozer.—*Saw-Mill Dog.*—Patented June 12, 1855. (Plates, p. 240.)

The teeth *c* of the head B take into the log. In each end of the movable head B are chambers in which the sectional nuts N move up and down (by means of the screw-shafts K taking into screw-threads in the nuts). These nuts, one at a time, are let down into their corresponding threads in the screw A; and as the carriage is gigged back, the wheel D on the end of the screw takes into corresponding cogs lying on a movable piece of timber by the side of the carriage, and turns the screw, moving the head B, together with the log, on towards the saw.

The inventors say: We do not claim the cog-wheel D on the screw A, or the cogs by which the wheel is moved;

But we do *claim* the combination of right and left hand screw A in connection with the movable head B, together with the sectional nut N, and the manner of operating the same, substantially in the manner specified.

No. 13,055.—Selden Warner.—*Curvilinear Sawing-Machine.*—Patented June 12, 1855. (Plates, p. 240.)

The nature of this improvement will be understood from the claims and engravings.

Figure 4 represents the stuff Y half cut.

Claim.—1st. Placing the saws J J in frames E F, which are allowed to move laterally in the saw-frame or sash B, and having said saws so attached to the frames F F as to be allowed to turn therein, said saws being turned by means of the rods *k n* attached to transverse bars *l* on the vertical rods *f*, these rods being turned by the rods *n* and levers *p*, the outer ends of the rods bearing against the pattern D, and operated as said pattern moves, as shown and described.

2d. I claim straining and attaching the saws J J to the frames E F, by having bows *h* attached to the ends of the saws and screws *i* passing through them, the ends of said screws resting upon the top and bottom strips *c* of the frames E F, as described.

No. 13,227.—F. A. Parker.—*Improvement in Saw-Sets.*—Patented July 10, 1855. (Plates, p. 240.)

The bed B is to be confined to a bench, and has a bevel *b* over which the saw-teeth are to be bent, the angle being regulated by the end of spring C, which comes down upon the saw with the hammer or jaw J when it strikes the tooth. By a pressure upon the handle of lever L the hammer is brought down upon the tooth by means of arms R and piece A, with sufficient force to fix it at the angle desired. S are stops of soft metal, which may be adjusted to accommodate teeth of different length. A can be moved by a set-screw, so as to adjust the space between the hammer and bevel to the thickness of the saw.

The inventor says: I make no specific claims on the several parts of the saw-set;

But I *claim* the arrangement of the circular spring C, the adjustable bar A, and the connecting-rods R, substantially as specified.

No. 13,351.—Frederick Field.—*Cross-Cut Sawing-Machine.*—Patented July 31, 1855. (Plates, p. 241.)

The walking-beam *b*, which supports the two saws *e d*, has its fulcrum in axle *c*. Motion is transmitted from pulley *k* to pulley *l*, and from *l* to the saws. *g* is a balancing weight. The feed-shaft *p* has a loose pulley *q* and clutch, and receives motion (when the pulley is in gear with the clutch on the shaft) from the driving-shaft by means of the belt on the pulley. The clutch and belt are not shown in the engravings. When the saws have descended below the under surface of the log, arm a^1, projecting downwards from the walking-beam, comes in contact with one side of lever c^1 pivoted at b^1, and connected at the other end with the fork that moves the clutch on shaft *s*, withdraws this clutch from pulley *y*, (which latter, by means of the clutch, revolved together with shaft *s*, and drew the walking-beam down by means of a rope r^1,) and the walking-beam and saws rise by the action of weight g. When the walking-beam rises, the lever w is caught in a notch in spring-bar j^1, (attached to the walking-beam,) and, in being lifted and removed from upright bar v, allows the clutch q to throw the feed-pulley into gear. The log then advances until it strikes an adjustable lever in front of it, which throws the spring-bar j^1 back, by means of which movement y will be thrown into and q out of gear.

Claim.—The arrangement of the two circular-saws, hung in a vibrating frame, and operated substantially in the manner set forth, in combination with the mode, substantially as described, of throwing the feed-motion in and out of gear.

No. 13,277.—Matthew Ludwig.—*Machine for Sawing Down Trees.*—Patented July 17, 1855. (Plates, p. 241.)

In order to adjust the jaw for standing trees, the bar H is withdrawn from socket N, turned, and replaced in the socket. In this case the bar I and saw J rest upon bar H, (see fig. 2,) but the friction-roller of the lever L still bears against the bar I, and keeps the saw to its work, the saw cutting in a horizontal direction.

Claim.—Attaching the connecting-rod F to a sleeve G, which works upon a bar H attached to the framing A, the bar I of the saw J being attached to the sleeve G, and the bar H having an arm K attached to it, which arm has a lever L attached, one end of which is provided with a friction-roller, which bears against the bar I and keeps the saw to its work, in consequence of the cord e and weight M attached to the opposite end. The bar H being arranged as herein shown, so that it may be turned and allow the saw J to cut in a vertical or horizontal position.

No. 13,305.—Benj. Fulghum.—*Sawing-Machine.*—Patented July 24, 1855. (Plates, p. 241.)

The circular saws being set in motion, and the saw-frame G being drawn along between the guides of frame G^1, the two saws cut a strip out of the log, leaving a vertical shoulder a^1 on the log. (See fig. 4.) When the saws have reached the end of the log, the clutch i is closed and the saw-frame is moved back by the operation of shaft V. Just previous to this, the plate N^1 is moved in consequence of the frame G operating rod O^1, and lever M^1 is moved thereby, and the bar s turns wheel K^1, and the log Y is turned a suitable distance, and frame G^1 is also lowered by the bevel-gear B^1 D^1 E^1 H^1 I^1 and screw-rods C^1 I^1. These movements of the log and frame G^1 prepare the log for the succeeding cut, and cause the log to be cut in a spiral form (see dotted lines in fig. 4). As the saws approach the centre of the log, the feed-motion that turns the log must be increased, in order to have the stuff sawed of an equal thickness. This is effected by having lever M^1 pass through the slotted plate N^1, which plate is attached to frame G^1, and, of course, at every stroke of the saw-frame and depression of the frame

G^1, is brought nearer the pin *w*, and thus the length of movement of the arm L^1 is gradually increased.

The inventor says: I do not claim the arrangement of swinging frames shown, for giving a horizontal reciprocating motion to the saw-frame G, for that has been previously invented, and was formerly patented by me.

But I *claim*, 1st. Placing the reciprocating saw-frame G, provided with a vertical circular saw H and a horizontal circular saw J, within a frame G^1, and placing or centering the log Y to be sawed between shafts X X^1, the frame G^1 being lowered at every stroke or vibration of the saw-frame G, and the log Y turning simultaneously therewith a gradually increasing distance at every stroke or vibration of the saw-frame G, for the purpose of cutting the stuff an equal thickness direct from the log and in a spiral manner, substantially as shown and described.

2d. I claim operating the log Y and frame G by means of the bar *s* attached to the wheel K^1, as shown, said bar being attached to the arm L^1, which is secured to the lever M^1, when said lever M^1 is passed through a slotted plate N^1 to the frame G, attached for the purpose of gradually increasing the feed-motion of the log, and causing the stuff to be sawed of an equal thickness, as described.

No. 13,354.—Liveras Hull.—*Machine for Sawing Ratan.*—Patented July 31, 1855. (Plates, p. 241.)

The screws C serve to set the bearer D at an angle with B, which of course will cause the stick of ratan R to be sawed in a diagonal line, so as to enable it to be used in the manufacture of whip-stocks. After a stick of ratan has been cut, the broadest ends of the two pieces are laid together, which will form a whip-stock of regular taper. (See fig. 3.)

The inventor says: I am aware that machines have been contrived for splitting a ratan longitudinally with one or more knives, the ratan having been supported between a series of rollers. I am also aware that timber, attached to a rectilinear moving carriage by dogging contrivances applied to its end, has been cut diagonally by means of a saw. I am aware, also, that it is not new to use an adjustable gauge-bar in connection with a movable carriage and saw. I am also aware that pressure-rollers are used in planing-machines, for maintaining a board against a movable carriage or bed during the operation of planing or dressing it.

The employment of such parts in a machine for sawing ratan requires a specific arrangement of them, or one which differs essentially from their arrangement in various other kinds of mechanism, such an arrangement having been before explained.

I therefore *claim* the above described arrangement of the rectilinear moving-carriage B, the adjustable holding-bearer D, the groove pressure-roller, and the saw, whereby, when a stick of ratan is clamped to the adjustable bearer, and the carriage B is moved forward so as to carry the said stick endwise against the saw while the latter is in evolution, such stick shall be sawed in a diagonal direction, in manner and for the purpose as specified.

No. 13,357.—Fielding H. Keeney.—*Circular-Saw Mandrel.*—Patented July 31, 1855. (Plates, p. 241.)

The saw B can be set at any angle by means of the set-screws *a a*, as will be understood from an inspection of the engravings.

Claim.—The mode of making a mandrel, as set forth, not confining myself to exact size or shape as described, but to the principle of the machine as herein set forth, or any other equivalent device to produce the same effect.

No. 13,390.—A. F. Gray and J. C. Fincher.—*Gauge Attachment for Hand-Saws.*—Patented August 7, 1855. (Plates, p. 241.)

This gauge serves to regulate the depth of the saw-cut.

Claim.—Attaching to one side of the blade of a hand-saw, a gauge formed of two strips *a b* and lugs *c*, having slots *d* made in them, through which slots set-screws *e* pass, the screws also passing through the saw-blade, substantially as shown and for the purpose set forth.

No. 13,399.—B. E. Parkhurst.—*Machine for Sawing Lumber.*—Patented August 7, 1855. (Plates, p. 242.)

The teeth of the pinion and rack moving the lumber-carriage I, when the latter were rigidly attached, would be liable to be stripped when suddenly thrown into gear. To prevent this, the rack is attached to the carriage by means of projections passing from the under side of the carriage through mortises in the rack-bar, the mortises being somewhat longer than the projections, so as to allow the rack-bar a slight longitudinal motion before it starts the carriage. Dogs P R serve as bearings for one end of each of the transverse shafts *a b*, which carry the pinions for moving the carriage back and forth. The dogs are pressed upward by springs; by shipping-handle N and bar O, one of the buttons *s* and *t* upon the stops will be brought beneath one of the recesses *u v*, when the stop will rise and the pinion engage with the rack-bar. The dogs B^1 C^1 slide upon bar F^1, and are provided with bolts f^2 drawn up into one of the notches g^1 of bar F^1, by springs e^1. When it is desired to move the dog, the bolt b^2 is depressed out of notch g^1, when the dog may be moved and secured to another notch of the bar. The dog G^1 is connected with guide P^1 by joint f^1, which permits the dog to be thrown out and in by hand-wheel Q^1 and gear d^1, without disengaging the rack H^1 from wheel I^1. Thus the dogs may be extended to a greater or less distance from each other, to accommodate different lengths of logs, by operating the hand-wheels L^1, at the same time that they may be moved out and in by hand-wheels Q^1. When the logs are short, an additional central dog S^1 is used, which can be made to protrude by means of hand-wheel U^1 and screw T^1. To hold the logs, when long, with sufficient steadiness near the centre, the points of screws V^1 are forced into the rear of the log after it is dogged to the carriage.

The saw-guide H^2 is pivcted at q^2 to a standard G^2 projecting from the main frame, so that when the log should be of more than usual size it will strike the guide and move it around pivot q^2 out of the way. (See dotted position of H^2 in figure 2.)

Claim.—1st. The method described of connecting the rack-bar to the carriage, so that the bar may have a slight motion, independent of the carriage, for the purpose set forth.

2d. The dogs P R, constructed and operated as described, in combination with the notched bar F^1, whereby they may be instantly moved, and set to accommodate them to different lengths of log, as set forth.

3d. The described method of connecting the dogs with their sliding guides P^1, whereby they may be operated longitudinally and transversely, in the manner set forth.

4th. The pointed screw-dogs V^1 V^1, operating in the manner substantially as set forth.

5th. The saw-guide H^2, so constructed as to be thrown out of the way by the log, in the manner set forth.

6th. The double dog S^1, which, when out of use, may be sunk flush with the surface of the head-block, and may be run in and out in the manner described, for the purpose of sawing the butt and point of shingles.

No. 13,422.—A. Brown and Abel Coffin, Jr.—*Mode of Straining Saws by Atmospheric Pressure.*—Patented August 14, 1855. (Plates, p. 242.)

The valves *a* a^1 do not operate with every reciprocating action of the saw, but only when it is necessary to clear the cylinders of air; and as these valves operate freely, and when no cut is on the saw, and only reciprocating the saw slowly to clear the cylinders, the saw will not be buckled by the operation of clearing the cylinders, while, when once cleared, the pistons in them will, by the atmospheric pressure on their exposed faces, strain the saw.

The inventors say: We claim nothing new of itself, in the arrangement of the cylinders and straining-pistons in relation to the saw, and straining the latter by atmospheric pressure, produced by a vacuum maintained between the pistons and cylinder-head, as such has before been done; and steam and compressed air have likewise been similarly employed.

But we *claim* providing the closed ends or heads of the cylinders B B^1 with snifting-valves *a* a^1 opening outwardly, as described, and freely hung or operating when the said cylinders, with their pistons, are arranged in relation to the saw, and the pistons and valves operate together as specified, whereby the reciprocating action of the saw is made to clear the cylinders without the aid of a separate exhaust-pump, as set forth.

No. 13,444.—Samuel R. Wilmot.—*Portable Steam Sawing-Machine.*—Patented August 14, 1855. (Plates, p. 242.)

The cylinder B is pivoted to the main frame in b^1 b^1; F is the boiler, and G a flexible steam feed-pipe; the rear dogs *a* of the frame are inserted into the ground, and the front dogs *b* are inserted into the log. The weight of the cylinder and saw are sufficient to feed the latter.

Claim.—1st. The attachment of the opposite ends of the steam cylinder to that at which the saw works, by a pivot, to the main frame or bed-piece A of the engine; so that the cylinder may swing to and from the said frame or bed-piece, for the purpose of allowing the saw a proper range and feeding it, substantially as described.

2d. Furnishing the main frame or bed-piece A of the engine with dogs *b b*, to drive into the tree or log on opposite sides of the saw, substantially as described, whereby greater stability is given to the engine and saw when in operation, without making the engine or its frame of great weight.

No. 13,531.—Elias Strange and Thos. B. Smith.—*Machine for Sawing Hoops.*—Patented September 4, 1855. (Plates, p. 242.)

The arrangement of saws serves to saw strips of equal thicknesses throughout at opposite sides of the poles in the process of sawing barrel-hoops from poles.

Vertical reciprocating motion is given (by means ot pitmans N) to the sashes F E and the saws G G, and the pole S fed in between friction-rollers Q H. As the pole increases in thickness from its point towards the butt, the movable sash F is forced outward in consequence of the pressure against the feed-roller H attached to the movable sash. The spring K causes the roller H attached to F to bear at all times against the pole.

We *claim* the employment of two reciprocating saws G G, arranged as shown, viz: one saw being secured in a laterally sliding sash, and the other in a permanent sash, or one which only has a reciprocating motion in a vertical direction.

No. 13,544.—Robert S. Eastham.—*Improvement in Saw-mill Carriages.*—Patented September 11, 1855. (Plates, p. 243.)

The nature of this improvement consists in providing the carriage of a saw-mill with steadying strips 4, which are forced up against the log 22 by means of wedge-blocks 3 and rods 5, and held in place by a catch 8 and ratchet 6, thereby preventing the log from springing. Each wedge-block, being clamped to the rod 5 by spring 7, will move with the rod until the strip 4 has been sufficiently raised to come into contact with the log, when the block will remain in place and the rod 4 will slide through the block until all the strips have been raised into contact with the log. Then the face-side of the ratchets 6 is turned up, when the catches 8 will fall into the teeth of the ratchet and hold the strips up to the log.

I *claim* the wedge-blocks 3 3 3 3, worked by the rods 5 5 and springs

7 7, for elevating the steadying strips 4 4 up against the log, for holding it steady while being sawed; the whole being operated by the machinery described and represented, for the purposes stated.

I also claim the combination of the catches 8 8 attached to the bottom of the wedge-blocks and ratchets 6 6, for holding the wedge-blocks to their place after elevating the strips against the log, for the purposes stated.

No. 13,554.—O. S. Woodcock.—*Improved Method of Operating Reciprocating Saws.*—Patented September 11, 1855. (Plates, p. 243.)

The upper end of the saw B is attached by a stirrup C to the upper rail *a* of the sash. The lower rail *b* is connected permanently to the pitman D a short distance below its upper end, and the ends of the rail *b* are connected to the lower ends of the stiles *c* of the sash by journals *d*. The lower part of saw B is attached by a pin *e* to the upper end of the pitman; and as the pitman extends a short distance above the rail, the pin *e* that connects the saw and pitman will be a short distance above the pins *d*. Consequently, as the sash is moved up and down by the pitman, the saw will be thrown in and out from its work, (in during the descent, and out during the ascent,) so as to afford ample room for the saw-dust to escape during the ascent of the saw.

I *claim* attaching the lower end of the saw B directly to the upper end of the pitman D by a pin *e*, which forms a joint connection, the pitman working on a suitable fulcrum or bearing *d* below the pin *e*, substantially as shown, for the purpose specified.

No. 13,573.—Dean S. Howard.—*Improved Sawing-mill.*—Patented September 18, 1855. (Plates, p. 243.)

On the carriage A are attached two setter-blocks L, which support the log C to be sawed. Fig. 4 is a side view of one of the setter-blocks. Fig. 2 is a front view of the same in two positions on the inclined plane, same scale as fig. 4. *f* is the foundation secured to the carriage, having four vertical racks *g*, one at each corner, over which the base-frame *h* slides. On the upper side of this frame *h* is fitted the vertical piece *k*, whose base k^1 has a sliding motion in a dovetail groove transversely to the carriage, and is constructed with two vertical sliding-dogs n^1, which may be moved up or down by pinions and racks, so that when the log is laid upon the base-frame *h*, these dogs secure it above and below; the under side of the frame *h* has a friction-roller *l* attached to it in such position, that that on one of the setter blocks shall pass over the inclined ways *m*, whilst that on the other passes over the inclines *n*; these inclines are pivoted to the stationary framing of the machine at one end, and rest on movable supports *o* at the other when elevated; the friction-rollers ascend their respective inclines immediately after a board has been severed from the log, and during the passage of the end of the log by the saw the distance of its diameter, thus elevating the setter-blocks, and with them the log, to any given height, which may be regulated by the pitch of the inclines; during its rise the elbow-lever p, which is attached at its angle, by a link or jointed-rod q, to the under side of the frame h, and at one of its arms, by a bracket r and link r^1, to the foundation f, changes its position and assumes that represented in dotted lines in fig. 4; the vertical arm slipping one or more notches in the under side of the base k^1 of the sliding vertical piece k, according to the height it is raised by the pitch of the inclined plane; when at the top of which, the roller l touches support o and forces it back, and, liberating the top of the inclined plane, allows it to drop, when the setter-blocks, by the weight of the log upon them, settle down to their original position, in doing which the elbow-lever p re-assumes its former position, its vertical arm carrying forward the dogs by being locked in a notch further back in the under side of the base k^1 than that occupied by it before rising, and thus carries the log forward the thickness of the board to be cut, both of the blocks operating as above, simultaneously. The circular saw D is so hung that its position can be changed from that shown in the engravings to that represented by dotted lines in fig. 3, for the purpose referred to in the second part of the claim.

I *claim* the method of setting the log forward after each board is severed, by mechanical devices operated by the weight of the log, substantially as specified.

2d. The method described of cutting from either end of the log with a circular saw, by hanging the saw in a vibrating frame, or its equivalent, so that the axis of the saw may be above the log when cutting from one end, and beneath it when cutting from the other end, so as to cut either way against the grain of the wood.

I claim the self-setting arrangement described, whether in connexion with the circular saw, or the single or double edged reciprocating saw, as equally applicable to either.

No. 13,670.—George W. Worden.—*Gauge Attachment for Sawing-Machines.*—Patented October 9, 1855. (Plates, p. 243.)

The stuff to be sawed (represented by dotted lines in figure 2) is placed upon platform A, one side of the stuff bearing against the face of gauge E. The stuff, by being held against gauge E, will press within its slot e the arm d at the outer end of lever G, and force out from its slot e the other arm d at the inner end of the lever. The stuff is pressed towards the saw C, and the arm d at the inner end of lever G will throw the stuff out from the face of the guide E, and present it obliquely to the saw. When the saw has entered the stuff a short distance, the back end of the stuff will be past the arm d at the outer end of lever G, and the lever then swings free, and the stuff is pressed towards the saw, and a strip of wedge-form sawed from the block.

Claim.—The vibrating gauge, formed of the lever G, with arms d d attached to its ends, the lever working on a pivot c attached to one of the arms F of the sliding-gauge E, and the arms of the lever G working horizontally through the gauge E, substantially as shown, for the purpose specified.

No. 13,715.—Luther B. Fisher.—*Improved Device for Gauging and Setting Saw-Mill Dogs.*—Patented October 30, 1855. (Plates, p. 244.)

The stop *i* is first placed upon serrated plate *l* fastened to the carriage, in the notch indicating the required feed, lever *f* being in position of figures 1 and 2, resting on stud *q*, bar *h* being at that time tightly held by ratchet-rim *m*. The lever *f* is lifted, causing the rotation of the ratchet R until the bar *h* strikes the stop *i*; the lever *f* then falls, causing the tapering stud *n* to press on bar *h*, and remove the said bar from the teeth of the ratchet-rim *m*, when the bar is carried round to stop *p*. The progress of the lever being stayed by stud *q*, the bar *h* being no longer acted on by the stud *n*, spring *g* carries it between the teeth of the ratchet-rim *m*, and the ratchet is ready for the next feed. The object of this construction is, making the limit of the ratchet movement independent of the movement of the operating lever *f*; for when the amplitude of motion is governed by the striking of a stop by the lever, the slipping of the pawl entirely destroys the accuracy of the feed; and as, by wear on pawl and teeth, this accident is liable to obtain, it becomes necessary to guard against it. This is done by spring-bar *h*; for if the pawlslip several teeth, the lever will move on until bar *h* strikes the stop, and the amount of the revolution of ratchet R will be the same, no matter how often the pawl may have slipped.

The inventor says: Making no claim to ratchet, pawl, and lever for giving the feed, I *claim* the spring-bar *h* in combination with the ratchet-rim *m* and lever-stud *n*, constructed, arranged, and operating substantially as and for the purposes specified.

No. 13,755.—Horace Lane.—*Saw-Horse.*—Patented November 6, 1855. (Plates, p. 244.)

The block of wood having been placed upon the saw-horse, the lever C is to be pulled back until the cord L has drawn the slide H (to which dog I is attached) down sufficiently, so that the block will be held tight by the dog. By throwing the pawl D out of the teeth of ratchet B, the dog will be caused to rise from the block of wood by the action of the spiral spring K.

The inventor says: I *claim* the use of the spur I to hold the wood or timber in its place on the saw-horse, while the sawyer is sawing the wood or timber into fire-wood, or into short pieces.

Also, the use of the roller, the ratchet-wheel, the lever, the dog and spring, the cord, the pulley-wheel, the slide, the arbor, and the groove and spiral spring, combined with the common saw-horse, substantially as set forth, and for the purposes stated.

I do not make any claim on the common saw-horse, but for the improvements on the same, as set forth.

No. 13,716.—Isaac N. Forrester.—*Improved Method of Hanging Mulley-Saws.*—Patented October 30, 1855. (Plates, p. 244.)

The flanges or ear-guides *n n* are riveted to the saw-blade, and are so arranged that the point of connection is in a vertical line with the base of the teeth, as indicated by the arrow *z*. The lower end of the blade may be hung in the same way, or merely a hole punched in a line with the base of the teeth, and by this hole attached to the end of the pitman. The ears *n n* slide in vertical ways attached to the frame of the machine. By this means the back edge and all the blade part of the saw is free from rigidity or stiffness. The ways (called grooves or guide-places in the claim) are not represented in the engraving.

The inventor says: I *claim* the manner or mode of hanging saw-blades, by forming thereon, or attaching to the front edge only of one or both ends, devices which I term saws or guide-flanges *n n* (figure 2); and the working or applying the same in grooves or guide-places *g g*, whereby the back edge and principal part of the saw-blade is free and unrestrained, and without any rigidity or stiffness other than that of the blade itself, substantially as set forth, and for the purpose specified.

No. 13,932.—Amos D. Highfield, assignor to Himself and Wm. H. Harrison.—*Method of Adjusting Circular Saws Obliquely to their Shafts.*—Patented December 11, 1855. (Plates, p. 244.)

By changing the relative positions of the two washers E F, the saw G can be set more or less oblique. When the thinnest portion of E coincides with the thickest portion of F, the saw will be at right-angles to the spindle A.

The inventor says: I do not claim the exclusive use of oblique circular saws for cutting grooves, as such are well known; but I *claim* the employment of two bevelled washers between a fixed collar on the spindle and the circular saw, in the manner and for the purpose specified.

No. 13,938.—T. C. Bush.—*Improved Saw-Set.*—Patented December 18, 1855. (Plates, p. 244.)

One of the scores *n* (adapted to the thickness of the saw) has to be set between the arms E J, and fastened by set-screw H. The operator then applies the score *n* to the teeth of the saw, and vibrates the handle B, and bends the teeth in each direction until the gauge-screws K F strike the plate of the saw and stop it from vibrating any further, so as to make the set of the teeth uniform.

Claim.—The additional guard or stop J, so constructed and arranged as to enable the operator to set the teeth of a saw alternately in each direction, without reversing the instrument or the saw, substantially as described.

No. 13,944.—Henry C. Green.—*Improved Automatic Feed-Motion for Saw-Mills.*—Patented December 18, 1855. (Plates, p. 244.)

The object of this improvement is to regulate, automatically, the feed of the log to the saw. As the saw-shaft B revolves, motion is given to the governor D by belt E, and the cones L L by belt *d*, and motion is given to the log-carriage by means of shaft Q and belt T. When the thick end of the log is being sawed, the saw has considerable work to perform; and consequently, the carriage moves moderately along, and gradually increases in speed as the thickness of the log diminishes, in consequence of the arm K moving the belt N along on the cones L L, the arm K being operated by means of the rack-bar J and pinion *c*. If the saw C binds and rotates very slowly in consequence, the collar G on the governor-shaft will be depressed, and the upper wheel F will gear into pinion H, and a reverse motion will be given to the arm K, so that belt N will cause the inner cone to rotate slower than the outer one and diminish the feed; and if the saw still continues to bind, so as to bring it down to less than its required number of revolutions, the arm K will cast the belt N off on a loose pulley at the large end of the inner cone L, and the feed will then be stopped until the carriage is gigged back and the saw relieved.

Claim.—The combination of the cones L L, governor D, and pulleys S S, arranged and operating substantially as shown, for the purposes specified.

No. 13,964.—Isaac Spaulding.—*Improved Saw-Set.*—Patented December 18, 1855.) Plates, p. 244.)

The saw is placed so that it rests on the elevating screws D D, by which the back of the saw is raised or lowered, so as to give the required set to the tooth with the teeth upon the anvil H, the tooth to be set directly under the punch F and in the angle of the slide E. A blow being struck on the punch at F, the operation is completed.

Claim.—The construction of the slides E, substantially as set forth, and their arrangement with screws D D and H, and punch F, operating in the manner described.

No. 13,982.—Soranus Dunham.—*Improved Method of Hanging Saws.*—Patented December 25, 1855. (Plates, p. 245.)

The object of this improvement is to get rid of the strain which occurs in the usual mode of hanging the saw in a saw-frame. *a c* are the beams of the saw-frames, hung on rocker-shafts *b* and *d*. At each end of the stiffening-bar *h* is fitted a wedge-shaped step *l m*, the edge of which sets and plays in a groove in bearing-plates *n o*, secured to said beams *a c*. The upper step serves as a nut, having a female screw cut in it, in which the male screw *p* (set firmly in the upper end of bar *h*) works, and by turning this bar round in one direction or the other, (provision being made for the turning in the lower step *m*,) the bars *a* and *c* will be spread apart or allowed to come together, so as to strain the saw properly. The saw *k* is set in forked plates *r*; screw-rods *s t* project from said plates, and pass through the ends of beams *a c* and through wedge-shaped steps *u v*, and are confined by nuts *x w*. These several wedge-shaped steps tilt in their respective grooves, and by so doing, when the saw-frame is moving with a reciprocating curvilinear motion, they provide for the necessary play of the saw and stiffening-bar h, at the same time keeping the parts properly strained. The saw receives motion by means of crank o^1 and pitman d^1.

Claim.—1st. The improved mode described of hanging the saw, when the frame in which it is hung has a reciprocating curvilinear motion, so as to provide for the necessary play of the same at its ends, said improved mode consisting in supporting and confining the saw at one end, or both its ends, in wedge-shaped steps arranged to tilt in proper grooves, in the manner and for the purpose explained.

2d. The vertical stiffening and regulating-bar, with its ends arranged in the wedge-shaped steps, and with one end made susceptible of the adjustment, as explained.

No. 13,989.—Westel W. Hurlbut.—*Improved Method of Hanging Circular Saws.*—Patented December 25, 1855. (Plates, p. 245.)

The saw D is hung upon the shaft C, which has its bearings F attached to the movable slides G ; the arm H extends from the bearing F to the cross-bar of the frame, passing around the edge of the saw, and forms the guides for the edges of the saw, as seen at L L^1. J is another arm, connected with the bearing F and arm H, and extends in the opposite direction to the back end of the machine, and is there attached to the knife, or opening-wedge K. The arm J has a flange similar to the one on H, which flanges serve to secure both said arms in their proper position, respectively, by means of the screws O and n.

The object of this arrangement is that the saw-guides L L^1 may assume their proper positions, when the saw D is set or regulated for splitting or re-sawing the board, either square or diagonally, thus permitting the guides for the board A A^1 and B B^1 to remain in all cases in an upright position.

I *claim* the arms H I, as connected with the saw-guides L L^1, the bearing F, and the opening-wedge K, in such manner as to adjust with the movement of the saw D.

No. 12,137.—Adrian V. B. Orr.—*Shingle-Machine.*—Patented January 2, 1855. (Plates, p. 245.)

The object of this invention is, first, to split the shingle from the bolt in such a way that the side of the bolt from which it is taken shall always be left smooth, thus leaving one side of the shingle smooth after the operation of splitting; and, secondly, to place that smooth side down on the platform over which the shaving-knife works, so that the piece shall be tapered and finished at a single stroke of the knife.

The frame D which holds the splitting-knife d slides in grooves g g in main-frame C, and it has two faces a and b, the face b being in a plane with the splitting-knife, the other with the free side of the shingle, the object of which is to allow the piece, when s
through without bending.

The piece being placed with the smooth side down on the platform R R, over which the shaving-knife f works, and kept from sliding by the hold-fast s (which latter projects through an opening in platform R), is now pressed close to the platform by roller *n*; the knife-frame P is put in motion through crank *o*, and, guided by the lower part of grooves T T, follows the roller and cuts the piece to the proper shape and thickness; then in its back motion along the upper part of grooves T T, rising from the platform, lifts lever A, in consequence of which hold-fast *s* is depressed below the surface of platform R, allowing the finished shingle to slide down the somewhat inclined platform.

Claim.—1st. Constructing the frame or slide D of the splitting-knife with two faces *a* and *b*, the one being the thickness of a shingle in advance of the other, as described and for the purpose set forth.

2d. The device of raising the shaving-knife *f* from the platform R R, during its back-motion, for the purpose of allowing room for the introduction of the piece to be shaved, substantially as described.

3d. Moving the hold-fast *s* up and down by the means described, so as to have its use, when wanted; and then removing it out of the way of the finished shingle, in its descent, as already set forth.

No. 12,206.—J. W. Hatcher.—*Rotary Shingle-Machine.*—Patented January 9, 1855. (Plates, p. 246)

There is attached to the under side of each of the cross-bars B B (at opposite ends) a knife C. The cells G, in the wheel F, have the length of a shingle, and little more than double its width, the depth of one end of each cell being equal to the thickness of the thick end of a shingle, and the depth of the other end of the inner cell something deeper than the thin end of a shingle, and of the outer cell the thickness precisely of the thin end of a shingle. Each cell has a revolving cylinder or axle H; from the under side of which extend two metal strips I I, in grooves cut for the same in the cells. Cords K and *m* pass through holes in the wheel, and connect each of the axles with a spring *n* and a lever L. A hole is bored through the wheel in each cell, outside the axle H, in which plays a bolt *b*, attached to a spring acted upon by levers O, which levers are worked by stationary beams S S. P is the feeder. As the wheel revolves, the inner half of each cell in turn passes under the feeder and receives a shingle. This is shaved on one side as it passes under C. The lever L then passes over S, thereby turning the cylinder and strips, and depositing the shingle, with the unshaved side up, in the outer half of the cell, while spring *n* draws the cylinder back in its old position. The shingle then passes under the other knife and is thrown off from the wheel by the bolt acted upon by the lever O, which has been set in motion by passing over S.

Claim.—Taking the shingles singly from an oblong feeder, open at the top and bottom, and partially so in front, by cells cut in the wheel; turning the shingle, after one side has been shaved, by means of a cylinder with bars attached, acted upon by a lever, and returned to its place by a spring, and throwing the shingle off the wheel by means of a spring-lever, after both sides have been shaved. The machine itself, when fed with rifted shingles, shaving both sides and turning out the shingles complete.

No. 12,600.—CHAS. LEAVITT.—*Shingle-Machine.*—Patented March 27, 1855. (Plates, p. 246.)

In operation the handle *q* is placed in notch 8, thus lowering the table *k* (by means of shaft *p*, pinion *m*, and rack *l*) to its lowest point; the bolt is placed on *k* and the machine is put in motion, and a piece sufficient to make eight shingles is split off by froe *h*, and sapped by the froe *s*; the upper part is thrown aside, and the handle being placed in notch 4, the froe subdivides the remaining piece into equal parts, and so on in the following rotation: 6, 7, 5, 2, 3, 1.

The shingle-holder consists of a wooden tail-block *t*, and two pieces *u* and *v*, placed horizontally between the guides *w*, *v* being fixed in mortises, and *u* in slots which admit of horizontal motion. Between these two pieces is a spring *x*, keeping *u* and *v* apart, but yielding to inequalities in the length of the shingles. The knives d^1, pivoted in *z*, have long lever arms, which, overbalancing the forward portions, rest upon bar a^1. When the plane-stock is forced forward, the levers are necessarily elevated by the bar a^1, and the cutting-edges of the knives describe arcs, which produce drawing-cuts on the edges of shingle e^1.

Claim.—1st. The elastic table *k*, capable of being elevated and depressed by the means described, or their equivalents, in combination with the froe or splitting-knife *h*, substantially in the manner set forth, and for the purposes specified.

2d. The elastic shingle-holder, constructed and arranged substantially as described, and for the purposes specified.

3d. The jointing-knives d^1, pivoted to the plane-stocks, in combination with the bar a^1, substantially as described, for the purpose of jointing the edges of the shingles with a drawing-cut.

No. 12,842.—ANDREW P. WILSON.—*Shingle-Machine.*—Patented May 8, 1855. (Plates, p. 246.)

The gate *s s s s*, which carries on its under side the two-edged knife *e*, being drawn to one end of the slides *f f*, the block in the box *o n o* falls on the slats *l l* on the door *t*. These doors *t t* are supported by slats *q* underneath, which are fastened to the frame *a* by means of bolts, which pass through oblong slots in the slats *q*, so that one end of one door can be lowered and the other end of it elevated, in order to give the shingle the desired thickness and taper. The springs *m* on the under side of the door extend a short distance under the knife, to support the shingle while it is being cut. The cleaners *g g* are attached to the under side of the cross-ties *c c*, and, extending from the ties towards the knife, reach down to the doors, and are provided with grooves (figures 5 and 6) for the slats *l* to pass back and forth with the gate.

When the gate passes back, it brings the knife in contact with the block, which rests on the slats *l*. The knife commences cutting the shingle, which passes under the knife, and is supported by the springs until the knife passes under the cross-tie *b* and over the cleaner, which cleans the shingle from between the knife and springs. The knife, having passed under the tie *b*, permits the block again to fall upon the slats on the door, on the opposite side of the knife. The gate, in passing back again, brings the knife in contact with the block again, which, having fallen on the door opposite to the first, reverses the new shingle to be cut, taking the butt of the second shingle from the end of the block from which the top of the first was taken, and so on alternately.

Claim.—The mode of adjusting the block out of which the shingles are to be cut, as set forth in the above specification, by means of the adjusting slats *q q q q* (figure 4) underneath the doors *t t*, and also the springs for supporting the shingle, as set forth above; also, the cleaners *g g*, for the purpose of cleaning the shingle from between the knife and springs, and for the purpose of keeping the doors *t t* clean from all substance that may fall on them.

No. 12,848.—JOHN TAGGART, assignor to HIMSELF and NEHEMIAH HUNT.—*Machine for Sawing Wedges or Shingles.*—Patented May 8, 1855. (Plates, p. 247.)

The main carriage C has a reciprocating longitudinal motion. The stationary lifters consist of rollers L M attached to the cross-bars P; so that when the carriage carries a lifter-catch G or H into contact with either of the rollers, it raises the catch out of its notched rack D or D^1, and this before the adjacent lever I or K is moved into contact with the rail next to it, N or O. When the friction-roller *y*, at the end of the lever I or K, comes into contact with the rail N or O, its lever will be turned on its fulcrum so as to draw towards it the arm *h*, and thereby turn the bearer F, so as to move the raised lifter-catch directly over the next notch of its notched rack, while the other lifter-catch remains in a notch in the other notched rack. This operation produces a slight forward movement of the carriage E, or feeds it up to the saw, at the same time causing the bearer F to be turned horizontally.

The block, from which the shingles are to be sawed, being made to rest against the front edge of the bearer, and being fastened to it, will not only be turned or vibrated with the bearer, but will be moved forward with it, so as to cause the saw, during each horizontal movement of the carriage, to cut from the block a shingle, commencing the cut at the thinner edge and terminating it at the butt end of the shingle.

Claim.—The peculiar combination of mechanism employed for moving the bolt forward, and changing its position so that a shingle or wedge shall be removed from it by the saw during each longitudinal movement of the bolt produced by the main carriage; the said combination consisting of the carriage E, the turning-bearer F, its lifting-catches G H, the notched racks D D^1, the two levers I K, the stationary lifters L M, and the stationary rails N O, the whole being combined with the main reciprocating carriage and the frame of the machine, and made to operate together and with the circular saw, substantially as mecified.

No. 13,155.—A. C. Billings and B. H. Ruggles.—*Machine for Riving Shingles.*—Patented July 3, 1855. (Plates, p. 247.)

As the wheel C rotates, the pin *c* thereon, in consequence of being fitted in the curved slot *b* of lever G, gives said lever a vibratory movement, and the gate I is consequently operated with a vertical reciprocating movement. The block A, from which the shingles are cut, (represented by broken lines,) is secured to the part *f* of the head-block N by clamp O. When the gate I has nearly reached the length of its upward stroke, the wedge-shaped projections *h* will act against the end of lever Q, and the pawl *n* will act against the rack M and feed the carriage L along, so that the edge of the block will be underneath the knife J in the gate. The head block and carriage are shoved back by hand by operating lever *p*, and throwing the pawl from the rack. The spring P allows the head-block N and block A to yield in case of any unusual resistance offered to the vertical stroke of the knife, such as knots, &c.

Claim.—1st. Giving the proper feed-motion to the carriage L by means of the inclined or wedge-shaped projection *k* on the gate I, lever Q, and pawl *n*, which acts against a rack on the under side of the carriage L, as shown and described.

2d. Connecting or attaching the head-block N to the carriage I by means of the spiral spring P, for the purpose of allowing the head-block to yield or give when necessary, as set forth.

No. 13,475.—Chas. Ketcham, assignor to Chas. G. Judd and Andrew Oliver.—*Machine for Sawing Shingles.*—Patented August 21, 1855. (Plates, p. 247.)

Ratchet-wheel M is driven by a lever attached to the gate, and screw-shaft G revolves with the ratchet-wheel; the screw-shaft moves the half-nut K, and the nut moves the block E by means of arms I, the arms I terminating in a clasp which holds the block E; the clasp has at its lower part a ring I, which slides on guide-rod H. From a bolt in the lower end of the clasp extends a rod connecting it with the inclined plane O, for which it serves as a guide. Brace P extends from the nut K to the same bolt, which fastens the rod above mentioned to the inclined plane. These two rods or braces communicate the feed-motion to O. The block-holder C is hinged, and supported by braces *s*. The receiver D supports the shingles when sawed, and receives the shingles, and, being movable, may be drawn back with its contents by lever B clear of the saws. The inclined plane O during its motion raises roller *o*, which, being attached to a coupling uniting the two wedges S S, raises them. The wedges are adjustable by lengthening or shortening the coupling, and are secured by clasps *q*; they (the wedges) communicate a lateral motion to rollers *n*, which, being attached to guide R, move it towards the centre of the machine. This lateral motion is communicated to the movable parts of the gate by means of friction rollers *m*. The gate holds the movable saws *x* hung in the lines they are designed to cut, and the saws of course move laterally together with the gate. By lengthening or shortening the coupling of wedges S, the movable saws can be set to cut lines in any desired direction in respect to the stationary saws *z*, moving the rollers *n* to another place on guide R, and changing the inclination of the plane O, as may be necessary. X X are binding rollers which hold the material while being sawed, and they are moved up and down in ways by racks *d* (connected to the boxes of the roller-shafts) and pinions on shaft *e*, which latter may be turned by means of a hand-wheel or crank.

Claim.—1st. The feeding-trough C for containing the shingle or stock-block E, constructed as described and arranged, in relation to the means for feeding and the means for cutting, as set forth.

2d. The receiving-trough D, having the grooves in it to receive each shingle while being cut, and holding them sufficiently to permit their easy and ready removal from the saws in compact and orderly condition.

3d. The arrangement of the adjustable inclined levers O O and adjustable wedges S S in relation to each other, and to the means for feeding, and to the saws for the more perfect and accurate adjustment of the movable saws, as herein set forth.

4th. Holding the stock-block within the troughs C and D by pressure exerted in the line of the edges of the shingle, being cut by means of the rollers X X, or their equivalents, held and moved substantially as stated, in contradistinction to the holding of the block by lateral and end pressure, as is usual in shingle-making machines, so that the shingle being cut is neither pressed upon the sides of the saw, as must occur when the lateral pressure is used, nor the block upon the teeth of the saws, as must occur when end-pressure is made.

No. 13,555.—Henry J. Weston.—*Improved Construction of Beds for Shingle-Machines.*—Patented September 11, 1855. (Plates, p. 248.)

The block of wood T being placed on the bed-plate R R^1, which is supported by elliptical springs, and is divided into two parts R and R^1 to allow the part R (when the part R^1 is forced below its natural position by the casually increased thickness of the piece V separated from the block) to rise to its proper position to form a gauge for the next succeeding piece which is to be split from the block by the next succeeding forward motion of the driver P.

I do not claim the general principle of splitting off a piece from the block thick enough to make two or more shingles, and then subdividing it; neither do I claim the combination of two or more riving knives for that purpose.

But I *claim* making the yielding-bed R R^1 in two parts, and arranging those parts in the manner described and represented.

No. 13,666.—William J. Scott.—*Improved Method of Feeding the Shingle-Bolt to Knives.*—Patented October 9, 1855. (Plates, p. 248.)

H is the knife-frame. The carriage F, for holding the block, is provided with racks *a a*, which are fed forward alternately by the arms *d d*, which are attached to each end of shaft D, and so adjusted that they operate alternately upon the two racks, for the purpose of feeding the carriage, in such a manner that at each stroke of the knife its face is at the required angle with said knife *h* in the ratchet attached to shaft D; said ratchet is rotated by means of the hand *g*, which is attached to one end of the lever *i*, whilst lever *i* receives a vibratory motion from the knife-frame H, to which it is attached by the rod Q passing through the stand *t*. *r r* are two adjustable stops, against which the stand *p* (which is attached to the knife-frame) strikes alternately in its upward and downward motion, thus giving the vibratory motion to the lever *i*, which is communicated by the hand *g* to the ratchet *h*, from thence through the shaft to the cams *v v*, which operate upon two racks *a a*, and thus feed the carriage; bearing the block to the knife E E are the handle cams for throwing down the arm M, for the purpose of allowing the carriage to be run back to receive another block, and repeat the operation already described. L L are two bars, to which the arms M M are pivoted, said bars being adjustable for the purpose of more readily adjusting the arms M M. *k k* are two springs for keeping the arms M M in gear with the racks *a a* when the cams are thrown up. *l* is a spring attached to the lever *f* to throw the hand *g* over the tooth of the ratchet *h*.

Claim.—1st. The application and construction of the two-handled cams *v v;* also the adjustability of the arms M M, by means of the bars L L, as described.

2d. I claim the combination of the rocking lever *i*, clutch or hand *g*, lever *f*, and spring *l*, with the knife-frame, for the purpose of feeding intermittingly the block to the knives in the manner described.

No. 13,958.—Joel Tiffany and Milo Harris.—*Shingle-Machine.*—Patented December 18, 1855. (Plates, p. 249.)

The shingle (represented in thick broken lines) is placed with its point against the head-block of the carriage; the operator draws the driver forward, and thereby causes the primary carriage to force the shingle far enough forward to place the butt of the shingle between the grippers F F. At this moment the pins *g g*, on the driver, act upon the outer ends of the grippers to firmly grip the butt of the shingle, and consequently to draw it between the knives B B and B^1 B^1. As the shingle passes between the knives, they gradually approximate each other by reason of the projection *k*, on the knife-carriage, sliding in and being acted upon by the oblique grooves I in the driver, and thus it receives the proper taper form; the knives B taking off the rough, and B^1 smoothing it from butt to tip.

Claim.—Providing a primary and secondary set of knives B B^1; a primary and secondary set of feed-rollers C C^1, and obliquely grooved driver D D^1 D^2; a primary feed-carriage E, and a pair of secondary feed-grippers F F; and arranging and combining the whole in the manner and for the purpose specified and shown.

No. 13,084.—George Fetter and Jos. L. Pennock.—*Machine for Cutting the inside Hold of Shovel-Handles.*—Patented June 19, 1855. (Plates, p. 249.)

The handle V is clamped upon carriage P, the outer edge of the hold *e* having been previously rounded in any proper manner. Motion is then given to shaft F, and cam G first acts upon the upper side of frame H and elevates the frame and sector I; and as rack K, on the sector, gears into rack N on the half cylinder M, said cylinder will be turned or moved in the grooves *b* in the projections L L, and cutter O will sweep over the upper half of the inside of hold *e* at one end, it being understood that one end of the hold is placed in contact with the cutter at the commencement of the operation. The cutter, therefore, during the movement above described, cuts a transverse groove on the inner side of the hold *e* at one end. The cutter now remains stationary as the cam G passes over the upper edge of frame H, and cam T acts against projection *d* of rack S and moves it; and as this rack gears into segment rack R of carriage P, the carriage is also moved, and the whole length of hold *e* is forced past the cutter O, which cuts or rounds the upper part of its inner side. As soon as projection *d* is relieved from cam T, the carriage is thrown back to its original position by spring U. Handle V is then reversed upon the carriage P, and the opposite side of the inside hold is cut by repeating the operation above described.

Claim.—1st. Cutting the inside hold of D-shaped handles for shovels, spades, etc., by means of a curved cutter O, so operated as to pass one half way round one end of the hold, the cutter smoothing or rounding a portion of said hold equal to its width, and then remaining stationary, while the handle is moved to force the remaining uncut portion of the hold past the cutter.

2d. We claim operating the cutter O, by means of the cam G, frame H, geared sector I, and half-cylinder M, to which the cutter is attached; also, operating the carriage P, to which the handle V is attached, by means of the cam T and racks R S. The above parts being arranged and operating conjointly as shown and described.

No. 12,424.—Samuel R. Smith and Elijah Cowles.—*Machine for Cutting Wood into Slivers.*—Patented February 20, 1855. (Plates, p. 249.)

The cutters *t* are attached to cutter-stocks X (of which the engravings represent only one). The cutter-stock fits over an arm V, which is fast on shaft B, and rotates with it. The slot *x* in the arm *w* of the cutter-stock fits over the ledge *y*. This ledge has an irregular curved portion *z*, which moves the cutter-stock so as to cause the cutters to pass over the stuff a^1 in a straight line. Thus the shavings will be cut in the direction of the grain.

As the hollow shaft F rotates, screw L communicates motion to worm-wheel K, and shaft J rotates, and cord *e* is wound upon one part of clutch *g*, weighted lever H is raised, and (as rack G gears into *c*) shaft B is gradually depressed, and the cutters are fed to the stuff. When the cutters have descended a distance equal to the thickness of the stuff, the clutch *g* is disconnected by the automatic operation of parts C M O, etc., the feed-motion ceases, and the cutters are raised and adjusted for a new cut.

Claim.—1st. Giving the necessary feed-motion to the cutters by means of the lever H, with the segmental rack G attached to one end, which rack gears into the recesses *c* cut in the lower end of the shaft B, the opposite or weighted end of the lever being raised by the cord *e*, which is wound around a clutch *g* on the shaft J, motion being given the shaft J by the worm-wheel K and screw L, substantially as shown.

2d. Giving the necessary direction to the cutters while passing over the stuff a^1, by means of the rim or ledge *y* on the disc Y, said rim or ledge having a bent or curved portion *z*, which, in consequence of the arm *w* working upon it, communicates the proper motion to the cutter-stocks, so that the cutters will pass over the stuff in a right or straight line, as described.

No. 12,459.—ASA LANDPHERE and SAMUEL REMINGTON.—*Spoke-Machine.* —Patented February 27, 1855. (Plates, p. 250.)

When the stuff F first arrives under the cutters L L, both these sets of cutters are forced away from the centre of the machine by the pins *m* m^1, (travelling along the properly curved profile-plates *g g*,) and only the straight parts *n o* of the cutters are allowed to come in contact with the stuff (see figure 4) and plane the top flat, as the first part which is acted upon is to form the shoulder of the spoke; as the carriage moves on, the pins *m* m^1 are allowed by the form of the profile-plates to approach the centre of the machine, and the curved parts of the cutters come into operation. The profile-plates *g g* direct the movements of the cutters laterally to the stuff, and the ways g^1 g^1 (against which the screws *o* o^1 rest which support the cutter-frames) direct their vertical movement, and thus regulate the form of the longitudinal profile of the stuff. One side of the shoulder is planed by the cutters *p p* on disc P^1, whose position is regulated by arm *r* and profile-plate *s*, which also serves to guide the disc and cutters out of the way of the wheels *b* b^1. When the stuff has passed through, the carriage is run back, the stuff is half-way turned round, and then again subjected to the planing operation, as above described.

Claim.—The dressing of spokes by means of a series of revolving cutters, whose edges present an oblique profile in part of the spoke, when said cutters are so arranged on their shafts as to reduce the spoke op narrow longitudinal sections, by which means much more smooth work is obtained than when the cutters reduce the spoke at one single ineration, as set forth.

No. 13,408.—WILLIAM VAN ANDEN.—*Spoke-Machine.*—Patented August 7, 1855. (Plates, p. 250.)

F^1 and F^2 are the cutter-holders, arranged on axles B^1 and B^2 in pairs, so as to cut, by means of the stationary and adjustable cutters G and H, the four sides of the spoke simultaneously and longitudinally, as the spoke is carried through the space between their curved cutting-edges. These holders are secured by pins *f* entering slots *g* on the axles, so as to allow of a lateral adjustment of the cutters at the same time that they are rotating, so as to vary or shape the sides of the spoke to the required pattern. Cutters G are fixed in the face of the cutter-holders. The cutters H have an adjusting lateral motion in the cutter-holders. When the spoke is nearly finished, the side-edges of it are squared; and as the lateral motion of the holders is uniform in its action upon all the cutters, it therefore becomes necessary to give the side-cutters an independent lateral motion to that of the holders. As the holders are separated, the shanks of the cutters H come in contact with fixed collars I, which forces the cutters out beyond the faces of the holders, and thereby flattens or squares the sides of the end of the spoke. Arms K are secured to the middle of yokes J, to which arms pins L are secured, which straddle guides M attached to bed N of the spoke-rest carriage, which guides govern the width of the spoke as well as the shape of it, as they move forward by the forward motion of the spoke-rest carriage. To the centre of the middle one of each set of double-acting levers V are secured the ends of pawls W, the tail-ends of which rest upon cams X. These cams govern the shape of the spoke, acting by means of the pawls upon the levers and upon the boxes of the cutter-holder axles, thereby increasing or decreasing the distance between the upper and lower sets of cutters, according to the irregularities in the shape of the cams.

Claim.—1st. The use of the upper and lower adjustable cutter-holders, made adjustable laterally on their axis, substantially as described, in combination with the curved stationary cutters G, and adjustable cutters H, and collars for adjusting the same, or their equivalents, for the purpose substantially as set forth.

2d. The use of the adjusting yoke and the attachments thereto for adjusting the cutter-holders, or their equivalents, in combination with the cutter-holders and guide-ways on the spoke-rest carriage, or their equivalent, substantially as set forth.

3d. The use of the double-acting adjusting levers, or their equivalents, for the purposes set forth, in combination with the cutter-holders and their axles, and their combination with the pawls attached to the double-acting adjusting levers, and cams for operating the same, or their equivalents, for the purposes substantially as set forth.

No. 13,731.—OWEN REDMOND.—*Spoke and Axle-Helve Machine.*—Patented October 30, 1855. (Plates, p. 251.)

The bed C is supported by journals *c* and *d*, the journal *d* being wrapped with a spring *i*, fastened at its outer extremity to the bottom of table B, and its inner end secured to the journal *d*. Two pieces of stuff are secured to the two faces *n* and *m* of bed C by means of dogs *l k* and l^1 k^1. The movements of the cutter-heads, in combination with the forward motion of the bed, cause the cutters to impart to the upper portion of the stuff the form of one side of a spoke. By this time the forward movement of table B has brought the lower end of catch *j* in contact with stud *b** on the side of frame A, moving it in the direction of the arrow and releasing arm *f*, (attached to journal *d*,) which by action of spring *i* flies up until it strikes shoulder *c**, bringing face *n* of bed C flush with the upper face of table B, so that the stuff secured thereto occupies the place of that secured to face *m*, and previously operated upon as above mentioned. The motion of table B will be reversed by means of driving-band L being shifted to pulley J, and the stuff on face *n* of bed C will now be submitted to the action of the cutters during the backward motion of the table.

Axe-handles may be shaped upon this machine by removing bed C and inserting in the table B either of the beds X, Y Z, instead thereof; the stuff to be first sawed in the curved form required for the axe-handle and secured upon one of its curved edges, when bed X is employed; when bed Y Z is used, the stuff is secured on a flat side to the swinging plate X^1. This plate is acted upon by a spring *d** so as to have a tendency in the direction of the arrow, and when moving with table B, has its edge *f** pressed upon by the adjustable guide *m**, so that the stuff shall pass under the cutters in a serpentine direction. The beds here represented will fashion but one side of the handle; others, the reverse in form, being required for forming the opposite side.

Claim.—1st. The partially revolving bed C, constructed, arranged, and operating substantially as described, so as to submit different pieces of wood to the action of the cutters at its forward and backward movements, substantially in the manner set forth.

2d. The bed Y Z, having a laterally swinging spring-plate X^1 in combination with the adjustable guide *m**, for submitting curved timber to the action of rotary cutters, in the direction of its curve, substantially as specified.

3d. The curved bed X, traversing with an undulating movement, for submitting curved timber to the action of rotary cutters, as set forth.

No. 13,882.—Thos. R. Markillie.—*Improvement in Spoke-Machines.*—Patented December 4, 1855. (Plates, p. 251.)

On the end of the rails of the carriage S a bed-plate *m* is firmly bolted. From this plate rise the standards *p*, from which standards two arms *q* are suspended by a bolt; these support a plate *r*, which carries the block in which the adjustable tracer *s* is fastened. At one end of this plate are two standards which support the shaft of the cutter-wheel M. At the other end of plate *r*, which slides in guides *t*, is a spring *u*, having its motion in the direction of the tracer. This spring, when the cutters are thrown out of gear by the cam U on the end of pattern V, acting on the tracer *s*, holds them in that position until the carriage has been again drawn back, when the spring is freed from its catch *t* by its curved point, which projects over the plate *m*, striking against beam N, when the plate carrying the cutter is drawn forward towards the blank W until the traces come in contact with the pattern, and held in that position by a weight attached to a cord passing over the sheave *n*, having its other end made fast to plate *r*. By this arrangement of the cutter and tracer on the same bed-plate, having a reciprocating motion, a duplicate spoke is cut instead of a reverse one, as in other machines.

When the spoke has been completed, the cam U on the end of the pattern, by pressing against the tracer *s*, throws the cutter out of the line of cutting, in which position it is held by spring *u*, leaving a portion of the blank undressed to form the tenon.

Claim.—The arrangement of the cam U on the pattern V, in combination with a tracer *s* and a spring *u*, in the manner and for the purpose described.

Also, the particular arrangement of the rotary cutter and tracer in combination with the plate that supports them, suspended in the manner and for the purpose described.

No. 12,221.—James Wilson Treadway.—*Stave-Jointer.*—Patented January 9, 1855. (Plates, p. 252.)

Under the circular saw E passes a carriage which supports the bed-piece A on two pivots *a a*. The stave is placed on the bed-piece and fastened closely to it by the clamps *b b*. The distance from the axis of the pivots to the ends of the upper side of the bed (which side of the bed is rounded as represented in fig. 3, so as to fit the curve of the stave) is always to be equal to the radius of the head of the cask for which the stave is intended. The axis of the pivots is in the same plane in which the saw-plate moves. By turning the bed-plate with the stave on said axis, the edge of a stave of any width can be brought in contact with the saw, and the cut will always be towards the axis of the pivots, and consequently towards the centre of the cask. The bed-piece is fixed in proper position by means of a serrated hold-fast *c* and catch *d*.

The inventor says: I do not claim the curved bed-plate upon which the stave is bent and held by the clamps, except in combination with suitable devices to allow it to rotate partially about a fixed axis, for the purpose of giving any degree of bevel to the joints, and for jointing both sides of the stave without its change of position on the bed-plate, as fully specified, and all of which I *claim.*

No. 12,900.—Daniel Drawbaugh.—*Stave-Machine.*—Patented May 22, 1855. (Plates, p. 252.)

The block from which the staves are to be cut is placed upon the vibrating bed C, one edge of the block bearing against the concave D; the bed is first moved downward by raising the outer end of lever H, in order to allow the edge of the block to bear against the concave. The bed is then raised by depressing the outer end of lever H, and the block is forced against the knife I, which cuts off a stave corresponding in thickness to the space between the knife and concave. The pressure-roller J, which keeps the stave against the knife, causes it to be cut in perfect shape corresponding to the concave.

Note.—The slides P P and the brace R, which are contained in the claim, are not referred to or represented in the original specification and drawings of the inventor.

The inventor says: I do not claim separately any of the parts of the above machine; but I do *claim* the combination and arrangement of the following parts, to wit: making the frame F to vibrate on the shaft G, and hinging it (the frame F) to the bed C, so as to operate it with a vibrating motion on the pivots *a a* and slides P P with the brace R, which renders the bed C rigid to traverse parallel, so as to cut the lumber either curved or straight, as required for staves, heading, or other purposes, substantially as herein described; the bar D being changed or reversed as described.

No. 13,035.—M. T. Kennedy.—*Machine for Planing Staves.*—Patented June 12, 1855. (Plates, p. 252.)

The disc and clamp revolving around their axes, the uppermost rod K will be acted upon by cam M, which bears against its end and forces it forward, so that its lip *c* will be forced outward from the disc I¹; the stave is then inserted between the lip and the edge of band J, and when the uppermost rod K passes the cam M, its spiral spring L will draw the lip *c* of said rod firmly against the edge of the stave, and the stave will then be secured between the edge of the lip and the edge of the band J. In this way all the staves are inserted, the cam M permitting this by forcing out the lip. The clamp and disc rotate against each other: and as the staves in the clamp come in contact with the cutters *b* they are dressed, as they pass around on the clamp; and when the ends of the rods come in contact with the cam M¹, the lips *c* are again forced forward or out from the disc I¹, and the staves fall from the clamp.

The inventor says: I do not claim, separately, the rotating disc C, with the cutters *b* attached, for that has been used for analogous purposes; but I *claim* the combination of the disc C and clamp, the clamp being formed of a series of rods K passing through the discs I I¹, and provided with springs L and lips *c*, arranged and operating in the manner and for the purpose shown and described.

No. 13,036.—M. T. Kennedy.—*Machine for Jointing Staves.*—Patented June 12, 1855. (Plates, p. 252.)

The stave is placed between the plates *e* and *g*, and there secured by turning cam K. The lower end of frame L is then moved towards the back part of frame A, so as to give clamp I an inclined position (see fig. 1). Pitman E is then set in motion; the clamp by its own gravity settles down between the two planes F F, and the edges of the stave are brought in contact with the cutters *a a*, which together with the planes F F are reciprocated by means of the pitman. The stave is then planed to the form A^1. The clamp is then elevated, by means of treadle N, above the planes, (see dotted lines in fig. 1,) and the position of the stave reversed by turning the clamp half-way round. The clamp is then again let down, when it will be cut to the form A^2. If the staves require to have rounded edges, (see A^3,) the staves are bent or sprung upwards at their centres in the clamp I.

The inventor says: I do not claim, separately, the reciprocating planers, for they have been previously used; but I *claim* the combination of the reciprocating planers F F and clamp I, constructed, arranged, and operated in the manner and for the purpose shown and described.

No. 13,045.—Geo. H. Swan.—*Stave-Machine.*—Patented June 12, 1855. (Plates, p. 252.)

C slides, and D cross-sills, for the reception of the cask; K and L levers, K carrying the croze, chamfer, and howelling knives, and a knife cutting in towards the outside of the staves at the extreme top of the chamfer, exactly opposite the levelling saw on lever L, cutting towards the inside of the staves, and meeting, forming the top of cask-levers K and L; M spiral spring pressing apart the rear ends of the levers, and consequently pressing the knives together to any depth that the gauging or connecting-rod N will admit. O short slide, making fast to arm I by means of a thumb-screw, with flange attached to receive spiral spring P, which also connects to lever L, and pressing the end of the lever carrying the levelling saw tight to the outside of the barrel, and governing lever K by the rod N, and consequently giving a uniform depth of cut to the work in every part, being round or not, as the levers are balanced off by putting a set of knives on each end, or the equivalent in weight, consequently getting the same pressure of spring at all points of the work.

Claim.—The chamfering, levelling, and bowelling barrels, kegs, &c., in fact, anything requiring a croze, chamfer, howel, or level, by means of the two levers operated upon by springs and gauging or connecting rod, in such a manner as to gauge the work from the outside of the cask, being round or not, working exactly to the shape of the head truss-hoop with any kind of croze knives or saws necessary to give the desired croze or groove, chamfer or level, howel and level; and the two levers are attached to the slide by pin or bolt acting on one or separate centres, and that slide made fast to the arm of the thumb or set screw at any part from the shaft, or doing away with the slide, and putting the pin or bolt directly through the arm at any point from the shaft required, consequently adjustable to any size.

No. 13,230.—WILLIAM ROBINSON.—*Stave-Machine.*—Patented July 10, 1855. (Plates, p. 253.)

The plate C and knife D have a reciprocating movement. The piece of wood fastened to head-blocks E F is dropped down to the plate as soon as the latter arrives under the piece of wood, and the knife D will enter the wood and split off a stave. The stave will remain under the piece of wood after being split off, but will fall upon the lower plate C^1 so as to be clear of it, the knife D sliding over the end of the stave to prevent it from springing up; as the return stroke of plate C is made, it will push the detatched stave towards the dressing-knives.

The inventor says: I make no claim to the mere fastening of the riving-knife to stationary springs, as shown in Stoddard's shingle-machine, patented December 7, 1852; neither do I claim the vertical slotted movable knife-bars shown in the patented shingle-machine of Stevens and Kidder.

But I *claim* the horizontally slotted spring knife-holders *a a*, combined as specified with the guides h h^1 on the driver C, so as to prevent the longitudinal movement of said holders during the riving operation, and cause the knife, at the completion of the cut, to force the stave into the lower bed and there hold it during the return stroke of the driver, as set forth.

No. 13,631.—LARKIN T. ATKINS.—*Machine for Gauging, Measuring, &c., Staves.*—Patented October 9, 1855. (Plates, p. 253.)

The two jaws A A attached to one end of the frame are shod with two plates D, having the proper bevel of the stave. The outer jaw is provided with two springs to allow the jaws to be opened for the reception of the staves E. A leather strap attached to a treadle F, and passing over a friction-roller *r*, connects with the outer jaw, which causes the jaws to close when the treadle is pressed down by the foot of the operator. The treadle is held down by a ratchet *s*. Vibratory arms C, worked by a weight, pass into the jaws and receive the staves from the jaws when they are opened. Similar jaws E E at the other end of the frame are provided with straight-edged plates D for jointing heading and shingles.

Claim.—The combination of treadles, clamps, and vibratory arms or levers, operating in the manner and for the purpose set forth.

No. 12,416.—JOEL HASTINGS, JAMES RAMSAY, and HENRY G. CHAMBERLAIN.—*Machine for Cutting Tenons.*—Patented February 20, 1855. (Plates, p. 253.)

The fixed plate L and the clamping-plate *m* are so adjusted that when the said clamping-plate and the dog *o* have advanced nearest to each other, the distance between them will be such as to grasp a slat tightly in the proper position with the parts to form its tenons opposite the axes of the cutter-stocks. Rotary motion is imparted to shaft J and shafts C C. The slats are inserted one at a time by the operator, by pushing them, during the retirement of the tenon-heads and opening of the clamping apparatus, between the gauge-plates V V on to the table K, being always pushed back against the stops *e e*. The clamping piece *m* and dog *o* advance and seize it, and the tenon-heads immediately approaching each other, (by the action of the cam-grooves *j* and *q*,) with their cutters in rapid revolution, cut down the tenons at both ends simultaneously. The retirement of the tenon-head, and subsequent opening of the clamps, (effected by the same cam-grooves *j* and *q*,) then leave the slat free; and the fly N coming immediately afterwards into operation (by the action of cam-groove *w*) throws it from the machine (see dotted lines in figure 2). The fly immediately afterwards, resuming its position below the surface of the table K, leaves the machine in condition to receive the next slat.

Claim.—1st. The arrangement of the two advancing and retiring tenon-heads, the clamping-piece *m*, the dog *o*, and the fly N, substantially as described.

2d. The construction of the cutter-stocks and arrangement of cutters, substantially as described, to wit: the cutter-stocks being composed each of an open flanch D, attached by a yoke or arms *d d* to its shaft, with a disc a^1 bolted to the said flanch, and having the cutters *b c* secured, one to its face and the other to its back.

3d. The fly N, arranged and operating in any way, substantially as described.

No. 12,449.—William Steele.—*Tenoning-Machine.*—Patented February 27, 1855. (Plates, p. 253.)

Frame A (secured to upright Y) receives gate C, in which incision-cutters *l l* and tenoning-cutters *j j* are secured, the sides of said frame serving as fender-posts for gate C to play upon. The gate C is moved downwards by means of pitman D and a treadle (not shown in the engravings). A sliding-box *d*, having a wing *g* secured to one of its sides, is connected to the innermost side of frame A by means of set-screw *e* (that passes through a slot in A) and nut L. Thus *d* can be moved up or down, and secured in place. The base *p* of gauge-box H, and of rest *s*, has a segmental wing *k* secured to its inner end, which wing is secured to wing *g* of box *d* by means of pin *n* and set-screw *f*, passing through a curved slot in *k*. Thus base *p* can be secured to box *d* in a position at right-angles to the sides of frame A, or in an oblique position thereto. The piece of joist is placed in gauge-box H; its forward end is then moved forward until it touches the tenoning-cutters; then the gauge-box is moved outwards from rest *s* a distance exactly equal to the required length of tenon; the joist is then secured in the box by set-screw J, and is fed forward to the cutters by means of lever *a*, pawl *b*, and rack *c*. The height of base *p*, box H, and rest *s* is adapted to the depth of the joist to be operated upon, so that there need not be an unnecessary amount of motion imparted to the tenoning-chisels. Figure 3 represents a horizontal section through line *x x*, in figure 1; figure 1 side view, and figure 2 front view.

Claim.—The arrangement of the feeding-box H, the rest *s*, and their base *p*, with each other and with the gate C which carries the tenoning-cutters, in such a manner that the said feeding-box may be moved from the said rest upon the base *p* the desired length of a tenon, and then be fed forward again to bring the joist to be operated upon in contact with the cutters, substantially as set forth.

Also, combining the base *p* of the feeding-box H and of the rest *s* with the frame A, in such a manner that the said base, box, and rest can e secured in an oblique position to the sides of said frame, and to the direction of the movements of the tenoning-cutters, whenever it may be desired to form tenons with oblique shoulders, in the manner set forth, or its equivalent.

Also, the combination of the incision-cutters *l l* with the angular-edged cutters *j j* in such a manner that the said incision-cutters will penetrate into the surface of the wood, in advance of the tenoning-cutters, a sufficient distance to prevent the said edges of the tenoning-cutters from tearing out splinters from the sides of the timber operated upon.

No. 12,864.—THOMAS J. KNAPP.—*Adjustable Tenoning-Tool.*—Patented May 15, 1855. (Plates, p. 254.)

The ring I, of a diameter equal to the diameter of the tenons to be cut, is placed around hub C. The two segments are then adjusted by operating screws *b c d*, the inner ends of the segments resting against the periphery of the ring I, thus leaving the aperture between the segments of a size equal to the desired diameter of the tenon. The shank A is placed in a rotating mandrel, and the end of the spoke pressed into the aperture formed by the segments, the cutters H G cutting tenon and shoulder. Thus by using various sized rings I, tenons of various diameters can be cut with the same tool.

The inventor says: I do not claim the cutters G H, for they have been previously used.

But I do *claim* the construction of the tool, as herein shown and described, viz: having two segments D D of a cylinder secured to the flanch B of the shank A by screws *b b*, and having screws *c c*, *d d*, pass through the flanches E E and projections *e e* of both segments, for the purpose of allowing the segments to be placed nearer together or further apart, as desired, and having a hub or boss C at the centre of the face of the flanch B to receive the rings I of different sizes, corresponding to the diameter of the tenons to be cut, and by which rings the segments are properly adjusted the required distance apart, for the purpose of cutting tenons of various sizes, as herein shown and described.

No. 13,049.—CHRISTOPHER SHARPS and GEORGE E. ADRIANCE.—*Tenoning-Machine.*—Patented June 12, 1855. (Plates, p. 254.)

The plane-stocks C C[1] (to which the planes and cutters D D[1] are attached) play freely up and down in box B, and in and out laterally, and are kept the required distance apart by a tongue E, which is made of the same width as the required width of the tenon to be cut. The cutters D^1 cut at right-angles to the planes, and form the shoulder on the spoke. The mouth F receives the end of spoke *a*. The chips, etc., pass out at G. The drop H (hinged at *h* to the box B) holds down the spoke. Springs I I are to throw the planes apart for the purpose of inserting a spoke. K K are followers, crowding the plane-stocks up to the tongue E and stuff to be tenoned, by means of springs L L. The slots *f* in the springs L are to allow the plane-stocks to be moved farther apart by the treadle and wedge J, when it is desired to tenon stouter spokes.

The operator places his foot on the treadle, throwing the planes apart, then the spoke is inserted and secured by the drop; the operator then removes his foot, thus letting the planes work up and down gradually un il the tenon is finished.

Claim.—1st. The arrangement and combination of the two planes and cutters for cutting the tenon-rest for supporting the spoke, drop for keeping it down upon the rest while the tenon is being cut, and guide-box B, having a mouth F and discharge G, substantially as and for the purposes set forth.

2d. The combination and arrangement of the plane-stocks, with planes arranged so as to move in and out; slotted springs to allow for said movements; springs I I and treadle J J, for causing said expansion, substantially as and for the purposes set forth.

No. 13,735.—C. P. S. Wardwell.—*Machine for Cutting Double Tenons.*—Patented October 30, 1855. (Plates, p. 254.)

The timber is represented in dotted lines. The operation of the machine will be understood from the claim and engravings.

Claim.—The combination and arrangement, substantially as shown and described, of the intermediate obliquely set or drunken saw F, with the clearing or finishing true circular saw M, for operation together in the manner specified, and whereby the drunken saw F not only serves to largely reduce the wood between the tenons, as required, for the completion of the tenons, but to form a wide kerf or pathway for the axle L of the finishing saw M, to admit of the deep insertion of the latter into the wood, and of its operation as a clearer between the double tenons during the continuous progress or feed of the timber as described.

No. 12,134.—Hazard Knowles.—*Cutters for Tonguing and Grooving.*—Patented January 2, 1855. (Plates, p. 254.)

The figures represent a cutter for grooving. It consists of three steel discs, the cutting-edges being formed by cutting oblique notches d^1 in their peripheries, and throwing outwards the angles c^1 at the rear side of said notches, and sharpening them. As the cutting-edges wear away, the rear sides of the notches are filed away sufficiently to form properly shaped points, which are again thrown outward. Consequently the diameter of the cutting-discs can be preserved in renewing the cutting-edges.

Tonguing-cutters are formed in a similar manner.

Claim.—Forming tonguing and grooving cutter-heads, of combined discs of steel, which have cutting-edges formed on their peripheries of such a shape that new cutting-edges can be formed upon them as the old wear away, without reducing their diameter, substantially as herein set forth.

No. 12,180.—Pulaski S. Cahoon and Samuel F. Ross.—*Improved Chuck for Turning Elliptical Cylinders.*—Patented January 2, 1855. (Plates, p. 255.)

The piece P to be turned is secured in the box *a* which projects from the slide E. The sliding-standard F, which supports the ring f^1 *f*, is then moved by means of screw *e* until sufficiently out of centre with the lathe-spindle B to give the required oval to the piece to be turned. By thus moving the standard, the slide E is caused to assume the position represented in dotted lines in figure 1, and the piece to stand eccentric to the lathe-spindle, and concentric (before the operation commences) with the hole *f* of the standard-ring. When the spindle is now revolved it carries the slide and box with it, and owing to the ring or flanch f^1, to which the slide is connected by the lips *b b*, being eecentric to the axis of the spindle, the axis of the slide is caused gradually to change its position, and in doing so to describe an oval, as represented in dotted lines in figure 1, and the piece, consequently, as it comes in contact with the cutter, to be shaped accordingly. By having the axis of the slide more or less eccentric to the spindle, the oval will more or less approach a circle.

Claim.—1st. The arranging of the ring f^1 upon the sliding standard F, and combining it with the lathe-spindle B, by means of the slide E, substantially as and for the purpose set forth.

2d. Arranging the slide between the face-plate and the standard F, instead of attaching it to a ring situated back of the face-plate, substantially as and for the purposes set forth.

No. 13,104.—John W. Russell.—*Improved Chuck for Turning Eccentrics.*—Patented June 19, 1855. (Plates, p. 255.)

The chuck is applied to irregular objects by starting one of the nuts, and then the other to follow in the same direction. Or one may be taken out, and then the screw and jaws follow, when the jaws can be revolved on the screw to any point of deviation desired.

Claim.—The application of a chuck to irregular objects, and points eccentric from the centre, using for that purpose the jaws C C and the screw E, in combination with the nuts or collars D D, all for the purposes substantially as set forth.

No. 13,208.—Wm. Blackburn.—*Automatic Machine for Turning Ship Spars, etc.*—Patented July 10, 1855. (Plates, p. 255.)

A is the spar to be turned; B is the driving-pulley, adjusted on the spar A; C, friction-rollers; E E the tool-carriages, worked by feed-screws H H; J J J rests, (for supporting the spar while turning,) adjustable by their screw-shafts and gearing N M, to be worked from shaft K; two pairs of bevel-wheels L *d* and L *d* serve, one to adjust the lower rest J, and the other the tool-holder O, by means of a vertical screw Q; P are the chisels. The engraving represents only about one-half of the machine, the other half being exactly alike; *x x* is the centre line.

Claim.—1st. The combination and arrangement of the gears N N and *d*, with the self-adjusting rests J J J, or mechanism substantially the same, for holding the stick, whether straight or tapering, always firm and steady during the operation of turning.

2d. The combination of the chisel-holder O, screw Q, and gear *d*, or mechanism equivalent thereto, for working the chisel-holder simultaneously with the self-adjusting rests, substantially as set forth.

No. 13,301.—Matthew F. Connet.—*Machine for Turning Cylinders of Wood, &c.*—Patented July 24, 1855. (Plates, p. 255.)

Before placing the material P between the centres on the swinging-frame, the latter is set free from catch F to be thrown forward by spring D against rail *m*, and the tightening pulley *h* is thrown off the belt *f* so as to stop the revolutions of the said centres. The material is then secured between the centres and the screws G G, adjusted so that when the swinging-frame is drawn up against them, (and secured by catch F,) the material P is reduced at the point opposite the cutters to the necessary depth. The tightening pulley *h* is now thrown on belt *f*, when the material will slowly revolve and be properly turned off by the rapidly revolving cutters.

The inventor says: I do not claim the revolving of the cutters at a rapid and the block at a slow motion, as this has been done before; but I *claim* so combining a swing-frame, which carries the block to be cut, with a cutter or cutters revolving around a fixed centre, as that the block may be swung up to the cutters and first cut to the required depth or gauge without revolving, and then be revolved slowly on its centre against the action of the cutters, to complete the turning or cutting at a single revolution of the block, substantially as described.

No. 13,899.—Israel Amies.—*Improved Application of Embossed Veneers.*—Patented December 11, 1855.

The veneers are perfectly polished on one side, and the rear side partially smoothed with sand-paper. Paper is then pasted over this rear side, and the whole is left a sufficient time to allow the wood to partially absorb the moisture of the paste. The veneer thus prepared is introduced between two dies, correspondingly carved—one convex, the other concave; both dies being moderately heated. The dies and veneer are then submitted to considerable pressure. On removing the veneer, its face represents in relief the pattern on the dies, and has all the appearance of an elaborate wood carving.

The inventor says: I wish it to be understood that, although I have described one particular process of treating veneers, before my improved art of embossing is practised thereon, I do not desire to confine myself to that process in every minutia, as the same may be modified, or equivalents substituted; but I *claim* the employment of embossing veneers, in the construction of furniture, and for other ornamental purposes, in the manner set forth.

No. 13,099.—David Pierce.—*Machine for Manufacturing Wooden Ware.*—Patented June 19, 1855. (Plates, p. 256.)

Notches are cut in a horizontal direction in the post, into which the cross-plank E may easily be moved, for the purpose of adapting the tail-spindle, which passes up through it, to the length of the work to be done. The notches are cut so deep as to permit the tail-spindle *a* to be moved to the right or left, sufficiently to secure any desired taper for the work to be done. F is the sliding-rest, holding the cutting-tools and sliding in grooves *g*. The gang of cutting-tools 1, 2, 3, 4 are placed at distances from each other to correspond to the desired thickness of the cylinder. To finish the inside of the work, a core is attached to the pulley-spindle of the lathes of the form of the interior of the work. The semi-circular longitudinal excavation in the core is to receive the chips; G is the cutter. Knobs *h*, cutter *n*, and spurs *m* (attached to springs H) serve to mark and cut the croze. The cutter-discs Q serve to turn the heads or bottoms. The slot in the piece *p* serves to admit the head to be finished, and the piece is hinged so as to bring it nearer to or further off the cutter according to the size of the head. Two centres pass perpendicularly one through the upper and one through the lower portion of the piece to the slot, and the unfinished head is adjusted to the plane of the cutter. The lower centre is fitted into a disc V. For the purpose of cutting elliptical heads, a pin *u* is placed in disc V. If then the band *x* be passed around the stationary pulley and the centre enclosing the pin, and the head be turned against the cutter, the pin operating against the band will retract the head from the cutter to the extent of the eccentricity of the pin, and thus the elliptical shape will be produced.

Claim.—1st. The application and use of a cutter or cutters in gangs attached to a sliding-rest, as described, or their equivalents.

2d. The apparatus for forming and finishing circular or elliptical heads or bottoms, as described, or its equivalent.

3d. The apparatus for turning out the inside of the cylinders and cutting the croze, as described.

No. 12,857.—Jacob A. Conover.—*Machine for Splitting Wood.*—Patented May 15, 1855. (Plates, p. 256.)

The blocks of wood are placed upright on the endless bed *c*. The plate *p*, with its elastic pad *v*, holds the block down to the bed during the descent and the rising of the knife *w*. The knife *w* has four blade forming a cross (see fig. 4), which work through a corresponding slot in plate *p*. As soon as the knife has cut through the block, and has again been lifted above the block and plate *p*, this plate *p* itself is raised by the operation of cam *u*, and the feed-motion of the bed takes place to advance the next block.

Claim.—The movable bed or carriage for carrying and advancing the blocks of wood in combination with the reciprocating cutters, operating at right-angles with the surface of the bed or carriage, substantially as and for the purpose specified.

Also, in combination with the bed or carriage and reciprocating cutters, substantially as specified, the employment of the clearing-plate through which the cutters pass, substantially as and for the purpose specified.

Finally, providing the said clearing-plate with an elastic pad, and imparting to it an up-and-down motion, substantially as specified, in its combination with the bed or carriage and reciprocating cutter, as specified; by means of which the said plate, under the combination specified, performs the double office of holding the blocks and clearing the cutters, as specified.

No. 13,485.—Wm. O. Bisbee.—*Machine for Splitting Fire-Wood.*—Patented August 28, 1855. (Plates, p. 256.)

The block of wood N being placed within box D, the horizontal knife C is forced in (by crank and pitman motion) until its front edge enters notch *d*, thereby splitting the block horizontally. By this time that part of frame B (frame B carries the knife C) which, in fig. 1, is immediately underneath the block of wood, has receded behind the box D, so as to allow the piece which has been rifted from the block to fall on table E. During the return-motion of the frame B, the shoulder *x* will push the piece before it, and draw it through the rifting-knives G. These knives are arranged as represented in the top view, fig. 2, because if the cutting of the knives were in a line, and the knives parallel with each other, it would be impossible to rift the piece into strips, as the latter would become wedged between the knives. Fig. 1 represents a vertical section.

The inventor says: I wish it to be distinctly understood that I do not wish to confine myself to the exact form or method described of operating the machine, or to the exact number of vertical knives shown.

But I *claim* the vertical knives G as arranged, with their edges a distance in advance of each other, and their sides at different angles, so as to act effectually as a means of rifting wood, as described.

No. 13,492.—G. W. B. Gedney.—*Rotary Wood-splitting Machine.*—Patented August 28, 1855. (Plates, p. 256.)

The splitting-knives *e* are attached to the revolving disc *a*, by means of the groove e^1 (fitting the edge of the opening in the rim) and the flanch and single bolt e^2. The V-shaped knife-edges e^3 project a proper distance above the outer perimeter of the rim to split the wood, which is put into hopper *f*, and rests by its own gravity upon the revolving rim; as the knives come around just in front of the hopper, there is a set of thin plates *s* that project down, one opposite each depression in the knife-edges, and others *r* opposite the small grooves in the apex of the V-knives. These serve to clear the knives and allow them to work freely. The small catches *h* on axis *o* (to which lever *i* is attached) come in contact with the wood and prevent its rising while under the action of the knife.

I *claim* the machine described for splitting wood, consisting of the V-grooved knives, acting upon the wood as described, having openings in their apex to receive the clearers, substantially as specified, and in combination therewith the fingers for holding down the rear end of the wood to be split.

I also claim the mode of attaching the knives by the groove e^1 at their back, and an overreaching flanch l^2, by which the resisting strain tends to hold the knife in place, as specified, without bringing the strain upon the bolt by which it is fastened.

XV.—STONE AND CLAY.

No 2,264.—John A. Messinger, deceased, by his administratrix and administrator, assignors to Ambrose Foster.—*Improved Building-Block.*—Patented January 16, 1855. (Plates, p. 257.)

This building-block is composed of sand and lime in the proportions of twelve parts of sand to one of lime, mixed together and pressed in moulds. For the purpose of facilitating the ripening of the block, it should, where this form is admissible, be perforated with holes as represented in the engraving. These bring the air into contact with the central parts of the block.

Claim.—The building-block described is claimed as a new manufacture.

No. 12,558.—Loomis E. Ransom.—*Improvement in the Manufacture of Bricks.*—Patented March 21, 1855. (Plates, p. 257.)

The material is placed between guide-bars *a b*, which are of the thickness of a brick, and the scraper B is passed over it.

The tool P^1 has a smooth surface *c*, equal in width to the thickness of a brick, bounded on one side by the flat plate *d*, and on the other by the knife *i*. The surface *c*, as the tool is drawn towards the operator, compresses and smooths the edge of the brick, while plate *d* and knife *i* sharpen the edges and trim off any superfluous material.

The tool P^1 is represented on a larger scale than the rest of the figure.

The inventor says: I make no claim to any portion of the processes of manufacturing bricks set forth in the French patents of Capgras and Chanon, June 21, 1843, and Charles Henry Maigret, May 22, 1840.

But I *claim* the manufacture of bricks substantially as described; that is to say, by first spreading the tempered mortar or clay at once upon the ground, where the bricks will be left to dry, and in beds of certain desired length, width, and thickness, and then while the mortar is in a soft state, or before it shall crack by too much drying, producing therein lines of weakening or separation, defining the dimensions of the bricks without regard to their smoothness or final finish; and after the bricks in drying shall have separated from each other along the lines thus formed, turning them on edge, and squaring and polishing their edges, and defining the thickness of the same, by rubbing over them the metallic tool P^1, or otherwise, substantially as set forth; the desired thickness of the bed being produced by means of guide-bars or moulds, and scraper or lute, substantially as specified, whereby I am enabled to dispense with off-bearers, and otherwise to simplify the manufacture of bricks.

No. 12,898.—JOHN CHASE, Jr.—*Improvement in Brick-Presses.*—Patented May 22, 1855. (Plates, p. 257.)

The frame B (which is vibrated by the action of the crank *r* on the pins *f*) operates alike on both sides of the frame A, one rod *c* ascending while the other descends; and there are two platforms F, one at each side of the frame A, the moulds being fitted at both sides of the frame. The figure represents only one of the platforms. When a rod *c* is raised, the empty mould B^1 is placed in between the grate I and the friction-rollers *j j*, and the spring O keeps the mould against grate I during the operation of pressing. The rod *c* in descending presses an empty mould in between the rollers and grate, and at the same time forces down a filled mould, which falls upon the step *h* on lever D. When *c* rises, the motion of lever D (which is imparted by means of rod C) will move the filled mould forward upon platform F.

The inventor says: I do *claim*, 1st. The swinging-frame B, constructed, arranged, and operating substantially as herein shown, for the purpose of feeding the empty moulds B^1 to the press-boxes H, and discharging the filled moulds therefrom.

I am aware that devices have been previously employed for preventing stones and hard substances from being pressed against and with the moulds, by affording facilities for removing said stones and hard substances from the moulds before the clay is pressed into them. I therefore confine myself specifically to the device herein shown and described for that purpose, viz: further, I claim the employment or use of the levers or pendents *l*, constructed with projections *n*, and placed between the bars of the grates I, as herein shown, so that as the filled moulds B^1 are forced downward from between the grates I and friction-rollers *j*, the stones, should any be in the moulds, will act against said projections and turn the pendants, so that a passage or opening is allowed for the stones to drop through.

No. 12,997.—Wm. H. Degges.—*Improvement in the Soak-Pits of Brick-Machines.*—Patented June 5, 1855. (Plates, p. 257.)

The clay, after it has been pulverized and soaked in the revolving-pit A, is shovelled on to the endless apron *h*; the pit being in motion, keeps the edge of the clay-bank therein all the time at the same distance from the endless apron, so that the hands supplying the machine need never change their position.

Claim.—The revolving soak-pit, or its equivalent, for uniformly soaking the clay when pulverized, and conveying it to a convenient position to be fed into the brick-machine, substantially as specified.

No. 12,998.—Wm. H. Degges.—*Improvement in Brick-Machines.*—Patented June 5, 1855. (Plates, p. 257.)

The crank *s* on the pug-mill shaft F gives a reciprocating motion to the sliding-frame H, the cross-head of which is connected with it by the pitman *t*; the ends of this cross-head extend on each side of the main frame D through slot 17 in the connecting-pieces *u*, (the slots allowing the cross-bar *z* to rest at each end of its travel,) which are jointed to the upper ends of two vertical levers *v*, one on either side, having a fulcrum common to both at *w*, and connected at their lower ends by links *x* and slides *y* with the cross-bar *z* under the tub E, extending from side to side through aperture I; the width of this aperture is equal to twice the thickness of bar *z* and the width of the mould-frame, so that at whichever side the bar may be, the mould-frames are inserted immediately under the centre of the tub, and moved thence by said bar exactly under the apertures for filling, thus economizing space and reducing the diameter of the pug-mill. The brushes, fig. 4, correspond in size and number with the mould-frames, which are fixed in any convenient position near the pug-mill, and are composed of alternate layers of bristles and sponge, for the purpose of cleaning and damping the moulds ready for use. The coal-dust with which the apron *h* is supplied, in order to prevent the clay from sticking thereto, mingles with the clay, and assists in burning the brick to great advantage.

The inventor says: I do not claim inserting the moulds under the pug-mill on either side of the centre, alternately, to be moved thence under the apertures to be filled; neither do I claim dusting the parts of a brick-machine generally for the single purpose of preventing the clay from adhering, nor do I claim mixing coal-dust with the clay in any other manner than that described.

But I *claim*, 1st. Causing the reciprocating bar *z* to rest at each end of its travel, whereby ample time is afforded for inserting the moulds under the bottom of the pug-mill, and for pressing the clay into them while in their rest position, substantially as set forth.

2d. Inserting the moulds immediately under the centre of the pug-mill, whence they are moved alternately to the right and left previous to being filled, whereby the size of the pug-mill and the power required to work it are economized to the fullest extent.

3d. Dusting the endless apron, or other device, for conveying the clay to the pug-mill with a mixture of coal-dust and sand, whereby the clay is prevented from adhering thereto, and at the same time the coal-dust is evenly mixed with the clay during its preparation, for purposes specified.

4th. I claim the series of brushes, or other device, substantially the same, that will hold water, whereby the moulds are both cleaned and damped at the same time, as set forth.

No. 13,042.—JNO. PLUMBE.—*Improvement in Cutting Clay into Bricks.*—Patented June 12, 1855. (Plates, p. 257.)

Wires F, the width and length of a brick apart, are stretched in contact with the inside bottom of the mould-frame D D; each end of every wire passing through a vertical transverse slit in the sides of the frame, and being fastened by a pin on the outside. When the clay has been filled in, and as soon as it has become sufficiently dry, the wires are pulled up, one or more, thereby cutting the clay into single bricks.

Claim.—The mode of forming bricks, or their equivalent, by means of wires, or their equivalents, cutting upwards substantially as described, irrespective of the manner in which the clay is prepared and placed in the mould-frame.

No. 13,110.—LEVI TILL.—*Improved Brick.*—Patented June 19, 1855. (Plates, p. 257.)

The object of this improvement is to secure greater strength by locking or binding the walls together, by means of grooves *a* and projections *b* in the bricks.

Claim.—The making of bricks with channels or grooves, and with spurs or conical projections, for the purposes and substantially in the manner set forth.

No. 13,123.—HENRY CLAYTON.—*Improvement in Brick and Tile Machines.*—Patented June 26, 1855; patented in England, December 13, 1852. (Plates, p. 258.)

The material as it comes from the pug-mill is forced against and cut by wires c^2 set to the desired width. c^3 are rods which slide through sockets fixed to the front of disc c^1, and have arms c^4 which embrace the wires c^2, so that when they are raised, the arms c^4 may scrape the wires and keep them clean. Next the clay is received upon a frame of carriers *e*, and is cut transversely into bricks by a movable frame *d* of parallel wires d^1. In order to separate the bricks one from the other, after having been cut by the wires, the outermost brick when it comes to the end of the carrying-rollers is received upon and by the last roller *e*, the surface of which is caused to move faster than the preceding rollers, by which means a brick is separated from the rest and laid upon a pallet or board which is placed on a movable or tilting platform.

Claim.—Combining with the wires c^2 c^2 their sliding-scrapers or cleansing mechanism, made to operate essentially as explained.

Also, the combination of the accelerating roller *e* and the tilting-board *f*, with the delivery rollers or their equivalent.

No. 13,129.—A. V. Hough.—*Improvement in Brick-Machines.*—Patented June 26, 1855. (Plates, p. 258.)

Claim.—The sides *ff* placed at the bottom of a pug-mill L, for the purpose of enabling the operator to regulate the rapidity of the egress of the clay, according as it requires to be subjected to the operation of the cylinder for a longer or shorter time.

Also, placing the shaft D, with its blades *c*, in a horizontal position within the cylindrical case C, as shown, whereby the machine is rendered extremely simple, the journals kept free from clay, and all the parts of the machine operated by the rotation of a single shaft.

No. 13,239.—Stephen Ustick.—*Improvement in Brick-Presses.*—Patented July 10, 1855. (Plates, p. 258.)

The longitudinal plates 12 are separated by thin plates 14, inserted between their edges, so as to leave a slit between them for the escape of condensed air, during the operation of pressing the brick. These spaces, widening towards the inside of the piston, form the air-channels 15.

I *claim*, 1st. Combining the inner and outer peripheries of the rim of the revolving casting C, shaped in segmental curves, eccentric with each other and with the centre of the shaft B on which the casting is secured, in such relation to the upper and lower pistons for pressing the brick frames to which they are attached, and to the friction-wheels E H in said frames, as to cause the said segmental surfaces to operate on the friction-wheels in their revolutions betwen the same, after the manner of a wedge, and thus avoid all liability of strain on the shaft B, arising from the resistance of the pressure exerted in pressing the bricks, by confining it to the body of metal between the two surfaces, substantially in the manner set forth.

2d. I claim forming the faces of the pistons of longitudinal and transverse plates 12 and 13, secured to the blocks or main body of the pistons by dove-tailed tongues or grooves, and wedges or gibs, and capable of being moved outward, sidewise and endwise, so as to increase the area of the face of the pistons, in case of wear, as before described.

3d. I claim forming a narrow slit in the centre or other part of the face of the piston, widening as it extends from the face, or not, as desired, and communicating with the outside of the pistons, through their ends, for allowing the air confined in the moulds to escape during the pressing of the clay into bricks, as described.

No. 13,502.—LEVI TILL.—*Improvement in Brick-Machines.*—Patented August 28, 1855. (Plates, p. 258.)

The pulverized clay passes from hopper E into the moulds *a*, and is, during the revolution of the moulds, pressed against the pressing surfaces *l* of the wheel D, which are perforated. Each of the surfaces *l*, as it comes into contact with the clay in the mould, comes also into connexion with an air-pump U by means of the channel *n* and tube O. Thus the air is exhausted from the clay while being pressed. The slots *m* serve for the discharge of all the excess of clay, and also, in combination with the teeth *h* on the mould-wheel, insure the accurate meeting of the moulds and pressers.

I *claim*, 1st. The use of the air-pump, in combination with the perforated pressers, by which the air is exhausted from the clay while under pressure, as stated, and not otherwise.

2d. I claim the device of the diagonal slots *m*, figure 3, in combination with alternating with the pressers, by which all the excess of clay may escape, and is discharged on one side, and not on both, of the machine.

No. 13,533.—JOS. ALEXANDER VICTOR.—*Improvements in Brick-Machines.*—Patented September 4, 1855. (Plates, p. 258.)

A cylinder, provided with spiral knives, cuts and mixes the clay and works it to the end of the cylinder into a reservoir, under which the endless chain of brick-moulds X X passes between two sets of rollers F and G; the upper rollers not only aid in carrying the moulds through the machine, but compress the clay in the moulds.

I *claim* the combination of the endless chain of moulds, connected substantially as described with the two sets of rollers, one of the upper of which, in addition to aiding in drawing the mould through, at the same time compresses the clay in the mould.

No. 13,572.—G. W. B. GEDNEY.—*Improvement in Brick-Machines.*—Patented September 18, 1855. (Plates, p. 258.)

To move the moulds *c* along, a ratchet-wheel is affixed to the axis of carrying-cylinder *b*, into which ratchet-wheel a pawl *m* works, which pawl is jointed to a lever having said axis for a fulcrum, and connected by a rod m^1 with a wrist *n* on the cam *o*, which cam operates the rod *p*. This rod extends down to a frame-work of two projecting-fingers p^1, that stand horizontally on each side of the lower portion of the endless chain of moulds, in such position as to receive the off-bearing boards *r* and deposite them on the off-bearing apron. The fingers p^1

are returned by a spring. The said boards r are placed in hopper q in front of the press-box, and as each mould is fitted and comes out from under the hopper, one of these boards drops upon the mould and is caught by two projections r^1 on said mould, and carried forward until it comes under the cords s, which keep said boards in contact with the moulds while they pass down around the cylinder b, and are thus reversed, bringing the board under the open mouth of the mould. At this instant the board is brought upon the ends of the fingers p^1, which then descend and deposite the board, with the bricks upon it, on an endless apron t below, which apron carries off the bricks.

I *claim* the off-bearing boards, applied and arranged as specified. I also claim the fingers for placing the board from the mould on to the endless apron.

No. 13,747.—Alexander H. Brown.—*Improvement in Brick-Machines.* —Patented November 6, 1855. (Plates, p. 259.)

The skeleton-wheel C, upon shaft X, has eight moulds, each open at top and bottom. E are the inside plungers, acted upon by the cams F upon shaft U^1 through the hooked ends of the compression bars G. The upper cog-wheel U transfers motion to upper horn of the ratchet-stock K by means of a crank-pin and connecting-rod J. The connecting-rod L extends from the lower horn of K to lever M, upon rock-shaft N. The two short levers O O are fastened to rock-shaft N, and the rods P P are attached to levers O by a pin passing through a horizontal slot in their lower end to allow for the vibration of levers O. The upper ends of bars P form a rectangular hook to act upon the cross-bars Q, fastened to the inside plungers E. These bars Q are taken hold of by the hooks of bars G and bars P alternately. S S are the outside plungers, provided with guide-rods and spiral springs, and acted upon by the two cams R R upon the shafts of the two cog-wheels U U. T is a fork attached to the inner end of the lower plunger S, which acts as a regulator of the skeleton-wheel by taking hold of a pin inserted into the ends of the moulds, so as to insure the plungers entering the moulds at all times. W is one of the pulleys of the delivery-band; x is the hopper; A^1 are two quadrants, one on each side of the skeleton-wheel, and attached to two levers for the purpose of regulating the amount of clay for making the brick, by causing the lugs Q to rise or fall as the surface of the quadrants are raised or depressed through the agency of the crank-shaft b and levers C^1.

The prepared clay passes into the hopper upon the top of the machine from the pulverizer, which is placed above the machine: as the moulds in the face of the wheel pass under the bottomless hopper, they receive the necessary quantity of clay to produce a brick; and when the mould arrives opposite the top plunger, the clay receives a sufficient amount of compression from the plunger to expel a portion of the air and consolidate the clay sufficiently to carry it to the second plunger, when the said plunger and the corresponding inside plunger approach each other, through the action of the reversed cams, upon the lower compression-shaft, making the bricks uniform in size and weight.

After the bricks have thus received three compressions (by the rotary motion of the wheel) the mould takes a vertical position with the skeleton-wheel shaft; and while the plunging is going on to compress the succeeding brick, and while the mould-wheel is perfectly at rest, the brick is expelled from the mould through the action of the vertical bars P P, which take hold of the cross-bars of the inside plungers, and force said plungers half an inch beyond the surface of the mould, and thus prevent the corners of the brick from becoming injured, as the bricks are deposited perfectly flat upon the carrying-belt and carried to the kiln door.

Claim.—1st. The combination of the outside plungers with the skeleton-wheel, inside plunger, and moulds, when arranged and operated as set forth, and not otherwise.

2d. Discharging the bricks by means of the ratchet-stock K, vertical bars P P, and inside plungers E, when arranged and operated as described, and not otherwise.

3d. The mode of regulating the amount of feed through the action of the quadrants upon the inside plungers, when arranged as described.

4th. Regulating the movement of the skeleton-wheel C, figure 1, by means of the fork T upon the lower plunger, when arranged as described.

No. 13,017.—Harlow H. Thayer.—*Improvement in Machines for Kneading Clay.*—Patented June 5, 1855. (Plates, p. 259.)

It will be understood from the engravings how a reciprocating motion is imparted to the fork M by means of a spring P and cam L revolved by a band N, and acting on the rear end *a* of the fork.

The plunger I forces the clay through the screen K, which, on account of its taper orifices, compresses the clay as it is forced through the lower end of chamber G.

Claim.—The employment of the inverted conical chamber G, provided with a screen or perforated plate K, which has its apertures of taper form, said chamber having a plunger I working in its upper cylindrical portion, for the purpose set forth.

Also, the press-chamber G, constructed as described, in combination with the reciprocating fork M, for cutting off the clay discharged from the press-chamber, the fork being operated substantially as shown and described.

No. 13,224.—John O'Neil.—*Improvement in Machines for Pulverizing Clay.*—Patented July 10, 1855. (Plates, p. 259.)

The clay is first pulverized by the yielding cutters *i*, and at the same time forced by their inclined surfaces into the lower section, where it is subjected to the friction produced by the plain or corrugated blades P, whose surfaces bear upon the inclined planes formed on the inner periphery of the hopper A, and thence permitted to escape, in such quantities as desired, through the perforated door B and slide C.

Claim.—The combination of the spring-blades with the ridged surface of the cylinder against which they act, substantially as and for the purpose set forth.

The combination of the aperture J in the depressed part of one or more of the ridges *f* with the spring-blades, which eject or force out the stones, substantially as set forth.

The combination of the cutting or pulverizing blades with the ridged surface of the cylinder, substantially as and for the purpose described.

The perforated or grated door and slide, for the purpose of regulating the discharge of tempered clay, as set forth.

No. 13,245.—William P. Walter.—*Improvement in Manufacturing Plate-Glass from Cylinders.*—Patented July 10, 1855. (Plates, p. 259.)

By moving the slide *e* upon the rod *a*, the rods *d* are caused to expand. The instrument is introduced into the cylinder C while exposed to the fire, and the rods expanded as above mentioned; at the same time the instrument is revolved at the will of the operator, thus causing all parts of the cylinder to be equally heated. The expansion of the instrument will of course produce two flat sides in the cylinder.

The inventor says: I am aware that cylinders of glass have been fashioned into an oblong shape, with two flat sides, by forcing them between two pieces of wood, in a heated state; also by holding them upon bars or rods in the kiln or furnace, and by "*percelars*" in the hands of the workmen, as in flint-glass works. The putting of glass cylinders into an oblong shape is not new. This I do not claim. The stretching of glass is not new. Cylinder-glass is stretched by the blower when he swings the hot glass in the operation of making the cylinder. Glass is also stretched in the operation of making glass tubes. Stretching glass I do not claim.

But I *claim* the forming of cylinder-glass into an oblong shape, with two flat sides, by my improved flattening instrument.

No. 13,411.—Phillippe Stenger, assignor to Pascal Yearsley.—*Improvement in the Manufacture of Plate-Glass.*—Patented August 7, 1855. (Plates, p. 259.)

The cylinder of glass, when blown, and still heated and in a plastic state of consistency, is placed over the vertical standards C, which latter are then moved from each other, thereby stretching the cylinder into the form of two parallel sheets of glass, which are finally separated from each other by the usual means.

Claim.—The application of tractile force to the manufacture of sheet-glass, by means of the mechanical arrangement described, or its substantial equivalent.

No. 12,820.—BENJAMIN HARDINGE.—*Improvement in Facing-Beds for Grinding Artificial Granite, &c.*—Patented May 8, 1855. (Plates, p. 260.)

The saucer-shaped dish A is filled with concrete cement of silici-calcareous and ferruginous trit, ground and mixed with coarse crude emery, steel turnings, sharp sand, and hydrous silicates, with which it is wet up into a plastic state, put into said dish, and levelled with a straight-edged spatula while in very slow motion, so that when hard it turns in a perfect level. The shaft is arranged with heavy steel bearings F and J, thereby giving it a slight vertical elasticity, which prevents all sudden jerking.

Claim.—The above described artificial grindstone or facing-bed, consisting of a saucer-shaped dish filled with a concrete of the materials specified, or others substantially the same.

Also, the suspension of the shaft which carries the rotary facing-bed upon steel supporters, substantially as and for the purpose above set forth and described.

No. 12,242.—SAML. H. ROBINSON.—*Improvement in Lime-Kilns.*—Patented January 16, 1855. (Plates, p. 260.)

Four side-kilns B (of which the sectional figure exhibits only two) surround a central kiln A. The waste heat and gases pass from the limestone in the side-kilns, through flue *o*, and (by means of dampers *p p*) through either of flues *q q*, into the central kiln.

The inventor says: I am aware that a series of kilns have been built in one stack, one placed over the other, and burned from the same or separate fires. These I do not claim; but I *claim* the arranging of a series of side-kilns around a central kiln, so that the waste heat from the former may be used for burning the limestone in the latter, substantially as described, whereby a great saving of fuel and labor is attained, a more regular disposition of the heat made available, and either of the surrounding kilns stopped off, cooled, and drawn, without interfering in the least with the others of the series, as set forth.

No. 12,521.—JESSE RUSSELL.—*Improvement in Brick-Kilns.*—Patented March 14, 1855. (Plates, p. 260.)

Fig. 1 is a front view, and fig. 2 a central section parallel to it; fig. 3 is a side view, and fig. 4 a cross-section parallel to it.

Claim.—The arranging of the fire-chambers outside of the kiln, and introducing the products of combustion to the brick to be burnt through avenues or passages extending from the fire-chambers entirely across the kiln, when said fires are placed and used on one side of the kiln only, substantially as described. (See engravings.)

No. 12,991.—Danl. Blocher and Geo. M. Blocher.—*Improvements in Burning Brick.*—Patented June 5, 1855. (Plates, p. 260.)

The object of this invention is to obviate the necessity for the long grates running entirely through the arches, which are now used, the burning being performed with greater regularity as regards solidity and color, a much larger quantity of brick being placed in the kiln than can be done when wood is used or coal employed, in the manner above stated. G are the grates running through the casement; openings *b* serve for removing the cinders; holes *c* for stirring the coal; *d* feed-opening; H are the arches of the kiln, and I the main body of the brick set close together, except the three outer courses m m^1 m^2 of each interior bench and the entire exterior supports P, which have the interstices common to ordinary setting.

We *claim* the furnaces F, entirely within the casement, and fed at the top, in combination with the close setting of the interior benches, as described, by which placing the fuel within the arches is avoided, and the burning of the kiln improved, as set forth.

No. 13,494.—Daniel Herr (Pequea).—*Improvement in Lime-Kilns.*—Patented August 28, 1855. (Plates, p. 260.)

a is the grate; *b* and *c* are the arched ribs.

I *claim* making the arch of the kiln with two series of arched channelled ribs, so arranged that the outer ribs shall extend over and across the spaces left between the inner ribs, and at the same time leave sufficient space between the outer and inner ribs for the fire and heat to pass into the limestone, whereby the fragments and loose lime are all prevented from falling into the fire, and are conducted down the channels into the proper receptacle below.

No. 12,813.—John T. Bruen.—*Improvement in Stone and Marble Saws.*—Patented May 8, 1855. (Plates, p. 260.)

The body of the saw-plate *a* is formed of woven wire or metal strips, the cross-wires of which are made to sustain a metal strip *b*, with a corrugated edge. The warp is so protected by means of the corrugated filling, or cross-wires, as to be out of reach of wear, and the kerf formed by the edge on the corrugated strip *b* is sufficiently wide to prevent undue friction on the sides of the saw-plate. a^1 a^1 is a band of lead with indentations c^1 c^1 attached to the edge of the woven blade, so as to freely admit the sand and bring it down upon the point to be cut.

The inventor says: I do not claim the using of sand and water, or other grit, with a plain metal plate, as new, nor the blade with grooves cut in it; but I do *claim* the making of the body of the saw-plate of woven wire, or strips of metal, or any analogous device, for the purpose of admitting the free passage of the grit, in the operation of sawing stone, substantially and for the purpose as herein described.

Also, in combination with the above, the waved cutting-edge, or any analogous device, substantially as and for the purpose herein described.

Also, forming the edges of the saw-blade thicker than the central portion, so as to admit the free passage of the grit on both its sides through the indentations, as before set forth.

No. 13,074.—JOSEPH ADAMS.—*Improvement in Stone-Sawing Machines.*—Patented June 19, 1855. (Plates, p. 260.)

The saw, (suspended as shown in the engravings,) when put in motion, strikes the stone, and is relieved by its segmental motion, at which time it again preponderates and comes back in the segment of a circle, with nearly its whole weight and momentum, when it again strikes and passes up, after making its cut, and so on, causing the counter weight *o* to have a short vibratory motion up and down as the saw strikes the stone and rebounds, gradually rising at intervals, as the saws are depressed, and allowing them to cut equally, however varying the density of the material may be. To guide the saw-gate *a*, fender-posts *p* are employed; outside of these are other posts p^1, and between these a box *q* slides up and down. The saw-gate is connected with these boxes by a guide-rod *r*, which slides through the box. Each end of rods *r* is attached to the frame by a portion of it bent at right-angles. The cover *s* is to prevent the guide from being clogged with sand, &c.

Claim.—The application to the saw-frame hanging from cords, so as to move of necessity in the arc of a circle of the counter balancing weight, which, at the same time, permits it to feed itself at all parts of its motion, substantially as described; and combined therewith the guides, constructed and operating as specified.

No. 13,540.—JOHN COCHRANE.—*Improvement in Machines for Sawing Marble.*—Patented September 11, 1855. (Plates, p. 261.)

By means of this arrangement of the saw-blades *g e*, two bevel sides of a taper block of stone can be cut at one operation. To prevent the warping or twisting of the saws, the latter are attached to sliders *k*, (connected to the reciprocating frame A by accommodation-links *a b c d*,) which work along stationary guides D E fast to the bed of the machine.

I *claim* the hanging of two saws in one gate, at any required angle with each other, in combination with the angular guides D and E, the slides *k l* and *m n*, and the accommodation-links *a b* and *c d*, or their equivalents, for the purpose of sawing two inclined or tapering sides of a block of marble or stone at one operation.

No. 13,591.—C. G. BIETEL and H. J. BRUNNER.—*Improvement in Machines for Sawing Stone.*—Patented September 25, 1855. (Plates, p. 261.)

The nature of this invention will be understood from the claim and engravings.

The inventors say: We do not claim flexible saws, radial and curved ways, or guiding-rollers, separately.

But we *claim* the combination of the flexible saws D D, rollers E E, adjustable radial ways G G, and concentric grooves or ways M M, whereby the saws are enabled to run at different angles, and their open ends to approach and separate, without affecting the degree of their tension, substantially as described.

No. 13,742.—Henry Burt.—*Improvement in Machines for Sawing Marble.*—Patented November 6, 1855. (Plates, p. 261.)

The frames B B^1, to which are attached the saw-blades c c^1, have their bearings or guides on frames A A^1. Each of the frames A and A^1 is provided with a thumb or set screw f^4; and by means of these thumb-screws f^4 and slots z the frames and saws can be set at various angles, so as to perform simultaneously two cuts forming a certain angle with each other. The connecting-rods d, which are to impart reciprocating motion to the frames B B^1 and saws c c^1, are connected to the crank-shaft h by means of ball-joints, so as to allow of their being swung out or in, together with the frames, as above described.

The inventor says: I am aware that different adjustable apparatus have been used for sawing wood, and that horizontal saw-frames have been used for sawing stone into square blocks, parallelograms, thin slabs, &c.; also, connections of various kinds have been used. I do not, therefore, claim the above devices separately.

But I *claim* the combination of the saw-frame B B^1, pivoted swinging adjustable guide-frames A A^1, and connection-rods d d^1, arranged and operated in the manner and for the purpose set forth.

No. 13,762.—Robert G. Pine.—*Improvement in Marble-Sawing Machine.*—Patented November 6, 1855. (Plates, p. 261.)

e are the saws; the guides g j cause the frame B, when set in a reciprocating motion, to work in a right line: and by turning the rods a in proper position, the rods d may be made to work parallel with each other; or by turning the two front rods a inwards or towards each other, the rods d will work in straight lines obliquely with each other, the jointed arms f n compensating for the obliquity of rods d.

The sockets k b g^1 allow the saw-frame to be raised and the marble block D to be placed underneath, after which the saws will feed themselves by their own gravity.

Claim.—The frame B, connected by jointed rods f n to rods d d, working in sockets b b, which are fitted loosely on rods a, when the above parts are all constructed and arranged in the manner and for the purpose set forth.

No. 13,773.—GEORGE W. BISHUP.—*Improved Marble-Sawing Machines.*—Patented November 13, 1855. (Plates, p. 261.)

The tube *s* can be turned by means of hand-wheel *u*. The nature of the improvement will be understood from the claim and engravings.

Claim.—Operating the saws of stone-sawing machines by placing the lever H on a tube *s*, which has an internal screw-thread cut in it, and is fitted on a screw-rod *r*, substantially as shown, whereby the lever H may, by turning the tube *s*, be raised or lowered to suit the height of the sashes, as set forth.

No. 13,777.—WILLIAM C. CHIPMAN.—*Improved Marble-Sawing Machine.*—Patented November 13, 1855. (Plates, p. 162.)

Each end of the saw-frames M M extends through the slots *a a* in the suspension-frame D, and are securely held lengthwise by the rollers and bolts *b b*, and at the same time may work laterally with freedom within the slots *a a*, this motion being made easier by a series of friction-rollers *d d*. *m* are the saws.

Claim.—The rollers *d d d d*, when arranged in the manner and for the purpose stated.

No. 13,784.—LUTHER B. FISHER.—*Improved Marble-Sawing Machines.*—Patented November 13, 1855. (Plates, p. 262.)

h is a perpendicular rock-shaft; J segments notched down on arm *k*, which latter is secured to the rock-shaft. The segments are secured to hub *n*, (which is keyed to the rock-shaft,) by means of braces *o*. It will be seen that by setting the hub higher or lower, the segments will be crowded further out or in, thereby widening or reducing the space between the ends of the two rods *g;* by this means the saw-blades *c* may be set more or less obliquely, so as to cut the marble block *b* more or less bevelling. The saw-frames will receive a reciprocating motion by means of chains or straps *s s*, which are secured to each end of segments J J, and crossed and secured to rods *g g*.

Claim.—Operating the frames of stone-sawing machines, by means of the segments J J, in combination with braces *o o o*, arm *k*, and chains or straps *s s s s*, for the purposes and in the manner set forth.

No. 13,788.—GEO. W. HUBBARD.—*Improved Marble-Sawing Machine.*—Patented November 13, 1855. (Plates, p. 262.)

E are the saw-blades; the sash D receives a reciprocating motion. The centre guides F^1 and centre cleats *d* keep the saw-sash D working in a right line, the centre cleats being permanently attached to the end pieces *f* of the sash D, while the oblique guides will move laterally the saws E, and cause them to cut the two sides of the block G in taper form, the degree of taper of course corresponding to the position in which the guides F are set, by means of set-screws or otherwise.

Claim —Constructing the adjustable guides F and permanent guides F^1 of stone-sawing machines, as shown and set forth.

No. 13,791.—WILLIAM B. KIMBALL.—*Improved Marble-Sawing Machine.*—Patented November 13, 1855. (Plates, p. 262.)

It will be seen that the two saws B^1 B^1 (set at an angle with each other) are driven from one crank K. The bar G slides in ways, and the connecting-rods F F are linked to the saw-frames and to said bar G at *d* and *f*.

Claim.—Driving the saw-frames or gates B B, by means of the jointed pitmans F F, and bar G, in combination with the geared segment H and crank K, when arranged as shown, and for the purpose set forth, and not otherwise.

No. 13,805.—CHARLES T. WARREN.—*Improved Marble-Sawing Machine.*—Patented November 13, 1855. (Plates, p. 262.)

As the saw-frame C is moved back and forth by means of pitman D, and as the sectors *f* gear into racks F, the sectors will, as the frame C moves, be turned or moved in a vibratory manner, and the pinions *d* will move the racks *c* in and out; the saws G (which are attached to these racks) will consequently be moved laterally in the saw-frame, while they are moving longitudinally, and two sides of the block will be cut at the same time in taper form.

Claim.—Giving the saws G G a lateral vibrating movement in the saw-frame C, as the saw-frame and saws work longitudinally, by means of the rack *c*, pinion *d*, grooved sectors *f*, and racks F, when arranged as described for the purpose specified, and not otherwise.

No. 13,829.—F. NOETTE and A. SCHMIDT.—*Improved Marble-Sawing Machine.*—Patented November 20, 1855. (Plates, p. 263.)

The block to be sawed is placed upright on the wheel Z, (see dotted lines, fig. 3,) and is properly secured within frame X. The lower or cutting edges of the saws O O are then set outwards, to correspond to the taper designed to be given the sides of the block. A reciprocating motion is then communicated to the saw-frame C by pitman D. As the saw-frame C vibrates, the saws cut the block from its top end downward; and at each stroke of the saw-frame the screw-rods G^1 are turned in consequence of the projection g^1, at the lower end of lever M, striking the pin *h*; and the nuts *i* will be moved further apart on their screw-rods G^1 at every stroke of the saw-frame, so that the block will be sawed in taper form. The block is fed upward as the saws cut, by means of worm-wheel *n* and screw-rod W. Worm-wheel *n* is turned by screw *m* on shaft T, the shaft being turned at each stroke of the saw-frame, by means of ratchet U, pawl *l*, and eccentric-rod R. When the sides of the block are sawed, the block may be run down below the saws by turning wheel Z, and the remaining sides may be sawed square, hexagonal, or in any other polygonal form.

Claim.—The combination and arrangement of the above described devices, when the same are all arranged and operated in the precise manner and for the purpose described, and not otherwise.

No. 13,866.—JOHN A. COLE.—*Improvement in Machines for Sawing out Tapering Blocks of Marble.*—Patented December 4, 1855. (Plates, p. 263.)

The nature of this improvement will be understood from the claim and engravings.

Claim.—Attaching the saws to swinging-frames *f f* by pivots at each end, which will admit the shoes *b b* turning to any angle to follow guides *e e*, the whole being arranged in the manner and for the purpose set forth.

No. 13,916.—GEORGE T. PEARSALL.—*Improvement in Machines for Sawing Marble, &c., in Taper Form.*—Patented December 11, 1855. (Plates, p. 263.)

The lower end of the levers C, at two opposite sides of the frame B, are attached to its sides, one lever being attached by a pivot *a* and the other by a button *b*, which works in a groove b^1 in the side of the frame. The upper ends of the levers C are connected to bars D of the framing A in a similar manner. The levers C C, at each side of the frame B, cross each other, and are connected to each other by a pivot *c*. The front and back ends of the frame B and the cross-pieces E have other levers C C attached to them in the same manner. The vertical bars H are fitted in longitudinal bars G, which are placed on the saw-frame F and slide freely on the end-pieces. The vertical bars H are attached to sockets I, which slide on cross-pieces J J, and to these sockets and to the uprights *f* of the framing A there are attached levers K by pivots *g* and buttons *h*, which work in the grooves *i*. The marble block to be sawed is placed underneath the saws L L, and the screw-rods N N are turned so as to operate the vertical bars H and the bars G, and thereby the saws can attain the intended oblique position. The cross-levers C C^1 allow the frame B to move up and down in a perfectly horizontal position, and the cross-levers K K cause the vertical bars H to be moved laterally in a perfectly vertical position.

The inventor says: I do not claim the adjustable bars H, irrespective of the mode of operating them; nor do I claim the laterally moving saws placed within a reciprocating saw-frame, for they have been previously used.

But I *claim* the employment or use of the levers C K; the levers K being connected to the sockets I of the bars H, and the uprights *f* of the framing A, and the levers C being attached to the frame B and the framing A, substantially as shown, for the purpose specified.

No. 13,164.—Geo. Finley.—*Improvement in Machines for Washing Sand.*—Patented July 3, 1855. (Plates, p. 263.)

The object of this improvement is to separate from sand, as it is dug from the earth, all such substances as roots, clay, pebbles, and any soluble matter, to make it suitable for the purpose of manufacturing glass.

The water is supplied to the cistern *a* from reservoir R through pipe *d*. The water-line in the cistern is kept at a uniform height by means of adjustable valves or gates *e* and *g* at both ends of pipe *d*. One-half of the cistern is occupied by hopper *b*, which is shaped like an inverted pyramid, terminating in the spout *c*. The water enters the hopper through small holes in its side; these holes entering with a downward inclination, so that the sand thrown into the hopper cannot fall through them. The spout terminates in a jet-block *h*, from which is suspended the riddle *k*, which vibrates horizontally. The lumps of clay, stones, &c., will remain on the surface of the coarse sieve *m*, and the sand will pass through it and the fine sieve *l*. The stones, &c., which will not pass through *l*, pass out of the riddle over an inclined apron at the side of it. The sand is shovelled from the water-trough *o* into another trough, where the sand receives its final washing with clean water.

I *claim* the use of the jet-block, or its equivalent, for the purpose of distributing the sand and water over the surface of the sieve, in the manner and for the purpose set forth.

No. 12,164.—Asa Keyes.—*Improvement in Machines for Cutting and Trimming Slate.*—Patented January 2, 1855. (Plates, p. 263.)

The improvement consists in applying a rapid succession of stone-hammer blows; each of which beats off a minute piece of the slate, while the latter is carried along for the purpose by a carriage. The under surface of the slate rests on the dog D, opposite to and near the point where each blow is given. The dog, which revolves freely on its axis, rises a little higher than the upper surface of the carriage, to insure its contact with the slate. By this arrangement the inventor proposes to keep the slate, however uneven, always in contact with its rest (the circular dog), and thus to prevent the frequent breaking of the plate which attends the use of a level-rest.

Claim.—The inventor says: I do not claim as my invention the cutting of roofing-slate by means of a revolving-knife and level-rest, after the manner of shears, as patented by James Carter, of Delabole, in the county of Cornwall, England, in the year eighteen hundred and forty-five, whose entire discovery, as recited in his specification and exhibited in his drawings, I wholly disclaim.

But I *claim* the combination of the cutters or hammers C on the fly-wheel A with the circular dog D, in direct contact with which each successive portion of the slate rests to receive the blows of the cutter, while the slate is fed up by a carriage on ways F; said combination operating substantially as and for the purpose above set forth and described.

No. 12,270.—HENRY J. BRUNNER.—*Improved Instrument for Cutting out Stone.*—Patented January 23, 1855. (Plates, pp. 264, 265.)

The axes of pinions E E pass through the cutter-stock, and have ratchets E^1 E^1 on their inner ends. Motion being given to shafts S S^1, the two pinions O N are alternately revolved, and pass the cutter-stock and cutters to and fro. The racks C C are moved downwards at each forward movement of the stock B, by means of the lever G striking against the stop or pin H on the bar H^1; said lever in consequence moving the pawls F F and turning the ratchets E^1 E^1. At the end of the forward movement of the stock B, the lever G strikes against the stop or pin H^2, and the pawls F F are moved back to their original position, so as to act upon the ratchets, as before stated, at the forward movement of the stock.

Claim.—Cutting out slate or other stone from quarries by means of a cutter-stock B provided with cutters D D, and having a reciprocating motion given it by means of a toothed wheel P, in which pinions O N are made to gear alternately in consequence of the arrangement of the teeth on the periphery of said wheel P, as shown; said cutters D D having the proper feed-motion given them by the pawls F F, ratchets E^1 E^1, pinions E E, and racks C C, or other substantially equivalent device, operating as set forth.

No. 12,531.—JAMES SMITH.—*Improvement in the Manufacture of Stone Paste-Board.*—Patented March 13, 1855.

The inventor says in his specification: I pulverize a stone found in Gibsonville, Livingston county, New York, composed of silica, alumine, oxyd of iron, sulphate and carbonate of lime, and sulphate of magnesia. I then mix it with paper-pulp, and water is applied until the compound becomes about the consistency of pulp prepared for making paper, but somewhat stiffer. I use about three-fourths in weight of the pulverized stone to one-fourth of the paper-pulp. This mixture, when about the thickness of very thick pulp, I mould into sheets, as brown wrapping hand-made paper was formerly moulded. When moulded I dry the sheets, and afterwards saturate them with drying-oil, so far as they readily absorb the same.

The inventor further says: I do not claim the use of bole of any kind, or chalk of any kind, or Spanish white, or glue, or paper-pulp, or linseed oil, either separately or the whole combined; nor do I in any manner use bole, chalk, Spanish white, or glue, or a compound of which they form a part; but I *claim* sheets for roofing, boarding, and other purposes, made or constructed in the manner described, or other equivalent manner, by combining said stone when pulverized with paper-pulp. I also claim the application of and combining drying-oil with said pulverized stone and paper-pulp, combined in sheets, as aforesaid, in the manner described, or in any other equivalent manner, so as to produce the results specified, or others substantially the same.

No. 12,666.—Solomon E. Bolles.—*Improved Machine for Raising and Transporting Stones.*—Patented April 10, 1855. (Plates, p. 265.)

The derrick B is firmly connected with bed-frame C by means of braces c^1 c^1 and C^1 C^1. The braces C^1 C^1 (more distinctly shown in figures 2, 3, and 4) support also the axle-trees B^1 B^1, the inner ends of which axle-trees consist of flanges *c c*, which are secured to the bed-frame by means of bolts *d*. On these axle-trees the wheels A revolve. With the derrick is connected a windlass E for operating the lifting-chain H. Cog-wheel Y, on the axis of the chain-drum, gears into a smaller cog-wheel on a shaft F, which shaft has its supports on the bed-frame C, and is operated by crank F^1. Disc K is fast to shaft F, and a drum K^1 with a ratchet-wheel *b* revolves freely on said shaft. A pin *a*, which can be passed through a hole in the disc into a corresponding cavity in the ratchet-wheel and drum K^1, serves to cause the said drum to revolve with shaft F, if so desired. Rope J is coiled round drum K^1. By setting free ratchet *b*, applying pin *a* as aforesaid, and attaching the team to the front end of rope J, the driving-power of the team can be made to raise the stone in conjunction with the power applied at the crank F^1.

The inventor says: I do not confine myself, in the aforesaid invention and specification, to any particular size of bed-frame; to any particular height of derrick; to any particular size of tackle-block, single or compound; to any size of axle-tree, or windlass, or even to the raising of stones as aforesaid, and removing them from their beds, without digging and blasting in the ordinary manner; said invention may be used for removing stones in any practical locality and position for the building of wall, and for other useful purposes.

Claim.—The construction of an axle-tree for "stone-digger," in combination with the bed-frame and derrick, substantially and for the purposes as set forth in the specification.

No. 12,766.—Louis S. Robbins.—*Improvement in Machines for Polishing Stone.*—Patented April 24, 1855. (Plates, p. 265.)

The driving-shaft A, the bearing of the upper end of which is not represented in the figure, rests with its lower end in a socket in the bed-piece B. The derrick C can be freely moved around shaft A. The arm *c* of the derrick, and a brace E attached to it, support shaft D, around which the disc F can be freely moved. The chuck L, which carries the rubber, is attached to shaft I by universal joint. Shaft I has its bearing in sleeve H, and can freely move up and down, there being a groove in the sleeve into which a feather on the shaft I fits. The sleeve, to which is attached the pulley *k*, has its bearings in the disc F, and in the outer end of arm G, through the other end of which the shaft D plays. The pulleys and bands keep the shaft I and rubber in rapid revolution, and by moving the derrick in arcs which have their centre in axis A and the disc round shaft D, the attendant can bring the rubber to bear upon any portion of the surface of stone M.

The inventor says: I do not claim the use of a revolving self-adjusting polisher or grinder, or the manipulating apparatus, separately considered; but I *claim* the manipulating apparatus (consisting of the shaft A, crane C, radial arm G, and wheel F, as above described) in combination with the revolving and self-adjusting rubber, or polisher, constructed and arranged substantially in the manner set forth, and for the purpose specified.

No. 13,116.—THOS. HODGSON, assignor to ROBT. LANSING WRIGHT.—*Improvement in the Manufacture of Artificial Stone.*—Patented June 19, 1855; patented in England May 9, 1854.

The inventor mixes together 90 parts (by measure) of sand or pulverized stone, 30 parts of plaster of Paris, and 24 parts of beast's blood. The mould is to be varnished, and immediately before the casting is performed it is to be coated with oil. The composition is to remain in the mould for about 16 hours.

The inventor says: I do not claim the admixture of blood with sand, or other earthy or mineral matter, except in the manner and for the purpose herein specified, as I am aware that such mixtures have been used for mortar, cement, stucco, &c.; but I *claim* the composition formed by the admixture of sand or pulverized stone, plaster of Paris, and beast's blood, when these ingredients are mixed in the manner and in about (without limiting myself precisely to) the proportions herein set forth, to be moulded or cast while in a plastic state, substantially as herein described, into blocks, architectural ornaments or devices, statuary, or ornamental or other forms or figures, and in such conditions used as a substitute for stone for building, architectural or other ornamental purposes.

No. 13,343.—ELIAS A. SWAN & DE WITT C. SMILEY.—*Improvements in Machinery for Dressing and Carving Stone.*—Patented July 24, 1855. (Plates, p. 266.)

The stone, while it travels with bed B, is operated upon by the twisted cutters on shaft *g*, which are so arranged that one end of the succeeding cutter shall take the stone before the end of the preceding cutter has left the surface; thereby there is no blow or vibration in the machine, and the cutting is a gradual shearing operation. *h* is a cross-slide, which can be adjusted higher or lower, as in planing-machines.

In carving out arches, circles, or ellipses for mantels, jointing the corners of beads that come together at an angle by a segment of a circle, and in various other kinds of carved work, the axis of the horizontal cutter should be parallel to and over the radial line from which said curves are described, or else in its rotation a perfect mould or cut could not be produced, in consequence of the change of direction of the cut for this purpose; we have therefore provided an attachment to the slide-rest *r* (see figures, which are an elevation and plan from below of the same). On the slide-rest the vertical shaft of the pulley 43, which is driven by a belt, is fitted with a pinion 44, which gears to and rotates a wheel 45 on the upper end of a vertical shaft, that is set in bearings on the slide-rest *r*, and carries near its lower end a disc 46 and strong sleeves in which the shaft rotates freely; and 47 is a lever, the inner end of which sets around the sleeve under the disc 46, and a pin 48 passing through holes in both the disc and lever gives the facility for turning the disc 46 completely around, by successively shifting the pin 48 into another hole in the disc. On the sleeves attached to the disc 46 that sets around the vertical shaft, a yoke *s* is attached and provided with journals. Carrying the shaft of a horizontal cutter *t* and s^1 are mitre gears, communicating motion from the shaft of the wheel 45 to the horizontal cutter *t*, which cutter is to be twisted and formed with the desired mould; and it will now be seen that by means of the lever 41, the cutters can be turned around so as always to stand on the radial line of any curved figure that said cutter may be forming. And in cases where the cutter *t* is required to be inclined out of a horizontal plane, the same may be done by forming the yoke *s* double, and providing holes by which it may be bolted with the cutter, and at any required angle to sleeve or disc 46, and in this case the bevel gear s^1 will have to be formed with rounding teeth to prevent their jamming.

In carving stones it is very often more expedient to use a circular rotating bed, carrying the material to be operated on; for where such a bed is used beneath the slide *h*, the mere turning of the screws 21 or 52, as the circular bed revolves, will cut all around a square, circle, oval, or similar shape for table-tops or other articles. To give this facility, a pinion 49 is provided on a cross-shaft, (see figures,) which receives motion from a pinion 10, before referred to, and, by means of mitre gears 50, rotates a vertical shaft 51, set in suitable bearings, the upper end of which is provided with a square or key seat, receiving the socket of a circular bed *u*, shown by dotted line. (See figures.)

The inventors say: We are well aware that spiral cutters have been used for shearing cloth and similar purposes, and also that cutters for planing-machines have been fitted spirally, or on an incline, and also that picks for stone-dressing machines have been arranged around a cylinder in rows that are not parallel to the shaft, but inclined thereto; therefore we make no claim to any of these devices. But we are not aware that cutting-tools for dressing stone have ever before been formed as a gradual or constant twist or incline, so that no point of the cutting-edges is parallel to the axis of motion of said cutter; thereby the cut on the stone or similar rigid substances is a shearing cut at all times, and no blow is given by said cutter on the stone, and all vibration of the machine is consequently prevented, and the cut partakes more of the character of a continuous boring operation, that both preserves the cutters and machines, at the same time the best character of work is cut, and that on very thin stone that would be broken to pieces if the cutter acted with any blow.

And we are also aware that the cutting-tools of various characters of machines have been directed in their motion by patterns and screws; but we are not aware that a drawing moved with the bed, and so fitted that the cutting-tool is regulated by keeping a tracing-point over said drawing, has ever before been used, whereby the most delicate work can be cut to a drawing without the expense of a pattern, the motion given to said cutter being, from the combined operation of the cross-slide rest acting with the main bed, to produce any desired shape of cut, according to the drawing the machine is made to follow.

1st. We *claim* the method described and shown, of dressing, carving or cutting stone, or similar rigid and unyielding substances, by the use of a rotating cutter, whose cutting-edge is spiral or at an incline with the axis of said rotating cutter; thereby the cutting edge always operates obliquely on the material under treatment, and effectually prevents any blow on the stone that would produce a vibration of the machine, or break the cutter or stone for the purposes as specified.

2d. We claim the method described of fitting a horizontal or inclined rotating cutter, so that it can be kept radially with any curved mould or cut it may be forming, viz: by the use of the yoke *s* and parts attached, substantially as specified.

No. 13,896.—JOHN TAGGART, asignor to HIMSELF and VERNON BROWN.—*Improved Machine for Channelling Stone.*—Patented December 4, 1855. (Plates, p. 266.)

B is the sliding-carriage; F F are the drill-gates, which carry a series of saws *b b*, which, together with the standards U, reach into the grooves cut into the block V.

The saw-gates having a quick vertical motion imparted to them, while they are moved horizontally by and with the carriage B, the drills of each will be made to cut the stone in a straight groove. As the mechanism for elevating the saw-gates is supported by standards extending down therefrom, and into the said grooves, and resting on the bottom thereof, the said operating mechanism will be caused to descend in proportion to the descent of the drills into the stone.

I *claim* supporting the operative machinery of the drill-saws by means of standards U U, extending down therefrom, and resting upon the bottom of the grooves made in the stone by said drills, the same enabling said operative parts or machinery to move downward with the drills in proportion as they may cut into the stone.

No. 13,952.—OLDIN NICHOLS and AMMI M. GEORGE.—*Improvement in Stone-Dressing Machines.*—Patented December 18, 1855. (Plates, p. 266.)

The iron platen B B slides in ways on the top of bed A, and is driven by means of rack C and gear D on transverse shaft E. The frame F is fitted to, and swings on, shaft E; and a secondary frame G is adjusted to, and slides on, frame F, to accommodate the various thicknesses of stones. This frame G is elevated or lowered by the screw H, operated by balance-wheel I, which is attached to the frame F. To cross-bar H^2, on frame F, is fastened the lower end of screw I^2, which passes through the centre of balance-wheel J^2. This balance-wheel is connected so as to revolve in the swinging-stand K^2 on the curved supports L^2. By this means any desired inclination can be given to frame F. The toggle-joints J are suspended to the top of the sliding-frame, and at their lower ends there are the tool-holders K, which receive the picks in holes L, and the finishing tools in slots N.

On the rear side of frame G there are two stands, seen at R R, which support the main driving-shaft S with pulley T. Two eccentrics V V are adjusted on said shaft by means of grooves therein seen at W, into which the ends of set-screws enter to hold firmly the eccentrics; both set-screws, of course, entering the same groove when the eccentrics are set so as to operate the finishing-tool; and one of the set-screws will enter one of the grooves on one side of the shaft, and the other set-screw will enter the other groove, which is formed on the opposite side of the shaft one quarter way around it, when the eccentrics are so set as to operate the picks alternately. The eccentric rods seen at Y drive the toggle-joints. The bevel-gears Z and A^2, (on shaft S,) and worm-shaft B^2, with the worm F^2 in connection with gear G^2 on shaft E, serve to feed the stone to be cut, and set-screws M^2 prevent the lateral movement of frame F.

The inventors say: We do not claim the use of fixed cranks or eccentrics to work the tools of stone-dressing machines, for they have been used before; but we *claim* the combination of the movable and adjustable eccentrics with the toggle-joints, for operating or driving stone-dressing tools, arranged and operated substantially in the manner and for the purposes fully set forth.

XVI.—LEATHER.

No. 12,128.—Jesse W. Hatch and Henry Churchill.—*Improvement in Machines for Cutting out Boot and Shoe Soles.*—Patented January 2, 1855. (Plates, p. 267.)

Shaft A carries at its lower end shoe B, to which punch C is secured. Shaft A is provided with journals which fit in boxes secured to the front of slide P, which, fitting in vertical guides in the framing of the machine, receives a vertical reciprocating motion from eccentric pin *p* on the end of driving-shaft Q. (See horizontal section, figure 2.)

When shaft A reaches its highest position, spur-wheel E comes to gear with toothed-segment F^1 on front end of lever F; lever F has its fulcrum in vertical pin *a*; and at the time of shaft A being in its highest position, cam I strikes the rear end of lever F and moves it round its fulcrum *a* the proper distance to cause spur-wheel E, and with it shaft A, to describe half a revolution. When shaft A is about reaching its highest position, the square part *b* of shaft A rises above guide J, and the round part *d* of the shaft comes into contact with the guide, thereby permitting the shaft to describe its half revolution.

Claim.—Giving the cutting-knife or punch C half a revolution on its axis after every cutting operation, substantially as described, by any suitable mechanical means, for the purpose of reversing its position for the next cut, and thereby, when its ends are of unequal width, preventing the waste which, without some such provision, would be unavoidable.

No. 12,607.—Henry G. Tyer and John Helm.—*Improvement in the Manufacture of Boots and Shoes.*—Patented March 27, 1855. (Plates, p. 267.)

The object of this improvement is, to produce a boot or shoe combining the qualities of India-rubber with the lightness and elegance attainable by the use of other materials, and capable of being resoled when necessary.

The upper *a* and in-sole *b* having been cemented together, perforations are made around and through the upper *a*, until the cemented surface of the in-sole *b* has been pierced; then these holes are filled up with cement; and then the outer sole *c*, properly cemented, is pressed upon the in-sole. If sufficient strength be used slightly to separate these several parts of the sole, (see figure 2,) the rubber-filament can be seen in threads *h*, as represented in said figure 2.

The inventors say: We disclaim the use or application of this our device or invention to any other matter or thing other than is described and set forth. We *claim* the uniting of the outer sole and upper, manufactured wholly or in part of vulcanized India-rubber, with the in-sole of boots and shoes, by means of cement, the cement passing through perforations made for that purpose in the upper, in the manner substantially and for the purposes described.

No. 12,670.—John Chilcott and Robert Snell.—*Improvement in Boot-Forms.*—Patented April 10, 1855. (Plates, p. 267.)

Clamp E consists of a strip of sufficient length to reach from the nick at *f* to the top of the front-piece A. This clamp E is prevented from being pulled out laterally by entering a recess *g* in the nick, and is secured at top by latch *h* catching pin *i* on the top of clamp E. The inside of the strip is furrowed from end to end, and the recess in which it is received is correspondingly furrowed to hold the material securely. The clamps I I, fitting to the outside of the front-piece and partly over the clamp E, are attached to screws K K, which fit in female screws in rod L. This rod L is fastened at its lower end to the bottom part of the front-piece. Its upper end is attached to the front-piece by plate P, in the manner shown in the figure. Plate P is fastened to the front-piece by screw N, which can be taken out. Pins *j j* serve to hold clamps I I in proper position.

Claim.—1st. The inner clamp E fitted to a recess in the front-piece A, substantially as described, so that its exterior presents the desired surface for a part of the front-piece; whereby, after having held the first edge of the front seam of the leg secure to the front-piece A, till the whole piece of material is lapped to the proper form, it may be drawn out lengthwise and re-inserted in the front-piece inside the edge it previously held, and thus throw out the said edge, and part of the material immediately behind it, into contact with that part of the material which overlaps and is to be united to it to make the seam.

2d. The exterior clamps I I, attached to screws K K, working in an upright L, which is attached to the front-piece A, substantially as described, in a suitable position for the clamps to hold the two parts of the front seam together, and in such a manner that the top part can be easily detached from the front-piece A, to allow the said front-piece to be taken from the "upper."

No. 12,793.—WARREN HOLDEN.—*Improvement in Boot and Shoe Stretchers.*—Patented May 1, 1855. (Plates, p. 267.)

If one side of the boot is to be stretched, a knob *n* is attached to one of the parts *e*, and the corresponding side of the boot is properly moistened. Then the last is placed within the boot, and the screw *l* is turned, thereby forcing apart the sections of the last and stretching the moistened side of the boot.

If the toe of the boot requires stretching, the link *d* is disconnected, by removing the screw d^1. The turning of screw *l* (in position shown in figs. 2 and 3) will expand the front ends of the parts.

If the instep requires stretching, the levers *j j* are placed in the groove *i*, as shown in fig. 1, and the levers then expand vertically. When in groove *g*, the levers are placed horizontally.

Claim.—Dividing the last A into a number of parts *a b c c*, connected by rods *e e f f* and a link *d*, and forcing said parts outwards so as to stretch the boot or shoe at any desired part, or at all parts, by means of the device composed of the jointed levers *j j*, nuts *k k*, and rod *l*, as herein shown and described.

No. 12,794.—HOSEA B. HORTON.—*Improvement in Boot-Crimping Machines.*—Patented May 1, 1855. (Plates, p. 267.)

A piece of wet leather having been laid across the jaws, it is, by the raising of them through the lever D, forced upon the crimp form F; by the repeated sliding up and down of the jaws on the leather, it is slicked smooth and embraces the crimp form. Should there be any thin place in the leather and the wrinkle not perfectly removed, the nearest set-screw *d* is turned, and the wire *b b* made to project beyond the face of the jaw opposite the screw. The angle-iron G, clamps I, and screw-rod H, serve to draw tight the leather on the crimp form.

Claim.—The adjustable wires *b b* (made so by set-screws *d d*) on the face of the jaws B B, arranged substantially in the manner and for the purpose set forth.

No. 12,816.—THOMAS DAUGHERTY.—*Improvement in Boot-Crimps.*—Patented May 8, 1855. (Plates, p. 268.)

The slides L M fit loosely to the arms of elbow G. The nut I is provided with projections *a a*, which extend up each side of the elbow so as to form two inclined planes which correspond with the inside of the clasp K, which clasp is perforated so as to traverse freely upon screw H, and the inside of the arms are scored so as to gripe the leather upon the projections *a a* of the nut. The edges of the leather having been inserted between the slides L M and clasps P Q, and between clasp K and nut I, the screw H is turned, by which means the elbow G is moved outward, and with it the clasps, thereby stretching the leather. As the screw is turned it slips a little on the nut, and slides and draws the clasps on, so that the scored part of the clasps gripes the leather tight. The elbow, etc., are received in a groove in the crimp-board A, formed by projections E and F.

Claim.—1st. (Substantially as above described and shown,) the projections E and F, to which the leather may be tacked after it is stretched, thereby permitting the stretching apparatus to be removed and applied to another crimping-board.

2d. The nut I, as constructed with projections fitting upon both sides of the elbow, applied and operating substantially as described.

No. 12,877.—JOHN H. THOMPSON, JAS. M. THOMPSON, and H. Q. THOMPSON.—*Improved Machine for Polishing the Soles of Boots and Shoes.*—Patented May 15, 1855. (Plates, p. 268.)

No description required.

Claim.—A machine for polishing the soles of boots or shoes, having a polisher or polishers *g h* made of bone or other proper material attached to a shaft *f*, which has a reciprocating motion imparted to it in any desirable manner.

No. 12,985.—ALFRED SWINGLE, assignor to ELMER TOWNSEND.—*Improvements in Hand Pegging-Machines.*—Patented May 29, 1855. (Plates, p. 268.)

By striking with a hammer upon the top of awl-driver C, the spring slider D is driven up into handle A, thereby splitting a peg from the block which is arranged in the space L between the said slider and the peg-wood driver M; to the latter is imparted a constant tendency to press the block forward by means of an elastic band N. The awl E and peg-driver F work through holes I and K in the said slider. G is an India-rubber spring attached to the lower end of the awl-driver.

The inventor says: I do not claim combining with an awl-holder or haft, and its handle, a spring slider independent of or separate from the handle, and made to play within it and to slide on the awl; the object of such slider being to draw or force the awl out of the leather sole, or other articles, immediately after having been driven into the same.

Nor do I claim a sliding peg-receiver or spout, applied to a peg-driver and made to move therewith, and to operate as described in the patent of William Kielder and Nehemiah Hunt, dated August 15, 1854, my invention differing essentially therefrom.

But I *claim* so combining the peg chisel or cutter with the spring slider and the peg receiving and discharging passage thereof, that such peg-cutter shall be moved upwards with and by the slider, so as to separate a peg from a strip of peg-wood, as specified, the same rendering it unnecessary to employ a spring bottom to the magazine, as is required when the peg-wood is moved against the knife.

I also claim the above specified manner of applying the spring to the peg-wood driver M and magazine, viz: by employing an elastic band spring, fastening it at its two ends to the magazine and the driver respectively, and making it to play around a grooved pulley applied to the handle or magazine, as stated; such a method of applying the spring having advantages, as set forth.

No. 13,003.—LUTHER HILL.—*Improved Machine for Skiving Boot and Shoe Counters.*—Patented June 5, 1855. (Plates, p. 268.)

Lever A^1 (turning horizontally) has its fulcrum upon the upper end of lever B^1. Studs *e f* projecting from lever A^1 rest respectively against the middle of the rocker-frame S and the end of shaft F. B^1 turns upon a fulcrum at *g*, and is operated by a foot-treadle below. By means of levers A^1 B^1, the clamp-plate and the rocker-frame S are simultaneously moved towards the rotary bed; the clamp-plate being carried into contact with the leather to be skived, while the knife is borne against the thin edge of the clamp-plate and against the edge of the rotary cylinder, so that whatever may be the thickness of the piece of leather, the knife will be made to adapt itself to that thickness. At the same time that the face-clamp is moved towards the cylinder-bed, the peripheral clamp will also be borne down upon a counter placed on the periphery of the cylinder and between the same and the said clamp.

I *claim* combining with the rotary cylindric bed or carrier B, its face-clamp E, and arc-cutter M, the peripheral clamp I and its cutter; the whole being arranged and made to operate so as to perform the function of bevelling along the arc of one counter and the chord of another during one revolution of the said cylinder-bed, as specified.

I also claim combining the arc-knife or cutter M, the cylindrical bed or carrier, and the clamp thereof, by means of mechanism, substantially as described, in order that such arc-knife or cutter may adapt itself to a leather counter of any ordinary thickness, held between the clamp and the plane surface of the cylindric bed; such means or mechanism being the rocker-frame S and the adjusting lever A^1, supported and made to operate substantially as specified.

I also claim supporting the bifurcated frame T of the rocker-frame S by means of a rotary journal and clamp, or the equivalent thereof, so that the angular position of the knife M, with respect to the plane surface of the rotary carrier B, may be changed as circumstances may require.

No. 13,023.—GEO. W. ZEIGLER.—*Improvements in Boot-crimping Machines.*—Patented June 5, 1855. (Plates, p. 269.)

The jaws are corrugated, as shown in the engravings, so as to work the leather away from the angle of the crimping-iron and into the foot and leg of the boot. The nuts *m* serve to stretch the leather upon the crimping-iron. The slots in plate K are so constructed as to move the foot of the crimping-iron into or between the jaws twice as fast as it does the leg, and to give it a compound motion which, in combination with the corrugations in the jaws, work the leather in the required direction to crimp it properly.

The inventor says: 1st. I *claim* the segment-gear M and rack J, or their equivalents, in combination with the slots K K, or their equivalents, for the purpose of giving to the plate H the described motion, for the purposes set forth, substantially as described.

2d. I am aware that the jaws of boot-crimping machines have been corrugated, and the ribs and grooves made parallel with their edges; therefore I make no claim to such corrugations.

But I claim corrugating them, substantially as described, for the purpose set forth.

No, 13,063.—Jean Pierre Molliere.—*Improved Machine for Cuting the Edges of Boot and Shoe Soles.*—Patented June 12, 1855. (Plates, p. 269.)

This arrangement consists in arranging two revolving tools G and *h*, both of the same construction, only *h* being of larger size than G. The side of the shoe is first held againt tool G, placing the edge of the guard-plate *g* between the upper and the sole. The shoe being thus slipped along to the point, and the same way with the other side of the shoe, the edge of the sole is finished with the exception of the heel. The heel is cut by the larger tool *k* (it being provided with a similar guard-plate *k*) in the same manner as the sole.

Claim.—The cutting down or paring and sawing the edges of the soles and heels of boots and shoes by means of the circular tools G and *h*, revolving upon horizontal, vertical, or angular axes; the whole constructed and operated substantially as described.

No. 13,072.—Caleb H. Griffin, assignor to Caleb H. Griffin and George W. Otis.—*Improvement in Machines for Cutting out Boot and Shoe Soles.*—Patented June 12, 1855. (Plates, p. 269.)

The operator places his foot on treadle H and moves it up and down, so as to impart to the shaft *c* a reciprocating rotary motion, he at the same time holding a strip of leather on the top surface of the table, and against guide-bar *z*. After each cut, he moves the strip forward against said guide or stop-bar. The carriage I has (by the means illustrated in the engravings) an intermittent reciprocating motion imparted to it, which carries each of the knife-bars K K in succession directly underneath the depresser-bar S, which, descending thereupon, forces the knife-bar and its knife P or R downward towards the cutter-block, and cuts through the leather. Figure 4 represents the soles as cut out by the machine. Each knife is elevated from the leather by the springs *a* on the rods that guide its bar K.

The inventor says: I am aware of the machine of Richard Richards, patented on the 16th December, 1854, the two cutting-knives of such machine having been applied to opposite sides of a revolving shaft, such shaft being partially revolved at suitable periods of time in order to bring alternately each cutting-knife in succession into the required position for it to cut through the leather, when depressed by the frame or mechanism by which it was made to act upon the same. I therefore do not lay claim to any such method of applying and operating the knives, it being attended with an uncertainty of action to which my improvement is not liable.

But I *claim* the combination of the depresser-bar with the reciprocating knife-frame, its two movable knives, and their elevating springs, or equivalent machinery, such being arranged and made to operate together, substantially as specified.

I also claim the combination of mechanism for imparting to the knife-frame its intermittent reciprocating movements, as described, the said combination consisting of the two momentum-levers W X, the barrel E on the driving-shaft, and the two sets of connecting-straps *k l*, *m n*, the same being applied to the carriage and shaft, and made to operate substantially as set forth.

No. 13,073.—John M. Wimley, assignor to J. A. B. Shaw.—*Improvement in Attaching Gutta-Percha Soles to Boots and Shoes.*—Patented June 12, 1855. (Plates, p. 269.)

The melted gutta-percha is filled into mould A up to the rim *e*, and the shoe placed thereon and pressed firmly down until the sides of the shoe come in contact, all round, with the rim *e*. The gutta-percha is thus forced into and through the holes previously made through the in-sole and edge of the upper.

The inventor says: I am aware that India-rubber, after being cut out and shaped like a sole, has been united to the in-sole and upper by means of an intermediating cement, caused to penetrate holes made in the in-sole and upper; but I do not claim this, nor do I claim uniting gutta-percha and leather for any other purpose than that of manufacturing boots and shoes, as described.

But I *claim* manufacturing or making boots and shoes with the outer sole made entirely of gutta-percha, when the said outer soles are simultaneously formed and united to the upper and in-sole, by means of heat and pressure, in a mould, substantially as described and set forth.

No. 13,095.—Jean Pierre Molliere.—*Improved Machine for Cutting Leather into Strips for Boot and Shoe Soles and Heels.*—Patented June 19, 1855. (Plates, p. 269.)

The leather being placed upon table A, it is slipped under the frame; for the traveller K, being at one of the extremities of the apparatus, does not hinder this being done. To cut the edge of the leather straight, the leather is kept in position by means of ruler M, which is a piece of iron with a longitudinal groove in it, in which plays the round steel chisel O, (when the ruler is kept down,) and at each end of this groove it is provided with an opening into which the traveller K can pass. The ruler is governed by two spring-rods N buried in the thickness of the table, and whose springs always tend to keep it raised; these rods are connected under the table by yoke P attached to a treadle (not shown in the engravings). When the leather is pushed under the rule the treadle is depressed, thereby pressing down the rule and keeping the leather firmly in its place. The handle R is then moved, throwing the parts (for moving the chain) into gear, and thus setting K in motion and cutting the leather. Arrived at the end, the traveller encounters one of the fingers S attached by levers T to rod 2, forces it forward, and throws the chain out of gear, thereby arresting the traveller. The treadle is then let up, which will allow the springs N to throw up the outer M so as to let the leather pass, which is cut straight, and which the workman pulls towards him until its smooth edge comes up to the two pins 3, and then the operation above described is to be repeated, and so forth.

Claim.—The cutting up of the sides of leather into sole and heel-strips of any required breadth, by means of the self-arresting curved knife-blade O, driven alternately to the right and to the left by the Vaucanson chain H, while the leather is held in its place against the adjustable pins 3 by the spring-ruler M, the whole constructed and operated substantially as described.

No. 13,144.—Reuben H. Thompson.—*Improvement in Hand-Machines for Pegging Boots and Shoes.*—Patented June 26, 1855. (Plates, p. 270.)

Motion is given to slide H by striking upon head K. The slide H carries the awl *i* and peg-driver *j*. The peg-driver is intended to drive one peg at the same time that the awl makes the hole for the next peg, the machine being moved the distance from one to the next by a slight pressure of the hand in that direction, this distance being measured, and this motion of the machine restricted to correspond, by the spring spacer T. The point at the lower end of T (it being held down by spring W) pierces the leather sufficiently to keep the machine from being easily moved out of place if a slight downward pressure is exerted upon the machine. Just as the slider H completes its descent, the arm O on the slider operates the tumbler P, the latter taking into a notch in the spacer T, and thus withdrawing the spacer from the leather at the time the awl is sunk deepest into it. T has also a slight vibratory motion in a groove in the spacer-plate Y, limited in one direction by the side of the groove, and in the opposite direction by the eccentric adjusting pin X, by the turning of which the travel of the spacer may be accurately graduated. The spacer bears against this pin; the pressure of the hand in the direction the machine is to move, will, however, move the spacer to the opposite side of the groove; then as it is raised by each descent of the awl and peg-driver, it will spring forward against pin X, and as the awl begins to rise will descend in the same position upon the leather; and when the awl rises so as to clear the leather, the pressure of the hand in that direction overcomes the spring of the spacer, and moves forward the machine for the next peg. The peg-wood is placed between the part G of the feeder and the spring D. This spring feeder is operated by the head of the screw V, which holds the awl and peg-driver to the lower end of the slider, and which projects through a slot in the front of the case, striking, as it descends, against a bend in the feeder, and pushing it back so as to give it new hold upon the peg-wood, at the time the wood is held firmly by the cutters and a clamp E.

Claim.—1st. The spring spacer or stepping-instrument T, constructed, arranged, and operating substantially as described.

2d. I claim the spring feeder F G, constructed, arranged, and operated by the driver-slide H, substantially as described.

No. 13,272.—S. T. Parmelee.—*Improvement in Attaching Metallic Heels to India-rubber Soles.*—Patented July 17, 1855. (Plates, p. 270.)

The India-rubber is vulcanized after having been fitted within the metallic casing A. It could not be done after the vulcanizing, as vulcanized rubber cannot be rendered sufficiently soft.

The inventor says: I do not claim the mere insertion of India-rubber within metallic rims or casings, for the purpose of forming the heels of boots and shoes, for that has been previously done; but I *claim* having the metallic rims or casings A formed with recesses *a*, arranged in any proper way; so that the soft or plastic India-rubber B, mixed with the proper vulcanizing materials, may be fitted therein, and the rubber and rims or casings be permanently locked together, by subjecting the rubber to steam heat, and vulcanizing it, when fitted within the rims or casings, for the purpose as set forth.

No. 13,296.—John Arthur and Evan Arthur.—*Improvement in Machines for Cutting Boot and Shoe Uppers, Soles, &c., from Sheets of India-rubber.*—Patented July 24, 1855. (Plates, p. 270.)

The endless apron B, which has an intermittent motion, receives the sheet of India-rubber *a*; the cloth J, between the India-rubber and apron, being properly wetted by its passage through a water-trough F. Two endless chains K carry the die-frames *d* with the dies *f*. The die-frames are pivoted to the chains at *g*, and are carried by the chains through the stove M and over the roller D. When over the roller D, the sides of the die-frames come under stationary plates *h*, and are, in their onward motion, firmly pressed upon the India-rubber, which is thereby drawn over roller D at the same speed with the dies. The dies cut, or rather melt, through the rubber, taking out pieces according to the shape of the dies. The pieces are conducted to apron G, (by means of thin fingers 20 secured to a swinging-frame *r*,) while the waste, still remaining attached to the piece *a*, passes over rollers *l* and *m*, and between *m* and *n*, on to the apron *o*. As soon as the die-frame, after having performed the cutting, comes around chain-roller *b*, the extensions v of the frame will strike and pass under pin w, which will throw up the front part of the die-frame so as to make room for the die-frame, which passes below.

Claim.—1st. The cutting or separation of India-rubber, by placing it on a wet cloth or other suitable moistened surface, and submitting it to the pressure of a heated die, having an edge of the form of the article to be cut, substantially as set forth.

2d. The combination of one or more reciprocating die-frames, each carrying a set of dies with a stove, and with carrying and pressing apparatus to carry the sheet or piece of rubber, substantially as herein described; so that the dies, by their reciprocating movement, may be carried into the stove to be heated, and then returned to cut or stamp out the pattern or article from the piece, as set forth.

3d. The method of raising the die-frames to carry the dies, on their return movement towards the stove, far enough above the roller D for other die-frames to pass below them, by extending the ends $v\ v$ of the side-pieces $d\ d$ of the said frame some distance beyond the pivots $g\ g$, which connect them with the chain; and providing pins $w\ w$ for the ends $v\ v$ of the die-frames to strike against, to throw up the opposite ends carrying the dies, substantially as described.

4th. The swinging-frame r, with its fingers 20, arranged and operating substantially as described, to conduct the points, or ends of the patterns, as soon as they are cut or separated between the roller D, upon which the cutting is performed; and another roller l, by which they are at once prevented curling it, and are conveyed along towards where they are delivered from the machines.

No. 13,796.—Jean Pierre Molliere.—*Improvement in Machines for Cutting Boot and Shoe Uppers.*—Patented November 13, 1855. (Plates, p. 270.)

G is the piston of the steam-cylinder operating the bed I. The skins are laid upon said bed I, and the cutters (consisting of pieces of bent steel) placed upon the skins, and then steam-power applied to lift and press the table, skins, and cutters, against the head-block C, so as to cut pieces corresponding to the shape of the cutters.

I *claim*, by letters patent of even date with the French letters patent for the same invention, the use of a large bed-plate pressed up against a resisting body by steam or other equivalent power, by which a large number of different parts are cut out from a skin by different cutters at the same time, substantially as described.

No. 13,798.—Charles Rice and Sylvanus H. Whorf.—*Improvement in Lasting and Applying Soles to Shoes.*—Patented November 13, 1855. (Plates, p. 270.)

The sole and the upper are first placed together upon a last A, the upper being made to overlap the outer surface of the in-sole, and affixed thereto by cement. The whole being thus prepared, it is next placed within the clamping-bed B, and the parts of the latter closed. Next the platen of the press is to be depressed, so as to carry the punches on the under side of die *w* into contact with the parts of the upper which overlap the inner sole, so as to make perforations through them, and either into or through the inner sole. The platen C is then to be elevated, and cement applied to the outer surface of the inner sole and overlapping parts of the upper. Next the outer sole is to be laid upon the cemented surfaces, and the slide G moved so as to bring the die *u* directly over the last A. Next the platen is to be forced down upon the outer sole, so as to press it closely in contact with the shoe, and expel from between the soles the superfluous cement.

Claim.—The holding-clamp B and last A, as used together, and in connexion with the pressing or puncturing mechanism, or both, for the purpose of fixing soles to shoes by cement, substantially as set forth.

No. 13,827.—Charles Rice and Sylvanus H. Whorf.—*Improved Machine for Preparing Leather, for the Manufacture of Boots and Shoes.*—Patented November 20, 1855. (Plates, p. 270.)

The skin is first introduced between the feed-rollers A B of the splitting mechanism, and by them is forced against the knife C; the upper portion of the skin split by said knife C passes towards and between the draught-rollers D E, which move it forward between the roller F and rasper G, which latter roughens the under surface of it. From the rasper, the leather moves over guide-bar W, between the pressure-roller I and the perforating-roller H; and finally it passes between the roller X and brush K, which latter revolves through a vessel L, containing cement, and applies the latter to the roughened surface.

Caim.—The above described mechanism, or machine, for preparing leather for the manufacture of shoes and boots, the whole being arranged and made to operate substantially in the manner and for the purpose set forth.

No. 13,852.—John S. Lewis.—*Improvement in the Mode of Cutting the "Uppers" of Boots.*—Patented November 27, 1855. (Plates, p. 271.)

The full lines in fig. 1 represent the form of the fronts cut on this improved plan, whereas the dotted lines represent the old way of cutting fronts. It will be seen that the improved plan saves a piece of leather included between the lines *a a* and point *f*.

Claim.—Cutting the boot-front all the way down, no wider, or but little wider, than the usual width of the part which is to form the front of the leg, and filling the vacancy *b* which is thus caused when the front is crimped by a separate piece, by extending the back or counter, or by a piece produced in any other way, whereby the saving is effected in cutting the fronts, substantially as described.

No. 13,854.—Jean Pierre Molliere.—*Improvement in Machines for Rasping and Dressing the Heels and Soles of Boots and Shoes.*—Patented November 27, 1855. (Plates, p. 271.)

The operator applies the bottom of the sole first to the tool *e*, and the bottom of the heel first to the tool T[1], (all the tools turning very rapidly,) for the purpose of having the rough parts of the leather and the jags of the nails taken off; the sole is then applied to tool *f*, and the heel to tool U, to finish the dressing.

Claim.—The circular tools *e f* T[1] and U, with rasping and dressing faces, revolved on horizontal, vertical, or angular axes, for the purpose of rasping and dressing the bottoms of heels and soles of boots and shoes, the whole constructed and operated substantially as described.

No. 13,867.—Alonzo R. Dinsmoor and Levi J. Bartlett.—*Improved Instrument for Chamfering the Edges of Shoe-Soles, &c.*—Patented December 4, 1855. (Plates, p. 271.)

The piece of leather is laid upon a flat table, and the blade A and spring-presser H are borne down simultaneously upon the leather, while the end of the lever-gauge G (turning upon fulcrum *d*) is made to rest upon said table. This done, the cutting-edge *a* of the tool is to be maintained at such an angle from the table as occasion may require, and the tool is to be pushed forward upon and around that portion of the leather which is to be chamfered.

The frame B is fastened to the knife by means of a wedge *c*.

Claim.—The combination and arrangement of the lever-gauge and spring-presser with the knife-blade or chisel, substantially as specified, the same being used in manner and for the purpose essentially as explained.

No. 13,875.—Jesse W. Hatch.—*Improvement in the Machine for cutting out Boot and Shoe Soles.*—Patented December 4, 1855. (Plates, p. 271.)

The cutter is attached to a vertical shaft, which is provided with journals to work in bearings in slide P, moving in vertical guides in framing C, and receives its vertical reciprocating motion from eccentric pin *a*, at the end of the horizontal-shaft D. The shaft in this machine makes only about three-fourths of a revolution in opposite directions alternately, the said movement being produced by a treadle F, which is a lever of the first order, with its fulcrum *g* at the rear end, secured to the floor or to a suitable bed-plate; said treadle being connected by a rod *o* with a wrist *b*, at the back of the fly-wheel G of the shaft D.

While the machine is at rest, the wrist is always held in one of these two positions by a weight *r* attached to or cast on the fly-wheel, the said weight resting upon one of two fixed standards *f f*[1]. In either of these conditions of the wheel and wrist, the treadle is of course raised. The operator stands in front of the machine, in a convenient position for placing the pieces of leather or other material in a proper manner upon the table H for the action of the cutter; and when he depresses the treadle by his foot, he moves the wheel far enough to bring the weight *r* over the centre of the shaft D; but the momentum the weight has acquired in moving to that point carries it past the centre, and then the pressure of the foot being taken from the treadle, it descends by the force of gravity until it reaches the other standard, thus completing the movement of the wheel. This movement of the wheel brings down the cutter and raises it again, and, just before its termination, it moves the lever E to reverse the position of the cutter by the action of one of two projections $d\ d^1$ upon one of the prongs $e\ e^1$ of a fork on the rear-end of the lever. The prongs $e\ e^1$ of the fork are at different elevations, and the projections $d\ d^1$ at different distances from the shaft D, to correspond with the elevation of the prongs $e\ e^1$; so that when the wheel moves in the direction of the full arrow shown in figure 1, the projection *d* may pass under the higher prong *e* and strike the lower prong e^1, thus throwing the lever to the position shown in figure 2; but when the wheel moves in the direction of the dotted arrow, the projection *d* may pass over the lower prong e^1 and strike the higher prong *e*, thus throwing the lever to the position the reverse of that shown in figure 2. The movements thus given to the lever take place just before the weight comes in contact with the standards $f\ f^1$, and is just sufficient to give half a revolution to the punch-shaft. To keep the cutter-shaft from turning back too soon, the friction of a bar *h* is applied in a yoke I; on the top of the framing, between the top of the yoke I and the bar *h*, the lever E works snugly, and the bar is forced up against the lever to produce the necessary friction by two India-rubber springs *i i*, one at each end. Stop-screws *j j* are also applied to the yoke I, to stop and regulate the movement of the lever.

Claim.—1st. The projections $d\ d^1$, at different distances on the face of the wheel, and the fork *e e* on the sector-lever E, having its prongs at different elevations, combined and operating substantially as set forth.

2d. The application of the spring friction-bar *h* to the yoke I for preventing the return of the sector-lever before the proper time, in the manner specified.

No. 13,886.—James Henry Sampson.—*Improvement in Boot-Trees.*—Patented December 4, 1855. (Plates, p. 271.)

In using this tree the foot portion is put in the boot; next the leg parts are passed into the leg of the boot, and connected to the foot C by means of dove-tail D. Then the lever E is to be turned down until the sections A B of the leg are pressed apart sufficient to mould the boot into shape. When the parts A B are closed upon one another, the pressure of the part A upon the upper arm of the lever-catch will prevent the stud *c* of said catch from entering recess *d*; but whenever the parts A and B are separated or forced asunder, the spring-lever catch will be free to be moved by its spring, and so as to lock the foot C to the leg portion B.

I *claim* the combination and arrangement of the three levers, and the screw M applied to the section A and B of the tree, and made to operate together, substantially as specified.

I also claim the combination of the spring-lever catch with the foot C and front section B in the manner described, and for the purpose specified.

No. 13,901.—HENRY E. CHAPMAN.—*Improvement in Boot and Shoe Peg Cutters.*—Patented December 11, 1855. (Plates, p. 271.)

This implement is shaped like the ordinary float used by boot-makers, but, instead of being faced upon its lower surface by rough teeth forming a rasp, this float has rectangular openings *c c* at right-angles to its length, cut obliquely backward from the bottom upwards, the solid portions *s* forming cutters like plane-irons for the purpose of cutting away the pegs within a boot.

I *claim* the making of shoe-makers' floats, or peg-cutters, with planing-cutters, substantially as the same is set forth and described.

No. 13,914.—JEAN PIERRE MOLLIERE.—*Improvement in Machines for Cutting out, Punching, and Stamping the Soles of Boots and Shoes.*—Patented December 11, 1855; patented in France July 22, 1853. (Plates, p. 271.)

The cutting out, pricking and stamping, (or numbering of the size of the shoe,) is done at one blow by means of punchers *a b* of the shape of the sole and heel, provided with prickers *d* and stamps *e* and P. *m m* and *o* are the detaching rods, the rods *m* working through holes *h* in the punching-frames; the punches are operated by eccentrics upon shaft L, which eccentrics are so set that the punches will successively, one after the other, arrive at their lowest or punching position, so as to distribute more equally power and resistance. When the punchers pass upward, the pieces of leather encounter the rods *m*, which are stationary, and are detached by them from the punches. The rod *o* serves as stamp, and also as detaching-rod for the heel-pieces. The workman holds the strip and puts it under the puncher; but, as this must be done with great rapidity, the leather finds itself guided and stopped in such a way that it can be instantly moved into its place. The guides to effect this consist of two pieces *r* forming the sides of a square—one of which is graduated and has on it the numbers of the sizes. As the puncher is always in the centre, these guides must be governed by a single movement in their approach to and retreat from it. This is done by the following mechanism: Two pieces S slide on a guide-bed, and have an oblique groove in which plays a pin fixed to the guides *r;* these two pieces S are connected by the cross-tie *t*, and controlled by the screw *u*, which, being attached to a bracket, causes the traverse *t*, of the pieces S, to move forward or backward, so that the pieces S, by their oblique grooves, push forward the guides. The stopping-piece *x* is supported upon a small axis *v;* it is kept down upon the leather on the flat by the counter-weight *y* attached to the axis, while the puncher is cutting; a spring-catch hitches under the piece *z* and lifts up the stopper as fast as the puncher rises, which leaves time enough to remove the leather that has been cut; when it gets up as far as it can, the piece *z*, which has described the arc of a circle, lets the catch go, and being drawn down by the counter-weight *y*, the stopper falls back to its place.

Claim.—The cutting out of soles and heels by the blades *a a* and *b b*, from strips of hammered or other leather, sliding between the guide-pieces *r r*, and held in place by the stoppers *x x*, the pricking and stamping of the heels and soles, so cut out by the awls *d* and the stamp at the same time; the three operations being performed at one stroke, the detaching from the blades and awls of the pieces cut out, pricked, and stamped by the detaching-rods *m m* and *o*, and the adjustment of the eccentrics upon the shaft L in such manner that no two of the punchers can operate at one and the same time; the whole constructed and operated substantially as described.

No. 13,950.—Jean Pierre Molliere.—*Improvement in Machines for Polishing or Burnishing the Edges of Soles and Heels of Boots and Shoes.*—Patented December 18, 1855; patented in France January 5, 1855. (Plates, p. 272.)

The steam passes from the boiler through pipe *x* and into branch-pipes x^1, one for each burnishing-tool *p*, and can be let into the hollow shaft *l* and the chamber *p* within the tool *p* by opening the cock U. A number of these tools, (each one rapidly revolving,) of slightly varying sizes, are arranged in line. The polishing of the edge of a sole or heel is performed by presenting said edge to the revolving polishing-tool, and pressing it gently upon it.

Claim.—Of even date with the French patent, the rotary hollow-tools, capable of being heated to any degree by the admission of steam or other heating medium, into the chambers through the hollow shafts on which they turn, from the regulating valve-cocks U, for the purpose of polishing and burnishing the edges of soles and heels of boots and shoes, the whole constructed and operated substantially as described.

No. 13,951.—Jean Pierre Molliere.—*Improvement in Machines for Mounting the Uppers of Boots and Shoes on Lasts.*—Patented December 18, 1855; patented in France August 19, 1854. (Plates, p. 272.)

The nature of this improvement will be understood from the claim and engravings.

I *claim* the arrangement of the adjustable frame I and thumb-screw G, armed with its tooth-clamp H, which, pressing vertically upon the inner portion only of the heel, holds the last securely in its position, and gives free access to the parts thereof on which any work is to be done by the apparatus, the whole substantially as described.

No. 13,119.—Wm. Boyd and Wm. F. Boyd.—*Improvement in Bridle-Winkers.*—Patented June 26, 1855. (Plates, p. 272.)

The object of forming the winkers on metallic plates is to keep them in proper shape when exposed to rain and when rubbed by the horse.

Claim.—Forming the flaring or projecting portions *a* of the winkers A of horse-bridles on metallic plates *b*, as shown, and for the purpose set forth.

No. 13,306.—Kingston Goddard.—*Improvement in Bridle-Reins.*—Patented July 24, 1855. (Plates, p. 272.)

The driver for ordinary purposes has only the snaffle-rein *c* in hand; but in case of accident, the loop *i* of the curb-rein is always convenient at hand for the control of the horse.

Claim.—The arrangement of the reins, substantially as described, by making the snaffle-rein tubular, in part, to receive the curb-rein, substantially as and for the purpose set forth.

No. 12,397.—Wm. D. Titus and Robt. W. Fenwick.—*Improvement in Bridle Bits.*—Patented February 13, 1855. (Plates, p. 272.)

Claim.—The described improvement in bits for stopping runaway horses, consisting in the application of pads, so arranged and controlled by a rein that, at the pleasure of the rider or driver, they may be made to close the horse's nostrils, and thereby check respiration, as set forth. (See engraving.)

No. 12,494.—Wm. L. Whittaker.—*Improvement in Machines for Stuffing Horse-Collars.*—Patented March 6, 1855. (Plates, p. 272.)

On table A are arranged two hoppers C sliding in ways, so that they may be moved towards or from each other; they are funnel-shaped, so as to contain but little straw at the point where the stuffing-rods pass through them, and not clog the rods. The outer sides of the hoppers are inclined and are composed of rods *b*, between which pass an inverted rack, as it were, its base *c* projecting out and resting on the straw. This inverted rack is weighted by a bar *d*, which causes it at all times to press equally on the straw and carry it down to where the stuffing-rods pass through the hopper, and thus keep a regular supply of straw at that point to take the place of that carried into the collar.

The inventor says: I am aware that a hinged rack inside of a hopper has been used, which, the inventor states, can be moved up or down to change the quantity which the stuffing-rods are to carry into the collar. It is not clear how it was done; but it differs from my plan, which keeps a regular and unvarying quantity at the spot which the rods pass through. It is deemed new, therefore, in its special application.

I am also aware that a collar has been stretched while it was being filled from one end only. This is not any part of my invention, because the same difficulty arises as to its susceptibility of having the straw lapped.

I *claim* in combination with the hoppers the weighted racks for bringing down a regulated quantity of straw to take the place of that carried into the collar by the stuffing-rods, as set forth.

I also claim stuffing the collar simultaneously from both ends, by means of stuffing-rods which travel past each other at the centre of the collar, by which means the straw is evenly lapped at the centre as at the ends, substantially as described.

No. 13,087.—ROBERT R. GRAY.—*Improved Expanding Block for Horse Collars.*—Patented June 19, 1855. (Plates, p. 272.)

This block is arranged for expansion, so as to lengthen and widen a horse collar.

The inventor says: I do not claim any device in the machine separately and alone considered.

But I *claim* the arrangement of the two sets of jaws B B, C C, by which the inner jaws C C alone are actuated directly by the screw G, while the outer B B are actuated by the expansion of the inner C C, and the pressure of the springs H H, both jaws being guided by the slots D D, E E, in the manner and for the purposes set forth and described.

No. 13,132.—PETER MOODEY.—*Improvement in Horse-Collar Blocks.*—Patented June 26, 1855. (Plates, p. 272.)

This collar-block is to be placed upon a wooden frame, jointed or split at its centre, which frame is curved like the under section of a collar; and to this frame is to be fastened the block securely, by inserting the supporters H H, V V, into holes bored for their reception. The block is to be opened until its extremities are thrown back to the desired position; then the collar is to be secured at each end to the holders F F, with the rim up; and then slides are to be thrown out by means of levers E E, and secured there by the regulators G. While in this position, the collar is to be stuffed. After stuffing, the slides are to be drawn in and fastened by means of the regulators; then the extremities of the block are to be be brought together, and the holders F secured by slipping a ring over them.

Claim.—The combination and arrangement of the hinge, the slides D D, levers E E, and regulators G G, or severally the equivalents thereof, so as to secure the stretching, stuffing, and blocking of a horse-collar of leather, cloth, India-rubber, or other material, without removal from the block, in the manner substantially as described.

No. 13,189.—T. J. VAN BENSCHOTEN.—*Improvement in Horse-Collar Blocks.*—Patented July 3, 1855. (Plates, p. 273.)

The collar G is placed over the three jaws C C B; the screw-rod D is then turned, and the two jaws C C are moved in consequence backwards and outwards, owing to the oblique position of the ways *c d*. The nut E has two spring arms *f*, the outer ends of which spring arms are attached to the front ends of the jaws C C.

The inventor says: I *claim* nothing new in the arrangement of the one screw-rod between the two jaws, nor yet in the mere arrangement of the three jaws or blocks, and the effecting of a longitudinal and lateral stretch to the collar, by the one operation of the screw, when these parts have been differently operated, as such have before been used.

But I claim the arrangement and operation together, substantially as shown and described, of the obliquely sliding back jaws C, along either side or edge of the collar, on its interior, with the stationary front jaw, to the narrow or upper end of the collar, as and for the purposes set forth, and whereby the advantages specified are obtained.

And I further claim giving to the obliquely sliding jaws C, acting separately, but in concert, increased freedom of action or play on their ways, to effect simultaneously the lateral and longitudinal stretch of the collar, by connecting the freely supported operating screw-nut E with the obliquely sliding jaws, by spring arms *f*, as set forth.

No. 13,949.—S. B. McCorkle.—*Improvement in Machines for Stuffing Horse-Collars.*—Patented December 18, 1855. (Plates, p. 273.)

It will be understood by reference to the engravings, how the lever Q is operated by means of roller K, and how the front arm of lever Q (represented in dotted lines) operates the ratchet N and roller L, the pins *h* of which, during the intermittent revolutions of the roller, draw the straw downwards from the box I and present it to the stuffing-rod C.

Claim.—The cylinder L provided with teeth or rods *h*, and operated by the roller K, lever Q, arm O, and ratchet N, for the purpose of feeding the straw to the plunger C, substantially as shown and described.

No. 13,965.—Samuel Shattuc.—*Improved Horse-Collar.*—Patented December 18, 1855. (Plates, p. 273.)

The tapering washer C turns about screw E, which latter fastens the two projections D D^1 and sections A A^1 together. The washer can be turned one quarter-circle, and the screw F secured into one of the holes C^1 so as to spread the sections A A^1 more or less by bringing the thickest part of the washer more or less towards that part of the horse-collar which rests upon the animal's neck, so as to adapt the horse-collar to the size of the neck. The slot H (in projection D), through which the screw F passes, allows the sections A A^1 to vibrate in relation to each other, so as to yield to the movements of the animal.

The inventor says: I am aware that horse-collars in one unjointed piece have been known and used.

I *claim* the key F, screw-key E, and sections A A^1, provided with the projections D D^1, arranged as set forth, and combined with the washer C, constituting a jointed collar, for the purpose described.

No. 13,858.—SAMUEL E. TOMPKINS.—*Improvement in Metallic Saddle-Trees for Harnesses.*—Patented November 27, 1855. (Plates, p. 273.)

By elevating the seat C, as shown by bridge B and ribs *c c*, a groove *e* is formed on each side of it from the head to the tail of the tree, and thus provision is made for the insertion of the ends *f f* of the leather seat G and the ends *g g* of the skirts H H. This gives the saddle a neat and finished appearance even when simply riveted together.

Claim.—Providing metallic harness saddle-trees with an elevated bridge B, substantially as and for the purpose set forth.

And in combination with the same, employing ribs *c c* on the front portion of the under seat, substantially as and for the purpose set forth.

No. 12,171.—CUNO WERNER.—*Improvement in Compositions for Dressing Leather.*—Patented January 2, 1855.

This composition consists of sixty pounds of saturated infusion of oak-bark, forty pounds of common train oil, sixteen pounds of rosin, (previously melted and incorporated with four pounds of hog's lard), and one ounce of creosote.

The advantages of this improvement the inventor alleges to be cheapness and a quicker and more intense action upon the leather.

Claim.—The use or employment of a compound for dressing leather, composed of a saturated infusion of oak-bark, (or other substances affording tannin), train or fish-oil, rosin, hog's lard, and creosote, compounded and combined as herein set forth, and for the purposes specified.

No. 12,226.—HIRAM L. HALL, assignor to JAS. C. STIMPSON.—*Improvement in Processes for Making Japanned Leather.*—Patented January 9, 1855.

The composition used in this process consists of two ounces of sulphur, (freed from its acid), half an ounce of alum, and half an ounce of borax. These substances are dissolved in one quart of water, and the leather thoroughly saturated therein.

Claim.—The improvement in the process of manufacturing patent or japanned leather, which consists in applying to the leather the composition herein above described, (prepared either with or without borax), and then submitting it with the varnish coatings thereon to a high degree of heat, whereby the surface of the leather is so matured as not to be affected by any temperature or change of climate.

No. 12,392.—MATTHEW H. MERRIAM and JOSEPH B. CROSBY.—*Improvement in Leather-Splitting Machines.*—Patented February 13, 1855. (Plates, p. 274.)

The cutter-frame E slides on ways B, and receives a reciprocating motion by means of connecting-rod G and crank F on driving-shaft D^1. Thus the cutter I has a reciprocating motion, and is, at the same time, revolved around its vertical axis by means of the endless band *a*. By this means the cutter is made to act on the leather with a drawing cut. The roller *o* consists of a cylindrical part *o* and a conical part *q*, both parts being connected by a universal joint; the part *q* has its outer bearing in a lever *s*, which is so connected with hand-lever N that said conical part *q* can at any time be brought closer to or further off the roller *i*. The function of this conical part is to draw out the puckers which are found on the belly-edges of hides; when any puckers occur, the operator presses the conical part harder against the leather, not allowing it to slip as much as before until the puckers are all drawn out in consequence of the increased surface speed being made to act more or less efficiently. K is a gauge-bed, its upper surface being parallel to the plane of motion of the revolving and reciprocating cutter I. The distance between the top of this bed, or the apron L on it, and the edge of the cutter, determines the thickness of the leather when split. The cutter is vertically adjustable by a set-screw *b*. The endless apron L stretches over the rollers *j* and *i* and bed K. The spring-plate *k* presses the leather against the apron and bed K. The rollers *o i* are so geared together that the surface speed of *o* is somewhat greater than that of the apron L or roller *i*, by which means the leather is drawn "taut" between the "bite" of the rollers and the spring-plate.

The inventors say: We *claim*, 1st. The disc-cutter having a simultaneous rotary and reciprocating movement, as applied to machines for splitting leather, and other analogous purposes.

2d. We do not claim the broad device of constructing a draught-roller so that it shall have a greater circumferential velocity in one part than in another.

But we claim constructing the draught-roller *o*, so that its increased circumferential velocity may be made to act more or less efficiently as desired, substantially in the manner described.

3d. The combination of the apron L, bed K, and draught-rollers *i* and *o*, when the roller *o* is constructed substantially in the manner and for the purpose set forth.

No. 12,444.—CHARLES MORRIS.—*Improved Machine for Shaving Leather Straps.*—Patented February 27, 1855. (Plates, p. 274.)

The inventor says in his specification: I pass the two legs K into holes in the bench, with the back of the knife towards me, apply my thumb to the thumb-piece L, force back the roller H, (see fig 1,) and insert the end of the strap downward between the roller and edge of the knife, when the spring I I will force the roller forward, so as to press the leather against the edge of the knife, when, by drawing the strap through, the knife will take off a shaving of the exact thickness of the space between the edge of the knife *a* and the gauge B, while the spring E E will yield, so as to allow the roller to conform to the thickness of the strap.

Claim.—The combination of the knife *a* and gauge B with the swinging-frame F, when this frame is supported by the frame *c c*, or its equivalent, to allow this frame F to yield to the varying thickness of the strap, and the whole is constructed, combined, and made to operate as described.

No. 12,806.—NATHAN AMES, assignor to SAMEL GREEN.—*Improvement in Machines for Polishing Leather and Morocco.*—Patented May 1, 1855. (Plates, p. 274.)

By turning the wheel C, it is apparent from the figure that in one half of its revolution the tool R will be raised up from the ellipsoidal table T, while during the other half the tool will pass as near to the table as may be desired without touching it.

Claim—The above described method of raising the figuring or polishing-tool R, while passing back over the table T T; *i. e.*, by making the tool-holding hand in effect a fixed part of the connecting-arm F F, constructed and combined substantially as described, so that the machine, partaking of the nature both of a reciprocal and rotary motion, may operate without joint, noise, or friction, as easily and silently as a wheel revolving on its axle, and as rapidly as may be desired, and at the same time moving in a uniform ellipsoidal orbit over the table without touching it.

No. 12,878.—WILLIAM MCK. THORNTON.—*Improvement in Machines for Creasing the Edges of Leather Straps.*—Patented May 15, 1855. (Plates, p. 274.)

The collars are adjusted singly, or two, three, or more of them, according to the width of the strap to be creased, are brought together; by which arrangement a strap can be creased to any desired width. During this adjustment of the collars, the creasers on shaft H are lifted out of the way by bringing the lever D, which supports said shaft, into position represented in dotted lines in fig. 2.

Claim.—The movable collars 1, 2, 3, &c., of different sizes on shaft F, in combination with the movable metallic creasers I I on shaft H, having flanges extending below the periphery of the collars, and shoulders with grooves *b b*, operating in the manner and for the purposes herein set forth.

No. 13,071.—JOSHUA TURNER, Jr, assignor to ASA BENNETT and WARREN COVEL.—*Machine for Ruling Leather.*—Patented June 12, 1855. (Plates, p. 275.)

A, frame of the machine; directly over the reciprocating bed B there a shaft *a* sustained by two arms *b*, affixed to and projecting from shaft *c*, whose journals are supported by standards *d* on frame *a*; shaft *a* being free to be moved up or down. Shaft *a* carries the wheels *e*, the central eyes of which wheels are of somewhat greater diameter than the diameter of the shaft, so as to enable each wheel not only to play freely up or down, with respect to bed B, or a skin placed thereon, but also to maintain the wheel in contact with an inking-roller *g*. For the purpose of stripping narrow as well as wide skins, some one or more of the stripping-wheels *e* can be thrown out of action by means of bar S and hooks *t*, as will be understood from an inspection of the engravings. Shaft *a* carries at each end a rocker-arm *u*, which has a catch-pin *v*. which pin operates in connexion with lever-catch *w*, that turns upon fulcrum *x*, the said catch having an arm *y* carrying a stud *z*. The lifter a^1, (applied to the top surface of bed B,) when carried against stud *z*, during the forward motion of the bed, raises said stud so as to lift catch *w* above pin *v*, and allow weight b^1 to so act upon arm *u* as to move it on frame A, and to cause it to elevate the series of stripping-wheels, and keep the same elevated some distance above the bed, until stud c^1 is carried by the bed, and during its backward motion, against arm *u*, or a pin projecting therefrom, so as to move said arm out of its upright into an inclined position, and thereby cause the stripping-wheels to descend towards and upon the surface of the bed; catch *w*, in the mean time, being elevated by and dropped upon pin *v*.

The top and front view in the engravings represent only one half of the machine, the other half being exactly alike.

The inventor says: I do not claim the combination of a printing-roller with a movable bed; but I *claim* my improved manner of combining each of the two stripping-wheels of the described machine with its movable bed and inking or closing roller, the same consisting in applying said stripping-wheel to its shaft in such a manner as to allow it to rotate on the same, and to play freely thereon in any direction, in a plane perpendicularly thereto, and to such extent as to accommodate itself to the changes in the surface of the leather, or article to be stripped, as specified, and at the same time keep in close contact with its coloring-roller, or cylinder.

I also claim combining with the series of stripping-wheels, arranged together and on a shaft, as described, a mechanism for raising and maintaining either one or more of them out of action with the surface to be stripped, and the coloring-roller, during the movements of the remainder of such wheels, and the bed and surface to be stripped, under them, as specified; such mechanism being a cross-bar or frame S and its set of suspension-hooks *t t t*, arranged, supported, and made to operate substantially as described.

I also claim combining with the series of stripping-wheels and the bed, a mechanism not only for raising said wheels from a skin resting on the bed, and after it has been stripped by them, but for maintaining them entirely off the skin during the backward movement of it and the bed, as specified; the said mechanism being a rocker-arm *u*, with its weight b^1, the catch or latch *w*, the lifter a^1, and the projection c^1, as applied to either one or both ends of the shaft *a*, and made to operate as explained.

No. 13,407.—John B. Tay.—*Improvement in the Bed-Spring of Leather-Splitting Machines.*—Patented August 7, 1855. (Plates, p. 275.)

As the hide or sheet of leather usually varies more or less in thickness, the spring bed will accommodate itself to its inequalities without producing creases in the lower surface of it, the spring plate or guard *a* operating to prevent such. The screws *e* are for the purpose of adjusting the top surface of the plate *a* to its proper distance from the roller, which is usually placed over it, and between which and the back spring the leather to be split passes, in its course to the edge of the splitting-knife.

Figures 1, 2, and 4 represent only one end of the plate.

Claim.—The improved bed or back spring, as composed of a thin guard or spring-sheet of metal *a*, and a series of separate springs *b b b*, &c., united to or forming part of a plate B, as described.

No. 13,605.—Chas. Weston, T. F. Weston, and Jno. W. Weston.—*Improvement in Leather-Finishing Machines.*—Patented September 25, 1835. (Plates, p. 275.)

The bed *b* is made elastic by means of spiral springs *e*. A piece of rubber is inserted between the tool *h* and the bottom of the tool-stock *i*; *g* represents the piece of leather to be finished.

Claim.—In a machine for finishing leather, in combination with the soft elastic bed and elastic finishing-tool, the cord *p*, secured to the tool-stock for the purpose of keeping the tool clear of the leather, during its retrograde movement over the bed, as set forth.

No. 13,756.—Jeremiah A. Marden and Henry A. Butters.—*Improvement in Machines for Splitting Leather.*—Patented November 6, 1855. (Plates, p. 276.)

When a hide is passed into the machine, it is carried under and against both the roller R and the wire S, it being held against the roller R by an attendant. Should there be any cockles in the hide, such as would be liable to produce wrinkles under the action of the feed-rollers, they will be carried against the wire or rod S, and will force it upward, and cause the arms *h* to be elevated so as to throw or permit the pawl T to fall down upon the ratchet, and thereby arrest the movement of the roller R, and in consequence of the same cause friction to be generated upon the skin, so as to retain it sufficiently to enable the feed to draw out the cockles. As soon as they are taken out, the wire S will fall, and cause the pawl to be thrown off the ratchet (this movement being insured by the action of cam U on the tilting-lever *i*, when said cam is carried against said leaver by the knife-carriage, which it will be when the tilting-lever is depressed into the path of the cam).

Claim.—Combining with the feeding apparatus a mechanism, substantially as described, by which the leather may be restrained in its delivery, so as to effect the reduction of "cockles," as specified.

No. 13,819.—Theodore P. Howell and Noah F. Blanchard.—*Improvements in Treating Leather for Enamelling.*—Patented November 20, 1855. (Plates, p. 276.)

The leather is placed inside the cylinder, which consists of smaller and wider ribs *k* L; and a number of iron balls *n* covered with India-rubber, to prevent their cutting the leather, are also placed inside said cylinder. The cylinder being revolved, the leather will be softened by the falling of the balls. The dust passes out between the ribs of cylinder.

The inventors say: What we claim, in our machine for softening tanned and dry leather, is not the details thereof, separately and apart from their use in combined action.

We *claim* the combination of the cylinder as constructed with the elastic shotted-bags, as constructed and used for softening tanned and dry leather for japanning purposes.

No. 13,920.—Charles Rice and Sylvanus H. Whorf.—*Improvement in Machines for Cutting Articles from Leather.*—Patented December 11, 1855. (Plates, p. 276.)

A pack of sheets of leather is laid on the bed A, and directly under the goring and forming die P; the pack-clamp R is then to be forced downwards, so as to confine the pack of leather to the bed. Next, the platen F is to be depressed by means of lever K, so as to carry the cutting-die in contact with, and cause it to cut entirely through, the pack. The plate H, which is in contact with lever K, is connected to the platen F by rods G, extending through the shelves E and F. Against the under surface of the platen there is a rotary disc L, which is furnished with a series of screw-clamps M M M, for the purpose of confining it to the platen. The said plate turns freely upon a vertical pitman N, which projects downwards through the platen; and the shaft E slides vertically and freely through them, and is supported on an elevating spring *o*, arranged and resting on shaft E.

Affixed to the under side of the rotary plate L is a cutting and goring die P, which extends downward from the plate, and has a pack-clamp R disposed within it, and so applied to the pitman N as not only to be capable of rotating freely thereon in a horizontal plane, but of being immovable, in other respects, relatively to the pitman, being moved either upward or downward by the pitman, which is operated or forced downward by a second cam-lever *r s*, which, when in use, turns in a fulcrum C, sustained by the middle shelf D. From the above it will be seen that whenever the adjustable plate L is revolved horizontally, it will produce a corresponding rotary movement of the pack-clamp R.

In order that the cutting and goring die may be accommodated to the pack of skins, or sheets of leather or cloth, or such as may be, and so as to cut up the same to the best advantage, or with the least waste of material, the cutting-die is applied to the platen by means of the rotary plate, as described, and the pack-clamps are also applied to its shaft or pitman, in manner as specified.

The inventors say: We lay no claim to any of the devices or combinations contained in the machines described in patents Nos. 6,095 and 12.128; but we *claim* combining the cutting-die with the platen by means of a rotary and adjustable plate L, in combination with so applying the pack-clamp to its pitman that it may turn thereon when the die or cutter is revolved, in the manner and for the purpose as specified.

We also claim the described arrangement of the operative mechanism of the pack-clamp, and that by which the cutter is either depressed or elevated.

No. 12,481.—Julius C. Dickey.—*Improvement in Harness Saddle-Trees.* —Patented March 6, 1855. (Plates, p. 276.)

The inventor says: I *claim* the shank-piece or prolongation of the nut C. for the purpose of enabling me to place the turrets higher up on the yoke, where they properly belong, and to prevent the reins passing through them from being too much spread at that point, as they would be if the turrets were placed at the joint, which is limited in its position, substantially as set forth. (See engravings.)

No. 13,213.—Daniel Campbell.—*Improvement in Saddle-Trees.*—Patented July 10, 1855. (Plates, p. 276.)

The act of girting this saddle to a horse will cause the side-bars to adapt themselves to the back of the animal, whether he may be fat or lean.

The inventor says: I do not claim uniting the side-bars of a saddle-tree to the pommel and cantle by means of joints; but what I *claim* is the combination of the side-bars *a a* to the pommel and cantle by means of the springs *b b*, substantially in the manner and for the purpose set forth.

No. 13,864.—Daniel Campbell.—*Improvement in Military Saddles.*—Patented December 4, 1855. (Plates, p. 276.)

The advantage of placing the connecting-strap of the holsters below the pommel of the saddle is, that the holsters can be confined much more securely, and also that a blanket or overcoat can be slung in the strap *b* which passes through the point of the pommel, without encumbering the rider. By dividing the valise into two receptacles, and securing them immediately in rear of the legs of the cantle, the cantle is not required to rise so high as usual by several inches, which enables the rider to throw himself into the saddle much more easily, and the double valise can be secured much more firmly and closely than a valise of the usual form.

Claim.—1st. Placing the arch of the connecting-strap of the holsters E below the pommel of the saddle, and supporting the holsters upon projections from the forward ends of the side-bars of the saddle-tree, or their equivalent, substantially as set forth.

2d. Covering the holsters by means of the roof-piece B, attached to the connecting-strap of the holsters, and the covers C C, which are pivoted to the sides of the holsters, substantially as set forth.

Also, constructing the valise of two connected receptacles D D, which are supported immediately in the rear of the legs of the cantle, substantially in the manner and for the purpose set forth.

No. 13,240.—ORRIN D. VOSMUS.—*Improvement in Open Stirrups.*—Patented July 10, 1855. (Plates, p. 276.)

By arranging the loop of the stirrup-strap D, the opening *a* in the stirrup-shank C, and the sliding-loop E, as represented in the engravings, the stirrup will always be held in proper position for placing the foot therein.

The inventor says: I do not claim an open stirrup, as it is not new; but I *claim* in combination with an open stirrup a shank-piece or arm, which passes between the stirrup-straps and is held in place by a loop, or its equivalent, substantially in the manner and for the purpose set forth.

No. 12,148.—R. KEELER, assignor to L. C. ENGLAND.—*Improvement in Tanning Processes.*—Patented January 2, 1855.

The stock is prepared in the tan-pits, as usual; also the tan-liquor. Into this the oil which is to be incorporated with the stock is poured little by little, the liquor and the skins being at the same time kept in constant agitation. In consequence of this process, the inventor says, a large quantity of oil is gradually absorbed by the leather, and a larger proportion of tannin is taken up by the skins than would be if there were no oil; and hence results the greater weight, quality, and value of the article.

(No illustration.)

Claim.—The herein described improved method of tanning leather, by introducing oil into the tanning-liquor and effecting its incorporation with the leather, in combination with the tannin, substantially in the manner and for the purposes as set forth.

No. 13,403.—GEO. W. SMITH.—*Improvement in Tanning Apparatus.*—Patented August 7, 1855. (Plates, p. 276.)

The outer surface of these leaches will always be saturated with water, and thus the planks of which they are composed will be better preserved than if intermittently wet and dry, as was heretofore the case.

A is one of the tan-leaches, and L the surrounding space filled with water.

Claim.—Surrounding the ordinary tan-leaches with a water-chamber, constructed in the manner and for the purposes herein set forth; not intending to limit myself to a particular form or mode of structure, but comprising any form by which the leaches are surrounded by water-spaces, substantially as described.

No. 13,443.—Otis B. Wattles.—*Improvement in Tanning-Compounds.*—Patented August 14, 1855.

Fifty sides, or the same bulk of skins, are placed in the vat. A composition of strong lime-water, 25 pounds of salt, and 3 gallons of soap, is then put into the vat, and the sides are worked in the vat until the hair is loose. When the hair is removed, the sides are placed in a vat containing soap-suds, (from 5 to 10 gallons of soap being used,) and the sides are well handled for two or three days. The sides are worked until the lime is entirely out of them, and are then placed in a strong tannin liquor made of any vegetable tannin and thoroughly mixed with from 5 to 8 gallons of soap. The sides are worked until the grain is set, the strength of the liquors being kept up until the sides are tanned, the same proportion of soap being used.

1st. I *claim* the employment or use of soap combined with salt and lime for unhairing or depilating the hides.

2d. I claim the employment or use of soap combined with the tan-liquor for tanning the hides, substantially as described.

XVII.—HOUSEHOLD FURNITURE AND IMPLEMENTS.

No. 12,689.—Samuel N. Maxam.—*Improvement in Machine for Paring Apples.*—Patented April 10, 1855. (Plates, p. 277.)

The scroll C attached to the driving-wheel A increases from its commencing point *a* to the height *b c* at its outer end. A rod D, guided by a standard K, slides with its inner end along the edge of the scroll as the latter revolves together with the driving-wheel; spring L always presses rod D against the scroll. The rod D is hinged to the lower end of a lever which has its fulcrum in F, and the upper end of which forming a toothed sector, gears into a similar sector on the end of a slotted arm F, which latter is fast on shaft G. The rod carrying the knife is hinged to the end J of shaft G, and passes through the slot of arm F. The prongs for supporting and revolving the apple are on the shaft of a pinion gearing into the cogs of wheel A. It is apparent that arm F will describe the semi-circular movement necessary to convey the knife to the surface of the apple from one pole to the other while the apple is revolved.

The inventor says: I do not confine myself to the precise form and exact arrangement of the several parts hereinbefore described, as I may choose to vary the same, while I attain the same results by means substantially the same.

Nor do I claim any particular form or construction of the knife to be used upon the machine set forth, but may use any of the known forms of straight, curved, circular, or cylindrical knives which have long been in common use.

But I do *claim* the inclined plane-scroll C, combined with the movers

D, E, and F, or their mechanical equivalents, combined, arranged, and operating substantially in the manner and for the purposes herein set forth.

No. 13,891.—Levi Van Hoesen.—*Improvement in Machines for Paring and Slicing Apples.*—Patented December 4, 1855. (Plates, p. 277.)

A is the cutter-wheel for slicing the fruit; I are the prongs for the reception of the apple, and they are revolved by the machine when C and G are in gear, or they are revolved by applying the hand to the crank H when C and G are out of gear. The rod M of the paring-knife L is connected to a wire spring *c*, so as to allow the knife to fit the inequalities of the apple. The knife is made to revolve round the apple one-half of a circle (while the apple is being pared) by turning rod M by hand.

Claim.—The combination of a paring with a slicing-machine, when constructed and combined substantially as described; that is, with an arrangement whereby the fork carrying the apple may be turned so as alternately to be brought into play with the paring-knife and with the slicing-wheel, being at the same time thrown into gear in the former case, and out of gear in the latter.

No. 12,656.—Wm. Stoddard.—*Improvement in Folding-Bedsteads.*—Patented April 3, 1855. (Plates, p. 277.)

The nature of this invention will be understood from the claim and engravings.

Claim.—The making the side-rails C of bedsteads in sections hinged together, and also to the head and foot posts, so that they will fold together substantially as specified, in combination with the slats E which support the bedding, which slats are so constructed and provided with pins or projections F near the ends of them as to brace or hold the folding-rails firm when the bedstead is extended, essentially in the manner and for the purposes set forth.

No. 12,693.—Joseph Rodefer.—*Improvement in Bedstead-Fastenings.*—Patented April 10, 1855. (Plates, p. 277.)

C is a ring open at *d*, and inserted into a circular groove in rail *a*. The circular open space *h* in post *b* contains a pin *g*, which passes between the rail *a* and the hook-end *e* of ring *c*; the post-mortise pressing firmly against the shoulder end *f* of the ring, so that the rail will resist any force to turn it in the post. The ring is easily adjusted by a tap given either against point *e* or shoulder *f*. The mortises can all be cut by circular saws.

Claim.—The circular split ring let into a segmental annular mortise in the rail, from which its upper end projects in the form of a hook, and its lower end in form substantially as described, permitting the passage of the catch-pin in the act of insertion, affording an additiona lateral bearing and a means of adjustment, as described.

No. 12,905.—Henry Gross.—*Machine for Cutting Screws on Bedstead-Rails.*—Patented May 22, 1855. (Plates, p. 277.)

The clamp *l*, holding the end of the rail, is opened and closed by the double-screw *m n;* the lever *h* works on a pivot in the middle, and is attached at each end to the nut-blocks *e;* by throwing wedge *f* to the left or right, one of the nut-blocks will be raised, and thus the left or the right screw on shaft *c* will be acted upon. The cutter is pivoted to the cylinder-head *d*, and works through a slot in the same, as seen in figure 3, and is kept in place, when cutting, by one of the screw-heads *j*. One side of each screw-head *j* is cut off, so that when the flat side is turned towards the cutter it is released, and can then be changed by pressing down the other end, which is secured in its place in the same way by the other screw-head *j*, which has only to be turned far enough to bring the rounded side of the head over the cutter at that point. The outer end of the rail rests upon a wooden block (figure 4) with a pointed iron set in the centre.

The inventor says: I do *claim* the mechanical combination and arrangement of the double-V cutter, and the manner it is secured in the cylinder-head when used by the screw-head *j j*. Also, the changes from right to left, by which changes right and left handed screws are cut with the same cylinder-head and double-V cutter. Also, the mechanical arrangement and combination of the lever *h*, the wedge *f*, and nut-blocks *e e*, when working together in the manner described, viz: the nut-blocks being right and left longitudinal sections of female screws, and used to direct the course of the shaft to which the cylinder-head is attached, the wedge by which the nut-blocks are changed, and the lever which guides the blocks and keeps them in their proper place when the change is made. Also, in combination as above, the double clamp *l* and nut *m n*, as shown in figure 1, the screw by which the clamp is opened and closed, and the rest-block, figure 4, for supporting the rail while being cut. All the foregoing I claim as a combination and arrangement for the purposes aforesaid, new, different, and far surpassing any machine now in use for cutting these screws. All else, and all other parts of machines designed for the same purpose, I disclaim.

No. 12,940.—Thomas Arnold, assignor to W. & J. S. Arnold & Co.—*Improvement in Invalid Bedsteads.*—Patented May 29, 1855. (Plates, p. 277.)

The object of this arrangement of invalid bedsteads is, to facilitate the sitting up and stepping out of bed. The dotted lines represent the inclined positions of the movable parts of the bedstead.

The inventor says: I do not claim, separately, any of the within described parts irrespective of the general construction of the bedstead; but I do *claim* forming the main frame, or bottom A of the bedstead, in three parts *a b c*, the middle part being stationary, and the head part *a* rendered capable of being elevated, and the foot part depressed, in the manner and for the purpose set forth.

No. 12,944.—E. Daniels.—*Improvement in Invalid Bedsteads.*—Patented May 29, 1855. (Plates, p. 278.)

The nature and purpose of this improvement are evident from the claim and engravings.

The inventor says: I do not claim the flap F, neither do I claim separately a hinged or swinging-body plane, for invalid bedsteads have been previously constructed so that their bottoms may be inclined at different angles in order to vary the position of the patient; but I do *claim* the combination of the body-plane B, thigh-planes formed of two parts C C, D D, the parts D D being movable, and leg-planes E E, provided with adjustable bars *n*, the movable parts D of the thigh-planes and the bars *n* of the leg-planes being operated by the racks *i m* and pinions *h p*, as herein shown, and for the purpose as set forth.

No. 13,034.—Florian Hesz.—*Improvement in Bedsteads.*—Patented June 12, 1855. (Plates, p. 278.)

The levers 12 are operated by the diagonal rods 4 and the central nut 11. When the nut is screwed down, the upper ends of the levers are pressed on the upper edges of the rails, and thereby hold them firmly to the posts.

Claim.—The four levers 12 attached to the inside of the posts, in combination with the diagonal rods 4 4 4 4, and plate 6 connected with the screw 7; all for the purpose of holding the side and end rails of the bedstead firmly to the posts, as set forth.

No. 13,188.—Hiram Tucker.—*Improvement in Spring-Bed Bottoms.*—Patented July 3, 1855. (Plates, p. 278.)

Figure 1 is part of a plan; section figure 3 represents the bearers B, when depressed by the mattress or other weight laid upon it.

The inventor says: I am aware that an inflexible bar, extended entirely across the frame, and supported on spiral springs, has been used in a bed foundation. I do not claim such, nor consider such as making part of my invention, as it is liable to sag in the middle, and is objectionable in other respects.

I do not claim an arrangement of helical springs and horizontal and transverse bands, wherein not only is each of the longitudinal bands connected to its supporting-frame by a helical spring extended from each end of such band to the adjacent end or bar of such frame, but each transverse band or strip of spring-steel is connected to the frame by four helical springs, two of which are fastened to each end of such strip.

But I *claim* arranging and connecting the inflexible bar B and lifter, and counter-sway springs, or their equivalents, together and with the frame A, so that each, when a mattress is laid upon it, shall extend under and give support to such mattress, and operate as an elastic foundation therefor, substantially in the manner stated.

No. 13,225.—Orson Parkhurst and Daniel Bullock.—*Machine for Cutting Screws on Bedsteads.*—Patented July 10, 1855. (Plates, p. 278.)

The cutters C, the head-stocks *d d* of which are arranged at an angle of about 45°, cut a score in the tenon V of the rail R, forming a screw as the cutter-stocks are traversed by the right and left screw-shaft *g*.

We *claim* arranging and operating the rotary cutters at an angle, so that we can traverse them parallel to the rail, and cut under the concave shoulder, substantially as described.

No. 13,253.—Benjamin Eastman.—*Improvement in Invalid Bedsteads.*—Patented July 17, 1855. (Plates, p. 278.)

The patient having been previously placed upon the sacking, the frame is lowered sufficient to admit of the attachment of the sacking to the hooks at the extremities of cords Q. The frame A is then raised, elevating the sacking and patient from the bed; then, by depressing the series of arms P (to which the cords Q are attached) on one side of the shaft H, the said shaft will turn in its bearings R R, and elevate the series of arms P on the other side, thereby inclining the sacking and turning the patient on one side. The lowering of the frame will bring the patient down upon the bed on the side on which he had been turned while elevated.

The inventor says: I make no claim to elevating the patient by means of frame A with cords and pulleys, as such has been done before.

I *claim* the apparatus described, composed of a shaft H, arms P, hooked cords Q, in combination with the detachable sacking and vertically moving shaft-bearings, arranged and operating substantially as set forth for the purposes specified.

No. 13,263.—J Carroll House.—*Alarm-Bedstead.*—Patented July 17, 1855. (Plates, p. 278.)

After the clock has sounded the alarm, the hour-hand strikes an arm at the upper end of shaft *o*, rotates the latter, and thereby withdraws arm o^1 from beneath pawl F, which of its own gravity drops, and thereby relieves lever E of its support; and this in turn descends, which descent rotates shaft S, unlocking hooks *a* from catches *b*, and thus relieving frame G H of its retention, which immediately assumes an inclined position.

The inventor says: Not confining myself to any particular style or pattern of bedstead, I *claim* the employment of the tilting-frame or bed-bottom in combination with a suitable catch or series of catches, connecting it with a clock in such manner as to be tilted at any required hour by the action of the clock, the whole constructed and arranged substantially as set forth.

No. 13,265.—TYLER HOWE —*mIprovement in Bedsteads.*—Patented July 17, 1855. (Plates, p. 278.)

The slats E, of which this bed-bottom is composed, being supported by spring bearings near the middle ot their length, do not sag in the centre, as is the case with sacking and other yielding bed-bottoms.

Claim.—The described bed-bottom, consisting essentially of the slats E and the springs *f*, constructed and operating in the manner substantially as set forth.

No. 13,607.—WILLIAM WHITE.—*Improvement in Bedsteads.*—Patented September 25, 1855. (Plates, p. 279.)

The rods *a*, forming a parallelogram, are pivoted to each other at *b*. It will be seen that by tightening the wedges *i i* the four posts A will be drawn towards the centre, by means of rods *d*, with an equal force in every direction.

Claim.—The jointed parallelogram of bars, provided with rods, or their equivalents, which extend to the bedstead at several points, and are secured thereto and tightened, substantially in the manner and for the purposes set forth.

No. 13,736.—JOHN W. YOTHERS.—*Improvement in Bedstead Fastenings.*—Patented October 30, 1855. (Plates, p. 279.)

The nature of this improvement will be understood from the claim and engravings.

Claim.—In connection with the screw-collars or tenons on the ends of the rails, the combination of the tubular segmental nuts G and the buttons H with each other and with the posts B, in such a manner that by turning the said buttons the rails A can be secured with their knobs in any desired position, and by the act of thus securing the rails the buttons themselves will be drawn so closely against the outer surfaces of the posts as to make perfectly tight and insect-proof joints between the buttons and the posts, substantially as set forth.

No. 13,803.—H. N. SHERMAN.—*Improvement in Forming Heads on ed-stead Screws.*—Patented November 13, 1855. (Plates, p. 279.)

The rod of which the screw is to be formed is placed between the jaws Q R, the end of the rod bearing against the end of the rod J, which serves as a stop; motion being given the shaft B, the jaw Q is moved towards the jaw R, and a piece of the rod of proper length to form the screw is cut off by the cutter *u* and firmly secured between the two jaws Q R. The levers H H are moved simultaneously with the jaw Q, and the dies I are brought towards each other in consequence of the movement of the levers K by the cams E E, and grasp the end of the bolt, the narrow part of the dies at their inner ends coming in contact with it. The two levers H are now moved towards the ends of the jaws Q R in consequence of the prominent part of cam D acting against the end of the sliding-plate G, and the collar O of the bedstead-screw is formed in the concaves *c c*, the inner side of the collar being pressed against the ends of the jaws, the square *p* is also partially formed in the dies I. The dies are now brought closely in contact in consequence of being forced between the friction rollers *k* on levers K, the ends of levers H being inclined; and the friction rod J is then forced by projection L on cam D against the end of square *p*, and the square *p* and collar O are completed.

The inventor says: I do not claim the jaws Q R, neither do I claim the cams upon the shaft B for operating the several levers as shown, for they have been previously used for the same or for analogous purposes.

But I *claim* giving the levers H H, to which the dies I I are attached, three separate or distinct movements in succession; the first lateral movement causing the smaller parts of the dies at their inner ends to grasp the end of the rod; the second and sliding movement causing the collar *c* to be formed and the square *p* partially formed; and the third lateral movement perfecting the square and completing the head, substantially as shown and described.

I further claim the combination of the dies I I, when operated as shown, with the rod J, operated as described, the rod J serving as a stop at the commencement of the operation, and also serving subsequently as a plunger to compress the metal forming the square *p* snugly within the dies I I.

No. 13,909.—Benjamin Hinkley.—*Improvement in Bedsteads.*—Patented December 11, 1855. (Plates, p. 279.)

The bedstead rests upon an X-shaped frame, the arms A of which may have some spring. The centre of this frame rests upon a short pillar F, which stands upon the middle of another X-shaped frame B B, supported by four feet E.

The inventor says: I do not claim the cross-springs as the means of support; but I *claim* the cross bars, whether springs or not, for the support of a bedstead-frame, when the same are mounted upon a pedestal and stand, as set forth.

No. 12,479.—Chas. Crum.—*Improvement in Processes for Making Bread.*—Patented March 6, 1855.

I *claim* the suffering the dough to pass into the acetous state, then reviving it by working and breaking into it fresh, dry, unfermented flour, and the subsequent process of cutting, piercing, raising in the open air, and baking in an open oven, or an oven freely ventilated; and I claim this invention in its application to wheat-flour or any other flour of which bread is made.

No. 12,182.—Wm. Hicks.—*Improved Paint-Brush.*—Patented January 2, 1855. (Plates, p. 279.)

The guide-roller *k* is attached to an inclined adjusting-bar *l*, which fits through a socket-piece *m* on handle A, and is held in the desired position (in relation to the brush *a*) by means of a set-screw *n*. The barrel A contains the color. The other features of the implement are sufficiently explained in the following claims of the inventor.

Claim.—1st. The arrangement, substantially as herein shown and described, of the brush-holder *b* with the reservoir-handle A, regulating-valve *g*, and branch or feed *f* conveying the color on to the top of the brush, essentially as specified.

2d. In combination with the self-feeding paint-brush or pencil the adjustable guide-roller *k*, to facilitate the run of the brush in its desired course and at a speed corresponding with a free flow and the continuous supply of color to and on the brush, and serving to form a rest at various angles of operation or hold of the brush; the said roller being arranged to run on the outside of the lateral spread or splay of the brush and on the off-side of it, as and for the purposes herein set forth.

No. 12,309.—Dexter H. Chamberlain and Jno. Hartshorn.—*Fountain Brush.*—Patented January 30, 1855. (Plates, p. 279.)

By drawing the valve C, with the brush D, more or less within the tube B A, there will be a more or less free flowage of the paint. By turning the valve C within socket B, the valve may be relieved from such liquid as may at any time become dried. When the fountain is to be supplied with paint, the brush end may be inserted in the paint, while the valve is to be drawn backwards into its socket. The person applies his mouth to the outer end of tube F, and exhausts said tube and the fountain of air. The external atmospheric pressure will cause the liquid to rush into the fountain, until float H is elevated by it so as to close the opening *a*.

The inventors say: We do not claim the combination of a fountain or reservoir with a brush or marking implement. Nor do we claim a tapering valve applied to a long rod, and working in a socket or tapering hole made through the bottom of a fountain pen-holder, the long rod extending through the fountain thereof. Nor do we claim a movable pin inserted in a conical tube extending into the body of a brush, and arranged at the lower end of a fountain-tube or reservoir, such pin, in order to increase the flow of the marking fluid into the brush, being raised by pressing the brush downwards against an object.

What we *claim* is arranging or applying the brush D, the valve C, its rod E, and the socket tube B, together as described, so that not only shall the brush be fixed directly to the valve, and be movable backward and forward and around, with, and by it, but the socket be made to so encompass the valve and brush that the marking fluid may flow down around the external surface of the brush before penetrating into its interior, the same affording important advantages in cleansing the valve and maintaining the flow of marking fluid.

We do not claim the application of a piston to the reservoir, so that by the movement of such piston, the reservoir may be filled with or emptied of marking fluid.

We claim so combining with the slide E and the fountain A a mouth-tube F, open at both ends, that such tube may not only serve to enable a person to supply the reservoir with paint or marking fluid, as described, but also to enable him to move longitudinally or rotate the rod E and its valve and brush.

And we claim the float H in combination with the opening at the inner end of the tube F, and as arranged to move on the slide-rod E and within the tube A, and to operate therewith substantially in the manner and for the purposes as stated.

No. 12,876.—James H. Stimpson.—*Improvement in Butter-Coolers.*—Patented May 15, 1855. (Plates, p. 279.)

a is the ice chamber, *b* the diaphragm containing the butter; when the lid *f* is closed, the ice chamber, diaphram, and lid are then entirely surrounded and kept from contact with the external air by a stratum of still air in the surrounding casing *d* and *g*.

Claim.—The improved butter-cooler, the same consisting in the double wall cover and reservoir, with a diaphragm or shelf between them, in the manner and for the purpose set forth.

Also, making the support for the butter knife upon the cover or handle, one or both, so that the knife cannot be put in place without closing the lid, thereby securing the economy of ice and the hardness of butter.

No. 13,291.—W. H. Elliot.—*Improvement in Devices for Sealing Preserve-Cans.*—Patented July 17, 1855. (Plates, p. 280.)

The bottle or can B, packed full, is filled with water nearly to the shoulder within the mouth; a metallic tubular stopper *p k* is placed in the bottle and sealed around its edges with wax; the bottle is then completely filled with water so that no air remains in it; then the lever *d* of the diaphragm pump *e a f* is raised, and the remaining space in the cylinder *a* and neck *r* is also filled with water. Both the pump and bottle being now perfectly filled, the tube of the stopper is placed in the mouth of the pump, and pressed gently down upon the India-rubber lining *n*, so as to insure an air-tight connection. Then, by depressing lever *d*, a large vacuum being created within the pump-cylinder, the water runs from the bottle by its own weight, leaving a vacuum in its place. When the bottle is properly drained, the lever is suddenly raised, when the liquid rushes up the neck *r*, striking the plunger *h* with considerable force, driving the plug *i* into the tube of the stopper, thus forming a perfect temporary seal, which is afterwards made permanent by wax or solder; the steel tube *j* and the tube *k* of the stopper form a perfectly continuous opening, which is insured by the steel tube being received a little way into the mouth of *k* and held there by a spiral spring.

Claim.—The use of the plug *i*, or its equivalent, in sealing exhausted vessels, with or without the tube *j* in connection with the plunger *h*, or its equivalent, operating in the manner set forth.

No. 13,667.—ELLIOT SAVAGE and NOAH C. SMITH.—*Improvement in Machines for Double-seaming Cans.*—Patented October 9, 1855. (Plates, p. 280.)

A is a metallic frame; B and C are two shafts connected by gears D and E. The driving-shaft B can be raised or depressed by means of boxes *a* and *b* and screw F. A third shaft G is connected to shaft C, by the universal joint H. The shafts B and C respectively carry rollers I and K; the latter is conical at *c*, and has a shoulder *d*, and is cylindrical at *e;* the upper roller being intended to work within, while the other operates without the pan, which, when it is completing the first seaming of the bottom, is arranged as seen at P, figure 2. The position of the pan, when the second overlap is being performed—that is, its position with respect to the conical surface of the lower roller—is shown in figure 3. Directly above the cylindrical portion *e* of the lower roller, and so as to bear against the outside of the bottom *f* of the pan, there is arranged a bearing-roller L, carried by a lever M, which turns on a fulcrum at *g*, and is pressed towards the pan by means of a screw N applied to a movable latch O. In order to guide the pan by its rim, two sets of conical rollers S S^1 and R R^1 are employed, the two rollers of each set being arranged with respect to one another as seen in figure. They should be attached to a screw *h*, extended through a slotted arm U jointed to a slide V, and in such manner as to enable it to be turned transversely off the slide fastened in position by the screw, as seen at *i* i^1.

The inventors say: By our improved arrangement of the bearing-roller and the shoulder of the roller K, together and with respect to the rollers I K, we not only bring the periphery of the bearing-roller to operate against the outside surface of the bottom of the pan, but we are enabled to dispense with the operation of bevelling or partially bending the two edges of the body and the bottom of the pan, before their last bending for completing the operation of double-seaming.

We *claim* the arrangement of the periphery of the bearing-roller L, that of the roller I, the cylindrical portion, shoulder, and conical part of the roller K, substantially as specified, and so as to operate together, in manner and to effect advantages as stated.

We also claim the arrangement and application of two sets of conical rollers, so as to receive and work against the rim of a pan or vessel, and support it, as explained.

No. 13,707.—STIMMEL LUTZ.—*Improvement in Sealing Preserve-Cans.*—Patented October 23, 1855. (Plates, p. 280.)

The nature of this invention consists in providing the can *c* with an external covering E, which is slid over it so as to make it double-sided, and consequently stronger, and also in providing a cement chamber D at the bottom so that the covering E is pressed down over the can; its lower end entering within the cement chamber, and this containing soft cement, the vessel becomes hermetically sealed.

The inventor says: I do not claim broadly a self-sealing can with a groove prepared with cement, nor do I claim a ground stopper and seat, nor a screw-cap and mouth, made air-tight, whether cement be used or not.

But I *claim* sealing a double-sided can or jar at the outside at or near the bottom, in the manner and for the purpose set forth.

No. 13,009.—Wm. S. Loughborough.—*Improvement in Fastenings for Carpets.*—Patented June 5, 1855. (Plates, p. 280.)

a^1 is a bed-piece in which the clamp-wire c^1 has an inclined axis, whereby the outer end of said clamp is made to press down upon the carpet C; and when it is turned back towards the base B, it is raised from the carpet. F is the floor.

Claim.—Securing the edge of the carpet by means of a button fixed to the floor and turning upon an inclined axis, so as to be self-clamping, as described.

No. 13,094.—Felix Miller.—*Improvement in Fastenings for Carpets.*—Patented June 19, 1855. (Plates, p. 280.)

The eyelet B is secured to the floor, and the hook D engaged into the eyelet, leaving the claws *d* in an upright position. The selvage of the carpet is then stretched over the points of the claws and down upon them, and as they appear through the carpet, the points are pressed down so as to hold the carpet to its place.

Claim.—The method of laying and securing carpets down upon floors by means and use of the claw-hook, substantially as described, operating and combining with the carpets and eyelets on the floor, substantially in the manner and for the purposes set forth.

No. 13,220.—Enoch Jackman and Edwin G. Dunham.—*Improvement in Fastenings for Carpets.* Patented July 10, 1855. (Plates, p. 280.)

The nature of this improvement will be understood from the claim and engraving.

Claim.—The method of securing carpets to floors by the arrangement and application of a socket and pin, or a plate and pin so applied that the friction which is caused by the contraction of the carpet in canting the pin in the socket may prevent the pin from slipping out of the socket, all in a manner substantially as set forth, so that the carpets may be put down and taken up at pleasure with nothing but the hand.

No. 12,578.—GILBERT L. BAILEY, assignor to GILBERT L. BAILEY and MIGHILL NUTTING.—*Improvement in Castors for Furniture.*—Patented March 21, 1855. (Plates, p. 280.)

The pin or oval guide B is set in a line with the axle of the truck F, which by its action on the inside of the hole in socket E causes the truck to keep always in a right position to be moved, and also throw it from the centre.

Claim—The pin B or oval guide put through or applied to the spindle A in any manner, or its equivalent, and attached to a straight truck-frame G with a socket hole E larger than the spindle A, in the manner and for the purpose substantially as described.

No. 12,357.—WM. B. CARPENTER.—*Improvement in Combined Chair and Crib for Children.*—Patented February 6, 1855. (Plates, p. 280.)

Claim.—The chair B in combination with the standards *c c*, and hinged thereto at A, when constructed and arranged so that by the reversal of the chair, as described, the whole forms a high and low chair and crib for children, substantially in the manner set forth. (See engraving.)

No. 12,587.—LEMUEL W. FERRISS.—*Improvement in Chairs.*—Patented March 27, 1855. (Plates, p. 280.)

The nature of this invention will be understood from the engravings.

The inventor says: I do not claim a chair wherein the parallelism of the back and foot-rest rails is maintained by the arms and seat.

But I *claim* hinging the seat at its back to the back of the chair only, in combination with hinging the rails of the foot-rest to the lower end of the pieces forming the back, so that the seat shall partake of the inclination of the back and foot-rest rails, and said foot-rest rails move on a changing centre, as set forth.

No. 13,479.—JNO. CRAM, assignor to HIMSELF and JNO. S. CRAM.—*Improvement in Folding-Chairs.*—Patented August 21, 1855. (Plates, p. 281.)

Claim.—Combining the platform or seat A with the back legs C C^1 by means of the turning or front legs D D^1 and the connecting links or bars E E^1, and so that said seat or platform may be either turned down horizontally so as to be supported on both sets of legs, or they and the seat be folded together.

No. 13,765.—JAMES SADGEBURY.—*Improvement in Clothes-Clamps.*—Patented November 6, 1855. (Plates, p. 281.)

The nature of this improvement will be understood from the claim and engravings.

The inventor says: I do not claim the mechanical principle involved in the operation of this clothes-clamp, as it is well known; nor do I claim a clothes-clamp that is made to string upon the line, there being a hole made in the clamp through which the line is passed; nor do I claim a clamp made to clasp the line by means of springs.

But I *claim* the grooved button D in connexion with the grooved protuberances A B, substantially as set forth.

No. 12,775.—H. and M. Blake.—*Clothes-Pin Machine.*—Patented May 1, 1855. (Plates, p. 281.)

The clothes-pins N are placed by hand in the holders E, as the cylinder D rotates, and during the time the inner ends of the pins *n* of the clamps F are upon the smaller portion of the rim *m*. When the inner ends of the pins *n* pass upon the larger portion of the rim *m*, the inner ends of the clamps F bind upon the clothes-pins N, and secure them firmly in the holders E; and the clothes-pins thus secured are forced against the saw B, which cuts the slot in them, the side-cutters *a* smoothing the sides of the slot, the clothes-pins then passing downwards, and the cutters K K bevelling the lower edges of the groove or slot cut by the saw. Just after the clothes-pins have passed the cutters K K, the inner ends of the pins *n* pass on the smaller portion of the rim *m*, and the inner ends of the clamps F are relieved from the clothes-pins N, which then fall from the holders E by their own gravity.

The inventors say: We do not claim the holding-cylinder D, irrespective of its construction and arrangement, and the manner in which it operates in connexion with the saw B, as herein shown; neither do we claim the saw B separately, nor the cutters K K, for they have been used for analogous purposes. But we do *claim*, 1st. The employment or use of the holding-cylinder D, and circular saw B, when both are hung on permanent shafts, and operating as herein shown, so that the cylinder rotates with a comparatively slow motion, compared with the saw, and conveys, by a continuous rotary motion, the clothes-pins over or against the saw, for the purpose of forming the grooves or slots therein.

2d. Securing the clothes-pins in the holders E of the cylinder D by means of the clamps F, secured to the periphery of the cylinder D, as herein shown, and operated by the rim or ledge *m* and flanch *j*, as herein shown, so that the clothes-pins will be firmly clutched in the holders E, while being operated upon by the saw B and cutters K K, and allowed to fall therefrom when the grooves or slots are finished.

3d. The combination of the cylinder D, saw B, and cutters K K, constructed, arranged, and operating as herein shown and described.

No. 13,447.—Saml. Pierce, assignor to Curtis B. Pierce.—*Improvement in Coffee-Roasters.*—Patented August 14, 1855. (Plates, p. 281.)

The object of the peculiar form of coffee-roaster referred to in the claim is, to completely stir and mix the coffee as the vessel is revolved in the operation of roasting the coffee.

The inventor says: I am aware that it is not new to combine a cylindrical or a spherical roasting vessel with a portable furnace or other heater, nor to make the journals of a cylindrical roasting vessel hollow; and I do not claim any such combinations or modes of construction.

I *claim* constructing the roasting vessel of a series of alernate longitudinal angular parts projecting inward, and co-extensive concave portions swelling outward, substantially as described, for the purpose specified.

No. 13,595.—JOSHUA E. HALL.—*Improvement in Coffee-Pots.*—Patented September 25, 1855. (Plates, p. 281.)

The coffee being placed in the coffee-pot and the water for boiling poured upon it, the cap B is placed over the spout, and the reservoir D placed in the top and filled with cold water and covered with the lid. As fast as steam is generated, it ascends to the floor of the reservoir and up into the tube E, and is condensed by the coldness of those surfaces, and dropped back into the coffee below. All the steam being condensed and restored to the coffee, all the aroma is preserved; H acts as a safety-valve.

The inventor says: I *claim* the conical tube E, with the knob F and aperture H, which serve as its continuation; this I claim in combination with the reservoir D D, as set forth.

No. 12,383.—PHINEAS EMMONS.—*Improvement in Cracker-Machines.*—Patented February 13, 1855. (Plates, p. 281.)

When the dough is placed upon the feed-board B and motion given to the shaft I, it rotates the eccentric K, and moves (by means of rod M, lever L, and pawl N) the wheel J one tooth (the wheel J having as many teeth or notches as the polygonal bed H has sides). The feed-rollers *c c* are moved at the same time by means of proper gearing, and carry in the requisite feed of dough. Then the eccentrics R R force down the cutters *t* upon the dough for cutting it. The cutters are then withdrawn, (a clearer-plate U clearing the cutters,) and the next motion of the revolving bed-plate carries off the crackers cut, by means of the endless band.

The inventor says: I *claim* the revolving intermittent bed-plate H, operated by means of an eccentric K on a driving-shaft I, and the connecting-rod M, lever L, pawls N, and notched wheel J, in combination; and this I claim, whether the said intermittent bed-plate be or be not combined with the endless band *o* surrounding it, for the purpose of conveying away the crackers substantially as set forth; it being understood that I do not claim in general the making of the machine so as to convey the dough beneath the cutters *t*, with an intermittent motion, that having been before done in other machines by passing the dough upon an endless band, carried with an intermittent motion over a fixed table, upon which the cutters work.

No. 12,150.—P. H. NILES, assignor to NILES and RICHARDS.—*Improvement in Curtain-Fixtures.*—Patented January 2, 1855. (Plates, p. 281.)

The barrel *c*, rotating in chamber *e*, is bored out to form a chamber *g*; which chamber *g*, in addition to chamber *e*, allows the insertion of a spring F, longer than it could be if barrel *c* were solid.

Another object of the arrangement of the operative parts, as shown in the figure, is to combine them so that they may be entirely separate from the stick or roller A, and yet hold together when applied to the same.

The inventor says: I am aware that it is not new to support the pulley on a pin fixed in and projecting from a bracket; nor to make the pulley with but one flange or head *d*, and so as to have its barrel *c* work in a chamber *e*; I therefore do not claim such. But what I do *claim* is arranging the spring in that chamber of the bracket in which the body of the pulley slides.

I do not claim making the pulley-body with a chamber for the reception of a spring or for any other purpose; but what I do claim is so arranging the secondary or lesser chamber *g* with respect to the chamber *e*, that the spring F may extend into both chambers, as specified, whereby an advantage as above stated may be attained.

No. 12,271.—DEXTER H. CHAMBERLAIN and JOHN HARTSHORN.—*Improvement in Rollers for Curtains.*—Patented January 23, 1855. (Plates, p. 282.)

The spring G, mentioned in the claim, consists of a doubled (elastic) cord, the ends of which cord are tied together. The doubled ends pass through the journals E, on which the curtain-roller A revolves, and through two holes *x* and *y* in each of the heads *v* of the journal. About midway between the two journals a pin I passes transversely through the roller and between the two parts of the cord. The revolutions of the roller will of course increase the twist of the cord, and the roller will be revolved backwards as soon as left free.

The inventors say: We do not claim the application of a torsion-spring to one end only of a curtain-roller. But we *claim* our improved manner of applying the spring to the curtain-roller; that is, extending it axially entirely through the roller and its two journals, and affixing it to the roller and both its brackets, (or journals extended from and fastened to them,) substantially as specified, such not only affording the advantages which a long spring has over a short one, but also important facilities in applying the spring, or modifying its tension as occasion may require.

No. 12,661.—JOHN HARTSHORN and DEXTER H. CHAMBERLAIN, assignors to JOHN HARTSHORN.—*Improvement in Fixtures for Curtain-Rollers.*—Patented April 3, 1855. (Plates, p. 282.)

The claim and engravings sufficiently explain the nature of this invention.

The inventors say: We do not claim balancing and supporting window-curtains by means of friction upon the ends of their rods; but we *claim* the bent socket E, pivoting upon the window-jamb, and secured thereto by means of screws, as described, whereby sufficient friction may be placed upon the rod, to balance a curtain of any size or weight, without the use of springs or other contrivances for the purpose.

No. 12,524.—Frederick W. Urann.—*Improvement in Curtain-Rollers.* —Patented March 14, 1855. (Plates, p. 282.)

A is the curtain-roller; E the pulley, and *d* the pulley-head, extended into bracket C; *c* the cord.

The inventor says: I do not claim the insertion of the end or journal of the curtain-roller in a chamber or bearing in the socket that supports it.

But I *claim* extending the pulley-head into the bracket, substantially in manner as described, and for the purpose of protecting the cord of the pulley from getting between the said head and the bracket during the process of rolling up or unrolling the curtain. (See engravings.)

No. 12,540.—Wm. Z. W. Chapman and John W. Chapman.—*Improvement in Knobs for Fastening Curtains and for other like purposes.* —Patented March 21, 1855. (Plates, p. 282.)

This improvement consists in securing carriage or other curtains in their place by means of an eyelet of metal or other appropriate material, which is fastened to the curtain, being placed so as to encircle a metallic knob that is covered with India-rubber, which, being elastic, holds it, the same being easily buttoned or unbuttoned from either side.

F, eyelet; B, India-rubber covering; A, knob.

Claim.—The combination of the eyelet mentioned, or its equivalent, with a shank or knob of metal, or other material that is covered, capped, encircled, or so connected with India-rubber, or the equivalent thereof, that by its elastic nature the said eyelet may be secured to it, as and for the purposes fully set forth.

No. 12,792.—John and Jacob Hartshorn.—*Improvement in Spring-Rollers for Curtains.*—Patented May 1, 1855. (Plates, p. 282.)

The blocks K K^1 are allowed to slide freely on rod G. The projecting trough *b* of the surrounding curtain-cylinder enters into the grooves *d* of the blocks, thereby causing them to revolve, together with the cylinder. M M^1 are right and left hand spiral springs, one end fastened to the corresponding block and the other to the rod. When the springs are allowed to run down, the blocks approach the ends of the rod and the springs expand freely, whereby a great range of expansion is attained. When the springs are wound up, the length of the coil increases and the blocks approach each other; and thus the twisting of the coil, as represented at M[1], is avoided, which twisting is apt to destroy the operation of the spring.

Claim.—Attaching one end of the springs to the sliding-block K K[1], for the purpose of enabling them to increase and diminish in length, as they are wound up or expanded, in the manner and for the purpose set forth.

No. 12,866.—Purches Miles.—*Improvement in Curtain-Fixtures.*—Patented May 15, 1855. (Plates, p. 282.)

To raise or let down the curtain, the tassel C is taken in one hand, depressing the lever *f* with the thumb, thus relieving the stationary guide-cord *g* from the grasp of the arm *e* of said lever. The tassel D and curtain can then be allowed to rise or be drawn down with the other hand, the tassel C at the same time running up on the cord *g* by the winding up of cord *d*. When the hands are removed, the weight of the curtain, through the cord *d*, draws the lever-arm *e* against the cord *g*, thereby securing the tassel C, cord *d*, and curtain in their position.

Claim.—The combination of the cords *d g* with the tassel C and cam-lever, the whole being arranged and operating in the manner set forth, for the purpose of making a simple, neat, and effective curtain-fixture, as described.

No. 12,881.—Benjamin B. Webster.—*Improvement in Spring Curtain-Rollers.*—Patented May 15, 1855. (Plates, p. 282.)

By drawing cord *d* the friction bracket c^1 is moved from the roller *b*, allowing it to be operated upon by drawing the curtain or by the recoiling of the spring *a*.

The inventor says: I do not claim stopping the curtain when rolled up, or partly so, by means of friction merely, as that is done in ways not new.

Neither do I claim a spring *a*, to be or which is compressed, by unrolling the curtain so as to cause the same to roll up again when the spring *a* is let into action. But I do *claim* in combination with spring *a* the friction-spring *c*, or its equivalent, combined with and operated by a cord *b*, so as to stop the rolling up of the curtain, or to cause the same to roll, or to hold the same at any point required, substantially as set forth.

No. 13,481.—Dexter H. Chamberlain.—*Improvement in Curtain-Rollers.*—Patented August 21, 1855. (Plates, p. 282.)

As the spool E is forced towards the jamb by spring *g*, the curtain will always be balanced, and none of the available width between the jambs B is used up by the motion of the spool, as is the case in other curtain-rollers.

Claim.—Attaching the spool directly to the spindle, and causing it to revolve with the curtain-rod, when the spool is forced towards the jamb by the spring *g*, as described.

No. 13,588.—PETER H. NILES, assignor to RALPH C. WEBSTER.—*Improvement in Curtain-Fixtures.*—Patented September 18, 1855.—Ante-dated March 18, 1855. (Plates, p. 282.)

This arrangement of curtain-roller saves the space of one-half of the spool on one end of the roller, and the whole of the knob or cap on the other end. F is the bracket.

Claim.—The combination of the bracket, having a hole of double diameter, with the spring-pin and the roller-end, either with or without a spool thereon, fitted to correspond to said hole, and dispensing with the knob or cap on the other end of the roller, substantially as described.

No. 13,931.—JOHN S. MARTIN.—*Improvement in Mosquito Curtains.*—Patented December 11, 1855. (Plates, p. 282.)

The upper bar *b* is fastened to the window-frame; the lower bar *a* can be fastened to catch *e* (see fig. 1). The curtain when freed from the catch will contract by reason of the elastic bands, and access can then be had to either of the sashes (see fig. 2).

Claim.—The mosquito curtain as made of two bars *a b*, a sheet of cloth or netting *c*, and a series of elastic bands *d*, arranged and applied, and so as to operate together, substantially as set forth.

No. 12,217.—JOHN LOUIS ROLLAND.—*Improvement in Machines for Kneading Dough.* Patented January 9, 1855. (Plates, p. 283.)

Two frames, consisting of outside bars I and alternating long and short arms H and G, are (with their long arms) attached to an axle E, and are revolved within a suitable trough B. The horizontal bar I serves to lift the dough from the bottom of the trough, and in carrying it upwards the dough passes through the spaces between the arms, and is drawn out in thin blades until the frame rises gradually to a horizontal position, when the dough falls back into the trough again, and so forth.

Claim.—The use of open frames for kneading dough, composed alternately of long and short blades projecting inwardly from the cross-bars, and operating in the manner substantially as set forth.

No. 12,640.—ANDREW MURTAUGH.—*Improvement in Pulley Arrangements for Dumb Waiters.*—Patented April 3, 1855. (Plates, p. 283.)

By suspending the waiter and weight between two cords, the use of a crane in raising the waiter is dispensed with; and the arranging of the cords double over pulleys has the advantage of giving a fixed end by which they can conveniently be drawn taut. C and D are the two cords.

Claim.—The manner shown of arranging and suspending the waiter A and weight B between the cords C D, arranged double over pulleys, substantially as and for the purpose set forth.

No. 13,834.—Hiram Carsley, assignor to Himself and Edmund Brown.—*Improvement in Nutmeg-Graters.*—Patented November 20, 1855. (Plates, p. 283.)

The piston E having been drawn back by means of wire F, the nutmeg is inserted through hole *c*. When the piston is let go, it will press the nutmeg against the rasping disc B, which is to be revolved by means of crank C.

I *claim* the combination of the box and holder, and the pressure-spring or contrivance with the rasping surface of the grater, the whole being applied and made to operate together, substantially as specified.

No. 12,419.—Carington Wilson.—*Improved Griddle.*—Patented February 20, 1855. (Plates, p. 283.)

The object of this invention is to prevent the heating of the centre part before the marginal parts, incident with the ordinary griddles when used over coal fires which are smaller than the griddle.

Claim.—Constructing the griddle by attaching to the under side of the main plate, by casting or otherwise, the rim or flange *c*, into which is fitted the supplementary plate *a*, in such manner as to form an intervening space or cell *b*, substantially as and for the purpose set forth.

No. 12,860.—Ezra Fahrney.—*Improved Hominy Machine.*—Patented May 15, 1855. (Plates, p. 283.)

This improvement (in machines which employ a revolving shaft of radial beaters within a stationary cylinder) is to render the machine self-feeding and self-discharging, and capable of retaining within it for a definite time a certain quantity of corn to be crushed or beaten into grains of a uniform size. The cylinder A in which the corn is cracked is provided with two self-adjusting slides, one F over the feed-passage G, and the other under the discharge-passage H; said slides being arranged in such relation to the two cams K K^1 on wheel L that at every revolution of said wheel they are both operated and caused to open said passages one after the other, the discharge-passage being opened first by the forward cam, and owing to the slow motion of the wheel L being kept open until the contents of the cylinder have escaped. As soon as the cam escapes by the slide, it will by the action of spring J be closed; whereupon the feed-passage will be opened by the rear cam to admit a new supply of corn, and afterwards will be closed as soon as the cam escapes by this slide, and so on.

The inventor says: I do not claim the self-acting slides F F separately, nor the wheel L, with cams for operating the same, as these devices are well known as the feed-movement of grain-mills, seed-planters, &c., &c.; but I do *claim* the employment of the two self-adjusting slides F F with the two cams K K, arranged a short distance apart on a wheel L, having a slower motion than the beater shaft B, essentially as shown and described, and for the purpose as set forth.

No. 13,883.—G. M. Morris and J. Newton.—*Improvement in Machines for Scouring Knives.*—Patented December 4, 1855. (Plates, p. 283.)

The knife to be scoured is held between the two rollers. By screwing up the discs F F^1 more or less tight, the roller-surfaces can be made to bind more or less tight upon the surfaces of the knife.

Claim.—The machine described and shown for scouring knives, the same consisting of two scouring-rollers D D^1, and a trough C containing the cleansing material, said rollers being arranged over each other above the trough, and each of them formed of a series of woollen or other absorbent elastic discs G, arranged on a screw-shaft E, and forced and confined compactly together by two movable metallic discs F F^1, substantially as and for the purpose set forth.

No. 12,373.—Wm. H. Allen.—*Improvement in Machines for Chopping Meat and other Substances.*—Patented February 13, 1855. (Plates, p. 283.)

The inventor says: I do not claim the use of chopping-knives on vertical sliding-heads, playing upon a block or receptacle. Nor do I claim the cam acting upon a circular corrugated disc, as a means of combining the lifting motion with a gradual rotary one, this having been done before in machines for drilling rocks.

But I *claim* the forming a machine for chopping meat and other similar substances, by attaching the chopping-knives H H H H to a central rotary spindle F, when this is operated by the combinations of the cam M and corrugated disc K, as described. (See engraving.)

No. 12,530.—John C. Schooley.—*Improvement in processes for Curing Meats.*—Patented March 14, 1855. (Plates, p. 283.)

The air passes down through the ice in cellar A, and enters the curing-room B through opening *k*. The object of circulating artificially dried air in the room where the curing takes place is to free the apartment of any warm or moist air, (caused by the introduction of meats before they are deprived of their animal heat, &c.,) and maintain a uniform temperature from freezing-point up to any desired point sufficiently cold for the purpose of curing provisions, so that they will keep when exposed to ordinary summer weather.

Claim.—The process of curing meat, and preserving fruit and provisions, by means of circulating currents of air artificially dried by ice, or its equivalent, through the room wherein the curing takes place, substantially as and for the purposes set forth.

No. 13,990.—Alexander Lightheiser.—*Improvement in Machines for Mincing Meat.*—Patented December 25, 1855. (Plates, p. 283.)

The inventor says: I do not claim any particular shape for the cutting-edge of the knives or blades K K K.

But I *claim* the placing of the knives or blades K K K in an inclined position on the surface of the cylinder, for the purpose of propelling the meat through the machine.

No. 12,365.—James A. Taylor.—*Improvement in Mop-Heads.*—Patented February 6, 1855. (Plates, p. 283.)

To introduce a mop into the stick, the handle A is to be turned in such a manner as to allow the bars B and D to separate a short distance to admit the mop between them, which may be fastened firmly in its place, and the parts held snugly in position, by turning the handle in such a way as to wind up the cord C. The handle A is then slightly driven into the bar D, so as to fasten it securely.

Claim.—The combination of the handle A and the bars B D with the cord C, or its equivalent, the whole being constructed and combined, and operating substantially as set forth, or in any other manner substantially the same.

No. 13,347.—Oliver D. Barrett.—*Improvement in Washing-Machines.*—Patented July 31, 1855. (Plates, p. 284.)

Claim.—Providing a pail with a foot-piece and treddle, in combination with the connecting-rods, lever, and sectors, operating the rollers; by which combination the rollers are thrown apart by their own weight, and brought together by means of the foot, and the action of the mop in being pulled out between them.

No. 13,812.—Alexander Barnes.—*Improvement in Mop-Heads.*—Patented November 20, 1855. (Plates, p. 284.)

The cross-piece being withdrawn the required distance from the bow by means of the screw, the cloth is inserted between the cross-piece and bow; and when it is adjusted, the cross-piece is screwed down upon it, and firmly holds it by pressing it upon the bow.

Claim.—Attaching the screw C to the cross-piece *d*, and riveting it in such a manner that its revolution is not obstructed; combined with the bow *a* and nut B, as described, for the purpose specified.

No. 13,125.—Samuel Eakins.—*Improvement in Ice-Pitchers.*—Patented June 26, 1855. (Plates, p. 284.)

The cover P of the spout is made self-acting by means of weight S. When the pitcher is inclined, the cover will be open; and when restored to a vertical position, the cover will be closed.

Claim.—The arrangement of the spout, lid, arm, and weight, in the manner and for the purpose described.

No. 13,400.—Edward Page.—*Improvement in Molasses-Pitchers.*—Patented August 7, 1855. (Plates, p. 284.)

v is the vessel referred to in the claim, hinged at H.

Claim.—The application to molasses-cups of a vessel to catch the molasses which drips from the cup, and the vessel to swing, as described.

No. 12,210.—HUGH L. MCAVOY.—*Improvement in Refrigerators.*—Patented January 9, 1855. (Plates, p. 284.)

A represents the panes of glass or porcelain.

Claim.—The application of glass to the purpose of lining refrigerators. Glass in any form or thickness, enamelled porcelain, or anything substantially the same.

No. 13,329.—WM. MOOTRY.—*Improvement in Refrigerators.*—Patented July 24, 1855. (Plates, p. 284.)

Ice being placed in the ice-box C, the lids D and E are shut down. The register J in the ceiling K being open, air from the outside passes through the holes *a* in the outer casing through the holes *b* into the space H surrounding the ice-box, thence through the register holes into the interior A of the refrigerator, and through vent-holes L into the open air. Ice-water can be drawn off by means of cock G.

The principal object of this improvement is a thorough ventilation of the refrigerator.

Claim.—1st. The application and employment of a vessel or chamber, substantially such as described, for containing both the ice and water together, serving the double purpose of a water-cooler and frigorific body, in combination with the ordinary refrigerating box, chamber, or other insulated non-conducting cooling apparatus; the whole operating in the manner substantially as set forth, and for the use and purposes mentioned.

2d. The mode of ventilation described, by compelling the outer atmospheric air, before entering the interior, to come into mediate contact with the ice or other frigorific body, operating in the manner substantially as and for the uses and purposes mentioned.

3d. The mode described of ventilating, in combination with the cooler and refrigerating chambers, as described.

No. 14,004.—JESSE D. WHEELOCK.—*Improvement in Sad-iron Heaters.*—Patented December 25, 1855. (Plates, p. 284.)

The flat-iron is placed upon table E, the legs of which fit into tubes *d*, and rest upon spiral springs inside said tubes; by closing the lids *c* over the table, the legs of the latter will be pressed down against the spiral springs, and the heat will be retained within the chamber. As soon as the catch fastening the two lids is withdrawn, the springs will lift the table and throw open the lids.

Claim.—The use or application of spiral springs within the tubes *d*, in combination with the tube E and lids *c c*, in the manner substantially and for the purposes specified.

No. 12,660.—Ozro A. Crane and Henry J. Lewis, assignors to Ozro A. Crane.—*Improvement in Scrapers for Removing Dirt from Boots and Shoes.*—Patented April 3, 1855. (Plates, p. 284.)

The nature of this invention will be understood from the claim and engravings.

We *claim* the method described and shown of causing the brushes *c* to accommodate themselves to any size or shape of boot or shoe, and brush off both sides at once, by so attaching said brushes that they shall be forced together by springs, substantially as specified.

No. 13,566.—Newell Cleveland and Jas. J. Johnston.—*Improvement in Heaters for Smoothing-Irons.*—Patented September 18, 1855. (Plates, p. 284.)

This improvement consists in making the heater grated. Between the bars P of this heater the fire can readily pass, and at the same time the scales forming upon the surface of the heater are thrown off better than in a solid heater.

We *claim* the grated or lattice-work heater for box smoothing-irons, substantially as described and represented.

No. 12,548.—Joel Haines.—*Improvement in Extension-Tables.*—Patented March 20, 1855. (Plates, p. 284.)

B C is a case to contain the tables when closed up. The table-frames E E E having been moved out, the slatted table-top I can be unwound and drawn out.

The inventor says: I *claim* the construction and arrangement of the top so as to wind up in the case, substantially as described; it being understood that I do not claim, in general, the device of the chain of slats to wind up, as that has already been used in window-blinds and shutters, but only the peculiar purposes for which it is applied to the table-top, as set forth.

No. 12,588.—Henry A. Frost.—*Improvement in the Mode of Supporting Table-Leaves.*—Patented March 27, 1855. (Plates, p. 285.)

B is the leaf; C swing-brace attached to the leaf by loops E, which allows the brace to swing freely when the leaf is raised; when the leaf is raised, the brace swings down to the shoulder-piece D. In order to lower the leaf, the brace must be swung up under the leaf; when it is dropped, the brace occupies the space between the leaf and the bed-piece of the table.

The inventor says: I do not claim the idea of using a brace to support table-leaves, as such.

But I *claim* the application to table-leaves of a self-acting swing-brace or support, which shall operate by its own weight when the leaves are raised, substantially as set forth.

No. 12,765.—Lucius Paige.—*Improvement in Combined Table and Writing-Desk.*—Patented April 24, 1855. (Plates, p. 285.)

The box C can be raised out of recess B, from position figure 1, into position figure 2, it being hinged at *a*. The desk-board H can be either slid into horizontal grooves *g*, as in figure 1, or into inclined grooves *f*, as in figure 2. A paper-rack E is supported on central journals *d*, around which it can be swung to reverse its position, so as to bring the papers or the openings of the recesses of the rack directly against either one or the other of the two sets of doors *b* or *c* of the rack-case, in order that access can be readily obtained, either when case C is horizontal, as in figure 1, or when it is vertical, as in figure 2.

Claim.—The combination of the desk-recess B and the hinged box or case C with the table, so as to operate therewith as specified.

Also the combination of a reversible paper-rack with the hinged box or case, provided with two sets of doors on its opposite sides, as specified, and adapted to a table, so as to fold into and out of the same, in manner as described.

No. 12,518.—Elijah Morgan.—*Improvement in Washing-Machines.*—Patented March 14, 1855. (Plates, p. 285.)

The full and dotted lines represent positions of the parts of the machine, during the process of washing; the broken lines represent the wings c c^1 of the rubber open for the reception of the clothes.

Claim.—Suspending a reciprocating rubber C between the yielding-bar D and wash-board B, in such manner that said wash-board and bar may both have a vertical motion during the action of the rubber, and at the same time an expansive action or motion, due to an over-accumulation of the clothes between the rubber and wash-board, as described. (See engravings.)

No. 12,658.—Geo. W. Edgecomb.—*Improvement in Washing-Machines.*—Patented April 3, 1855. (Plates, p. 285.)

When the disc G is rotated, and when the face of the large end of one of its ribs *o* passes over the small end of a rib I on the bottom of the tub, the clothes will be forced towards the smaller ends of the two ribs, at the same time that they are forced forwards; the action of the next two ribs (when the smaller end of rib *o* passes over the large end of rib I) will force the clothes in an opposite direction, and so forth.

The inventor says: I am aware that a wash-board has been made of a conical form, having its surface higher above the bottom of the tub at the circumference than at the centre, with radial ribs of the form of a half-cone attached to it and to the bottom of the tub, with their broadest end outwards, and with spaces between them of such width and depth as to receive the clothes between them in such manner as to turn them over, and has been used before, in the washing-machine of Joel Wisner, patented November 8, 1853; and therefore I do not claim such as my invention, for the reason that I have made a marked improvement thereon.

I *claim* the alternating radial arrangement of the tapering rubbers O and I, upon the under side of the actuating disc G, and upon the bottom H of the tub, substantially in the manner and for the purpose set forth.

No. 12,757.—Jason W. Corey.—*Improvement in Spring Connecting-Rods for Washing-Machines.*—Patented April 24, 1855. (Plates, p. 285.)

The eccentric E on shaft F being revolved, the coiled spring C imparts a reciprocating motion to dasher B, which is made to slide on guides D D. The dasher agitates the water and beats the clothes against the slats L L. The binder K, being movable to any point around the eccentric, regulates the coiled spring, increasing or diminishing its stiffness. This arrangement of the coiled spring prevents, by reason of its pliancy, too great pressure and strain upon the clothes.

Claim.—The coiled spring C, combined with the eccentric E, or its equivalent, for the purposes specified.

No 12,996.—Lewis W. Colver.—*Improvement in Washing-Machines.*—Patented June 5, 1855. (Plates, p. 286.)

The connecting-rod E imparts to the crate C a reciprocating motion, which at the same time moves the crate around its axis. The rod E is connected with a pitman, crank, fly-wheel, etc.

The inventor says: I am fully aware that clothes have been washed in machines by being put in a crate or basket and rotated in a suds-box; but in these the clothes scarcely ever change their positions, and there is no way of driving the water through them (which really performs the washing). I make no claims to such means or method.

But I *claim* the washing of clothes by placing them in a crate, basket, or creel, which has a reciprocating motion through the ends or wash-box, and at the same time a rotary motion around its own axis, by which means the position of the clothes is constantly changing while they are forced through the water, and are washed without being rubbed or injured, the motion of said crate, basket, or creel being obtained by means substantially as described.

No. 13,222.—Wright Lancaster.—*Improvement in Washing-Machines.*—Patented July 10, 1855.—(Plates, p. 286.)

The rods B are intended to float when the box is properly filled with water. The box is represented in the engraving with the front part broken out so as to show the floating rods.

The inventor says: I *claim* the floating-rods marked B, which are claimed to be an improvement on any former machine, for their freedom of action, their adaptation to the washing and cleansing of clothes by friction without damage to the finest fabric, and the facility and speed with which this otherwise laborious process is performed by the use and application of this improvement. And to all other parts of the above machine I hereby disclaim all and any right to letters patent, and confine my claim especially to the aforesaid improvement B by the use of floating rods.

No. 13,372.—Samuel M. Yost.—*Improvement in Washing-Machines* —Patented July 31, 1855. (Plates, p. 286.)

I *claim* the arrangement of two corrugated rollers, one above and mashing into the other without coming in contact with the lowest lines, and each being tightly covered with canvass or other strong material, the whole combined and operating in such a manner as to effectually wash any cloth submitted to it, and without breaking the buttons or other hard substances upon the linen or cloth.

No. 13,344.—John H. Atwater—*Improvement in Washing-Machines.*—Patented July 31, 1855. Plates, p. 286.)

By turning crank-wheel *t*, motion is imparted to the washing-frame, and at the same time the platform of slats is made to move forwards or backwards by bringing pawl *x* or *y* in gear with its ratchet-wheel *e* or *f*. The pawls are operated by hand-levers *a* and *b*.

Claim.—The arrangement of the washing-frame *m n o v*, and the endless platform of slats *h h*, together with the respective parts combined therewith in such a manner that the same first mover will, at the option of the operator, simultaneously impart a reciprocating movement to the washing-frame, and a forward or a rearward movement to the said platform, or operate the said washing-frame substantially in the manner and for the purpose set forth.

No. 13,356.—Josee Johnson.—*Improvement in Washing-Machines.*—Patented July 31, 1855. (Plates, p. 286.)

The pounder is suspended upon a cord C, each end of which is secured to an upright spring B. The pounder is composed of a disc perforated with a number of holes, through which a number of pestles *d* are inserted, which by small pins through the upper portion are prevented from falling through the holes. Spiral springs J allow the pestles to accommodate themselves to the uneven surface of the clothes in the tub A, when the pounder is worked up and down.

Claim.—The arrangement and combination of disc D, pestles *d*, and spiral-springs J, or their equivalents, which form the pounder, as described and set forth.

No. 13,635.—John A. Bills.—*Improvement in Washing-Machines.*—Patented October 9, 1855. (Plates, p. 286.)

The partition *i o* is stationary, but the other two partitions *i i* swing on the central axis for the purpose of distributing the clothes (which have been introduced through the door *d*) around the inside of the machine in such a manner as to balance the machine and cause it to revolve with ease. The partitions are perforated to let the water pass through.

The inventor says: All I *claim* is the movable partitions *i i*.

No. 13,682.—JOSEPH KEECH.—*Improvement in Wash-Boards.*—Patented October 16, 1855. (Plates, p. 287.)

The face of the wash-board *f* is elevated at *m*, and depressed at *n n*. The hand will naturally rest in the valleys *n*, and draw the article over the protruding middle *m*, pressing it upon the said middle portion, and rendering the corrugations on that part as effectual in cleaning the article as if the hands were directly applied at said middle portion.

The inventor says: I disclaim expressly the curving of the corrugation, as patented by Lester Butler in 1852; but I *claim* constructing the operating-face of wash-boards of a laterally depressed and centrally elevated corrugated surface, substantially as specified, for increasing the effective operation of the board, in the manner set forth.

No. 13,692.—CHAS. LOVE.—*Improvement in Washing-Machines.*—Patented October 16, 1855. (Plates, p. 287.)

The operation of this machine will be understood from the claim and engravings.

The inventor says: I make no claim to rollers or brushes, as applied to washing-machines and separately considered; but I *claim* the construction within the tube and above its bottom of a rack composed of radial fluted cones *r*, each capable of an independent rotation, arranged and supported as described, and operating as set forth, for facilitating the washing operation.

No. 13,751.—DANIEL HALDEMAN.—*Improvement in Washing-Machines.* Patented November 6, 1855. (Plates, p. 287.)

The position of the parts of this machine, represented in figure 1 in dotted lines, sufficiently elucidates the nature of this improvement, as set forth in the claim.

Claim.—The combination of the hinged arms F, crank-shaft *i*, restraining-hooks *m*, and rubbing-board E, for the purpose of holding and operating said rubbing-board in its proper position whilst washing, and to enable the operator to raise it out of the machine, to replace the clothes by simply throwing back the restraining-hooks, and drawing the shaft (still pivoted to the arms) towards the end of the machine, as set forth.

No. 12,340.—GEORGE A. MEACHAM.—*Window-Washer.*—Patented January 30, 1855. (Plates, p. 287.)

Claim.—The arrangement of a sponge or brush E at the end of a hollow-handle or tube D, connected by a hose or pipe C to a body of water B higher than the object to be washed, so that the water flows through the said sponge or brush at the very time it is rubbed or scrubbed against the window.

No. 13,482.—Jno. J. Crooke.—*Improvement in Window-Shades.*—Patented August 21, 1855. (Plates, p. 287.)

The nature of this improvement will be understood from the claim and engraving.

Claim.—So constructing and hanging a window-shade that the roller thereof shall be capable of being raised and lowered, and at the same time shall roll or unroll the shade, and this without interfering with the fixtures for raising the bottom of the shade in the ordinary manner, as described.

No. 13,684.—John McLaughlin.—*Improvement in Wringers for Clothes.*—Patented October 16, 1855. (Plates, p. 287.)

The clothing is passed over hook H and confined in drum B between the serrations *f h*, arm D, and lever C, being pressed and held together by engaging one of the serrations *l* of lever E with pin *i* on lever C. The drum is then revolved so as to wring the clothes.

Claim.—The serrated rotary drum in combination with the ratchet levers C and E, constructed, arranged, and operating as and for the purposes specified.

XVIII.—ARTS POLITE, FINE, &c.

No. 13,044.—Edgar A. Robbins.—*Method of Tuning Accordeons.*—Patented June 12, 1855. (Plates, p. 288.)

The inventor says, in his specification: I construct an accordeon or flutina after the plan of the French semi-toned instruments, without the double semi-tone key, and apply my invention by simply tuning the reeds, as shown by the figure. The figure shows that as the scale of C fingers, the scales of E and A*b* finger; that as D fingers, so do the scales of F♯, F, B*b*; as E*b* fingers, so do those of G and B; and as the scale of F fingers, those of A and D*b* also finger; thus requiring but four modes or forms of fingering to perform the twelve major scales.

Claim.—Such mode of tuning the reeds of accordeons and flutinas as will require but four modes or forms of fingering to perform twelve scales any number of octaves within the compass of the instrument, as described.

No. 12,279.—William Ives.—*Book-Brace.*—Patented January 23, 1855. (Plates, p. 288.)

A A is the brace in the form of a book on one side; and extending its whole length is an adjustable slide B, which may be secured in its proper place by the thumb-screw C. To the upper end of the slide is attached a pointed spring-bolt D, into the lateral slot E of which the shank of the bolt may be drawn when the brace is to be placed on or taken from a shelf. To secure the foot of the brace to the shelf, there are two spurs F F. If the brace be placed at the end of a row of books on a shelf, with the slide adjusted so as to bring the shelf above within reach of the bolt, it will be seen that by detaching the shank of the book from the rest, the shelf will be pierced by the point of the bolt, which with the two spurs will hold the brace firmly in its place. The object of this invention is to confine books in an erect position on the shelves of a library, &c., thereby preventing them from falling or attaining an inclined position. The slide serves to make the brace available on transferring it to various shelves having spaces between them of greater or less width.

Claim.—The combining with the brace the pointed spring-bolt and spurs, substantially in the manner and for the purpose described.

Also, the application of the adjustable slide to the brace, substantially in the manner and for the purpose set forth.

No. 12,954.—GABRIEL LEVERICH.—*Apparatus for Paging Books.*—Patented May 29, 1855. (Plates, p. 288.)

To the slide B there are attached two frames B^1, one above and one below said slide; and to these frames levers *b* are attached, two at the front and two at the back edges of frame B. To the ends of these levers platforms C C are attached to support the book to be paged. The levers are so arranged that when one platform is depressed the other will be elevated in a corresponding degree, (the levers and platforms to be raised and lowered by operating thumb-screw *c*,) for the purpose of compensating for the unequal heights of the two sides of the book, which, of course, are of equal height only at the middle of the book. *j* are the type-wheels, which at proper times are brought down on the pages.

The inventor says: I do not claim simply placing type on the periphery of a rotating wheel for the purpose of printing, for that has been previously done in machines termed mechanical typographers; but I *claim* in combination with the type-wheels the adjustable platforms C C, for the purpose as set forth.

No. 13,166.—CHAS. FOLSOM.—*Improved Book-Clasp.*—Patented July 3, 1855. (Plates, p. 288.)

This clasp can be applied to books already bound, and may be adjusted to those of any size or thickness. *b* is a tightening-screw.

Claim.—The described book-clasp, constructed and operating in the manner as substantially set forth.

No. 13,302.—F. O. Degener.—*Improved Paging-Machine.*—Patented July 24, 1855. (Plates, p. 288.)

A number of cylinders *a* are fitted into each other, each cylinder having on its circumference a stop-flange *c*, a guide-flange *g*, (except the last cylinder, which needs none,) a pawl *f* to operate it, and a stop-spring to hold it in its place. The face of the cylinders I divide into ten equal parts, and place on those ten divisions of each cylinder the figures 1, 2, 3, 4, 5, 6, 7, 8, 9, 0; each figure to correspond with a notch in the stop-flange, and a notch in the ratchet. One of these ten divisions in the stop-flange I divide again into two equal parts *t*, into which the stop-spring *d* catches, when the machine is being set, before commencing operation, so that no other figures will show except those which count, as shown at K, where the first cylinder has moved from the commencing place and counted one, 1. When the first cylinder has counted nine, 9, then by the next move the second pawl, which has been resting on the guide-flange of the first cylinder, will drop into the ratchet of the second cylinder or wheel, which counts the tens; and when the odd of the first cylinder comes opposite the figure one, 1, of the second cylinder, then both will move together to the proper place and count ten, 10. In this manner the first cylinder, by each revolution, will move the second cylinder one figure till they count ninety-nine, 99, when at the next move the third pawl, which thus far has been resting on the guide-flange of the second cylinder, or has been supported by the second pawl by means of a stop-finger *l*, during the tenth revolution of the unit cylinder, will drop into the ratchet of the third cylinder, and then move the three cylinders or wheels and count hundred, 100. In this manner it will operate until the last cylinder or wheel has moved around, when the machine will have to be set again. The third pawl has a stop-finger *l* which rests on the second pawl, thus causing the third pawl to be supported by the second pawl, when the notch in the guide-flange of the second cylinder is brought under the pawls, when it counts 90; the second pawl supporting the third pawl by means of said stop-finger on the third pawl, during the tenth revolution of the unit cylinder, or unit series of figures. Each succeeding pawl after the second pawl must have a stop-finger, to rest on the succeeding pawl, for the reason already explained. The pawls are attached to a rotating, reciprocating, or vibrating arm D; said arm D is operated by a cam-shaped piece E, which cam E gives to the arm D and paws *ffff* their requisite motion and time of rest, when the machine is being used.

A round disk or inking-table *m*, figures 1 and 2, is placed in the type-cylinders or wheels, so that the inking-table shall be surrounded by the figures or type. A spring *n*, figure 2, is attached to the shaft or inking-table *m*, to keep said inking table in its proper position, and allow it to be raised when ink is being applied to it. An inking-roller is affixed to a jointed-arm *p*, to be operated by a cam F, to take ink from the inking-table and ink the type. A face-plate *g*, figures 1 and 2, is attached to the machine in such a manner that it shall allow the inking-roller *a* to pass under it, to take ink from the table, and then ink the type. Said face-plate has an opening *r*. Upon this face-plate and directly over this opening *r* is that part of the article laid which is to be numbered, when the platen *s* will carry it down with the face-plate and print it, and then relieve it from the type after it has been printed. The joint *p* in the roller-arm will allow the roller *o* to be pressed down when the impression is given, so as not to interfere with the face-plate.

The inventor says: I lay no claim to producing combinations by means of movable figures, in whatever position they may be placed; but,

1st. I *claim* the guide-flange or guide-flanges, in combination with the pawls, stop-finger or stop-fingers, and ratchet, for the purpose as described.

2d. I claim the cam or cam-shaped piece E, in combination with a rotating, reciprocating, or vibrating arm D and pawls *f*, for the purpose of operating movable series of figures, so as to produce different combinations; and giving to them, by means of the cam E, the requisite motion and time of rest, when the machine is in operation.

3d. I claim the combination of the inking-table being surrounded by the figures or characters to be printed.

4th. I claim the combination of the inking-table and spring, for the purpose described.

5th. I claim the combination of the spring-hinge, or joint, with the roller-arm, for the purpose described.

6th. I claim the mode of adjustment, by raising or lowering the machine, in the manner and for the purpose described.

No. 13,477.—Wm. C. Demain, assignor to A. B. Ely.—*Machine for Paging Books, &c.*—Patented August 21, 1855. (Plates, p. 289.)

The gate F, which works in ways E, carries the wheel-shaft G and other operating parts of the machine. When F is raised, the ratchet-wheel L on shaft G engages with pawl M attached to the frame, and shaft G is turned a portion of a revolution. Drum N is loose on shaft G, and is held by a carrying-ring P attached to a bracket Q projecting from the gate. R, S, S^1, etc., are the numbering-wheels, each of them having ten arms *x* with the numbers from 0 to 9 on. R is fast to the shaft. The wheels S S^1, etc., are carried by the drum; each of the latter wheels has a spring-bolt *d* which enters one or other of the grooves *f g* in the surface of the drum. When the bolts are in one of grooves *f*, the wheels are in position to be operated; when in groove *g*, they are not. *h* is a spring upon the side of wheel R, which, on arriving opposite cam T upon gate F, is pressed in, and thus as R is again revolved by the rise of the gate, the spring strikes one of the arms *x* of wheel S, by which means S is carried a short distance (the length of the cam-face); the spring, having passed the cam, flies out and disengages wheel S until R and the spring have again made another revolution, when the same operation will be repeated. Wheel S thus continues to move one notch, or one number, until, when S has made an entire revolution, spring *k* (see fig. 4) is brought opposite to spring *h*, and then the spring *k*, actuated by means of spring *h*, will operate upon the second wheel S^1 in the same manner as *h* operated upon wheel S, as above described; and R and S, moving together, will actuate S^1 until the latter has also completed one entire revolution, when the springs upon R S and S^1 are brought in a line, when S^2 will commence to be actuated by its spring, in the same manner as was previously done with S^1 and S, and so on with any number of wheels. For the purpose of repeating, if desired, the same number a given number of times, before the wheels are changed to give the next number, the repeating-wheel L^1 is attached to shaft G, so as to allow it to revolve with slight friction upon the shaft; the broad-footed pawl X is inserted in place of pawl M; the notches *t* are deep, to allow the pawl to enter deep enough to catch the teeth of L, and turn the shaft. When, however, the pawl falls into the shallow notches *r*, it is held off from L; and when the gate rises the repeating-wheel is carried round upon the shaft without turning it, and the numbering-wheels remain unchanged. Thus, the same number will be printed twice in succession. Figure 6 represents a repeating-wheel with nine shallow and only one deep notch, by means of which the same number will be printed nine times in succession.

The printing apparatus is not exhibited in the engravings, as it does not form part of the claims.

Claim.—Operating the numbering-wheels, by means of the springs *h k*, whereby the first wheel is caused to actuate all the others, and the operation of the machine is rendered automatic, in the manner set forth.

2d. The repeating-wheel operating according to the form and frequency of the notches thereon, substantially as described.

3d. The drum, with its notches *f g*, in combination with the numbering wheels.

4th. The gate F, in combination with the numbering wheels, and the parts which set them in motion.

No. 13,501.—M. Riehl.—*Machine for Trimming Books.*—Patented August 28, 1855. (Plates, p. 289.)

The nature of this invention will be understood from the claim and engravings.

I *claim* hanging or attaching the knife H to the cross-piece B of the uprights A A by the arms *a a*, as shown, whereby a drawing or oblique cut of the knife is obtained, and operating the knife by means of the worm-wheel E, screw F, and connecting-bar D.

No. 12,477.—E. C. Benyaurd.—*Self-Catch for Breast-Pins, &c.*—Patented March 6, 1855. (Plates, p. 289.)

This invention consists in combining together two slotted cylinders, the one within the other, the outer one being soldered to the back of a piece of jewelry with its slotted side up, and the inner cylinder fitting nicely within the outer one, so as to be capable of being rotated therein, by means of a thumb and finger piece fixed thereto; so that the point-end of the pin proper may be admitted or released at pleasure through the slot when they are matched or brought together, and held securely when the inner cylinder is so as to close the slot in the outer cylinder.

The inventor says: I do not claim the application of a safe-catch generally, for the purpose of holding the point-end of the pin or a piece of jewelry.

But I *claim* the application and use of a safe-catch constructed substantially as described, for the purpose of holding safely and securely the point-end of the pin of breast-pins, cuff-pins, chatelaines, or any other piece of jewelry requiring a catch and pin.

No. 12,700.—ALBERT S. SOUTHWORTH.—*Plate-Holder for Cameras.*—Patented April 10, 1855. (Plates, p. 290.)

This plate-holder consists of a stationary casing A B E F. B is a zinc plate in front of the daguerreotype-plate Y, and contains a square opening C equal to one-fourth of the plate Y. The hollow square space within part E of the casing is of proper dimensions, so that when the frame G, holding plate Y, is successively slid into the four corners of said hollow space, the parts 1^1 2^1 3^1 4^1 of plate Y will be successively exhibited opposite the opening C, ready to receive the picture. The plate-holder G is brought into said four positions by moving the square knob I into the four corners of opening *k* in the rear part F of the casing. This motion can be made so quickly that the four pictures can be taken without covering the aperture of the camera from first to last. When exhibiting 3^1 and 4^1 the knob rests with its under surface on the top surface of the hinged block L, which block is then in a vertical position close to frame F. To exhibit 1^1 after 4^1 the block L is brought into position shown in fig. 4, the knob is slid down along edge *a* till its corner rests in groove *c*, and then passing the knob to the other end of said groove, 3^1 is exhibited. The object of this arrangement is to obtain rapidly a succession of pictures, timing them differently in order to select the best, and also to take stereoscopic pictures with one camera.

Claim.—The within described plate-holder in combination with the frame in which it moves, constructed and operating in the manner and for the purpose substantially as herein set forth.

No. 12,344.—DAVID N. B. COFFIN, Jr.—*Improved Daguerreotype-Plate Holder.*—Patented February 6, 1855. (Plates, p. 290.)

b is the block, *a* the frame, and *c* the bed-piece. The block *b* is of about the same length and width as the plate which is to be held, and is provided with a pin *m* at its centre, about which the bed-piece turns. The plate is held upon this block by the frame *a*, as follows: The projecting part within the corners of the frame *a* overlaps the corners of the plate, and draws them against the depressed corners of the block; the frame is made to act on the corners of the plate and force them to conform to the space between the projections within the corners of the frame and the depressed corners of the block by the force of the hand or by the action of the bed-piece, which is as follows: the plate is placed on the block, and the frame slipped on and over it; the bed-piece is slipped on to the pin *m*, so that its ends may pass within the frame; then the bed-piece is turned, so that the ends of the bed-piece pass between the frame and the block, and the surface of that part of the frame at p and q, with which the ends of the bed-piece come in contact, is inclined to the surface of the block, so that as the bed-piece turns, its ends pass up the inclines p and q, and the block is pressed further into the frame, and so held there.

Claim—The peculiar combination and arrangement, substantially as described, of the block-frame and bed-piece, for the purposes specified, the same being constructed and operated substantially as set forth.

No. 12,560.—DAVID SHIVE.—*Machine for Polishing Daguerreotype-Plates.*—Patented March 20, 1855. (Plates, p. 290.)

Spur-wheel I is attached to the bottom of eccentric H, the shaft E passing loosely through both. Spur-wheel K is fixed to shaft E. Spur-wheels M N are of one piece. The circular pieces C D are carried round by shaft E and arms F, they being connected therewith by cranks G, and (the eccentric H being confined in the centre of the lower circular piece D, and driven at a different speed from that of the shaft E, upon which it turns) the circular piece C, with the fixed-block P, to which the plate is attached, is necessarily caused to gyrate, and the polishing-pad B being held down upon it, the plate will be polished.

The inventor says: I do not claim effecting a gyratory motion of the pad for polishing the surfaces of daguerreotype plates, or other like surfaces, by means of machinery, as such has been so effected before for similar purposes.

But I *claim* the shaft E, with its arms F, cranks G, the pieces C and D, or their equivalents, and the eccentric H, with its spur-wheel I, in combination with the united spur-wheels M and N, and the spur-wheel K, when constructed and arranged substantially and for the purposes as described.

No. 13,196.—EDWARD BROWN, assignor to THE SCOVILLE MANUFACTURING COMPANY.—*Machine for Bevelling and Polishing the Inner Edges of Daguerreotype Face-Plates or "Mats."*—Patented July 3, 1855. (Plates, p. 291.)

The face-plate O^1 is placed over frame O, and motion is given to shaft J and arm K and burnishing-roller f; the plate is pressed against the burnishing-roller by means of lever P, and the roller forms a bevel on the edge of the aperture of the plate, and also polishes the bevel. The lever P is operated by hand, and the frame O is allowed to yield in order to allow for the difference between the diameters of the aperture, or, in other words, for the variation in distance from the centre of the plate of the several points forming the shape or pattern of the aperture.

Claim.—The combination of the rotary burnisher f and vibrating or yielding frame O, arranged and operating as shown, and for the purpose set forth.

No. 13,410.—HALVOR HALVORSON, assignor to HORACE BARNES.—*Improvement in the Manufacture of Daguerreotype Cases.*—Patented August 7, 1855. (Plates, p. 291.)

The composition referred to in the first part of the claim consists of gum shellac and some fibrous material dyed to the color required, and ground with shellac and between hot rollers, so as to be converted into a mass which, when heated, becomes plastic. The other features of this improvement will be understood from the claims and engraving.

The inventor says: I am aware that boxes have had sheets of paper or pasteboard glued or cemented to their surfaces; I therefore do not claim the mere application of paper by such means.

I *claim* the improvement in the manufacture of picture-cases or other articles of like character from a composition of shellac and fibrous material, as described; the same consisting in making said case or article of the said composition and one or more sheets of paper, and pressing and combining the whole together in a press or between dies, as described, so that the paper shall combine or connect itself directly with the composition without the aid of cement interposed between them, and sure to add great strength to the article so made.

And I claim the improvement of ornamenting the surfaces of the impression of the die with burnished gold, substantially as set forth; the same consisting in applying the gold to the surface of the sheet of paper, or its equivalent, burnishing it while on said surface, and laying the said burnished surface in contact with the surface of the die, and pressing said paper and the plastic composition together and into the die, so as to force the burnished gilding paper and composition upon it, and produce the result as specified.

I also claim the extension of the paper up the inner surface of the sides of the case, and by means of pressure in the mould; the same being for the purpose of enabling me to affix to the side the velvet-covered frame for the support of the picture, the mat, and the glass thereof.

No. 13,665.—DAVID SHIVE.—*Improved Daguerreotype-Plate Holder.*—Patented October 9, 1855. (Plates, p. 291.)

The part a, with the hooks b b, is attached to the part A, and the part a^1, with the hooks b^1 b^1, is attached to the part A^1; so that when the two parts A A^1 (which are held apart by a spiral spring between them, which latter is not represented in the engravings) are compressed, the parts a a^1 will recede from each other (see position fig. 1) so as to make room for the insertion of the plate.

The inventor says: I do not claim a two-part daguerreotype plate-holder, nor do I claim actuating the two parts by means of springs and the force of the hands.

But I claim a daguerreotype plate-holder, so constructed that when its under side is compressed by the hand of the operator, as described, its upper side shall expand so as to admit of the plate being placed between the hooks b b and b^1 b^1 thereon; and so that when the pressure of the hand is relaxed, the said upper side shall contract, causing the hooks *b b* and b^1 b^1 to catch upon the outer edges of the plate, and hold it firmly upon the face of the holder, substantially as described and set forth.

No. 13,701.—SAMUEL S. DAY.—*Improved Daguerreotype-Plate Vise.*—Patented October 23, 1855. (Plates, p. 291.)

The clamp *e* is actuated by the combined operation of the screw-rod *c* and cam-piece *f*, whereby the screw on the rod furnishes the means for holding plates slightly varying in size; while the cam-piece acting on the screw-rod, as the latter is turned, becomes a ready means for clamping the plate 5 or plate 6 to the holder, or releasing the same therefrom.

Claim.—The combination of the clamp *e* with the screw-rod *c*, bow *d*, and cam-piece *f*, to hold the daguerreotype-plate between and beneath the lips 1 and 3, or 4, and *g*, in the manner as specified.

No. 13,516.—JOS. ALEXANDER ADAMS.—*Improved Machine for Electrotyping.*—Patented September 4, 1855. (Plates, p. 291.)

The sliding-carriage P being at the front end of the machine, the mould to be coated with the material is placed indented side upwards on the turn-table Q, and the powdered plumbago is sprinkled over it; the crank-wheel D is then turned. As the brush M vibrates, the carriage and mould move slowly under the brush. When the mould has entirely passed under the brush, the motion of the crank-wheel is reversed, and the mould re-passes under the brush. The turn-table and mould are then turned at right-angles to their former position, and again passed forward and backward under the vibrating-brush as before, and the mould is sufficiently coated. As the mould advances at the same time that the brush descends, the advancing sides of the indentations of the mould are acted upon more efficiently by the brush, and are thus better coated than in case the mould remained stationary.

By the above operation, each of the four sides of the mould is in turn the advancing side; thus all the sides, as also the bottoms of the indentations and tops of the projections of the mould, are acted upon by the brush.

I *claim* the reciprocating or vibrating brush, operated as shown, or in an equivalent way, for the purpose of covering or coating the moulds for electrotyping purposes with any proper powdered substance; the said vibrating brush being combined, when necessary, with a carriage N, arranged as shown, or in an equivalent way, so that the whole surface of the moulds may be presented gradually or successively to the action of the brush, as the moulds pass underneath.

No. 13,521.—AARON D. FARMER and RANSOM RATHBONE.—*Improved Mould for Backing Electrotype-Shells.*—Patented September 4, 1855. (Plates, p. 291.)

The shell is laid face down upon the bed-plate A, the top-plate C is then put on, and, by means of the clamps, the whole tightly secured together. The mould is then placed on end, having the handle thrown backwards and downwards, which inclination prevents the metal from coming in contact with the tinned surface of the shell while running down into the mould.

We *claim* the use of the mould-frame B, or its equivalent, in combination with the bed-plate A to plate C, and clamps and handle G, or their equivalents, for the purpose substantially as described for backing electrotype-shells.

No. 12,841.—HENRY WHITNEY, Jr.—*Improved Inkstand.*—Patented May 8, 1855. (Plates, p. 291.)

The piston *c* being depressed so that its circular ribs *d* will enter the corresponding cavities *b* in the bottom of the ink-stand, the ink, being prevented from flowing outwards by the air between the ribs, rises through *m* into the dipping-cup *f*. The cap C, it is apparent, is not required to fit with an air-tight joint.

Claim.—The well *k* and the cylinders *b b b*, in combination with the piston *c* and the cylinders *d d d*, for the purpose of raising and sustaining the ink above its level in the inkstand, without the necessity of using the tight-packed joints heretofore required.

No. 13,515.—ALBERT BINGHAM, assignor to HIMSELF and A. J. BAILEY. —*Improvement in Inkstands.*—Patented August 28, 1855. (Plates, p. 292.)

By moving the hand against the part of the bow-lever which is in front of and above the cover, the latter may be raised up and a pen held in the hand may be inserted into the ink-port B. As soon as the pen is withdrawn and removed from the wire or bow-lever, the cover will fall by its gravity and cover the ink-port.

I *claim* arranging and combining with the hinged-cover G of the pen-port B, substantially as described, the bow-lever H, whereby the cover may be raised under circumstances and in the manner specified.

No. 13,902.—CHARLES T. CLOSE.—*Improved Fountain-Inkstand.*—Patented December 11, 1855. (Plates, p. 292.)

Figure 1 represents the inkstand when being filled, and figure 2 when filled and ready for use.

Claim.—The arrangement and combination, substantially as specified, of the upper tube or passage *b*, connecting the top or air-space of the reservoir with the pen-cup, at or immediately below the level; the ink is designed to stand in said cup, the latter being connected with the reservoir in manner shown, or equivalently thereto, and the ink in the pen-cup forming a fluid-valve, that, upon the insertion of the pen and the withdrawal thereof, alternately opens and closes the lower end of the upper connecting tube, for the free, rapid, and certain admission of fresh air at intervals in the reservoir as required.

No. 12,713.—JEREMIAH CARHART.—*Improvement in Melodeons.*—Patented April 17, 1855. (Plates, p. 292.)

Space B communicates with the air-receiving chamber of the bellows through apertures E, the air following the direction of the arrows in the figure. The key N being pressed downwards, rod *a* acts upon lever H and the valve D is raised; the air passes up through the reed *r*, which is arranged on the *under* side of the reed-board C, and out at E. The hammer I is arranged *underneath* the reed, and not between it and the bellows. In its upward motion it strikes the buff M (a strip of soft leather, one end of which is secured and the other is left free) and intermediately through this buff the reed *r*, by which arrangement the percussion of the hammer is prevented from affecting the tone of the reed. The whole of the striking action, with the exception of the hammer-rail J and hammer, can be removed at once through the front of the case, all except those parts attached to a rail K, which (its ends running in dovetail guide-grooves *d* in the side-board L of the case) can be drawn forward by hand when the front of the case is open.

The inventor says: I am aware that a similar arrangement of the reed has before been adopted, and the air forced upwards through it to produce the tone by the bellows from below, and the hammer caused to strike the reed from beneath; but this has only been done in instruments employing a forced current of air produced by the blowing action of the bellows, and the hammer has necessarily been arranged within the wind-chest, or between the wind-chest and reed, and intermediate of the current passing from the bellows to the reed, whereby much inconvenience arises in the removal of the hammer and adjustment of the reed, and to remove the hammer destroys or stops the operation of the reed for tuning or playing; this I do not claim.

But what I do *claim* is the arrangement herein shown and described, in instruments operated by exhaustion of the air, of the reeds, valves, and hammers, in relation to the exhausting bellows or passage, so that the hammer is caused to operate outside of the influence of the bellows, and not between the bellows and the reed, and whereby the hammer may be readily detached and taken out of the instrument for repairs, tuning, or adjustment of the reed, without destroying the capability of the reed to speak or play.

Without claiming the application of buffs, consisting of strips of leather, to musical instruments generally, or for any other purpose than that which I have specified, I claim their application to reed instruments in connection with hammers, substantially as and for the purposes herein fully set forth.

No. 12,938.—Thomas Foster Thornton, assignor to George A. Prince and Thomas Stephenson.—*Improved Swell for Melodeons, &c.*—Patented May 22, 1855. (Plates, p. 292.)

The swell C *c* is divided into two halves C and *c*, the hand-lever E acting only upon the half C. Figure 1 represents the swell when closed, figure 2 when open, and figure 3 when half open and half closed.

The inventor says: I do not claim the whole swell when acted upon by the foot pedal J, the rod *g*, and levers *h h*, when the swell is not divided. as I am aware that it is now so constructed and used.

But I do *claim* the *divided swell*, constructed substantially as set forth, so that one-half (or part) may be used separately when desired, for producing more variety in the tone of reed instruments, such as melodeons, melo-peans, seraphines, and reed-organs, all of which are very similar in construction.

No. 13,021.—William C. Whipple and William C. Bowe.—*Improved Melodeon.*—Patented June 5, 1855. (Plates, p. 292.)

By pressing down one of the keys, for instance C, the pitman *h* will force down the extreme end of lever *c*, raise the contiguous ends of *c* and *d* at *g*, and thus open the valve at B, which is under the key C^1, so as to sound the octave. Thus the instrument will have the same power as if made with two sets of reeds.

Claim.—The use of two sets of levers, located in the wind-chest under the valves, and so connected as to enable us to play any desired note and its octave, when the whole is constructed, arranged, and made to operate substantially as described.

No. 13,048.—Henry W. Smith.—*Improved Coupling for Organs and Melodeons.*—Patented June 12, 1855. (Plates, p. 292.)

a a are the two banks of keys of a melodeon, with two sets of reeds *c c*; *b b*, two sets of valves. One end of lever *f* is attached by a hinge to the upper surface of a key of the lower bank, and has a slot *s* through which the valve-rod *e* plays; the slot, however, is too narrow to let the valve-rod shoulder *h* pass through. The other ends of all the levers *f* rest in a groove in the cross-rail *g*, which latter is hinged (by means of two arms *t*) at both its extreme ends to the key-frame, so as to allow the rail to rise and fall, and thereby elevate or depress the ends of the levers *f*. When the coupling-rod *k* is pulled towards the performer, the end of arm I, which rests on the rear extension *u* of rail *g*, forces down the rail and all the levers until they rest upon the surface of the keys. When the rod *k* is pushed back towards the action, the rail and levers are elevated by the springs *l*. When a key on the lower bank is depressed, the shoulder will rise above the surface of the key if the coupling apparatus is not adjusted. If the coupling-rod be pulled out, and the levers are forced down on a line with the lower keys, and the lower key be then depressed, both sets of valves must be opened by means of the action of shoulder *h* on lever *f*. Thus both banks of keys are coupled in such a manner that the lower bank will play both sets of reeds.

Claim.—The combination of the lever with the shoulder on the valve-rod, operating in the manner and for the purpose described.

No. 13,704.—George G. Hunt.—*Improvement in Melodeons.*—Patented October 23, 1855. (Plates, p. 292.)

A reed-board of ordinary construction is divided along its central line, and the two parts are removed a short distance apart, as shown at *a a*. Upon these is placed another ordinary reed-board *b*, but having no valves of its own, and in which the reeds are placed at the same distance apart as are those in the first board, whereby each pair of reeds of the upper board will be in the same vertical plane with a pair of the lower board. The divisions between the reeds of board *b* must be continued through space *c* to the lower face of board *a*, (as indicated at *d*,) and in each of these spaces a block *e* is put as a guide for valve-rod *f*. One valve *g* covers the slot in the lower reed-board, and this slot is just so much longer as the two parts *a a* have been removed asunder. Two other sets of reeds may be added by removing the parts *a a* still further and dividing board *b* as before described for *a*, when another reed-board may be placed upon that.

Claim.—The described construction whereby two, four, or more sets of reeds may be operated by one and the same valve, in the manner set forth.

No. 13,959.—Thomas F. Thornton.—*Improvement in Organ Melodeons.*—Patented December 18, 1855. (Plates, p. 292.)

The nature of this improvement will be understood from the claim and engravings.

Claim.—Providing an additional set of valves F and one or more additional sets of reeds E E, arranged as described, in a position the reverse of the usual arrangement of valves and reeds, and extending the keys backwards in rear of the fulcrum to actuate the additional set of valves through push-up pins, to play on the additional set or sets of reeds at the same time as they actuate the other sets of valves D, through the push-down pins to play the reeds C C, which are below them, substantially as described.

No. 12,628.—Gustavus Hammer.—*Improved Valve for Wind Musical Instruments.*—Patented April 3, 1855. (Plates, p. 293.)

1 is the valve which works in the cylindrical case 18, and is operated by means of bow 15 and string 16, which passes around the top of the valve and is prevented from slipping by being held with the screw 19. By arranging the attachment for working the valve between the bearing and the valve, the uneven wear of the latter is avoided.

Claim.—The combination of the bow 15, string 16, and the screw

19, for working themselves and preventing the string from slipping, all for purposes set forth.

Also, the privilege of applying the improved valve to all musical instruments to which such valves are commonly attached.

No. 12,724.—Isaac Gallup.—*Improved Apparatus for Turning the Leaves of Music-Books.*—Patented April 17, 1855. (Plates, p. 293.)

When the leaves are placed between the prongs of the fingers C and the first leaf is to be turned, the performer depresses the knob 1, and thereby forces the key G and lever-bar F down. The lever-bar in descending causes the barrel-pulley D to turn, thereby extending the spring H and allowing spring I to contract. Spring H is thus prepared to bring back the finger and leaf, and the contraction of I aids in turning the barrel-pulley which carries the finger. As soon as the finger C occupies the position C^1, (shown in broken lines in fig. 1,) the spring-catch K has caught the stop J of the key, and thus prevents its rising and the finger returning with the leaf. When all the keys (or as many as there are leaves to be used) have been operated, the keys are moved laterally to a small extent, which the oblong form of the key-holes *h* permits. Thereby the keys are freed from their catches, and they are elevated to their original position by the springs H, in doing which they cause the fingers, with the leaves, to return to their former place. As the fingers are returned by the force of the springs H, they and the leaves would be injured were not the force of said springs counteracted by springs I; and after the fingers are at rest, they would be subjected to strain, if it were not for the counterbalancing action of springs I.

The extension *a b* of the fingers is intended to facilitate their supporting the leaves, and to turn them more evenly.

The inventor says: I do not claim the revolving, self-adjusting pulleys or finger-carriers; but I do *claim*—

1st. The employment and arrangement of the swinging-bars F F F F and keys G G G G, in combination with said revolving self-adjusting pulleys or finger-carriers D, substantially as and for the purpose set forth.

2d. The employment, substantially as herein shown, of the spring I in combination with the spring H, for the purpose set forth.

3d. Providing a stop J on each of the keys G, and a spring-catch K on the under side of the top A^1 of the case A, to fit against said stop, substantially as and for the purpose set forth.

4th. Providing each of the fingers C with an extension from *a* to *b*, for the purpose herein specified.

No. 13,001.—Wm. Fischer.—*Method of Composing Music.*—Patented June 5, 1855. (Plates, p. 293.)

Divisions *a b c*, &c. contain three cards each A B C, each of which has four numbered musical phrases a^1 b^1 c^1 d^1 for the composition of a musical piece. Of these twelve phrases, one may be chosen from several or from all of the divisions, and they will become a coherent and rounded musical composition, perfect in itself.

Claim.—So arranging a certain number of musical phrases on each of a number of cards, that by selecting and combining one from each card, as described, a number of musical and perfectly melodious pieces can be composed, all of them differing from each other.

No. 13,365.—George S. Shepard.—*Improvement in Musical Reed Instruments.*—Patented July 31, 1855. (Plates, p. 293.)

The air to act upon the reeds is made to enter each end of the reed-chamber J through the oblong openings *g g* from the bellows. Sounding-posts *n n* are inserted between the socket-board *c* and the bottom *s* of the valve-chamber I, which serve a purpose similar to what they do in stringed instruments. By opening to their full extent the valves *e e*, and thereby allowing the air to escape freely from the valve-chamber into the sounding-chamber A, tones of increased fullness and power will be produced; and by nearly closing said valves, the softest tones will be produced. The auxiliary sounding-chamber B adds to the vibrating surface of the instrument, and prevents the air in the valve-chamber from reacting against the reeds when the valves *e e* are nearly closed, thereby adding to the fullness and evenness of the tones.

Claim.—The combination of the auxiliary sounding-chamber B and the swell-chamber A with the valve-chamber I, substantially in the manner and for the purpose set forth.

No. 13,642.—Daniel George.—*Wind Regulator for Organ-Pipes.*—Patented October 9, 1855. (Plates, p. 293.)

The nature of this improvement will be understood from the claim and engravings.

Claim.—Constructing the lower part of each or any of the pipes of an organ with a transverse seat fitted with a plug *b*, like that of a faucet, having a suitable passage or passages, the area of which is regulated by turning the plug, for the purpose of regulating the tone of the tube and tuning the instrument, substantially as described.

No. 13,668.—J. C. Stoddard.—*Apparatus for Producing Music by Steam or Compressed Air.*—Patented October 9, 1855. (Plates, p. 293.)

A is the steam-chest, (constantly supplied from a boiler.) B are the whistles attached to it. The valves are placed in valve-boxes C between the whistles and the steam-chests, with their stems *a* protruding from the said boxes to enable the valves to be opened by the keys. The valve has two puppets b b^1, one end of its stem *a* passing through stuffing-box *c*, and the other end through guide *d*, being exposed to the steam. b^1 is a trifle smaller than *b*, so that the pressure of the steam tends to keep the valve tight when closed. The steam escapes at *g*.

The valve remains open only as long as the finger is on the key, and is closed as soon as the finger is off by the steam-pressure on the stem.

Claim.—1st. The musical instrument described, consisting of a number of what are commonly known as steam-whistles, of such tones as to produce a musical scale, arranged in a convenient manner, upon a steam-chest, chamber-pipe, or generator, and furnished with valves and a rotating studded-barrel, finger-keys, or other suitable mechanical means of opening the said valves to allow the escape of steam or air to the whistles, substantially as set forth.

2d. As a part of the said musical instrument, the described valve, with its two puppets and seats of unequal size, and with one end of its stem exposed to the atmosphere.

No. 13,946.—H. B. Horton.—*Machine for Registering Music.*—Patented December 18, 1855. (Plates, p. 294.)

This apparatus is intended as an aid to music composers, by registering the notes played upon a sheet of paper D, which has a slow motion. The notes played are marked, as the keys H, when depressed, allow the supporting rods *i* to descend and the markers *h* to descend upon the paper.

Claim.—1st. Attaching the markers *h h*, by which the notes are registered, to high springs or flexible-bars *g g*, which are so supported by the keys H, when the latter are raised or not in operation, as to hold the points out of contact with the roll D, or other travelling sheet, upon which the notes are registered, until their respective keys are depressed, when, losing that support, the points fall or are gently pressed upon the surface of the sheet, substantially as set forth.

2d. The within described method of operating the bar-marker *o m*, by which the bars are registered, by making it sufficiently elastic to hold its point off the sheet while it is left free, and striking it down in contact with the sheet at intervals of time bearing a proper relation to the movement of the sheet, by means of a hammer applied substantially as described, and operated by a cam *s* on one of the rollers, which supports and moves, or is moved by, the sheet.

3d. The vibrating indicator *u*, arranged so as to be visible by the player, and operated substantially as described, and the cam *c* on the axle of one of the rollers, which drives or is driven by the sheet, for the purpose of marking the time, to lead or guide the player.

4th. Attaching all the note-markers and the bar-markers, and the upper guide of the rods through which the keys support the note-markers, to a frame *k k*, so that the whole can be moved simultaneously in a lateral direction to mark in different lines, substantially as and for the purpose set forth.

No. 12,255.—R. L. Hawes.—*Machine for Making Boxes of Paper.*—Patented January 16, 1855. (Plates, pp. 294, 295, 296, and 297.)

Almost all the features of this machine can be well understood from the engravings and claims. A more detailed description would occupy too much space.

Claim.—1st. The pasting apparatus, consisting of rollers 20 21, working in the open bottoms of vessels containing paste or other adhesive material, said rollers having cavities or cells to receive paste from the paste-vessels, and transmit it to such parts of the roll or piece of paper, or other material, as may be necessary, as the paper passes between them before entering the machine, substantially as described.

2d. The employment of a series, consisting of any suitable number, of moulds G 1 to G 6 of proper form for the boxes, arranged so as to work radially, or nearly so, upon or within a revolving mould-wheel I J outside a series of tables N N, in such a way that a piece of paper or other thin material, to form a box, is taken between each of the several moulds and their respective tables, and drawn between the edge of a projection *b* and of a clamp *t*, or other edges attached to or forming part of the wheel, for the purpose of bending the paper up the sides of the mould, substantially as described, and thus forming three sides of the box.

3d. Attaching to the mould-wheel, between the moulds, a number of blades *c c*, corresponding with the number of moulds, so arranged at equal distances apart, and at equal distances from the axis E of the wheel, that the distance between their cutting-edges shall be equal to the required length of paper to form the box, and that they are severally and successively caused, by the revolution of the mould-wheel, to act in combination with a fixed knife *d*, suitably arranged in any way substantially as described, and cut the paper or material from the roll in proper lengths to form the boxes.

4th. The clamps *t t*, arranged one at the side of each mould, and actuated and operating substantially as described, for the purpose of lapping the part 37 of the paper or material over the mould, to form part of the fourth side of the box.

5th. The presser P, arranged relatively to the mould-wheel, and operating substantially as described, for laying down the edge of the part 37 of the paper or material, and confining it till the part 38 is lapped over it.

6th. The presser R, arranged relatively to the mould-wheel, and operating substantially as described, to lap the part 38 of the paper or material over the part 37, to complete the fourth side of the box, and at the same time lap the part 40 over the end of the mould to commence the end of the box.

7th. The levers V V, working as described, on or within a wheel, or its equivalent, which rotates with the mould-wheel, and successively so operated as described, by coming in contact with a fixed tongue 33, or other fixed part of the machine, during the revolution, that each at the proper time folds the part 41 of the paper or material, and laps it over one side of the part 40, as represented.

8th. The arm U, arranged as described, carrying the plate 15, by coming in contact with which, the part 42 of the paper or material is folded and lapped over the opposite side of the part 40 to that covered by the part 41, also carrying the plate 14, which is caused by the action of studs *l l* on the wheel H, or its equivalent, to be moved from the central shaft to lap the part 43 over the part 40 41 42, and thus complete the formation of the bottom of the box.

9th. The stationary smoothing and pressing-plate 16, arranged as described, to smooth and finish the bottom of the box by the revolution of the latter in contact with it.

10th. The arrangement and mode of operating the roller 44, to take the glue or adhesive material from a fixed trough 47, and distribute it over the bottom of the box, to prepare the same for sanding.

11th. The shears 62 69, arranged and operating as described, at the end of the table 61, along which the printed string of labels passes, from the printing apparatus, in combination with the intermittent motion of the said printing apparatus, for the purpose of moving forward the labels at proper intervals, and cutting them off one by one, as required, to be presented to the boxes.

12. The table 71, attached to or supported above the shear-blade 69, so as to move with it for the purpose of receiving the cut label, and applying it to the box, as set forth.

13th. The arrangement of and mode of operating the tongs 74 75, substantially as described; that is to say, the said tongs being arranged so that each box will pass them after being completely formed, and being operated to move towards the mould-wheels with open jaws, and to close upon the box when it is between them, so that the mould, by a movement in the direction of its length, may be withdrawn from the box, and then to move with closed jaws from the mould-wheel to carry the box to a trough or box of sand 76.

14th. The hook 88 attached to one jaw of the nippers, and operated upon substantially as described, so that as the said jaws descend with the box towards the sand-box, it throws up the mouth of the box to bring it to a nearly upright position, to dip the bottom in the sand, substantially as set forth.

15th. The rod 92, arranged and operated upon as described, to knock over the finished box from the sand-box.

16th. The general arrangement, and combination of the several portions of the machine, as described, either with or without the labelling and sanding apparatuses and their appendages.

No. 12,786.—JOHN A. SMITH and S. E. PETTEE.—*Machine for Making Paper Bags and Envelopes.*—Patented May 1, 1855. (Plates, p. 298.)

The paper, cut into the proper form, is placed in the box H, and paste or gum in the trough P, and motion given to shaft U (the bars I and L being in position figure 1); the bar I turns up in the direction of arrow, figure 1; and after it passes the bar L, the box is lowered a little to allow the bar K to take the weight of the pile partly off the end of the under sheet, and to insure the bar I to draw or slide back the under sheet, as shown in figure 2. The bar L, then following, lifts the box and paper away from it, and the bar L passes over the end of the sheet, as shown in figure 3; the bar I then returns to former position, and the bar L, continuing its motion, brings the paper between them, as shown in figure 4, ready for the jaws E to take it.

Now the table (which consists of two parts, x and x^2, connected together by springs W W) moves back far enough to allow the former A to be lifted; it then moves forward, as also do the jaws E, carrying the lifter M into the box H, lifting the weight off the under sheet, to allow it to be drawn out without disturbing the others; the former A lifting enough to clear the fore part of the jaws, and to open them by pressing against the roll on the back end of the lower jaw, keeping them open until they pass over the paper, the jaws stopping the former A by dropping a little, thereby allowing the spring to close the jaws on the paper. The jaws now, moving back, draw the paper out of the box H, the bar L turning a little to be out of its way, and afterwards turning to place in time for bar I to pass it to take another sheet; the table moving back with the jaws receives the paper, and the former A is closed on the paper with its edges in the proper position for the folds, and part C^2 (figure 6) of the paper lying on the bar B, and the sides B^2 B^2 lying on the guards G G; the former A, pressing the lower jaw down, removes the paper from the points in the upper one. The table x now moves forward, carrying the former A under the bar B, thereby folding the part C^2, as in figure 7, the guards G G guiding the sides B^2 B^2 over the folders D D, and to pass under the sponges and pasters, which (having been dipped into the paste whilst the table is receiving the sheet) pass outward as the paper passes under them, so as to spread the paste along the straight edges of B^2 B^2; the side-folders D D now move on simultaneously, whilst the part C^2 is held smooth by bar B, as in figure 8, and press the sides on the pieces F F, casting off the springs W W; the part x of the table now moves back, the folders D D holding the paper on to x^2, when the former A is drawn out the width of the envelope, or so as to cover only D^2; the whole table then moves back, carrying the folded part beyond the bar B, when the springs T T lift it, and the table moving forward, again it is folded over the former A and lap D^2, finishing the envelope; the table, moving far enough forward to bring the whole work by the bar B, delivers the work on its return, the bar B serving to hold whilst the table is moved back for another sheet; the springs T T in passing under the bar B are pressed down sufficiently to allow the former A to slide over them, allowing x and x^2 to come together before they get clear forward, the finished work falling on to a band, which, passing over a roll V, carries it into a suitable receptacle.

The inventors say: We do not claim the exact form or arrangement of any of the parts, but only the following points:

Claim.—1st. The bar K, to relieve the end of the under sheet of the weight of the pile, partially or wholly.

2d. The friction-bar I, to separate the under sheet.

3d. The guide-bar L in connexion with bar I, to hold the sheet in place for the jaws.

4th. The lifter M, to relieve the sheet from the weight of the pile.

5th. The feeding from the bottom of the pile.

6th. The combination of the weight-bar, friction-bar, guide-bar, and lifter, constituting a feeding apparatus.

7th. The jaws, to place the paper in position.

8th. The former A, to fold the paper over or around.

9th. The pasters and side-folders.

10th. The combination of the table, the bar B, the side-folders and pasters, all constructed as herein set forth, or any other substantially the same.

No. 12,736.—T. J. BALDWIN.—*Improved Paper-Ruling Machine.*—Patented April 17, 1855. (Plates, pp. 298 and 299.)

Power being applied to the roller C, the apron B moves around rollers C L D, and the sheets of paper a^1 are fed in between the apron and the cords *b*. The front edges of the sheets strike the pendant *f* and rotate the shaft N a trifle, so as to throw the recess *e* in the pulley O off the pulley M; and as the peripheries of the two pulleys are then in contact, the shaft N rotates, and the prominent parts of the cam Q raise the lever T, and lever T in rising throws the spring-catch S out of the collar *l* and hub *i*, and spring *m* consequently forces the loose pulley R against the fast pulley K, sufficiently tight to cause the loose pulley to rotate, and the projections V (on pulley R) tilt the pen-beam W, and the pens are raised from the paper.

Projections V are to be of proper length, and to be so arranged as to operate the pen-beam as desired. The pulley R may make one revolution during the passage of each sheet of paper underneath the pens, and the pulley O also makes one revolution in the same time; the recess *e* fitting over the pulley M, and preventing any casual movement of the shaft N. The stop *g* and lever U serve to stop the motion of the shaft N at the termination of every revolution.

Claim.—Lifting the pens X from the sheets of paper a^1 at the proper intervals by means of the mechanism herein shown and described, viz: having the front edges of the sheets a^1, as they move along on the endless apron B, strike against a pendant *f* attached to a pulley D^1 on a transverse shaft N on the frame A; said shaft being provided with a pulley O at one end, having a recess or groove *e* cut in its periphery and a cam Q; the pulley O working on bearings on a pulley M underneath it; said pulley M being driven by a belt *d* from the driving-shaft C. The cam Q operating a lever T, by which the spring-clutch is allowed to act and connect the pulleys K R at one end of the drum L; the projections V on the lower pulley R raising the pen-stock so as to leave the blanks or spaces at the desired parts of the sheets, as herein set forth.

No. 12,945.—E. W. GOODALE.—*Improved Machine for Making Paper Bags.*—Patented May 29, 1855. (Plates, p. 299.)

The feed-roll G receives an intermittent motion to feed at proper intervals a suitable length of paper; I, stationary blade; I^1, vibrating blade for cutting off the paper after it has been delivered to the endless apron L. They are so bent as to cut the paper after the line 3 3

3^1 3^1 (see the form of paper represented in fig. 4). J J are the stationary and J^1 J^1 the movable blades of two pairs of shears for cutting out the slides 3 1, so that when the piece is folded in the line 1 1 (fig. 4) to bring part 3 on top of part 4, the side parts 2 2 of the other half may lap over it to close the sides of the bag. The stationary blades are all attached to bridge-piece I^2. Shear-blade I^1 is pivoted in i to a fixed stand i^1, and its other end is operated by a cam and rod i^2. Blades J^1 J^1 are on shafts j^1 sliding in guides j^2, and are operated up and down by cams; they (blades J^1) are set slightly out of parallel with the blades J, and are connected by spring j^3, which tends to draw them together, and while cutting to throw their edges slightly across those of J J; but their downward movement in contact with J J forces them gradually apart. Apron L receives the cut piece, and while on the apron the under faces of the side margins are pasted ready to receive the lapping-pieces 2 2; they receive the paste from rollers *c* revolving in a trough, and alternately raised and lowered to take up paste and give it to the paper.

The intermittently revolving folder consists of two plates M secured to a shaft M^1, the whole length of the folder being nearly equal to the length of the bag-piece, but considerably narrower. O^1 is the intermittently revolving creaser-shaft; the creasers consisting of arms O, of such length and at such distances apart that each pair forms a mould upon which to fold the bag. As soon as the piece is deposited on the folder, the creaser-shaft O^1 makes a portion of a revolution (see arrow), and brings down a pair of creasers upon part 3 of the bag-piece; then the folder makes half a revolution (see arrow), and folds part 4 over on to the part 3. During the folding operation, the operating pair of creasers has been supported by the stands *d d*, and then remains below the creasers after the folding operation and during the operation of lapping the sides, which is performed by means of lappers N N hinged to the folding-stands, which, during the preceding operations, have been thrown back, (see dotted lines fig. 2). As soon as the side-lap has been completed, the lappers N N are again thrown back, and the movement of the creaser-shaft to bring another pair of creasers into operation takes place; but previous to this, the stands *d d* move farther apart to allow the creasers to pass. This is done by means of a cam acting on lever V^2 and elbow-levers *u u*. The removal of the bag (before entering between the pressing and delivery rollers S S) is commenced by the sharp teeth of roller T.

The bag-piece is represented in the figures in strong broken lines.

Claim.—1st. Giving the blades of the side-shears J J^1 J J^1 a curved angular or irregular form near the point, for the purpose of cutting out each by a single cut the whole piece necessary to leave the lap on that side of the bag.

2d. Hanging the movable blades J^1 J^1 of the side shears on shafts or pivots j^1 j^1 perpendicular to their faces, for the purpose of allowing them to cut slightly across the fixed blades by a slight lateral movement, which they receive simultaneously with the movement usually given to shears, substantially as described.

3d. The intermittently rotating folders, arranged and operating substantially as described, to receive the bag-pieces from the feed-apron

L after the cutting and pasting operations, and to support or partly support them until the creasers come into their operative positions, and afterwards to fold them over the creasers.

4th. The combination of the side lappers N N and the laterally moving folding-stands *d d*, operating in conjunction with the creasers, substantially as and for the purposes herein set forth.

5th. The toothed roller T, hung in a frame T^1 from the axle of one of the pressing-rollers, and operating substantially as described, to commence the removal of the bags from the creasers.

6th. The general arrangement and combination of the several working parts of the machine, substantially as herein described.

No. 12,982.—Francis Wolle.—*Machine for Making Paper Bags.*—Patented May 29, 1855. (Plates, p. 300.)

The working parts of the machine, with the exception of the printing and drying apparatuses, are erected upon a bed A, and the movements of all the parts are derived from the main shaft B. The paper from which the bags are made is on a roll C, of a width suitable to the depth of the bags. One end is led between feeding-rollers D, caused to revolve and unwind a sufficient length of material to form a bag, and then remain stationary for a time. The paper which has been drawn between the feed-rollers lies upon an inclined plane, and is cut off by a shear-blade; other blades also cut away part of one edge, so that the remaining part of the same edge can, when the piece is doubled, be lapped over it to form the bottom of the bag. The blade *r* creases the sheet, and folds it by driving the crease through a slot far enough to be seized by a pair of rollers *s*, which draw the folded piece through the slot, and, assisted by two other rollers s^1, carry it into a swinging-frame Q, whence it passes upon an endless apron U. The apron has an intermittent motion, and receives one movement for every revolution of of the driving-shaft B. One of these movements of the apron serves to carry the folded paper from the conveyor Q to the bottom-pasting and lapping apparatus 38, where the bottom of the bag is closed; and the next movement to carry it from the bottom pasting and lapping apparatus to the side-pasting and lapping apparatus, where the side of the bag is closed. Each of these parts consists of paster, paste-feeder, lapper, and creaser. The bag is printed by means of a type-cylinder revolving suitably with the velocity of the bag to be operated on and inked by rollers. The bags b^5 b^6 b^7 are dried in a chamber 73 at the end of the machine, through which a current of air circulates, and into which they are carried by an apron or series of cords 74 and 99.

Claim.—1st. The conveyor for conveying the folded paper to the apron, by which it is carried to the folding and lapping apparatus, substantially as set forth.

2d. The construction of the lappers 20 and 38, and their connexion with their respective lapping-tables 21 and 68, as shown and described.

3d. The arrangement of the drying-chamber, and the aprons which convey the bags through it, as described, so that the bags are severally delivered to the aprons with their sides in a positive oblique to the direction in which the aprons move, and thus, as they are successively deposited, have the net-laps of their sides and ends left uncovered by their successors.

4th. The general arrangement and construction of the whole of the machinery described, whereby a piece of paper of suitable length is cut from a roll, cut out to the proper shape, folded, pasted, lapped, printed in any desirable manner, and dried, at one continuous operation.

No. 13,349.—Jno. A. Elder and Jno. Richardson.—*Improved Machine for Ruling and Paging Paper.*—Patented July 31, 1855. (Plates, p. 300.)

The railways M are fastened to the frame A, as also the rails G G and F F; the rails H H and J J are jointed to the rails G, and the rails L L and K K to F F; the circular rails X and 24 are jointed to frame A. The tables B travel on said rails by means of gear-wheels T taking into rack-teeth 12; I are the pens. The dogs Z, at each end of the tables, raise at proper times the arms *c* of the pen-clamp D and the pens from the table. 16 and 17 are rocker-shafts, to which the nippers 14 and 15 are attached, the latter being held together by spring 27 applied to arms 4 of said rocker-shafts. Lever *o* jointed to frame A in 29; lever 6 7 jointed to arm 13, projecting from frame A; 9, inking-roller; 11, weight. The roller 9, by passing backwards, takes the ink from the under side of plate 23, and in passing forward transfers it to the type at the lower end of type-rod 5. Springs 19 press against rails L. When the wheels *u* of the lower table B stand on H, and wheels *v* on K, then, by moving crank P forward, T will be turned, carrying along B; the front end of B will depress rails L till they rest on M; and when *u* passes over J, pin 26 will lift lever Q extending from K, and thus raise K; *u* will then pass under K and down on 24. Pin 25 lifts R, and lifts at the same time rail X and lets *u* pass under it and along on rail M; *u* will then lift H until the latter sinks down, (and rests again on M,) when *u* passes the end of M; *u* will then pass up over rail H; the weight of the table depresses L. Spring 19 raises L, for wheels *v* to pass under rails L and along on M; wheels *v* lift 24 and J, and travel up the rail X. Cam N works O, so as to press the type-rods down on the paper while B stands at rest, the rails 24 being so shaped at their ends that the table will not move during the depression of the type-rods. At the same time the bar 2 moves forward and opens arms 4, and the nippers receive the edge of the paper; the table moves back, and the paper falls down on the lines Y.

Claim.—1st. The arrangement of machinery for the ruling, printing, and paging paper for the manufacture of blanks, books, or other like purposes; when the ruling, printing, and paging are done before the paper is removed from the car or table where it is ruled, as specified.

2d. The combination of a car or table B, and ratchet-bar, with its type-rods 5, or their equivalents, for the purpose described.

3d. The pliers or nippers for the purpose of removing the paper from the car or table, when operated as described.

No. 13,647.—E. W. GOODALE.—*Machine for Making Envelopes, &c.*—Patented October 9, 1855. (Plates, p. 301.)

a are the blanks on table C, from which the envelopes are to be ormed. Table C is attached to shaft C^1, which works in guides *c c*, and is operated by cam C^2, which latter has not a complete rotary motion, but only turns far enough to enable the cam to raise the table up to bring the top blank in contact with the under side of the gluten dish *b*, said cam being operated upon for the above purpose, by spring *d* applied to cord d^1. The head E is carried by two cranks d^1 d^1 on shafts *d d*, and has attached to it the paste-box *e*, which carries the paste to stick together the three flaps of the envelope; the gluten die *f*, which takes the gluten from the gluten dish *b* and puts it on the seal-flap; and a screw *g*, which gives pressure to the stamp which stamps the seal on the seal-flap. The cranks are geared together, and operated by lever F^1 and cam F, so as to give, at proper times, one-third of a revolution to shafts *d* and cranks d^1, so as to carry the head E from position figure 3 (where the gluten die is in the gluten-dish) to position figure 1, where the gluten die rests on the top sheet of the pile of blanks, the head being returned again by the action of spring *i*. The bottom of box *e* (the horizontal form of which corresponds with the margin of the flaps) is closed, except at the proper time to apply the paste, by means of a spring-valve *j*; this valve is opened, at proper times, by lever j^1 striking stand j^2, attached to gluten-trough *b*. The gluten die *f* corresponds in form with the shape of the seal-flap of the envelope. The seal-stamp consists of two dies, one covex *h* and one concave h^1. *h* is attached to an arm *i* secured to slider F*, which reeeives a movement back and forth by elbow-lever *j** operated upon by cam G and spring j^{1*}. The wedge i^1 at the front end of arm *i* passes under the top blank, to make room for die *h* to get under; said top blank being, at the time the wedge arrives, momentarily raised a little by the gluten die, as the latter commenees rising after putting on the gluten; and the table with the remainder of the pile being, at the same time, lowered a little by the action of cam H upon arm H^1, and pawl *k* upon ratchet-wheel I, which latter is fast on the shaft D of cam C^2. The top die h^1 is attached to arm l^1 working in bearings l^2 attached to slider F*.

During the return of head E, the nippers *o* o^1 come into operation to remove the top blank to the folding-stand J. The nippers are opened and closed by the movement back and forth which they receive, for the purpose of carrying the blank. The nippers go backwards open to rereceive the blank, and at the end of this movement are closed by the release of lever r^1 from spring-catch *r*, the release being effected by the catch being thrown inwards by passing inside a fixed incline r^2. After the nippers have advanced with the blank far enough to place it on folding-stand J, lever r^1 comes in contact with fixed stop r^3, and the concluding portion of the movement of slider K^1 throws back lever r^1 till it is again caught by catch *r*. These movements are effected by means of cam L and elbow-lever L^1. Attached to the four sides of the stand J are the lappers t^1 t^2 t^3 t^4, which crease and fold the envelopes, by means of levers u^1 u^2 u^3 u^4 pivoted to the stand, and operated by means of rods v^1 v^2, etc., levers M^1 M^2, etc., and cams N^1 N^2, etc. The lappers are thrown back after the creasing and folding operations, by springs applied to their hinges. When the blank has arrived on the stand, the plunger P descends upon it, the lappers rise to a vertical position, the plunger withdraws again, and the lappers close upon the envelope. The seal-flap receives no preparatory creasing operation like the other three flaps; and as, when the folding operation takes place, the plunger P is raised, two creasers W (the engraving represents only one) are employed, which are attached to lever R; which latter works transversely of the machine on a fixed pivot w^1, and is operated by lever M^1, by which lapper t^1 is operated to throw the fingers w down upon the envelope at the time the lapper t^4 commences to operate. The finished envelope, after the lappers have been thrown back, is lifted up and removed by nippers y y^1. The lifting up of the envelope from the stand high enough to be caught by nippers y y^1, is effected by lifter 10; which latter is flush with the stand until, at proper times, it is raised by stud 8 striking lever g.

I *claim*, 1st. The employment, in a machine for making envelopes or bags, to support the blanks during either or all of the operations of pasting, stamping, and applying the gluten, of a self-adjusting table C, supported by a cam, whose position is so controlled by a spring or its equivalent, applied to its shaft, that as the blanks are removed one by one, the table is caused to rise, to bring the next one to the proper height or position to be pasted, stamped, or have the gluten applied, substantially as set forth.

2d. Giving the self-adjusting table a drop movement, substantially as described, by means of the cam H, the lever H^1, pawl k^1, ratchet-wheel I, or their equivalents, acting on the shaft of the supporting cam C^2.

3d. Applying the gluten which makes the envelope or bag self-sealing to that part of the blank which is to form the seal flap or closing flap of the envelope or bag, by a die, while in the machine, at the commencement of the process, substantially as described; whereby the said die serves the two purposes of applying the gluten and of lifting the blanks one at a time from the pile, or retaining the top one while the remainder of the pile is lowered away from it.

4th. Applying the two dies h h to the arms or jaws i l, which are connected together by a hinge or its equivalent, arranged at the rear of the table C, and have a sliding motion back and forth, substantially as described, to move the said dies out of the way of every successive blank, till the latter has had the gluten applied, and been separated from the pile, and then to bring them forward again, to receive the separated blank, and to receive the pressure of the screw g, or its equivalent.

5th. Attaching the paste-box, the gluten die, and the screw g, or other equivalent device, which gives pressure to the stamp which produces the seal, to a head E, receiving such a motion as is described from a pair of cranks, or their equivalent.

6th. The employment of a pair of nippers o o^1, having a motion of a positive length in the line parallel with the line in which the blank is required to move, from the pasting to the folding apparatus, either to take a cut blank from a table, or to draw the material before it is cut from a roll, and measure off the proper length to be cut, substantially as set forth.

7th. The method of giving the necessary movement to the lappers t^1 t^2 t^3 t^4, by means of the bent levers u^1 u^2 u^3 u^4, and the springs t^* applied to their hinges, substantially as described.

8th. The creasing-fingers W, arranged and operating substantially as described, to hold the blank in position and crease it in the line for folding the seal flap, substantially as described.

9th. The nippers *y y*, arranged and operating in a lateral direction, substantially as described, to remove the printed envelopes or bags at one side of the folding stand.

10th. The lifter 10, applied, substantially as described, to the folding stand, and operated by the lever which carries the nippers *y y*, for the purpose of lifting the finished envelope or bag at one side thereof from the stand, to enable it to be taken by the nippers.

11th. Applying a stamp V, to work through the table C, substantially as described, for the purpose of stamping a card, &c., on a bag during the process of manufacture.

12th. The general arrangement and combination of the several working parts of the machine, substantially as set forth.

No. 13,838.—EMANUEL HARMON.—*Improved Envelopes.*—Patented November 20, 1855. (Plates, p. 301.)

This improvement consists in preparing envelopes with lines *a* ruled or otherwise made in such manner that the lines shall not appear externally, and shall become visible when the face and back of the envelope are brought together, so as to guide the hand in writing the address.

Claim.—The manufacture or preparation of envelopes with parallel lines on the interior of the back, as set forth.

No. 12,301.—NEWELL A. PRINCE.—*Improved Fountain-Pen.*—Patented January 23, 1855. (Plates, p. 301)

Claim.—1st. The elastic spring C unfixed in the feeding-tube, whether the said spring be placed under or above the pen, it being so placed that it is made to vibrate by the action of the pen in writing, substantially as described.

2d. The under recess, formed by inserting the feeding-tube in the lower end of the main reservoir tube, the said under recess acting as a receptacle of the ink, which re-flows when the point of the pen is turned upward, substantially the same as described.

3d. The combination of the conical part of the piston-rod with a conical seat for the same in the screw-cap, so that when the piston-rod is drawn outward in charging the main reservoir tube with ink, the hole in the screw-cap is closed ink and air tight, substantially as described. (See engraving.)

No. 12,727.—Hugh K. McClelland.—*Improved Fountain-Pen.*—Patented April 17, 1855. (Plates, p. 301.)

The valve *e* is operated during the writing by tapping the key *h* with the fore-finger.

The inventor says: I do not claim separately any of the within described parts; but I do *claim* the construction of the implement as herein shown and described, viz: Having a bag or receptacle B placed within a tubular handle A, the lower end of said bag having a tube C attached to it, which tube is provided with a valve *e* and button or spur *c*; the tube, valve, and button or spur being enclosed by the pen-holder D, which contains a sponge G, and is provided with openings or channels *j*, through which the pen is supplied with ink as the valve *e* is operated, as herein shown and described.

No. 12,734.—William H. Towers.—*Improved Pen-Holder.*—Patented April 17, 1855. (Plates, p. 302)

The writer by dipping this pen in the ink has to insert the nibs sufficiently far below the surface of the ink to enable the sponge to become saturated with ink. By pressing the fore-finger upon the knob C the lower end of lever A will press against the sponge D, and the nibs will be supplied with ink, &c., till the sponge is exhausted.

Claim.—The combination of the sponge with the lever and pen, arranged and operated in the manner and for the purpose herein described.

No. 13,534.—Geo. W. White.—*Improved Fountain-Pen.*—Patented September 4, 1855. (Plates, p. 302.)

B is the inner tube and C the outer tube, with the pen-holder E attached to it.

I *claim* the manner of constructing the holder by having two small tubes, one fitting close over the other; the inner tube joined to the main band, and the outer tube having the holder for the pen attached, and having a hole drilled through both tubes on the side that the pen is attached to, so that the ink may flow out into the pen *r*. When the outer tube is turned or revolved around on the inner tube, the holes are turned away from each other and the holder closed; this outer tube to be turned and regulated by means of a small projection on each tube to the place desired.

No. 13,995.—Newell A. Prince.—*Improvement in Fountain-Pens.*—Patented December 25, 1855. (Plates, p. 302.)

The pen is provided with a bead C, to prevent it from lifting too much in writing; the pen is notched at A B so as to interlock with the notched part A^1 B^1 of the feeding-tube T, and prevent the moving laterally or slipping out of the pen.

The inventor says: The claims I now make are for improvements in addition to those already made and patented January 23, 1855.

I *claim*, 1st. The elevation or bead on the back part of the pen, near its heel, being designed to keep the pen, by coming in contact with the inside of the main reservoir tube, from lifting too much, substantially the same as set forth, as described and shown.

2d. I claim the pen, notched near its heel, and the combination of the same with the feeding-tube, correspondingly notched; so that, the two placed together and infixed in the main reservoir tube, the pen cannot get out of its position, substantially the same as shown and described.

No. 12,722.—Walter K. Foster.—*Improvement in Moulds for Casting Pencil-Sharpeners.*—Patented April 17, 1855. (Plates, p. 302.)

The conical core of the mould is seen at D; the shape of the pencil-cutter at E and F. A small spring-holder G is inserted through an opening in the side of the mould for holding in place a blade or cutter H. The side of the mouth I is furnished with a flat sliding-core J. The spring-holder is pushed in sufficiently to allow the blade to be slipped in between the spring ends of the holder. The holder is then drawn back until the flat sliding-core J will come against the side of the blade so as to stop the metal composition from coming to the bevel edge or back of the blade, and at the same time hold the edge of the blade to its place in the groove of the conical core D, the object of this groove being to drop the edge of the blade to make it take the wood in cutting. The mould is then placed upon the base A, with the sliding-core at a sufficient distance from the gauge K to admit the flat core of the other part of the mould L to come between it and the gauge K, which gauge is for the purpose of varying the depth to which the edge of the blade shall be placed in the groove of core D in order to make a thicker or thinner shaving. The mould-piece L is placed on the base, with the sliding-core P touching the gauge. This core is for the purpose of shutting off the composition from the front edge of the blade, in order to form a throat for the passage of shavings. The ring *m* is placed on the mould, and the composition is poured in. The finished pencil-cutter is seen at *o*.

Claim.—The arrangement of the spring-holder G, sliding-plates J and P in relation to the grooved core D and gauge K, for the purpose of adjusting and holding the blade H in the mould, and the forming of the slot in the pencil-sharpener, as herein set forth.

No. 12,391.—William S. Mac Laurin.—*Method of Teaching Penmanship.*—Patented February 13, 1855. (Plates, p. 302.)

This method consists in having a series of diagrams, figure 1, figure 2, and figure 3, etc., marked or engraved on a slate or tablet, each diagram forming a continuous or endless curve. The pupil, with a pencil, stilus, or other hard point, follows the mark or groove in the tablet as rapidly as possible without ever raising the stilus from the groove, increasing in speed until the hand becomes perfectly accustomed to the form.

Claim.—The employment of figures such as described, marked on or formed in the surface of a tablet, slate, or other surface, for the purpose of aiding the hand in guiding the point of a pen, pencil, or stilus in retracing therewith the lines of the said figures an indefinite number of times, as described, to train the hands of pupils in teaching them the art of writing.

No. 13,885.—Isaac Rehn.—*Improved Photographic Bath.*—Patented December 4, 1855. (Plates, p. 302.

The reservoir 1 being filled with solution, and supposing it to have been used, is covered with a film. By pouring some additional solution into the receiver 2, the current passes through opening 3 into receiver 1, causing the fluid to flow over the lower wall of the reservoir 1 into a conducting-trough 4, along which it is carried into a proper receptacle, from which it is to be again used as before.

The overflowing of the fluid carries with it all the scum or film, leaving the surface entirely clean; which scum, when allowed to come into contact with the plate, is apt to adhere to it, and cause a stain.

Claim.—The overflowing bath with the conducting-trough and receiving-chamber, or their equivalents, as set forth.

No. 12,188.—Dwight Gibbons, assignor to Fred. Starr.—*Improved Brace for Piano-Frames.*—Patented January 2, 1855. (Plates, p. 302.)

The part of the plate marked F shows the suspended point of the plate. The part H of the plate forms the back of the frame. L M is a brace which extends from the right-hand front corner of the frame or plate to near the middle of the back, which are the two weakest points in the instrument. Thereby the instrument is to be better kept in tune than by any of the means heretofore in use.

The inventor says: I do not claim any method of bracing from the suspended point of the plate F to the iron frame, rest-plank, or back, knowing them to be in use.

But I do *claim* making use of a diagonal brace extending above the plate and strings, operating in the manner and for the purpose substantially as herein described and set forth.

No. 12,315.—Alexander Hall.—*Improvement in Piano-Fortes.*—Patented January 30, 1855. (Plates, p. 302.)

This improvement relates to the celestial piano patented in April, 1854, which is provided with octave strings in addition to the normal strings.

The octave string d^3 passes very near to the normal string d^2 at the commencement, where the motion from vibration is very slight. Further on, however, the string recedes from this normal string and descends to the bridge *a*, which is low enough to allow the octave string to vibrate clear of the normal string. From this bridge the octave string passes through perforations *e* in the usual bridge, and to the depressed hitch-plate *c*, the face of which is on a level with the perforations *e*. Thus the octave string requires but little extra room. Pieces of leather $n\ n^1$ are attached to the curved strip *m*, which project over and are struck by the hammers up against the strings in such way as to imitate the thrumming upon the harp. In order to adjust the distance between the octave and normal strings, the normal strings are carried over a bridge-pin *p*, which is vertically adjustable by means of its screw shank. The top of it has channels *r r* on its sides, and at the rear apex *s* there are notches *t t* for confining the strings. The octave strings pass round or against the common form of bridge-pin *o*. When the bridge-pin is turned to the right or left it carries both normal strings with it, and thus increases or diminishes the distance between them and the octave strings.

Claim.—1st. Sinking the middle octave bridge *a* below the level of the normal strings, so as to be clear of their vibrations as set forth.

2d. In combination with the depressed bridge *a*, the perforations in the bridge *b* on the level with the top of bridge *a*, for the purposes set forth.

3d. The extra hitch-plate *c*, in combination with the depressed *a* and perforated bridge *b*, as set forth.

4th. The adjustable bridge-pin for the normal strings, furnished with a screw, and the notches and channels on its two sides; so that the normal strings can be regulated in their relative distances from the octave strings, either vertically or laterally, as set forth.

5th. Making the buff stops of two qualities of leather, a hard and a soft, for producing the harp effect, as set forth.

No. 12,362.—James A. Gray.—*Improved Sounding-Board for Piano-Fortes*.—Patented February 6, 1855. (Plates, p. 302.)

The inventor says: I do not confine myself to any particular form or number of corrugations, but any number that may be necessary.

But what I *claim* is the improvement of the sounding-board of the piano-forte by corrugating its surface, thereby increasing its sounding surface, and giving it sufficient stiffness or strength, without gluing cross-bars on either side.

No. 12,432.—Henry S. Ackerly.—*Improvement in Piano-Forte Frames*.—Patented February 27, 1855. (Plates, p. 302.)

A is the wrest-plank; *a* and *b* are the strings; the diagonal part *c* of the wrest-plank is more elevated than the part *d*, for the purpose of allowing the strings to be arranged in two tiers (*a* the upper and *b* the lower tier).

The inventor says: I do not claim the arrangement of the strings in two tiers crossing each other, nor the construction of the wrest-plank with one part elevated above the other.

But I *claim*, 1st. The arrangement of the wrest-plank of a square piano-forte along the front and across one of the front corners of the instrument, as described, to receive two tiers of strings, of which the tier comprising the longest strings is arranged nearly parallel with the front and back of the instrument, and the shorter ones diagonally across the same, said arrangement being for the purpose fully set forth.

2d. The construction of the metallic plate B B, with the straight brace *e* across the back, and the arched moulding or brace *f* running from the said straight brace to the front of the instrument, as and for the purpose set forth.

3d. Constructing the plate B B with a recess to receive the wrest-plank, so that it may be firmly secured against the tension of the strings, substantially as set forth.

No. 12,737.—Stephen P. Brooks.—*Improved Piano-Forte Action.*—Patented April 17, 1855. (Plates, 303.)

The object of this improvement is to bring the greater bulk of the action of an upright piano below the line of the keys, thereby economizing the space required for the upper portion of the instrument, and obviating the necessity of casing the top to protect the action. The front of the key-lever being depressed, the fly *c* strikes against butt *d* which projects from bar *e*, said bar being hinged to the frame in *h* and h^1; thereby the hammer *m* is raised and strikes the string *n*. As soon as the hammer falls back, the bar *e* in its descent moves somewhat towards the string, as apparent from the figure, and the damper *o* is brought against the string *n*, the damper being hinged at *r* to the bar.

The hammer can be held close to the string after the blow has been given, so as to be in readiness to strike the string again, without moving through its whole arc, by means of a projection S attached to bar *e*, against which projection, after the fly *c* has been thrown off the butt *d*, the end of the key-lever strikes and thereby partly lifts the bar and the hammer, keeping it close to the string as long as the key-lever is depressed.

Claim.—Transmitting the blow from the key-lever to the hammer, by means of the vertical bar, arranged and actuated substantially as described, whereby I am enabled to place the action below the level of the keys, as above set forth. I also claim attaching the damper-arm to the vertical bar in such a manner that the up-and-down movement of the said bar will alternately bring the damper against the string and relieve it from the same, as set forth; I also claim the means used for keeping the hammer close to the string after the blow has been given, the same consisting of a butt attached to the vertical bar, and actuated by the key-lever, as described.

No. 12,763.—William Munroe.—*Improved Piano-Forte Action.*—Patented April 24, 1855. (Plates, p. 303.)

When the finger-end of the key A is depressed, the jack D rises and throws the hammer into position figure 2. During this upward motion of the jack, the inclined escapement F comes in contact with the hammer-rail M, and is pushed therefrom, carrying with it the jack and check G. The hammer in rising acquires a momentum, and is projected beyond the positive action of the jack so as to strike the string; the reaction of the string causes the hammer to fly back, where it is caught by the check acting on the hammer-butt, which prevents the hammer from rebounding against the string. When the finger-end of the key is allowed to rise through a small portion of its movement, the jack descends proportionally, and the weighted check causes the end of the jack to move forward and engage the end of link E, whilst the hammer is held up by the top of the check in a position ready to repeat the blow. In repeating, the finger does not allow the key to rise its full stroke; consequently the hammer does not descend and bear on cushion N, but follows and rests on the top of the check G, whilst the surface *d* is disengaged from contact with the hammer-butt, and the jack is drawn by the weighted check partially under link E, and obtains sufficient engagement therewith to give the repeating blow without its being necessary for the key to traverse the whole of its stroke. The uniformity of the resistance of the inclined escapement is intended to avoid the inequality of touch occasioned by the abrupt resistance of the ordinary escapement

Claim.—1st. The combination of the escapement-jack and check, co-operating to sustain the hammer in position to repeat and to prevent its rebound, substantially in the manner set forth.

2d. The inclined escapement as applied to piano-forte and other similar actions, substantially in the manner herein set forth.

3d. The application of the toggle-joint to piano-forte and other actions, in combination with the jack and hammer, for the purpose herein set forth.

No. 13,091.—Robt. M. Kerrison.—*Improved Piano-Forte Action.*—Patented June 19, 1855. (Plates, p. 303.)

The hammer-butt has an angular notch C, at the apex of which the impulse is applied to drive the hammer against the string; A is the impeller, with two branches *b a*, the action of which will be understood from the engravings; H is the key; figure 2 represents the key when depressed.

The inventor says: I *claim* solely the means I employ to check and hold the hammer after the blow. I claim, as part of those means, the impeller made of the form described, substantially in the manner and for the uses set forth.

I also claim the hammer-butt, of the form described, suitable and proper to receive the action or pressure of two parts or branches of the impeller at two different points, in the manner and for the purpose substantially as set forth.

I claim the impeller and the hammer-butt as described, or their equivalents, acting conjointly in the manner substantially as set forth.

I claim them conjointly, as a means of checking and holding the hammer after the blow.

No. 13,236.—ANDREW STOECKEL.—*Machine for Cutting Legs for Pianos, Tables, &c.*—Patented July 10, 1855. (Plates, p. 303.)

The frame C with the stuff Y is first moved to the right as far as necessary to bring the disc Q over the end of the stuff. Motion is then given to the shaft T, and the cutter-disc Q, drum V, and the stuff are rotated. Carriage B is then moved along by turning shaft L by hand, and the frame C and stuff are raised and lowered by the patterns, or rather by means of projections attached to frame N, against which the patterns bear, frame C being kept elevated by operating lever I by hand, and the stuff will consequently be turned in form corresponding to the patterns. When the stuff is turned, the frame N is raised in order to free the stuff from the cutter-disc, and the drum V is also raised, or the swinging frame to which the drum is attached. Carriage B is then moved back to its original position, and the stuff secured in any proper manner so that it cannot turn. The frame C is then again raised by means of lever I, and motion given to shaft U, and shaft *o* and cutters *i* rotate, and the carriage is moved as before, and the cutters *i* will plane a flat surface on the stuff. The carriage is then again moved back, the stuff turned a proper distance, and second flat surface is planed, and so forth.

The inventor says: I do not claim turning the stuff the required forms by means of patterns, for that has been previously done; neither do I claim the cutter-disc Q, nor the rotating cutters *i i;* but I *claim* the arrangement of the carriage B, with the vibrating frame C attached, cutter-disc Q, and cutters *i i* on the shaft O, and drum V placed in the swinging-frame W, substantially as shown, for the purpose set forth.

No. 13,924.—FRANCIS TAYLOR.—*Improved Piano-Forte Action.*—Patented December 11, 1855. (Plates, p. 303.)

b represents the key; *a* the key-board.

The inventor says: I do not claim the button *m* taking the second knuckle of the hammer-butt, as this has been used as an attachment to the key. But I am not aware that this button *m* has ever before been made as a permanent attachment to and moving with the fly of the jack, in the manner as specified, whereby the fly of the jack is held to the knuckle 3 by the button *m*, until the hammer is sufficiently raised for said button *m* to clear the knuckle 2, and also replaces the said fly of the jack beneath said knuckle 3 immediately that the key is released, and the hammer descends but a short distance, producing an instantaneous and uniform repeating action.

Therefore I *claim* the regulating-button *m*, permanently connected to, moving with, and governing the fly of the jack in its action on the butt of the hammer, the whole arranged and operating substantially as specified.

No. 13,942.—Spencer B. Driggs.—*Improvements in Piano-Fortes.*—Patented December 18, 1855; patented in England November 1, 1855. (Plates, p. 304.)

The object of the various features of this improvement is to produce more free and prolonged tones in all parts, particularly the treble of the instrument.

The inventor says: I *claim*, 1st. Securing the sounding-board H within a metallic frame I, or its equivalent, substantially in the manner and for the purpose set forth.

2d. Combining the sounding-board and its enclosing frame with upward projections from an open metallic base frame B, and with a wrest-plank and an upper metallic frame or hitch-plate D F; by which I am enabled to make a piano-forte without using wooden blocks or other wooden supports for the wrest-plank, sounding-board, and upper metallic frame, substantially as set forth.

3d. In connexion with the combination of the upward projections from the open metallic base frame with the metallic sounding-board frame, the wrest-plank C and the upper metallic frame, combining a thin bottom-board *b* with a shallow wooden frame A, which encloses the said open metallic base frame, substantially as set forth.

4th. In connexion with the enclosure of the thin sounding-board within a metallic frame, and the combination of said frame with the upper metallic frame, the wrest-plank and the open metallic base frame; also the combination of the said enclosed sounding-board with the thin bottom-board of the instrument by means of a sounding-post *l*, for the purpose of adding additional stiffness and vibratory power to both of said boards, substantially as herein set forth.

5th. Supporting the strings upon metallic saddles *n*, which stride the sounding-board bridge, and are combined with said bridge *k* and with the sounding-board, substantially in the manner set forth.

No. 13,580.—John S. Morton.—*Piano-Forte Action.*—Patented September 18, 1855. (Plates, p. 304.)

The movement of the key A is transmitted to the hammer (which latter is attached to the end of hammer-rod *i*, and is not represented in the engraving) by means of jack C applied to the key operating on the hammer-butt *a*. The lever *b* is hinged to C by pin *d;* at about the middle of its length it rests upon the cushion-headed pin *e*, which works freely in the lower arm of the jack, and rests upon the jack-bottom D which is secured to the key. As the front end of the key is depressed, to throw up the hammer to strike the string, the movement of the jack in escaping carries the lever under the block *c;* and when the hammer falls, the block *c* falls on the end of the lever, and causes the hammer to be arrested a very short distance below the string (see fig. 2). While in this position the hammer is ready for repeating the blow, to effect which it is only necessary previously to allow the front end of the key to rise about a sixteenth of an inch. As the hammer falls on the raising of the front end of the key, the rounded face of block *c* acts in such a way on the end of lever *b* as to have the effect of returning the point of the jack into the notch of the hammer-butt. The spring *g* prevents the jack flying too far out of the notch when it is let off.

I *claim* the arrangement and operation together, shown and described, of the lever *b* pivoted to the jack, post, or cushion *e*, and block *c*, with the jack and hammer, to effect the repeat; and whereby, while the use of an additional spring or upright is dispensed with, the weight of the hammer operating on the lever returns the jack to its notch in, or position under, the butt, essentially as set forth.

No. 13,960.—HUBERT SCHONACKER.—*Improved Piano-Fortes.*—Patented December 18, 1855. (Plates, p. 305.)

i is the sounding-board; *b* the bridge upon the sounding-board; and *g* the bridge upon the wrest-plank *h*; *a a* are the hitch pins, and *c* a metallic fret. In order that none of the vibrations of the strings be wasted, the screws *e e* are employed; each one of which has fixed bearings, and passes through a movable nut *f*, to which nut the string is attached, or the two ends of the string where a double string is employed for each note.

I *claim* constructing the instrument so that the strings shall rest on a fret at the nodal or octave points, or substantially similar rest, upon the bridge of the sounding-board, whereby free vibration is allowed to the whole length of string between the hitch-pins and bridge on the wrest-plank, substantially as described.

2d. Though I do not of itself claim connecting the two strings of a note with a single horizontal turning-screw, I claim the connexion of the two strings with the same screw, when that is combined with the employment of fret *c* or other rest, merely supporting the string on the sounding-board at single points, and not confining it, substantially as set forth.

No. 13,014.—SAMUEL J. H. SMITH.—*Improved Portfolio.*—Patented June 5, 1855. (Plates, p. 305.)

The bars A B are parallel to each other, and united at their two adjacent ends by elastic band springs *c c*, which will permit these bars to be moved apart from one another, and to return towards one another when the separating force is withdrawn from them. To each of the bars is connected a sheet of pasteboard C, which may be hinged to the bar; there is applied to both bars an elastic back D, which should be made of a width as great as it may be desirable for the bars to be separated. The letters are clamped between the bars, and the elastic back will be extended in proportion as the mass of paper filed is increased; the back serving to protect the papers from being injured or forced from between the bars.

The inventor says: I do not claim a bill or paper file composed of two straight bars or plates, and elastic bands uniting them together at their ends, when such bars or plates are so arranged that one may be parallel and above the other.

But I *claim* combining such with two covers and a flexible, elastic, or extension back, so as to constitute the file portfolio for the retention and preservation of papers or letters, and protecting the mass or file of them on both sides and the back of the same.

No. 12,118.—DAVID BALDWIN.—*Improved Apparatus for Feeding Paper to Printing-Presses and Ruling-Machines.*—Patented January 2, 1855. (Plates, p. 305.)

Frame B consists of two arms *e e* suspended loosely upon shaft C, and arms *e e* carry horizontal tube E, provided with a series of tubes *f* projecting downwards. Connected with tube E is air-pump F, which has its bearings *g g* on shaft C, which turns loosely through them. Piston-head G of pump, and piston-rod H, are worked (through a slot in the cylinder) by pinion I, which is firmly attached to shaft C. Through slot *j*, in connecting-rod K, passes shaft L of cylinder A; pin K, on connecting-rod K, works against cam M on shaft L; shaft L passes also through slot *m* in connecting-rod *o;* pin m^1, on connecting-rod *o*, works against cam o^1 on shaft L; feed-table Q is pivoted in *p p;* arm U, which carries roller-wheel T, is pivoted in *s* to arm V projecting from rock-shaft X; arm Y projecting downwards from shaft X fits into end of rod Z, which works through bearings *u u* attached to main frame; pin *w*, on rod Z, fits into slot *v* of arm A^1.

Operation: motion communicated to shaft L; and if tubes *f* are over the sheets *x* (see figure 1), cam o^1, operating against pin m^1, causes rod *o* to turn shaft C, and pinion I draws up the piston; thereby a vacuum is formed in tubes E and *f*, and the sheet of paper underneath is forced against the tubes by the external atmospheric pressure. Cam M now operates against pin *k* and vibrates frame B, which moves the tubes *f* with the sheet of paper towards cylinder A until the sheet is caught by cords *a a.* At this point, pin m^1 is released from the cam, and spiral spring P turns shaft C back; the piston descends, destroying the vacuum, and releases the sheet. Pin *k* is now freed from its cam, and frame B falls back to commence the above operation again.

When frame B returns to its original position, arm g^1 strikes the end of lever U, thereby moving rock-shaft X and arm Y. This moves forward rod Z, and moves pin *w* off the tooth against which it rested in slot *v;* spring R throws up the front end of table Q the distance of one notch in slot *v*. This upward motion of the table serves to keep the sheets always close to the tubes.

Claim.—1st. Feeding sheets of paper, singly or one at a time, to a printing press, paper-ruling, or other machine requiring the feed of a single sheet at a time, by means of a vibrating frame B having at its lower end a series of tubes *f* which, as the frame B vibrates, pass over the sheet to be fed to the machine, and also over a portion of the cylinder or other device for receiving the sheet; a vacuum being formed and destroyed in said tubes *f*, by means of an air-pump F attached to the frame B, and operating as herein shown, for the purpose of causing the tubes *f* to convey the sheets from the feed-table to the receiving device of the machine.

2d. The self-adjusting feed-table Q, constructed and arranged as herein shown, or in an equivalent way, so as to be operated by the vibrations of the frame B, and keep the sheets *x* close to the ends of the tubes.

3d. The tubes *f*, arranged as herein shown, on a vibrating frame, when said tubes, with the aid of an air-pump, are employed for conveying the sheets to the press or machine.

No. 12,178.—STEPHEN BROWN.—*Improved Press for Printing different Colors.*—Patented January 2, 1855. (Plates, p. 305.)

The type being placed upon the platens P P P, and motion being given to the shaft B, the cam C is made to rotate, and first acts upon friction-roller *d* and depresses the ends of the horizontal arms of the bent levers K K. This movement throws the ink-rollers I (each of them being charged with a distinct color) across the platens, inking all of them simultaneously. This is done by means of lever K drawing lever F towards the centre of the machine, thereby pushing the arms H the same way, to the ends of which arms the boxes J are attached which support the ink-rollers; projecting-pins on the boxes and the horizontal grooves *a* (in the side of the frame of the machine) serve to confine the rollers to a horizontal motion across the platens, which latter slide in vertical grooves *f*, and are supported at the proper height for the operation of the inking-rollers by small pins *h*. The cam C now acts against the segment M, and throws the rollers back to their original position; and the cam then raises the movable bed N, and the bed raises the sectional platens P, and forces them simultaneously against the paper, which is between the platens and the upper bed-plate U, producing a simultaneous impression in different colors.

The sheets of paper are fed in by the fingers R, attached to bar Q on one of the upper boxes J (see figure 4). As the fingers are moved forward with the ink-roller, the lower end of the screw *t* depresses the back end of the upper jaw *k*, which is caught by the spring-catch *n*, leaving the jaws *j k* open (see dotted lines, figure 2). The fingers move over and beyond the platens in an open state till they pass under the bar T, at which point the fingers reach the sheet, the edge of which passes between the jaws and the spring-pin *r*, throwing the catch *n* off from jaw *k*, which closes by means of spring *m*, and the fingers return with the sheet, the bar T rolling backwards to allow them to pass. The sheet, being thus brought over the platens, receives the impression. and is then conveyed back with the return motion of the fingers and caught by the fly-rollers.

Claim.—1st. The employment or use of a series of platens P P P, (two or more,) arranged as herein shown, or in an equivalent way, so that the form or type on said platens may be inked simultaneously with separate and distinct colors, and operated or pressed simultaneously against the sheet, as set forth.

2d. The arrangement herein shown of the cam C, bent levers K K, and arms F F, which are connected to the bent levers, for the purpose of operating the movable bed or platen N, rollers I, and fingers R, as herein shown and described.

3d. Operating the fingers R by means of the screw *t* and rolling-bar T, whereby the fingers are made to convey the sheet to and from the platens P, as herein described.

No. 12,183.—Sidney Kelsey.—*Improved Printing-Press.*—Patented January 2, 1855. (Plates, p. 306.)

The sheet of paper e^2 is passed over the platform A^1 and through the slot d^1, so that its front edge reaches into the space between the projection *i* on the part *g* of the carriage and the edge of the part *h*, this space being held open by means of small springs *j j* between said part *h* and projection *i*. As the pulley L rotates, it winds the band *t* around it, and the carriage H is drawn underneath the platen B and over the form on the bed G; and as the band *t* is attached to the projection K on the part *h*, the edge of said part *h*, adjoining the projection *i*, will overcome the power of springs *j j* and bear against projection *i*, and thereby the edge of the sheet will be grasped between them. The sheet is thus carried between the platen B and bed G. Just before the carriage gets underneath the platen B, the cord c^1 acts upon the pulley b^1 and turns the shaft X, the rods *z* being turned over in a horizontal position and occupying the place the carriage left, the pulley being retained in proper position by the pawl Y, which catches against the notch d^2. When the carriage is between the platen and bed, it is retained there by a spring-catch; the platen descends and presses the sheet against the form on the bed. The pulley I is now rotated, and the carriage is drawn back by means of cord *l*, the rods *z* of fly W fitting in the notches *s*, in the outer end of part *h* of the carriage, the sheet of paper passing over the rods *z*. Just as the carriage reaches again its original position, the pawl Y is thrown up, the spring a^1 throws the shaft X around, and also the rods *z*, to their original position, the sheet being deposited thereby upon a proper fly-board.

Claim.—1st. Feeding or conveying the sheets to the form by having the carriage H formed of two parts *g h*, and arranged substantially as herein shown, so that the edge of the sheet may be grasped between the two parts of the carriage as it is moved between the platen and bed.

2d. The fly W, operated as herein shown, viz: by means of the pulley b^1 attached to the carriage H by a cord c^1, said pulley being hung on the shaft X, which is provided with a spring a^1, as set forth.

No. 12,185.—James Lewis.—*Improved Printing-Press.*—Patented January 2, 1855. (Plates, p. 306.)

The paper is placed upon feed-board indicated at a^1. The cylinder B is then rotated; and when the bent end *y* of the rod H comes in contact with the bent rod *v* attached to the upright *a*, the fingers *w* on the rod H are raised from the cylinder, and the edge of the paper is slipped underneath them, the fingers falling upon the paper when the bent end y has passed the bent rod v. The sheet, being thus carried round with the cylinder, is pressed between it and the form b^1 on bed C, which is moved by cog-gearing c e. When the fingers w get near the point at which they receive the next sheet, the pins u, at the ends of the cylinder, act upon bent-levers r, the inner edge of bar G is depressed, and the edge of the printed sheet c^1 is caught between the bar G and lips h of the bar F; the buttons p p now come in contact with the bar E of the fly D, force it back till the lower ends of the upright f strike against pins d^1 projecting from frame A, and the uprights f assume the position shown in dotted lines in figure 2, the rod l passing over a bar e^1, which causes the inner end of bar G to be depressed, and the printed sheet falls upon fly-board I. Spring k serves to press the edge of bar G close against the lips h, when the bar is left free to the action of said spring.

At this time the cogs c have passed the rack e, and bed and fly are brought back to their original position by the action of a weighted cord A^1.

Claim.—1st. The employment or use of the fly D, constructed, arranged and operating in the manner and for the purpose as herein set forth.

2d. The combination of the fly D, cylinder B, and bed C, when the above parts are arranged and operated as herein shown and described.

No. 12,213.—Robert Neale.—*Machine for Printing from Engraved Plates.*—Patented January 9, 1855. (Plates, p. 307.)

The engraved plate, with its bevelled edges, is secured between the cleats K I, which project from the plate-holder a^1, and which are correspondingly bevelled on their inner sides, so as to fit the bevel of the engraved plate. The plate-holder is movable around a central pivot v^1, which projects from the bed-plate H, which latter is so attached to an endless chain G that it travels with it and around the drums A B. M and D are the press-cylinder, and E E are cog-wheels on the axle of the lower press-cylinder, which gear into the links of the chain and move it along. Immediately after having passed through the press-cylinders, the projection b^1 on the plate-holder a^1 strikes against a stud on the frame of the machine; so that the plate-holder, together with the engraved plate, will assume an oblique position, as represented in figure 3. This is for the purpose that the ink may not be wiped out from the engraved lines, which are parallel with the edge of the plate, when afterwards passing over the wiping and polishing belts and rollers; which wiping out will be apt to be the case with all the lines which lie in the direction of the motion of the bed-plate, the chains, and the wiping and polishing belts and rollers.

The plate, after having travelled around drum A, arrives in front of the inking-roller X. This roller receives the ink from the distributing-roller W. By employing inking-rollers of various lengths, various widths of the plate can be inked. The length to which the plate is to be inked is regulated by the strips or bearers d^1 d^1, which are made to correspond to the length of the portion of the plate which is to be inked. As soon as the strips d^1 strike against the end e^1 of the levers Y Y, which support the axle of the inking-roller, the end e^1 will be depressed and the inking-roller will be elevated and pressed against the surface of the plate. When the strips have passed over the end e^1, the inking-roller will fall back from the surface of the plate. The plate then arrives over the wiping-belt *g g*, which passes around two rollers *f e*, and is destined by its contact with the surface of the plate to wipe off the ink from said surface. The pressure with which the belt operates upon the surface of the plate can be regulated by pressure-roller i^1 and by lever f^1 (to which is suspended the axle-box of belt-roller *f*) and a weight h^1. The belt is to be kept clean by means of a scraper *h*, which is kept in contact with its surface. The plate then passes over the polishing-rollers *i i*. These rollers *i i* are to be kept clean by their being in contact with an endless belt *q*, which is impregnated with whiting or similar substance, which it receives by travelling over a roller o^1, which is arranged within a suitable box filled with said substance. The belt passes from roller o^1 over a revolving brush P^1, which removes the grit, &c., from the surface of the belt. The plate then travels around roller B and arrives between the press cylinders, where the sheet of paper is fed in over the belt u^1. Of course, all the above-mentioned belts, rollers, &c., have a width corresponding to the greatest width of plates to be printed on this press.

Claim.—The combined apparatus herein explained for inking, wiping, and polishing engraved plates used in copper and other plate-print; the same consisting—

1st. In the attachment of the engraved plate to an endless chain, with which it revolves while undergoing the several processes of inking, wiping, polishing, and printing, substantially in the manner herein before described.

2d. In the bed-plate H, with its movable plate-holder a^1 and its strips or bearers, as constructed and operating substantially in the manner described.

3d. In the mode of inking the plate, so as to confine the ink to the engraved portion, substantially as described.

4th. In the mode of regulating the pressure of the wiping belt upon the plate, substantially as described.

5th. In the mode of keeping the polishers clean by an endless belt of cotton or other proper cloth, itself kept in proper order by the application of whiting or other suitable drying powder, and preserved from dust and grit by the action of the revolving brush.

No. 12,401.—A. B. Childs and Henry W. Dickinson.—*Machine for Feeding Paper to Printing-Presses.*—Patented February 20, 1855. (Plates, p. 307.)

When the fan in *a* is revolved, the air enters trunk *c* (the valve c^2 being open) through the slit in the lowermost part of trunk c^1, and escapes from the fan-case into the tube *f*, and thence into trunk *i*, and out from the slit in said trunk *i*. These currents of air (an inward current through *d*, and an outward one through *i*) raise the uppermost sheet s^1 from the pile *s*. The sheet adheres to trunk *d* over the slit, and the outward blast from *i* serves to raise up and separate the whole length of the sheet from the pile. As c^1 revolves, (see arrow *x*,) it carries the sheet around. When it has made a half revolution, the valve c^2 is closed, and hole *m* covers over the mouth of pipe *g*, and the hole connected with trunk *e* comes opposite the mouth of pipe *f* (the trunk *d* is attached to trunk c^1, extending nearly the whole length of the latter; within trunk c^1, and extending nearly its whole length, is a trunk *e*, having a slit e^1, extending nearly its whole length; just beyond the end of slit e^1 is a hole through trunk c^1, which, as the trunk revolves, comes opposite to the open end of blast-pipe *f*, and at other times is open to the air). The inward draught through *d* is now succeeded by an outward blast through the same, and also an outward blast issues from *e*, thus throwing off the sheet on to the table ready to be seized by the fingers of the press. The force of the blast is regulated by valve *h*.

Claim.—The raising and delivering the sheets by means of the inward and outward currents, said currents being produced and operating in one and the same trunk, through one and the same slit or opening, by means of the fan or its equivalent, trunk *c*, valve c^2, revolving-trunk c^1, pipe *g*, and aperture *m*, and in combination therewith the outward blast produced through the trunk *e*, by the means set forth.

Also, in combination with the inward blast through the revolving-trunk c^1, for raising the paper, the outward blast through revolving-trunk *i*, for separating the sheets, as described.

Also, the projecting trunk *d*, in combination with the main trunk c^1, in the manner and for the purposes set forth.

Also, the combination of the regulating and supply-valve *h* with the shut-off valve c^2, in the manner and for the two-fold purpose as set forth.

No. 12,553.—Chas. Keniston.—*Improved Hand-Press for Printing.*—Patented March 21, 1855. (Plates, p. 307.)

The arm D, to which are attached a piston and type, operates upon a swivel. There are two pads H H attached to the bed F of the press; one for inking the type, the other for the article to be placed on when printing. The press is operated by turning the arm over the pads, and striking a light blow upon the top of the piston alternately.

Claim.—The arrangement and construction of the press described.

No. 12,568.—Lemuel T. Wells.—*Improved Printing-Press.*—Patented March 21, 1855. (Plates, p. 308.)

The object of this invention is to facilitate the feeding and accurate placing of the sheets. The platen U, instead of having a uniform vibration with the arms V, is capacitated for a distinct rotation upon pivot *x*, which distinct rotation is effected by the reaction of the stationary pin *n*, (projecting from the frame,) when the lower portion of the platen impinges against it; the platen at first moving with the arm a sufficient distance to lift the sheet square off of the type, then impinging against the pin, the lower edge is held nearly stationary, while the outer or upper edge immediately sinks, so as to present the platen in any approach to a horizontal position that may be deemed to best facilitate the action of feeding.

Claim.—The platen U, hinged or pivoted to vibrating arms V, in combination with the stationary pins or pin *n* and retracting springs X, or equivalent devices, for the purposes explained.

No. 12,634.—Ebenezer Mathews and Wm. D. Siegfried.—*Apparatus for Feeding Paper to Hand Printing-Presses.*—Patented April 3, 1855. (Plates, p. 308.)

c c are sliding clamps, moving freely upon the guide-rods *d*, and moved by cords attached to the cam-levers *f f* The cords pass from *f*, under the pulleys *e e* on the platen, and from thence to the clamp to which they are fastened; other cords, attached to the back edge of the clamp, pass over the pulleys e^1 to the weights *j*. The mode of operation is as follows: Supposing the guide-rods raised up and resting against the top of the press, the frame with feed-board, etc., slid back on its ways, clear of the form, the paper is now placed on the board *a* and the inking apparatus properly adjusted; the frame is then slid forward into the position represented in the figure; the guide-rods, with clamp, weights, etc., are then lowered into their place; the top sheet is now fed into clamp *b* and its front end properly clamped; the form is next run in, and the impression made; the clamp *c* has followed the sheet, until it arrived at the edge of the platen, when the additional pull given by the movement of the cam-lever *f* striking against *l*, the jaws of it pass beyond the edge of the sheet; at the same instant the teeth *x x* on the edge of the platen, striking catches *y y*, close the clamp on the margin of the sheet, the carriage is run back, bringing the printed sheet fast in the clamp *c*, until it arrives at the stationary bar *g*, which presses it open, the sheet is dropped, and the catches *y* are hooked at the same moment. The clamp *b* had been opened (by the fingers *k* striking and pushing back the spring-catches *r*), by means of which the clamp was closed down, when the form was run in, and thus returns open, ready for another sheet.

Claim.—The feeding hand-presses automatically, by means of the operation of the clamps, guide-rods, cords, weights, pulleys, catches and springs, arranged in the manner and for the purpose set forth.

No. 12,702.—John Bishop Hall.—*Apparatus for Feeding Paper to Printing-Presses.*—Patented April 10, 1855. (Plates, pp. 308, 309, and 310.)

The finger-shaft has its bearings in sockets E. It consists of a solid shaft *f* within a tubular shaft *e*. To this shaft *e* there is secured a series of plates *g*, and also to the inner shaft *f*, one set of plates *g* being directly above the other, their ends being in contact and forming fingers kept together by spring h, which passes around two arms i j secured into each shaft e f, the two arms being close side by side. The figures 1 and 2 exhibit two of the finger-shafts F. Driving-shaft B transmits motion to cords D D around pulleys C on shafts B B, their slipping being prevented by collars a on the cords fitting in recesses b in said pulleys. The ends of the cords are fastened to the sockets E of the finger-shafts, the sockets also fitting into corresponding recesses in the pulleys. The cords drawing the finger-shafts along, the grooved hubs q at one end of the shafts run on ways H, and the outer friction-rollers l on the arms i run on ways G G^1, keeping the fingers steady. Friction-roller l in its way ascending the curved front end of way G, the finger-shaft is gradually turned in its sockets, and roller n at the end of arm m is depressed and catches against the curved end of way H, and falls into recess p. The finger-shaft moving still forward, it is turned about one-half of a revolution, the other roller l following guide G, which is so curved that the fingers are properly presented to the edge of the sheet of paper, which has been raised up by nippers Z. Now the inner roller l strikes projection j^2, and arm j is kept nearly horizontal, while outer roller l follows the curve of the guide projecting nearly horizontally a short distance, forming a jog; the two arms i j are consequently expanded, the fingers g are opened, and when the roller has passed off projection j^2, the fingers grasp the sheet in consequence of spring h and carry it down and around pulleys C, the sheet receiving its impression during its travel from the front to the back pulleys underneath. When travelling further on around and above the back pulleys, the inner roller l strikes projection z on way H, the roller passes underneath the same, and depressing arm j opens the fingers g. Now, cam N having passed and having left free the end of rod M, spring L throws down frame K, and the sheet is pressed into receptacle I. Cam N then causes frame K to return to its original position. The nippers Z operate as follows: Stocks S S^1 slide in vertical ways R; and both being alike, a description of one will be sufficient. Stock S is moved upward by cam W acting on rod V, crank g^1, shaft T, and suspending links b^1 d^1. When the cam frees the rod V, the stock S descends by its own weight, assisted by spiral spring g on shaft T. The stock being elevated, the nippers Z present a sheet of paper (represented in the illustration in thick lines) to the fingers g, (see fig. 2,) and while in this position arm 10 (projecting from pulley C) strikes and raises roller m^2 on the outer end of lever H^1, and causes thereby the inclined projection n^3, at the inner end of lever H^1, to act against roller x^1 on bar A^1, thereby forcing this bar (which is linked to the stock by arms t^1 t^1) obliquely downwards. The bar A^1, being arranged between the jaws p^1 p^1 of the nippers Z, forces, in its descent, these jaws apart and the sheet of paper from them, just after it has been grasped by the fingers g, which are now commencing to withdraw the sheet of paper. The stock descends now, and roller i i on bar A^1 strikes projection f^3, which is screwed to the main frame A of the machine, whereby bar A^1 is moved a little back. The jaws p^1 p^1 (still in a distended state) are pressed upon the uppermost sheet of paper of the pile upon platform I^1. The jaws are pressed upon the paper by the weight of the stock, and with a force equal to the power of springs o^1. The pressing down of the jaws causes the portion of the sheet between them to rise to a certain extent. This being done, cam C^1 acts against roller b^2 of bar B^1, and the outward bent front end of bar B^1 striking roller w^1 moves arm v^1 (projecting from bar A^1) inwards, thereby elevating bar A^1 to its original position. The jaws, being freed from the distending action of bar A^1, grasp firmly the ridge of the sheet of paper between them. Cam W now again acting on rod V, causes stock S to ascend till the edge of the paper is brought against rod E^1 and properly presented to the fingers g. The springs 8, slightly depressing the sheet of paper just behind rod E, insure its outer edge bending upward. The jaws are more or less extended according to the thickness of the paper, by adjusting projection f^3. The extension of the jaws being properly adjusted, they cannot grasp more than one sheet at a time. The arrangement hereafter described serves to prevent two sheets being drawn off the pile, on account of their sticking together by moisture, &c. When the nippers press upon the uppermost sheet, the points of hooks w^2 w^2, which are attached to eccentrics v^2 v^2 on shaft K^1, just bear upon the back part of said sheet, and also rollers u^2 u^2 on the same shaft. They remain so till the nippers elevate the paper, when the upper inclined parts of projections 6 on bars 5 (which are secured to the stock) touch rollers 7 on both sides of platform I^1, (which slides in horizontal grooves,) and move said platform outward, (see arrow 1, fig. 5,) and shaft K^1 is turned simultaneously by means of rod z^2, so that the points of the hooks pass through the upper sheet, which is grasped by the nippers, and consequently does not move outward with the pile. After having passed through the upper sheet, the ends of the hooks, owing to their curved form, rest upon the second sheet without passing through this, as it moves with the platform and in the same direction with the hooks. The second sheet and those underneath it have the weight of shaft K^1, with its rollers w^2, &c., upon it; for the hooks in turning elevate the shaft K^1 with its appendages (see fig. 5). Now projection 6 leaving rollers 7, the platform moves inward, (see arrow 2, fig. 5,) actuated by spring 4, shaft K^1 is turned back, and the hooks pass out from the perforations their points made in the uppermost sheet, leaving the rollers upon the second sheet and the weight of shaft K^1, just previous to the withdrawal of the raised sheet by the fingers g.

Claim.—1st. Lifting or picking up the sheets of paper from the feed-boards, or piles of paper to be printed, by means of nippers, pincers, or tweezers Z, constructed as herein shown, or in an equivalent way, and operated by any suitable mechanism, so that the jaws of said nippers may, when slightly open, press upon the sheets of paper in such a manner that when closing they may grasp or nip the upper surface of the top sheet, or the ridge of the sheet formed between them by their pressure, as herein described, and for the purpose as set forth.

2d. Separating or detaching the uppermost sheet of paper on the feed-board, or pile of paper to be printed, so as to prevent the removal therefrom of more than a single sheet at a time by means of the hooks w^2 w^2, or by pins or points so constructed and operated as to answer the same purpose; and rollers u^2 u^2 attached to a shaft K^1, or other suitable fixtures, and in connexion with said hooks and rollers, by means of a reciprocating motion of the platform I^1, feed-board, and pile of paper, the said hooks, pins, or points acting conjointly with the said movable platform, substantially as herein shown and described.

3d. Conveying the raised sheets of paper from the feed-boards to the form to be printed, and also from the form when printed to the proper boxes or receptacles, by means of the fingers *g* attached to the tube and shaft F, which is secured to endless bands or cords D D, which are provided with collars or teeth to prevent them slipping in passing around suitable pulleys C, or by means of other fixtures which would be equivalent to the above named parts, when they are used in connexion with a vibrating or movable frame K, or its equivalent, for properly adjusting the printed sheets in the boxes or receptacles I when released from the fingers, as set forth in the body of the specification.

No. 12,733.—Caleb A. Thompson.—*Improvement in Making Printers' Ink.*—Patented April 17, 1855.

The object of this improvement is to produce ink which is tenacious and adhesive, and which, when dried, adheres firmly to the surface of the paper. The inventor takes for ordinary inks four pounds of litharge, two pounds of acetate of lead, and forty gallons of linseed oil. Before using the linseed oil it is subjected to heat of six hundred degrees Fahr., after adding litharge and lead as dryers. It is subjected to the action of heat from forty-eight to sixty-five hours, as to the desired quality of the ink. Thereby the inventor intends to eradicate the greasy nature of the linseed oil. The oil having been prepared as above, gum copal is added—four pounds of it to one gallon of the oil (as above prepared). For news-ink the following ingredients are used: fifteen pounds of the above oil, ten pounds of resin (common), two pounds brown resin soap, and five and a half pounds lamp-black. These proportions vary for other inks. (No illustration.)

Claim.—The composition of oil-varnish, made in the manner herein set forth, to be mixed with resin-soap, lamp-black, &c., for printers' ink.

No. 12,761.—Isaac B. Livingston and Miles Waterhouse.—*Machine for Feeding Paper to Printing-Presses.*—Patented April 24, 1855. (Plates, p. 309.)

Motion is transferred from driving-shaft N to cog-wheels W S and K^1, with the latter of which the crank-shaft K is revolved. This crank-shaft works between arms *a a* attached to levers G G. When the crank-shaft in its revolution strikes the upper part of arms *a*, it raises said arms and the front end of levers G and the cross-bar E, to which the ends of the pliable tubes D are attached. Continuing its revolution, it presses against the front side of arms *a*, pushing the levers G with cross-bar and tubes forward along the guide-ways F. A crank-pin on wheel K^1 and pitman *c* are so arranged that when the tubes have been raised and carried forward, the exhausting apparatus C shall blow off the paper that was attached to the tubes; and when crank K has pushed back the tubes, and allowed them to drop upon the paper, the crank K will commence to operate the exhausting-apparatus so as to exhaust the air from the tubes, which causes the paper to adhere to them by suction; the crank K in the mean time has passed the top of arm *a*, raising levers G G, cross-bar E, and tubes, and carrying them forward as described.

Claim.—The use of the angular guide-ways in combination with the cross-bar, or its equivalent.

The use of a crank, or its equivalent, working between the arms of levers, as described, in combination with levers and cross-bar, as described.

The raising-table in combination with the cams, shafting, and gearing, (forming said table), or their equivalents, as described.

The inventors add: We do not claim the raising of paper by atmospheric pressure, as that has been before used for that purpose.

But we do *claim* the combination of machinery, as described, for carrying forward the paper, a sheet at a time, and feeding it to printing-presses.

No. 12,989.—James Albro.—*Improvement in Registering Blocks for Printing Oil-Cloths.*—Patented June 5, 1855. (Plates, p. 310.)

The cloth D (represented in the engravings in strong broken lines) is placed upon table A, and the straight edge C pressed down upon it; said straight edge being exactly parallel with the ends of the cloth. The heads of the screws *d d* are then placed against the back flanges of the stops E, and the head of screw *e* is placed against the side of the projection of the left-hand stop. If the screws are properly adjusted, the figure on the block F will be printed in proper position on the cloth, as by means of said adjustment the parallelism of the transverse and longitudinal lines of the block is preserved. The block is elevated after each impression, and placed between the adjoining two stops. The recess *c* affords a space to receive the right-hand stop E, so that no adjustment of the stops is required. The matching of the rows is at a point on a line with the ends of the projections of the stops E, because the blank surface *b* of the face of the block occupies the space between the stops; consequently, in adjusting the cloth D, the edge of the printed surface is brought to this line.

The inventor says: I do not claim a transverse bar or straight edge C applied to a printing-table A separately, for it has been previously used; but what I *claim* is the bar or straight edge C with T-shaped stops E attached permanently to it, in combination with the guide-screws *d d e*, in blocks, and the recess *c* in the right-hand corners of the blocks, and smooth surface *b* on the faces of the blocks, as shown, and for the purpose as set forth.

No. 13,069.—Joel G. Northrup, assignor to James D. Mather.—*Improvement in Printing-Presses.*—Patented June 12, 1855. (Plates, p. 310.)

The nature of this improvement will be easily understood from the claims and engravings.

R, bed; Q, stand; P, connecting-rod; *a*, wrist-pin; O, chain.

The inventor says: I *claim* the manner by which I give motion to the bed of a printing-press by means of a vibrating connecting-rod; one end of which is attached by a stud-pin, or wrist, to a stud that projects from the lower side of the bed; the other end being connected by a similar wrist to an endless chain which passes around two wheels which revolve, thus producing an elongated crank-motion.

I do not claim the application of a chain for this purpose; but I *claim* the manner of attaching, combining, and communicating its motion to the bed, substantially as described.

No. 13,183.—A. H. Rowand.—*Machine for Feeding Sheets of Paper to Printing-Presses.*—Patented July 3, 1855. (Plates, p. 310.)

The sheets of paper W to be fed to the press are placed in a pile upon bed B, and springs force the paper up against the rollers G G on shaft F of swinging-frame E. Motion being given to driving shaft K, the rollers G G first move the uppermost sheet of paper backwards against guides T T, and projection Q on arm N then strikes against shaft F on swinging-frame E and raises it, and consequently rollers G G are freed from the uppermost sheet of paper, and arm P is also moved back, and the lip *q* is depressed or forced downward in consequence of bar S being operated by eccentric O, the slot *s* moving segment *r* at the proper time. Arm N and bar P are now moved forward, and lip *q* catches the back edge of the uppermost sheet, which was previously moved backward, and the lip closes and grasps the edge of the sheet, which is shoved forward (see figure 5) and caught by nipper or other devices, by which it is passed to the press.

The plan figure 2 represents only one half of the machine, the other half being similar; *x x* is the centre line.

Claim.—The employment of the swinging-frame E, provided with the rollers G G, in combination with the vibrating arm N and clamp or lip *q*, operated by the slotted bar S. The parts being arranged as shown, and operating in the manner and for the purpose as set forth.

No. 13,197.—Wm. McDonald, assignor to R. Hoe & Co.—*Machine for Mitreing Printers' Rules.*—Patented July 3, 1855. (Plates, p. 311.)

The bed H can be raised and lowered to the required angle, and causes the rule R, which rests on said adjustable bed, to be properly presented to the plane D, so that the end of the rule may be bevelled as desired.

The inventor says: I do not claim the plate A and plane D, for they are well known and in general use.

But I *claim* attaching to the bed A and framing B a sector guide-plate E, to which plate the bed F is secured by a set-screw G, the bed F having a curved projection *e* attached to it, which projection works in a recess or groove *a* in the plate E, by which the bed F may be adjusted at the desired angle with the plate A, and the ends of the rules bevelled or cut, as shown, and for the purpose set forth.

No 13,333.—Andrew Campbell.—*Machine for Feeding Paper to Printing-Presses.*—Patented July 24, 1855.

A description and engravings to illustrate this invention would necessarily be too extensive to be given in this Report.

The inventor says: I do not wish to be understood as making claim to the method of lifting sheets of paper from a pile by means of an exhaust bar or surface acting on the principle of suction, as I am aware that this was first done many years ago by M. Remond, in Europe, in a machine for making envelopes.

But I *claim*, 1st. The mode of operation, substantially as described, by which the pile or piles of sheets is or are moved up and down, and kept in position as the sheets are taken from or laid on the pile, that the upper sheet of the pile may at all times be in the proper position to be acted upon by the mechanism which removes it or which deposits it; which mode of operation consists in lifting or depressing the table or tables on which the pile is placed, by the mechanism described, or the equivalent thereof, in combination with the gauge and holding-bar, or any equivalent therefor, which rests on the top sheet to hold it down, and which, by its position on the pile and its connection, governs and controls the movements of the lifting or depressing mechanism, substantially as described, the said mode of operation being employed to govern the position of the pile on the feeding and receiving tables, or either, as set forth.

2d. The mode of operation for holding the upper sheet of the pile on the receiving table on which the sheets are in succession deposited, so that as the sheets are drawn over the pile in succession, the sheet previously deposited shall not be moved from the position in which it was deposited, the said mode of operation consisting in the use of a wing or plate, which rests on the upper edge of the pile, at that end of it over which the sheets move as they are drawn over the pile, and which then rises to permit the newly deposited sheet to fall, and again descends to hold it preparatory to the introduction of another sheet, the said wing or plate being operated in the manner substantially as described, or any equivalent therefor. And this I claim whether the said wing or plate be employed for the sole purpose of holding the upper sheet of the pile on the receiving table, or for this and for the purpose of gauging and controlling the mechanism which depresses the receiving table.

3d. Making pressure on the surface of the pile of sheets on the feeding-table, and within a short distance of the edge which is to be lifted, for the purpose of forcing the sheets at that end of the pile to open fan-like, that the separation may be more effectually made, substantially as described. And this I claim, whether the required pressure be made by the rounded surface on the exhaust-bar, or by a separate bar having the like mode of operation.

4th. Making the surface of the exhaust-bar concave, substantially as described, so that the upper sheet which is to be removed from the pile shall be drawn into the said concavity, and thereby draw its edge within the edge of the pile, the more effectually to insure the separation, as set forth.

5th. Placing the concave surface of the exhaust-bar so that it shall overlap or extend beyond the edge of the pile, substantially as described, that the whole of the portion of the sheet which is to be acted upon by the exhaust may be within the concavity of the exhaust-bar, thus causing the inward current to pass upwards under the projecting concave surface of the exhaust-bar, and across the edges of the pile, to insure the separation of the edges of the sheets, as described.

6th. Blowing in a current of air at one or both ends of the concavity of the exhaust-bar, substantially as described, to insure the separation of the top sheet from the rest of the pile preparatory to removing it from the pile, as set forth.

7th. The employment of the transferring or holding-bar, connected with the exhaust and blowing bellows, or equivalents therefor, substantially as described, in combination with the exhaust-bar, substantially as described, by means of which the end of the top sheet, which has been separated from the pile, is transferred from the exhaust-bar to and held by the said transfer-bar until it is taken by the gripers, or any equivalent therefor, that the exhaust-bar may be removed preparatory to carrying off the top sheet as set forth.

8th. The method of thumbing the edge of the pile to insure the separation of the edges of the sheets where they are liable to adhere, by means of the rotating spring flippers, substantially as described, or any equivalents therefor.

9th. The mode of operation, substantially as described, by which the griping ends of the jaws which gripe the sheet of paper are made to maintain a fixed position relatively to the edge of the sheet held by the transferring and holding-bar during the entire operation of griping the sheet, and during a portion of the operation of reversing the jaws, by the continued motion of the endless chain or band with which the rear parts of the jaws are connected, as set forth.

10th. Removing the sheets in succession from the pile by griping one end of each sheet in succession with griping-jaws, after the end has been separated and lifted from the pile by other means, and rolling it off the pile by the motion of the griping-jaws towards the other end of the pile, substantially as specified, by which all tendency to disturb the sheets below is avoided, as set forth. I do not claim broadly the fact of rolling off the sheet from one end of the pile towards the other, but limit my claim to the mode of operation specified, by which I attain this end in a practical and efficient manner.

11th. Making friction on both surfaces of the sheet when brought over the pile on the receiving table, substantially as described, in combination with the gripers, so that the moment it is liberated by the jaws the friction on both surfaces shall hold and leave it in place on the pile, substantially as described, whereby the accurate deposit of the successive sheets is made to depend solely on the period of opening the jaws, as set forth.

12th. Making pressure on and packing the pile of sheets on the receiving table as each sheet in succession is deposited, by means of a perforated plate, substantially as specified, the perforations in the plate giving free access to air, that the plate may be lifted without drawing up the sheets with it, as set forth. I am aware that pressure has been made on the top sheet of a pile in the operation of depositing the sheets, by means of bars called a fly; and therefore I do not claim broadly making pressure on the pile, except by means of a perforated plate extending over the entire pile to pack down the entire surface, as specified, the small perforations preventing the sheet from being lifted, while they do not prevent the packing down of the pile. And finally, the employment of pointed punches to punch holes in the sheets as they are piled on the receiving table, substantially as described, so that the pile may be transferred and moved without shifting the position of the sheets relatively to each other in the pile, as set forth.

No. 13,335.—MERWIN DAVIS.—*Improved Printing-Press.*—Patented July 24, 1855. (Plates, p. 311.)

Taking the distance between axes b and d^2, and dividing it into a number of equal parts, apportioning a certain number to the length b and the bed b^2, the others to the radius of motion of the platen d, and then dividing the radius of the platen into the same number of parts, and taking the same proportional distance from the axis to the periphery that the platen bears to the bed, the point x is determined on which the circle is struck which the segment platen is formed on; this is the curve that meets the straight line of the oscillating bed b^2. The distributing-rollers f deliver the ink on the distributing-plate b^3, from which it is transferred, at each forward vibration, on the inking-rollers e e, under which plate b^3 passes. The leaf of paper is placed at g, against little gauges h^1 at the front edge and under fingers h^2 on the slide h; when the sheet is placed, the fingers are let down upon it and hold it steadily; slide h is connected with arm h^3 on an axis just below, from which axis projects downwards arm h^5, so as to be struck by the front part of the bed, by which the slide is thrust out towards the platen to deliver the sheet; the slide at the same time turns the gauges h^1 down out of the way, and the fingers i on the platen seize the paper.

Claim.—1st. The oscillating bed, having a plane surface, and operating substantially as described, and contradistinguished from a rotary bed or sliding bed by oscillating between two given points.

Also, the construction and operation of the platen in combination with said bed, as specified.

Also, the adoption and combination of the well known fly n, for laying the sheets, the slide h for the reception and the delivery of the paper, and the inking apparatus b^3, e e, and f, to the new construction and arrangement of the bed b^2 and platen d, substantially as described.

No. 13,376.—JAMES MELVILLE and JOS. BURCH.—*Machine for Printing Textile Fabrics.*—Patented August 7, 1855. (Plates, p. 311.)

Q is a longitudinal slide-rest bed, having upon it carrier S, with adjustable head-stocks T U for the support of the printing-roller V; *g* is the bottom color-supplying roller, and *i* the color-box; the roller *g* supplies the color to the endless feed-band *j*, which latter passes over rollers *g* and *k* and presses against the printing-roller. The carrier S, with the printing-roller and coloring apparatus upon it, can be traversed by rack-gear *t u*, and set by pawl *x* and serrated plate *v*; the serrations correspond to the particular repeats in use at any given time in printing, a new serrated plate being attached at each change in the repeat.* The shaft *o* is grooved longitudinally, and wheel *c* (fitted upon it with a key) revolves with it, at the same time being free to traverse on said shaft. The carrier-wheel *b* also gears with a wheel *e*, set loose on a stud in the head-stock, and made to drive wheel *f* fast on the overhanging end of the shaft of *g*. This shaft is carried in bearings in the standards *h* of a separate adjustable carrier, the base of which is dove-tailed to traverse upon slide-rest carrier S at right-angles to the traverse of such carrier. The piece to be printed is stretched over the periphery of main cylinder D, and the printing-roller, previously prepared, is then shipped upon the head-stock centres of the slide-rest apparatus. The printing-roller is then set up to the end of the slide-rest bed, as in figure 2, and the end portion of the pattern is printed as at 1. For the next repeat of the pattern, the rack-gear *t u* is operated to traverse the entire printing apparatus along the bed to a distance equal to the width of the repeat, the main cylinder being moved from printing contact by proper means. The serrated plate serves as an accurate set for the printing-roller at the termination of the shift for the repeat. The main cylinder is then again moved into printing contact, and the above operation repeated, and so on until the whole width of the pattern has been printed, and the same process is repeated with the rest of the colors, inserting for each new color a new coloring apparatus.

In printing lengths of goods, as in carpeting, the different colors are laid on by a corresponding set of surface rollers, set one in advance of the other, as regards the direction of revolution of the main cylinder, each roller and color apparatus being carried upon a slide-rest, as above described.

Claim.—1st. The mode of printing in two or more colors by means of a movable color apparatus, in connexion with the pattern printing-roller traversing laterally on a slide-rest, by means of which the colors in a repeat are printed without shifting the printing-roller, as described.

2d. The application of a slide-rest guide apparatus for guiding and regulating the action of the pattern printing-roller, when the mode of connecting such slide-rest apparatus with the impression cylinder is by means of a grooved shaft traverse movement, as described.

3d. The mode of adjusting the position of the printing-roller at the

* *Repeat* means the traversing of the printing-roller for repeating the length of the pattern comprised within the longitudinal limits of the roller.

repeat shifts, by means of notched or serrated plates set to correspond to the different repeats.

4th. The mode of printing carpeting and other fabrics, by means of printing-rollers corresponding to the several colors in the repeat, set, one in advance of the other, in the direction of the main cylinder's revolution, that revolution being continued until the colors are duly impressed upon or into the fabric.

No. 13,423.—Dexter H. Chamberlain.—*Improved Hand-Press for Printing.*—Patented August 14, 1855. (Plates, p. 312.)

In order to adjust the bed I to the surface of the types, the nut L is loosened, and the block E is forced down upon the bed (see figure 2). The bed then assumes the required position, and may be fastened so by tightening nut L; and if the plate at any time be replaced by another of a different thickness, the adjustment is to be repeated.

Claim.—1st. The described percussion hand-press, consisting essentially of the block or type-carrier E sliding upon the lever C, the ink-roller G and bed I, operating in the manner and for the purpose substantially as described.

2d. The method, substantially as described, of adjusting the head to the surface of the types, whatever may be the thickness of the plate which carries them, for the purpose set forth.

No. 13,462.—John Hope and Thomas Hope.—*Machine for Engraving Calico Printers' Rolls.*—Patented August 21, 1855. (Plates, p. 312.)

One end of the cylinder S (to be engraved) is supported by rollers *g h*, and the other end by rollers *e f*, extending into the groove i^1, formed around the surface of the cylinder. From the carriage Y (resting and sliding on carriage K) projects the tracer V, its point projecting over the table and up on the design carried thereby. When the trace is moved transversely, the carriage K will be put in motion, and, by means of band L, will communicate motion to wheel M, so as to rotate shaft *o* and cylinder S. When the tracer is moved longitudinally it will move carriage Y, and with it the grooved wheel X and carriage C^1, so as to move the several gravers *t* in a longitudinal direction on the surface of the cylinder.

In arranging the machine for operation, the two index arms Q and R are disposed at an angle to each other equal to that measured by the arc through which the surface of the cylinder is moved while a section of the pattern is being engraved. Having effected the engraving of such section, it becomes necessary to draw forward the pattern-sheet P a distance equal to another such section, which distance corresponds with that between either two marking points *i i* of each set thereof. The pattern-cloth is then moved forward so as to bring those points which were previously under the rear-markers directly underneath the front markers. Next, with a piece of chalk, marks are made on the side of the rim of M, where the index points Q R cross it; next, the set-screw of wheel M is loosened, so that it can revolve freely upon shaft

N, and the tracer is drawn forward to that part of the design upon which it rested previously to the pattern being moved. While doing this, although M will be rotated upon its shaft, it will not rotate the shaft *o*; next, taking the foot off the treadle which pressed the connecting-rod P[1] upwards, the gravers are thrown off the cylinder. The shaft *o* is then turned by means of arm R until the arm comes up to the chalk-mark, which was made as above mentioned. This movement of the shaft rotates the cylinder the distance necessary for the engraving corresponding to the section of the pattern over which the tracer is next to be carried. After clamping M to shaft N, and letting the gravers down upon the cylinder, the machine is ready for the tracer to be moved over a fresh section of the pattern. While the markers serve to determine the distance to which the pattern is to be moved, the rods E F operate as hold-backs to keep the pattern-cloth properly strained upon the top surface of the table B. By means of this machine a number of similar figures (this number corresponding to the number of gravers employed) can be engraved on a cylinder by the use of a single pattern.

Claim.—The combination and arrangement of the two sets of measuring-markers *i i i i*, the hold-back rods F F, and roller with plane surface table, the same being not only to enable the design to be transferred, it being brought forward in regular sections, but to be maintained flatly upon the table; also, the two measuring indices, in combination with the large pulley and the shaft of the driving-roller of the cylinder to be engraved; also, the means of holding and moving the cylinder, so that it shall not only be rotated by pressure against its external surface, but may be readily either removed from or applied to its supports, the same consisting in employing a driving-roller and a bearing-roller at one end of the cylinder, in combination with two sets of bearing-rollers, made to extend into a groove around the cylinder, and to support such cylinder, both laterally and longitudinally; also, the arrangement of the pattern-table, the tracer and its carriage, the several other carriages, the mechanism for operating each, the wheel, the shafts, and the supports of the roller to be engraved; the whole constituting an improvement in engraving machinery, and securing to it important advantages in operation as well as in construction, as set forth.

No. 13,576.—Daniel K. Winder.—*Improved Card-Printing Press.*—Patented September 18, 1855. (Plates, p. 313.)

The nature of this improvement consists in connecting the two chambers D C at the ends of the platen B (in which chambers the cards *e* are deposited) by a groove in the under side of the platen, so that a card may pass from one chamber to the other by the action of a spring-driver E secured to the bed of the press, which, as the platen is drawn back for the inking of the form, catches the lower card in the blank card-chamber C, forces it into the groove beneath the platen, and causes the last-printed card to be delivered at the bottom of the pack in the other chamber, while a card *p* is left in position to receive the impression when the platen shall return.

I *claim* the combination of the connected chambers C and D of the platen with the spring-driver E of the bed, constructed, arranged, and operating substantially as specified, for the automatic feed and delivery of cards.

No. 13,579.—Edmund Morris.—*Improved Seal and Stamping Press.*—Patented September 18, 1855. (Plates, p. 313.)

When the die is to be inked, the frame A is to be thrown over by means of hinge F.

I *claim* the causing of the frame which contains the die or plate to work to and fro on a joint or hinge, so that the latter may be turned over, with its face upward as described, in a convenient position to receive a supply of ink.

No. 13,587.—Samuel W. Lowe, assignor to Himself and Jacob M. Beck.—*Preparation of Metallic Plates for Printers.*—Patented September 18, 1855.

The copper plate (or steel plate coated with copper by the galvanic process) has its surface amalgamated with mercury, by rubbing the latter on the said surface; after that the plate is ready for engraving or etching. In a plate already engraved, the engraved lines are first to be filled up with resinous matter, and then the surface to be covered with the amalgam, as above described. The resinous matter is then removed, and the plate is ready for the press.

The inventor says: I do not claim engraving or etching designs, or figures of any kind, upon metallic plates, or surfaces of any material, for the purpose of printing therefrom, as these processes have been known and practised for a long time.

I *claim* coating the plane or engraved face or surface of the plate (which is intended for leaving the white or unprinted surface of the paper) with a mercurial amalgam, that will have the effect of preventing the ink used in printing therefrom from adhering to or soiling the same, whilst the figures engraved or etched thereon readily receive the ink, and thus admit of printing from the plate, by a letter or any other press, either from the plate alone, or from the plate in the same "form" with the type, without the "wiping" heretofore required in printing from steel or copper plates, substantially as described.

I also *claim* the coating the plane surfaces of etched or engraved steel plates with an alloy of tin and mercury, substantially and for the purposes as described; and also the coating of etched or engraved copper plates, in the same manner and for the same purposes, and the coating of the plane surface of metallic embossing plates, in the same manner, and for the more special purpose of using the sunken parts, when filled up with a resinous substance, as a plate to print from, thus saving an extra color-plate, when it is desired to have the parts to be embossed first printed in any color.

No. 13,671.—DANIEL K. WINDER.—*Improved Card Printing Press.*—Patented October 9, 1855. (Plates, p. 313.)

The press being in position figure 1, the card x is inserted under the holders i i. Lever L, then being moved in the direction of the arrow, draws form F from under roller R, and to the top of face a of the bed, and gradually produces the descent of platen P. The pressure still continuing, the card under the platen is drawn with sufficient force against the form to effect an impression. During which operation the lever L, pressing on short arm of lever C, raises its long arm, and causes roller R to revolve in contact with surface g, the several parts being now in position figure 2. On removal of pressure from lever L, the reverse operation will obtain. The springs D, keeping roller R in contact with supply surface g, cause the roller to rotate over it in passing off, and insure a full supply of ink.

Claim.—The double inclined bed B, traversing form F, and inking surface g, in combination with the lever C, spring roller supports t, and operating lever L, constructed, arranged, and operating substantially as and for the purpose specified.

No. 13,689.—DANIEL K. WINDER.—*Inking Apparatus for Card-Printing Presses.*—Patented October 16, 1855. (Plates, p. 313.)

The backward movement of platen P throws roller R between the platen and bed, and the outward motion of arm m, which takes place as the platen moves forward, withdraws said roller, the springs p keeping the roller upon the form F during this movement, and causing it to impart ink to the type. On the withdrawal of roller R from between the platen and bed, which is accomplished before the platen has half completed its forward course, it immediately comes in contact with the supply-roller E, then revolving by the action of spring T; this contact produces a rotation of roller R, and enables supply-roller E to impart to it a sufficient quantity of ink before the platen reaches the termination of its course, when inking-roller R again passes over the form as the platen is withdrawn, and the operation proceeds as before. L is the handle to operate the platen with.

Claim.—1st. The double-armed rock-shaft S, and outward-pressing roller-frame G, or their equivalents, in combination with the platen and the springs actuating the arm m of said rock-shaft, constructed, arranged, and operating substantially as and for the purposes specified.

2d. The above mechanism for operating the inking-roller, combined with the supply-roller E, actuated by the movement of the platen, substantially as specified.

No. 13,703.—THOMAS HARSHA.—*Card-Printing Press.*—Patented October 23, 1855. (Plates, p. 314.)

The strip of paper x x having been inserted between rollers M l, and ink having been placed in box I, the lever C is brought directly over the ink-bed J; by which movement the rack o is moved by means of levers E G, and frame H is moved by levers E and F. The rack in moving turns rollers M l, and feeds the ends of the paper over bed Q; and as frame H is moved the roller d passes over ink-bed J, distributes the ink thereon, and passes into recess f to receive an additional supply of ink. The form being now over ink-bed J, lever C is depressed so as to depress and ink the form. Lever C is then relieved, and rises by the action of spring W. The lever is then moved back and box S brought over bed Q; by this movement frame H is moved back again over ink-bed J, which is thereby charged with ink, and rack O is moved back without turning rollers M l as the pawls slip over the ratchets. The form is then pressed down upon the card-paper on bed Q, and a projection a^1 at the side of box S depresses knife U and cuts off the card printed at the previous movement of lever C.

Claim.—Attaching the box S, which contains the form, to the lever C, and connecting said lever with the ink-rollers c d and feed-rollers M l, as shown, so that when the lever C is moved and the form brought over the ink-bed J, the paper will be fed over the bed Q on which the paper is printed; and when the lever is moved over the bed Q, in order to print the paper or cards, the charged ink-rollers will pass over the ink-bed J, whereby the ink-bed is kept properly charged with ink, and the paper fed over the bed on which the paper is printed, and the paper or cards printed by simply moving or operating the lever C, as described.

Also, passing the paper between two knives T U, arranged substantially as shown, in relation with the lever C, so that the printed paper or cards will be cut off in proper lengths as the form is pressed upon the paper, the knives cutting off a previous impression at each depression of the lever.

No. 13,737.—Henry W. Dickinson, assignor to Lansing B. Swan.—*Machine for Feeding Paper for Printing-Presses.*—Patented October 30, 1855. (Plates, p. 314.)

The frame-work, consisting of air-pumps B, air-chambers A, valve cylinders J, and system of horizontal tubes I, and short vertical tubes 1, 2, 3, 4, 5, 6, (the latter open at bottom,) slides on ways N, and receives a reciprocating motion by crank or other mechanism. The piston-rod P being firmly attached to the main framing of the machine at o, a vacuum will be maintained within cylinder B and chamber A as the frame-work above described travels back and forth. As the frame arrives over the paper on table W, the bars K (to which are attached the valve-rods a of the valve-cylinders J) strike the projections M and open the valves, when the suction from the exhausted chambers A causes the uppermost sheet of paper to adhere to the orifices of the vertical tubes 1, 2, 3, &c. As soon as the frame arrives at the other end of its way, the bars K strike the projections L, and the valves in the cylinders J are again closed. The suction ceasing, the sheet of paper is ready to be caught by the nippers, and to be delivered to the frisket-frame of the press.

Claim.—The general arrangement of the devices described, and for the purposes set forth.

No. 13,818.—Thomas Henderson.—*Machine for Printing Yarns and Cloths.*—Patented November 20, 1855. (Plates, p. 314.)

The printing-types B are moved up and down by a crank and pitman motion. G are the lower set of color-distributors, revolving in boxes H. The upper color-distributors revolve in boxes J. The boxes H are fixed to the bottom of frame A; the upper ones J are fixed to the movable frame I. The journals of the color-distributors turn in the ends of their boxes. As they are turned one-quarter way around at each advance of the yarn or cloth being printed, a fresh supply of color is imparted to the cloth on its under side, and a fresh supply of color is imparted to the lower ends of the types B by the angular upward movement of the frame I, carrying the boxes and distributors J and K, so as to impart the color first to the types B, then by them to the cloth on its upper side, exactly over the color-distributor G, by the types descending perpendicularly after the frame I and color-boxes K are moved angularly down so as to let the types B pass by them.

The cords F are attached to the tops of the types B and to the top-piece E, they passing up through slots in plate *e*, and through the eyes of the needles *f*; these needles are pressed back against and into the surface of the jacquard cylinder X by the spiral springs *g*. The knots *h* in the type-cords F are for the purpose of keeping up the type which print the desired figure; the knots are drawn up through the large part of the slot *d*, and are carried longitudinally over the narrow parts of these slots by means of the eyes in the needles *f*, these needles being driven longitudinally forward by the jacquard cylinder. Thus it will be seen the types will not descend until the next revolution of the crank, while the other needles enter the holes in the surface of the jacquard cylinder, and leave the types which they govern free to descend.

I *claim*, 1st. The printing or coloring types B, arranged and operated essentially and for the purposes set forth.

2d. I claim the coloring-distributors, and boxes in which they revolve, when they are constructed and operated substantially as described, for the purposes set forth.

3d. I claim the types B, in combination with the jacquard operation for printing and coloring figured goods, when they are arranged and operated substantially and essentially as set forth.

4th. I claim the types B in combination with the color-distributors and boxes, arranged and operated essentially as set forth.

No. 13,857.—Cyrus A. Swett.—*Improved Printing-Press.*—Patented November 27, 1855. (Plates, p. 314.)

The rollers *k* each receive its proper colored ink from a fountain; and as the cylinder C rotates, the rollers *k* will distribute upon the periphery of the cylinder belts of different colored inks, the width of the belts being equal to the width of the blocks X. The shaft of the rollers *k* has a longitudinal vibrating movement given it as it rotates, by the screw S and fork T, so that the rollers *k* will distribute evenly the ink upon the cylinder. The different colored belts of ink will be taken from the cylinder C by the rollers V, which ink the plates X as they pass underneath them, each plate receiving its appropriate color. The sheet of paper to be printed is placed upon the plate J, and as the cylinder C rotates, the cam F will operate the lever G and arm H, and the bed J will be drawn underneath the cylinder C and plates X, which will leave their impressions upon the sheet of paper. At the first impression, there will be as many colored impressions as there are blocks and belts of color (three being represented in the engraving). After the first impression, the operator moves the belt L, (by means of handle N,) and consequently the sheet of paper on bed J, the distance of the width of one block to the left, and at the next impression two of the previous impressions will receive a different color; for instance, the previous impression from block 3 will now receive an impression from block 2, and so forth.

Claim.—Inking the blocks X with their appropriate colored inks by means of the rollers *k* V, arranged as described; further, the cylinder C, vibrating bed or plate J, with endless apron L attached, when the above parts are constructed, arranged, and operated as shown for the purpose specified.

No. 13,915.—Robert Prince and Ambrose Lovis.—*Improvement in Processes for Calico-Printing.*—Patented December 11, 1855.

An even mixture of 220 pounds of carbonate of soda, 40 pounds of common salt, 41 pounds of sulphate of soda, and 280 pounds of white sand, is to be introduced into a reverberatory furnace previously heated to a white heat. The materials are quickly fluxed to a stiff transparent glass, and rapidly withdrawn and cooled, and subsequently ground. For the purpose of dunging. a standard solution is made, containing one pound of the silicate in one gallon of the solution, by boiling the powdered salt in clean water. To the hot solution is to be added one pound to each gallon of a mixture of carbonate of soda, containing about twenty-five per cent. of common salt and sulphate of soda. Twelve gallons of this standard solution may be added to a fly-dung cistern, containing from 600 to 700 gallons of water, already heated to about 190 Fahr.; and when 20 pieces of ordinary calico have been run, an addition of one gallon to each twenty pieces succeeding must be made, until 300 or 400 pieces have passed, when the cistern ought to be emptied and refilled as above.

The inventors say: We are aware that pure silicate of soda alone, or with pure carbonate of soda, has been used heretofore in dunging; and we are also aware that silicate of lime has been used for the same purpose.

We disclaim the use of these substances, confining ourselves to the use of the silicate in mixture with neutral and alkaline salts. We *claim* the manufacture of silicate of soda, or potash, containing foreign neutral salts, and the use of this compound with carbonate of soda and neutral salts in dunging operations, substantially as set forth.

No. 12,468.—DANIEL W. MESSER, assignor to DANIEL W. MESSER, R. B. FITTS, and ALBERT JAMES.—*Improved Hand-Stamp.*—Patented February 27, 1855. (Plates, p. 314.)

To produce a hand-stamp where the type-plate *a* accommodates itself to the surface of the paper, the inventor secures the type-plate to the handle B by means of an India-rubber connexion A.

The inventor says: I do not claim uniting the type-plate to the handle by means of a ball and socket-joint, as that has been done before; but I *claim* the India-rubber connexion between the plate and the handle, operating in the manner and for the purpose set forth.

No. 13,308.—JOS. HARRIS and ELBRIDGE HARRIS.—*Improved Hand-Stamp.*—Patented July 24, 1855. (Plates, p. 315.)

The cylinder B, through which the die-piston A C works, swings on pivots D D. When swung back, the die C will come into contact with, and be inked by, roller G. When brought back to the vertical position and depressed, the impression will be made. J J are adjusters, being attached to the pivots D D, and are operated by the same motion with the die, and come in contact with the ink-roll boxes H, for the purpose of adjusting the roll to meet the die to be inked.

Claim.—The peculiar oscillating motion of the stamp; also the same in combination with the inking-roll and its adjusters, for the purpose described.

No. 13,495.—HORACE HOLT.—*Improved Hand-Stamp.*—Patented August 28, 1855. (Plates, p. 315.)

The cover or platen *l* having been turned out of the way, the stamp *c* is brought down upon the surface of the ink *h*; the stamp thus inked is allowed to rise back; the platen is turned inward so as to cover the ink-pan; the article to be stamped is laid on the platen, and the stamp brought down upon it.

Claim.—A hand-stamp in which the stamp is inked and the impression effected by the movement of the stamp in one vertical plane; the ink being arranged directly under the stamp, and provided with a cover which can be moved away from or upon the said pan or fountain; thereby serving both as a cover to the ink-receptacle and as a platen to the article to be printed.

No. 13,470.—STEPHEN P. RUGGLES.—*Hand-Stamp.*—Patented August 21, 1855. (Plates, p. 315.)

The nature of this invention will be understood from the claims and engravings.

Claim.—In a hand-stamp the connecting of an electrotype plate to the handle of the stamp by means of a screw-cap, as described, for the purpose of facilitating the removing and replacing of the electrotype, or portions thereof, as set forth.

Also, the combination of devices for holding the bed-plate E to the shank B, so as to preserve the ball and socket or yielding joint, prevent them from being separated, and to keep the coiled spring in place; the same consisting of the flanges *a f* on the bed-plate with the holes therein, the large opening *c* in the shank-piece and the pin *e* passing respectively through them, as set forth and described.

No. 12,157.—Willard Cowles.—*Improvement in Apparatus for Stereotyping.*—Patented January 2, 1855. (Plates, p. 315.)

The material for the moulds is composed of treacle and plaster of Paris.

The frame *a* (with its upper and lower surface perfectly parallel) surrounds the types to be moulded, for the purpose of forming a bed *b* around the mould. Upon this bed is placed a metallic casting-frame (fig. 2) with its aperture exactly corresponding to the sides and bevel edges of the required stereotype-plate, and its thickness such as, together with the difference between the height of the moulding furniture *a* and that of the type, shall be equal to the required thickness of the stereotype-plate. Or bars (fig. 3) may be used of a length to suit the page, which laid, previous to the last impression, upon furniture, as the type, less the required thickness of the stereotype-plate, become imbedded in and adhere to the moulds, and in place of the frame determine the size for and bevel of the stereotype-plates.

The material is placed upon the mould and within the frame or bars—all parts being in position shown in fig. 6; a metallic plate is placed upon it, and the whole submitted to pressure until the plate comes into contact with the upper side of the frame or bars.

When a page is to be cast in several pieces, leads *d* as high as the type, less the required thickness of the plate, are placed between the lines of type (wherever the plate has to be divided), thus forming narrow grooves. Slips *e* are prepared of equal thickness with these leads, and of a breadth which, together with that of the leads, shall correspond to the height of the type. Before making the last of the successive impressions (of the material upon the type), these lips are inserted in said grooves, and after the impression they remain imbedded in the mould. Upon making the cast, these lips remain in it, dividing it to a sufficient depth to render it easily separable.

Claim.—The inventor says: I do not claim the use of plaster of Paris or clay for the making of moulds for stereotyping, nor do I claim the making of moulds by successive impressions; nor do I claim simply the use of furniture, nor the determination of thickness of the stereotype-plate by means of bars or a frame placed upon the mould-plate.

But what I do claim, is the use of furniture of an exact and proper height, for the purpose of forming a bed in and around the mould on which to place the metallic casting-frame, which gauges the size, form, and bevel of the stereotype-plate, or which shall itself form a support, on which bars are placed for the purpose of being pressed into the mould and imbedded therein, thus taking the place of the frame.

I also claim the use of a frame placed upon a bed in and around the mould, or in its place the use of bars pressed upon and into the mould, substantially for the purpose and in the manner described.

I also claim the use of slips of metal or other substance for the purpose of dividing the plate into two or more pieces, used substantially in the manner described or any similar manner of accomplishing the same object, whether these be used in combination or separately, or whether the material used in making the stereotype-plates be common type-metal or a compound, in which a principal ingredient is gutta percha, India-rubber, shellac, or any other substance or composition of which stereotype-plates are now or may hereafter be made.

No. 12,257.—Jno. F. Mascher.—*Stereoscopic Medallion.*—Patented January 16, 1855. (Plates, p. 316.)

No description required.

The inventor says: I am aware that a daguerreotype-case has been converted into a stereoscope, by constructing it with a supplementary flap or lid containing two lenses, a patent having been granted to me for such an arrangement in 1853; therefore I do not claim the use of a supplementary lid, as used in a daguerreotype-case for such a purpose. But I *claim* constructing a medallion or locket A B B, with two supplementary lids C C, containing each a lens D, and arranged so as to fold within the picture-lids B B, and in such relation to the same that, upon being opened and properly adjusted, they shall cause the lenses to stand opposite said lids, and thereby convert the medallion into a stereoscope; said arrangement also rendering the medallion useful as a microscope and sun-glass, substantially as described. (See engraving.)

No. 12,451.—John Stull.—*Improved Stereoscope-Case.*—Patented February 27, 1855. (Plates, p. 316.)

The inventor says: I do not claim constructing a stereoscope-case with a single adjustable flap or supplementary lid within the case, as such invention has been made and used before in daguerreotype-cases; but I *claim* constructing a stereoscope-case, with the three jointed pieces E E E, or their equivalents, so applied as to preserve at all times a perfect parallelism between that part of the case containing the lenses and the part which contains the figures, so that a perfect stereoscope is formed of the whole as described, and the two figures B B, by binocular vision, are apparently formed into a solid figure, the whole being at the same time adapted to fold or close into a small flat case, (resembling the common daguerreotype-case,) that may be conveniently carried about the person, if so required, substantially as described. (See engravings.)

No. 13,093.—Jos. H. Marston.—*Apparatus for taking Stereoscopic Photographs.*—Patented June 19, 1855. (Plates, p. 316.)

The nature of this invention will be understood from the claim and engravings. C is the camera-box.

The inventor says: I do not claim the using of guides or frames or angles on a camera-board, as such have been used before; but I *claim* the particular arrangement of two frames or guides or boards, that shall work on and be fastened or pivoted on the camera-board A at a a^1, the frames or guides to be at right-angles, so that when closed together they form the raised ledge *b b* parallel with each other, by which to adjust and centre the angle-board A previous to setting the angles for stereoscoping; also, the application of the spring *c* to hold the frames together, the right and left screws e e^1, and nuts d d^1, and eye-plates f f^1, to force apart and hold the frames B B^1 in any desired angle. I claim the described apparatus for moving the camera-box and giving the true stereoscopic angle at one and the same time.

No. 13,106.—Abel S. Southworth and Josiah J. Hawes.—*Apparatus for Moving Stereoscopic Pictures.*—Patented June 19, 1855. (Plates, p. 316.)

In figure 1 the plates 10 10^1 are in the fields of vision; as the crank revolves, the toes *g* having passed the arms *h* upon the vertical rod, the latter are left free to revolve when necessary; the cogged segments now engage with the racks upon the bottom of the plates 4 9^1 4 9^1, and drive them in the direction of the arrows, moving forward (as they advance) the plates 5 10 5^1 10^1 into spaces S, the plates depressing the fingers *k* as they advance into recesses in the sides of the box revolving the rods *i* into position figure 8; the teeth of segment P having now become disengaged from the rack on the bottom of plates 9 9^1, the said plates are allowed to remain stationary in the fields of vision during a semi-revolution of shaft *o*. While this is taking place the toes *g* come again in contact with the arms *h*, and the fingers *k* are again thrown into position figure 1; by which means the plates are moved forward in the direction of arrows *f*, thereby again vacating the spaces S, &c.

The inventors say: We *claim* giving to the pictures of a stereoscope, or other analogous instrument, a panoramic motion into and out of the fields of vision by means of mechanism substantially as described, or by any other equivalent means.

No. 13,609.—Samuel S. Weed.—*Machine for Making Printers' Types.*—Patented September 25, 1855. (Plates, p. 316.)

This machine separates copper or other metal wires R into short pieces, and forges each with a body and a letter of right shape. The die L is for forming the letter on one end of the type, the die K forms the body, and the die M shapes the opposite end of the type. The receiving orifice N has the rod extended through it from the feeding mechanism; and during its movement it carries the rod against the edge of the body-die I, by which a blank will be severed from the rod, and crowded into the body-die. The lever R is connected to the arms pivoted to the dies K L, so that during the time that the dies K L are moved towards each other, the lever R will be moved so as to force the type-rod inward between the dies, it being held to the lever by the nipper, which previously has been forced against it by the action of the outer shoulder U of the rod T, during the previous return movement of the lever R. During the forward movement of the lever R, the feeding nipper is carried against shoulder V, and moved out of contact with the type-rod, and so as to enable the lever to be moved backward without carrying the type-rod with it.

Claim.—For making type, the described combination and arrangement of the stationary body, or bed-die I, and bottoming-die M, and receiving orifice N.

Also, the combination of the feeding-lever R, the nipper S, and the rod T, provided with shoulders U V, as set forth, the whole being for the purpose of feeding the type-rod into the mechanism or its dies, as specified.

No. 13,710.—Wm. S. Loughborough.—*Machine for Composing and Setting Types.*—Patented October 23, 1855. (Plates, p. 317.)

Figure 1 represents a plan of the machine; figure 2, a transverse section; and figure 3, a perspective view. W is the composing-wheel on shaft A, inclining towards its centre, as seen at a^2 a^2; B, base-plate of the type-cells U. The column Z supports drum V, to which the base-plate is attached. A segment of B and V is removed from the front side, so as to bring the galley N up to wheel W. The bands F and E pass round the type-cells, which are constructed of thin sheet-metal, with an edge open as seen at U. Directly under each cell is a slide R, operating in a groove formed in the plate B; and each slide should be so fitted to its respective cell as to deliver but one type at once, which is done by making the blank elevation *r* of the slide to correspond to the thickness of the types in its cell, and the lower edge of the front-band F must also agree with the thickness of the types contained in the cell opposite each. A spring S extends from the ring A^1 to each slide, and these springs are distended when the slides are in their proper position, where they are retained by the detents *w*, until said detents are drawn out by touching the keys.

There are cam-grooves in the lower face of the composing-wheel W, seen in dotted lines at L and K, which open from and into its periphery, and receive the tooth *o* of the slide R.

The type-bearers, or transits, are constructed with two jaws I and H, the jaw H being constantly pressed out by spring b^1; but the stop J permits a change, as at the instant of delivering a type. There is a recess across the under side of H, through which I passes, and one shoulder of this recess limits the opening of I. The spring e^1 is constructed for the purpose of securing the opening or closing of the jaw I at either extreme. *n* is a tappet fixed to the jaw I, and hangs in the angle of the groove, which brings it in a line with the jaws of the transit and the slides R, and is drawn out by the tooth *o* bringing the jaw with it and closing it upon the type. The tappet *m* is fixed to the upper side of the same jaw, which opens it for the delivery of a type, by striking the adjuster *k*, which also changes the pressure of the spring e^1 to the opposite bevel, and holds it open until the tooth of a slide shall close it again. The other jaw H is opened at the same instant by the adjuster *j* striking the tappet *v*. The cam *n* is turned upon its axis by the pressure from the tooth *o* as it passes out, and when thus turned it forces the slide back to its place. The bridge D is bolted to the base-plate B, and supports the vibrating lever G, which has a uniform vertical motion, received from the tappets T, the spring Y giving the counter motion, and as often as a transit passes the point of delivery, said point being at the guard-plate P. The spring s^1 keeps the lower face of the head *i* parallel with the types, as it presses them down into the line by forcing the point at the bottom against the plate Q, as seen in figure 2.

The bed-plate Q extends across the machine in front, and directly over the key-board C; it hangs on the pivots P^1, and is kept in its upright position by catches, which are withdrawn when the galley N is to be removed, when the bed-plate may be swung down to its horizontal position. The arm C^1 is connected with this plate, and supports the lining register *a*. The rod E^1 carries the index N^1 over the index-plate K^1. The spring B^1 is attached to cross-head *f* of register *a*. The cam-ratch *b* is constructed, as seen in figure 2, so as to allow the register *a* to move down or recede, according to the thickness of each type deposited by the transits, but to prevent it from returning until the line shall be filled, when the point *c* strikes the ratch *b*, thus throwing spring *y* into the notch q^2, which holds it there until the bar *a* is thrown up by the spring B^1, which is done as soon as the galley N is moved forward, so as to clear the end of the bar *a* and the set-screw *e*, when the bar rises, strikes the ratch *b* and throws it back again, when the spring f^1, through the detent *g*, presses it down on to the bar. The lever O has a vertical and horizontal motion, and receives the strokes from the tappet *t* at H^1, and through said lever the change of the justifier M is gauged. The depressing-face of the tappets *t* throws the opposite end of lever O up immediately after making the horizontal change, when the tooth of the detent *g* catches under the projection J^1, thereby preventing another vibration until the next line shall be filled, when the arm *d* throws the detent *g* back, thus relieving the lever O, and the spring e^1 brings it up horizontally, and the tappet strikes the screw H^1, forcing it out from the shaft, and the opposite end of the lever carries the justifier M forward, and the depressing face of the tappet forces the lever back to its present position, where it is caught and held by the detent *g* as before shown.

The summary of the operation is as follows: When a key is touched, the detent *w* (by means of lever h^1 and spring g^1) is drawn out, and the tooth of the slide enters the groove K, and the type is delivered to the jaws of transit, when the discharge groove L and the cam *u* throw the slide R back to its place, and the type is conveyed to the point of delivery, where it is deposited in the angle of the line-register *a* and guard-plate P. The head *i* presses it down into the line, the bar *a* recedes according to its thickness, and so on, until the line is filled, when the galley is thrown forward by the justifier M, the depth of the line relieving the bar *a*, and it flies back to its place N^1 carrying the index up to the top of the plate K^1, ready to commence the next line. When the galley is filled it may be transferred, and when the cell is emptied it may be filled.

Claim.—1st. The presentation of the type-cells in the machine, those of each case in the front forming the arc or segment of a circle in the manner specified.

2d. The means above described, or their equivalents, which shall deliver the types from the various cells into the jaws of the transits, fixed to a wheel, or other rotary motion.

3d. The construction and application of the transits as described, or their equivalents, attached to a wheel or other rotary motion, for conveying the types from the slides, or their equivalents, to the galley or composing chamber.

4th. The combination of the lever G, head i, tappets T, and springs y and s^1, with the line-register a and its appurtenances. The lever O, rule or justifier M, detent g, and the index N^1, and index-plate K^1, whereby the operator is enabled, simply by touching the keys, to do the entire business of composing types, and without a transfer of each line separately.

No. 13,935.—Daniel Moore, assignor to George S. Cameron, James H. McWilliams, and Himself.—*Improved Machine for Rubbing Types.*—Patented December 11, 1855. (Plates, p. 318.)

The nature of my invention consists in so constructing the parts that the types are taken one by one into a suitably constructed slice i, and carried with the letter-end first through between cutters o o^1, which remove the projection on the base of the letter and the burrs on the type, by cutting from the face of the letter down towards the base of the type; thereby the hair-lines cannot be broken, there being no crossways cut on the same. And the parts that act on the base are such as not to damage the type. The main vertical shaft f receives motion by means of horizontal shaft 1 and the cog-wheels e e^1. The slices i i are attached between the plate g, permanently secured to the shaft f and the movable cap h. The metallic box k is attached to the bed c, and the cutters o o^1 are adjusted by means of a set-screw m passing through the bridge l on said box.

The types are fed into the machine in the following manner: p p are slides receiving the ends of the follower-block q, that is drawn towards the centre of the machine by an India-rubber spring 7. 8 8 are slideways for the types, passing beneath the plate 9, which is adjustable by means of set-screw 10. To carry the types on the race-way u, the lifting-block 12 is fitted so as to slide up and down on screwed guide-rods 13, and is actuated by means of a cam r on the shaft 1, which causes the spring S to lift and momentarily retain the block 12 in its elevated position, after which the spring S draws said block 12 down again. t is a gauge-block, having fingers extending out on the level of the race-way u in slots therein, and said gauge t is attached by a set-screw 14, and the lifting-block 12 is formed with grooves receiving said gauge-fingers. These gauge-fingers t are to be so adjusted that the space between their ends and the end of the plate 9 shall be slightly more than the width of the type to be rubbed, so that when the lifter 12 is depressed, the end type in the line will, by the follower *q*, be pressed up to these gauge-fingers *t;* and therefore when the lifter 12 is elevated, this one type only is raised to the surface of the race-way *u*, and the motions are so timed, that the lifter holds the type in that position until carried off by one of the slices *i*. The corners of the bottom of the slice are taken away, (as seen in fig. 2,) so that the power to force the type through the cutters is applied near the centre of the lower end of the type, at the point where the sprue is broken off, or at each side of said point, and where the groove is cut in the usual manner after the types are set up in a line; thus the corners on which the type stand remain untouched.

I *claim*, 1st. Constructing the slice *i* with openings to receive the type at such an angle, relatively with the direction which said slices move, that the cutters shall commence to act at the latter end of the type carried by said slices, and that the cutting operation shall tend to force the type into the bottom of said angle, and thereby retain the type in place in said slice, in the manner and for the purposes specified.

2d. I claim constructing the slice *i* in such a manner as at *v*, that the power to force the type in an end-ways direction, or nearly so, through the cutters, shall be applied to or near the middle of the bottom end of the type, in the manner as specified.

3d. I claim the follower *q*, slides 8, and holding-plate 9, to supply the machine with a line of type, in the manner as specified.

4th. I claim the lifter 12 combined with the gauge-fingers *t* and end of the plate 9, or other stop, for the purpose of elevating one type at a time, to be taken by the slices, as specified.

No. 13,878.—MATTHIAS KELLER.—*Improvement in Cutting the Fronts and Backs of Violins.*—Patented December 4, 1855. (Plates, p. 318.)

The nature of this improvement will be understood from the claims and engravings.

Claim.—1st. The slides B and C, with the pattern M and lever K, in combination with the connexions I and J, lever G, and spindle F, with its cutter *i*, the whole being arranged and constructed substantially in the manner set forth, for the purpose of forming any number of exactly similar backs and fronts of violins from one pattern.

2d. The supplementary lever Q, with its connexions R, S, and J, in combination with the levers K and G and slot *k*, for the purpose of forming the concave sides of backs and fronts of violins without changing the pattern used for forming the convex sides, and for the purpose of giving the said backs and fronts a gradual and correct tapering thickness.

No. 12,502.—JAMES M. BOTTUM.—*Polishing Apparatus for Watch-Makers' Lathes.*—Patented March 13, 1855. (Plates, p. 318.)

The polishing-wheel J is adjusted by unscrewing the screws F and

F^1; then, while it moves freely in any direction, bringing it up to the work, and tightening the two screws'

Claim.—The application of the polishing spindle to the lathe, in such a manner that it has a universal movement, substantially as described, for the purpose of adjusting the polishing-wheel to surfaces of various forms.

No. 12,538.—ELIHU BLISS.—*Improved Swivel for Watch-Chains.*—Patented March 20, 1855. (Plates, p. 318.)

The ends of the two pieces of the loop are matched exactly together, (as seen at *h*,) so that they set together the same as if the loop had been made whole in one piece, and had been cut in two at the joint; they are fastened together by pivot *f* When the loop is opened, the loose side turns over sideways into position represented in the engraving in full lines.

Claim.—The specific arrangement of the joint of the swivel, in the manner and for the purpose substantially as set forth.

XIX.—FIRE-ARMS.

No. 12,411.—JOHN S. KEITH and JOHN BROOKS.—*Improved Bullet-Mould.*—Patented February 20, 1855. (Plates, p. 319.)

The object of chamber C and tube F is to allow the molten metal to pass down into chamber C, and from thence to flow up through the several matrices, (which communicate with each other by means of their holes *c*,) and drive the air that is in them before it and into the chamber H, from whence it will escape through *f*. The space H receives the metal from the several holes *c* of the upper set of matrices, and holds a surplus of it sufficient to supply what metal may be needed in the several matrices during the time of its passing from a molten to a solid state therein, it being subject to contraction at such time. By making the thickness of the mould-plate equal to the radius of the matrix, the outer surface of each of the mould-plates becomes tangential to the matrices thereof, and so that when the lead or molten metal in contact with it is removed, the balls are left in the matrices, and without any sprues or projections. By passing a knife between the outer surface of a mould-plate and the lead adhering thereto, such lead may readily be separated from the balls so as to leave them in a fit state for use.

Claim.—Combining with the mould-plates A and B the air and lead chambers C and H, the passages *c c*, and a tube F leading out of the chamber C and terminating above the level of the receiving space above the upper mould-plate; the object of the said chambers G and H connected with the several matrices, as described, being as herein before specified.

Also, the arrangement of the outer surface of either of the mould-plates tangentially to the spherical or adjacent surfaces of its several matrices, or so that after the mould has been filled with metal, and the sheet of metal against the tangential surface removed therefrom, the balls shall be left for all practical purposes without sprues, or in a state fit for use, as specified.

Also, arranging two or more sets of mould-plates and their matrices together, so that the matrices of each set shall be made to respectively communicate with those of another set placed either above or below it, as specified.

No. 12,774.—William Ashton.—*Improved Bullet-Mould.*—Patented May 1, 1855. (Plates, p. 319.)

The nature of this improvement consists in forming the mould with a movable core, whereby hollow or Minie bullets are to be cast with facility and perfection.

Claim.—Constructing the mould as herein shown and described, viz: Having a conical aperture *a* made in a piece of metal, and having a projection or core E and flanch *d* attached to a metal strip D, which is secured to the shank or handle C of the mould by a pivot *c*, so that said projection or core may be inserted in and withdrawn from the aperture *a*, as herein shown and described.

No. 12,295.—Samuel Huffman, assignor to Himself and Dennis O. Hare.—*Improvement in Repeating-Cannon.*—Patented January 23, 1855. (Plates, p. 319.)

The circular flange *n* surrounding the ring *m*, is intended to relieve the pin *e* from any undue strain from the recoil of the gun when discharged. The revolving table *d* is kept in proper positions by a spring-pawl *v*, which catches into notches in the table exactly under the centre lines of the sections *f*. Rack and pinion *l g* serve to move the forward section *c* back, and its flange *h* over the bevelled front end of section *f*, until the flange *h* abuts against the flange *g* on section *f*. The water-jacket a^2 serves to keep the gun cool in rapid firing. The vent-closer, figures 3 and 4, is secured upon the gun in any manner, and consists of a cylinder a^1 fitting truly in a lower bearing b^1, and pressed down thereon by caps c^1. A vertical projection d^1 from the cylinder forms the nipple; and a hole bored down through it, the cylinder, and the bearing b^1 to the chamber of the gun, forms the vent. A lateral projection e^1 is a stop to prevent the tube from going beyond the perpendicular in one direction, while it is free to move about one-fourth of a circle in the other, when the crank-handle g^1 is elevated. In this position the upper portion of the bore is placed at right-angles with the lower portion, thus cutting off the communication effectually without the use of the thumb, which usual mode of stopping the vent would be impracticable with the revolving sections.

Claim.—1st. The movable forward section *c*, with its flange *g*, in combination with the revolving rear sections *f*, secured to the plate *d*, constructed and arranged substantially as described, and for the purposes specified.

2d. The flange n, in combination with the projection m on the plate a, substantially as described, and for the purpose specified.

3d. The jacket or cold-water tank a^2, substantially as described, and for the purpose specified.

4th. The vent-closer, constructed and arranged substantially as described, and for the purpose specified.

No. 12,629.—LUTHER HOUGHTON.—*Improved Mode of Loading Rifled Cannon.*—Patented April 3, 1855. (Plates, p. 319.)

At the moment of explosion the force of the expansive gases will, by acting on the base of the sabot S, drive it forward on the butt of the ball A, and into the grooves of the gun.

Figure 1 represents the ball and sabot before the explosion takes place, and figure 2 represents them at the moment of the explosion.

The inventor says: Disclaiming the sabot as now used, I *claim*, for loading rifled or grooved cannon, the employment of a deep sabot at the base of the projectile, so as to be driven thereon and into the grooves of the gun at the moment of discharge, for rendering said grooves effective in producing the rotation of the projectile, as specified.

No. 13,851.—ALFRED KRUPP.—*Improvement in Cannon.*—Patented November 27, 1855. (Plates, p. 319.)

These cannon are made of solid pieces of cast-steel shaped on the outside under the hammer, and finished by being turned in a lathe. The cannon is bored by placing it in a lathe in the usual manner.

The gun A, of cast-steel, may be surrounded by a casing B of cast-steel or gun-metal, to give the necessary weight. When either the gun or the casing has become injured or unfit for further use, the gun can be taken out by removing screw D, pin F, and vent C, and the defective piece replaced by a fresh one.

Claim.—1st. The manufacture of cannon from solid pieces of cast-steel, as described.

2d. The surrounding of cannon and other parts of artillery, when made of cast-steel, with an outer casing of cast-iron, steel, or wrought-iron, or gun-metal, in the manner and for the purposes described and represented.

No. 13,927.—DANIEL TREADWELL.—*Improved Manufacture of Cannon.*—Patented December 11, 1855. (Plates, p. 319.)

The inventor says: I do not claim using hoops generally in making cannon, as the earliest cannon known were formed in part by hoops brazed upon them; but my invention consists in constructing cannon with hoops around, and shrunk upon, a body in which the calibre is formed in the manner described. (See engraving.)

No. 13,984.—JOHN GRIFFEN.—*Improved Manufacture of Wrought-Iron Cannon.*—Patented December 25, 1855. (Plates, p. 319.)

Figure 1, mandrel on which the pile is to be formed; figure 2, mandrel with the two end rings fitted to it; figure 3 shows a series of longitudinal bars arranged around the mandrel to form the bore of the cannon; figure 4 shows the pile, consisting of a series of iron bands wrapped spirally around the central bars; figure 5 shows the pile turned off at its end, and having a plug M inserted to form the breech; figure 6, section of it.

The inventor says: Having discovered that the mode of preparing the pile or faggot, described in my specification, is specially adapted to being welded under the rollers, and that welding such a prepared mass by means of rolling is entirely practicable, and will secure a more homogeneous and perfect union of the parts without weakening or rupturing the fibre, I do not desire to claim the described mode of preparing the pile or faggot, when the faggot so prepared is welded by blows or under the hammer.

But I *claim* the manufacture of wrought-iron cannon, by forming the faggot or pile of longitudinal bars surrounded by a series of bands of iron, and then welding together the whole mass by passing it between rollers.

No. 12,545.—ABBOT R. DAVIS.—*Improved Shot-Cartridge.*—Patented March 20, 1855. (Plates, p. 319.)

The shot is mixed with wet clay or other plastic material, that when dry will readily crumble apart, using no more of such than will be sufficient to fill the cavities between the shot when in close contact. The mixture is rolled into balls or cylinders, and these are then rolled in contact with fibrous material, so as to form a coating which will serve as wadding.

The inventor says: I am aware that a shot-cartridge has been, with a woven wire frame, filled with shot and loose sand, and covered by paper pasted around it; I therefore do not claim such a mode of making a cartridge. But I *claim* an improved shot-cartridge, made by mixing the shot in a plastic material or compound, of the character described, subsequently reducing the mass to the shape required for the cartridge, and covering its external surface with fibres of wool, or other material, belted or applied thereto substantially as specified.

No. 12,556.—ABNER N. NEWTON.—*Improvement in Cartridges.*—Patented March 20, 1855. (Plates, p. 319.)

The percussion powder is arranged on the end of a metallic rod *a*, the other end of the rod being attached centrally to the base or rear end of the ball or cartridge; the percussion being thus arranged, secures certainty in the explosion when struck by the hammer, or its equivalent.

Claim.—The arrangement of the percussion priming with a metallic rod, in the manner specified, whereby said priming is ignited within the chamber of the gun between the ends of two metallic rods, as set forth.

No. 12,942.—Chas. F. Brown.—*Improvement in Cartridges.*—Patented May 29, 1855. (Plates, p. 319.)

By giving the ends of the case a hemispherical or similar form when the explosion takes place, they will be caused to expand in such form (see fig. 2) as to fill the bore of the piece tightly.

The water-chamber in rear of the case will be burst by the explosion of the charge, and the water will serve to cool the breech and prevent the escape of fire at the breech and the consequent loss of force.

Claim—1st. Making the ends of the metal cartridge-case of hemispherical or other convex form, substantially as and for the purpose herein set forth.

2d. Providing a water-chamber S in the rear end of the cartridge-case, to be filled with water previously to the insertion of the cartridge in the gun, substantially as and for the purpose set forth.

No. 12,189.—Joshua Stevens, assignor to the Massachusetts Arms Company.—*Improvement in Repeating Fire-Arms.*—Patented January 2, 1855. (Plates, p. 320.)

The hammer B is provided with a small notch *a* for the catch part *b* of the trigger to take into, in order to set said hammer to a full cock. By the side of the hammer is a tumbler O, which rotates freely on the same axle N as the hammer, and is moved thereon by means of an arm *d* projecting from trigger C, and made to work into notch *e* of the tumbler. To this tumbler the turning-lever L is jointed, fitting into a recess *f* which is formed in the tumbler, and is provided with a bearer *x* against which lever L rests, so that it may be elevated by the tumbler. The spring *g* presses down the front end of said lever. The tumbler has also a notch *h* and a cam-surface *i* for the purpose of operating upon and withdrawing the bolt M, which for this purpose is provided with a tail-hook *r;* the said bolt being moved forward by spring P. During the rotation of the tumbler by the trigger, the bolt will be withdrawn until the cam *i* shall force the hook *r* above the notch *h*, when the spring P will move the bolt forward. The rear end of the cylinder is formed with locking recesses and a ratchet or turning-cams such as usually employed. During the reciprocating rotary movement of the tumbler (caused by the action of trigger C and spring H) the turning-lever L will be moved forward and upward against the bearing face of a tooth of the ratchet, and be subsequently drawn backward away from the ratchet, while the spring *g* will force it downward so as to cause it to pass over the next succeeding tooth of the ratchet. While being moved by the tumbler, the bearer *x* comes in contact with the turning-lever, and forces it upward. From the rear side of the hammer there projects a cam-pin *w*, which, while the hammer is descending, elevates the tail-hook *r* above the notch *h*, so that the notch may pass by and not act on the hook, in case the movement of the trigger should be too rapid to give the proper time for the rotation of the cylinder. During the next backward pull, then, of the trigger, the hammer being down upon the nipple, the notch *h* will again act on the hook and withdraw the bolt, so as to allow the cylinder to be revolved.

The advantage of the above-described arrangement of the turning-level L is that there is no tendency to press the cylinder longitudinally against the barrel or its front support, which often attends the application of the usual springs or pawls, and produces considerable friction.

The inventor says: I *claim* so combining the trigger, the hammer, and the mechanism for rotating the cylinder, that, by a single pull on and during the back movement of the trigger, the hammer shall be discharged or set free from the trigger, (so as to fall on the nipple when the touch-hole of one charge-chamber of the cylinder is in connexion with it,) and the cylinder subsequently rotated, so as to bring up to the percussion-nipple, or its equivalent, the touch-hole of the next chamber of the series thereof. And in combination with the mechanism for turning the cylinder, and that for locking and unlocking it, I claim a cam-pin, (projecting from the hammer,) or its equivalent, for preventing the cylinder from being unlocked, or for locking it in case the movement of the trigger is so rapid as to render the cylinder liable to be rotated before the charge fired by the action of the hammer has left its chamber.

I do not claim jointing an impelling pawl directly to the lower part of a percussion-hammer, so that, by the reciprocating rotary movement of the hammer, the said pawl may be moved against and drawn away from the ratchet of the revolving cylinder; nor do I claim jointing a lever directly to the trigger, so that, by the movement of the trigger, such lever may be moved against one tooth of the ratchet or drawn back over the next succeeding tooth and against a spring acting upon the rear end of the lever.

But I do *claim* the hereinbefore described and represented arrangement and combination of the trigger and its spring, a rotary tumbler separate from the trigger and moving on a separate pin or fulcrum, a turning mechanism of the cylinder, and the locking and unlocking mechanism thereof; by which arrangement and combination, during and by a back and forward movement of the trigger, the cylinder will be locked or unlocked, and have an intermittent rotary motion imparted to it, and the cock or percussion-hammer be actuated essentially as specified.

Also, the bearer *x*, or its equivalent, in combination with the turning-lever L, and the part or tumbler to which it is connected or jointed, and by which motion is imparted to the said turning-lever, as specified; this combination attaining an important advantage, as hereinbefore explained.

No. 12,230.—Thos. H. Barlow.—*Improvement in Fire-Arms.*—Patented January 16, 1855. (Plates, p. 320.)

The engravings, in combination with the claim, fully explain the nature of this invention.

Claim.—1st. Constructing the chamber of the cannon nearly square and winding, so as to give the projectile the motion and accuracy of the rifle-ball.

No. 12,235.—EDMUND H. GRAHAM.—*Improvement in Fire-Arms.*—Patented January 16, 1855. (Plates, p. 320.)

The short barrels *e* which receive the charge are hinged by pivots *f* to a horizontal and cylindrical revolving plate *g*. The barrels hang all in a vertical position; the one opposite to the gun-barrel *d* is raised by bringing the lever *m n* from position figure 4 into position figure 1. This motion at the same time cocks the gun, by means of the arm *n* pressing against the cam projection *p* of the sliding-bar *q q*, to which the hammer *l* is attached. The projection *r* of this bar abuts against a notched pawl *s*, upon which a spring *t* presses. The bar is forced back until a notch in the same passes beyond the pawl *u* of trigger *v*, when pawl *u* is forced into the notch of the bar by a spring *w*, and thereby holds it and the hammer ready for action. The plate *g* is revolved by means of the knob and pin *x*, which catches into notches on the upper surface of the plate. The plate, together with the barrels, can be removed (by unscrewing cover *h*) for the purpose of charging the barrels. By this arrangement each of the barrels *e* is, at the moment of firing, removed a long distance from the remaining tubes, being in a horizontal position, while the others hang down; and hence the fire cannot be communicated from the tube discharged to the others.

Claim.—So attaching the tubes or short barrels, in which the charges are placed to a revolving plate, as to admit of their being separately and successively elevated into a horizontal position, in a line with and so as to form a continuation of the gun-barrel, while the others retain a vertical position as set forth, and for the purposes specified; also, the lever *m n*, arranged and operating as described, for elevating and lowering the tubes which hold the charges, and for cocking the gun, as set forth.

No. 12,244.—ALONZO D. PERRY.—*Improvement in Fire-Arms.*—Patented January 16, 1855. (Plates, p. 320.)

To insure the forcing of the cap down on the nipple, the rear part of the chamber, along the path of the nipple, is made in the form of an eccentric plane *w*, against which the top of the cap rubs, and by which it is forced down on the nipple.

Claim.—The arrangement of the tube in the stock for containing the caps, and a spring to force them forward in a line radiating from the axis of motion of the turning-breech, and placing the nipple also in a line radiating from the axis of motion of the breech, so that, when the breech is opened to receive a charge, the nipple will be brought into the same radial line with the cap-tube, so that the same spring which forces the caps forward in the tube may also force one of them upon the nipple, thus simplifying the mechanism for automatic capping; also, the use of an eccentric or its equivalent, as specified, in combination with the capping-tube and nipple on the movable breech, as specified, for the purpose of forcing the caps to their proper place on the nipple as the breech is brought in line for the discharge, as specified; also, pivoting the trigger *k* to the lever *h* for operating the breech, as described, so that the trigger shall be carried in and out by said lever, and shall not be brought into a position to act upon the lock until the breech is in a line with the barrel, as set forth.

No. 12,328.—Alexander O. H. P. Sehorn.—*Improvement in Portable Fire-Arms.*—Patented January 30, 1855. (Plates, p. 321.)

When the box B is within the cylinder *p*, the stud *t* of the hammer *m* enters the opening *r* in spring s^1, so that, as the box B is drawn from the receptacle, the cheeks *e* and *f* move forward and leave the hammer held by said spring s^1, the stud t^1 entering the opening r^1 in spring *s*, when the box moves sufficiently far forward, and resisting the spring *l*. A pressure on lever *q* depresses the spring s^1, releasing the hammer *m*, which flies forward and discharges the load. By lifting spring *s* from stud t^1 the box B is moved back; the stud *t* engages spring s^1, and, if the load be ready, the arm is prepared for a new discharge. C H is the handle of a cane enclosing the whole apparatus.

Claim.—The combination of the box, springs *s* and s^1, coiled spring *l*, hammer *m*, and casing *p*, constructed, arranged, and operating as set forth, when used in connexion with an external case C H, for the purposes specified.

No. 12,440.—Danl. B. Neal.—*Improvement in Repeating Single-Barrelled Fire-Arms.*—Patented February 27, 1855. (Plates, p. 321.)

The rod F is fitted into the stock close to the under side of the barrel, and has a mortise in each end: one at *o* receives the tenon of lever E; the other at *s* receives the tenon of the trigger G, which trigger G is used only for throwing the false hammer D forward. Two charges being inserted, both cones capped, and the false hammer placed on the front cone, the front charge is fired in the usual manner; after this the finger remains in the guard, and the hammer C is again cocked with the thumb of the hand used on the trigger, at the same time the finger presses forward against the front trigger G, which will throw the false hammer forward, so that the back charge may be instantly fired.

The inventor says: I *claim*, 1st. The combination of the elongated hammer with the false hammer, arranged as described.

2d. The arrangement of the lever E and rod F for throwing the false hammer forward, substantially as set forth.

And I hereby disclaim the original invention of the double-shooting one-barrel fire-arms, and of all and singular the parts and the combination and arrangement of the parts thereof, except the arrangement and the combination of the parts merely which I before claim.

No. 12,470.—JEHU HOLLINGSWORTH and RALPH S. MERSHON.—*Improvement in Fire-Arms.*—Patented February 27, 1855; patented in England August 1, 1854. (Plates, p. 321.)

In loading this arm, after the catch G is raised up, the chamber C and scape-wheel E are slipped forward on the spindle just far enough to release said scape-wheel from the anchor, and then the two (for they are connected together) are free to be turned round so as to meet the ramrod J and be charged; whilst the chambers and scape-wheel are thus run forward, the nipples can be capped; and when returned, the catch put down, and the handle turned to coil the spring, the fire-arm is ready for its series of discharges. On the rear of the scape-wheel E is a series of cam-planes *m* corresponding to the number of chambers in the revolving breech. Against these planes the end of bolt *n* works as follows: When the breech-piece or chamber is released, by means of the anchor and trigger, in turning, one of the planes *m* pushes back the bolt *n*, and brings a shoulder *o* on said bolt against the hammer D, pushing back or cocking said hammer, and in the act of cocking it compresses a spring *s*, the point of which presses against the arm *p* of said hammer, and in this position the arm stands after each discharge, it being caught by the anchor and escapement. Now, by pressing the trigger the anchor is released, the scape-wheel makes another partial rotation, bringing the nipple to the exact spot for receiving the blow of the hammer, and simultaneously letting off the hammer, which carries forward with it the bolt *n* for the next repeated discharge. It is, therefore, (after the arm is loaded and the reservoir of power filled,) but the simple act of pressing the trigger that cocks, rotates, and discharges the fire-arm, so that the whole series of discharges may be let off without taking down the arm. A more detailed description would take up too much room.

Fig. 2 is a section through 2 2; fig. 3, through 3 3; fig. 4, through 4 4; fig. 5, rear view of stock and spring box.

The inventors say: We do not claim a reservoir of power, simply for rotating the breach, as that has heretofore been done; but we *claim*, 1st. The application of a reservoir A of power to the rotating of the cylinder or breech C, in combination with the cocking and releasing of the hammer D in concert, so as to produce two or more discharges from a repeating fire-arm without replenishing said reservoir of power, substantially as described.

We also claim combining a reservoir of power with a rotating toothed scape-wheel E, anchor K, and trigger F, in such a manner that at each periodical releasement of said scape-wheel, by the operation of the trigger and anchor, or anchor escapement, the reservoir of power will rotate the chambered breech to the required distance, and simultaneously trip an independent hammer, substantially as described.

We also claim so combining a rotating chambered breech with a reservoir of power and cock or hammer, as that by the periodical releasement of said reservoir, by means of the scape-wheel, trigger, and anchor escapement, said chambers shall be caused to rotate to their required distance, and meet the blow of the hammer at the exact instant that each chamber in succession comes opposite the barrel, substantially as described.

We also claim combining a reservoir of power with an independent cock or hammer, so that, by the periodical releasement of said reservoir of power, said hammer shall be tripped at the exact moment that each chamber of the series comes opposite the barrel, substantially as described.

We also claim so combining the stock with the frame, as that by turning said stock a spring or springs shall be wound up, which shall be capable of actuating the fire-arm for a series of discharges, substantially as described.

We also claim the peculiar form of guard M, or protection to the hand on which the fire-arm may be supported, so as to guard the hand from any accidental discharge of the chambers when not opposite the barrel, while said accidental discharge may escape from the fire-arm without detriment to the user, substantially as set forth.

We also claim the conical plate *j* and ring i^1 as a means by which the stock and spring box are united to the frame so as to make a firm connexion, and at the same time allow the one to be turned upon the other for the purpose of coiling up or compressing the spring, substantially as described.

No. 12,471.—Ralph S. Mershon and Jehu Hollingsworth.—*Improvement in Repeating Fire-Arms.*—Patented February 27, 1855; patented in England August 1, 1854. (Plates, p. 321.)

When the barrel C is swung back around its fulcrum *a*, the cam projection *b* will strike the screw nut H, and (the nut being prevented from turning round by moving between guides) move the nut along on screw-shaft I, thereby compressing spring J, the compressed spring then forming the reservoir of power for discharging successively all the chambers. One end of I turns on a journal in frame F; to its other end is attached the toothed wheel *c*. Shaft I projects forward of wheel *c*, and forms the spindle upon which the revolving chambers B turn. Its front end is embraced by the latch E. A pin *e* on the toothed wheel projects into a hole in the rear of the breech-piece B, thereby securing its revolution together with the said wheel. By pulling trigger D, arm *i* of lever L recedes from between the teeth of wheel *c*, thereby allowing the rod 1 to be revolved by the action of spring J and nut H. At the time *i* recedes, the arm *f* of lever L enters between the nipples *s*, the uppermost nipple (the breech-piece revolving together with the screw-rod and wheel) strikes against the arm *f* thereby exploding the cap and discharging the chamber. The screw-rod is of sufficient length and the spring of sufficient power to revolve and discharge all the chambers without recocking the arm.

Claim.—1st. A reservoir of power capable of discharging two or more barrels or chambers of a repeating fire-arm, substantially as described.

Also, exploding the cap, or similar percussion priming for discharging the chambers, by means of the blow caused by the rotation of the chambers or barrels, bringing each nipple or cap in succession against a vibrating arm, or its equivalent, thus causing said rotation to perform the ordinary function of a cock or hammer, substantially as described.

Also, so hinging the barrel or frame which supports it to the stock, as that when said barrel is swung back or forward, either for removing or recharging the chambers, it will contract the spring to supply a reservoir of power capable of discharging two or more chambers or barrels successively without recocking or recharging at each discharge, substantially as described.

No. 12,528.—Rollin White.—*Improvement in Breech-Loading Fire-Arms.*—Patented March 13, 1855. (Plates, p. 321.)

The engravings fully explain the nature of this improvement.

Claim.—The connection of the breech with the hammer in such a manner that it may be withdrawn to open the chamber to receive the charge by the act of cocking the hammer, and replaced to close the chamber by the falling of the hammer, when the latter is set free to explode the charge, substantially as set forth.

No. 12,529.—Rollin White.—*Improvement in Breech-Loading Fire-Arms.*—Patented March 13, 1855. (Plates, p. 322.)

When the piece has just received its charge, the hammer being down and the breech withdrawn, (see fig. 2,) the raising of the trigger-guard to close the chamber will cause the tooth D to throw back the tumbler and hammer; and when these are thrown back, (see fig. 1,) their return is prevented by reason of the tooth D standing nearly at right-angles to the line of movement of the breech, and thus locking them. The drawing of the trigger moves the sliding-piece *d* upwards, and lifts the tooth D from the sere, thus setting free the hammer to cause the explosion. Near the termination of the descent of the hammer, the sliding-piece *d* comes in contact with pin *e*, and is thereby thrown back to be operative for the next discharge. Pin *e* also serves the purpose of preventing the hammer descending further than is desirable after the withdrawal of the breech for the next charge, by catching sere *b*. When the discharge is not to be immediately repeated, the cocking of the hammer (by the replacement of the breech to close the chamber) is prevented by throwing back eccentric *f* to keep tooth D disengaged from the sere. The hammer may be then cocked by hand. The eccentric *f* is controlled by a small crank outside the stock not shown in the engravings).

Claim.—1st. The connection of the breech or breech-piece with the hammer in such manner that the latter may be cocked by the act of moving the former into its place to close the chamber, substantially as set forth.

2d. The peculiar manner of effecting the cocking and setting force of the hammer by means of the spring-tooth D attached to the breech or breech-piece, and the sliding-piece *d*, working in the tumbler to be acted upon by the trigger for the purpose of disengaging the said tooth, substantially as set forth.

3d. The employment of a crank or eccentric *f*, arranged and operating substantially as described, for the purpose of disengaging the tooth D from the tumbler, and thereby disconnecting the hammer from the breech or breech-piece, when the immediate repetition of the discharge is not desired.

No. 12,555.—Frederick Newbury.—*Improvement in Fire-Arms.*—Patented March 20, 1855. (Plates, p. 322.)

On the front face of ratchet-plate R P (but not shown in the figures) is a slot part of an arc to receive pin 5 which projects from the rear of cylinder C, the slot being just long enough to permit the cylinder to revolve so far as to bring two chambers in succession in line with the barrel. When the trigger T is operated, its upper cam end will elevate lever R L, (pivoted in *l*,) and thereby raise pawl R K and operate the ratchet-plate (the upper end of the pawl projecting and working through a slot in face-plate F P). One of the stop-levers is shown behind the other in the figure, with a small part S K projecting downward, and has just above this projection, on its upper edge, a stop-catch, which, when the loaded cylinder is first put on, is pressed up into a cavity cut for the purpose in the cylinder, the proper position for the cylinder being indicated by a stop-pin marked S P, which pin must rest against the right-hand side of the stock. When in this position, the pin 5 will be on the left-hand corner of its slot. On drawing the trigger for the first fire, the lever R L causes the ratchet-plate to revolve (the cylinder being six-chambered) one-sixth of a revolution; but as the pin 5 is acting in a slot for that distance, the movement of the plate R P does not carry with it the cylinder, whose first chamber remains in line with the barrel as placed. At the moment of discharge the end of lever R L touches the stop-lever S K and throws the catch out of its notch, leaving the cylinder free to turn till every chamber has been fired and the stop-pin strikes the left-hand side of the stock. The second stop-lever C L has a similar projecting catch, which enters into cavities on the surface of the cylinder, (it being held up by a spring,) so that at the moment of firing the pawl holds it firmly in one direction, and the lever C L in the other. The lever is discharged from its hold by the pressure of the front end of the lever R L just as the pawl begins to act upon the ratchet-plate. A notch on the rear of the hammer and of the tumbler M L lies against the spring M S, which operates the hammer so that the drawing of the trigger pressing the tumbler down relieves the hammer from the pressure of the spring which holds it down against the cylinder. The spring presses against the hammer as well as against the tumbler, and the hammer moves round independently of the tumbler, on a common axis. The cocking-spring C S has just sufficient power to keep the hammer pressed against spring M S. The hammer lies flush with the outer surface of the stock, so that the pistol cannot be discharged by reason of an accidental catching of the projecting hammer, as heretofore in use. The hammer, when down, being near to a horizontal line, has forced the cylinder slightly forward, and presses it against the barrel.

Claim.—The ratchet-plate, with its ratchet indentations and its slot, in combination with the pin by which it connects with the cylinder.

Also, the two stop-levers below the cylinder to regulate and sceure the connection between the chambers of the cylinder and the barrel, substantially as set forth.

Also, the arrangement and combination of the tumbler with the hammer and cocking-spring, to enable the hammer to act independently of the tumbler in the act of firing.

Also, the arrangement of the hammer to lie within the stock and to act in such line of direction upon the nippers as to press and hold the cylinder firmly against the barrel in the act of firing; the whole substantially as set forth.

Also, the arrangement of the apparatus for disengaging and attaching the barrel with the cylinder to the stock, viz: the thumb connecting plate T P or detent with the spring S to hold it in place, and the notch *n* in the mandrel to receive the detent, substantially as set forth. (See engraving.)

No. 12,567.—Alexander T. Watson.—*Improvement in Breech-Loading Fire-Arms.*—Patented March 20, 1855. (Plates, p. 322.)

The magazine being filled with cartridges, (see figure,) and the breech-piece L being raised by means of handle M, the first cartridge is drawn forward through cylinder G into its place in the chamber of the barrel, by means of a string, which, however, is only necessary on the first charging of the gun. Any further supply of cartridges can be hooked to the loop *y* on the last cartridge of this series. The first cartridge 1 thus drawn into its place, the breech-piece L (the hand being removed) is pressed upon the end of the cartridge by means of spring *q*. Cylinder G, with its attachment below, is drawn back by the finger in the ring *k*, which also forces down the long end of lever H, taking off the pressure of plate *l*, and leaving cylinder G to pass freely backwards and on to the next cartridge 3 (the rear end of the cylinder being bell-mouthed for the purpose). Ring *k* is then released and plate *l* is pressed (by spring *i*) on the cartridge to hold it firmly, while the cylinder and cartridge within it is pressed forward by the action of spring K. The breech-piece is then forcibly shut down, cutting off the end of cartridge 1 and the connecting-string of cartridge 2. These pieces are pushed out through the opening below by the action of the part *m* of the breech-piece.

The touch-hole *r* passes through the breech-piece, ending in the centre of cavity *o*, directly at the bottom of the chamber. By the circular shape given to the breech-piece and its slot, it can be made so as always to press forward against the base of the barrel, thereby constituting a firm and tight breech-piece.

Claim.—The mechanical combination and arrangement of the cylinder G, the bent lever H, and the forked standard j^1 j^1, acted upon by the rod J and spiral spring K; also the spring *i i*, by which J being drawn back, the cartridge, constructed and arranged as described, is released from the pressure of *l*, and the cylinder is made to pass over the next succeeding cartridge, and the pressure of the finger being removed from J, the cartridge is firmly griped by *l*, and carried forward towards the chamber by the action of K and J, pushing before it also the next preceding cartridge ready to be deposited in the chamber upon the raising of the breech-piece; which operation being repeated after each discharge, in connexion with raising the breech-piece, secures a measured supply of charges from the magazine in the stock to the chamber, to an extent and with a facility not heretofore attained in breech-loading fire-arms.

Also, the breech-piece of a segment of a circle, having the concave space *o* for the bottom of the chamber, with its central point of depression in the line of the axis of the barrel.

Also, the forming the lower end of the breech-piece into two cutters—one front, the other back—with the rounded swell *m* between, operating as well to hold the cartridge in its place, as to cut off the end and remove the parts thus cut off, as described.

No. 12,638.—Rollin White.—*Improvement in Breech-Loading Fire-Arms.*—Patented April 3, 1855. (Plates, p. 322.)

A is the sliding-breech, with the tumbler *a* and sere *e*. The breech carries a nipple at *i*, which, by striking against the projection *j*, explodes the cap. The cartridge is placed upon a hinged plate B, forming a seat for it; and when the breech is raised, the plate (being sufficiently elastic) is pressed down by the cartridge with the fingers, in loading, far enough into the breech-space to enable the cartridge to be pushed from it by the fingers into the chamber. When the cartridge is in the chamber, and the fingers are removed, the plate, by its elasticity, stands across the rear of the chamber, (see full lines in figure 3,) and thus prevents the cartridge from falling out at the rear of the chamber. The breech is chamfered out on one side, so as to push the plate aside when it (the breech) is falling (see dotted lines, figure 3).

Claim.—1st. The application of the sliding-breech to operate in connexion with the trigger, through a tumbler, as described, in substantially the same manner as the hammer in ordinary fire-arms; thereby making the breech serve not only its proper purpose of closing the rear of the chamber, but as the hammer for effecting the explosion of the charge, as set forth.

2d. The spring-plate B, applied substantially as described, to serve as a guide to conduct the cartridge into the open chamber, and as a guard to prevent the cartridge from falling out at the rear of the chamber before the breech is liberated, as set forth.

No. 12,648.—Rollin White.—*Improvement in Repeating Fire-Arms.*—Patented April 3, 1855. (Plates, p. 323.)

The nature of this invention will be fully understood from the claims and accompanying engravings.

I *claim*, 1st. Extending the chambers *a a* of the rotating cylinder A right through the rear of the said cylinder, for the purpose of enabling the said chambers to be charged at the rear, either by hand or by a self-acting charger, substantially as described.

2d. The application of a guard to cover the front of all the chambers of the cylinder which are not in line with the barrel, or any number thereof which may have been loaded, combined with the provision of a proper space for the lateral escape of the exploded powder, substantially as described, whether the said space be between the cylinder and guard, or in rear of the cylinder, and whether the said guard be constructed with a recess to receive the balls, or be of such form as merely to stop the balls.

3d. Combining a charging piston G with the hammer, by means of gearing substantially as described, or by the equivalent thereof, in such a manner that, by raising the hammer to cock the lock, the piston is moved towards the chambered cylinder to force a cartridge from the magazine into one of the chambers thereof, and by the falling of the hammer the piston is withdrawn to allow a new cartridge to be supplied ready to be driven into the next chamber of the cylinder, as the hammer is again raised to cock the piece, as fully set forth.

4th. Furnishing the hammer with an attachment *m*, by which, in the act of falling, it may close the mouth of the magazine, substantially as described, before exploding the priming, and thus protect the charges within the magazine from ignition.

No. 12,649.—Rollin White.—*Improved Repeating Fire-Arm.*—Patented April 3, 1855. (Plates, p. 323.)

The cartridges are placed in magazine D, parallel with the barrel, and one upon another. The magazine opens at the bottom into charging-tube C, which stands parallel to the barrel, in such a position that one chamber of the cylinder is opposite to it when another is opposite the barrel. The charging-piston E is attached by a link *b* to lever F. When this lever is thrown forward, (by means of gearing G *g*,) a cartridge drops into the tube; and when the lever is pulled down, it drives the piston forward and the cartridge into the chamber. When F is thrown up to draw back the piston, the tooth *d* engages with a tooth *f*, and rotates the cylinder so as to bring opposite the barrel a chamber previously charged. As rack-rod *g* is moved back, tooth *k* engages with a tooth *l*, and raises the hammer till trigger *m* falls into sere *n*. M is a light steel springing plate secured to the recoil plate J; it covers nearly the whole of the exposed part of the rear of the cylinder, and one end *t* of it fits into one of notches *u*.

The inventor says: 1st. I do not claim the employment of a magazine to supply cartridges to the chambers, or caps to the nipples, when the cartridges are arranged therein end to end, to be fed into the chamber or chambers, or on to the nipple of the piece.

But I *claim* the method, substantially as described, of combining and applying the magazine and charging-tube, either for cartridges or priming, to wit: the charging-tube being arranged in line with the chamber or one of the chambers, or the nipple or one of the nipples, of the piece, and the magazine being arranged in such a manner, relatively thereto, that the cartridges or caps lie side by side, to be fed sideways one by one, as required, into the charging-tube, by gravitation, a spring, or other means, on the retraction of the said piston from opposite the magazine, and to be fed into the chamber or on to the nipple by the movement of the piston towards it.

2d. I claim combining the rotating chambered cylinder with the charging-piston, or its equivalent, in the manner substantially as set forth, so that by the operation of retracting the charger, after charging a chamber of the cylinder, the cylinder shall be rotated to the extent required to bring a new chamber in line with the barrel.

3d. I claim combining the hammer with the charging-piston, in the manner substantially as described, so that during the operation of moving the charging-piston, to drive a cartridge from the magazine into the chamber, the hammer shall be raised to cock the lock.

4th. I claim the spring protecting-plate M, applied substantially as described, to fall into notches *u u* in the rear of the rotating cylinder, to protect the other charged chambers from the effects of lateral fire from the discharge of the chamber which is in line with the barrel.

No. 12,655.—George H. Soule.—*Improved Breech-Loading Fire-Arm.* —Patented April 3, 1855. (Plates, p. 323.)

The nature of this invention will be fully understood from the engravings.

Claim.—The use of the double-acting cam, in combination with the charging-chamber and breech-piece, having an opening P in it, for cutting off the end of the cartridge and discharging the fragments thereof from the charging-chamber, the said parts made and operating in combination with the barrel and magazine of a gun, substantially as set forth.

No. 12,681.—Ferdinand Klein.—*Improvement in Fire-Arms.*—Patented April 10, 1855. (Plates, p. 323.)

A flexible spring *d* is attached to the lever *e* (that turns the faucet or cylinder) at one end, and to the valve *c* of the chamber that receives the charge at the other end; this spring is intended to open and close the valve as the lever is moved up or down by the force or action of the spring *a*, whereby the faucet is also turned. Spring *d* also assists in preventing the faucet from moving endwise, in case the screw *f*, which holds the faucet in place, should become loosened. This screw also holds cap *b*, which covers the spring *a*. The cap is intended to protect the spring, and to prevent it from loosening the screw, which would be the case if it were in immediate contact with the screw-head.

Claim.—The improvement in the manner of opening and closing the valve, or cover of the chamber which receives the charge, in the manner and for the purpose above set forth. Also, the use of a cap for the purpose of protecting the chamber, and the spring that moves it, as set forth in the foregoing specification.

No. 12,836.—John Stowell.—*Improvement in Fire-Arms.*—Patented May 8, 1855. (Plates, p. 324.)

By pulling down the lever D, the long arm of the lever *d* is depressed, and the short arm thereof thrown forward, throwing forward the bottom of the tumbler and raising the hammer to cock it. When the lever D is raised to drive up the sliding-crotch to wedge the rotating-breech up to the barrel, the pin *h* moves back in the slot *i* of the stirrup *f*, that being the purpose for which the slot is provided.

The inventor says: I hereby disclaim the invention of the combination of the hammer with the sliding-crotch, for the purpose of effecting the cocking of the lock simultaneously with and by the same movement as the rotation of the breech, in any other way than substantially described in the specification.

But I do *claim* the method herein described of effecting the connexion between the hammer and the lever D, by which the sliding-crotch is operated by means of a lever *d* and two stirrups *e* and *f*, applied and operating substantially as herein described.

No. 12,906.—Henry Gross.—*Improvement in Fire-Arms.*—Patented May 22, 1855. (Plates, p. 324.)

The nipples *p* are placed in the periphery of the cap-cylinder, with vents extending through to *f f*. This cap fits on the cylindrical roller *e* and on each change of lever *a*; in order to load the gun the cap-cylinder is turned forward, presenting one of the vents *f* opposite and over the vent *g* in the roller, which communicates with the charge in the gun.

Claim.—The combination and arrangement of the cap-cylinder with the cylindrical roller, as described and shown in figures 2 and 3. All else and all other parts of said gun are disclaimed.

No. 13,039.—Frederick Newbury.—*Improvements in Revolving Fire-Arms.*—Patented June 12, 1855. (Plates, p. 324.)

The nature of this improvement will be understood from the claim and engravings.

A is the cylinder; P the guard-ring.

Claim.—In fire-arms having revolving cylinders to carry their load and priming, the construction and use of a guard to prevent the fragments of exploded priming-caps from impeding the rotation of the cylinder by covering up the cones with a movable metallic ring, containing for each cone a chamber as large as can be conveniently made within the ring to enclose each cone; each chamber having an aperture to permit the priming-cap to pass through, and also having lateral openings to pass off the gas produced by the detonation of the cap; the whole substantially as set forth.

No. 13,154.—Ethan Allen.—*Improvement in Fire-Arms.*—Patented July 3, 1855. (Plates, p. 324.)

The nature of this improvement will be easily understood from the claims and engravings.

The inventor says: I do not claim a rotary charge-receiver placed within or applied to a gun-barrel, nor do I claim a sliding or movable breech so applied to the barrel as to constitute not only a breech thereto, and a means of uncovering the rear end of the bore of the barrel, and enabling a cartridge to be introduced into the same, but a contrivance for shearing or cutting off the rear part or end of the cartridge after the introduction of such cartridge into the barrel.

Nor do I claim a perforated rotary breech-cylinder so combined with or applied to a gun-barrel, and entirely in rear of its charge-chamber, that it (or its cylinder) may be capable of being rotated or turned in one direction so as not only to uncover the rear end of such charge-chamber, and present a passage or opening through which a charge may be introduced into it, but rotated back so as to cover the rear end of the charge-chamber, and constitute a breech thereto.

Nor do I claim a turning-lever and breech so applied to the rear end or part of a gun-barrel as to be capable of being turned up so as to uncover such end sufficiently to permit a charge or cartridge to be introduced into the barrel.

But I *claim* so combining a rotary or movable breech and a charge-chamber together, and with the barrel of a fire-arm, that not only when they (the said breech and chamber) are moved or rotated in one direction, shall the breech uncover the passage into the barrel, and such charge-chamber be brought into a position to permit a cartridge to be passed into it and the barrel; but when they (the said breech and chamber) are rotated in the opposite direction, such a breech shall be made to cover the passage into the barrel, and such charge-chamber, in conjunction with the barrel, be caused to bend, break, and hold such cartridge, as specified.

I also claim the charge-chamber and breech-block D, provided with journals and bearings as specified, extending the movable breech F beyond the block in manner as described, in order that not only may said breech serve as a scraper to the inside surface of the case C, but that the block may be protected from the injury and friction of the carbonaceous matter resulting from the explosion of a charge, as specified.

No. 13,292.—John A. Reynolds.—*Improvement in Fire-Arms.*—Patented July 17, 1855. (Plates, p. 324.)

The chambers *a*, with the nipples *b*, are arranged upon a hollow cylinder A, supported at its periphery by suitable rings and ribs (to receive the recoil), so that while they are radial therewith, the butt of the chamber shall permit the touch-hole to be reached from within the cylinder, while the discharge ends shall be in contact with the rear ends of the barrels B. Cylinder A is turned by means of hand-lever operating a pawl which takes into notches on the circumference of the cylinder. These same notches match with rack-teeth on the underside of a perforated block F, and cause the latter to move successively along as the cylinder is turned. The cartridges having been placed in the perforation of block F, (which correspond with the chambers in the cylinder,) the cross-head L is depressed by means of pedal G and connexion H H K K, when the ram-rods M, attached to the under side of the cross-head, will drive a row of cartridges from F into the chambers *a* underneath. At the same time the levers K act upon stirrups 5 and levers N and rods 6, thereby elevating a cross-bar *o* which carries a row of vertical plungers, which force the caps from the perforated hollow cylinder P, and on to the nipples above said row of plungers, the nipples belonging to those chambers which have been simultaneously loaded by the ram-rods from above. The cap cylinder moves together with cylinder A by means of cogs around their ends. Springs 3 thrôw the levers K and appurtenances up as soon as the foot is withdrawn from the pedal. The handle and levers R S serve to move the hammer T, and also the swabbers *n*, the end of the swabber-bar being guided in slots 10 in the frame of the machine. U is a back set, which, being connected with the bar S at its end by the projecting end of the swabbing-bar, is made to retreat from hammer T, when S is drawn by R, the movement of U being allowed by a slot 11; W is a shaft-spring, playing through a slot in S, and has two angular projections upon its inner face, the office of the one of which is to catch on ear 14 on the hammer-stock and draw it back, while the other projection, in passing post 15, throws the spring into the slot of S, and releases the front catch from ear 14, and permits the hammer to be driven by spring V against the capped nipples. The swabs in their withdrawal pass through sheets of sponge retaining water, by which the filth drawn with the swab may be removed therefrom. In figure 2 the swabbing-box is shown, with the lid (mounted with sponge) thrown open.

Claim.—Constructing fire-arms with a hollow cylinder A, containing chambers *a a*, as described, in connexion with barrels B B, substantially in the manner and for the purposes set forth; also, loading the chambers *a a*, by foot pedal G, straps H H, levers K K, operating the plungers M, in combination with the simultaneous capping of the nipples by lever N, straps 5 and 6, cross-bar O, and plungers thereon, for removing caps from cylinder P and placing them on the nipples, as set forth; also, drawing the hammer T by lever R and bar S, furnished with spring W and catch thereon, or its equivalent, in connexion with the annular liberating projection on spring W, and the liberating post 15, for the purposes set forth; also, drawing the hammer back, in the manner set forth, in combination with the simultaneous swabbing of the discharged chambers, in the manner substantially as described: also, likewise, the swabbing-box for containing the swabs *n n*, as described, furnished with sponge or its equivalent; the whole operating substantially in the manner and for the purposes set forth.

No. 13,293.—JOHN A. REYNOLDS.—*Improvement in Fire-Arms.*—Patented July 17, 1855. (Plates, p. 324.)

Spokes I^1 serve to turn the chain-wheel H^1, and by means of chain K^1 the chain-wheel E with its screw-shaft F, thereby elevating or depressing the frame Y of the machine, and thus adjusting the range. (The engravings belonging to this patent will be better understood by reference to engravings No. 13,292.) The shield L^1, in front of the machine, with the side wings M^1 and a roof, (not shown in the engravings,) serve to protect the machine and operator. (See engravings No. 13,292.)

Claim.—The elevating of the manifold fire-arm by the screw F^1, nut G^1, on swivelled arms a^1 a^1, as described, in connexion with pulleys H^1 and E^1, chain K^1, or their equivalents, substantially as set forth; also, the adaptation of the shield to the manifold fire-arm, or similar machine, substantially in the manner and for the purpose set forth.

No. 13,294.—JOHN A. REYNOLDS.—*Improved Apparatus for Cocking Repeating Fire-Arms.*—Patented July 17, 1855. (Plates, p. 324.)

Sheet-iron metal tubes F^2, closely surrounding the barrels B, are soldered into the box A^2, which is filled with water.

(This engraving will be better understood by reference to the engravings No. 13,292.)

Claim.—The application of a refrigerator, constructed as described, to the barrel or tubes of fire-arms, for the purpose of keeping said tubes from undue heating, substantially in the manner set forth.

No. 13,474.—JNO. SWYNEY, assignor to HIMSELF and JAMES DANDRIDGE.—*Improvement in Breech-Loading Magazine Fire-Arms.*—Patented August 21, 1855. (Plates, p. 325.)

To the cartridge-rammer P is attached a rod y, the rear end of which rod has a long slot a^1, in which one end of lever b^1 is placed, which lever works on a fulcrum pin, as shown in figure 3, the lower end of said lever entering a slot in the cap-rammer t; the length of slot a^1 is such that, as the rod y is moved backward with rammer P in loading, it will not move lever b^1 until the cartridge is almost driven home in chamber C; then striking lever b^1, it will force rammer t forward through the hole in the carrier R, and thereby force a cap on the percussion-nipple G, by the time the cartridge is driven fully home; then, by withdrawing the hand from the handle j of rammer P, said rammer will be drawn back to its original position by spiral spring i, and with it the rod y will be also drawn back, when the rear end of slot a^1 will hit lever b^1, and thereby move rammer t back from the nipple and into the position it previously occupied. A piece of rubber in the front end of t will prevent exploding a cap in the act of forcing it on the nipple. The rest of the features of this invention will be understood from the engravings.

Claim.—The carrier R, its spring S, in combination with the magazine or tube F, for the purpose of bringing a cap from the magazine F downwards, or into line with the rammer *t*.

Also, the rammer *t*, in combination with the rammer P, and the mechanism by which they are connected, so as to operate together, as described, such mechanism consisting, in part, of the rod *y* and the lever b^1.

Also, combining with the charge-chamber C and the magazine E, the intermediate chamber or carrier M, said charge-chamber C and carrier being connected with, and operated simultaneously by, the guard K.

No. 13,507.—Benjamin F. Joslyn.—*Improvement in Breech-Loading Fire-Arms.*—Patented August 28, 1855. (Plates, p. 325.)

A represents the barrel, and B the shaft of the fire-arm. The inventor says: The powder being ignited, the first instantaneous act of the expanded air and gas will be to drive the pin E towards the interior of the projection D on breech C; and, as the rings are prevented from moving backwards by the end of the said projection, it necessarily follows that the coned head of the pin must cause the rings to expand, and consequently to fit tight to both the inside of the barrel and outside of the cone-head of the pin, effectually preventing all escape of air or gas towards the breech.

I *claim* the combining of the cone-headed pin E, and two or more expanding rings G and H, with the radial breech C, of breech-loading fire-arms.

No. 13,547.—Josee Johnson.—*Charge for Fire-Arms, &c.*—Patented September 11, 1855. (Plates, p. 325.)

The charger A is filled by opening valve E (see figure 3); the patch is placed over the hole in cylinder A, and pressed down even with the end of the cylinder; the patch is then cut off smoothly. Then a short ramrod is applied to it, by which the ball is pressed through the cylinder into the projecting position represented in the figures. The powder is then put in, and the valve E is closed. The cap *c* is then inserted in the clamp B. To transfer the charge to the gun, the valve E is opened, the powder poured into the calibre of the gun, the charger inverted, and the projecting-ball applied to the muzzle of the gun. Then the ramrod is slipped through the cylinder, forcing the ball out of the same, and down to the powder at the breech of the gun. The charger is then slipped off over the end of the ramrod, the latter withdrawn, the cap (in clamp B) applied to the nipple of the gun, and the charger withdrawn. The chargers are designed to be loaded at home or in camp; and the inventor states, a rifle can be loaded with this charger in less time than with the rifle cartridge now in use.

I *claim*, 1st. The combination of the projecting-ball C, cylinder A, and cut-off valve E, arranged and combined in the manner and for the purposes described, as set forth.

2d. I claim using clamp B in combination with the charger, for the purpose of facilitating the rapid completion of the process of loading fire-arms, as described.

No. 13,581.—Wm. W. Marston.—*Improvement in Fire-Arms.*—Patented September 18, 1855. (Plates, p. 325.)

When the trigger *h* is pulled on, the sere K is thrown up, the end whereof takes the notch behind the projection 17, rotating the barrels until the stop 13 on the trigger comes against the stop 20 on the edge of the face-plate *e*, the act of doing which elevates the hammer, the tumbler 9 of which takes the half-cock notch 15. The trigger is then released to be thrown forward by spring 12, which causes the sere *k* to take the second projection 18, and on pulling again on the trigger, the barrels are moved nearly to the point to bring the succeeding barrel into the same place as that occupied at the previous discharge of the piece; the stop 13 now takes against the lowest notch of the double stops 21, and the hammer is raised to the full-cock notch 16. If now the trigger be released, the sere *k* cannot descend far enough to take the next projection 17; so that on pulling the said trigger, the sere slides over the back of said projection until it comes against the notch 19, thereby turning the barrels just enough to allow the tumbler 9 to slip off the end of the cam at 16, and strike the cap on the cone 4 and discharge the piece. In this case, the stop 13 comes in contact with the stop 21 on the edge of the face-wheel, stopping the farther rotation of the barrels, and the parts are again ready to be operated on, as before described, to discharge the next barrel.

I do not limit myself to the size or character of arm fitted with my improvements; neither do I make any claim for rotating and cocking a fire-arm simultaneously; neither do I claim the sere K to act upwards and rotate the barrels, as this is also well known.

But I *claim*, 1st. Elevating the hammer to cock and discharge the piece, by means of a cam *d* revolving with the barrels or chambers, and formed with as many points as there are barrels or chambers, so that the hammer shall be raised and discharged by simply rotating said barrels or chambers, as specified.

2d. I claim the revolving face-plate *e*, formed with projections on its face to take the sere K, and with notches on its edge taking the stop 13 on the trigger, the two acting to rotate and stop the barrels at the precise point requred, and prevent the strain on the trigger from turning the barrels too far, as specified.

3d. I claim the mode specified of constructing and fitting the parts of the cam *d*, face-plate *e*, trigger *k*, sere K, and stop 13, so that the hammer shall be cocked by one, two, or more pulls on the trigger, in the manner as specified.

No. 13,582.—Frederick Newbury.—*Improvement in Revolving Fire-Arms.*—Patented September 18, 1855. (Plates, p. 326.)

When the trigger is drawn, the limb *d* enters into the angular recess behind one of the ratchet teeth at *a*, and as it is pressed forward compels the wheel R to turn round until it has moved round one-sixth of its circuit. The projection *f* of the trigger locks into projection *k* of the hammer H; consequently, when the trigger is drawn and the cylinder C is revolving, the hammer gradually rises until the instant that the cylinder, reaching the last point of its movement, has placed a chamber in line with the barrel, when the points *f* and K pass each other, releasing the hammer, which is then thrown forward by the spring M S upon the cap. The shoulder *e* of the trigger is so adjusted that just at the moment of the discharge it presses against the edge of the wheel R, so as to hold the barrel immovable during the discharge. As soon as the trigger is released from the pressure of the finger, the spring B throws it back, the trigger-slot allowing K and *f* to repass each other. The lever-catch F is kept against the projection S at the lower end of a piece Q extending from the barrel downwards by means of a spring V, thus attaching the barrel Y to the face of the cylinder. The finger-piece X serves to detach the barrel.

I do not claim the use of an oblique toothed ratchet-wheel, nor the revolving mandrel attached to both cylinder and ratchet-wheel; but I *claim* the method of operating an oblique toothed ratchet-wheel by the direct action of the upper limb or cam end of the trigger, which trigger, also by the same action, cocks and discharges the hammer, and holds the cylinder firmly in place during the firing of the piece, substantially as set forth.

I also claim the employment and use of a slot in the trigger directly upon the hammer, in order to enable the trigger to replace itself behind the hammer as before the discharge of the same, substantially as set forth.

I claim the apparatus for attaching and detaching the barrel to the stock, to wit: the catch-lever lying in the stock underneath the cylinder, with its hook, finger-piece, and spring, together with the recess and stop in the block.

No. 13,592.—Frederick Beerstecher.—*Improvement in Fire-Arms.*—Patented September 25, 1855. (Plates, p. 326.)

When the longer part *d* of the head of the hammer is in the position shown in the engraving in full lines, and the hammer is let go, it explodes the forward cap C, and thus the forward charge only is discharged. The hammer is then again cocked, when with his finger the operator turns down the longer part *d* of the head, into the position shown by dotted lines, when the hammer is again let go and the rear charge discharged.

I do not claim the general arrangement whereby two loads may be discharged in succession from one barrel, without reloading, as such arrangement is not new.

But I *claim* constructing the head of the hammer of fire-arms of this description, so that the part of the head which discharges the forward load can be capable of being turned down for the purpose of allowing the shorter part of the head to strike the rear tube only, and so that when turned up it shall strike the forward tube only, without the use of the intermediate covering-lever heretofore required, for the purpose of preventing the explosion of the rear cap, in fire-arms of this description, the same being constructed, arranged, and operating substantially as described and set forth.

No. 13,691.—H. B. WEAVER.—*Improvement in Breech-Loading Fire-Arms.*—Patented October 16, 1855. (Plates, p. 326.)

The chamber A is hinged at *a*, and has an arm *c* extending from said hinge, which arm works in a slot in the sliding-piece *d*. The lever D is connected with hammer F, and pivoted at *e;* when the rear end of lever D is depressed, the hammer F will be drawn back so as to clear the pin *h*, and allow it to move back out of the way of slide *d*, which latter, by the time the pin *l* has arrived at the top of slot *n*, will lift the sliding-piece *d* to receive a cap from the repository E, and the piece *d* will simultaneously actuate arm *c* and move the chamber A laterally, (see dotted lines in figure 2,) ready to receive the charge.

Claim.—1st. Combining the hammer with the laterally swinging chamber, for the purpose of effecting the simultaneous opening of the chamber and cocking of the hammer, by means of the lever D, the pin *k*, slide *d*, and lever-arm *c*, all operating substantially as described, whether the said slide *d* be a priming-slide, or simply employed to connect the chamber A with the lever D.

2d. Combining the priming-slide *d* with the lever D and the hammer F, by means of a pin *l* attached to the lever, working in a slot *n*, in the slide or a link attached thereto, so that the lever D will draw back the hammer before moving the slide far enough to allow the pin *h*, or its equivalent, through which the hammer strikes the cap, to move out of the receiving-hole in the slide before the slide is acted upon by the lever, substantially as set forth.

No. 13,941.—JOSEPH C. DAY.—*Improvement in Fire-Arms.*—Patented December 18, 1855. (Plates, p. 326.)

The barrel A can be readily removed by unscrewing screw *e*, and then turning piece B^2 on its hinge *h*.

The grooves C^1 and A^2 afford a place for the fouling to collect, and also serve to decrease the surface upon which the sliding-collar D moves, thereby lessening the friction and securing a free and easy motion of the sliding-collar. The face of this collar is made on the same sweep as the rear of the barrel A, and its outer edge sharp, so as to serve for cutting the cartridge C^1, when the barrel A is forced from position figure 4 into position figure 1.

Claim.—In addition to my former claims, granted to me in letters patent—

1st. Connecting the two side-pieces B^1 and B^2, between which the barrel is hung by a hinge B^3, and locking them by the projection B^4 and a corresponding recess, substantially as set forth.

2d. Making the face of the sliding-collar D of the shape of an arc, with a cutting edge, so as to act in combination with the rear end of the cartridge as described.

3d. The grooves C^1 and A^2 in the breech, and the rear end of the barrel, for the purpose set forth.

No. 13,999.—E. K. Root.—*Improvement in Revolving Fire-Arms.*—Patented December 25, 1855. (Plates, p. 326.)

b are the usual longitudinal and diagonal grooves, in which the driving-pin *c* works to rotate the breech to bring the chambers successively in line with the barrel. The pin, provided with a spring *e* as usual, is fitted to a block *d*, which slides in slot *f* in the bottom-plate *g*, below the rotating breech. After the block has been drawn back sufficiently far to pass out of one of the diagonal grooves, to turn the breech and bring the next chamber in line with the barrel, and to enter one of the longitudinal grooves to hold the chamber in line, its rear end then strikes against the end of the spring-catch *i*, which holds the tumbler *j* of the hammer, and liberates it, that the hammer may be forced on to the cap.

Claim.—Combining the driving-pin that works in the grooves to rotate and hold the breech in line with a slide below, adapted to the reception of and to be operated by the trigger finger, and acting on the lock at the end of the back motion, to liberate the cock or hammer, to discharge the load, substantially as described.

No. 14,001.—Gilbert Smith.—*Improvement in Breech-Loading Fire-Arms.*—Patented December 25, 1855. (Plates, p. 326.)

The nature of this improvement consists in the opening cap A being eccentric to the axis of the barrel B, thereby opening in one position and closing in another, according as the aperture in the cap A is in position either to close or open the breech of the gun; said aperture being made to roll concentric with the barrel B, opening the same to receive the charge; the reverse motion combined with the traverse motion, obtained by the screw on the inside of the cap, and exterior of the eccentric portion attached to the barrel B for closing the breech. The closing of the breech is completed by the screw-pin C being turned, and forwarded into the seat, as soon as the eccentric has thrown it direct over the axis of the barrel, which position is insured by stops on the outside of the cap.

Claim.—The eccentric and traverse motions combined, for opening and closing apertures, by means of a cap perforated eccentric to itself, as described.

2d. Closing the aperture by means of an inserted screw-pin being screwed forward direct from the cap, when the eccentric throws it direct over the axis of aperture as described.

No. 12,810.—PHILIP BACON.—*Improved Tape-Fuse.*—Patented May 8, 1855. (Plates, p. 327.)

A common fuse (for blasting under water,) as represented at A, is run through a melted composition of pitch and coal-tar, then the tape B in a clean state is wound round it, and afterwards the external covering C is wound around the tape, then the fuse is again run through the melted composition to give it an external water-proof covering.

Claim.—The application, substantially as described, to tape-fuse, of an external winding of thread, whereby the loosening or cracking off of the tape and water-proofing substance is effectually prevented, and the manufacture of the fuse cheapened and simplified, as herein set forth.

No. 13,138.—ABRAHAM POWELL, Jr.—*Improved Fuse-Stock for Bomb-Shells.*—Patented June 26, 1855. (Plates, p. 327.)

This improvement consists in making a stock of two cylinders C C, one within the other; the outer one being screwed fast in the shell B, the inner can be turned at pleasure to any required point of a graduated scale on the ends of both. This scale is marked with the number of seconds, 2, 5, 10, &c., to the greatest extent of time, according to the driving of the fuse, that may be required for the bursting of the shell; and when the inner cylinder is set with, say S, opposite 2 on the outer cylinder, then the shell when fused will burst in 2 seconds. This is effected by apertures down the sides of both cylinders, so arranged that when any particular hole of the inner cylinder is opposite the corresponding hole of the outer, all the others are shut and safe from fusion; and when the fuse has burned down to that particular hole, (indicated by the scale on top,) the explosion will take place.

Claim.—A double-cylinder fuse-stock, so graduated as to burst shell-shot at any required number of seconds, as described.

No. 12,124.—J. S. BUTTERFIELD.—*Improvement in Locks for Fire-Arms.*—Patented January 2, 1855. (Plates, p. 327.)

In turning the hammer back (fig. 1) the cam *c* hooks into slot *h* of spring *d*, while connecting sear *e* (being attached to lower end of cam *c*) is drawn forward until notch e^1 fastens on catch *f*, the end of sear *e* being pressed down by spring *g*. When the hammer is cocked and it is desired to discharge, the after end of sear *e* is raised, and notch e^1 cleared of catch *f*; whereupon spring *d* operates cam *c*, the hammer falls, and the sear comes back (fig. 2).

Claim.—The combination of the sear *e* with the cam *c* (the latter operating in the slot *h* of the main spring *d*); the respective parts being arranged in the manner herein set forth.

No. 13,442.—Michael Tromly.—*Improvement in Gun-Locks.*—Patented August 14, 1855. (Plates, p. 327.)

The nature of this invention will be understood from the claim and engravings.

Claim.—Constructing the lock by having the lower part of the hammer B formed of two prongs *d e*, against one of which *d* the spring C bears; and having the upper end of the tumbler *f* attached to the other prong *e*, the lower end of said tumbler being attached to the upper end of the trigger D; the front end of the spring C having a small spring *l* attached to it, which spring *l* bears against the trigger D, as shown and described.

No. 13,825.—John Phin.—*Improvement in Gun-Locks.*—Patented November 20, 1855. (Plates, p. 327.)

On the trigger being pulled, it will elevate the hammer by means of link L; and as the point *r* of the trigger recedes from the sear *s*, the pressure of spring *i* will cause said sear to press against tumbler T, and to fall into the notch *n* when said tumbler has been rotated sufficiently. The trigger may now be allowed to move forward in obedience to the spring *a;* but until it touches the tail of sear *s*, the hammer will remain cocked. Instantly on the trigger pressing on this sear, however, it will release the hammer, which will descend in obedience to the main spring.

While the sear remains in the notch *n*, no discharge can be effected by pulling the trigger.

Claim.—Securing accuracy of aim and safety in the use of trigger-cocking fire-arms by means substantially as described, which consist, first, in the sear *s* and spring *i* to hold the hammer up; and, secondly, in the spring *a* acting on the trigger to release said hammer.

No. 12,522.—Christopher Wolter.—*Improvements in Ordnance.*—Patented March 13, 1855. (Plates, pp. 327 and 328.)

The two barrels are adjustable at different angles to each other, for the purpose of firing chain or rope shot at various distances, and always extending the chain or rope between them without danger of breaking it. By moving the slider F back and forth, the angle of the two barrels can be adjusted, (see broken lines in figure 2,) and they will always bear the same relation to a central line along which the sight may be taken. Each breech has, at its rear end, a pivot *b*, which works in a slot *c* in head C of frame C C, D E. The slots *c* are arcs described from the vertical stands of the universal joints *a*. The sectors D are described from the horizontal pivots of the universal joints *a*. The cross-piece, which connects the two sectors, carries the slide E through which slider F works. The hammers (not shown in the engravings) are driven by springs, and are made to strike the cap by being drawn back against the force of the spring and then suddenly let go. The hammers are connected with links *l*, which are connected with sliding-piece *v*. From piece *v* a cord *n* passes over pulley *o* at the end of slide E, and from thence under pulley *p*, attached to cross-piece *q*, and connects with rod *r*, to which rod is applied a spiral spring S, so as to throw it inwards, and outside the carriage it connects with lever *u*. The barrels are discharged simultaneously by pulling this lever outwards and suddenly letting it go; the lever draws the rod *r* out, and drawing-cord *n* draws forward the links *l*, thus drawing back the hammers against the springs, (above mentioned,) which apply to them the required percussive force. The effect of these springs is assisted by spring S, which quickly slackens the cord. To vary the length of cord *n*, as the relative positions of the barrels vary, it is passed through an eye in the end of rod *r*, and thence to a winch *w*, on which it can be wound or unwound by means of handle *x*.

The inventor says: 1st. Though I do not claim, of itself, the mounting of a gun-barrel, or piece of ordnance, upon a universal joint or pivot, I *claim* the connexion of two barrels or pieces thus mounted, in such a manner that they may be adjusted and held at any desirable angle relatively to each other, substantially as and for the purposes described.

2d. I claim the connexion of the barrels by means of the toggle-joints *j j* and the central slider F, working in a suitable slide supported by the carriage, substantially as described, for the purpose of adjusting the barrels at the desired angle.

3d. Supporting the breeches for the purpose of varying the elevation of the barrels, by means of a frame composed of sectors D D D D, and slotted heads C C attached thereto, as described, whereby the necessary changes of elevation, and of the angle of the two barrels, are provided for independently of each other; this I claim, irrespective of any mechanical devices that may be employed to raise and lower the frame.

4th. The connexion of the two hammers, or the triggers, or their equivalents, by means of two links, with a sliding-piece *v* operated upon by a cord or chain connected with a rod *r*, which passes through the side of the carriage, and has a spring S applied, substantially as and for the purpose set forth.

5th. Connecting the cord or chain with the rod *r*, or its equivalent, by merely passing it through an eye at the end thereof, and attaching it to a winch *w*, conveniently situated to keep it always wound up to the proper degree to give it the required length, as fully set forth.

No. 13,249.—Charles F. Brown.—*Improved Mode of Mounting Ordnance.*—Patented July 17, 1855. (Plates, p. 328.)

The nature of this improvement will be understood from the claim and engraving.

Claim.—Mounting a cannon or any other piece of ordnance, substantially as described, in a carriage A, of spherical, spheroidal, or other circular form externally, which carriage is arranged to close the port or embrazure through which the piece works, but to turn freely therein in a horizontal, or nearly horizontal, direction, and which has an opening *c* within it, of suitable size and form to receive the gun, and to allow it the necessary upward and downward swinging movement on its trunnions, whereby an efficient protection is afforded against the entrance of the enemy's shot or projectiles, and the smoke of the discharge is excluded, and at the same time a desirable range in a lateral and vertical direction is maintained.

No. 13,679.—ANDREW HOTCHKISS.—*Improved Projectile for Ordnance.* —Patented October 16, 1855. (Plates, p. 328.)

By placing the cap E over the tail-piece B, so that the corners *c* of the tail-piece will pass through and beyond the notches s^1 in the annular projection *s*, and then, turning the cap a little, the cap and tail become locked, and the parts A and E are held together.

In the act of firing, the ring C (made of lead or other soft material) is expanded so as to take a full impression of the grooves, in consequence of the force with which the cap E is driven towards the body of the shot, before momentum is communicated to the latter, and thus windage is prevented and a rotary motion communicated to the projectile.

Claim.—1st. Constructing a shot or projectile capable of being fired from a cannon having rifle grooves, said shot consisting of three parts, two of which parts, A and E, are of hard metal, and the other, C, of some flexible expansive material, in the form of a band or ring attached to one of the hard metal parts, and overlapping the edge of the other in such a manner that, either by the act of loading or of firing, or of both, the said ring shall be so expanded, or distended, that it shall take the impression of the grooves and be made to fit the bore, as described.

2d. The tail-piece B, for securing the cap E to the body of the shot, and as a guide to the cap in its forward motion, in the manner described.

No. 12,258.—HEZEKIAH CONANT.—*Improvement in Moulds for Casting Projectiles.*—Patented January 16, 1855. (Plates, p. 328.)

The metal is fused in G, and then the stop-cock *k* opened and the wheel D revolved, when the metal will fill the moulds c^1 as rapidly as they are presented, the air escaping between the wheel and band. The slugs escape underneath at E.

Claim.—The arrangement of moulds c^1 in the periphery of a wheel D, combined with a stationary band B, which also forms part of each mould, the whole operating as described.

No. 12,574.—WILLIAM M. B. HARTLEY.—*Improved Press for Making Cylindro-Conical Hollow Projectiles by Pressure.*—Patented March 21, 1855. (Plates, p. 329.)

The die A is made up of four sections, B B^1 C C^1. The solid ball, of a size to enter the die without difficulty, is inserted in the die, it passing clear of the ridges *c c*. The carriage H is run to stop *b*, and punch P permitted to descend, forming the cavity within the ball, (see fig. 4,) and spreading the ball so as to fill the die, the straight portion exactly filling the collar R. Punch P rises instantly, and the operator, after moving carriage H up to stop b^1, places the cap *i* on the butt of the ball and depresses punch P^1, which drives the cap to its place. When the operator releases the punch P^1, it flies up by the action of a spring. Punch P is then made again to descend, and by striking the ball frees it from the collar, and causes it to drop through opening T in bed K. The die is then closed by turning rim G, and the machine is ready for the reception of another ball. The collar R, by forming the butt, prevents the change which would take place in the calibre of the balls by the wear of the machine, if the butt were formed by the sectional die, as from the pressure there would necessarily be a variable expansion of the parts from the wear of the machine, which would render the calibre of the balls unequal.

The inventor says: I do not claim the manner of operating the die sections.

But I do *claim* the collar R, in combination with the sectional parts of the die, constructed, arranged, and operating substantially as and for the purposes set forth.

I also claim the arrangement relative to the punches P P^1 of the die A, with a horizontal motion of sufficient amplitude to admit of the successive action of the punches, substantially as and for the purposes set forth.

I further claim capping the ball while in its die, and while held firmly at its base, by a punch, which on the opening of the sections will, by a subsequent or continuous motion, discharge the ball capped and ready for use.

No. 12,795.—Eben Hoyt, Jr.—*Improved Projectile for Fire-Arms.*—Patented May 1, 1855. (Plates, p. 329.)

In order to impart a rotary motion to heavy shot and shells, the inventor furnishes the rear end of the ball with inclined planes B, against which the force of the discharge reacts, thereby rotating the ball.

Claim.—The employment of inclined surfaces upon the rear end of the ball, operating in the manner and for the purpose substantially as herein set forth.

No. 12,801.—W. J. Von Kammerhueber.—*Improvement in Projectiles.*—Patented May 1, 1855. (Plates, p. 329.)

This projectile is lens-shaped, its section in one direction being represented in figure 5; the section perpendicular to the plane of the section figure 5 is exhibited in figure I, the other figures being variations of the shape of figure 1.

The inventor throws this projectile from a cannon or gun-barrel, the cross section of which corresponds to the cross section figure 1 of the projectile. He further intends to impart to this projectile a revolving motion around its shorter axis, by means which form not a subject of these letters-patent.

This projectile, moving through the air in a plane corresponding with its circular section, will offer to the air the smallest possible resistance.

Claim.—The lens-shape of the projectile, made of any desirable material or combination of materials, solid or hollow, as above described, and which projectile is to be thrown by any exploding or expanding substance.

No. 13,469.—AUGUSTUS MCBURTH.—*Improvement in Percussion Projectiles.*—Patented August 21, 1855. (Plates, p. 329.)

f is the position of the discharging-rod while holding the hammer *e* previous to the point *h* striking the ship or any other substance; *g* is the percussion-cap, ready to be discharged by hammer *e* when set free by the collision of the point *h* of the discharging-rod with the surface at which the shot was directed. The spiral spring *d* prevents the hammer from striking the cap (which serves to explode the charge within the shell) previous to the point of the rod coming in contact with the said surface.

Claim.—The improvement in bomb-shells or missiles having four arms b, b^1, b^2, b^3, and eight flutes, with sharp edges 1, 2, 3, in the manner and for the purpose substantially as described; also, a rod to pass through the shell in a longitudinal course, for the purpose set forth; and also a hammer, with a flat spring attached, together with a spiral spring *d*, as shown and described.

No. 13,799.—SYLVANUS SAWYER.—*Improved Compound Projectile.*—Patented November 13, 1855. (Plates, p. 329.)

The coating of lead C is employed as anti-friction metal, and the force of the explosion, acting in conjunction with the hard metal at the butt or rear end of the shell, is intended to expand a portion of such anti-friction metal, so as to press it firmly against the inner surface of the bore of the piece and cut off the passage of the flame; which latter would otherwise tend to rip and melt off the coating. The explosion will force the surplus anti-friction metal at *b* up the tapering edge of the shell, so as to insure the closest fitting between shell and bore.

Claim.—Combining with the butt, or flat rear end of the cylindro-conical iron shell, a layer of lead or softer metal than that of which the body of the shell is composed, and united or not to a layer of such metal, extended around the sides of the shell, as described; the same operating in manner as specified, while the shell is being projected through the bore of a gun by a discharge of the powder therein.

2d. Making the rear part of the shell tapering or conical, as seen at *a a*, combining therewith a ring *b b* of lead or its equivalent; the same being substantially in manner and for the purpose as herein before stated.

Also, confining the explosive screw-cap to the body of the shell, by means of a softer or yielding metal or casing, which, when the cap or shell shall strike an object, shall give way under the force of the blow, and let the cap down with force so as to compress the percussion-wafer or priming in it, or on the main screw, stopper, or plug, and so as to create an explosion thereof.

No. 12,285.—E. K. Root.—*Improved Compound Rifling Machine.*—Patented January 23, 1855. (Plates, pp. 329 and 330.)

One end of the connecting-rod v is pivoted in w to an arm x of the frame of the machine; the other end fits over a pivot u, which passes up from rack t through a slot in the top of the carriage l. When the carriage is at the extreme of its range of motion in the direction of the arrow, the connecting-rod v is at right-angles to the line of motion of the carriage; so that, as the carriage recedes, the end of rod v, connected with the rack t, describes a circle (see dotted line, figure 2); and, as this circular line moves from the line of motion of the carriage, motion is imparted to the transverse rack t, and by it to the series of spindles r, which are thus caused to turn faster the more the carriage moves back. Each of the mandrels carries a wheel, with as many notches in its periphery and as many cogs on its face as the number of rifles to be cut in the barrel. The levers h^1 h^1 h^1 h^1 have their fulcra i^1 i^1 i^1 i^1 in the slide g^1, and are held down by springs j^1 j^1 j^1 j^1. The forward ends of these levers are wedge-formed, and pass under pins k^1 k^1 k^1 k^1 projecting from the faces of sliding-dogs c^1; so that, as the slide moves forward in the direction of the arrow, the dogs are lifted up out of the notches in the wheels d^1 of the mandrels, which liberate them so that they can be turned freely; and, as they are liberated, the heel l^1 of each lever strikes against the cog on the wheel, to turn it the required distance to determine the distance between any two of the rifles to be cut. The pins on the dogs then pass over the rear end of the inclined planes of the levers, and are forced down by the springs to re-engage the next set of notches in the wheels to hold the mandrels in place for the next cut. On the return of the slide, the levers rise and ride over the pins of the dogs. As the cutters leave the barrels, the small wedge-rods z in the tubular end of the cutter-rods y strike a series of stops x^1, by which the wedge-rods are forced in to wedge out the cutters. These stops are forced forward a little at every cutting operation, which is effected in the following manner: These stops slide in the head-block y^1, (figure 4 represents a horizontal section of the same,) and have rack-teeth on their upper surface, engaged each by a pinion z^1 on shaft a^2, provided with a spur-wheel b^2, engaged by vertical worm-arbor d^2, which carries a ratchet-wheel e^2, operated by pawl f^2 on slide g^2, which is forced back by a spring (not shown in the engraving) and moved forward to give the proper motion by bar q^1, which carries an adjusting-screw h^2, the end of which strikes the end of the slide. By means of this screw, the amount of motion can be adjusted at pleasure. The arbor d^2 is hung in bent-lever i^2, to which is jointed a hand-rod j^2, by which the worm can be disengaged from wheel b^2 to reset the stops at the end of each complete operation. The crank-pins g can be adjusted in radial slots in the crank-wheels e f so as to adjust the stroke of the connecting-rods *h*, and consequently the way of the carriage *l*, as the other ends of the rods *h* are connected to wrist-pins *i* on slides *j*, fitted to ways *k* on the opposite sides of the carriage, which in turn slides on ways *m*. The slides *j* are adjustable on the carriage by means of two screw-shafts *n n* and bevel-gearing and shaft *o p q*. Thus, by turning shaft *q*, the slides *j* can be shifted to regulate the position of the carriage with reference to the other parts of the machine, when the crank-pins *g* are shifted to adjust the range of motion of the carriage. Figure 5 is drawn on a larger scale than the rest of the figures.

Claim.—The method of giving the motion to the cutter-stocks for giving the increasing twist, by means of the connecting-rod *v*, or its equivalent, turning on a fixed centre *w*, and describing a circle at the point *u* of its connexion with the cutter-carriage *l*, which moves in a tangent line, substantially as specified; also, combining a series of cutter-spindles *r* with the said connecting-rod, or its equivalent, by means of a sliding-rack *t* connected with the said rod, and engaging pinions *s* on the said spindles, substantially as described; also, in combination with the mandrels b^1 b^1 b^1 b^1 that carry the barrels, the slide g^1 and its appendages, to act upon and turn the mandrels, in combination with the dogs for locking and holding the barrels during the rifling operation; the said dogs being operated by the said slide, substantially as specified; also, the mode of operating the series of stops to insure an accurate adjustment of the series of cutters, substantially as specified; and, finally, the adjustable crank-pins for operating the cutter-carriage, in combination with the mode of forming the connexion of the connecting-rods with the carriage, by means of slides governed by adjusting geared screws, substantially as specified, as a means of adapting the machine to the rifling of barrels of various lengths, without the necessity of changing the relations of the mandrels, and the stops for setting the cutters, as set forth.

XX.—SURGICAL INSTRUMENTS.

No. 13,467.—Louis H. Lefebvre.—*Improvement in Warm-Bath Apparatus.*—Patented August 21, 1855. (Plates, p. 331.)

The vapor generator A consists of two compartments, one of them B for generating sulphurous acid gas, and occupying about one-third of the capacity of A, and the other C for the generation of vapors of water, occupying the rest of A. H H boxes for inserting alcohol-lamps. By means of stop-cocks *e f* the vapors may be allowed to pass either mingled or separately through the pipe K and into the bag M, in which the patient receives the bath.

Claim.—The portable steam-bath apparatus, composed of a double generator, so arranged that the products generated in the two compartments may be conveyed to the bath mingled or separately, of a bag

M, and of a connecting-pipe K; each of said parts constructed and arranged as described.

No. 13,974.—Joseph Buhler.—*Improvement in the Pipes of a Vapor-Bath.*—Patented December 25, 1855. (Plates, p. 331.)

The pipe G is perforated with one or more rows of perforations *e*, extending straight down the front; the sleeve H, fitted over said pipe G so as to be capable of being turned, has one or more spiral rows of perforations *f f*. The patient sitting on C may, by pulling handle *k*, turn the sleeve until one of the holes *f* coincides with that one of the holes *e* which is at the proper height to direct the vapor (entering pipe G from a suitable retort) to any part of his back as required.

Claim.—The back distributing pipe G, with its sleeve H, operated by a cord with a handle and weight, or by any equivalent means; the said sleeve having perforations *f f* out of line with the perforations of the pipe, to allow the patient to direct the concentrated vapor to any part of his back, substantially as set forth.

No. 12,972.—Richard A. Stratton.—*Improvement in Chairs for Dentists' use.*—Patented May 29, 1855. (Plates, p. 331.)

The set-screw G, the end of which fits into a longitudinal groove in the screw-shaft D, prevents the latter from turning, and confines it to a vertical motion upwards or downwards, for the purpose of raising or lowering the seat B. N are friction rollers.

The inventor says: I do not claim screws and wheels for raising and lowering the seats of chairs, as they are old and well-known applications, nor do I claim the use of friction rollers exclusively.

But I *claim*, as a simple arrangement for raising and lowering, and steadying the seats of operating-chairs, the annular piece L, with its projecting pins and rollers, or their equivalents, in combination with the slotted screw D, its nut-wheel N, and bracket F.

No. 13,396.—D. W. Perkins.—*Improvement in Dental Chairs.*—Patented August 7, 1855. (Plates, p. 331.)

The nature of this improvement will be understood from the claims and engravings. The standard B can be raised or lowered in a socket (not shown in the engravings).

Claim.—1st. Tightening the ball and socket-joint so as to secure the body of the chair in the desired position, by means of the band F which encompasses the socket E, the band being operated upon by a clamp G, as shown, whereby the parts *e f g* of the socket may be pressed or bound snugly around the ball *d*, substantially as shown and described.

2d. Attaching the head-rest N to the inner edge of the plate O by hinges, and having the head-rest secured at the desired angle of inclination by a segment rack P, the plate O being allowed to slide laterally upon a plate Q at the upper part of the bar R, which bar works in an opening S in the back of the chair, and is secured at the desired point by the rack V and spring-catch W, for the purpose of rendering the head-rest capable of perfect adjustment, as set forth.

No. 12,951.—EDWARD G. HYDE.—*Improvement in the Construction of Ear-Trumpets.*—Patented May 29, 1855. (Plates, p. 331.)

This improvement consists in intersecting the tube of an ear-trumpet, near where it enters the ear, with a passage communicating with an artificial ear, which is intended to lead such vibrations as fall on it to unite with the vibration passing round through the tube, so as to enable the person to hear the utterances of others or other sounds, without the confusion of sounds that accompanies the instrument without the improvement.

Claim.—The artificial ear C applied to an acoustic auricle or ear-trumpet, substantially as and for the purpose described.

No. 13,953.—DANIEL PARRISH.—*Improvement in Instruments for Modifying Focal Length of the Eye.*—Patented December 18, 1855. (Plates, p. 331.)

By applying the mouth D of the tube A closely over the eye, and either drawing the piston back if the eye is too flat, or forcing it against the eye if the latter is too convex, the eye in the former case will be drawn into the mouth D by reason of the vacuum formed in the tube, so as to assume more than its ordinary convexity, and in the latter case it will be flattened by the pressure of the air in the tube. The instrument should be applied every evening for the space of about twelve minutes, and applying it to one and the same eye for one week, and then change to the other for the same period, and so on for the term of about three months.

Claim.—The improved optical instrument described, for the purpose of improving and restoring the sight by giving greater convexity to the eye when flattened, and also by depressing that organ when too convex, in the manner specified.

No. 12,636.—HENRY MELLISH.—*Improvement in Lancets.*—Patented April 3, 1855. (Plates, p. 331.)

A is the blade of the lancet; B, a slot along its centre, with ways D at its edges for the ways *b* of the charger G and the guide *c* of piston H to slide on; C slot, and E recess or channel in which the tubular portion G of the charger slides; F is also a channel, but in the opposite side of the blade in which the shaft *a* of the charger slides. When the instrument has been charged, (by bringing the point into contact with the matter and withdrawing the piston,) and when the puncture has been made with the point of the lancet, and while the latter is still in the puncture, the finger is pressed upon guide *c* of the piston, and it is slipped with the charger outward until one of the guides *b* of the charger comes in contact with the end *g* of slot B, when the guide *b* at the end of the charger will have run off its ways into space *h*, where the ways are discontinued; at which instant the guide *c* of the piston will act upon the bevelled lip *d* of the charger, and throw it out of line with the piston, (see fig. 4,) and the end of the piston will present at the end of the charging-tube at *i*, and the matter will be forced from the charger into the puncture.

The inventor says: I do not claim the combination of a piston and charger, as such, for the purpose of depositing vaccine or other matter.

But I *claim* the construction of a lancet, in combination with a charger and piston inside its blade, substantially as described, for the purpose of depositing vaccine or other matter in a puncture made for that pur pose, before the lancet is drawn.

No. 13,360.—Wm. H. Rhodes.—*Improvement in Artificial Legs.*—Patented July 31, 1855. (Plates, p. 332.)

The standard *f f i* is held firm by means of the hook at the end of brace *g* taking into the notch in the roller *h;* the spiral spring within roller *h* serves to crowd the hook to its notch. The standard and brace are turned back out of the way (into position fig. 2) by knocking the lower end of the brace outwardly, (with foot or cane,) which frees it from its notch and lets it slide in the roller and snugly turn back together. The leg-piece is fastened at the knee by a capsular band and buckles, thus avoiding all binding or unpleasant sensation to foot and leg.

I *claim*, 1st. The knee-joint, as described in specification and drawing, and ankle-joint as set forth.

2d. I also claim the standard *f f* and brace *g*, with their nge-joint connection to foot-plate; coiled spring, with roller to hold the same, which retains the brace and standard in position when walking, as set forth. These principles and improvements united forming the within apparatus, which is of great utility to the afflicted.

No. 13,404.—Addison Spaulding.—*Improvement in the Construction of Artificial Legs.*—Patented August 7, 1855. (Plates, p. 332.)

The nature of this invention will be understood from the claims and engravings.

The inventor says: I disclaim the knee-joints, as patented in France by Ferdinand Leopold John, November 11, 1835, wherein the central pins withstand all the wear and shock of the leg when in use. I also disclaim any part, device, or thing embraced in the patent granted to Jonathan Russell, August 17, 1852. I also disclaim the application and use of India-rubber as applied to move the leg, as in the patent granted to John L. Drake, August 31, 1852. I also and finally disclaim the surface of deer-skin stuffed with hair, and attached to the bottom of the foot, described with the invention patented by Frank Palmer, August 17, 1852, as such will not retain any elasticity when used, but will cake together as hard as the wood of which the leg is composed.

I *claim*, 1st. The knee-spring F, or its mechanical equivalent, for throwing forward the portion of the leg marked A at each step of the artificial leg, essentially in the manner and for the purposes set forth.

2d. I claim the ankle-spring K, or its mechanical equivalent, for swinging up the forward portion of the foot on the axis or pin M, or other turning-point, at each step of the operator, essentially in the manner and for the purposes set forth.

3d. I claim the chain or rod G connected and combined with the India-rubber J, or their mechanical equivalent, which is secured in the heel of the foot, to allow the leg A a slight elasticity when placed upon the ground and tipped forward by the operator, to prevent the shock upon the cords and nerves in the stump of the natural leg, essentially in the manner and for the purposes set forth.

No. 13,611.—Jno. Taggart, assignor to Himself and Theo. D Parker.—*Improvement in Artificial Legs.*—Patented September 25, 1855. (Plates, p. 332.)

By attaching the instep part of the foot firmly to the leg without any joint between them, and providing the foot with a joint at the instep, the straightening of the leg is facilitated, and the leg is caused during its further movement to turn upon the lower part of the front end of said instep part, so as to lift the heel off the ground. As soon as the weight of the body is taken from off the artificial leg, the action of the springs G and L is such as to bend the knee and throw the heel still higher above the ground, and so far clear of it as to admit of the leg being readily moved forward for taking another step.

Claim.—Making the leg and foot without any ankle-joint, as specified. Also, combining together and with the foot the part A, and the thigh-case B, the two springs G and L, so as to operate therewith, substantially and for the purpose as specified.

No. 13,318.—Jonas Moore and D. P. Adams.—*Improvement in Apparatus for Administering Pulverulent Medicines.*—Patented July 24, 1855. (Plates, p. 332.)

The nitrate of silver is ground (by means of a wheel H, which has its circumference coated with emery) to fine dust, to be inhaled by the patient.

Claim.—The combination of the machinery described for turning the emery-wheel with a discharge-pipe L, for the purpose of administering pulverulent substances, in cases of inflammation of the mouth and throat.

No. 13,453.—ALMOND C. BUFFUM.—*Improvement in Obstetrical Extractor.*—Patented August 21, 1855. (Plates, p. 332.)

To prepare the instrument for application, the cross-bands C and E are to be doubled at *h h*, and the folds drawn through the fenestræ *j j*, entering on the concave side of the fingers until the fingers are brought together, when the slide F should be placed upon the handles A of the fingers, taking care to keep the down-straps on the inner or concave side of the fingers, and the cross-bands drawn through and folded on the back or convex side of the fingers (see fig. 2).

Claim.—An obstetrical extractor which, from the peculiar form of its fingers, and by means of three cross-bands interlaced by down-straps, so clasps and supports the head of the child as that the force necessary to assist its delivery can be applied without injury to mother or child.

Also, that by means of the fenestræ, the instrument can be ready for application, so small and of such shape that it can be applied more readily and with less risk and pain to the patient than any forceps or other extractor in use.

No. 12,960.—ERASMUS A. POND.—*Improvement in Pill-Making Machine.* Patented May 29, 1855. (Plates, p. 332.)

In the faces of the cylinders *a a*, which are properly geared so as to revolve against each other, are a double row of hollow excavations equal to the size of half of one pill. The pill mass being fed in between the two cylinders, pills will be formed where the excavated faces of the two cylinders meet. The bent wires *w w* fit into holes extending from each excavation towards the centre of the cylinder. These wires resting on the stationary eccentric *b*, are thrust outwards as the cylinder revolves around the said eccentric, thereby expelling the finished pills. The ring *d* confines the bottom part of the wires to the face of the eccentric.

Claim.—The hollow working cylinder, with a stationary cylinder inside and eccentric to it, the pits on the outside cylinders being perforated, and the perforations supplied with wires bent as described, with a ring around the loops, so that the wires are thrust out and drawn in, as described, as the working cylinders revolve.

No. 13,975.—JOSEPH BUHLER.—*Improvement in the Combination of Injecting Syringes.*—Patented December 25, 1855. (Plates, p. 333.)

To prepare this instrument for use the cock *a* is opened, and cock *g* turned to close all communication from the receiver, either to the force-pump or to the atmosphere, as shown in the drawing, and the piston *c* requires to be pushed upward as far as possible. The glass receiver A is thus inserted into the vagina, and the piston *c* of the exhausting-pump is drawn out by hand to exhaust the air from the re-receiver, which causes the neck of the uterus and the surrounding parts to be drawn up so far into the receiver as to be visible to the eye of the medical attendant through those parts of the sides of the receiver which are outside the body. Medical application can then be made by drawing the preparation through suction-pipe *h* into the pump, (by drawing out piston *d*). After this, cock *g* is to be turned so as to open communication from the pump to the receiver and to close passage *f*; and then the injection can be made by forcing in piston *d*.

The inventor says: I do not claim any of the parts of this apparatus.

But I *claim* the combination of the receiver A and pumps C and D, provided with cocks *a* and *g*, in the manner and for the purpose set forth.

No. 12,156.—Sharpless Clayton.—*Improvement in Teeth.*—Patented January 2, 1855. (Plates, p. 333.)

Claim.—The dovetail grooves formed in the base of the teeth and the holes through the teeth; said groove and holes to be filled with pure tin, cadmium, or any other fusible metal, so as to form a dovetail flange and pins, as described, for the purpose of securing the teeth and gums to metallic plates, as herein set forth and described. (See illustration.)

No. 13,801.—Barclay Arney Satterthwait.—*Improvement in Preparing Artificial Teeth.*—Patented November 13, 1855. (Plates, p. 333.)

The nature of this improvement will be understood from the claim.

The inventor says: I do not claim the compounding the material or enamel, or modelling around teeth, to be connected to metallic plates in the form of block-work or continuous gum; neither do I claim the carving upper or lower sets out of one piece of porcelain material.

But I *claim* the described improvement in making whole or half sets of artificial teeth, gums, and plates entirely of porcelain by modelling porcelain material around porcelain teeth, or mounting porcelain teeth on porcelain plates, substantially as set forth.

No. 12,266.—Wm. M. Bonwill.—*Improvement in Hernial Trusses.*—Patented January 23, 1855. (Plates, p. 333.)

A A is the part of the hoop which passes around the waist of the patient; it bends down at B in front to the groins opposite to the ruptured parts or internal abdominal rings at C C. The strap I, with the pad K, forms the truss for umbilical hernia. The strap L passes horizontally across the groins of the patient at the pads M M to prevent the truss from opening too much in case of a violent strain.

The inventor says: I do not claim the hinges F F, the adjustability of the pads, or the form of the hoop, separately.

But I *claim* the combination of the peculiarly-formed hoop with the umbilical pad and strap, for the purpose of preventing the movements of the body from displacing the pad in either umbilical or inguinal hernia, as set forth.

No. 12,986.—LUCIEN E. HICKS, assignor to HIMSELF and HIRAM L. HALL.—*Improvement in Pads for Hernial Trusses.*—Patented May 29, 1855. (Plates, p. 333.)

Claim.—1st. An India-rubber pneumatic truss-pad without opening or valve, and filled with compressed air, as set forth. (See engravings.)

2d. Making the top of the pad thicker than the bottom or cushion, whereby the former is rendered sufficiently rigid to allow of the attachment of the shank while the cushion maintains its entire elasticity, as set forth. (See engravings.)

3d. Sinking the button beneath the surface of the pad within the cavity, and securing it therein by the flange *a;* by which means the button is surrounded by the annular cushion, and is prevented from coming in contact with the person, in the manner substantially as set forth. (See engravings.)

No. 13,430.—E. B. GRAYHAM.—*Improvement in Hernial Trusses.*—Patented August 14, 1855. (Plates, p. 333.)

The rack and catch prevent the shifting of the pad towards as well as from the body; and by means of the set-screw passing through slot *f*, an adjustment of the pad is admitted of between the adjustments allowed by the movement of the spring-catch from one tooth to the other of the ratchet.

The inventor says: I do not claim of itself the invention of a square-toothed rack and square-pointed spring-catch.

I *claim* the described means of adjusting and securing the pads in any desirable position, consisting of the square-toothed rack *c*, the adjustable square-toothed and slotted spring-catch *d*, and screw *e*, all combined and applied substantially as set forth.

No. 13,548.—FRANCIS GRACE MITCHELL, M. D.—*Improvement in Hernial Trusses.*—Patented September 11, 1855. (Plates, p. 333.)

The lever *b* presses on steel spring *d* into groove *e*. An elastic strap fastens on stud *f*, then passes under the thigh, and is buttoned on the supporter on the costa of the ilium; this produces an inward action, and increases or diminishes the pressure of the pad.

I do not claim the form or application of the pad; but I *claim* the mechanical arrangement on the back of the pad, on the metal plate, which consists of a lever, which presses a steel spring into a longitudinal groove formed in the centre of the metal plate, on the back of the pad.

No. 12,965.—HENRY A. ROSENTHAL.—*Improvement in Uterine Supporters.*—Patented May 29, 1855. (Plates, p. 333.)

The instrument, with the leaves *a a* closed, (see figure 2,) is to be inserted into the vagina. When inserted as far as the shank will allow, the leaves are to be extended (by means of drawing together the two shank-rods c c^1) into position figure 1, and as they are extended should be gradually pushed inward. When fully extended, (the two shank-rods being close together and confined in this position by screwing up the nut d,) the leaves will find a suitable resting place, and the shank will then be within the vagina, with but little more than the nut d protruding beyond the pubes. By this means the uterus will be efficiently supported, enabling the female to take all reasonable exercise.

Claim.—The uterine-supporter, composed of the two supporting-leaves a a, and the double shank c c^1, combined to be applied substantially as set forth.

XXI.—WEARING APPAREL.

No. 13,002.—Franklin J. French.—*Improvement in Boot-Jacks.*—Patented June 5, 1855. (Plates, p. 334.)

The nature of this improvement will be understood from the engravings.

Claim.—The spring-wedge D, in combination with the lever B and spring E, or their equivalents, and arranged substantially in the manner and for the purpose set forth.

No. 12,725.—Samuel H. Hopkins, assignor to W. C. Greene, &c. —*Improved Stud and Button Fastening.*—Patented April 17, 1855. (Plates, p. 334.)

The button A being on the outside of the garment, and the tubular shank B being passed through eyelet-hole a^1 of the garment, the shank c can be inserted on the inner side of the garment, as apparent from the engraving, thereby effectually securing the button to the garment.

Claim.—The construction of the fastening, as herein shown and described, viz: Having the shanks of the stud or button formed of a tube B, which contains a spiral-spring b, and having a bar c fitted in slots d, in the outer end of said tube, and between the outer end of the spiral-spring b and a pin a attached to the outer end of the tube, the outer side of the bar c being provided with a recess e, which, by means of the spring b, is kept over the pin a, and the bar c consequently secured in a transverse position with the tube B.

No. 13,926.—Amasa S. Thompson.—*Improvement in Cutting Cloaks.*—Patented December 11, 1855. (Plates, p. 334.)

The nature of this improvement will be understood from the claim and engravings.

I *claim* cutting a cloak from seamless cloth, without sleeves, but so that, by making four cuts of the proper length for the sleeves, the cloak may be worn as a sleeved sack or overcoat by merely changing the buttonings, substantially in the manner described.

No. 12,379.—John Dick.—*Improvement in Stays for Articles of Dress.*—Patented February 13, 1855. (Plates, p. 334.)

The India-rubber webbing or spring *b* is attached at one end to the upper end of the whalebone *a*, and at its other end to the lower end of the whalebone a^1, both the whalebones being placed within the part of the dress *c c* which is to be kept without wrinkles. As this stay is so placed within the garment that the India-rubber spring is a little compressed, its constant tendency to expand will keep the part of the garment stretched to which it is attached. Figure 1 represents the garment (designated by dotted lines) in a wrinkled, and figure 2 in a stretched state.

Claim.—The described improvement in stays, as applied to articles of wearing apparel, consisting of two or more supporting-pieces, with a spring or springs applied to extend them, substantially as set forth.

No. 13,011.—Daniel Minthorn.—*Improved Brace for Supporting Garments.*—Patented June 5, 1855. (Plates, p. 234.)

The straps F serve to hold up the stockings without elastics, in case of cold feet or a predisposition to varicose veins.

The inventor says: I do not claim to have invented shoulder-braces or suspenders, for they have long been known and used.

But I *claim*, substantially as described, the arrangement of straps forming a suspender for the use of ladies or gentlemen, by which the hips are relieved of the weight of the lower garments, which are sustained by the shoulder-straps, at the same time leaving the region of the lower ribs and viscera perfectly free in their action.

I claim, also, in combination with the above suspender, the short corset by which the weight of the skirts is employed to raise and adjust the breasts, and by tightening or loosening which, the said weight is more or less transferred from the shoulders to the breast.

I also claim the straps F, attached to the boot or stockings, substantially as and for the purposes set forth and described.

No. 12,173.—William F. Warburton.—*Improvement in Hats.*—Patented January 2, 1855. (Plates, p. 235.)

The inventor says: I do not mean to confine myself to the precise form of corrugations described and represented, as they may doubtless be varied in shape, and yet accomplish the end desired; but I do *claim* forming the rims of hats with corrugations, channels, or grooves, or other ridges of the form described, or other form substantially, thereto, for imparting strength, softness, and elasticity to the rim, with less weight of felt or other material, and a decreased quantity of stiffening substance of which it is composed, than is ordinarily employed in forming rims; the said rims being previously, or at the same time such corrugations, grooves, or channels are formed, slightly raised or arched at the front and back parts immediately next the body or crown of the hat, and depressed or slightly curved downward at the sides or not, as fancy or taste may dictate, to give them the proper brace or set, as herein set forth. (See engraving.)

No. 13,361.—Wm. Sellers.—*Improvement in Ventilating Hats.*—Patented July 31, 1855. (Plates, p. 335.)

The nature of this improvement will be understood from the claims and engravings. Figure 3 represents a section of the hat when closed; figure 2, a view of it when open for ventilation; which position of the parts of the hat is also represented in figure 3 in dotted lines.

I *claim*, 1st. Making the hat or other similar head covering to open at its side or sides, by dividing the body of the hat, and connecting or arranging the separated portions or sections of the body so that the one portion of the body may be adjusted to form an open or close connexion with the other portion of the body, substantially as and in the manner specified.

2d. Providing the divided body at the junction of the two sections with a gimp guiding-strip, or reticulated telescopic lining or casing D, arranged for operation in connexion with the movable section of the body, essentially as and for the purposes set forth, and whereby an ornamental and unbroken appearance is given to the hat all round when the body of it is open for ventilation, as described.

No. 12,313.—Hezekiah Griswold.—*Improvement in the Yoke of Shirts*—Patented January 30, 1855. (Plates, p. 335.)

This yoke permits a free motion of the neck, and obviates the necessity of cutting the top of the bosom in a curved line.

The inventor says: I do not claim the insertion of gores upon the shoulders of shirts or other garments, that being old.

I *claim*, in shirts, the compound yoke, substantially as and for the purpose set forth.

No. 12,899.—Rufus K. Chandler.—*Improvement in Wristbands of Shirts.*—Patented May 22, 1855. (Plates, p. 335.)

In wearing these wristbands, both flaps should be turned forward at first (see figures 1 and 2). When the outer flaps a are soiled, they can be turned back (as shown in figures 3 and 4) for the purpose of exposing the clean surface of the under flap b.

Claim.—Making wristbands with double flaps a b, substantially in the manner and for the purpose herein set forth.

No. 13,212.—S. N. Campbell.—*Improved Sun-shade.*—Patented July 10, 1855. (Plates, p. 335.)

This sun-shade is attached to a head-band, or a cap, to be worn by a person.

Claim.—Attaching a frame formed by the rods *a* and *e* to a head-band *f*, or to a cap; said frame being covered with any proper material, for the purpose set forth.

No. 13,850.—Joseph Kleemann.—*Improvement in the Preparation of Umbrella-Sticks, &c., of Ratan.*—Patented November 27, 1855.

The ratan sticks are soaked in logwood extract for about four days, and then immersed in a solution of any of the iron salts, which gives the ratan a deep black dye. Afterwards the ratan is exposed, in a close vessel, to steam of three or four atmospheres pressure for about one hour, and then dried. The sticks are then soaked in linseed, and again dried at a temperature of 110 to 125° Fahr., till the oil has become hard. The sticks are then placed into an iron cylinder (capable of standing the pressure of at least ten atmospheres), connected by a pipe with an open vessel, containing a varnish made by dissolving 120 parts of shellac and 200 parts of burgundy pitch in 90 parts of absolute alcohol. The air having been exhausted from the cylinder, the cock connecting it with the vessel containing the varnish is opened, when the atmospheric pressure will force the varnish into the cylinder and into the pores of the ratan. To make the impregnation more perfect, varnish may be pumped into the cylinder by a force-pump. The ratan, when taken out, has to be dried.

The inventor says: I disclaim impregnating woods in general; but I *claim* the preparation of ratan by impregnating it with drying-oils and varnishes, substantially as described, for the purpose of giving it flexibility, elasticity, tension, and an appearance similar to whalebone; the ratan so prepared to be used as a substitute for whalebone in the manufacture of umbrella and parasol frames, and for other purposes for which whalebone can be employed.

No. 12,903.—Wright Duryea.—*Improvement in Umbrellas.*—Patented May 22, 1855. (Plates, p. 235.)

The umbrella being closed, the spring *o* can be pressed in, and the slide M be pushed up, surrounding the umbrella with the hand at the same time to prevent the stretchers L from spreading. The outer parts of the ribs H will then slide along by the sides of the parts G, thereby reducing the length of the ribs nearly one-half. The length of the staff can be reduced correspondingly by unscrewing it at C.

Claim.—The improvement in umbrellas with ribs made in two parts, so constructed that one part slides on, or traverses in or beside, the other part, in connecting the stretchers to the outer or traversing portions or parts of the ribs, so as to hold the said portions out in their proper position when the umbrella is spread, substantially as described; and also, that those portions of the ribs to which the stretchers are connected may be traversed by moving the stretchers and slide, as described, when the umbrella is closed.

XXII.—MISCELLANEOUS.

No. 12,433.—John Bale.—*Improvement in Hotel Annunciators.*—Patented February 27, 1855. (Plates, p. 336.)

When one of the wires, leading from the rooms, and connected to the lifters G, is drawn, the hook on the lifter elevates the rear end of frame L; the stop M passes between stop P and arm O, crowding the latter back so as to allow stop M to pass over stop P, when it is forced forward by arm O, far enough to disengage the bar *i* from lifter G, and at the same time carry stop M forward of stop P, and the frame L is allowed to drop back to its original position, allowing the hammer to strike the gong S.

(The weighted end of frame L is hung on straps *k*, so that the frame has a tendency to move backward, and to bring stop M against arm O.)

The raising of the lifter G draws the catch E out of the notch in bar A, which is then thrown out by the action of spring H, and exposes to view the number on C.

Claim.—1st. Arranging the number-plate C upon the sliders A, or their equivalents, in combination with the screen-plate D, so that the number-plate shall be pushed forward, lifting the screen-plate, and thus exposing the number to view.

2d. The combination of the frame L, its hangings, and the stop M, with the stop P, arm O, and lifters G, or their equivalents, by which the wires are made to act independently of each other in striking the gong, unless the hammer shall, at the instant, be in active operation.

No. 13,600.—Jacob Nelson.—*Improved Awning for Horse and Dray.*—Patented September 25, 1855. (Plates, p. 336.)

This awning is designed as a shelter from sun and rain; the awning, when not required to protect the load, can be thrown (from the position shown in the engraving) over, so that the poles *f* of the awning will be supported by the brackets *h*, when it will serve to protect the team.

Claim.—The portable and reversible dray and horse canopy, whose poles *f* are hinged at one end to posts *b*, and supported in either the forward or backward position by branches or brackets *g h* projecting from the posts.

No. 12,222.—Fred. Tesh.—*Improved Beef-Spreader.*—Patented January 9, 1855. (Plates, p. 336.)

The operation of this instrument is apparent from the engraving.

Claim.—The construction of a spreader for beef, of a stick *a* and tongue *c*, operated by a cog-wheel and ratchet-work, substantially in he manner hereinbefore described.

No. 13,089.—Geo. W. Hildreth.—*Improved Mode of Hanging Bells.*—Patented June 19, 1855. (Plates, p. 336.)

To secure the swing of the tongue always at right-angles with the yoke, the cap *d* is dowelled to the yoke so that it cannot turn with the nut *e*, and the nut *c* is prevented from turning by a projection from the sides of the bolt sliding in grooves in the cap *d*. (See fig. 3.)

Claim.—The round tapering shank *b* and corresponding hole in the yoke *a*, in combination with the bolt *c* and cap *d*, to secure the bell firm into the yoke.

2d. The dowelling of the cap *d* to the yoke *a*; also, the manner of securing the bolt *c* from turning in the cap *d*, for the purpose set forth.

No. 12,604.—George W. Palmer.—*Improved Bill-holder.*—Patented March 27, 1855. (Plates, p. 336.)

N L is the box, D is the arm hinged at E, and is constantly pressed against the back N of the box by means of a flat spring H, which is bent at nearly a right-angle.

Claim.—An oblong box, of suitable size, for holding files of bills or papers, having upon one of its sides a hinged movable arm and attached spring, by which the papers are held in place, as fully described

---

No. 12,202.—James Hanley.—*Improvement in Devices for Stoppers of Bottles.*—Patented January 9, 1855. (Plates, p. 336.)

The inventor places the bridge B obliquely, as he considers this to be a more convenient arrangement than to have the cork exactly at right angles to the neck of the bottle.

Claim.—Making bottles so that the resistance of their contents shall bear laterally upon the cork or stopple; also the oblique position of the bridge B, for the purposes as above set forth, in the manner stated, or by its equivalent.

No. 12,501.—Theophilus A. Ashburner.—*Improvement in Bottle-Stopper Fastenings.*—Patented March 13, 1855. (Plates, p. 336.)

b b are the two stirrups, and *c* is a tin band, with which the cork may be permanently fixed to the bottle.

The inventor says: I am aware that many devices have been essayed for securing corks in bottles, but all of them involve expense, intricacy, or difficulties in placing or removing them from the bottle; and I do not claim any such contrivances, meaning to limit myself to what I have described and represented, relying mainly upon the hinging of the stirrups to the button, which greatly facilitates the placing or removing of the button from the cork, and securing the stirrups on one side to the neck of the bottle.

I claim the device described for securing corks in bottles, viz: a button provided with hinged stirrups for catching under the projection of the bottle, for the purpose of more readily placing it on or removing it from the cork; and this I claim, whether said device is a fixture on the bottle or separate therefrom, as described.

No. 13,266.—JULES JEANNOTAT.—*Improvement in Bottle-Fastenings.*—Patented July 17, 1855. (Plates, p. 336.)

The nature of this invention will be understood from the claim and engravings.

The dotted lines represent the position of parts when the cushion is removed from the mouth of the bottle.

Claim.—Forcing or pressing a cushion H of India-rubber, or other suitable material, over or upon the mouth of the bottle A, by means of a lever E inserted in a plate D, which plate D has flanches F F attached to it by rods or links C; the plate D also having attached to it a plate G to which the cushion H is secured; the above parts being arranged and applied to the bottle as shown, for the purpose set forth.

No. 13,338.—JOHN ALLENDER.—*Improvement in Bottle-Fastenings.*—Patented July 24, 1855. (Plates, p. 337.)

A strap of sheet-metal E is so arranged as to swing across the cork C after it is inserted into the mouth B of the bottle. E is connected to another strap F, which partially surrounds the neck A of the bottle, so as to form a pair of hinges by means of the wire G, which fastens both straps to the neck of the bottle.

Claim.—The strap E, constructed with hinge and hook for holding in and releasing the corks or stopples of bottles, jugs, &c., substantially as set forth.

No. 13,402.—AMASA STONE.—*Improvement in Forming Screw-Threads, &c., in the Necks of Glass Bottles and Similar Articles.*—Patented August 7, 1855. (Plates, p. 337.)

Figures 1 and 2 represent the tool used for forming the orifice of a bottle; fig. 3 represents a section of the orifice formed by said tool.

Claim.—In the construction of tools for forming screw-thread, angular, or other scores in the necks and orifices of glass, earthen, or other bottles, and other articles, making the plug which forms the interior of the orifice to turn with the bottle, jug, or other article, while the material of the orifice is worked around it, substantially as described.

No. 13,659.—JOHN SMYLIE.—*Improvement in Register Bottle-Fastenings.*—Patented October 9, 1855. (Plates, p. 337.)

The bottle as shown in the engraving (which latter represents only the neck N of the bottle) is supposed to be standing upright, the ball E resting on the lower collar *d* at the bottom of spindle F, and inclined projection L on arm H bearing with its upper end against the projection of lever J. On the bottle being partially inverted in order to pour out a portion of its contents, the ball E slides along the spindle, strikes the upper collar *d*, moves the spindle, and with it the arm H. The projection L on the latter moves the lever J on account of its inclination, and thereby moves spring-catch K and the dial to the extent of one tooth. The amount of contents required being now obtained from the bottle, it is restored to its upright position, and in doing this the ball slips down the spindle F, and, striking the lower collar *d*, restores the spindle and its arm H to its former position; at the same time, the inclined projection L being likewise depressed, the lever J and catch K are by means of a spring caused to move backwards the distance of a tooth. Thus, every time a portion of the contents is poured out, the dial is moved so as to register the number of times the bottle is used for pouring the contents therefrom. D is a casing which is hinged at *e*, and can be provided with lock and key.

Claim.—The spindle F, with its sliding-ball E, in combination with the arm H, projection L, lever J, spring-catch K, and dial G, or their equivalents, arranged and constructed substantially in the manner and for the purposes specified.

No. 13,782.—Jos. C. Day.—*Improvement in Ring and Gudgeons for Bottle-Fastenings.*—Patented November 13, 1855. (Plates, p. 337.)

B represents the neck of the bottle; *s*, the stopper.

Claim.—1st. The construction of the hoop *b* with two gudgeons thereon, one of which is made open or divided in halves, and which when in actual use are brought together, thereby forming one gudgeon, which, with that on the opposite side, are the recipients of the gudgeon-boxes represented by the letters c^2 c^2, fig. 4.

2d. I claim securing the two half-gudgeons c^1 c^1 together, and consequently the collar to the bottle, by means of one of the gudgeon-boxes of the bail.

No. 12,335.—Elisha Waters.—*Improvement in Cylindrical Boxes.*—Patented January 30, 1855. (Plates, p. 337.)

The paper sides of a box cost less than if made of wood, because the paper can be formed into cylinders on long mandrels, and be cut into sections of the proper length by machinery.

The inventor says: I do not claim in general the combination of wood and paper in the manufacture of all descriptions of boxes; but I *claim*, in the manufacture of cylindrical boxes, making the sides of said boxes of paper tubes, and the ends of wooden discs, substantially as and for the purpose set forth and described, whereby I am enabled to produce at once a better and a cheaper box, by making each part of the most suitable material and in the cheapest manner.

No. 12,511.—LOUIS KOCH.—*Machine for Making Paper Boxes.*—Patented March 13, 1855. (Plates, p. 337.)

The inventor says, in his specification: In the accompanying drawings the machine is represented in a position supposed that a box has just been completed. The cam 19 has already moved a little the lever P^1, and consequently the rod L^1; and as this motion has been just sufficient for the projection x on the latter to have acted on the catch L, so as to bring the same clear of the valve I, said valve is now only kept closed by its weight a^1, and can be pressed open by the mould h to allow the same, by the motion of the wheel H, to move out of the outer mould. By turning now at the handle C, setting thereby the machine in motion, the cam 19 will still act upon the lever P^1, and consequently on the rod L^1, which comes now in contact with one of the pins h^1 of the wheel H, and turns the same partly around, thereby bringing the inner mould h (where we suppose a finished box attached) clear of the outer mould, formed by the projections k k^1 and the valve I. In this operation the valve I has been pressed down to allow the mould h to pass, and is then closed again by its weight a^1, and the catch L, which has likewise been released from the projection x of the rod L^1, is pressed by its spring K so as to lock the valve I firmly to its place. The cam 19 moves the lever P^1, and consequently the rod L^1, so far, that by the action of the latter against the pin h^1 of the wheel H, this wheel H has turned so far around that its arms stand at an angle of about 45°. During this time the cam 18 has acted continually on the lever Z, keeping the same, and consequently the pincers P, steady, and the latter in the lower position.

As soon as the arms of the wheel H arrive at the above described position, namely, at an angle of about 45°, the cam 18 relieves the lever Z from its action, and the same is now acted upon by the spring z, thereby forcing the pincers P suddenly upwards. The pincers are at this top position pressed apart by the action of their spring K^1, as above described, so as to allow the arms or moulds h of the wheel H to pass within them.

At the same time the above mentioned operation has been going on, the toes 3, 4, 5, 6, and 7, have acted alternately upon the levers T and T^1 respectively, moving the same backwards and forwards, and which motion has been communicated through the rods m m^1 to the cranks f f^1 of one of the lower rollers D, producing thereby just one whole revolution of the same; and as all the lower rollers are connected together, and the circumference of each roller is just equal to the length of paper required to make one box, such a length of paper has by this operation been brought forward. As will appear by the following, there is on the table of the machine, formed by the plates H^1, and in the position the machine is now in, one required length of paper lying close to the forward end of the machine, ready cut and pasted for folding, close to and behind the same one piece of paper already cut or stamped out, only requiring to be pasted, and behind and close to this last, again, the paper drawn into the machine by the rollers D D^1 from an endless roll of paper R P, situated behind the machine.

By the operation of the rollers, the first piece of paper, ready cut and pasted, is now brought forward and pushed over the opening of the outer mould, resting upon the top slides M and M^1, ready to be folded, while the second piece of paper is brought forward and into the place which was occupied before by the first piece, to be pasted by the next operation, and a further piece from the endless roll of paper is pulled into the machine and into the place occupied before by the second piece, to be cut off by the stamp the required length and shape. The first projection of the cam 19 has brought the arms of the wheel H in the position above mentioned, and kept the same stationary in that position until the motion of the rollers, and consequently of the paper, is completed. The second projection of the cam 19 acts now upon the lever P^1; and consequently on the rod L^1, producing a further motion of the wheel H, sufficient to bring the arm or inner moulds h into the outer mould, thereby folding the paper which was moved over the opening of said outer mould around itself, and taken it with it into the same, leaving only those parts of the paper projecting which are to form the top and end of the box. This cam 19 keeps, then, the lever P^1, and consequently the rod L^1 and wheel H, for some little time stationary in that position, until one of the top-slides has been moved over the mould h, by which the same is secured, when the cam 19 leaves the lever P^1, and the spring q^1 at the end of the rod L^1 pulls said rod and lever back again to its original position, ready for the next action.

After the mould h has been moved into the outer mould forward by the projections k k^1 and the valve I, the toe 14 acts upon the lever X, which lever, by its connexion, through the bell-crank y, with the top-slide M, moves the latter over the mould h, folding thereby the projecting paper over the top of the same.

The toe 15 acts now upon the opposite side of the lever X, and brings, therefore, the top-slide M away again; but at the same time the toe 16 acts upon the lever X^1, which, by its connexion through the bell-crank y^1 with the top-slide M^1, moves the same over the mould h, folding, thereby, the projecting paper over the top of the same, and over that piece of paper which was folded by the slide M; and as this last folded piece of paper was pasted beforehand by this operation of folding the second over the first, the same will likewise be pasted together. The toe 12 acts now upon the lever V^1, which, by its connexion through the bell-crank W^1, with the side-slide N^1, moves the latter over the end of the mould h, folding thereby the projecting piece of paper over the same. The toe 13 acts then upon the opposite side of the lever V^1, and brings thereby the side-slide N^1 away again. When the side-slide N^1 is about half-way brought back again, the toes 9 act upon the lever U, which, by its connexion with the lower slide O, brings the same upward and over the end of the mould h, and at the same time the toe 1 acts upon the levers S and S^1, which, by their connexions through the levers R and R^1 and the rods n and n^1 respectively with the beams F and F^1, and the connexion of the latter with the upper slide O^1, move this upper slide downward and over the end of the mould h. By those two last operations, the top and bottom projecting pieces of paper have been folded over the end of the mould *h*, and over the piece of paper which was folded before by the side-slide N^1. The toes 2 act then upon the opposite sides of the levers S and S^1, and the toe 8 upon the opposite side of the lever U, bringing thereby the upper and lower slides O^1 and O back again. During the return motion of those slides, the toe 10 acts upon the lever V, which, by its connexion through the bell-crank W with the side-slide N, moves said side-slide over the end of the mould *h*, thereby folding the paper projecting on that side over the end of the mould, and likewise over the three other pieces which have been folded before over the end of the mould *h;* and as this last folded piece was pasted beforehand, the same has, by the last mentioned operation of folding, been pasted to the other pieces of paper, which were folded first. The toe 11 acts then upon the opposite side of the lever V, bringing thereby the side N back again, and the toe 17 acts upon the lever X^1, whereby the lap-slide M^1 is brought back, leaving the box finished, folded, and pasted together on the mould *h*. During the time that this operation of folding the paper around the mould *h* has been going on, the cam 18 has acted on the lever Z; and through the connexion of the latter with the pincers P, the same have been pulled downward by the action of this cam on said lever Z. In this downward motion of the pincers P, the projections *p* p^1 on the pincers P have passed over the elevations on the slides Q Q^1, and have pressed thereby the pincers P together, so that the same take hold of the finished paper box, which we have supposed to be on the lower mould *h*, and pull, therefore, said box off this mould *h*. As soon as the pincers have been moved sufficiently far downward to have pulled off the paper box clear of the mould, the projections *p* p^1 leave the elevations on the slides Q Q^1, and the pincers are then forced apart by the action of their spring K^1, thereby allowing the box to fall out from between said pincers. During the downward motion, the beams F F^1, the knife or stamp-frame E, and the pasting-frame G, have been pressed down and upon the paper situated upon the table H^1, and between the rollers D and D^1, and held stationary at that moment by said rollers. By this action, the stamps or knives attached to the under side of the frame E cut off a piece of paper the required shape necessary for one box, while the boxes, with paste, situated within the pasting-frame G, have (by being pressed upon the paper) pasted that piece of paper in the required places, which lies underneath the same, and which was cut out by the former action of the stamp-frame E, and has been pushed into that place by the former action of the rollers D D^1. The cam 19 begins now again to act upon the lever P, and the above described actions will be repeated.

Claim.—1st. The application of a series of rollers connected together, and worked by an arrangement of levers and toes, or cams, for the purpose of bringing paper from an endless roll, and of a required length, into the machine and pieces of paper previously shaped and pasted by the machine, to the place required, substantially as described.

2d. The application of a stamp-frame, with suitable knives or stamps attached, situated between the rollers, for the purpose of cutting off the paper the required size and shape, from the endless roll, necessary for one box.

3d. The application and construction of the pasting-frame, with paste-boxes situated between the rollers, and arranged in such a manner as to paste the already shaped paper in the required places, as set forth.

4th. The construction and application of a wheel with arms, having at their extremity the moulds attached, around which the boxes are to be made; said wheel, with mould, being moved by an arrangement of a rod and lever, actuated by a cam, in the manner described.

5th. The application and use of a series of slides for the purpose of folding the ends of the papers around the mould, said slides being worked by a combination of levers, &c., actuated by toes, in the manner set forth.

6th. The application and use of a pair of pincers, for the purpose of pulling the finished paper-box off the mould, constructed and worked in the manner set forth.

7th. The construction of the outer mould, formed by two projections attached to the frames, and a hinge-valve, and the operation and manner of working said valve; the various parts of the whole machine being combined and arranged for the purpose described.

No. 12,177.—Daniel Wells.—*Burglars' Alarm.*—Patented January 2, 1855. (Plates, p. 338.)

B is the door (end view); A the door-jamb. When the door is closed against the door-jamb, the projection *e* on arm *d* (which is pivoted to the door-plate D) will strike the upper part of latch *a*, and will be forced along the circular slot *h*, and against the spring *f*, until i has arrived opposite the recess *x* in latch *a*, when it will catch into aid recess, there being a small angular groove in the top of said recess, for the reception of the corner of projection *e*. The arm *d* is then in position d^1. When, after that, it is attempted to open the door, (the arm *d* being continually pressed towards a vertical position by means of spring *f*,) the projection *e* in ascending the slot *h* will raise the latch *a* (into position figure 2), thereby pulling rod K, turning lever *n* around its fulcrum, so that the notch on its shorter arm will clear the square head of rod O, which rod (being connected with any suitable alarm apparatus) will be suddenly protruded by means of spring *p* (see dotted lines figure 1), and give the alarm. When it is desired to restore the bolt *o* to its original position, the rod *l* is pulled so that the upright arm of bell-crank *m* will press rod O back until its square head catches again into lever *n* (see dotted lines figure 2). When it is desirable to render the apparatus inoperative, the cam *c* is turned back (see dotted lines figure 1), which will depress latch *a*, so that it will be clear of projection.

The inventor says: I am well aware that the giving of alarms from the opening of doors is old and well known. I do not, therefore, wish to claim that exclusively; but I do *claim* the radial arm *d*, with its projection *e*, and the latch *a*, with its notched recess *x*, for effecting the disengagement of the spring-bolt on the movement of a door, or its equivalent, in the manner described, and for the purpose specified.

No. 12,351.—DANIEL HALDEMAN.—*Improved Burglars' Alarm.*—Patented February 6, 1855. (Plates, p. 338.)

The alarm being charged and cocked, the lever G is turned over as seen in the engraving, and its end slipped under the door T. The opening of the door presses down the lever G, which carries down the end of lever F, releases the catch *o* from the hammer, and the hammer is thrown by the spring *c* on to the cap *a*.

The inventor says: I do not claim the letting off an alarm in the act of opening a door, nor do I claim an alarm which requires fastening of any kind, either to the door or floor, to insure its going off, as several of these are already known.

But I *claim* combining with the trigger, lever, or dog, which holds the hammer at a cock, a hinged inclined lever G, the end of which simply passes underneath the door, and requires no fastening other than it receives by being held by the door itself, as it is pushed open, as described.

No. 13,157.—EPHRAIM BROWN.—*Burglars' Alarm.*—Patented July 3, 1855. (Plates, p. 338.)

The alarm apparatus consists, as usual, of a bell F, hammer H, key *a*, coiled spring *c*, key-wheel *d*, ratchet *e*, lantern-pinion *g*, and escapement-wheel *i*, which operates the bell by means of an escapement *k*. The main-spring being set free, it will sound the bell. Pin *l* projects from the escapement-wheel, and operates in connection with stud *m*, extending from discharging-lever *n*, which plays on fulcrum-pin *o*, extending through the two arms a^1 b^1 of locking-lever *q*. When *n* is drawn forward it will cause *m* to be drawn away from *l* so as to permit wheel *i* to revolve. The return movement of the lever will carry the alarm-stud into the path of rotation with the pin *l*, and when *l* strikes against the stud its further rotation will be arrested. Springs *r* draw levers *n q* towards the alarm apparatus. Bolt-rod I is jointed to spring-bolt K, which works through the side of the drawer (to which the whole apparatus is applied) and into the rail of the drawer-case. Rod I has a projection r^1, which enters a notch in arm a^1. I is pivoted to bent lever L, which has its fulcrum *t*, its shorter arm resting against rod *u* of secondary knob M (*u* sliding freely within *v* in main knob B). *v* extends through opening W, (in lever *n*,) into which the stud *y* extends from arm a^1. Rod *v* has also a projection *z*, which (actuated by spring f^1) rests against projection c^1 on lever *n*. Thus rod *v* can be turned upon rod *u* so as to move projection *z* directly in rear of stud *y*. Then (after pulling the main knob) *q* will be moved so as to unlatch I, and to enable I to withdraw bolt K; the latter movement being effected by forcing M inwards against short arm of L. Levers *n q* are drawn by their springs against stationary block N, which is provided with two series of holes g^1 and h^1, the two external series of wicar aran aginstred

a^1 b^1, while the internal rows thereof are disposed against the discharging-lever. Each hole is to receive pins d^1, resting each against one of three key-levers O P Q on a common fulcrum k^1. By these holes, pins, and key-levers, either of levers *n* or *q* may be operated, the spring-bolt being always retracted by forcing in knob M (and the motion of the key-lever forcing the pin against the lever *n* or *q*). Lever Q may be operated by a knob R and lever-fork t^1. Affixed upon shaft *h* of pinion *g* (where *h* passes through counter-plate S^1) is a tooth m^1 operating in gear *n*, that carries index-pointer o^1, and turns on arbor p^1 projecting from S^1. Thus all improper attempts to open the drawer will be registered. There are three ways of opening the drawer without sounding the alarm: 1st. Seize the main knob and rotate it until *z* is directly in rear of *y*, pull the knob from the drawer, press the knob M inward with the palm of the hand, and seize and pull forward the drawer. 2d. Pull the right-hand key-lever, press inward M, and pull the drawer outward. 3d. Pull R, the wrist of the arm being borne against M, and pull forward the drawer. The other two key-levers may be arranged as decoys, or with respect to *n*, so that when one is moved in one direction it may move the discharging lever, and also when the other is moved in the opposite direction it shall produce such a movement of the discharging-lever as to sound the alarm.

The inventor says: I *claim* arranging the locking and discharging-levers, the main and secondary knob-rods, the unbolting lever and bolt-rods, substantially as specified and represented.

Also, arranging the key-levers with respect to the locking and discharging levers, as described, and combining such key-levers with the locking and discharging levers, by one or two series of sliding-pins, or their equivalents, made to operate through holes in a block arranged with respect to the locking and discharging levers, essentially as set forth.

Also, arranging the alarm apparatus with respect to the locking and discharging levers, as specified; also arranging the counter-wheel or apparatus, and combining it with the alarm apparatus, in manner as described, not intending to claim the use of a counting apparatus or register, or connection with the alarm apparatus, such having been claimed by me in my former patent.

Also, combining with the lever Q the third knob R, by which said lever may be operated under certain circumstances, as specified.

Also, arranging the main and secondary knob-rods so that the latter may slide through the former, or the former be made to slide on the latter, under circumstances as specified.

No. 13,478.—Albert Bingham, assignor to Himself and Andrew J. Bailey.—*Burglars' Alarm.*—Patented August 21, 1855. (Plates, p. 339.)

The friction-match is inserted between the clamping-cam *d* and the projection *c* of the plate I, which is capable of sliding on plate C, and is thrown forward when the spring-pawl P is thrown out of its notch *b*. The match when thrown forward rubs against the rough face of disc R, which latter is attached to its holder S so that it can be rotated, to bring a new portion of the roughened surface into contact with the match. The face of the disc is arranged at an angle with the path of the match, which latter is indicated by the arrow in the engravings. The rod of the bell-hammer F and the arm K attached to the match-holder are so arranged, that while the match-holder is retracted, its arm bears upon the rod of the hammer and forces it down so as to prevent the play of the escapement of the alarm apparatus. When the match-holder is moved forward the hammer is relieved, so that the escapement-pallet H will play freely, or be made to operate by the escapement-wheel. The cast-off lever L (on fulcrum *a*) will be operated by projection *m* on arm K, during the forward motion of the match-holder, so as to throw the extinguisher M off the wick-tube and expose it to the flame of the match.

Figure 1 represents a plan; figure 2, a side view; and figure 3, part of an end view of the apparatus.

The inventor says: I do not claim combining with the match-holder a roughened surface for the match to rub against.

But I *claim* arranging the friction-surface when applied to a spring, bent as set forth, at an angle with the match-holder or its path of movement, as described, in order to facilitate the ignition of the match when the holder is in movement.

I also claim making the friction-surface to revolve, as described, in order that a fresh portion of the surface may be exposed to the match whenever any part of the surface becomes worn or unfit for use.

I do not claim the combination of an alarm apparatus or movable match-holder, or friction-surface, and a lamp, nor the combination therewith of a contrivance for casting the extinguisher off the wick-tube of the lamp.

But I claim the described arrangement of the match-holder, cast-off lever, and hammer-rod of the escapement, whereby, while the match holder is retracted, the escapement apparatus will be controlled, as described; but during the forward motion of the match-holder, not only will the cast-off lever be tilted so as to throw the extinguisher off the wick-tube, but the escapement set free, so as to enable the alarm mechanism to operate and strike the hammer with repeated strokes upon the bell.

No. 13,738.—Daniel E. Eaton, assignor to Himself and Perley O. Eaton.—*Improved Burglars' Alarm.*—Patented October 30, 1855. (Plates, p. 339.)

When set, one arm of the trigger should project by the door in such manner that, while the door is being opened, it may move said trigger far enough to produce a complete withdrawal of the bolt from the hammer H; in order that said hammer, when thus relieved from retention in a vertical position by the bolt, may be thrown, by the spring F, upon the top of rod I. By the pressure of spring F, the hammer H is borne down upon the top of rod I, so as to depress the said rod and the spring *g* upon which it rests. This movement of rod I will set free an alarm apparatus of the usual construction (not shown in the engravings . The match-holder is stationary while the igniting surface, or sector G, is made to move against the match, the said sector operating to throw the extinguisher O off the wick at the same time it ignites the match.

The inventor says: I do not claim the combination of a lamp and an alarm apparatus, a match-holder and an igniting surface, nor employing therewith a contrivance for casting the extinguisher off the lamp; but I *claim* arranging both the igniting sector G and escapement hammer H on one rotary shaft, (controlled by a spring, as set forth,) in combination with so arranging the wick-tube *c*, the match-holder C, and the escapement-rod I, that immediately after the trigger-bolt N has been withdrawn from the escapement hammer, not only shall such hammer be thrown over upon the top of the escapement-rod, but the igniting sector be caused to discharge the extinguisher off the wick-tube, as described.

No. 13,874.—Samuel Hamilton, Jr.—*Improved Burglars' Alarm.*—Patented December 4, 1855. (Plates, p. 339.)

In figure 1, the drawer is pushed into its case, and is secured without the aid of a lock by the stop *t* and the stop *u* resting on or bearing against each other, and the hook part of the spring-catch *o* passing into the partition *s* of the casing *a;* and in addition thereto, the secret spring-slip *x x* presses up against the side of the drawer, and thereby keeps the two stops *t* and *u* locked against each other; and before this drawer can be opened, the pull-knob *b* must be turned to the left. This turns the block L, which actuates the links *n n*, which draws out the hook of the spring *o;* and in the act of doing this, the drawer must be also pressed to the left by the pull, and in the act of pressing to the left, hold on to the knob; and with drawing the drawer, no alarm is given, when operated by any one familiar with the devices.

The knob *b*, if pulled out directly, will, in the act of passing outward, cause the stop-pin *h** to pass from off the hammer-handle *h;* and this lets fly the alarm-spring *i*.

Again: if the knob *b* is thus pulled out, the spring-catch *c* catches on to the outside of the drawer, and the alarm cannot be stopped until the spring is again pushed into its place, and the knob pushed back also.

In order to increase the certainty of detection, the side alarm is inserted in the side of the casing instead of within the drawer; and in the act of depredation, as the drawer is being drawn out, the hook *o* carries along with it the sliding-block s^1, by the hook catching the pin t^1, and as this is done, the stop *y* is taken off the hammer axle X, when the side alarm is set going.

The inventor says: Disclaiming the clock devices and bells, I *claim* constructing an alarm with an actuating pull-knob *b*, having a catch-spring *c*, ratch-disc b^1 formed with a tube c^1, together with the rail *d d d*, the blocks K K^1 L, links *n n*, hook-springs *o*, sliding-block s^1 with catch devices t^1 t^1, the secret spring-strip *x x*, through all of which, in combination with clock-work devices, are actuated bells or alarms J, substantially in the manner described and for the purpose set forth.

No. 13,876.—Horace L. Hervey.—*Improved Burglars' Alarm.*—Patented December 4, 1855. (Plates, p. 340.)

A number of sets of levers B C D E F are arranged alongside of each other. The fulcrum levers F rest on springs *a*; the stop levers D and E are attached to levers F, and also to levers C, which are connected to key levers B; levers G G^2 are connected to the alarm L and bolt M. When it is desired to connect the bolt and alarm with levers B, the stops E are to be pressed down, which press levers C down on lever G^2 until it is firmly grasped in the slot in C; this conneets it with the alarm. When it is desired that an alarm be given by any one attempting to open the drawer, the stop levers C are to be pressed down until the slots fit over G^2, which connects the alarm with the key levers B. All the stop levers, except one, having been pressed down in this way, the one is to be connected with the bolt by pressing down its stop D, so as to cause the slot in C to fit lever G. If any one then attempts to open the drawer, and touches any one of the levers C, except the one connected to the bolt, it will lift the catch J and set the alarm free, and ring the bell L.

The springs *a* are so arranged that, as the stops D and E are pressed down alternately, the point of bearing of the fulcrum lever F is changed, yet the pressure of the spring is always against the point of bearing, keeping it always at the required position, and holding the levers C in the proper place.

The inventor says: I do not claim decoys of themselves, as decoys are not new; but I *claim* the combination and arrangement of levers F, springs *a*, stop levers D and E, slotted levers C, key levers B, flat-sided levers G G^2, connecting levers H H^2, and links I, varying the alarm at pleasure, by means of the stop levers D and E, and for drawing the bolt M, as described.

No. 12,674.—Wright Duryea.—*Card Exhibitor.*—Patented April 10, 1855. (Plates, p. 340.)

A strip J, with the name, etc., printed upon it, from end to end, at suitable intervals apart, is substituted for separate cards, and is wound upon roller D, its last end being held by spring-holder G, as apparent from the illustration. If a card is to be drawn, the hinged holder G is opened, as indicated in figure 2, by broken lines, the strip drawn out to the length of a card, the holder G closed again, so that it holds the strip firmly, and the card is torn off.

Claim.—1st. The within described improvement in card exhibitors and distributors, consisting in the application of the roller D, printed strip J, guide *d*, and self-closing spring-holder G, substantially as and for the purposes herein described.

2d. I claim the use of a printed strip, for the purpose herein described.

No. 12,599.—Samuel B. Knight.—*Method of Chalking Lines.*—Patented March 27, 1855. (Plates, p. 340.)

C is the chalk, and L the leather or rubber packing, which rubs off all the superfluous portions of chalk.

The inventor says: I claim the described method of chalking a line by drawing it through the cylinder or other vessel containing the fine chalk, and also through the rubber of leather or other compressible substance, for the purpose and in the manner substantially as set forth; and this I claim when used for chalk or other coloring material.

No. 12,605.—David Sholl.—*Improvement in Coffins.*—Patented March 27, 1855. (Plates, p. 340.)

Claim.—The production of a coffin composed of *terra cotta* or pottery ware.

No. 12,497.—Virgil Woodcock.—*Improvement in the Arrangement of Desks in School-rooms.*—Patented March 6, 1855. (Plates, p. 340.)

I *claim* the diagonal arrangement of the seats B and desks A, as described. (See engravings.)

No. 13,371.—Wm. G. Wolf.—*Improvement in Writing-Desks.*—Patented July 31, 1855. (Plates, p. 340.)

The nature of this improvement will be understood from the claim and engravings.

I *claim* the horizontal inclined levers E, and inclined and declined planes J, with the upright traveller H working thereon, which causes a graduation that of a desk to be formed, or else entirely concealed, at pleasure, as described, using for that purpose the aforesaid horizontal inclined levers, inclined planes, and upright traveller.

No. 13,414.—Francis Arnold.—*Improved Egg-holder.*—Patented August 14, 1855. (Plates, p. 340.)

By means of boxes fitted with a number of egg-holders, as the one represented in the engravings, eggs can be transported safely and without requiring the usual packing.

I *claim* securing or holding eggs within boxes, or on table castors and other articles, by means of the elastic clamps *b*, constructed as shown and described, or in any equivalent manner, for the purposes specified.

No. 12,643.—Stephen R. Roscoe.—*Fire-escape Ladder.*—Patented April 3, 1855. (Plates, p. 341.)

B C D are the sections of the ladder C, being in the progress of extension beyond B, and coming up from behind it. When the ladders are down, C lies behind B, and D behind C; the bottom of each being on the platform P, and B being hinged to the platform at H. The operation of the machine is as follows: When the sections lie in a tier, one behind the other, (the section D being in rear of the others and resting against wheel W,) and when the wheel W is turned, its teeth act on the rack-teeth with which the rear parts of the sections are provided, the section D mounts upward, hook *h* running along the groove under the iron *b*, and keeping the ladders together, so that when the bottom of D arrives at the sloping top of C, the bottom of D is guided along the top of C until the upper lines of the two sections range, and the dovetailed tenon and mortise fit each other, when the detent *k*, which has been kept down by the pressure of the ladder above it, springs out, and prevents D from sagging back along C. To lower the ladder, W is turned backwards; and when the points of junction of two sections have passed just below the wheel, a tooth of it depresses the detent, and allows the upper section to slide down under and along the lower one.

The roller R and roller G (pressed against the sections by means of springs S) serve to keep the sections in their places as they move up and down. Platform P slides a small distance back and forth to accommodate itself to the movement of the ladders as they rise and fall.

Claim.—In the described sectional ladder for fire-escapes and other purposes, the combination and arrangement of the mortise *m*, tenon *t*, spring-latch *k*, hook *h*, and the groove in which it traverses, the whole being constructed to operate as described, for the purposes set forth.

2d. In combination with the described sectional ladder, I claim the traversing platform P and traversing roller G, so constructed and arranged as to allow the sections of the ladder to be operated as set forth.

No. 13,068.—Charles De Saxe, assignor to Thomas H. Bate.—*Improved Serpentine Spinner to Catch Fish.*—Patented June 12, 1855. (Plates, p. 341.)

The object of this improvement is to give continual motion to the hook or tackle, even where there is but very little current.

Claim.—A spinner, substantially as described, constructed of a piece of metal *a* twisted or coiled upon its edge, and then attached to and winding about a hook, or other piece of tackle, for the purpose set forth.

No. 13,081.—Richard F. Cook.—*Improved Fish-Hook.*—Patented June 19, 1855. (Plates, p. 341.)

To make the hook (see figure 2) ready for fishing, the collar *e* is moved down (see figure 1) and held so by the elasticity of the strips A; this causes the lower ends of these strips to move close enough together to allow the ring *e* to be placed round them. When a fish takes the baited hook into his mouth sufficiently far to bring hooks *b* between his jaws, and then pulls slightly upon the bait-hook D, it will be sprung, (see figure 2,) and the hooks *b* caused by the action of spiral spring C to move laterally from each other.

Claim.—The combination and arrangement of the steel-strips A A A A, having barbs *b* formed on them, collar or plate *c*, rod B, spiral spring C, ring *e*, and bait-hook D, substantially as and for the purposes set forth.

No. 13,649.—Job Johnson.—*Improved Fish-Hook.*—Patented October 9, 1855. (Plates, p. 341.)

The nature of this invention will be understood from the claim and engravings.

The inventor says: I do not claim forming a spider of hooks in themselves, as the same have been used for meat, and a variety of other purposes; but I *claim* the method described and shown of catching fish by means of a cluster or spider of hooks, beneath and around suitable bait 3, so that said hook can be suddenly raised up, and catch the fish while nibbling at the bait, in the manner and for the purposes specified; I also claim the method set forth of attaching and hanging the hooks *e* from the ends of the spider-arms *c*, by means of the spring-throat 2, whereby said hooks can be raised or replenished, in the manner and for the purposes specified.

No. 12,395.—Jefferson Parker.—*Improvement in Machine for Slaughtering Hogs.*—Patented February 13, 1855. (Plates, p. 341.)

c are longitudinal bars for the hogs to rest upon while undergoing the scalding process. The hogs are removed from the scalding-vessel A to the scraping-table B by means of fingers *d*, being brought from position figure 2 into position figure 3, when the hog, elevated by the said fingers, will glide off upon table B.

Claim.—The arrangement of the elevating fingers *d d* and the chains *e e*, with the operating levers, and with the scalding-vessel A and the scraping-bench B, substantially in the manner and for the purpose set forth.

No. 13,817.—Edward Pierre Fraissinet and Henri Emile Reboul.—*Ticket-Holders.*—Patented November 20, 1855. (Plates, p. 341.

This instrument is intended for travellers, for preserving, securing, and exhibiting their tickets. The ticket is to be secured between the spring-hooks B, and the hook E may be inserted in the button-hole of a coat.

The inventors say: We do not confine ourselves to the forms described, as they may be varied without deviating from the principle described; but we *claim* the construction of an apparatus, or instrument, for carrying, securing, and exhibiting tickets, as described and erred to.

No. 12,748.—William H. Webb.—*Improvement in Metallic Hones.*—Patented April 17, 1855. (Plates, p. 341.)

The body A of the strap is made of metal of a softer nature than one or more metal strips *a b c* which are inserted within said body, so that the whole presents an even surface.

The softer metal is intended to seize and retain the sharpening material and to enable it wear down the blade, the strips of harder metal serving to smooth or polish the abraded edge.

Claim.—A hone constructed with its sharpening surface composed of a combination of metals of different degrees of density, and arranged together substantially as specified, and intending to claim the broad ground of constructing a hone of metal.

No. 13,409.—Addison Capron and Joseph S. Dennis, assignors to Themselves and Henry M. Richards.—*Improved Machine for attaching Hooks and Eyes to Cards.*—Patented August 7, 1855. (Plates, p. 342.)

Each link of the endless chain or feeding receiver G has a recess *b*, for the reception of a hook and eye; see fig. 3, which is a top view of a couple of the links on an enlarged scale.) The grooves *c c* in the wheels E F, and rail I, supporting the chain, allow each hook and eye placed in the chain recesses to extend out of said recesses and across the line in which they are to be sewed to the card placed upon plate H. The metallic band M prevents the hooks and eyes from falling out of their recesses while passing around the wheel, and until they (the hooks and eyes) come to lie with their faces on the card on plate H; this card is moved along upon plate H with the feeding receiver G, both being moved with the same intermittent motion; for this purpose, the feeding roller N is placed underneath wheel F, and works through a slot O, formed transversely through plate H. For the purpose of sewing the hooks and eyes to the card, any suitable sewing-machine is employed, so arranged that its needle or needles may work on either or both sides of that part of the feeding receiver resting upon the card; figure 2 of the engraving represents only part of the sewing-machine, to wit, the end of needle-carrier arm X, with needle U, and other parts of a common sewing-machine.

Claim.—The described combination, or other substantially the same, of a feeding-receiver, made to receive the articles and maintain them at proper distances asunder, a card or sheet-feeding mechanism, and sewing machinery on one or both sides of said receiver.

No. 13,098.—W. D. Parker.—*Improved Ice-House.*—Patented June 19, 1855. (Plates, p. 342.)

When the door O is open, the doors n^1 are closed; and when flooring *l* is covered, the door O is closed, and the doors n^1 opened and the articles lowered into the house. By not keeping the door O and the doors n^1 open at the same time, the lower chamber is kept free from atmospheric influence and change of temperature.

The inventor says: I do not claim making an ice-house with double sides, and packing a non-conducting substance between the sides, for that is well known; but I *claim* the construction of the ice-house as shown and described, viz: having the ice-house formed with double sides *a a*, and double roof *c*, with a suitable non-conducting substance packed between them, the house being provided with a slotted floor g having an ice-chamber underneath it, and also provided with a double inclined floor *i i* at its upper part, underneath which a screen *j* is secured, on which charcoal and other absorbents *k* are placed; a flooring *l* being placed on the flooring *i i*, and having holes or traps *n* provided with doors n^1 made through it, and also through the flooring *i i* and screen *j*, the flooring also containing ice, substantially as described and for the purpose set forth.

No. 13,809.—Cornelius R. Wortendyke.—*Improved Machinery for Raising Ice from Rivers, &c.*—Patented November 13, 1855. (Plates, p. 342.)

e is an inclined plane, formed of main string-pieces, supported by suitable braces, and receiving an open bottom composed of bars *t*, attached to a cross-piece 2, setting in notches 3 in the string-piece *e* of the incline, so as to run the ice off into the house *d;* and as the said house becomes gradually filled with ice, the section 1 can be replaced, commencing at the bottom, so that the ice will be higher elevated before being run off by the chutes *f* to the house. The ice is drawn up the inclined plane *e* by means of hooks 4 4 attached to an endless chain *i*, that passes around pulleys *g* and *h*, and to the pulley *g* the required rotary motion is given.

By the following arrangement this apparatus is applicable, notwithstanding any change of the level of the water: *k k* are side-beams, connected by joints to the ends of the string-pieces *e* at the edge of the dock *b*, carrying at their lower end the bottom-board *l*, connecting said ends together. The outer ends of these beams *k k* rest on a cross-piece *m*, between two floats *n n;* and the joints between the beams *k k* and string-piece *e e* are such as to allow the said outer end of the incline to rise and fall the required extent. *o o* are frames on the side-beams *k k*, carrying the journal-boxes 5 5 of the wheel *h*. These journal-boxes are attached by means of screws passing through a long slot in the frame *o o*, and into a plate 6 beneath. 7 7 are screws passing through fixed nuts 8 8, and through small holes at 9 in the journal-boxes, where they are secured by a pin, so that said journal-boxes can be adjusted to tighten the chain as the water rises, or slacken the same as it falls. To prevent the slipping away of the ice from the hook when passing the angle at the joints 10, formed by the side-pieces K K and the string-pieces *e e*, a movable bottom is applied, which bottom is attached to the beams *k k* by joints 12 12, and is formed of side-pieces 11 11 and cross-bars, carrying strips *p*.

The bottom *p* nearly coincides in its inclination with the chain *i* whatever the height of the water may be, and consequently the hook 4 will always have a firm hold on the ice to draw it up the incline.

Claim.—The method set forth of adapting the ice-elevating chain to work under changes in the level of the water by joining the frame-work carrying the lower wheel of said elevating-chain to the lower end of the fixed incline, and sustaining the said frame-work wheel and chain on scows or floats, substantially as specified.

Also, the movable bottom *p*, fitted and arranged as specified, to pass the ice from the movable to the stationary part of the incline, in the manner specified.

Also, in combination with the frame-work, jointed to the lower end of the fixed incline, carrying the lower wheel of the elevating-chain, the adjustable slide journal-boxes and screws, for regulating the tension of said elevating-chain, as the water rises or falls, as specified.

Also, making the rear part of the shell tapering or conical, as seen at *a a*, combining therewith a ring or annulus *b b* of lead, or its equivalent, the same being substantially in the manner and for the purpose specified.

Also, confining the explosive screw-cap to the body of a shell, by means of a softer or yielding metal, or casing, which, when the cap or shell strikes an object, shall give way under the force of the blow, and let the cap down with force, so as to compress the percussion wafer or priming in it, or on the main screw, stopper, or plug, and so as to create an explosion thereof, as stated.

No. 13,824.—Charles A. McEvoy.—*Improvement in Railroad-Station Indicators.*—Patented November 20, 1855. (Plates, p. 342.)

The signs *a*, with the names of stations written on both sides, are hinged to the angles *m* of a reel, which may be turned by the conductor so as to cause one of the signs to drop through the slot *s* in the roof R of the car, and exhibit the name to the passengers.

The inventor says: Disclaiming the use of an indicator pointing fixed signs, and also movable signs where but one side is visible, *claim* presenting a movable sign *a* or symbol to passengers of a railroad-car so that both sides of said sign shall be visible and utilized as annunciators, by hinging said signs *a* to the angles of a polygonal reel, in such manner as to make each sign in turn drop through a slot, substantially as set forth.

No. 12,759.—Norman C. Harris.—*Improvement in the Manufacture of Slate-Pencils.*—Patented April 24, 1855. (Plates, p. 342.)

When the carriage B, bearing a slab of slate, moves in the direction of the arrow, the cutter D, bearing against its carrier A, cuts a slight depth into the slate, making grooves at the proper distances apart for forming the individual pencils side by side. Then as the carriage recedes it swings the cutter away from its carrier, (the cutter swinging round pivots *b*,) and thereby frees it from the slate. In the mean time the cutter-carrier is moved down a little, and when the carriage again moves forward the cutter makes the grooves a little deeper in the slate.

This continues till the pencils are half formed in the face of the slab, which is then turned the other side up, and the same process repeated till the other halves are formed in like manner.

Claim.—Cutting the pencils, completely formed, from slabs of slate, by means of a cutter or series of cutters, grooved so as to half form the pencils on one side of each slab; and then reversing the slab and forming the other halves of the pencils, substantially as herein set forth.

No. 12,141.—SYLVANUS SAWYER.—*Machine for Splitting Ratans into Strips.*—Patented January 2, 1855. (Plates, p. 343.)

The stick of ratan, passing through a series of feed-rollers C C^1, is introduced into guide-stock Q, and by it guided to radial wing-cutter R. Guide-stock Q consists of four movable guides *e e e e* (see fig. 6), each two of which are jointed together by levers *g g*, and are forced towards one another by springs *f f*. Cutter R consists of a series of chisels *k* with their cutting-edges in radial direction. Between each two of these chisels there is a spring *l* fast to the tubular box *m*, which sustains the cutter R. Springs *l* serve to force the sectoral strips of cane close down into the angle of the cutters, and thereby insure the proper direction of the strips to the annular cutter S. Within this annular cutter S there is a series of tubular cutters *n*. (See fig. 7.)

Cutter R separates the ratan into sectoral strips *a* (represented on an enlarged scale in fig. 3); cutter S separates from each of these sections *a* what is termed the strand *b* (see fig. 4); the tubular cutters *n* round off the triangular portions *c* to the shape as represented in fig. 5.

Claim.—Combination of mechanism for splitting the ratan into sectoral strips, and a mechanism for removing annular or segmental strands therefrom, substantially as above specified.

Also, combination of mechanism for splitting a stick of ratan into sectoral or triangular parts or strips, and a mechanism for rounding and dressing or finishing either one or more such strips, substantially as specified.

Also, combination of mechanism for splitting a ratan into sectoral parts or strips, a mechanism for removing or separating from such parts annular or segmental strands, as specified, and mechanism for rounding, reducing, or finishing either one or more or all of the triangular strips or parts of the pith or inside portions of the ratan, as specified.

No. 13,627.—CHARLES C. REED, assignor to HIMSELF and WILLIAM S. REINERT.—*Machine for Preparing Ratans, &c.*—Patented October 2, 1855. (Plates, p. 343.)

The table G is provided with suitable openings to give room for the feed-rollers H, and allow them sufficient side-play; the table can be adjusted to the grooves *g* in the rollers by means of set-screws *s*, so as to suit different thicknesses of ratan. The vertical shafts E bear against the slide-rests S, which are inserted into the lower table D. The outer ends of these slide-rests bear against set-screws *k* and springs K, which allows the feed-rollers to yield sufficiently to receive and embrace the ratan between them. The side-bar L, which can be adjusted by set-screws l, prevents the springs and slide-rests from moving out too far.

The engraving represents only one set of the feed-rollers, and none of the cutters, as the latter form no part of the invention.

Claim.—1st. The combination of the adjustable table or plate G with the upright feeding and guide rollers H, for enabling the upper surface of said table or plate to be graduated to the grooves in the rollers, substantially in the manner and for the purpose set forth.

2d. Arranging the adjustable side-bars L in such relation to the upper and lower parts of the flexible portions of the springs K as to enable them to be graduated so as to arrest the outward movement of the lower flexible portions of said springs, at such points as to allow the rollers to yield sufficiently to receive and embrace the ratan between them, and yet prevent one of them from moving further from the centre than the other, so as to keep the ratan at all times in the centre groove, and at the same time allow a slight and stiff elastic movement to the upper portions of the springs above the bars, to allow either of the rollers to yield to the inequalities on either side of the ratan, as fully set forth.

No. 13,823.—Joseph McCord.—*Policemen's Rattles.*—Patented November 20, 1855. (Plates, p. 343.)

The nature of this improvement will be understood from the claim and engravings.

Claim.—In policemen's rattles the securing of the handle to the edge of the ratchet-wheel, and at right-angles to the axis of the latter, for the purpose of turning down the handle out of the way, thereby rendering the instrument more convenient to carry in the pocket, and for the further purpose of combining a mace and rattle in one instrument, substantially in the manner set forth.

No. 13,913.—Gilbert D. Jones.—*Improvement in Sand-paper Making Machines.*—Patented December 11, 1855. (Plates, p. 343.)

The sand is heated by hot air entering passage b for the purpose of effecting the immediate setting of the glue, by driving off the moisture by contact with the hot sand. The sand is thrown up by means of the wheel d against the glued surface of paper p arranged around the drum e. The face of the drum has two grooves e^1 near its edges, for receiving the sides of the sand-box, and thereby making a better joint. As the gluing-roller e^2 presses the paper against the drum, the extreme edges would not receive glue properly by reason of their sinking into the said grooves; each groove is therefore filled up for a short distance by a curved stationary piece O, attached to the side-framing.

Claim.—1st. Applying the sand or grit, in a heated state, to the glued surface of the paper, for the purpose set forth.

2d. The method of depositing the sand upon the glued surface; that is to say, by projecting it forcibly against said surface while in such reversed position, that the excess shall fall off by gravity, as described.

3d. The combination of the stationary pieces O, or their equivalent, with the moving drum, the paper, and the gluing roller, for the purpose set forth.

No. 13,449.—Edward Louis Seymour, assignor to William Oland Bourne.—*Improvement in Apparatus for Sifting.*—Patented August 14, 1855. (Plates, p. 343.)

This apparatus consists of several tubular sieves *a*, *b*, *c*, *d*, of various dimensions, placed one within the other, and so contrived as to be capable of being rotated around their common centre; and whilst thus rotating as one body, to sift out the different sizes of the material (sands, mineral earths, or the like) corresponding to sizes of sieve-cloth of which the tubes *a*, *b*, *c*, *d* are formed. The centre sieve is the coarsest.

Claim.—The described rotary sifter, in the several cylindrical sieves, one within the other, provided with conduits so arranged as to deliver the proceeds continuously, substantially as described.

No. 12,426.—N. C. Sanford.—*Improved Skates.*—Patented February 20, 1855. (Plates, p. 343.)

Claim.—Securing the runner B to the stock A by having disks *a a* on the upper ends of the knees C C, these disks being fitted within tubes or cylinders D D in the stock, the tubes or cylinders having a suitable elastic material within them, and their upper and lower ends covered by plates *b*, secured to the stock, whereby a requisite degree of elasticity is given the skate, as shown and described. (See engraving.)

No. 12,427.—N. C. Sanford.—*Improved Skates.*—Patented February 20, 1855. (Plates, p. 343.)

The back part of the skate will rise with the heel when the weight of the body is thrown upon the front part of the skate.

The inventor says: I do not claim merely forming the stock of two parts, for that has been previously done; but I *claim* having the stock A of the skate formed of two parts B C, and connected by a spring D, when said stock is combined or used in connexion with an elastic spring runner E, for the purpose set forth.

No. 12,338.—Jas. S. Ewbank, assignor to Wm. Everdell, Jr.—*Improvement in Spurs.*—Patented January 30, 1855. (Plates, p. 343.)

E is a screw-nut working on the screw *f*, its under face being concave so as to insure the outer edge coming in contact with the shoulders *e e* of the branches *a a* outside of the pivots *c c*, thus necessarily compressing the branches towards each other, and securely fastening the spur to the heel. In the same manner the conical screw F operates, by expanding the upper ends *g g* of the arms *a a*, and effects the same result.

Claim.—The construction of a spur having a divided hinge-branch *a a* for embracing the heel of the boot or shoe. Also, I claim the mode of sustaining the divided branches *a a* by means of the shoulder screw-nut, either as constructed by having said nut E with its bearing outside of the hinge of the jaws, or as sustained by means of the cone F, substantially as described.

No. 13,312.—John Jeune.—*Improvement in Stalls for Horses, &c.*—Patented July 24, 1855. (Plates, p. 343.)

The partitions yield readily to the pressure or kick of the animal, so as to prevent him from injuring himself. The partition can be swung in line so as to leave the space open and unobstructed, to make it available for other purposes when desired. The sleeves R surrounding the lower part of the chains turn freely, so that the animal when rubbing against it cannot be injured by coming in contact with the chain. The chain *c* can be hitched across when desirable. The doors D prevent the animals from getting their heads together; but they can be swung in line, whenever it is desirable to make the stable more airy or to expose the heads of the animals.

Claim.—In stables or stalls for horses or other animals the swinging partitions S S, constructed and arranged substantially as described, for the purposes set forth.

2d. The doors D D in combination with the planks A A, so constructed and arranged as to operate substantially as described.

3d. The sleeves R R in combination with the chains C C, to prevent the animal from injuring himself by rubbing against the chains.

No. 12,926.—George Turner.—*Mandrel for Cutting Tapering Sticks.*—Patented May 22, 1855. (Plates, p. 344.)

A, face-plate; B, hole through face-plate A and mandrel K for the passage of the stick when turned; the jaws E have projecting nuts G that pass through slots F in the face-plate, with a right and left-handed screw (cut on shaft H) working through said nuts; shaft H is attached to face-plate A by means of box I and said nuts G. On the outer end of shaft H is a cog-wheel L, and as this wheel is turned the two jaws E are made to approach each other; one of the jaws carries chisel M, which, as the plate revolves, cuts the stick down to a size which will let it pass through between the jaws. The cog-wheel receives its motion from the single screw-thread P on the stationary rim R, within which the face-plate revolves.

Claim.—The construction of the face-plate with the two jaws E E, with the cutter M on one of them, and made to close together by means of the right and left screw-shaft H H, moved by means of the cogged wheel L and the screw-thread P P P P on the rim R R, as herein described, or by any other construction substantially the same, and which will produce the intended effect.

No. 13,709.—Robert A. Smith.—*Improved Machine for Sweeping Gutters, &c.*—Patented October 23, 1855. (Plates, p. 344.)

The gutter-brush H is made adjustable, so that it can be readily removed and another put in its place. The pivot at the outer end of axle L, which the gutter-wheel I turns on, is inclined upwards, so that the wheel will also be a little inclined, for the purpose mentioned in the third part of the claim.

Claim.—1st. An adjustable gutter-brush, made to conform to or correspond with the shape of the gutter to be swept, so constructed and arranged that it may be removed from and applied to the end of the shaft which carries it, with facility, substantially as described.

2d. The guard or gauge-wheel *c*, arranged so as to prevent the gutter-brush from being carried too hard against or over the curb-stones so as to derange or injure it.

3d. So arranging the gutter-wheel by means of an angular axle, that the lowest portion of the tire, and the lower portion only, will come in contact with the curb-stones, substantially as described.

No. 13,455.—Dugald Campbell.—*Swimming-Glove.*—Patented August 21, 1855. (Plates, p. 344.)

In swimming with these gloves, the fingers and thumb are brought close together when the hands are pushed forward, and are expanded upon the backward and downward stroke.

Claim.—The use or employment of flexible webs, uniting the thumbs and fingers of gloves.

No 12,832.—M. M. and J. C. Rhodes.—*Machine for Leathering Tacks.*—Patented May 8, 1855. (Plates, p. 344.)

The links *c c* which connect the cylinder E and lever F have slotted the holes which receive the pin *f*, elongated to leave some free play for the pin, and to allow the final action of the cylinder, in driving the leather down upon the punch, to be effected by a shoulder *g* at the top of driver G, so that the leather will not be cut till the tack is driven through it. Each time the feeder H moves forward (by means of lever connection K L,) the tongue I passes behind the foremost tack, and, separating it from the rest, moves it towards the cylinder, but does not push it into the mouth *o*, as, while the divider moves forward, the cylinder and mouth have moved downwards. The tack is pushed into the mouth by the tongue I^1 as the divider moves back, while the cylinder rises. When the tongue I^1 begins to act, the head of the tack is in contact with the face of the cylinder, and the tongue acting near the point throws the point inwards, and gives the tack a slanting direction. The point, being thus carried in beyond the projecting-lip *k*, is prevented from slipping down outside the cylinder.

Claim.—1st. The employment, substantially as herein described, of a hollow cylinder with an opening in the side to receive the tacks, and a driver working within it in a suitable manner to expel the tacks at the end of the cylinder, and drive them into the leather or any material serving the same purpose, which is presented in a suitable manner to receive them.

2d. Operating the cylinder and driver, substantially as described, so that the former may receive a short and the latter a long movement, and that the final operation of the former to cut the leather or other material on the punch may not take place till after the termination of the operation of the latter in driving the tack through the said leather or material as herein fully set forth, but may be produced by a continued movement of the latter after it has driven the tack through.

3d. The divider, consisting of one or more tongues, similar to I I^1, having a straight edge working nearly close to and across the entrance of the receptacle into which the tacks are fed, to be submitted to the operation of the driver, and having a bevelled end terminating in a point to separate the tacks one by one as they are brought by the feeder contiguous to the aforesaid receptacle, and to conduct and push them as required into the said receptacle, substantially as herein described.

4. Forming the mouth *o* of the barrel with a projecting lip *k*, substantially as described, for the purpose of passing outside the point of the tack as the barrel rises and the tack is entering the mouth, and thereby preventing the point from going down the outside of the barrel and letting the tack fall head-foremost into the barrel.

No. 12,417.—Jos. G. Goshon and Saml. M. Eby.—*Preparation of Maize-Leaf as a Substitute for Tobacco.*—Patented February 20, 1855.

After cutting, the blades are thrown into heaps, and, when sufficiently wilted, are hung upon frames or lines to dry; after drying, the leaf is dipped in warm water, and its middle portion or string removed. Corn-stalks, gathered at any time before frost and cut fine, are then boiled with a sufficient quantity of water to reduce the central parts of the stalk to a pulp and form a syrup, to which are added quassia and capsicum, or other bitter botanical production. The blades, after preparation as above, are then placed in this syrup and allowed to simmer and soak until well filled with the mixture; they are then taken out and submitted to the same process as tobacco for manufacture of segars, &c. Four ounces of quassia and half an ounce of capsicum to a pound of the corn-leaf.

Claim.—Preparing the leaf of Indian corn substantially as set forth, for the purposes specified.

No. 12,125.—James Caffrey.—*Improved Trap for Catching Animals.*—Patented January 2, 1855. (Plates, p. 344.)

When the animal walks on the platform D, and pushes against the grating G, the back edge of frame H raises the spring L, which lifts the lever I, which frees the pin K, and the platform D falls downward with the weight of the animal, entrapping the latter in a box underneath. At the same time, pin K on the next platform D is caught by notch of spring I, and the trap is again set.

Claim.—The peculiar arrangement and combination of the lever I, spring L, and wire grating G, acting simultaneously with the revolving platform D, to cause the trap to act and set itself, substantially as herein described.

No. 12,892.—LUCIUS B. BRADLEY.—*Improved Rat-Trap.*—Patented May 22, 1855. (Plates, p. 345.)

The bait being placed on the tilting plate which is hinged at *d*, and a rat passing upon it, his weight depresses it. Thus the arm *e* of the tilting plate is thrown up, whereby, through the agency of the balancing spring F, the stop *c* is thrown up, and the spring D enabled to act and force the drop C down upon the rat. (Position shown in figure 3.)

Claim.—The employment of a tilting or swinging plate E, and balancing or counteracting spring F, in combination with the ordinary spring-drop or fall C, substantially as and for the purposes set forth.

No. 13,483.—LUCIUS B. BRADLEY.—*Improvement in Trap for Catching Animals.*—Patented August 28, 1855. (Plates, p. 345.)

After the animal has stepped upon the tilting bottom G, the fall D is thrown down by the action of springs E, thus catching the animal; the pawl J, bearing against one of the teeth of ratchet I, prevents the fall from being lifted by the animal.

Claim.—The application of the ratchet-bar I and pawl J to the trap, for the purpose of rendering it capable of confining the animal after being caught under the drop or fall, as set forth.

No. 13,853.—LEONARD S. MARING.—*Improvement in Attaching Casters to Trunks.*—Patented November 27, 1855. (Plates, p. 345.)

The inventor attaches the casters to brackets E, secured to the edges of the trunk A, so that it may roll on said casters, even when tilted, as represented by the dotted lines in the figure.

I *claim* constructing and arranging casters on trunks, in the manner substantially as described and shown, and for the purpose set forth.

No. 12,671.—THOMAS C. CONNOLLY.—*Improvement in Machines for Recording Votes in Legislative Bodies.*—Patented April 10, 1855. (Plates, p. 345.)

The frame D has three vertical divisions, the central one containing the names of the members; over the unoccupied divisions, to the right and left, are to be inscribed "yeas" over the one, and "nays" over the other. The tablets *b*, in the central division, which exhibit the names, are moved either to the right or left division under the yeas or nays by means of levers *i*, which levers have their fulcra in a vertical rod *g*, passing through the fulcrum holes *h* in said levers. Cords *f f* (represented in the figures by broken lines) pass from each lever *i*, as indicated in figure 1, below the floor of the apartment, to a T-shaped lever L. The end *d* of this lever can be moved to the right or left round pin *l* by means of an upright lever *d*, which is pivoted in *e*, and the upper end *c* of which, reaching above the desk of the respective member, can be moved by said member either to the right or to the left, whereby, in consequence of the just described arrangement of cords and levers, the tablet *b*, exhibiting his name, will be moved under the yeas or under the nays.

The rear end of each lever *i* is forked, so as to move the corresponding one of the plates *k* to the right or to the left, whenever the front end of lever *i* is moved to the left or to the right. These plates *k* are arranged, one above the other, in a galley E, and contain, on their front faces, the corresponding names of the members in types, stereotypes, or the like. When the names in frame D are recorded among the yeas and nays, the stereotyped names in the galley will also be accordingly arranged. The series of plates *k* in the galley are provided with three holes *l l l*, which, whether the plates be moved to the right or left, or be in the centre of the galley, will all come one immediately over the line of the other, so that a pin may be passed through the galley and the plates. The galley, so locked, can be removed to a printing-press and impressions taken from it.

The inventor says: The recording of votes by a system of knobs, bell-pulls, cranks, and wires, has been used, and is well known. These I do not claim, nor do I claim the working of slides on which the names of the members are printed or engraven; but I *claim* the moving of the slides, containing the members' names, into columns of yeas and nays; this arrangement being one that is well calculated for the convenient display of the vote to all the members of the body voting, substantially in the manner described.

I also claim the arranging of a series of types, stereotypes, or plates, in a galley, by a system of levers, cords, wires, etc., extending from each desk or seat to said galley, so that any number of impressions of the exact record of the vote taken may be instantly printed or struck off.

I also claim the so arranging, in a galley, of a series of types, stereotypes, or plates, as that they may be readily moved therein to the left or right, and instantly locked into a form, from which printed impressions may be taken by any of the well known means.

No. 12,886.—Samuel Huffman, assignor to Himself and C. D. Hay. —*Mode of Indicating the Numbers of the Yea and Nay Balls in Machines for Taking Votes in Legislative Bodies.*—Patented May 15, 1855. (Plates, p. 345.)

The slightly inclined conductor *t* or *u* (one for the negative and the other for the affirmative votes) guides the balls into inclined or vertical transparent tube x or w, which have an inner diameter to fit the balls loosely, and which are provided with a scale s or s^1. The divisions of the scale being equal to the diameter of one ball, and being numbered from the bottom upwards, the number of votes polled on either side can be read off the scales at a glance.

Claim.—The use of transparent tubes, or their equivalents, provided with index scales, for the purpose of showing the number of votes on either side of a question by indicating the number of balls which have passed into each tube, substantially in the manner herein set forth.

AGRICULTURE.—(Omitted.)

No. 12690.—A. H. MORREL.—*Improvement in Cultivators.*—Patented April 10, 1855. (Plates, p. 346.)

This improvement is particularly designed for the thinning out of superfluous cotton plants and the cultivation of the remaining plants.

The thinning point m, of the shape of a common plough-point, has its shank fast to a cross-beam L, which can be slid laterally on rods n by turning handle e^1 on shaft a, a cord f^1 (both ends of which are fastened to cross-beam L) passing over guide-pulleys q around pulley p on shaft a. The cultivator r is attached to a beam K, which can be made to swing in a vertical plane round rod k^1 by depressing the handle-lever J, which latter has its fulcrum in l^1. The front end of beam K projects a little, so as to fit into a recess in the plate m^1, which plate is screwed to the cross-bar F of the main frame A C D. The handle of lever J can also be turned to the right or left, its support l^1 turning on a vertical journal y; thereby beam K, with shear r, can be slid laterally on rod k^1, after having lifted the front end of K out of the recess in plate m^1. A frame k k has its rear ends hinged to the axle j of the wheels H. Its front is supported by a plate o projecting from the under side of the main frame A. This frame k k is confined by a catch d^1 entering a recess in cross-bar E of the main frame. Frame k k supports a shaft b carrying an arm c and knife d. Gearing h g serves to revolve shaft b and knife d. This knife, when in position shown in figure 2, cuts the plants below the surface of the ground; if not desired to cut, it can be elevated above the top of the plants by pulling back cord s, as indicated in figure 3.

The stationary knife e serves to clean the rotating knife d.

Claim.—1st. The combination of the adjustable thinning point (or points) m at the forward end of the cultivator, with the adjustable cultivating point (or points) r at the rear end of the cultivator, substantially as herein set forth.

2d. The combination of the rotating cutter d with the laterally adjustable thinning point (or points) m, and the cultivating point (or points) r, substantially in the manner and for the purpose set forth.

LIST OF RE-ISSUES AND CLAIMS FOR 1855.

No. 286.—*Improvement in Mowing-Machines and Harvesters.*

What I claim is the combination of the bar that supports the cutter with a diagonal lever, held down at its inner end substantially as described, and resting upon the axle of the carriage as a fulcrum, or upon some other equivalent support that will perform the function of a fulcrum, whereby the outer end of the cutter-bar is held up, substantially as herein set forth.

JOHN H. MANNY.

No. 287.—*Improvement in the arrangement of Joints for attaching Trucks to Harvester-Frames.*

What I claim is the arrangement of a flexible joint in the line of the cutter or thereabouts, in such manner that the machine will bend freely up and down along this line, to keep the cutter as nearly as may be at a uniform height from the surface of smooth or undulating ground.

JOHN H. MANNY.

No. 288.—*Improvement in Arrangements for Controlling Harvester-Cutters.*

What I claim is controlling the flexure of the machine, hinged so that it will bend in the line of the front edge of the cutting apparatus or thereabouts, by means of an adjustable stop and arm, or their equivalent, in such manner that the cutter will be kept at the proper elevation on smooth ground, will be free to rise and fall to conform to a gently undulating surface, and will be restrained from descending into furrows, and other sudden and narrow depressions, while it will be free to rise to any extent required for passing over boulders, stumps, or other like protuberances in its path, substantially as specified.

JOHN H. MANNY.

No. 289.—*Improvement in Harvesters having a Leading Truck.*

What I claim is the leading carriage to carry the driver in a position in advance of the cutter, where he can readily see obstructions, and observe the character of the surface of the ground, in time to adjust the machine properly for operating upon any given part of its path before reaching the same, in combination with a cutter-carriage joined to the leading carriage by a hinged bar or other flexible connexion, cutter-carriage the being provided with an adjusting lever or arm, and extending forward to the leading carriage, where it can be conveniently reached by the driver, to enable him to raise or lower the cutter as required.

JOHN H. MANNY.

No. 290.—*Improvement in the Frame Construction of Triangular Harvesters.*

What I claim is constructing the frame which supports the cutting apparatus of a triangular or trapezoidal form, one of its acute angles being at the end of the finger-bar next the standing grain, so that the frame will not bear against the standing grain back of the finger-bar, and will permit the wheel which supports the outer end of the platform to be placed a considerable distance within the end of the finger-bar, yet sufficiently far from the frame, and at the same time not too far back of the centre of weight to poise or balance the machine properly.

JOHN H. MANNY.

No. 291.—*Improvement in Cutter-Fingers of Harvesters.*

What I claim is constructing the lower part of the finger or the upper, or both, with a recess on either side in front of the finger-bar, whereby the clogging of the cutting apparatus is effectually prevented, as herein described.

I also claim constructing the finger so that the sides of its upper half will overhang those of its lo er half, the cutter playing between the two, substantially as herein set forth.

I also claim bevelling the upper corners of the shank of the lower part of the finger, so as to form a cutting-edge *e* thereon, in the position and for the purpose described.

JOHN H. MANNY.

No. 292.—*Improvement in Harvesters.*

What I claim is:

1st. The arrangement of the track-scraper at the outer end of the machine. and the wheel or wheels which support the opposite end of machine, whether driving-wheels or not, in such relative position that the wheels while the machine is cutting one swath will run in the track cleared by the former while the machine was cutting a previous swath, as herein set forth; but in this patent I make no claim to the track-scraper itself.

2d. The projection 7, on the under side of the upper bars 5 of the top *m* of the finger, in combination with the chamfer or recess on the lower inside corners of said bars, to counteract the tendency of wire-grass or other fibrous obstructions to pass in between the cutter-bar e^1 and the sides of the recess in the upper part of the finger in which it is guided.

3d. Forming the guard-finger *o* of two parts *m* and *n* interlocked at the point substantially as herein set forth, so that grass cannot lodge in the joint and form an impediment to its entering between the stalks of the standing grain.

4th. In combination with the raker's stand or seat, I claim the removable platform or raking bottom, constructed with a wing that extends from the outer end of the cutter over the frame, and holds up the butts of the straws above the stubble, which otherwise would obstruct the discharge of the grain from the platform, substantially as herein set forth.

JOHN H. MANNY.

No. 293.—*Improvement in Spark-Arresters.*

We claim:

1st. The arranging of a series of chambers and channels between two conically shaped plates, the channels being so formed as to cause the products of combustion to impinge against that side of each of the dirt-chambers which has the openings and caps, and thereby force the sparks, dirt, &c. into them in the manner described herein.

2d. We claim the piece *p* suspended in the central aperture at the top of the spark-arrester, arranged and operating in the manner and for the purpose substantially as herein before described.

3d. We claim the double cover or top, for the formation of a second series of dirt-passages, arranged and operating in the manner and for the purpose substantially as hereinbefore described.

JAMES RADLEY.
JOHN W. HUNTER.

No. 294.—*Improvement in Ploughs.*

What I claim is:

1st. Combining with the plough-beam between the plough and the forward end of the clevis, by means of a single shaft, two wheels, one on each side of the beam, and of different diameters—the one resting in the furrow, and the other on the land—for the purposes set forth and described.

2d. I also claim making the tread of the furrow-wheel narrow, for the purposes described.

3d. I also claim making the furrow-wheel bevelling outward on the side which presses against the land, as above described, and for the purposes hereinbefore set forth.

4th. I also claim making the small wheel adjustable with reference to the shaft or axle and the large wheel, as described.

I also claim the adjustable hangers in combination with the plough-beam and axle, for the combined purpose of bracing the axle and rendering the wheels simultaneously adjustable with reference to the beam, without disturbing their adjustment relatively to each other, as described.

CORNELIUS R. BRINCKERHOFF.

No. 295.—*Improvement in Fastening Lanterns.*

We claim attaching the lamp to the lantern by means of the combination of the catches *e* with the flanges *a* and *f*, and the ring to which the catches are hinged, or its equivalent; the purpose and object of the ring being to give the hinged ends of the catches a motion concentric or parallel, or nearly so, to the side of the lantern, or the flange through which the catches pass.

CHARLES MOUNIN.
WM. M. BOOTH.

No. 296.—*Improvement in the Mode of Constructing a combined Caldron and Furnace, for the use of Agriculturists and others.*

What I claim is, first, combining a caldron with a portable furnace having a fire-chamber of smaller size than the area of the caldron, by spreading out and extending the sides of the furnace to form an outer casing partly or wholly surrounding the caldron, and forming a flue space between the two leading to the exit pipe, substantially as and for the purpose specified.

I also claim making a casing to form a flue space around the caldron, by elevating and spreading out the plates of the furnace, and fitting to and combining therewith sectional side-pieces, substantially in the manner described and for the purpose specified.

JORDAN L. MOTT.

No. 297.—*Improvement in Candlesticks.*

What is claimed is the employment of elastic packing attached to the standard, bar, spring, or slide of a candlestick, substantially in the manner described, whereby I am enabled to support said part, prevent the leaking of the grease, and use a shorter sliding-socket than when the cork is inserted loose in the socket.

JOHN W. ROCKWELL.

No. 298.—*Improvement in Grain and Grass Harvesters.*

What I claim is making the outside or dividing-finger hollow, so that while it affords sufficient room for the play of the end of the brake, the bearing of the latter therein will not be so long as to afford a lodgment of grain, grass, &c., in sufficient quantity to clog it.

JOHN H. MANNY.

No. 299.—*Improvement in Grain and Grasss Harvesters.*

What I claim is, the combination of the reel for gathering the grain to the cutting apparatus, and depositing it on the platform, with the stand or position for the forker, arranged and located as described, or the equivalent thereof, to enable the forker to fork the grain from the platform, and deliver and lay it on the ground, at the rear of the machine, as described.

JOHN H. MANNY.

No. 300.—*Improvement in Grain and Grass Harvesters.*

What I claim is, the combination of the fence to compress the grain against, at the outer end of the machine, and guide it while sliding off the platform, and the position, stand, or seat for the forker at the inner end of the platform, with the platform, substantially as herein set forth.

JOHN H. MANNY.

No. 301.—*Design for Metallic Coffins.*

What is claimed is, the ornamental polygonal design for a metallic case or coffin, substantially as described and represented.

M. H. CRANE.
A. D. BREED.
JOHN MILLS.

No. 302.—*Improvement in Machinery for Separating Flour from Bran.*

I claim: 1st. The platform D (always at right-angles with the sides of the bolt when not made conical), or close horizontal bottom when used in connexion with upright stationary or revolving bolt for flouring purposes.

2d. The opening at D^5 for the admission of a counter current of air through the bottom and into the bolt, and the opening and bran-spout F, as described, in combination with the platform D.

3d. The upright stationary bolt, or bolt and scourer combined, with its closed-up top, except from air and material; or in combination with claims first, second, and fourth, or either of them, or their equivalents, to produce like results in the flouring process.

4th. The use of the revolving, distributing, scouring, and blowing cylinder of beaters and fans, by which the material is distributed, scoured, and the flour blown through the meshes of the bolting-cloth.

ISSACHAR FROST.

No. 303.—*Improvement in Vault Covers.*

What I claim in covers for openings to vaults in floors, decks, &c., is making them of a metallic grating or perforated metallic plate, with the apertures so small that persons or bodies passing over or falling on them may be entirely sustained by the metal, substantially as described; but this I only claim when the apertures are protected by glass, substantially as and for the purpose specified.

And I also claim, in combination with the grating or perforated cover and glass fitted thereto, the knobs or protuberances on the upper surface of the grating or perforated plate for preventing the abrasion or scratching of the glass, substantially as specified.

THADDEUS HYATT.

No. 304.—*Improvement in Grain and Grass Harvesters.*

What I claim is:

1st. The combination of the rake O, swinging or suspended from one rod of the reel, with the guides L L and ways S S, substantially as above set forth and described; for the purpose not only of delivering the grain at the rear of the platform, but also for better directing the standing crop to the cutters.

2d. I claim the guides for forcing the grain into the end of the reel as described, and for the purposes set forth.

3d. I claim the latch *f* and appendages, by which the operator is enabled to permit more or less grain to accumulate on the platform between the successive actions of the rake.

4th. I claim placing the vibrating knife-bar 3, and cutters thereon, between alternately placed fingers 4 and 5, for the purpose of dispensing with the slot-guards, and sustaining the line of cut by throwing the action of the alternate shear-edge of the blades of said cutters on the upper and lower sides of the said fingers.

5th. I claim the alternate edging of the same tooth, and so placing them together that the two adjacent edges of successive teeth which act against the same finger may be alike turned in one direction, while the next two edges acting against the next finger are alike turned in the contrary direction.

ABNER WHITELEY.

No. 305.—*Improvement in Harvesters.*

What we claim is, discharging the cut stalks and heads of grain from the main platform D, on which they first fall, by means of the combination of the rake C with the overhung lever B, moved by gearing located within the inner edge or circle of said platform, as herein set forth.

AARON PALMER.
STEPHEN G. WILLIAMS.

No. 306.—*Improved Winnowing Machine.*

What I claim is:

1st. The employment or use of a vertical blast-spout F, gradually enlarged from its lower to its upper end, so that the strength of the blast is decreased in the upper portion of the spout owing to the increased space or area of the spout, for the purpose of preventing any sound or perfect grain being carried with the light foreign matter over the upper edge of the spout; the blast being formed or generated in said spout in any proper manner.

2d. I claim the blast-spout F, either gradually enlarged from below upwards or of the same dimensions throughout, and communicating with the atmospheric current through the screen H, in combination with the hopper E^1 and the fan placed at the end of the opposite vertical spout D, to separate the chaff and other impurities from the grain, in the manner substantially as herein described.

3d. I claim the employment or use of a vertical blast-spout, either gradually enlarged from below upwards or of the same dimensions throughout, when said blast-spout is so arranged that the grain is cleaned or separated from impurities within said vertical spout.

B. D. Sanders.

No. 307.—*Improvement in the Hinge of Rolling Iron Shutters.*

What I claim is, constructing shutters of slats of sheet-metal with joints formed by curving the edges of the slats as described, and securing them in place in the manner specified, viz: either by turning down projections from or attachments to the ends of the slats, and thus forming an even edge to the shutters, or by means of wires inserted in the curves and bent or headed at the ends, the shutters sliding up and down in the grooves of the window-frame in which it is placed, the whole being constructed substantially as herein specified.

A. Livingston Johnson.

No. 308.—*Improvement in the Construction of Moulds for Pressing Glass.*

I claim so combining with a mould-fountain or reservoir, provided with a plunger, one or more matrices or moulds, that a liquid mass of glass, when pressed in said fountain or reservoir by the plunger, may be made to flow or pass therefrom, and into such matrix or matrices.

I claim combining with a series of matrices, and a press-chamber or reservoir surrounded by them, an auxiliary annular and concentric chamber, (as seen at *h h*, in fig. 2,) formed in the two mould-plates, and made to perform the function of preventing the plunger from clogging the mouths of the matrices under circumstances as above stated, and also to prevent the chilled glass from obstructing the downward movement of the plunger.

I also claim so combining with the lower mould-plate or movable bottom-block *f*, that the same may not only serve to form a bottom to the main and auxiliary mould-chambers or to the former, but also enable a person to detach the pressed glass or metal from the lower mould-plate, under circumstances and in manner as above set forth.

HIRAM DILLAWAY.

No. 309.—*Improvements in Machinery for Reducing Metal Bars.*

What I claim is, the method of rolling bars or rods on four sides by the combination arranged with the axis of two of them parallel, and the third at right-angles thereto, substantially as specified, whereby two opposite faces of the bar or rod are drawn between the two rollers on parallel axes, and the other faces between the periphery of the third roller and the face of a cavity formed in one or both of the other rollers, as specified.

And I also claim in combination with the three rollers, combined and arranged substantially as specified, the employment of the bolster, substantially as specified, to prevent the forming of a pin on the bar at the junction of the two parallel rollers, as set forth.

DEXTER H. CHAMBERLAIN.

No. 310.—*Improvement in Endless Chain Horse-Power.*

What is claimed are, the links *c* of the parallel endless chains, which carry the travelling bed (formed with cogs on their inner edge meshing into the side pinions K) on the driving shaft, when the latter is arranged back of the forward end of the power to receive motion by the straight run of the cog-links over the said pinions, as shown and described.

ALONZO WHEELER.
ALEXANDER F. WHEELER,
Executor of this last will and testament of
WM. C. WHEELER, *deceased.*

No. 311.—*Improvement in Bleaching Apparatus.*

What I claim is, the combination of one or more air-tight vats for receiving and containing the goods, an apparatus for exhausting the air therefrom, and the necessary vessels for containing the liquids used in the process of bleaching, whereby the various steps may be performed in a much shorter space of time than has heretofore been required, as set forth.

C. T. APPLETON.

No. 312.—*Improvement in Looms.*

What I claim is, the yielding-rest or support K, for the picker, arranged substantially as described, to break the sudden blow or concussion with which the shuttle infringes upon the picker, thereby preventing the filling of the cop from being jarred off and entangled, and relieving the picker from danger of being broken.

I also claim separating or freeing the lever K and the picker from the end of the shuttle by the same movement which shifts the shuttle-boxes, operating through a combination of levers, cams, and springs, substantially as herein set forth, or through levers, cams, or treadles, worked from any part of the loom.

BARTON H. JENKS.

No. 313.—*Improved Nut and Washer Machine.*

We claim the machine, substantially as herein described, for making nuts, by cutting the blank from a heated bar of iron, punching its eyes in a closed die-box, pressing it into shape while in the die-box and on the punch, and then discharging it as specified.

HENRY CARTER.
JAMES REES.

No. 314.—*Improvement in Machinery for Felting Hat-Bodies.*

What I claim is, giving the felting action by means of the moving apron, arranged substantially as specified, to receive the article or articles to be felted within the fold thereof, and there confined and compressed by the rollers or their equivalent, acting on the apron, and resulting in a mode of operation substantially as described.

WILLIAM FUZZARD.

No. 315.—*Improvement in Corn-Planters.*

What I claim is, the peculiar construction of the horizontal slide G, made reversible from end to end, for the purpose of varying the quantity of seed planted, in the manner set forth and specified.

SAMUEL MALONE.

No. 316.—*Improved Portable Grinding-Mill.*

What I claim is, the alternate deep and shallow sections of furrows upon the main grinding surface of the burr, for the purpose of distributing the material over said surface, and preventing a surfeit or clogging upon any one point of said grinding surface, substantially as described.

I claim the method of supporting the shell and adjusting the burr therein by means of the lower bridge-tree, grooved legs, sockets, and adjusting screw-rods, when said legs serve the double purpose of supports to the shell and guides to the bridge-tree, as described

I claim the arrangement of driver G, arms I, burr B, and shell A, constructed as herein shown and described; so that the several operations of breaking the ear, cracking the cob, and grinding into meal, may be all conducted without straining the mill or power applied, substantially as described.

LYMAN SCOTT.

No. 317.—*Endless Chain Horse-Power.*

What I claim is, the construction and attachment of the gearing, substantially as herein set forth, having a hub or pinion permanently affixed on the ends of each shaft, to either of which the centre caps or hubs of either the driving or band wheels fit and are fastened.

GEORGE WESTINGHOUSE.

No. 318.—*Improvement in Cotton-Presses.*

What I claim is:

1st. The use of the slats or guide-strips E, arranged and operating in the manner substantially as set forth.

2d. I claim the tenons upon the transverse bars of the doors, which, entering mortises in the frame-work, relieve the hinges from the strain which would otherwise come upon them.

3d. I claim hinging the doors of the press in the manner herein described, to prevent them from violently bursting open when the bar which confines them is removed.

C. J. FAY.

No. 319.—*Improvement in Machinery for Dressing Treenails.*

What I claim is, the use of the cutters *a*, in combination with the enlarging and heading apparatus, or apparatus analogous thereto, when used for the purposes substantially as herein before set forth; and this I claim, whether any one or more of the parts of the enlarging or heading apparatus, or apparatus analogous thereto, are used separately or collectively in combination with the said cutters, whereby treenails are cut and shaped by the use of such mechanical devices as herein before substantially described.

DELIA A. FITZGERALD, *Assignee of*
JESSE FITZGERALD.

No. 320.—*Improvement in Looms for Weaving Figured Fabrics.*

What I claim is:

1st. Operating the headle-frames by the application of a cylinder having two or more patterns, interchangable at pleasure, in the manner and for the purpose substantially as herein before set forth.

2d. The employment of two or more distinct patterns simultaneously in the same loom, interchangeable at pleasure, operating in the manner and for the purpose herein before described.

I also claim the apparatus for turning the cylinder, substantially as herein specified, whereby it can be moved through a greater or less space, as may be required.

RICHARD GARSED.

No. 321.—*Improvement in Hot Water Apparatus.*

What I claim is, connecting the ends of the horizontal or nearly horizontal water-pipes of hot water warming apparatus, by means of return bends or elbows of less caliber, and entering within the end or ends of such pipe or pipes, substantially as and for the purpose specified.

And I also claim making each horizontal or nearly horizontal pipe, having the bend or elbow at one end of reduced caliber, with the calibers at top in the same line, substantially as and for the purpose specified, whether made in one piece or the bend or elbow separate, and then united, the said elbow being connected with the next pipe above it by entering the end thereof, substantially as and for the purposes specified.

And I claim the construction and arrangement of the apparatus for the purposes and substantially as specified.

JOHN BROWN.

No. 322.—*Improvement in Bathing-Tubs.*

What I claim is, the before described mode of combining with a bathing-tub, or other like vessel, either one or both of the channel-ways substantially as described, and making, when constructed, part of the tub or vessel, one of which channel-ways connects the overflow and waste or discharge holes with the waste-pipe, and the other channel-way is adapted to the insertion of the hot and cold water pipes, and discharging the hot and cold water together, at or near the bottom of the vessel, and in a horizontal or nearly horizontal direction, substantially in the manner and for the purpose specified.

JORDAN L. MOTT.

No. 323.—*Method of Warming and Ventilating Buildings.*

What I claim is:

1st. The mode herein described of warming and ventilating buildings, railroad cars, and apartments of every known description, the same consisting in introducing the air from without by conducting it under the floor of the building or apartment, and directly under the air warmer or ventilator, for the purpose of being warmed for distribution; the air, after being thus warmed, rising in a central or otherwise convenient apartment or passage, and thence being admitted into the various rooms of the building, or into the apartment near the ceiling or roof, without the aid of pipes, and thence passing downwards, and through openings in the lower part of the rooms or apartment, and thence outwards through the various channels provided, connected with the foul-air shaft.

2d. I claim the arrangement of the radiating pipes or flues of the air warmer, in combination with the fire chamber situated within or between them, in the manner substantially as set forth.

3d. In combination with the elevated air chamber and flues of the air-warmer, I claim the arrangement of the openings for admitting heated air above the fire, to complete combustion, as herein set forth.

4th. I claim the construction of the fire-grate, as herein set forth, viz: with one or more grates of cylindrical or other form raised above the ordinary grate floor; said raised grates being capped or covered in such manner as to protect the vertical bars from the fuel substantially as herein set forth, and the principle of their action being substantially as herein set forth.

5th. I claim the mode of conducting the air into the pure-air shafts, whatever may be the direction of the wind or of the external currents of air, by placing a swinging valve or shutter at the mouth of said shafts, substantially in the manner herein set forth.

6th. I claim so constructing or placing the mouths of the pure-air shafts, for the ventilation of railroad cars, that by the motion of the car, the incoming pure air may be increased in quantity, as herein fully set forth.

H. RUTTAN.

No. 324.—*Method of Closing and Opening Gates.*

What I claim is, a double-span rotating gate, opening and closing by an intermittent rotating motion in one direction only; said motion being derived through lifting pieces or levers, cam planes, weights or cords, or their equivalent, substantially as herein set forth.

WM. G. PHILIPS.

No. 325.—*Improvement in Lanterns.*

We claim constructing and arranging the spring-catches I, in the manner described, or its equivalent, to cause the attachment of the lamp to the lantern by the operation of pressing the lantern down upon the spring-catches.

Also, arranging the thumb-pieces L, within the flange G, at the base of the lamp, by extending the springs I towards each other horizontally, as described, and thus forming the elbow-catch to rest against the shoulder on the flange E of the lantern, in the manner and for the purposes specified.

HUGH SANGSTER.
JAMES SANGSTER.

No. 326.—*Improvement in Hinges.*

I claim the bridge or inclined plane at the base of the pin, and the corresponding elongation of the eye, operating and in connection with the hook and catch, attached and connected in the manner described.

I also claim the elongated or enlarged eye independently of its combination with the bridge, for giving the lateral motion to the blind, to effect the disengagement of the lower catch, as described.

I also claim so placing the catches on the two parts of the hinge as to cause the strains produced by the wind or otherwise to act directly upon the screws, whereby the pin and eye are relieved, as described.

WM. BAKER.

No. 327.—*Improvement in Air-Heating Stoves.*

What I claim is, making the bottom plates of the flue-spaces of air-heating furnaces or stoves for the passage of the products of combustion outward or inward among or around the air-passages inclining inward and downward towards the fire-chamber, substantially as described, for the purpose of facilitating the increase of the heating surface, without the inconvenience of the accumulation of ashes, soot, and other solid matter on such plates, as set forth.

And I also claim the combination of the inverted domes or frustrums F I M and plate P, with the short tubes *b b*, *f f*, *i i*, *l l*, connecting them substantially in the manner herein before described, for the purpose of effecting the connection between the lower ends of the fire or draught-flues, and carrying the air through them to the spaces between the cylinders or tubes.

J. M. THATCHER.

No. 328.—*Improvement in Grain Dryers.*

What I claim in the method of kiln-drying grain is, the employment of an endless pan or apron, made of metal and passing around drums, or the equivalents thereof, substantially as specified, in combination with and operated within a heating chamber, substantially as set forth.

JOHN MASSEY.

No. 329.—*Improvement in Pumps.*

What we claim is, the combination of the cylinder or chamber A and the piston, constructed as herein described, and its valves and the induction and eduction passages; so that the water, all entering said cylinder under pressure alternately at its ends, is discharged under pressure through the opening at its side, producing a constant and direct stream through the piston-heads from the cylinder, substantially in the manner and for the purpose set forth, thus dispensing with chambers and partitions in the barrel and valves at the eduction port, preventing leakage and rendering the pump more simple and effective, and less liable to derangement.

LEVI P. DODGE
WM. F. DODGE.

No. 330.—*Machine for Sawing Lumber.*

What I claim is, the employment of two pairs of shifting-guides, substantially as herein described, in combination with a circular saw, alternately in opposite directions, substantially as and for the purpose specified.

I also claim setting the log or timber, by means of the two screw-shafts geared in the manner described, or the equivalent thereof, and operated by griping-pawls, which act against stops at the end of the motions of the carriage, in combination with the arms and adjusting slides, to determine the degree or extent of set intended to be given to the log, substantially as specified.

And, finally, I claim, in combination with the method of setting the log at the end of the several motions of the carriage, substantially as described, the method of throwing the setting apparatus out of gear by the bar which carries the log, substantially as described, to prevent the said bar with the holding-dogs from approaching too near to the saw, as set forth.

PINNEY YOUNGS.

No. 331.—*Improvement in Harvesting-Machines.*

What I claim is, in combination with a *frame* nearly balanced on its supporting-wheels, and a *tongue* hinged to said frame, a lever connected to one, and projecting towards the driver's stand or seat on the other; so that the driver, who is the sole conductor of the machine, may, from said stand or seat, raise or depress the cutters at pleasure, during the operation of the machine for cutting the grain or grass at any suitable height above the ground, or for passing over any intervening obstacle, substantially as described.

I also claim, in combination with the operative parts of a harvesting-machine, a conveyor, which first carries the cut grain horizontally across the machine, and then elevates it so as to discharge the grain into the bed of a wagon driven alongside of the machine, when the conveyor-frame is connected to the bed by a flexible joint, in manner and for the purpose described.

JONATHAN HAINES.

No. 332.—*Machine for Sticking Pins in Paper.*

What I claim is, separating the pins laterally, one by one, from the lower end of a pile sliding by gravity between guides; by combining with such channel-way between guides a grooved or notched instrument, substantially as described, so that when the groove or notch is brought in the line of the channel-way, the lowest pin of the pile shall enter the groove or notch, and, by the lateral motion, separate it from the pile without any conflict of the heads; the surface of the said instrument beyond the groove or notch acting as a stop to prevent the farther descent of the pile, until a groove or notch be again brought in line, substantially as set forth.

I also claim the channel-way between guides for the descent of the column of pins by gravity, substantially as described, in combination with one or more guide-grooves, and one or more followers working in the said guide-grooves, substantially as described, whereby the pins, after being separated, can be transferred, as described.

And I also claim, in combination with one or more guide-grooves and one follower, being connected therewith substantially as described, for guiding the pin or pins as pushed forward, the employment of a clamp or holder for clamping or holding the paper in the required position during the operation of inserting the pin or pins.

And, finally, I claim the combination of the channel-way for the column of pins between guides, the guide, groove or grooves, or notches, the follower or followers, and the clamp, or equivalent therefor, substantially as and for the purpose specified.

SAMUEL SLOCUM.

No. 333.—*Improvement in Mortising-Machines.*

What I claim is:

Firstly. The sliding-wrist *o*, connected with the chisel and also with the driving-power, in the manner described, in combination with the mechanism described, or its equivalent, for sliding said wrist; so that the operator can, during the motion of the machine, vary the depth of the cut of the chisel, or cause it to be suspended, without disconnecting the driving power.

Secondly. The combination in a mortising-machine, substantially as described, of treadle and opposing spring or weight, connected to a toggle, one end of which being pivoted to the frame, the other is pivoted to a sliding-wrist upon a vibrating arm, actuated by the power; the said wrist being slid out and in upon the arm with varying power and speed by the action of said toggle and its attached weight or spring and treadle as explained, or their equivalents.

JOSEPH GUILD.

No. 334.—*Improvement in Reaping and Mowing Machines.*

What I claim is, raising and depressing the finger-bar K, and consequently the cutters *e f*, by means of the vertical bars M M, having wheels *o o* at their lower ends, arm P attached to the cross-piece N of the bars M M, lever Q and shaft R with its arm S attached, the above parts being arranged substantially as herein shown and described.

I also claim supporting the ends of the stationary cutters *e* by means of the sockets, or their equivalents, in the knobs or projections *d* of the fingers *c*, substantially and for the purpose as set forth and described.

SAMUEL ROCKAFELLOW.

No. 335.—*Improvement in Spark-Arresters.*

What I claim is, the combination of the central chamber C, the series of tangential openings E E, the larger circular chamber A, furnished with a series of vertical openings J J leading into exterior chambers or channels, for separating sparks and other particles of matter from the gaseous current discharged from locomotive or other chimneys, substantially in the manner set forth.

WM. C. GRIMES.

No. 336.—*Improvement in Sofa-Bedsteads.*

What I claim is, drawing down or depressing the cushion at the joint between the back and seat by means of the cords *b*, or their equivalents, connected automatically with the seat A and back B, for the purpose herein set forth.

CHARLES F. MARTINE.

LIST OF DESIGNS AND CLAIMS FOR 1855.

No. 683.—*Design for Parlor Stoves.*

What we claim is, the ornamental design and configuration of parlor-stove plates, as herein described.

JAMES WAGER.
VOLNEY RICHMOND.
HARVEY SMITH.

No. 684.—*Design for Cooking-Stoves.*

What we claim is, the design, configuration, and arrangement of the ornaments, in bas-relief, on the plates of cook-stove "Emporium," as herein set forth.

GARRETTSON SMITH.
HENRY BROWN.
JOSEPH A. READ.

No. 685.—*Design for Parlor Cook-Stoves.*

What we claim is the design, configuration, and arrangement of the ornaments in bas-relief, as herein described, forming an ornamental design for a parlor cook-stove.

GARRETTSON SMITH.
HENRY BROWN.
JOSEPH A. READ.

No. 686.—*Design for Stoves.*

We claim the design, configuration, and arrangement of the ornaments in bas-relief of the stove "Fanny Fern," as set forth.

GARRETTSON SMITH.
HENRY BROWN.
JULIUS HOLZER.

No. 687.—*Design for Lanterns.*

I claim the ornamental figures c^1 c^2 on the sunken planes or faces B, the beads *a a* at the sides or edge of the planes or faces, and the ornamental figures D D on the upper parts of the sides of the lantern, when combined as herein shown, to form a new and original design for a lantern.

WM. D. TITUS.

No. 688.—*Design for Metallic Coffins.*

What I claim is, the ornamental design for a metallic burial-case, as herein described.

M. H. CRANE.

No. 689.—*Design for Parlor Open-Front Stoves.*

What I claim is, the ornamental design and configuration of parlor-stove plates, as described.

N. S. VEDDER.

No. 690.—*Design for Parlor Stoves.*

What we claim is, the ornamental design and configuration of parlor-stove plates, as described.

N. S. VEDDER.
EZRA RIPLEY.

No. 6ϛ *Design for Coal Stoves.*

What we claim is, the ornamental design for a coal parlor stove, as herein described, to be known and called the "Diadem."

CONRAD HARRIS.
PAUL W. ZOINER.

No. 692.—*Design for Stoves.*

What I claim is, the design and configuration of the ornaments and mouldings herein described, forming an ornamental design for stoves.

S. W. GIBBS.

No. 693.—*Design for Cooking-Stoves.*

What we claim is, the design and configuration of the ornamental flutings and mouldings herein described, forming an ornamental design for cooking-stoves.

S. W. GIBBS.

No. 694.—*Design for Daguerreotype-Cases.*

What I claim is, the construction of a case for daguerreotype or other purposes, rounded on its front and back edges, and banded as described.

HENRY A. EICKMEYER.

No. 695.—*Design for Stoves.*

What we claim is, the combination of the different ornaments as represented in the drawing.

JOHN HANFBAUER.
HENRY WAAS.

No. 696.—*Design for Table Forks.*

What I claim is, the peculiar configuration of the parts C, D, and E, substantially as described.

JOSEPH W. GARDNER.

No. 697.—*Design for Cooking-Stoves.*

What we claim is, the design for "Fanny Forester" stoves, as herein set forth.

JACOB BEESLEY.
EDWARD J. DELANY.

No. 698.—*Design for Forks' and Spoons' Handles.*

What I claim is, the use and combination of the ornamental designs herein before described, with the handles of forks, spoons, and other articles of table cutlery, for the purpose of ornamenting them, substantially as set forth.

HENRY BIGGINS.

No 699.—*Design for Spoons.*

I claim the ornaments *a b i j*, and the beads *c f*, with their scrolls *d e g h*, when the whole are arranged to form a new and ornamental design for a spoon, as herein shown.

JOHN GORHAM.

No. 700.—*Design for Clock-Fronts.*

What I claim is, the design and configuration of the casting, as herein described, forming the front and sides of a clock-case.

WILLIAM B. LORTON.

No. 701.—*Design for Cooking-Stoves.*

What we claim is, the ornamental design and configuration of cook-stove plates, as herein described.

GEO. WARREN.
S. H. SWETLAND.
E. C. LITTLE.

No. 702.—*Design for Water-Coolers.*

What we claim is, the design, configuration, and arrangement of the herein described castings, and the ornaments in bas-relief thereon, constituting a new and original design for water-coolers.

GARRETTSON SMITH.
HENRY BROWN.
JOSEPH A. READ.

No. 703.—*Design for Cooking-Stoves.*

We claim the ornamental design and configuration of the side of the stove, inclusive of its base and cornice mouldings.

And we particularly claim the ornamental design of the larger door I, and that of each of the doors K L.

We also claim the ornamental design of each door P or Q, that of the hearth-plate, and that of each of the legs, the whole being as exhibited in figure I.

BENJAMIN WARDWELL.
EPHRAIM R. BARSTOW.
GEORGE C. HARKNESS.

No. 704.—*Design for Equestrian Statues.*

What I claim is, a design for an equestrian statue, such as represented.

CLARK MILLS.

No. 705.—*Design for Water-Coolers.*

What I claim is, the design and configuration of the parts A B C, as herein described, forming a water-cooler.

GEO. HODGETTS.

No. 706.—*Design for Cooking-Stoves.*

What I claim is, the arrangement and combination of the several figures and mouldings, the whole forming an ornamental design for a cooking-stove.

SAMUEL D. VOSE.

No. 707.—*Design for Cooking-Stoves.*

What I claim is, the arrangement and combination of the several figures and mouldings, the whole forming an ornamental design for a cooking-stove.

SAMUEL D. VOSE

No. 708.—*Design for Cooking Stoves.*

What I claim is, the arrangement and combination of the several figures and mouldings, the whole forming an ornamental design for a cooking-stove.

SAMUEL D. VOSE.

No. 709.—*Design for Parlor Stoves.*

What I claim is, the arrangement and combination of the several figures and mouldings, the whole forming an ornamental design for a parlor stove.

SAMUEL D. VOSE.

No. 710.—*Design for Sewing-Birds.*

What I claim is, the design for sewing-birds as herein set forth.

JOHN NORTH.

No. 711.—*Design for Parlor Stoves.*

I claim the composition of mouldings and the ornaments in relief on the front plate, base plate, and cornice or top plate, and considered irrespectively of those of the feet C.

And I also claim the peculiar design or configuration of the urn K, and that of each of the blowers.

ABNER J. BLANCHARD.

No. 712.—*Design for Cooking-Stoves.*

What I claim is the design of each of the doors as well as that oı the side-plate, the larger hearth-plate, and the leg.

ABNER J. BLANCHARD.

No. 713.—*Design for Stove-Plates.*

What I claim is, the combination and arrangement of ornamental figures and forms represented, forming the ornamental design for the plates of a cooking-stove.

S. W. GIBBS.

No. 714.—*Design for Cooking-Stoves.*

What I claim is, the ornamental design and configuration of stove-plates, such as herein described.

RUSSELL MANN.

No. 715.—*Design for Spoons.*

What we claim is, the use and arrangement of the ornamental devices hereinbefore described for ornamenting the front of the spoon and back of the spoon, or either of them, in combination with the spoon or other similar articles of table cutlery.

H. HEBBARD.
JOHN POLHAMUS.

No. 716.—*Design for Franklin Fire-Places.*

What I claim is, the ornamental design of the stove, as represented in the drawings, and particularly I claim that of each of the members hereinbefore mentioned.

N. P. RICHARDSON.

No. 717.—*Design for Labels on Bottles and Jars.*

What I claim is, the construction of the design and the fashioning of the letters, and the marking out of the same, so as to leave them on the glass in a permanent form.

W. A. ROGERS.

No. 718.—*Design for Ornamenting Stove-Plates.*

What I claim is, the combination and arrangement of ornamental figures and forms represented in the accompanying drawings, forming together the ornamental designs for the plates of a stove.

SAMUEL W. GIBBS.

No. 719.—*Design for Cooking-Stoves.*

What I claim is, the new design for a cooking-stove, consisting of the ornamental, plain, and scolloped mouldings *d d* and *e e*, and figures of harps or ancient lyres, as herein described.

APOLLOS RICHMOND.

No. 720.—*Design for Iron Railings.*

We claim the ornamental design and configuration of looped bars, rosettes, balls, and pickets, substantially as herein described.

M. H. FOWLER.
ENOCH JACOBS.

No. 721.—*Design for Stoves and Fire-Places.*

What I claim is, the ornamental design as exhibited in the accompanying drawings; or more particularly, that of the front plate and the end plate, the hearth plate, the leg, the top plate, and the blower.

WINSLOW AMES.

No. 722.—*Designs for Cooking-Stoves.*

We claim the configuration of surface and combination of ornaments, constituting the exterior design of the stove, as set forth.

RUSSELL WHEELER.
S. A. BAILEY.

No. 723.—*Design for Portable Fire-Places.*

What I claim is, the design of the front of the stove; or, more particularly, that of the urn of the front plate, the upper blower, the fender, and the lower blower, as shown in the drawings.

WINSLOW AMES.

No. 724.—*Design for Cooking-Stoves.*

What we claim is, the design, configuration, and arrangement of the ornaments on the several parts of the cooking-stove, as herein described.

CONRAD HARRIS.
PAUL W. ZOINER.

No. 725.—*Design for Trade-Marks.*

I claim the above ornamental design or trade-mark, to be fixed on articles of manufacture, and formed without the padlock.

THOMAS LEWIS.

No. 726.—*Design for Cooking-Stoves.*

I claim the bas-relief upon each foot *a*, a part of the design, the whole combined with the mouldings and panels exhibited in the drawings.

WM. T. COGGESHALL.

No. 727.—*Design for Table-Casters.*

I claim the wreath on the circular projection *h*, and the embellishments on the doors B and feet C, when the whole are arranged and formed, as herein shown, to constitute an ornamental design for a table-caster.

EDWARD GLEASON.

No. 728.—*Design for Parlor Grates.*

What I claim is, the design for parlor grates, as above set forth.

JAMES ANDREWS.

No. 729.—*Design for Stoves.*

I claim the design, configuration, and arrangement of the ornaments in bas-relief on stove-plates A and F, as herein described.

JAMES H. CONKLIN.

No. 730.—*Design for Ovens of Cooking-Stoves.*

What I claim is, the ornamental design and configuration of the side and end plates of an elevated oven of a cook-stove, as described.

GEORGE W. CHAMBERS.

No. 731.—*Design for Cast-Iron Monuments.*

What I claim is, a design for a cast-iron monument for the head of graves, combining the figures of the harp and heart with a recess for the insertion of a miniature likeness and inscription, and a locket of hair.

J. H. WILSON.

No. 732.—*Design for Metallic Covers for Jugs.*

What I claim is, making metallic covers and pitchers of the shape and pattern shown in the drawings.

ORRIN NEWTON.

No. 733.—*Design for Ornamenting Daguerreotype and other Mats.*

I claim the design of the ornaments surrounding the opening in the matting, and also the design of the ornament formed on the surface of the matting.

HIRAM W. HAYDEN.

No. 734.—*Design for Burial-Cases.*

I claim the form and configuration of metallic burial-cases, substantially as described.

M. H. CRANE.

No. 735.—*Design for Cooking-Stoves.*

What I claim is, the ornamental design and configuration of the cook-stove plates, as described.

JAMES WAGER.

No. 736.—*Design for Cooking-Stoves.*

What I claim is, the ornamental design and configuration of cook-stove plates, as described.

JAMES WAGER.

No. 737.—*Design for Parlor-Stove Plates.*

What I claim is, the ornamental design and configuration of parlor-stove plates, as described.

JAMES WAGER.

No. 738.—*Design for Stove-Plates.*

What I claim is, the ornamental design of a stove-plate, as described.

CALVIN FULTON.

No. 739.—*Design for Coal-Stoves.*

What I claim is, the combination of ornaments herein described, the whole forming an original design for a coal-stove.

JAMES HORTON.

No. 740.—*Design for Stoves.*

I claim the design for doors, top, and side-plates of a stove, as described.

ANDREW O'VEILL.

No. 741.—*Design for Cooking-Stoves.*

We claim the combination and arrangement of the several ornaments and mouldings as herein described, the whole forming an ornamental design for a cook-stove.

EZRA RIPLEY.
N. S. VEDDER.

No. 742.—*Design for Coal-Stoves.*

What we claim is, the combination of ornaments herein set forth, the whole forming an original design for a coal-stove.

GARRETTSON SMITH.
HENRY BROWN.

No. 743.—*Design for Stoves.*

What I claim is, the border B on the edge of the base A, the borders F and other embellishments on the doors E E, the wreath *m* on the rim or projection *l* of the cylindrical portion C of the stove, and the embellishments on the panels *i* on the cylindrical portion C, when the above parts are arranged as shown, to form an ornamental design for a stove.

BENJAMIN WARDWELL.

No. 744.—*Design for Steam-Tube and Hot-Air Covers.*

What we claim is, the scrolls *j k*, and the figured ground-work *l* on the trusses, and the circular embellishment A, and figured fret-work B on the panel C, and also the fret-work in the frame or borders D E, when the whole are arranged as herein shown, to constitute a new and ornamental design for a hot-air and steam-tube cover.

JAMES O. MORSE.
JULIN W. ADAMS.

No. 745.—*Design for Parlor Stoves to Burn Wood.*

What we claim is, the design, configuration, and arrangement of the several ornaments as set forth, forming a new and original design for a wood parlor-stove, to be known and called *Parlor Gem.*

CONRAD HARRIS.
PAUL W. ZOINER.

No. 746.—*Design for Clock-Frames.*

What I claim is, the combination of the octagon form with the oval corner, as distinguished from some other form.

J. C. BROWN.

No. 747.—*Design for Parlor Stoves to Burn Coal.*

What we claim is, the design, configuration, and arrangement of the several ornaments, as set forth, forming a new and original design for a parlor coal-stove, to be known and called the "Carbon."

CONRAD HARRIS.
PAUL W. ZOINER.

No. 748.—*Design for Six-Plate Box-Stoves.*

What we claim is, the mouldings, flutes, and scrolls, arranged as set forth, to form an ornamental design for six-plate or box-stoves.

CONRAD HARRIS.
PAUL W. ZOINER.

No. 749.—*Design for Cooking-Stoves.*

What we claim is, the configurations and arrangement of the several ornaments as set forth, forming a new and original design for premium cook-stoves, to be known and called the "Kanzas."

CONRAD HARRIS.
PAUL W. ZOINER.

No. 750.—*Design for Strap-Hinges.*

I claim the design for strap-hinges, to be made of malleable cast-iron, ornamented by the straight, curved, and annular ribs, substantially in the manner set forth.

ENOCH WOOLMAN.

No. 751.—*Design for Ships' Caboose-Stoves.*

What I claim is, the new design for a cook-stove, consisting of the bar mouldings, rivets, and bead or scolloped moulding, as described.

A. A. LINCOLN, Jr.

No. 752.—*Design for Table Knives and Forks.*

What I claim is, the part marked C, substantially as set forth.

JOSEPH W. GARDINER.

ADDITIONAL IMPROVEMENTS.

No. 118, to original Letters Patent No. 10,655.—*For Improvement in Seed-Planters.*

What I claim is, the attaching the box or hopper to the beam and handles by means of holes left in casting the box, or any equivalent device.

Also, placing the bottom of the lime-box below the slide for the purpose of preventing the lime from choking the machine and impeding its action, substantially as described.

J. GRAHAM MACFARLANE.

No. 119, to original Letters Patent No. 11,024.—*For Improved Lubricator.* (Plates, p. 347.)

I claim: 1st. The division of the plug into two longitudinal chambers C and D, and the relative positions of the feed and discharge openings in said chambers, so that while one chamber is discharging, a simultaneous feed will take place in the other.

2d. I claim the insertion of the tubes f and f^1 relative to the feed openings of cup and plug, as described, whereby they perform the double function of vent and steam passages; the feed-openings of the plug passing under the tubes, and discharging the steam contained in the plug, clear of the oil in the cup, before communicating with the feed-channel of the cup.

ROBERT M. WADE.

No. 120, to original Letters Patent No. 10,294.—*For improvement in Pen-Holders.* (Plates, p. 347.)

I claim constructing the thumb and finger rests *a b*, or either of them, so that they shall extend along the sides of the cylinder a greater or less distance, substantially as described, whether the outer corners of the same be rounded off or left full, or whether the said cylinder be adapted for fitting over the pen-barrel or tube, or for attaching to the end of the stick, so as to dispense with the barrel or tube as described, or whether the said rests be soldered to the cylinder A, or cut out with the same.

E. W. HANSON.

No. 121, to original Letters Patent No. 11,068.—*For Improvement in Saddle-Trees.* (Plates, p. 347.)

What I claim is, the combination with the hinged pommel and cantle of the self-adjusting side-pieces for the purpose of preventing an unequal pressure upon the edges of the side-pieces, however much the saddle-tree may be expanded or contracted.

WM. E. JONES.

No. 122, to original Letters Patent No. 12,039.—*For Mode of Regulating the Furnace of Hot-Water Apparatus.* (Plates, p. 247.)

What I claim is, the arrangement of the three dampers or valves P O M, and their several connecting-rods, in combination with a single float p laced in the open tank above, as set forth.

THOMAS J. TASKER.

No. 123, to original Letters Patent No. 10,341.—*For Machine for Sawing and Planing Clapboards.* (Plates, p. 347.)

Lever M, to which cutter F^1 is hung, is pivoted at B, its other end N being pressed against the board to be cut by a spring D. The lever G, which guides the saw-arbor E, is pivoted at H, its other end being pivoted at I to a rod J, which bears at K against the board.

What I claim is, the peculiar mode of hanging the front cutter F^1, or plane, or its equivalent.

Also the manner of hanging the saw or its equivalent, so that they will adapt themselves to the different thicknesses of boards; I claim them in combination as above set forth, or either one when operating separately.

EPHRAIM PARKER.

No. 124, to original Letters Patent No. 10,365.—*For Improvement in Quartz-Crushing Machines.* (Plates, p. 347.)

The object of this improvement is to keep the pestle down with sufficient force to pulverize the material operated on, and still to prevent the pestle from grinding too finely, *i. e.*, simply to crack the ore instead of grinding to powder.

I claim the combination of the weighted levers 12, and adjusting screws 11, with the pestle *d*, set and moving on the shaft *c*, the whole constructed and operating in the manner described.

JAMES HAMILTON.

No. 125, to original Letters Patent No. 12,590.—*For Improvement in Screw-Wrenches.* (Plates, p. 347.)

What I claim is, adding the tube E, with the nut G, surrounding the rod *c* of the movable jaw, to the arrangement of the adjustable toothed plate with springs, the toothed shank, and the eccentric, with its strap attached to the toothed plate, as set forth.

LORENZO D. GILMAN.

No. 126, to original Letters Patent No. 12,555.—*For Improvement in Fire-Arms.*

claim the combination of the *hammer, trigger, ratchet action, and cylinder spring stop-lever*, to operate jointly in the process of firing.

I also claim the apparatus for detaching and re-attaching the barrel to the stock, viz: the *bent* lever lying in a recess, within a metal projection from the barrel, with its *catch* at its back end, fitted to hold into a notch in the stock and kept in place by a spring lying within said recess, in combination with the hinge-plate, this arrangement being a substitute for the *thumb connecting-plate.*

F. NEWBURY.

No. 127, to original Letters Patent 12,752.—*For Improvement in Propellers.* (Plates, p. 348.)

What I claim is, the elongation of the paddles, or buckets, outside, to the same length that they are on the inside, or nearly so, of each wheel, or disc, and presenting a nearly equal surface on each side of the said wheels, or discs, so as to equalize the power and prevent any lateral strain or drawing of the said wheels, or discs, together, and forming perfect balance paddles, or buckets, taking almost all strain from the connecting-ring of said paddles, or buckets, as described.

WILLIAM D. JONES.

No. 128, to original Letters Patent No. 5,958.—*For Improvement in Warming and Ventilating Buildings, &c.* (Plates, p. 348.)

What I claim is, the foul-air receptacle *a*, said receptacle being connected with the vertical passages and ventilating chimneys, substantially as set forth.

R. RUTTAN.

No. 129, to original Letters Patent No. 12,124.—*For Improvement in Locks for Fire-Arms.* (Plates, p. 348.)

What I claim is, the receiver *e* and the connexion of the carrier *c* to the peculiarly arranged tumbler *b* used in my former lock, patented January 2, 1855, wherein the action is direct, and, being on the inside of the lock, is secured from any and all exposures; also the putting of the charged charger *d*, containing the primers, into the chambers *a*, by which means the primers are readily applied, and with greater precision than by any other now in use.

J. S. BUTTERFIELD.

No. 130, to original Letters Patent No. 13,270.—*For Improvement in Apparatus for Heating Feed-Water to Locomotive-Engines.* (Plates, p. 348.)

M are sectional cones deflecting the current of steam, gases, &c., in contact with the tubular feed-water heaters *c*.

What I claim is, the arrangement of the tubular water-heater sectional cones in relation to each other and to the chimney, as set forth.

DAVID MATTHEW.

DISCLAIMERS.

Improvements in Pump

Your petitioner, therefore, hereby enters his disclaimer to that part of the claim in the aforementioned specification which is in the following words, to wit:

"Second. I claim the particular mode above described, of constructing a pump with an air-chamber below the lower box, into the bottom of which chamber the pipe communicating with the cistern, or well, is inserted or connected in any proper manner, and through which chamber another pipe passes, the lower end of which is situated immediately over the top of the induction pipe, while the upper end is joined or connected to the top of the chamber, the said pipe communicating at top with the pump-barrel, the lower valve of the same being immediately over and upon its ends, and at bottom by any sufficient number of holes or orifices bored through the same, with the said chamber, through which it passes; the whole being arranged substantially as described, and for the purpose of permitting the water from the cistern or well to rise into the chamber during the downward stroke of the piston or upper box, and otherwise operating in the manner as herein before explained and set forth, meaning in the above not to claim the addition of an air-chamber to the lifting-pump, but my particular mode of constructing and applying the same as described."

Which disclaimer is to operate to the extent of the whole interest in said letters patent vested in your petitioner, by the grant to him, as aforesaid, who has paid ten dollars into the Treasury of the United States, agreeably to the requirements of the act of Congress in that case made and provided.

JESSEE REED.

Improvements in Marine Steam-Engines.

Your petitioner, therefore, does, by these presents, disclaim so much of the second section of the before recited specification of claim as covers broadly the lifting of the valves by means of mechanism acting directly upon the valve-stems, hereby limiting the said second section of the claim to the lifting of the said steam-valves, one at each end of the cylinder, by toes on shafts arranged outside of the steam-chest, and acting on a foot attached to and projecting from each valve-stem, substantially as described, in the before recited letters patent.

CHARLES W. COPELAND.

Improvements in the Construction and Operation of Wagon-Brakes.

Your petitioner, therefore, hereby enters his disclaimer to that part of the claim in the aforenamed specification which is in the following words, to wit: "The application of wagon-brakes to the forward wheels of wagons, by using the hounds, sway-bar, block-tongue, or other appendages, running back from and firmly attached to the front axle, as the frame for the support and steadying of such brakes."

Which disclaimer is to operate to the extent of my interest in said letters patent so granted to your petitioner, who has paid ten dollars into the Treasury of the United States, agreeably to the requirements of the act of Congress in that case made and provided.

JEHIAL E. BLODGETT.

Improvement in Wrought-iron Beams and Girders.

Your petitioner, therefore, hereby enters his disclaimer to so much of the said specification as may be construed as claiming, as the invention of the said Pollak, a combination of top and bottom T-pieces, with plates disconnected from each other, and so arranged as to leave spaces between them in a longitudinal direction; which disclaimer is to operate to the extent of the interest in said letters patent vested in your petitioner, who has paid ten dollars into the Treasury of the United States, agreeably to the requirements of the act of Congress in that case made and provided.

A. POLLAK.

EXTENSIONS.

Improvement in the Method of Working the Steam-Valves of Steam-Engines, when the Steam is cut off and allowed to act expansively.

What we claim is, the combination of an additional and separate eccentric wheel, to work a rock-shaft to raise the steam-valves, in combination with any of the several methods hitherto used for working the exhaust valves.

We also claim the manner in which the toes are applied to the rock-shaft, so that the shaft is made to vibrate during a certain interval without either toe communicating motion to either valve.

We also claim the combination of the cog-wheel and rack in the manner set forth, for the more completely effecting our object.

ROBERT L. STEVENS.
FRANCIS B. STEVENS.

Improvement in the Mode of Applying Water to Fire-Engines, so as to render their Operation more Effective.

What we claim is, the employing the pressure of a column of falling water, or the tendency of the hydraulic pressure on water at rest, to assist in the working of fire-engines, by combining a hose or pipe conducting said water with the receiving tubes of an engine or pump, operated by animal or mechanical power, the same being constructed as set forth.

FRANKLIN RANSOM.
UZZIAH WENMAN.

Improvement in Seed-Planters.

What we claim consists:

1st. In the arrangement of the spur-wheels, for the purpose of connecting the seed-roller Y and hopper P to the shaft O, as before described.

2d. The combination of the rectangular staple i^2, the beam L, the slider S^2, the compound lever L M N, and the hopper frame V W, for the purpose of throwing the hoppers in and out of operation, when sowing point and other land, without stopping the horse, as well as for planting seeds that require caltivation in rows, and also for raising the drill U above or allowing it to sink into the ground a sufficient distance, as described.

MOSES PENNOCK.
SAMUEL PENNOCK.

Machine for Cutting Square-Joint Dovetails.

What I claim is, the arrangement of the carriages, one moving at right-angles to the motion of the other, in combination with cutters arranged with their axles inclined, as described.

WILLIAM PERRIN.

Improvement in the Construction of Iron-Truss Bridges.

What I claim is, the method of sustaining the flooring of bridges by iron trusses, containing cast-iron arches. formed in sections or segments, in combination with diagonal ties or braces, to sustain the form of the arch against the effects of unequal pressure, (with or without vertical posts or rods,) and wrought-iron arch strings, or thrust-ties, to sustain the thrust, and prevent the spreading of the arch in case the abutments and pins be not relied on for that purpose. Also, the divergence or horizontal expansion of the arch from the middle portion to the ends thereof, in wooden trusses or arches as well as in those composed of iron.

SQUIRE WHIPPLE.

Improvement in the Form of the Screw Propeller for Propelling Vessels.

I claim curving the wings of the screw-paddles, or propellers, in a direction perpendicular to the shaft, or their axis of revolution, as set forth.

EBENEZER BEARD.

Improvement in Pumps.

I claim, 1st. The method of confining the lower valve to its seat, so that it may be easily removed therefrom, for repair or other purpose, by means of a spring, to which the valve is connected, and which rests on the upper surface of the bottom of the pump-barrel, its end pressing against the interior circumference of the barrel, the same being arranged and constructed as herein described.

2d. I claim the particular mode of constructing a pump with an air-chamber below the lower box, into the bottom of which chamber the pipe communicating with the cistern or well is inserted or connected in any proper manner, and through which chamber another pipe passes, the lower end of which is situated immediately over the top of the induction-pipe, while the upper end is joined or connected to the top of the chamber, the said pipe communicating at top with the pump barrel, (the lower valve of the same being immediately over and upon its ends,) and at bottom (by any sufficient number of holes or orifices bored through the same) with the said chamber through which it passes, the whole being arranged substantially as above described, and for the purpose of permitting the water from the cistern or well to rise into the chamber during the downward stroke of the piston, or upper box, and otherwise, as set forth.

3d. I claim the method of adjusting the pump-handle, or lever, which raises or depresses the upper box, by attaching said lever to the top plate or cover of the pump-barrel, and arranging said plate or cover as before described, so that it may be turned around and fixed in position by a screw, or other similar or suitable contrivance, as described.

JESSE REED.

Improvement in the Method of Constructing Screw-Wrenches.

What I claim is, moving the sliding-jaw by a screw combined with, and placed by the side of, and parallel with the bar of the permanent jaw and handle, substantially as described, when the required rotation for sliding the jaw is given by the head or rosette, or its equivalent, which retains the same position relatively to the handle during the operation, substantially as described.

And I also claim moving the sliding-jaw by a screw combined with, and placed by the side of, and parallel with the bar of the permanent jaw and handle, substantially as described, in combination with the rosette, or its equivalent, retained in its position relatively to the handle, in the manner described.

LORING COES.

Improvement in the Manner of Constructing Railroad Carriages, so as to Ease the Lateral Motion of the Bodies thereof.

What we claim is, the connecting the turning-bearing to the truck-frame of the above described kind, resting on four wheels or more, by a mechanism substantially such as described, that shall not only allow such turning-bearing (independently of the wheels and axles) a lateral play, movement, or movements, in directions transversely of the carriage, but bring or move it back to its central position, after the lateral deflective force has ceased to act.

CHARLES DAVENPORT.
ALBERT BRIDGES.

Improvements in Machinery for Removing Bars and other Obstructions from Harbors, and for Forming and Cleaning out Docks.

What I claim is, the combination of the scrapers K K with the swivel-plate M M, for the purpose and in the manner described.

Also, the arrangement of ploughs in combination with the cutters J for the purpose and in the manner specified.

J. R. PUTNAM.

Improvements in the Machine for Riving and Dressing Shingles.

What I claim is, the manner in which I have combined and arranged the frame *b b* for holding the bolts with the vibrating frame *c c*, its panels and riving-knife, with their appendages, so as to rive shingles from the bolt, by an apparatus operating substantially as described.

I also claim the within described manner of constructing and combining the dressing and jointing apparatus, the dressing-knives being made to approach towards each other, and the jointing-frames being affixed and operating as set forth.

WM. S. GEORGE.

Improvement in the Manner of Arranging the Low-Pressure or Condensing Steam-Engine, so as to Adapt its Parts to be Used by Vessels for Ocean Service.

I do not claim the mere placing of the cylinder of a steam-engine obliquely, as this has been done for other purposes; but as I produce a new and useful effect by so placing the steam cylinder and its appendages in the combination above claimed, on board of vessels for navigating the ocean, I limit my claim to the so placing them under the said combination as to attain the objects herein fully made known.

Secondly. I claim the manner of arranging and working the steam and exhaust valves, as set forth, the same being effected by a direct action; that is to say, without the employment of the lifting-rods and lifters usually required for that purpose.

Thirdly. I claim the manner of combining and arranging the condensing apparatus, the air-pump being placed at the same angle, or nearly so, with the cylinder, and attached by its lower end to the channel-plate, the delivery valve being also placed on the upper part of said plate, the combination intended to be claimed under this last head consisting in the arranging of the several parts enumerated; that is to say, the air-pump, the channel-plate, and the delivering-valve substantially in the way herein described.

CHARLES W. COPELAND.

Improvement in the Endless Chain Horse-Power for Driving Machinery

What is claimed is, the links of the parallel endless chains which carry the travelling-bed, formed with cogs on their inner edge, meshing into the side pinions K on the driving-shaft, when the latter is arranged back of the forward end of the power to be received by the straight run of the cog-links over the said pinions, as shown and described.

ALONZO WHEELER.
ALEXANDER F. WHEELER,
Executor of this last will and testament of
WM. C. WHEELER, *deceased.*

Improvement in the Portable Circular Saw-Mill.

What we claim is, the employment of a revolving fluted or channelled cylinder like that herein represented and marked G, or of any analogous apparatus which will present, in rapid succession, a number of picking or cleaning edges, to operate upon the burrs, seeds, or other foreign matter contained in wool or cotton, in combination with the fine-comb cylinder and the picker cylinder, and other apparatus analogous thereto, by which wool is carried up and presented to the action of the revolving fluted or channelled cylinder, as herein represented.

WM. W. CALVERT.
ALANSON CRANE.

Improvement in the Manner of Constructing Gins for Ginning Cotton.

I claim the forming of a speculum ani, with tapering or conical blades, united at their larger ends to forcep handles, standing at a suitable angle with the blades to admit of the ready inspection of the parts, and furnished with a set-screw to regulate the opening of the blades, by which combination and arrangement of its parts the instrument is rendered more effective, and more convenient in use, than such as have been heretofore made for the same purpose.

JOSEPH T. PITNEY.

Improvement in Machines for Removing Buildings, &c.

What I claim is, the combination of the pivot and bolster, herein referred to, with the truck employed for removing buildings and other weights, consisting of an axle and wheels, having mortises on their periphery, to which the levers are adapted for moving them. Also, in combination with said truck, the screw figure, having a movable nut, with hooks attached to it, for raising buildings, all as herein described.

LEWIS PULLMAN.

Machine for Sticking Pins into Papers.

What I claim is:

1st. The plate, with grooves for receiving the pins.

2d. The sliding-hopper, which deposites the pins in the grooves as described.

3d. The sliding-plate or follower, with the wires attached thereto, in combination with the hopper, as described.

SAMUEL SLOCUM.

Improvement in Machinery for Making Pipes or Tubes of Lead, Tin, and other Metallic Substances.

What we claim is:

1st. The long core or core-holder, formed and held stationary with relation to the dies, as described.

2d. We claim the constructing of the piston B hollow, in the manner described; and the combination of the same with the long core or core-holder, upon which the piston slides.

3d. We claim as a modification of our invention, the arrangement and combination of the several parts above mentioned, as exhibited in what has been termed "*the reverse arrangement,*" shown at figure 7 in the accompanying drawings.

GEORGE N. TATHAM.
BENJ. TATHAM, Jr.

Improved Manufacture of Wire-Heddles for Weavers' Harness.

What we claim is, the making of such heddles in one continuous piece from end to end without a joint; and this we claim whether said heddles be made of one entire piece, recurved at one end to form the end-eye, or of two pieces of wire, each having the end-eyes formed by twisting the ends of the wire round in the ordinary manner.

ABRAHAM HOWE.
SIDNEY S. GRANNIS.

Improvement in the Saw-Mill for Re-sawing Boards and other Timber.

What I claim is, the method of presenting, gauging, or guiding the board by means of the rest and pressure-rollers, or their equivalents, substantially as herein described, in combination with the saw, substantially as described.

And I also claim the method of hanging and straining the saw by the combination of three stirrups at the ends of the saw, constructed and connected in manner substantially as herein described.

PEARSON CROSBY.

Improvement in Spark-Arresters.

What I claim is, the mode of separating sparks and other particles of matter from the gaseous current discharged from locomotive or other chimneys, by passing the current from a central chamber through tangential openings into a larger circular chamber around them, wherein the sparks and particles are retained by centrifugal force, and revolve till they are consumed, or are passed out through proper openings, as through J J, into exterior chambers made for that purpose.

WILLIAM C. GRIMES.

Improvement in Machines for Threshing and Winnowing Grain.

What I claim is, the combination of the stationary screen K, and spring-rakes U, their teeth projecting through the apertures in the screen with the straw-carriers arranged above said screen, as set forth.

Also, in combination with the foregoing arrangement, the inclined plane B^2 and returner R^2, the latter having cross-pieces r, for pushing the grain down the inclined plane B^2, to the screens Q^1, &c., all as set forth.

Also, constructing the movable shoe Q with a chaff-screen Q^1, as set forth, in combination with the screens Q^2 Q^3 and tail-screen Q^4 arranged below it, and the inclined board V, for separating the tailings from the clean grain, the whole being combined and operating as described; likewise the combination of the foregoing with the return-belt R^2 and fan E, and further combining these with the elevators X and Z, and the trunks W and Y.

ANDREW RALSTON.

Errata in Classified List of Patents issued (Vol. 1, *pp.* 131–208).

Page 132, Class I.—Agriculture, 12,690* (Cultivator, A. H. Morrell, Marlen, Texas, April 10, 1855) should be inserted between 12,653 and 12,744.
135, Class I. Agriculture........................instead of 13,480, read 12,480.
140, Class II.—Metallurgy........................instead of 13,052, read 13,050.
141, " " 13,295 should be inserted between 13,255 and 13,625.
141, " " instead of 13,355, read 13,255.
144, Class III.—Fibrous and Textile Manufactures, 13,698 should be inserted between 13,614 and 13,963.
145, " " " " instead of 13,740, read 13,750.
145, " " " instead of 12,563, read 12,565.
146, " " " instead of 13,620, read 13,629.
147, " instead of 12,557, read 12,577.
147, " " " instead of 12,734, read 12,754.
148, " " " " instead of 13,550, read 13,350.
150, Class IV.—Chemical Processes..............instead of 12,576, read 12,116.
156, Class V.—Calorifics........................instead of 13,535, read 13,525.
156, " "........................instead of 13,233, read 13,283.
158, " "........................instead of 13,686, read 13,868.
158, " " 13,728 should be inserted between 13,589 and 13,868.
161, Class VI.—Steam and Gas-Engines...........instead of 13,805, read 12,805, and insert it bet'n 12,495 and 13,238.
161, " " "...........instead of 13,622, read 13,652.
162, " " "...........instead of 13,350, read 13,359.
163, Class VII.—Navigation......................instead of 13,184, read 12,184.
164, " "......................instead of 12,940, read 13,940.
165, " "......................instead of 12,778, read 12,718.
166, Class VIII.—Mathematical Instruments.......instead of 13,062, read 13,162.
166, " ".......instead of 13,665, read 13,655.
166, " ".......instead of 12,121, read 12,129.
168, Class IX.—Engineering......................instead of 12,234, read 13,234.
169, " "......................instead of 12,126, read 12,136.
171, Class X.—Land Conveyance....................instead of 13,860, read 13,869.
173, " "....................instead of 12,252, read 12,251.
174, " " 13,297 should be inserted between 12,976 and 13,418.
174, Class XI.—Hydraulics and Pneumatics.........instead of 12,743, read 13,743.
175, " " ".........instead of 13,219, read 13,319.
175, " " ".........instead of 12,625, read 12,626.
176, " 13,559 should be inserted between 13,459 and 13,622.
176, ".........instead of 13,179, read 13,979, and insert it bet'n 13,967 and 13,590.
177, " " ".........instead of 12,246, read 12,346.
178, " " ".........instead of 13,069, read 13,067.
179, Class XII.—Lever and Screw, &c., 12,259, see Class IV.
179, " "...........instead of 13,191, read 13,199.
181, Class XIII.—Grinding Mills, &c., 13,415, see Class VI.
181, " "...........instead of 13,927, read 13,937.
184, Class XIV.—Lumber...........................instead of 13,102, read 13,101.
187, " "...........................instead of 13,251, read 13,351.
187, " "...........................instead of 13,719, read 13,354, and insert it bet'n 13,305 and 13,357.
187, " "...........................instead of 13,916, read 13,716.
188, " "...........................instead of 12,606, read 12,600.
191, Class XV.—Stone and Clay....................instead of 13,860, read 13,866.
192, Class XVI.—Leather..........................instead of 13,101, read 13,901, and insert it between 13,886 and 13,914.

---

* For Description and Claims No. 12,690, see Vol. I, p. 300.

Page 195, Class XVII.—Household Furniture, &c......instead of 13,088, read 13,188.
196, " " " ......instead of 12,361, read 12,661.
196, " " " ......instead of 13,150, read 12,150, and insert it bet'n 12,383 and 12,271.
197, " " " ......instead of 13,272, read 13,372.
198, Class XVIII.—Arts Polite, Fine, &c.........instead of 13,199, read 13,166.
198, " " " ....instead of 13,575, read 13,515.
199, " " " ....instead of 12,836, read 12,736.
200, ...instead of 13,956, read 13,580.
201, " .. " ....instead of 13,691, read 13,671.
201, " " " ....instead of 13,740, read 13,470.
203, Class XIX.—Fire-Arms, 13,294 should be inserted between 13,293 and 13,474.
203, " " instead of 12,539, read 12,529.
204, " " instead of 13,706, read 12,440, and insert it bet'n 12,328 and 12,470.
204, " " ....instead of 12,592, read 12,522.
204, " " ....instead of 13,678, read 13,679.
206, Class XXI.—Wearing Apparel. ....instead of 13,213, read 13,212.

ILLUSTRATIONS.

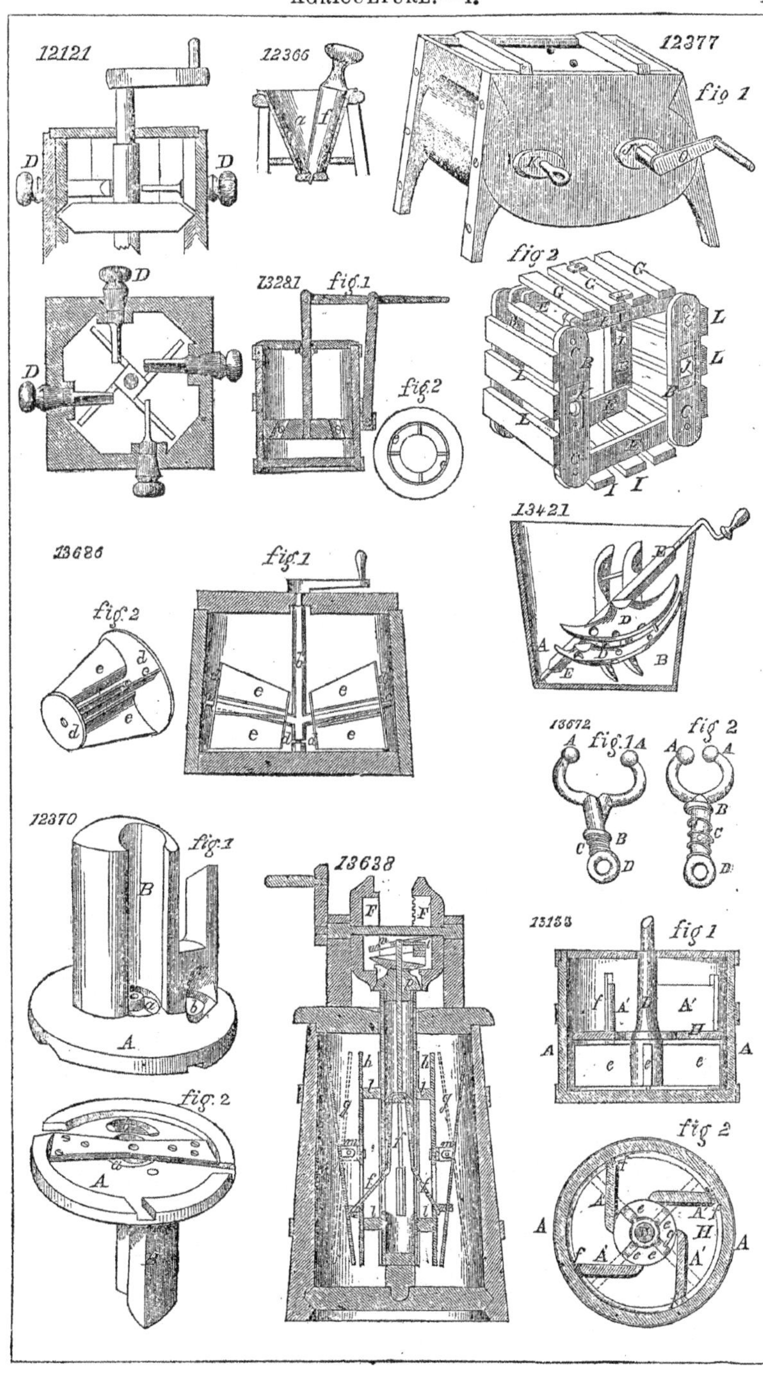
12121
12366
12377
fig 1
fig 2
13281 fig.1
fig.2
13421
13686
fig.2
fig.1
13672 fig.1A
fig 2
12370
fig.1
13638
13133
fig 1
fig.2
fig 2

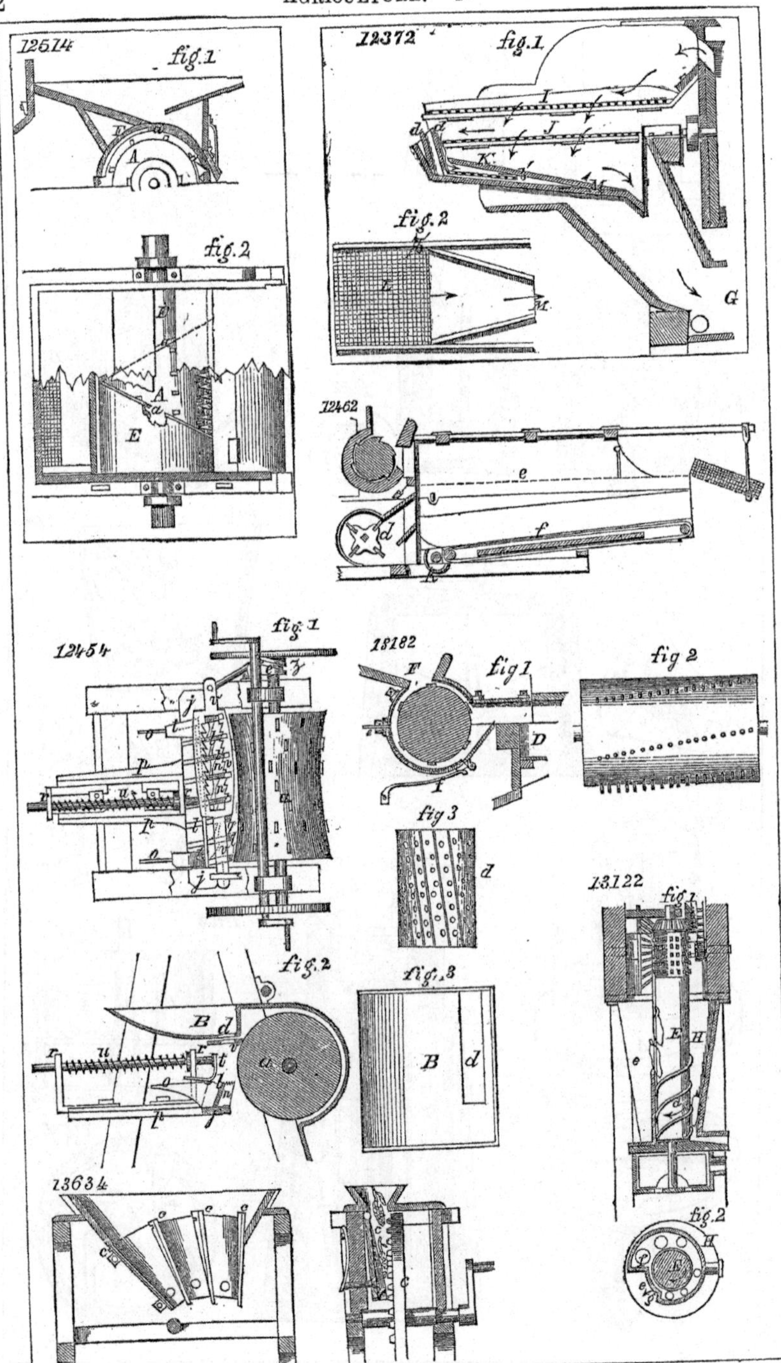
12514
fig.1
fig.2
12372
fig.1
fig.2
12462
12454
fig.1
fig.2
18182
fig 1
fig 2
fig 3
fig.3
13122
fig.2
13634

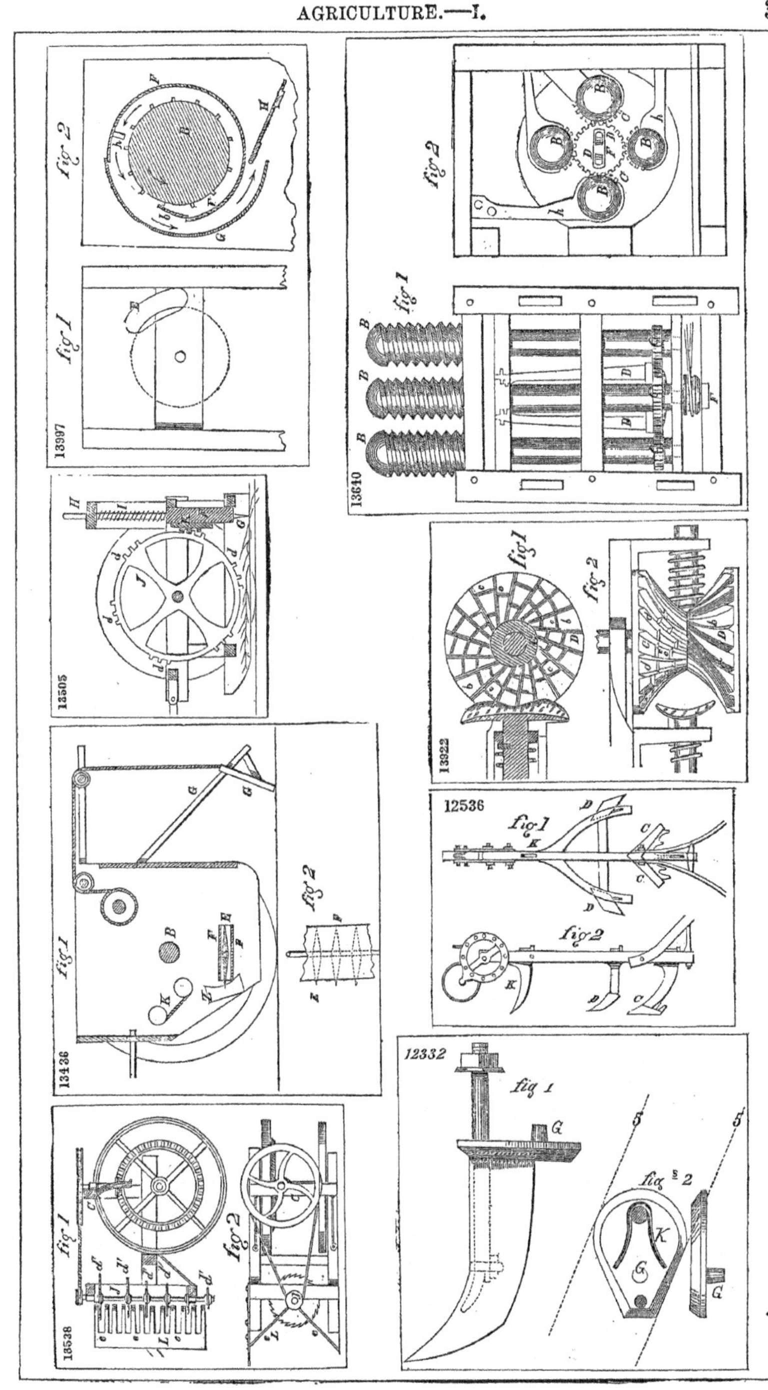
13997
13640
13505
13922
13436
12536
13538
12332

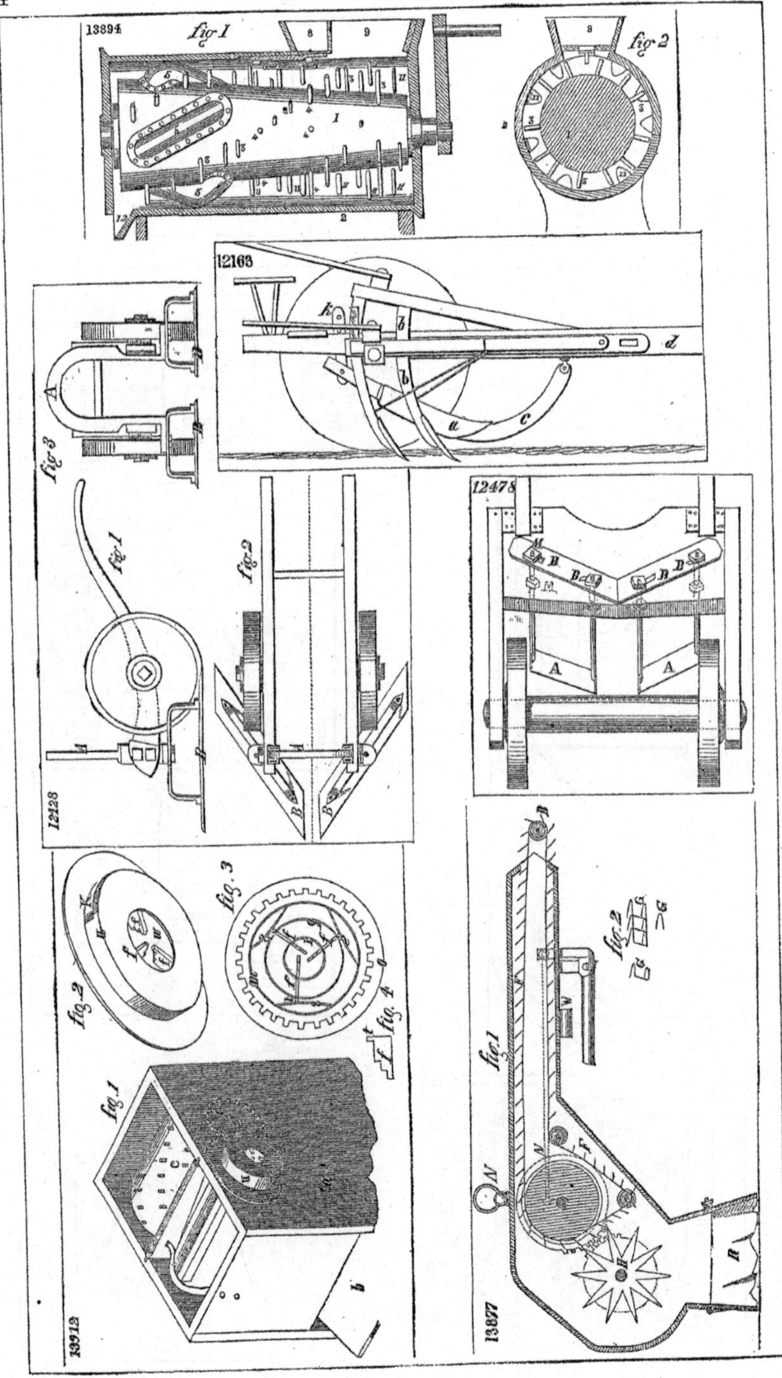
13894
fig 1
fig 2
12163
12128
12478
13912
13677

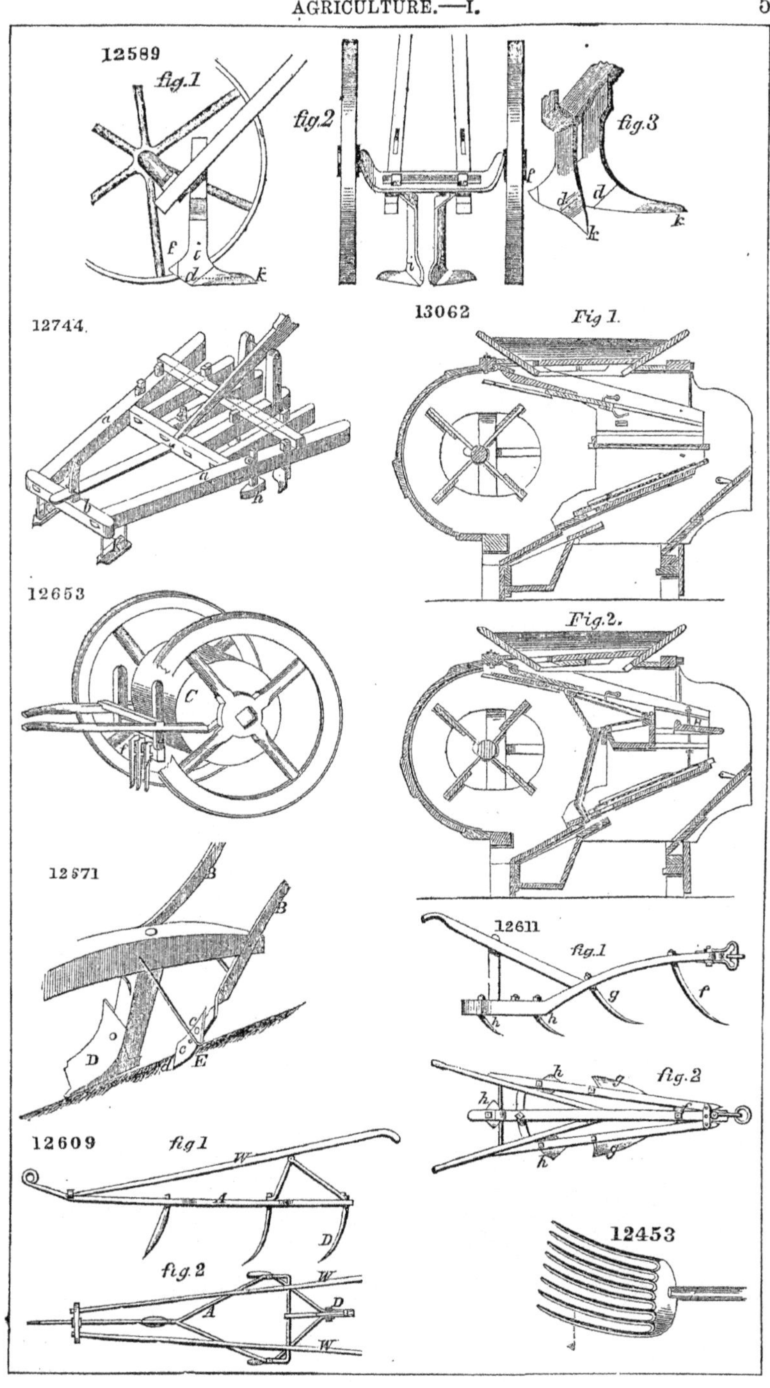
12589
fig.1
fig.2
fig.3
12744.
13062
Fig 1.
12653
Fig.2.
12571
12611
fig.1
fig.2
12609
fig 1
fig. 2
12453

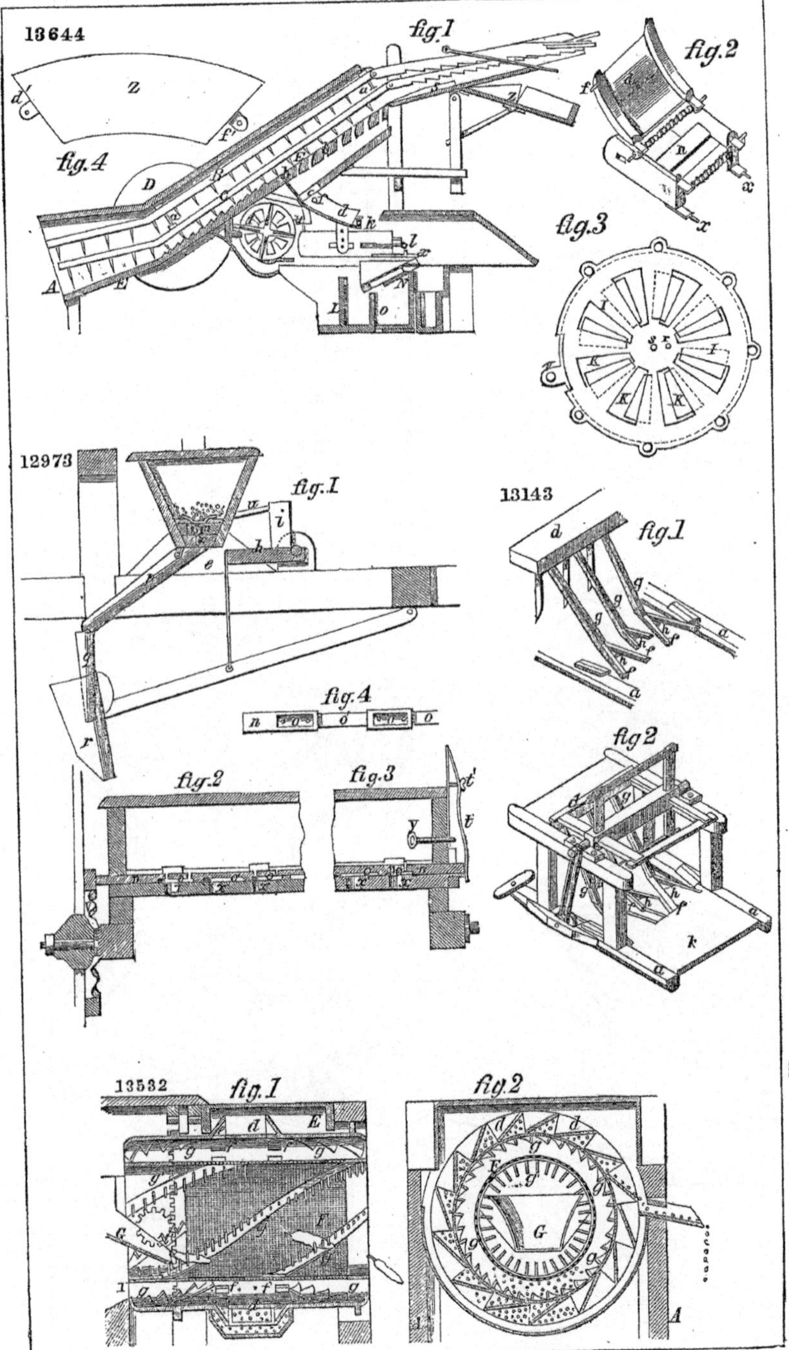
13644
fig.1
fig.2
fig.3
fig.4
12973
fig.1
fig.2
fig.3
fig.4
13143
fig.1
fig.2
13532
fig.1
fig.2

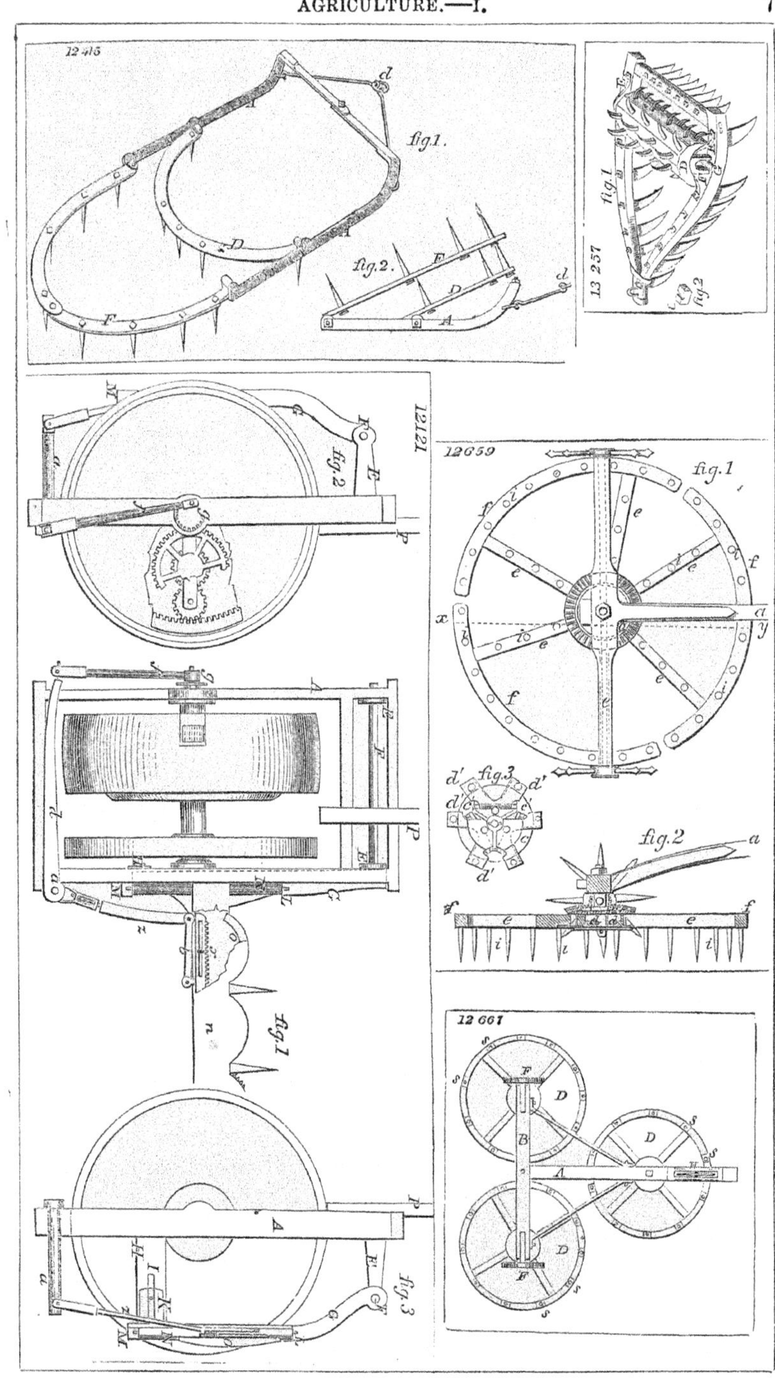
12 415
fig.1.
fig.2.
13 257
fig.1
fig.2
12121
fig.2
fig.1
fig.3
12659
fig.1
fig.3
fig.2
12 661

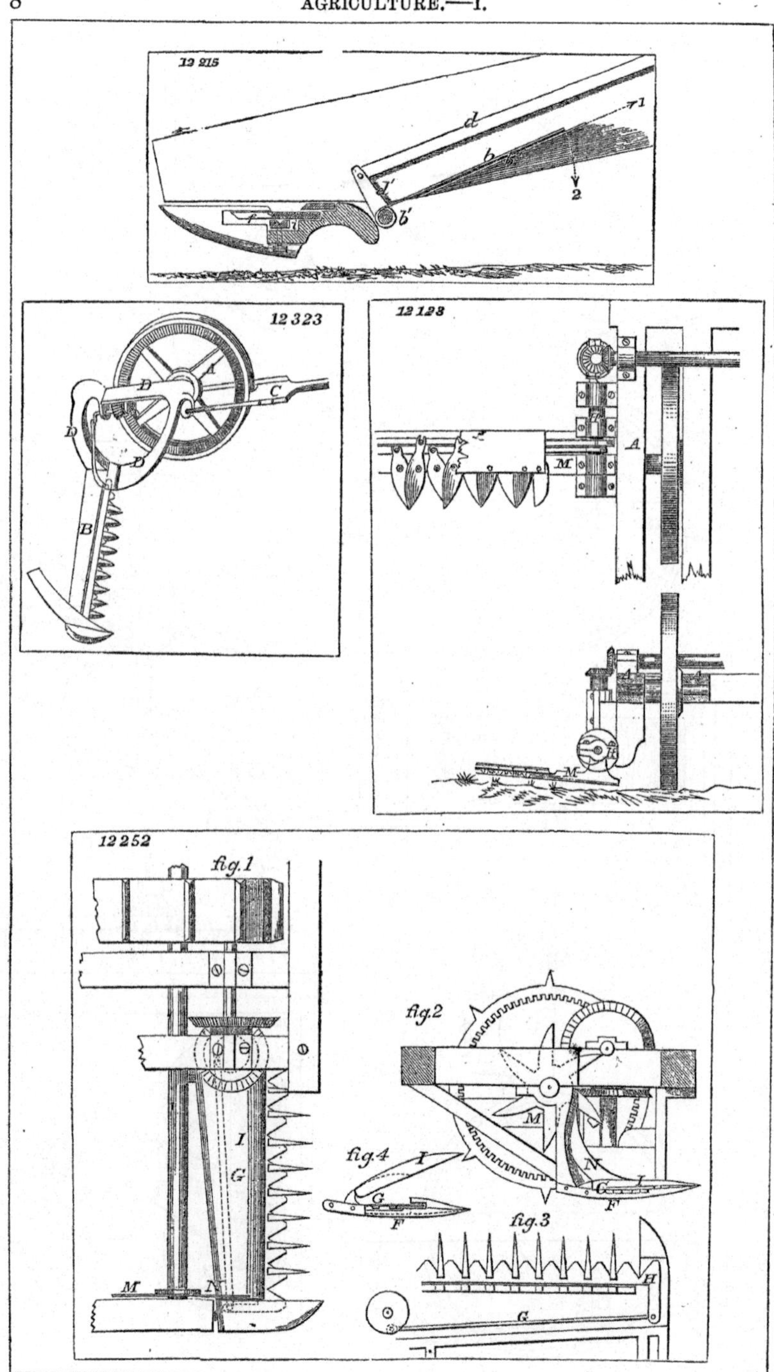
12 215
d
b
b'
1
2
12 323
A
D
C
B
12 128
A
M
12 252
fig.1
fig.2
fig.3
fig.4
I
G
M
N
F
H

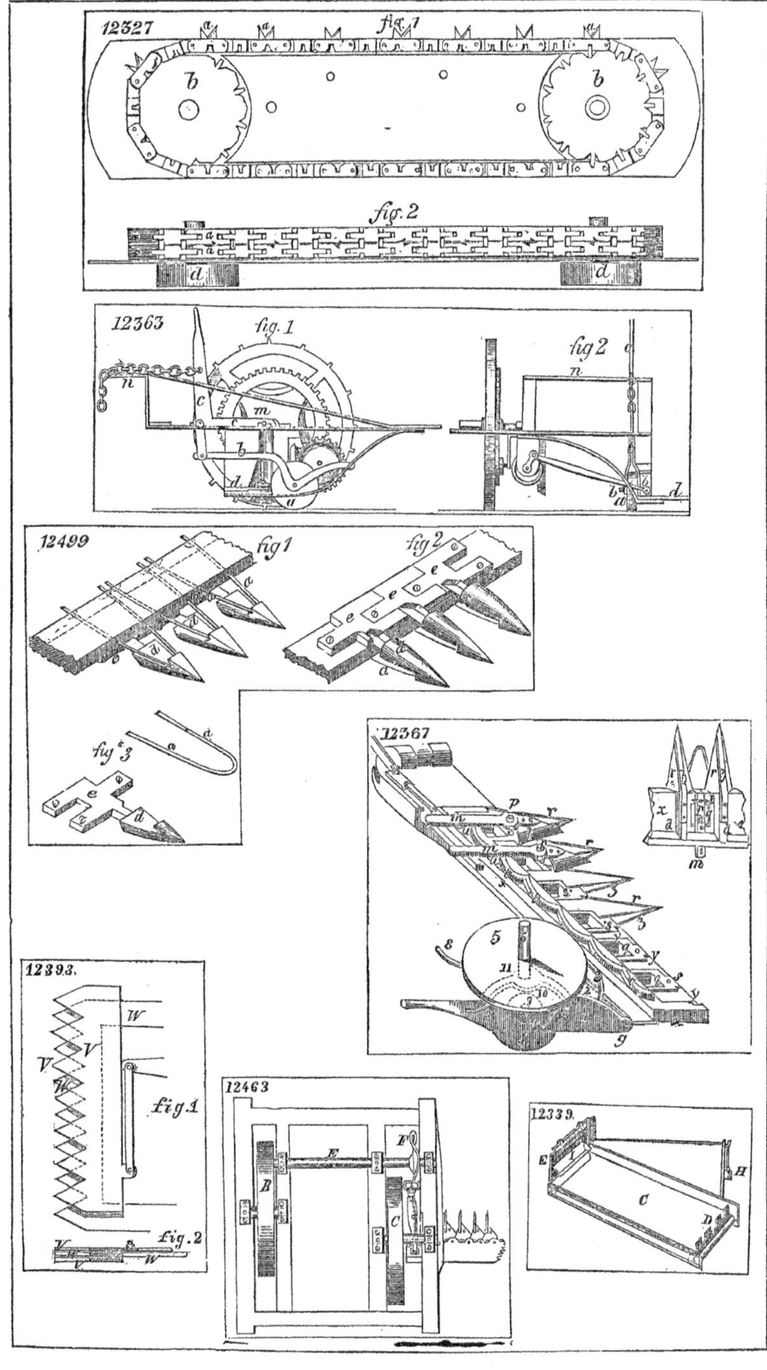
12327
fig. 1
fig. 2
12363
fig. 1
fig 2
12499
fig 1
fig 2
fig 3
12367
12393.
fig. 1
fig. 2
12463
12339.

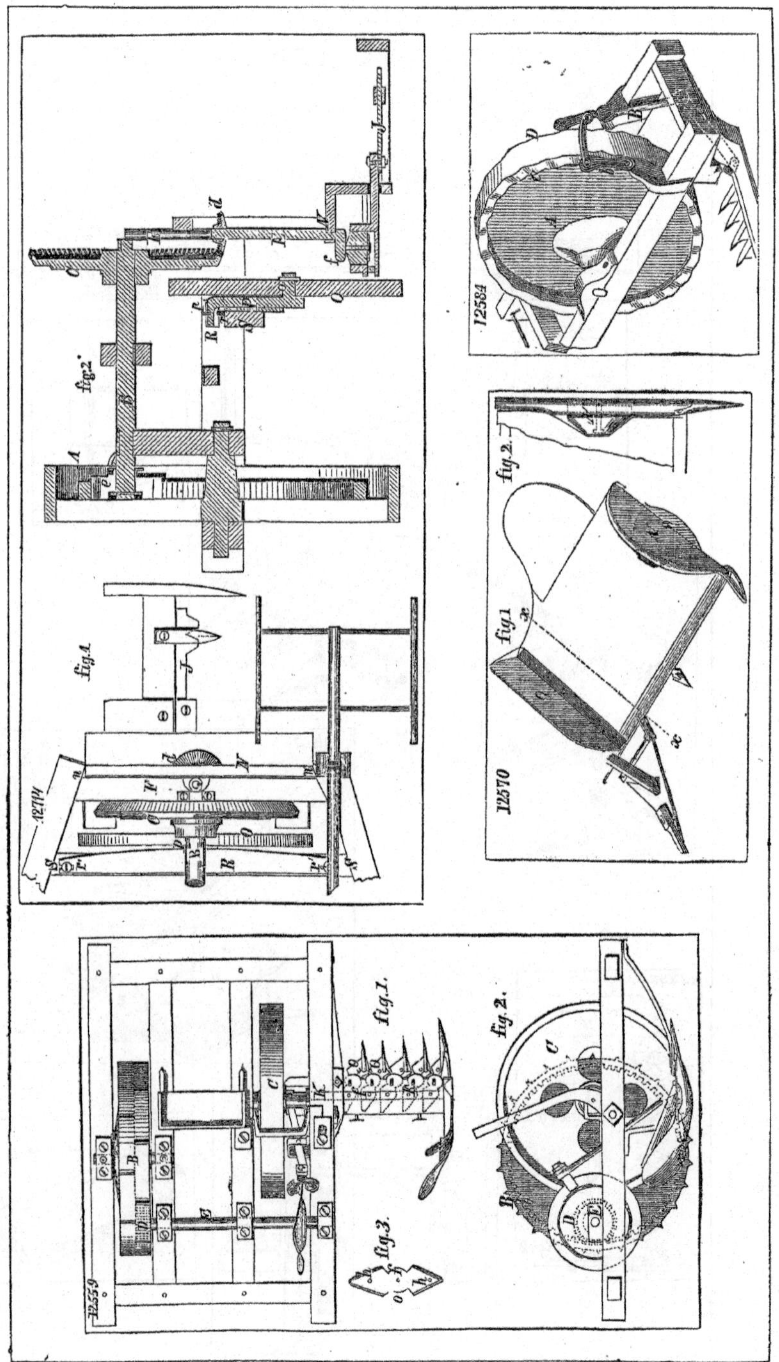
12714
fig.1
fig.2
12584
12570
fig.1
fig.2
12559
fig.1
fig.2
fig.3

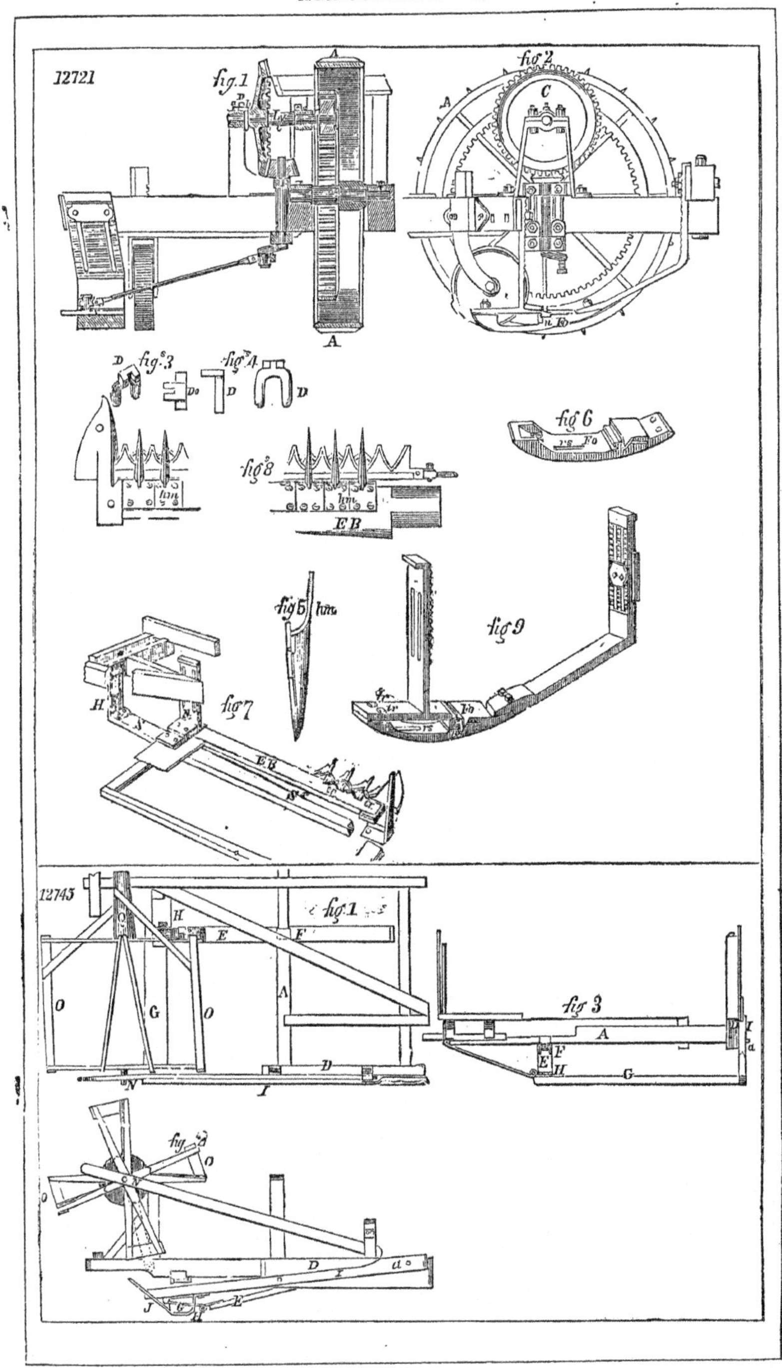
12721
fig. 1
A
fig. 2
C
fig. 3
fig. 4
fig. 6
fig. 8
EB
fig. 5
fig. 9
fig. 7
12745
fig. 1
H
E
F
O
G
A
D
I
N
fig. 3
A
G
fig. 2

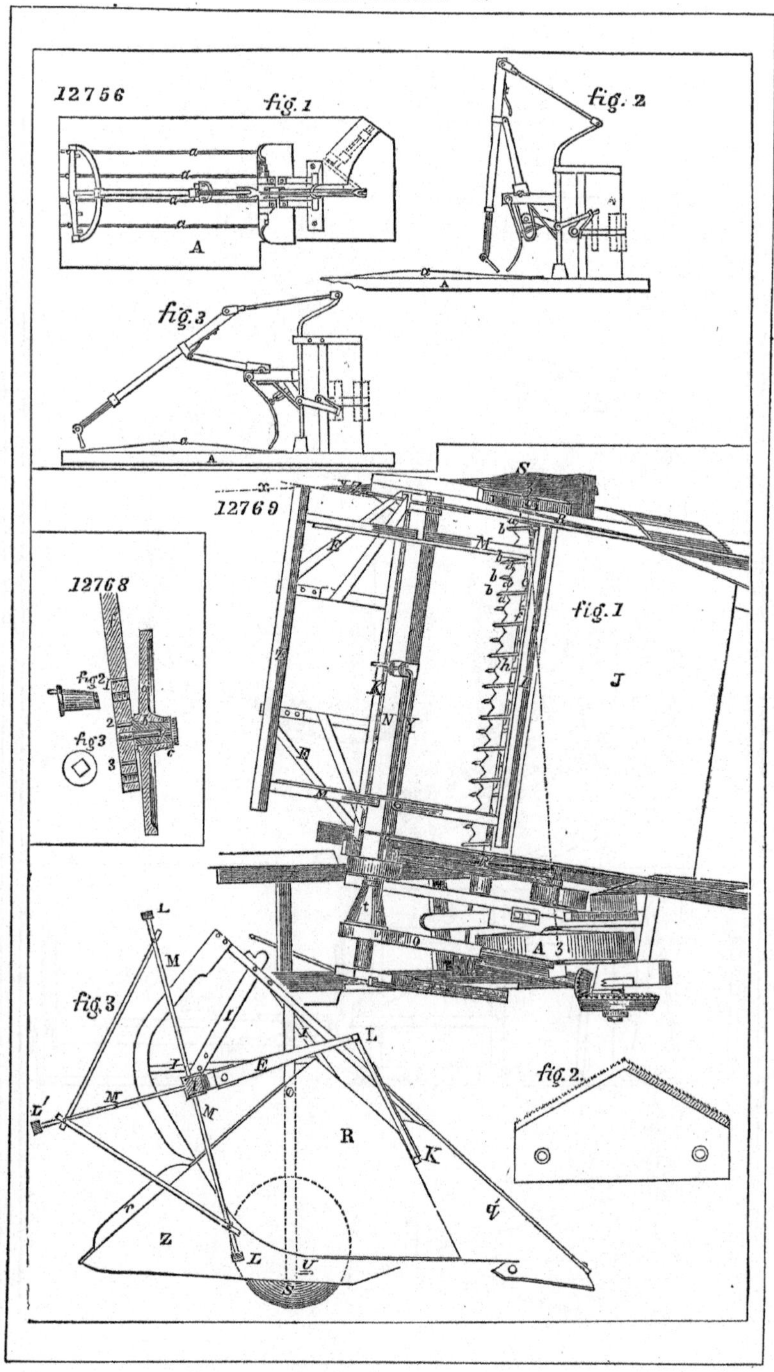
12756
fig. 1
fig. 2
fig. 3
12769
12768
fig. 2
fig. 3
fig. 1
fig. 3
fig. 2.

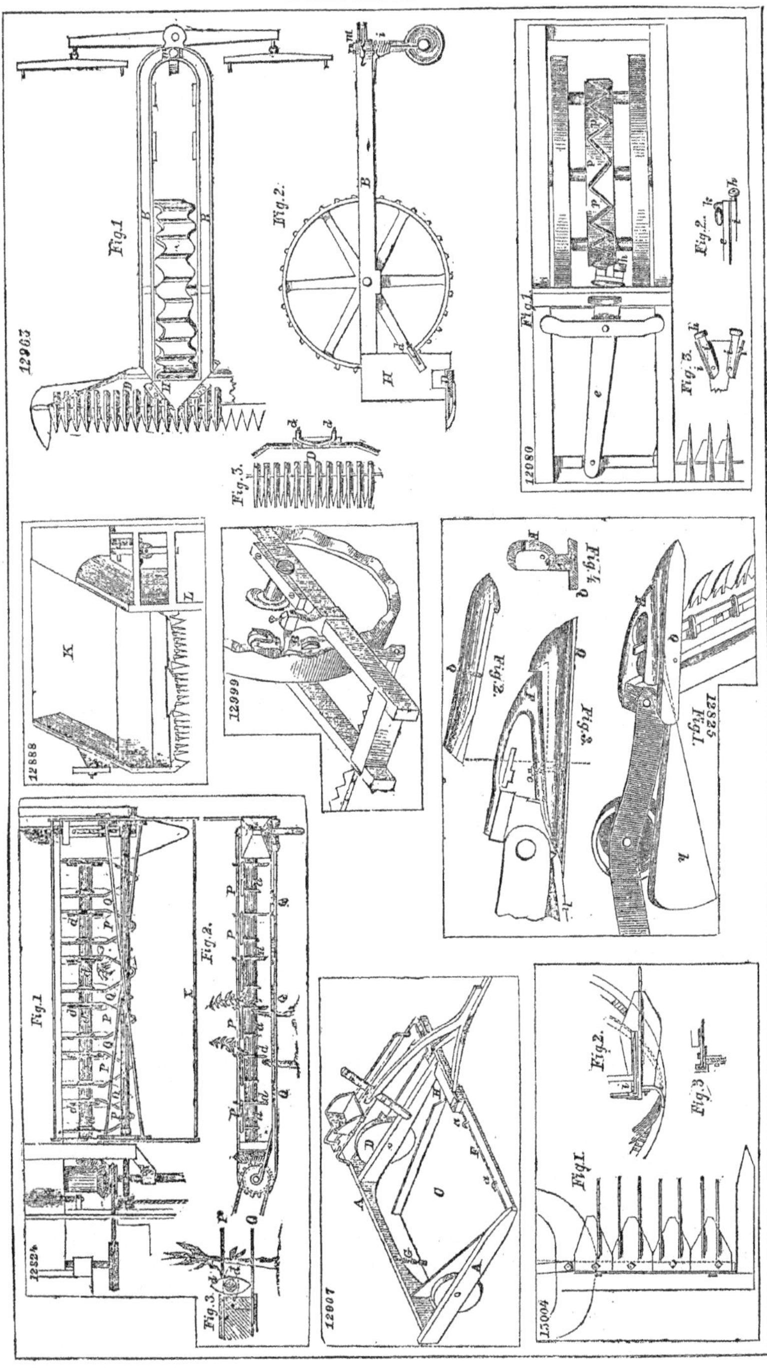
12963
12080
12888
12999
12825
12824
12807
13004

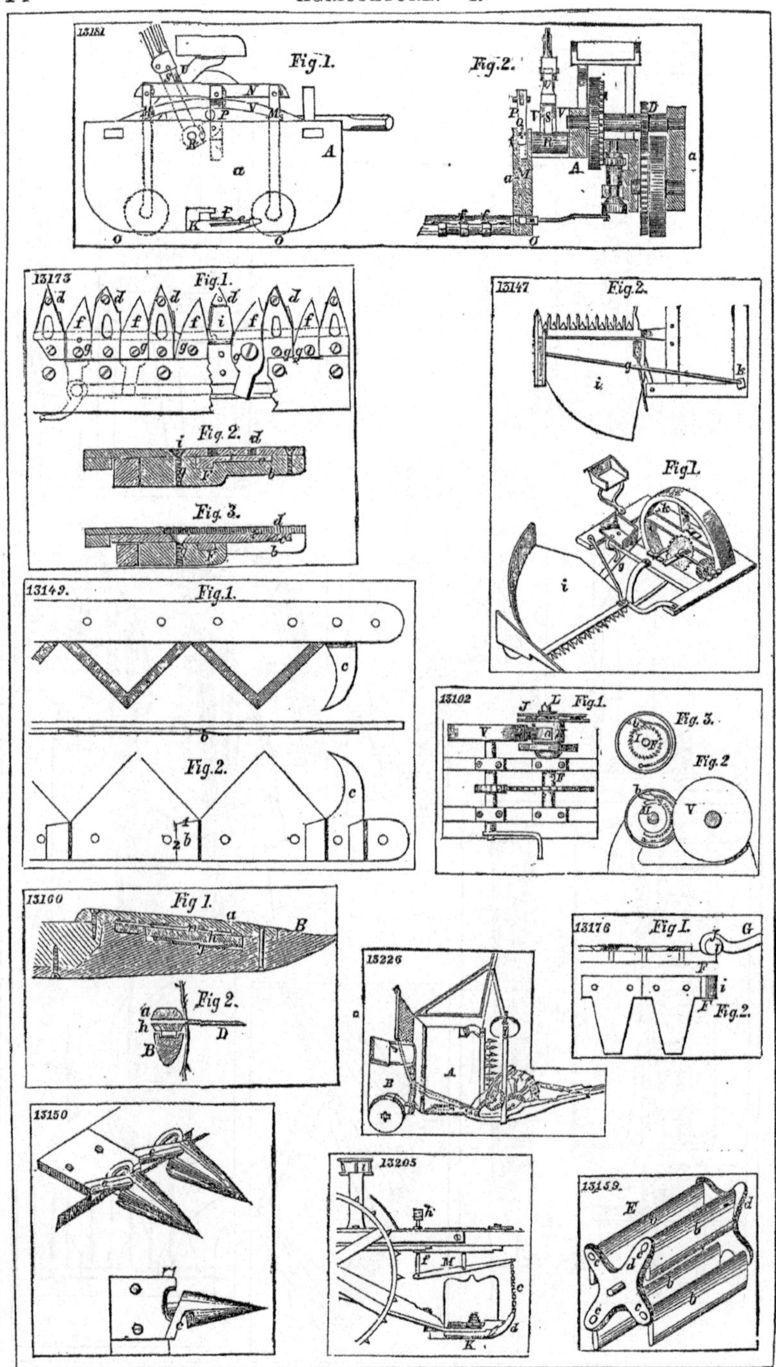
13181
Fig. 1.
Fig. 2.
13173
Fig. 1.
Fig. 2.
Fig. 3.
13147
Fig. 2.
Fig. 1.
13149.
Fig. 1.
Fig. 2.
13102
Fig. 1.
Fig. 3.
Fig. 2
13160
Fig 1.
Fig 2.
13176
Fig I.
Fig. 2.
13226
13150
13205
13159.

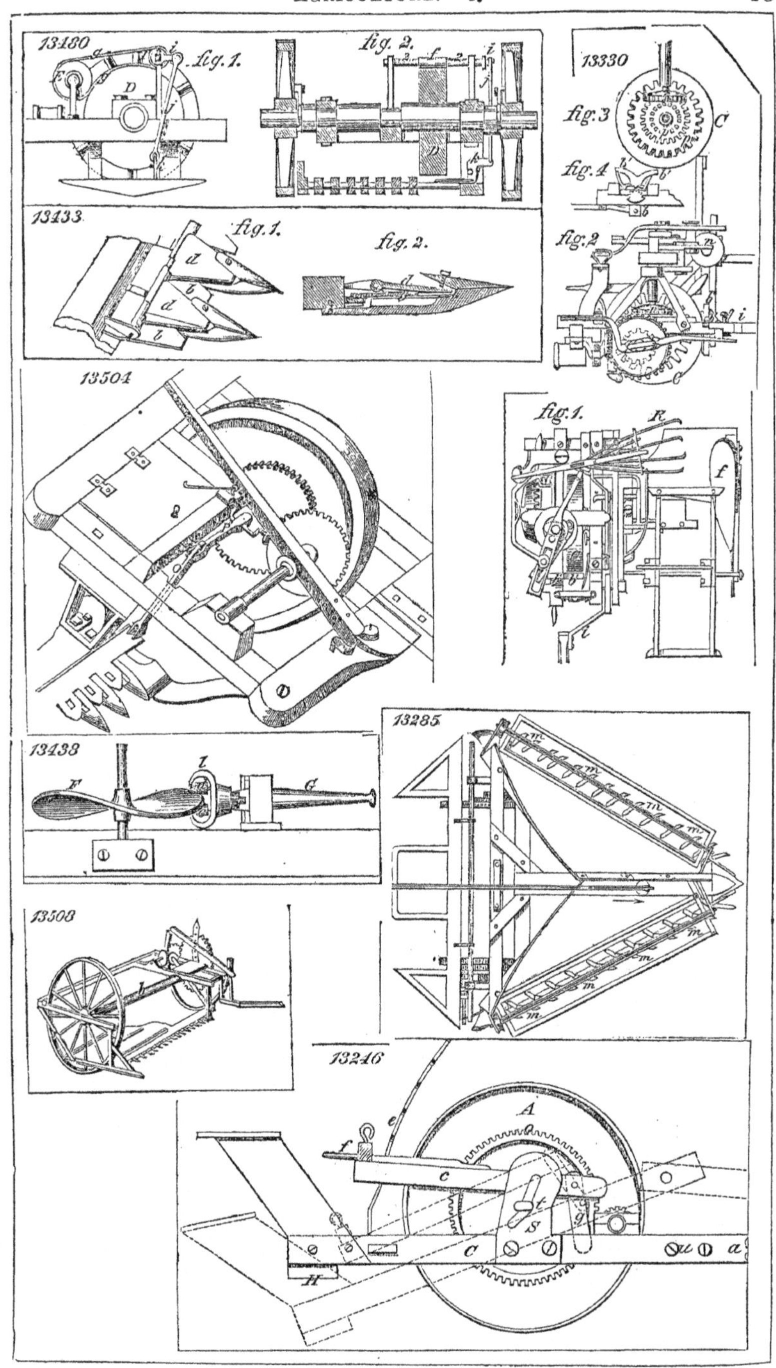
13480
fig. 1.
fig. 2.
13330
fig. 3
fig. 4
fig. 2
13433
fig. 1.
fig. 2.
13504
fig. 1.
13285
13438
13508
13246

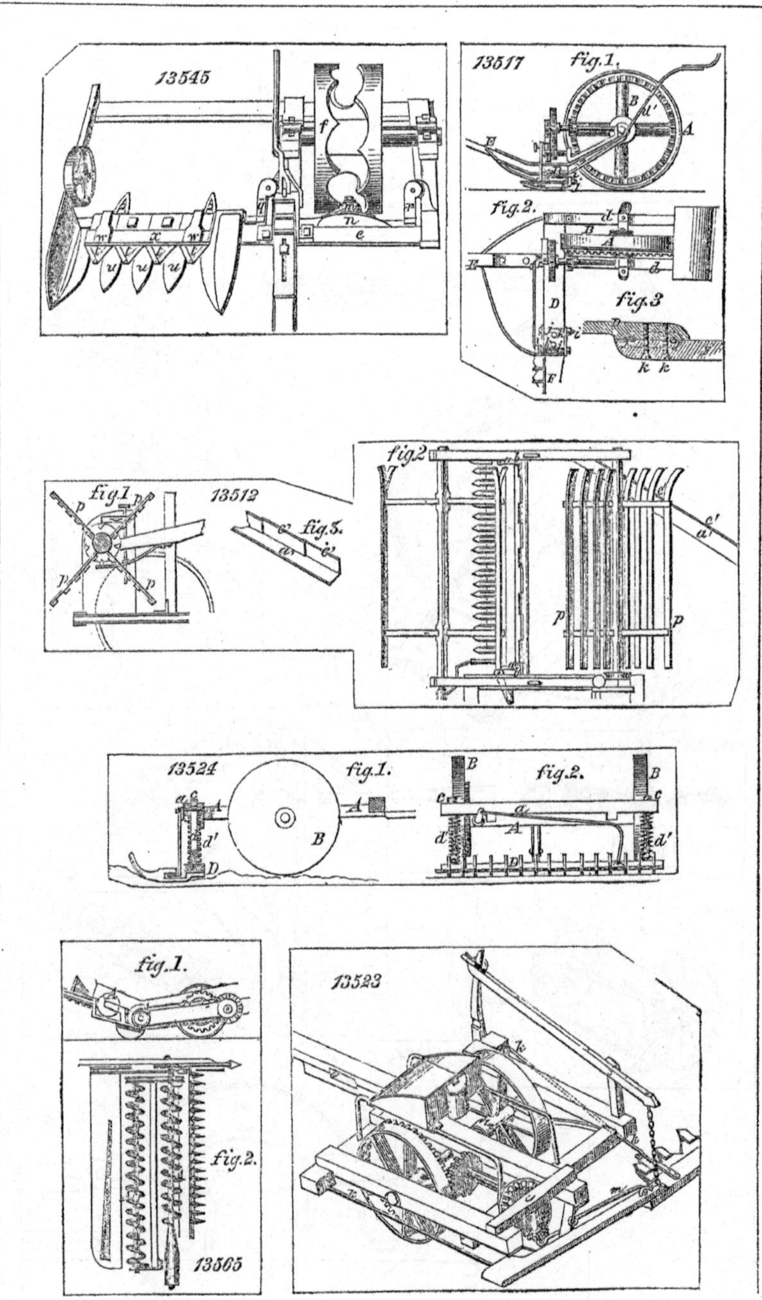
13545
13517
fig.1.
fig.2.
fig.3
fig.1
13512
fig.3.
fig.2
13524
fig.1.
fig.2.
fig.1.
fig.2.
13565
13523

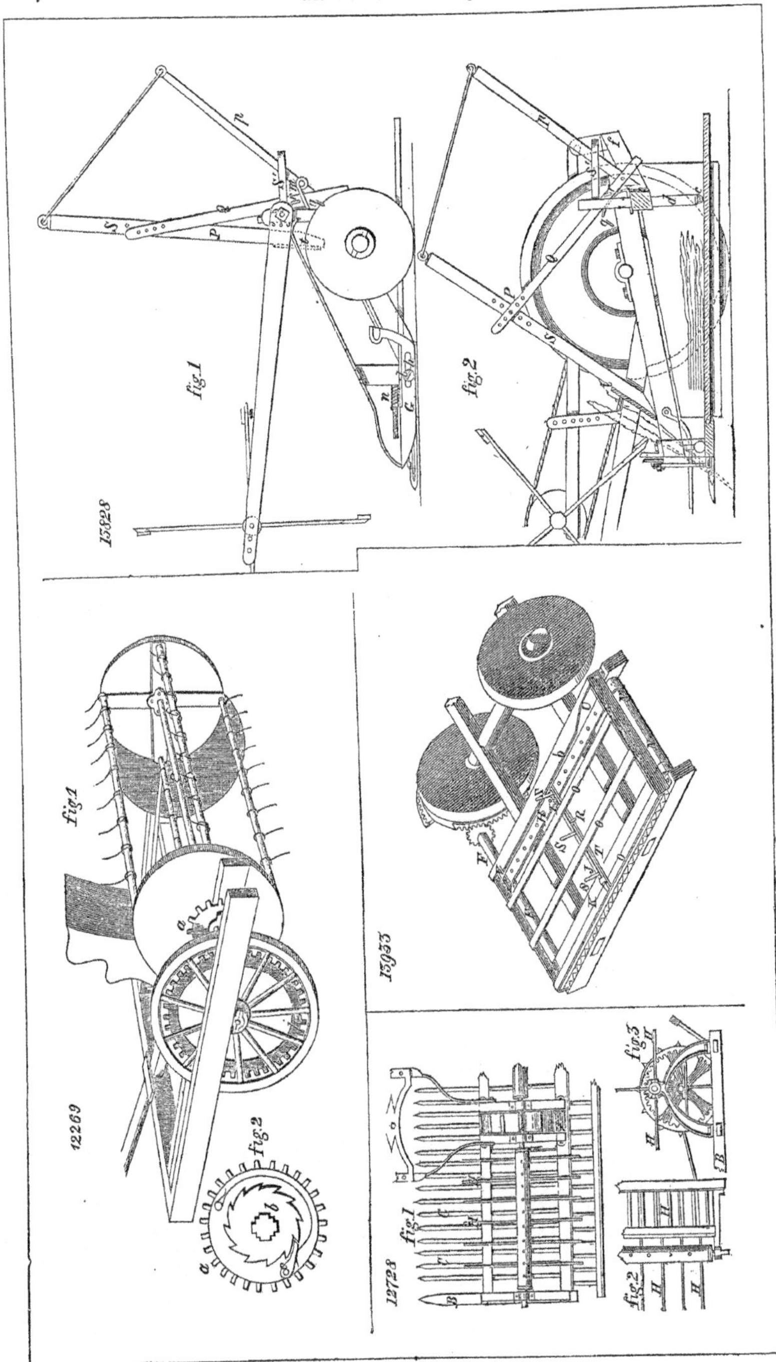
13828
fig.1
fig.2
12269
fig.1
fig.2
13933
12728
fig.1
fig.2
fig.3

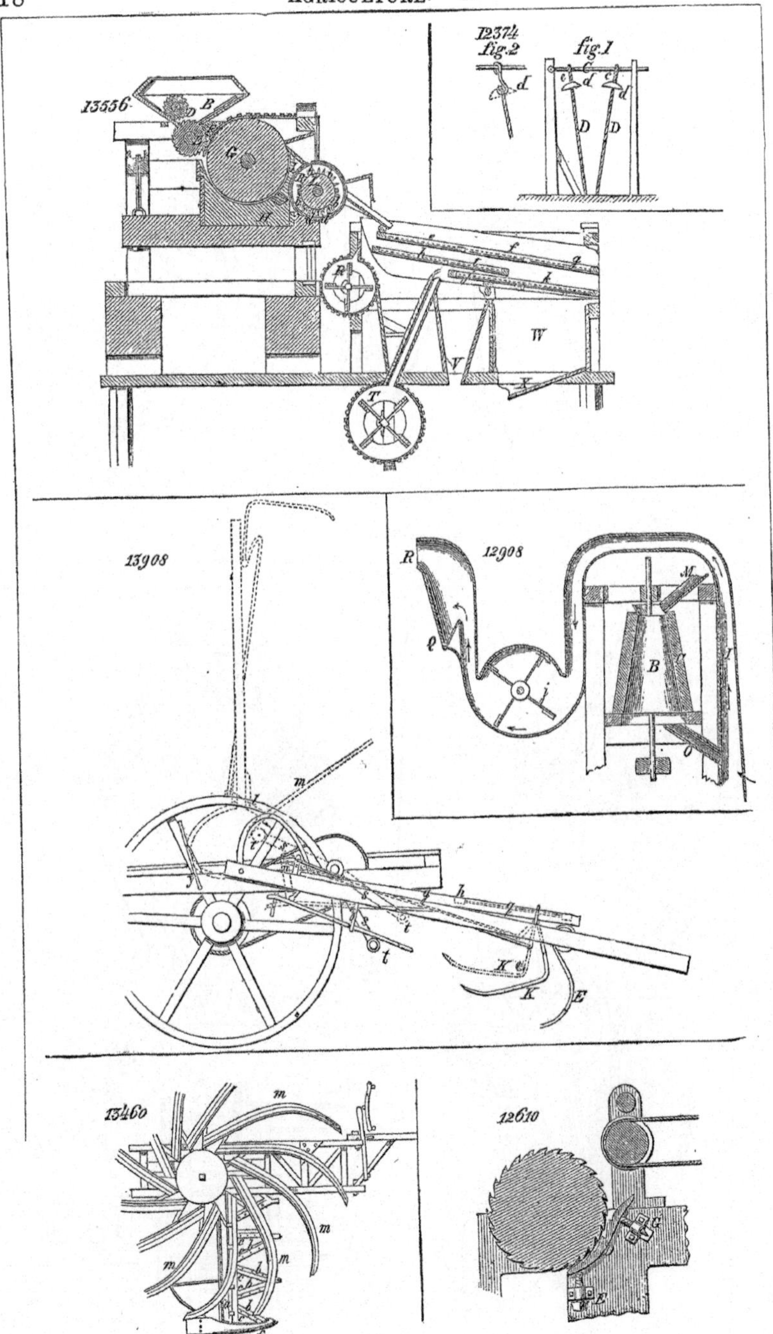
12374
fig.2
fig.1
13556
13908
12908
13460
12610

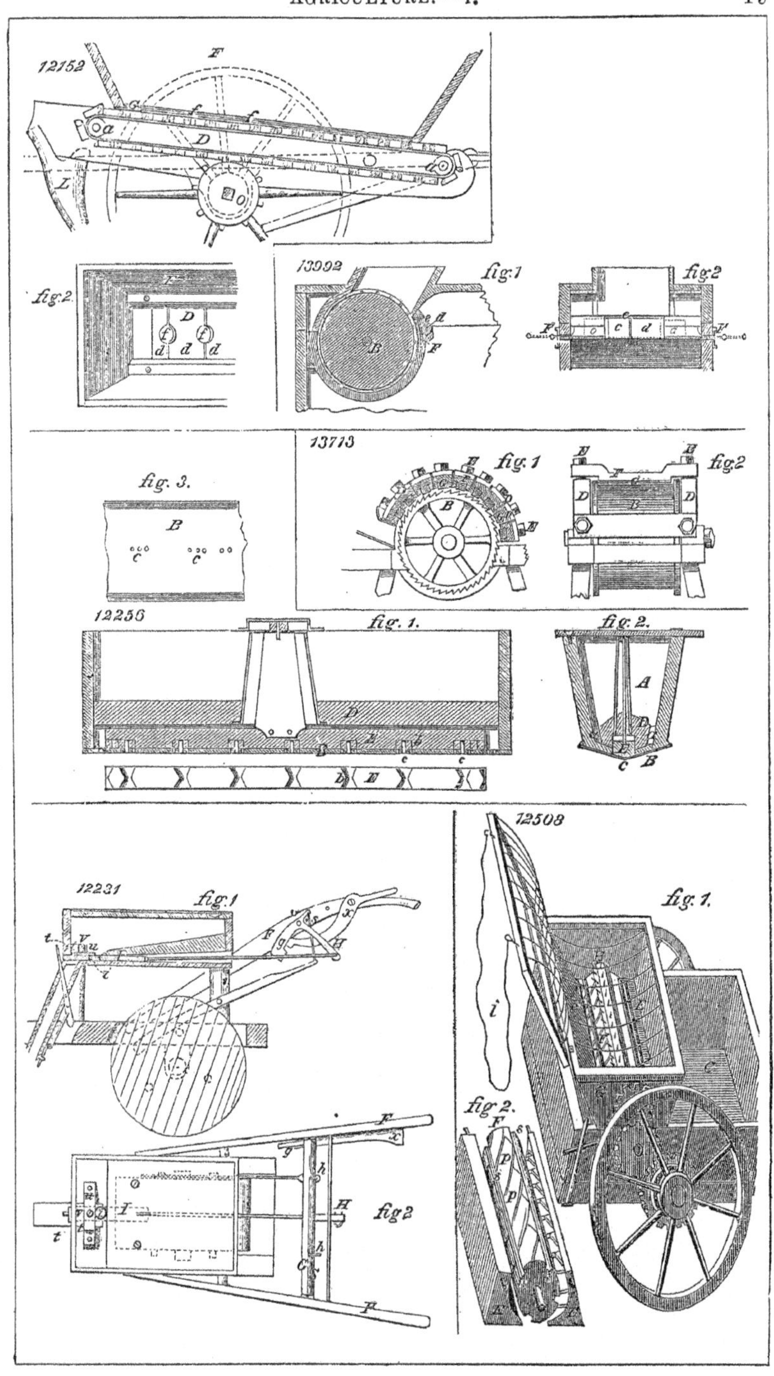
12152
fig.2.
13992
fig.1
fig.2
fig. 3.
13713
fig.1
fig.2
12256
fig. 1.
fig. 2.
12508
fig. 1.
12231
fig.1
fig2
fig 2.

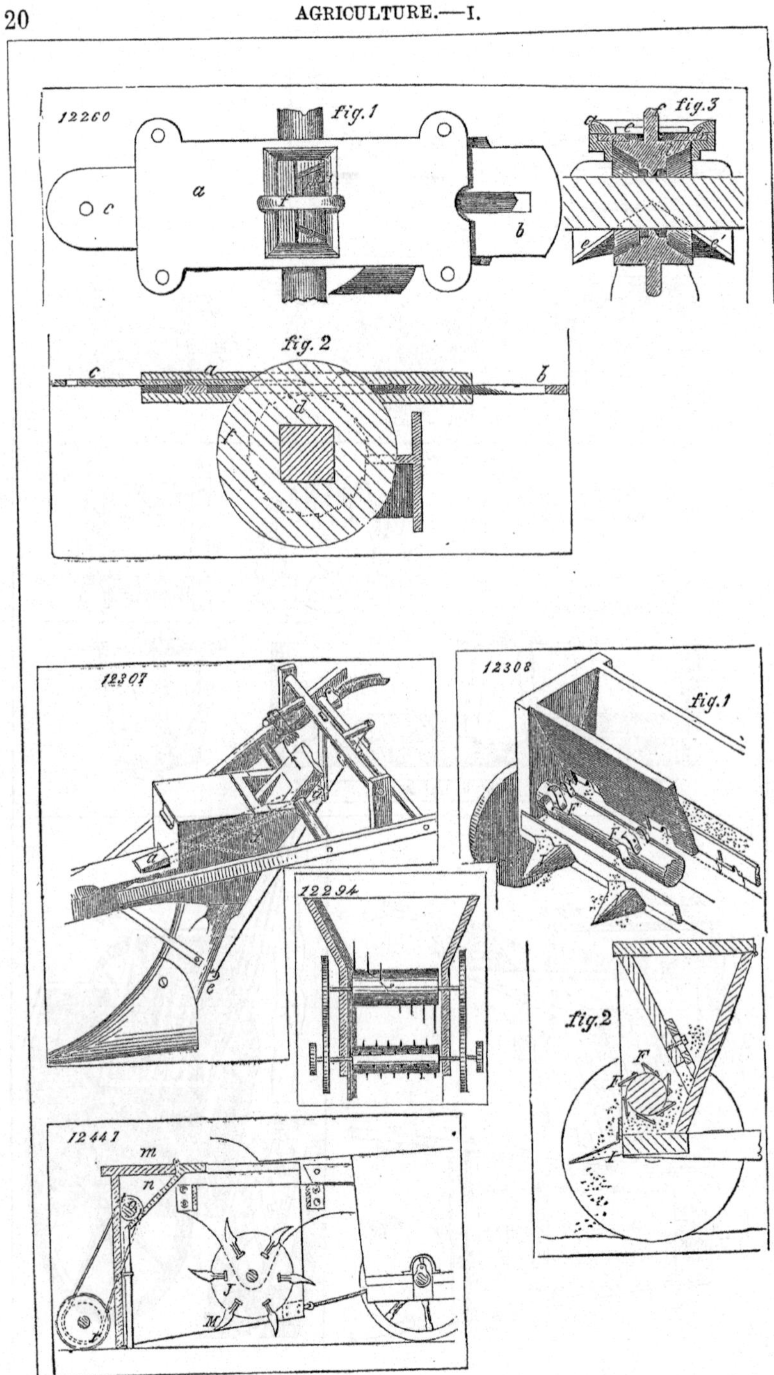
12260
fig.1
fig.3
fig.2
12307
12308
fig.1
12294
fig.2
12447

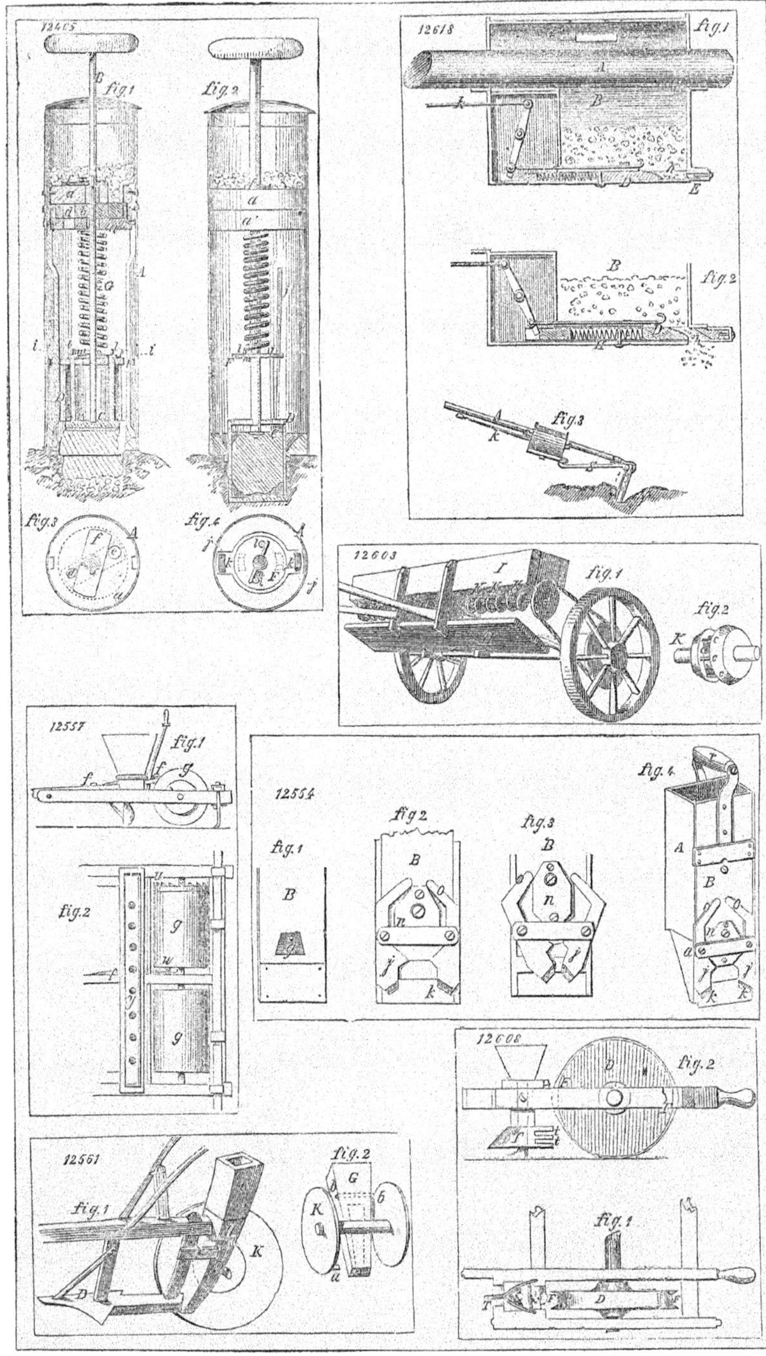
12405
fig.1
fig.2
fig.3
fig.4
12618
fig.1
fig.2
fig.3
12603
fig.1
fig.2
12557
fig.1
fig.2
12554
fig.1
fig.2
fig.3
fig.4
12608
fig.2
fig.1
12561
fig.1
fig.2

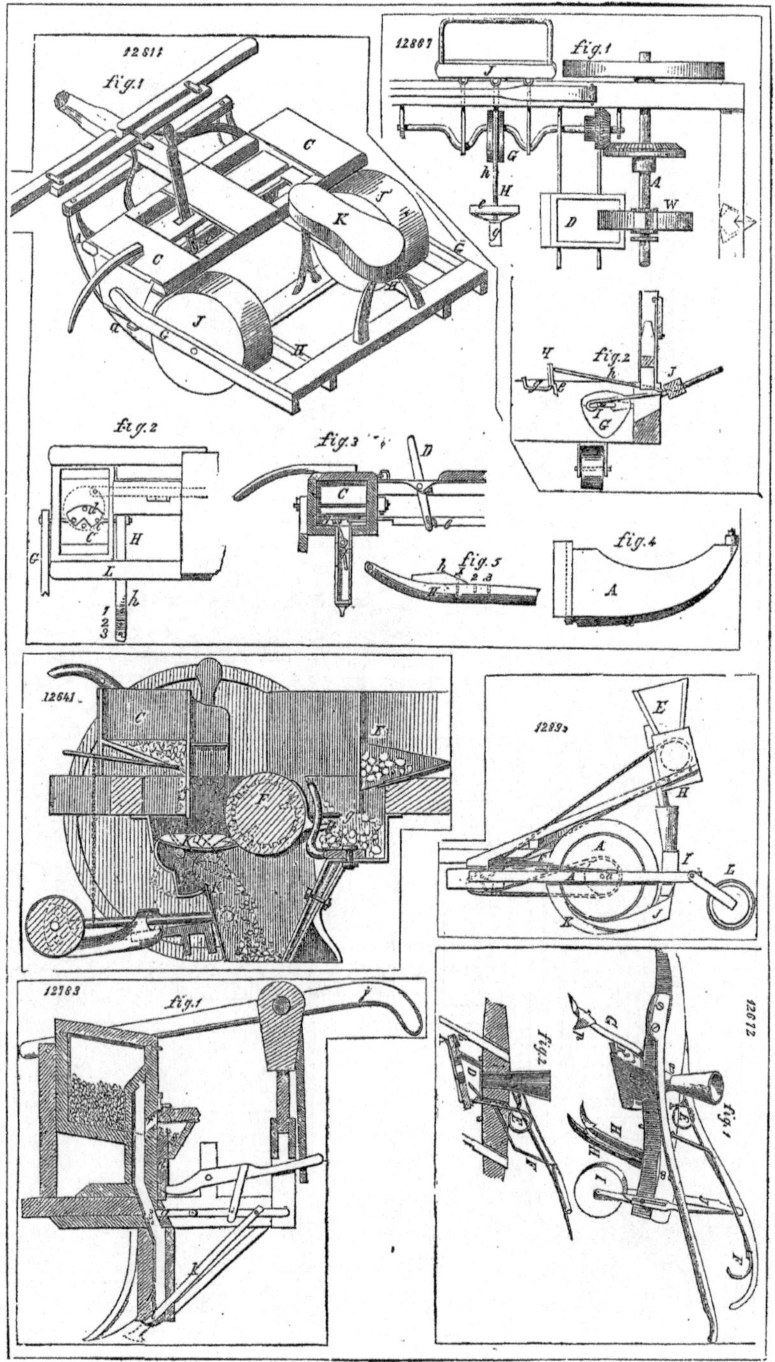

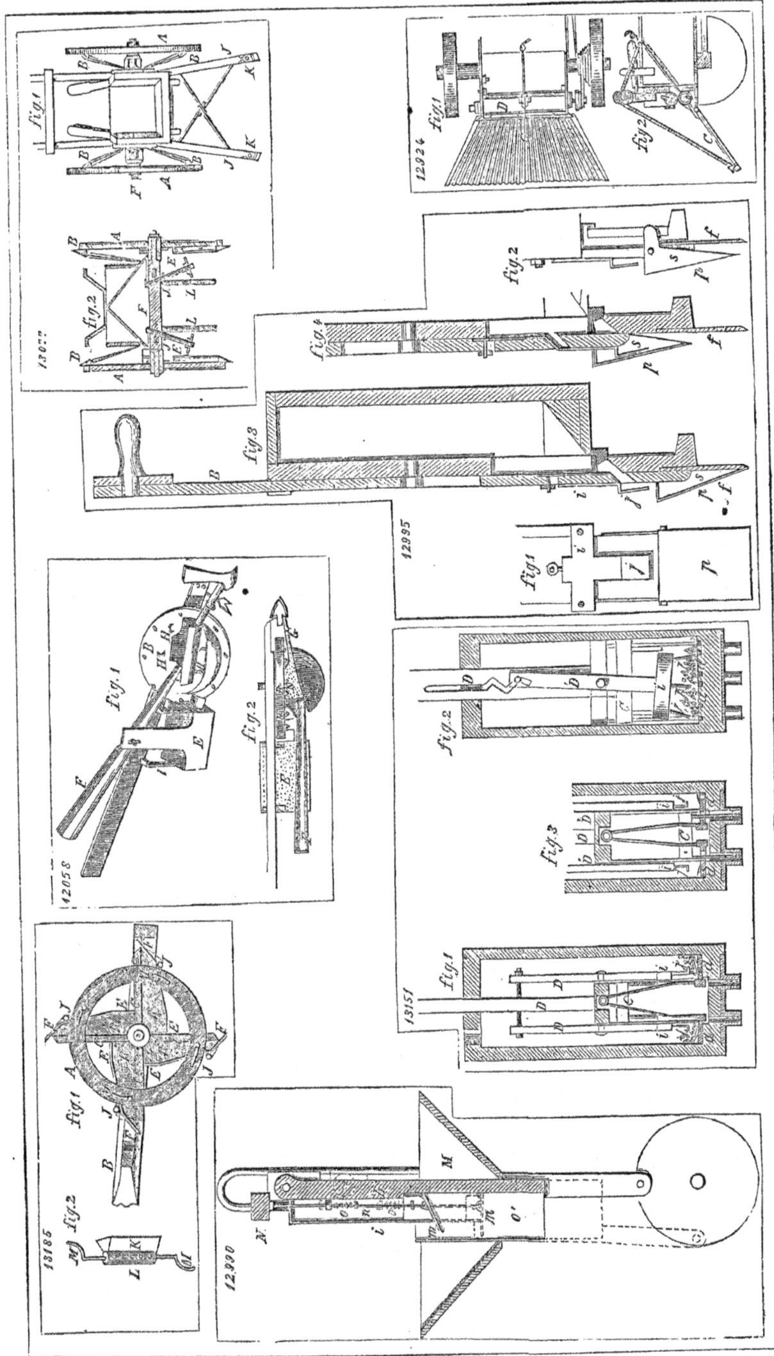

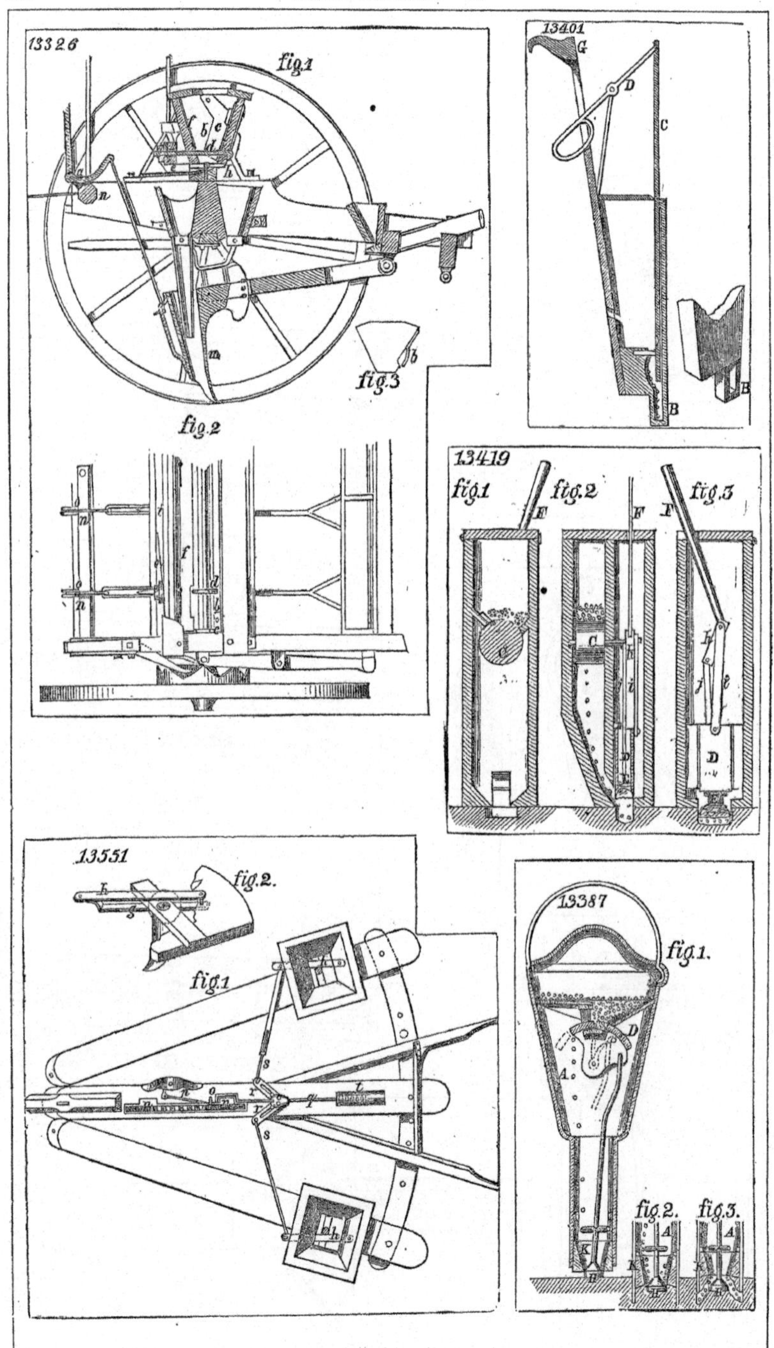
13326
fig.1
fig.2
fig.3
13401
13419
fig.1
fig.2
fig.3
13551
fig.1
fig.2.
13387
fig.1.
fig.2.
fig.3.

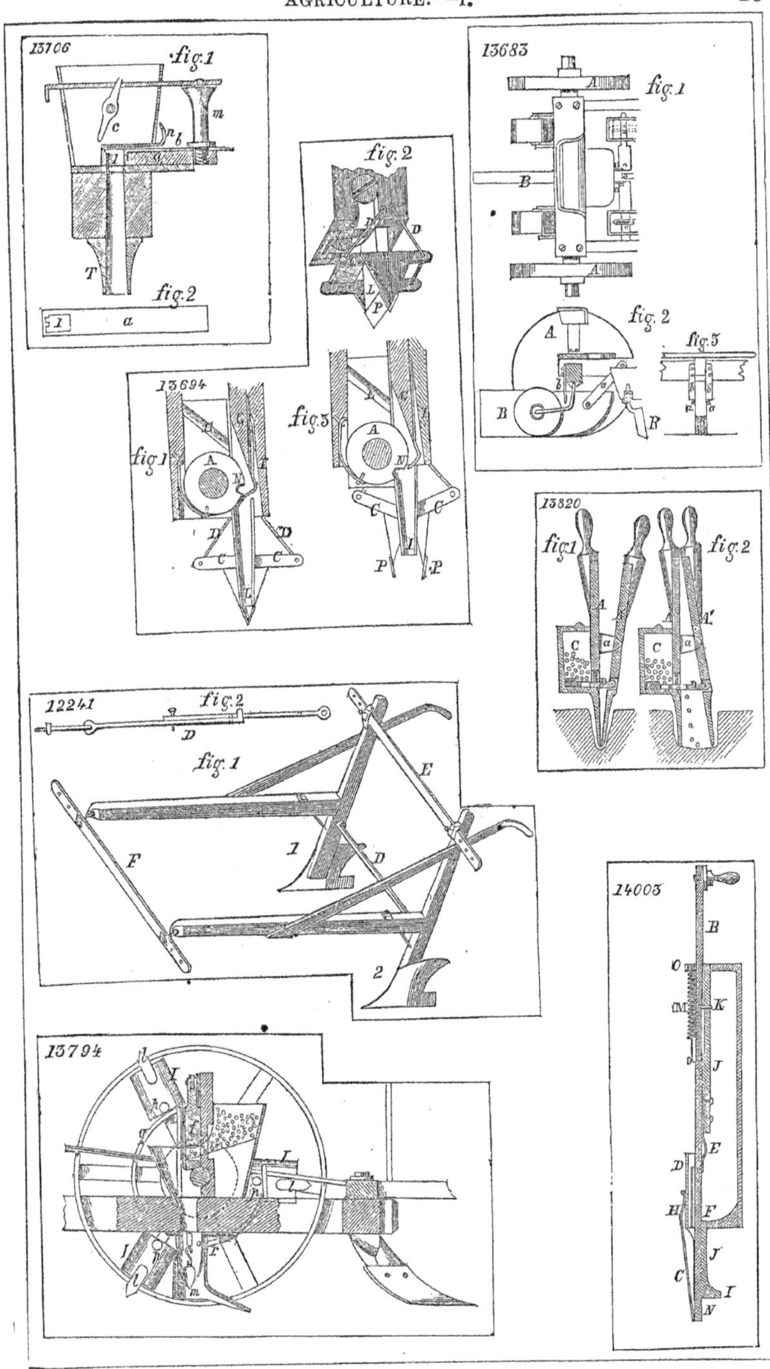
13106
fig. 1
fig. 2
13683
fig. 1
fig. 2
fig. 3
13694
fig. 1
fig. 2
fig. 3
13820
fig. 1
fig. 2
12241
fig. 2
fig. 1
14003
13794

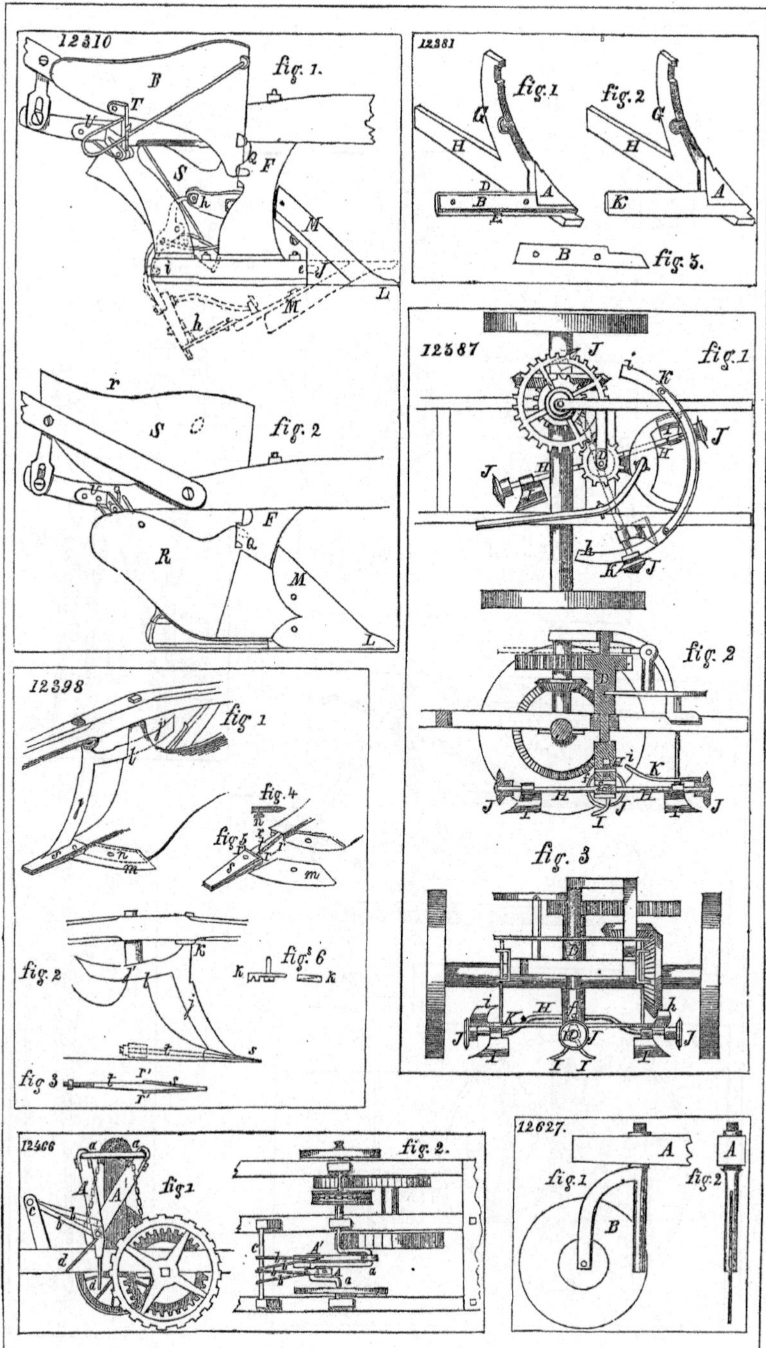
12310
fig. 1.
fig. 2
12381
fig. 1
fig. 2
fig. 3.
12387
fig. 1
fig. 2
fig. 3
12398
fig 1
fig. 4
fig. 5
fig. 2
figs 6
fig 3
12466
fig 1
fig. 2.
12627.
fig. 1
fig. 2

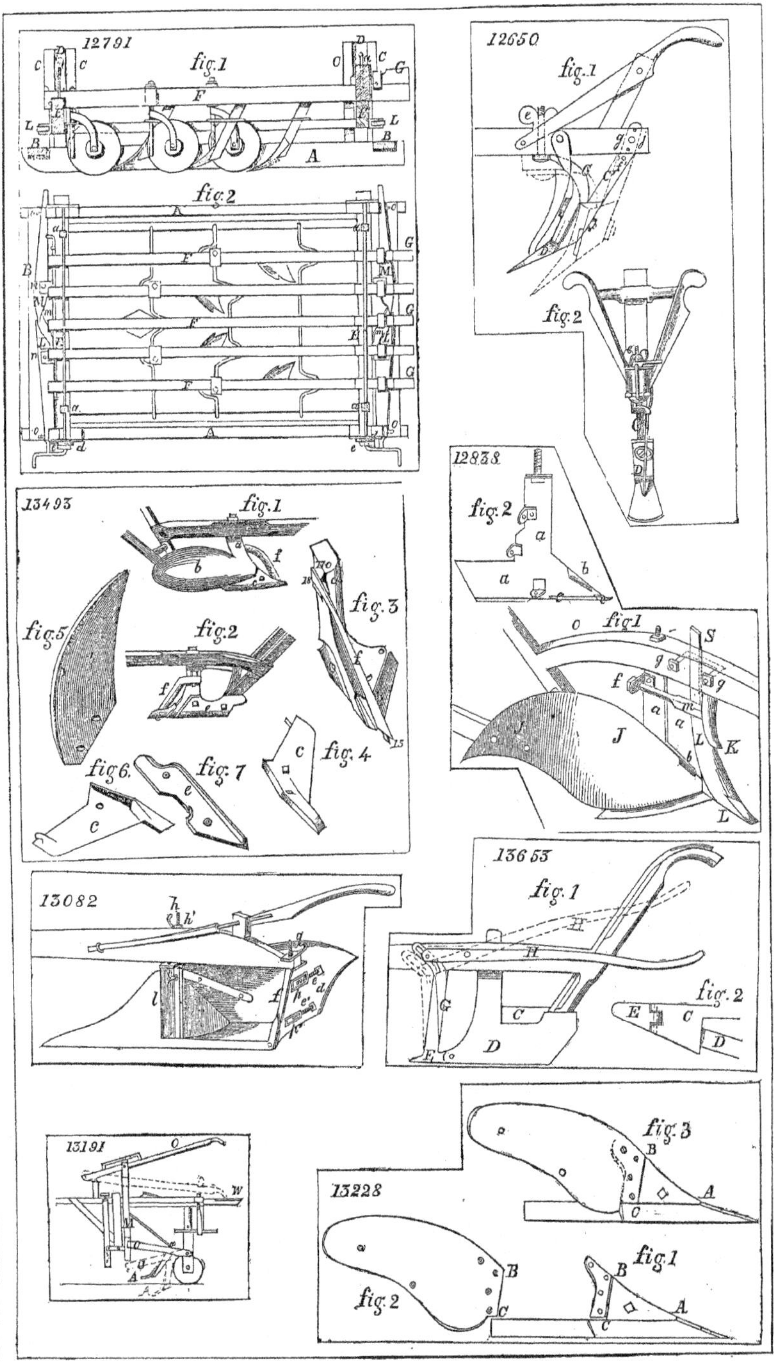
12791
fig. 1
fig. 2
12650
fig. 1
fig. 2
12838
fig. 2
fig. 1
13493
fig. 1
fig. 2
fig. 3
fig. 4
fig. 5
fig. 6
fig. 7
13082
13653
fig. 1
fig. 2
13191
13228
fig. 3
fig. 2
fig. 1

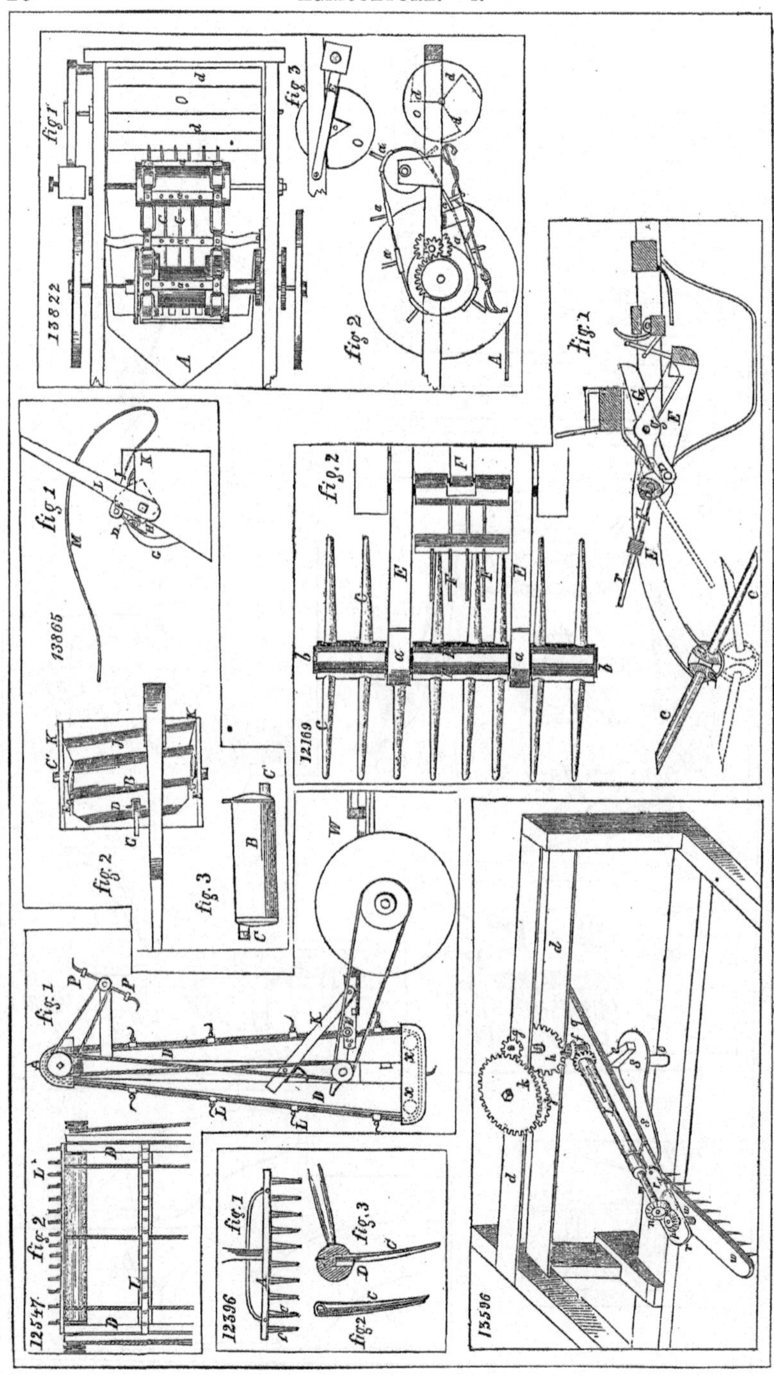
13822
13865
12169
12547
12396
13596

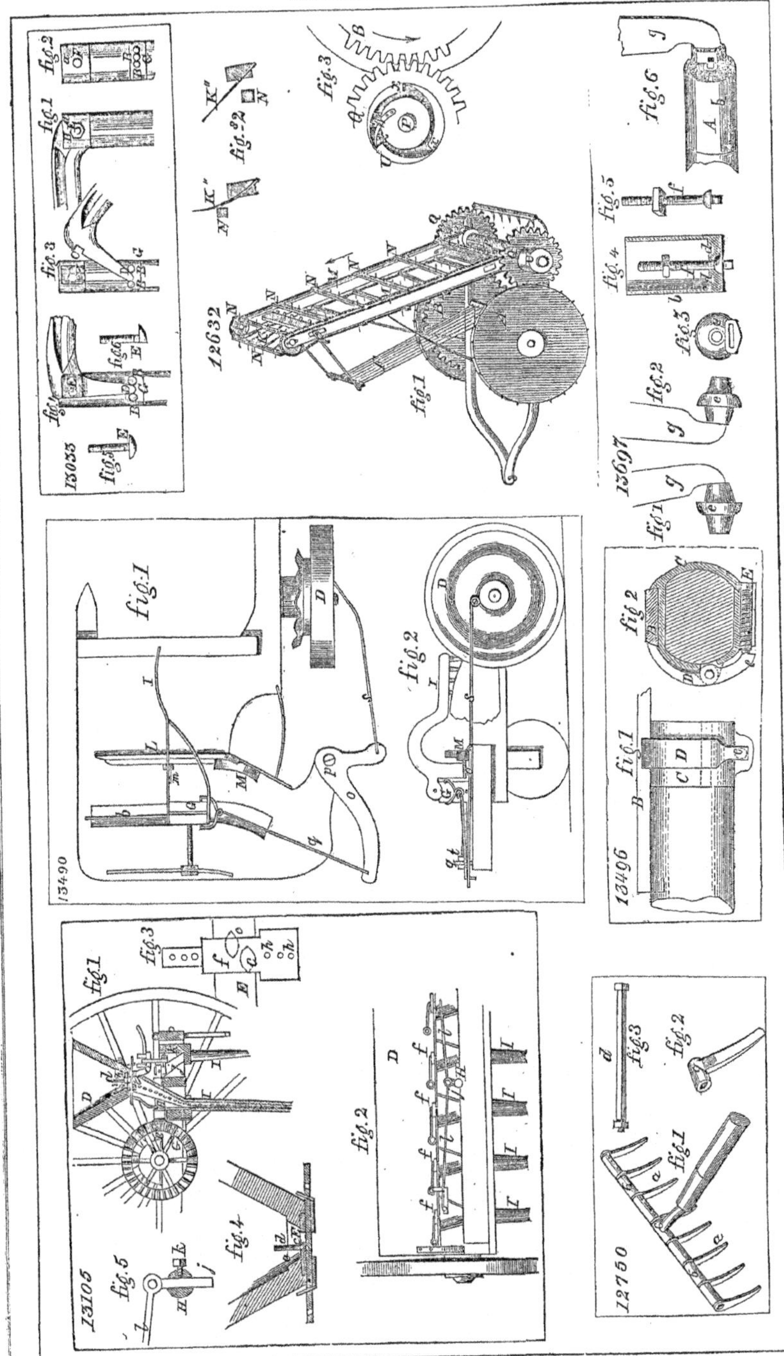
13035
12632
13490
13105
13496
13697
12750

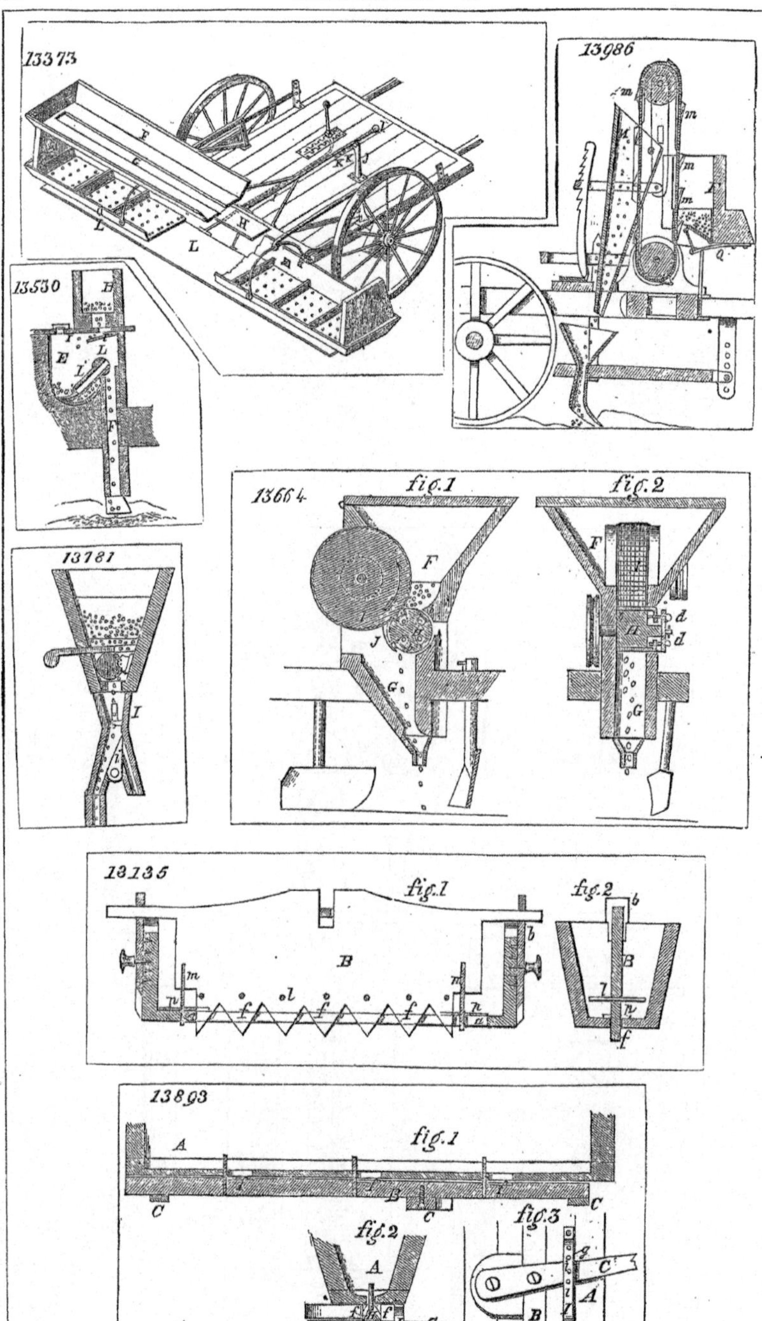
13373
13986
13530
13664
fig.1
fig.2
13781
13135
fig.1
fig.2
13893
fig.1
fig.2
fig.3

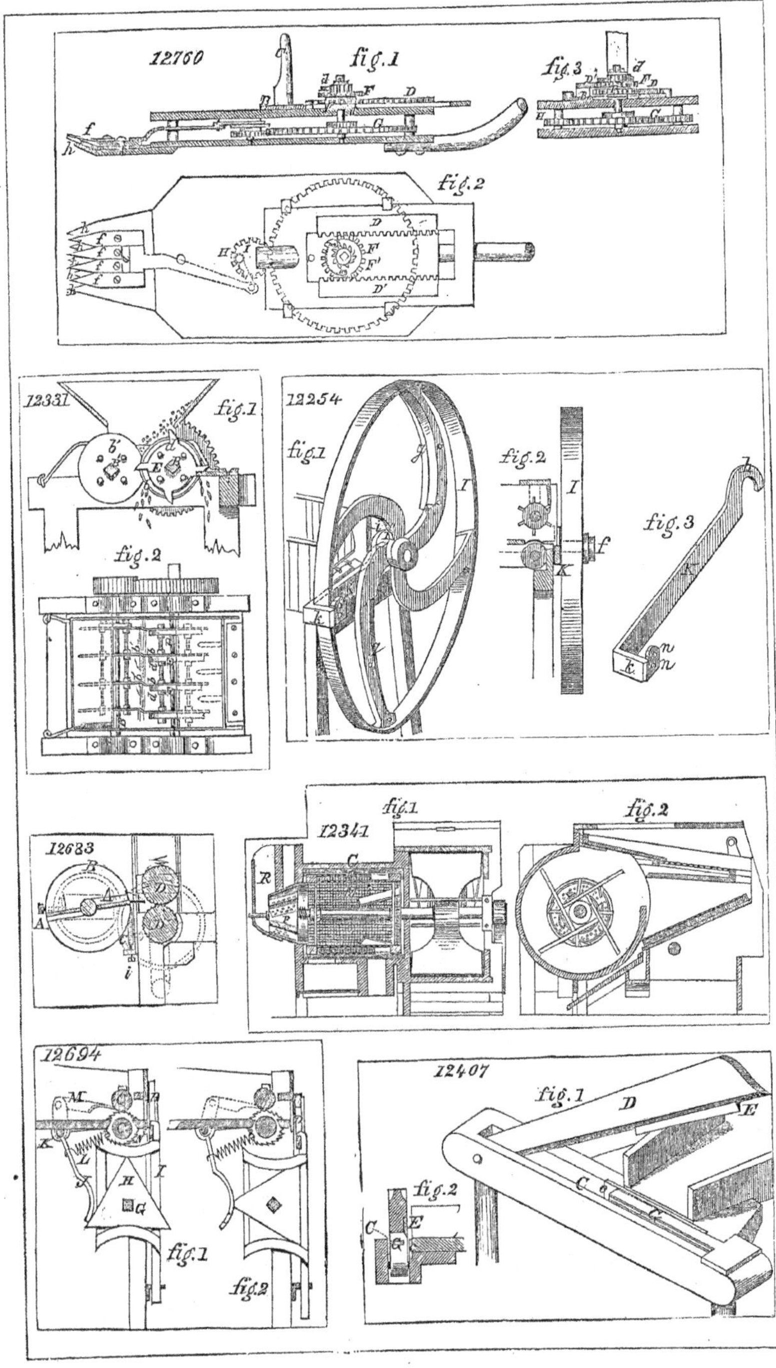
12760
fig. 1
fig. 3
fig. 2
12331
fig. 1
fig. 2
12254
fig. 1
fig. 2
fig. 3
12633
12341
fig. 1
fig. 2
12694
fig. 1
fig. 2
12407
fig. 1
fig. 2

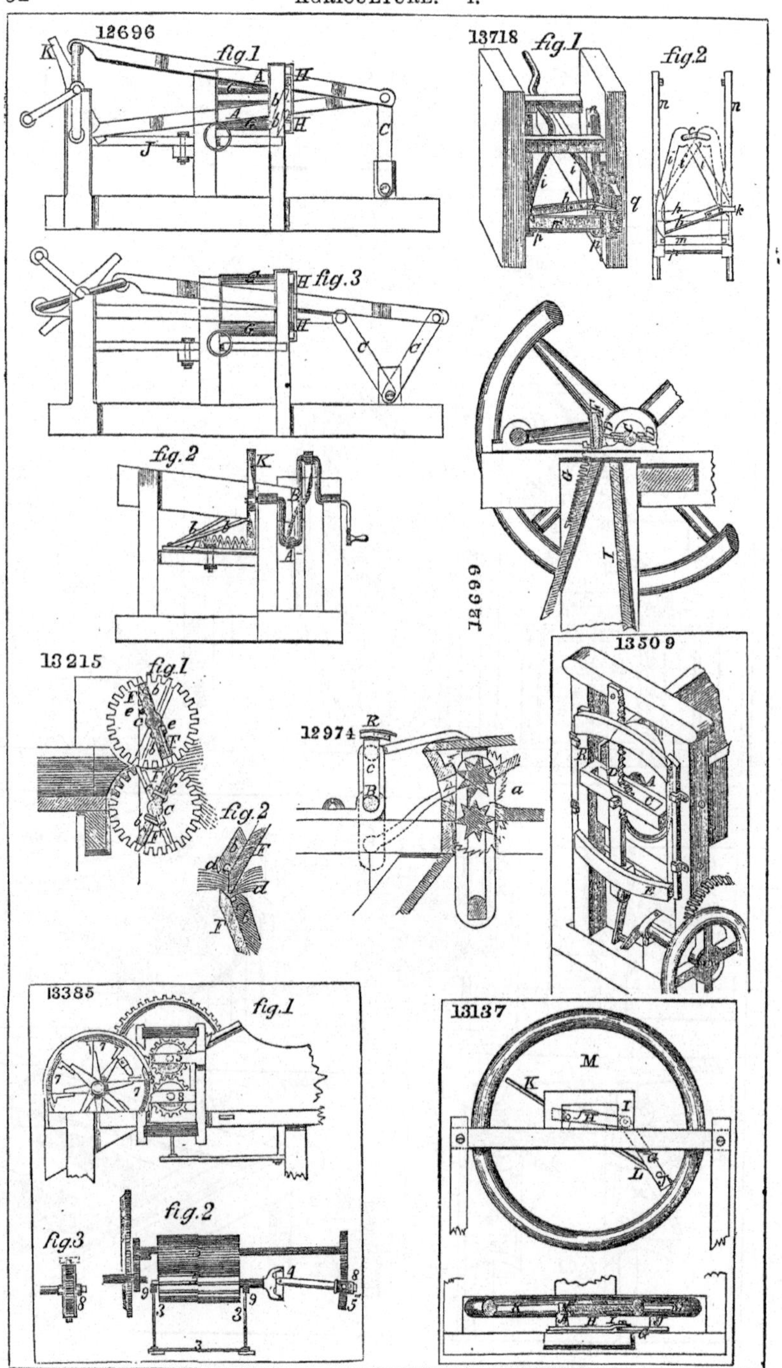
12696
fig.1
fig.3
fig.2
13718
fig.1
fig.2
12699
13215
fig.1
fig.2
12974
13509
13385
fig.1
fig.2
fig.3
13137
M

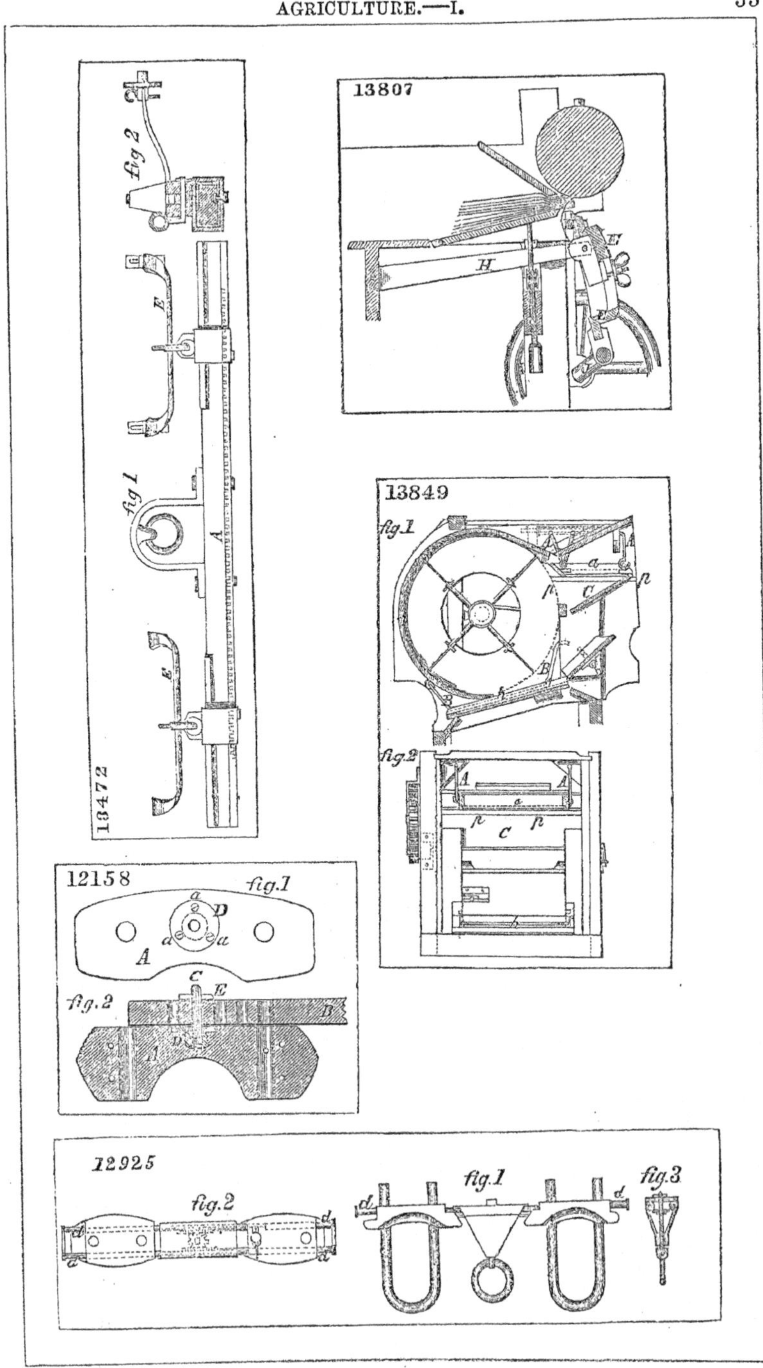
13807
H
E
13849
fig.1
fig.2
13472
fig 2
fig 1
A
E
12158
fig.1
fig.2
12925
fig.2
fig.1
fig.3

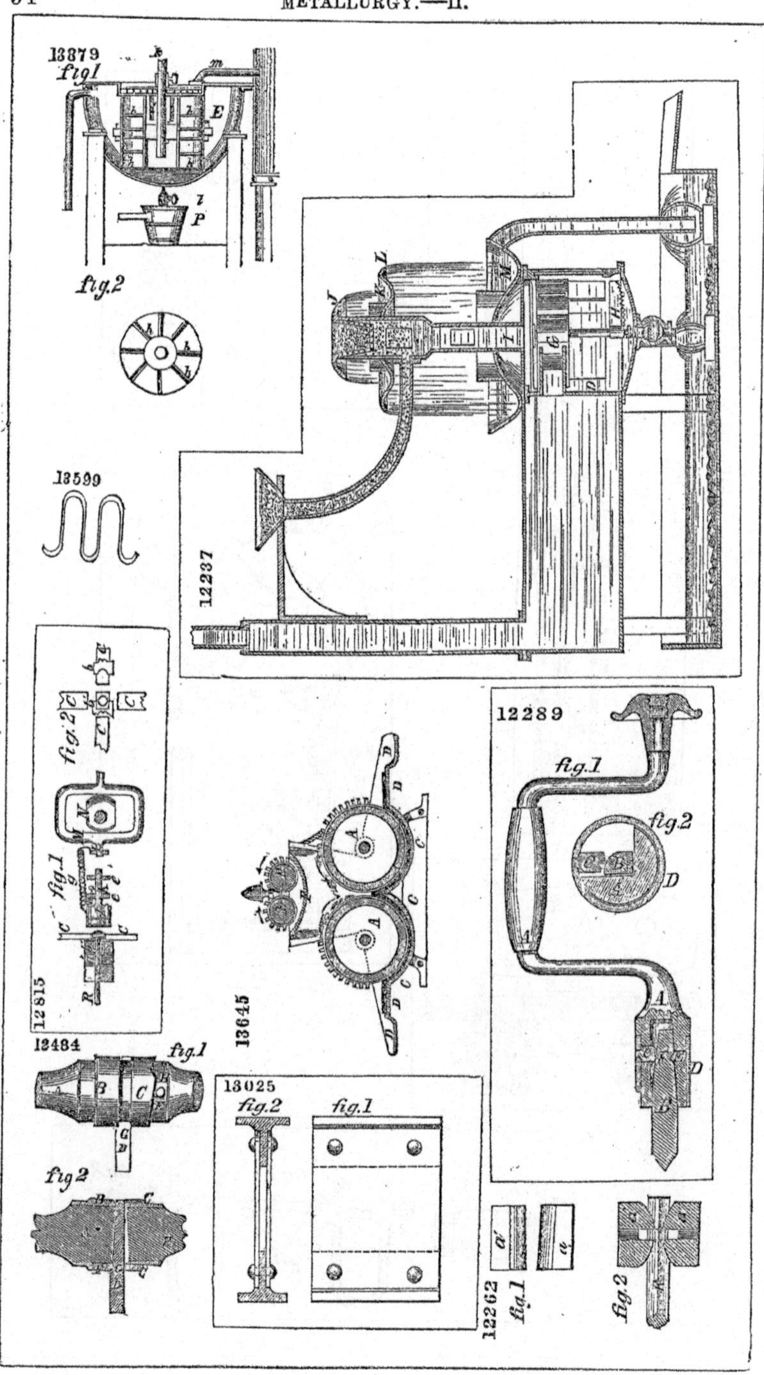
13879
fig1
fig.2
13599
12237
12289
fig.1
fig.2
12815
fig.1
fig.2
13645
12484
fig.1
fig 2
13025
fig.2
fig.1
12262
fig.1
fig.2

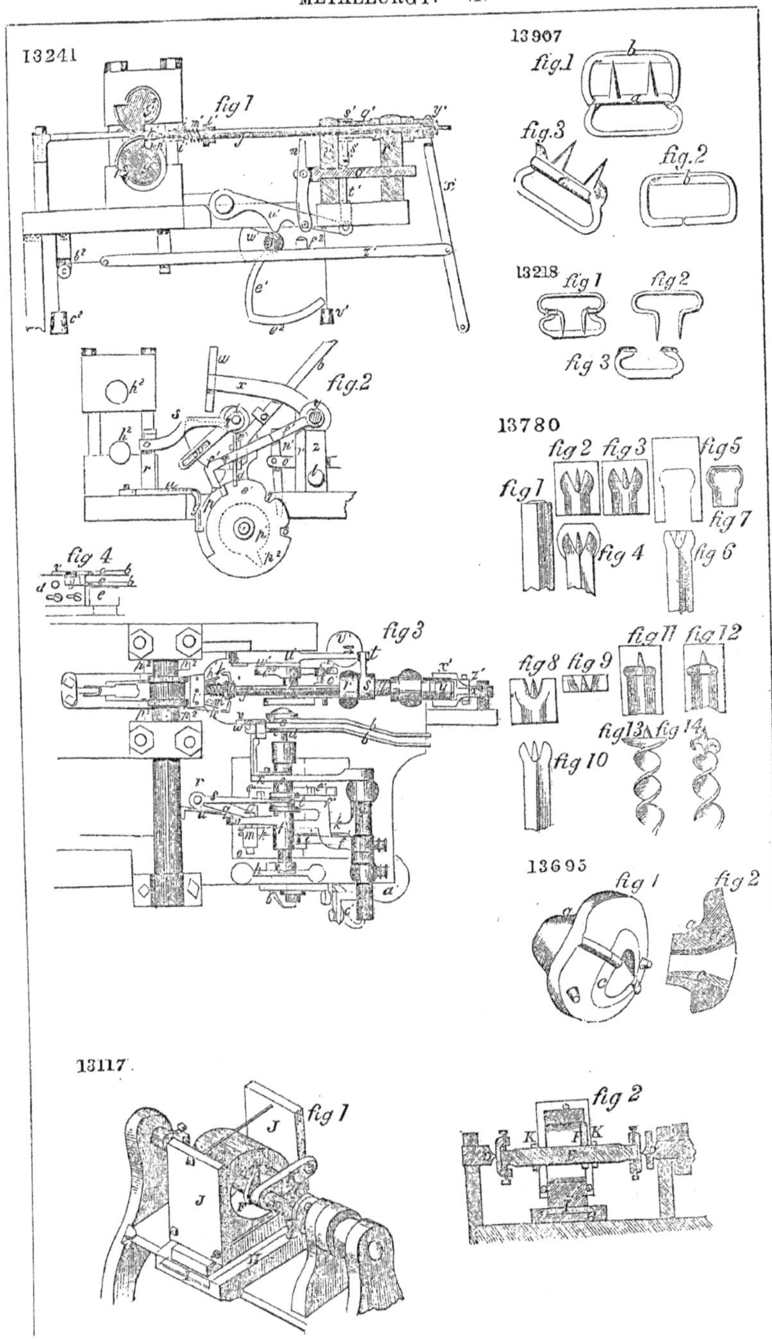
13241
fig 1
fig.2
fig 4
fig 3
13907
fig.1
fig.3
fig.2
13218
fig 1
fig 2
fig 3
13780
fig 1
fig 2
fig 3
fig 4
fig 5
fig 6
fig 7
fig 8
fig 9
fig 10
fig 11
fig 12
fig 13
fig 14
13695
fig 1
fig 2
13117
fig 1
fig 2

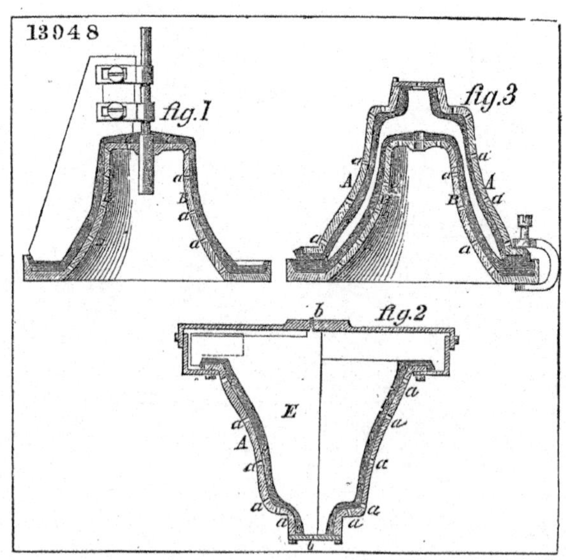
13048
fig.1
fig.3
fig.2
A
B
E
a
b

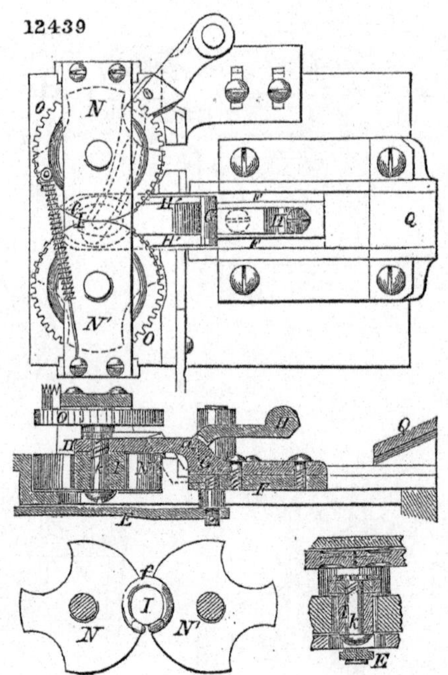
12439
N
N'
O
Q
I
H
F
E
f
k

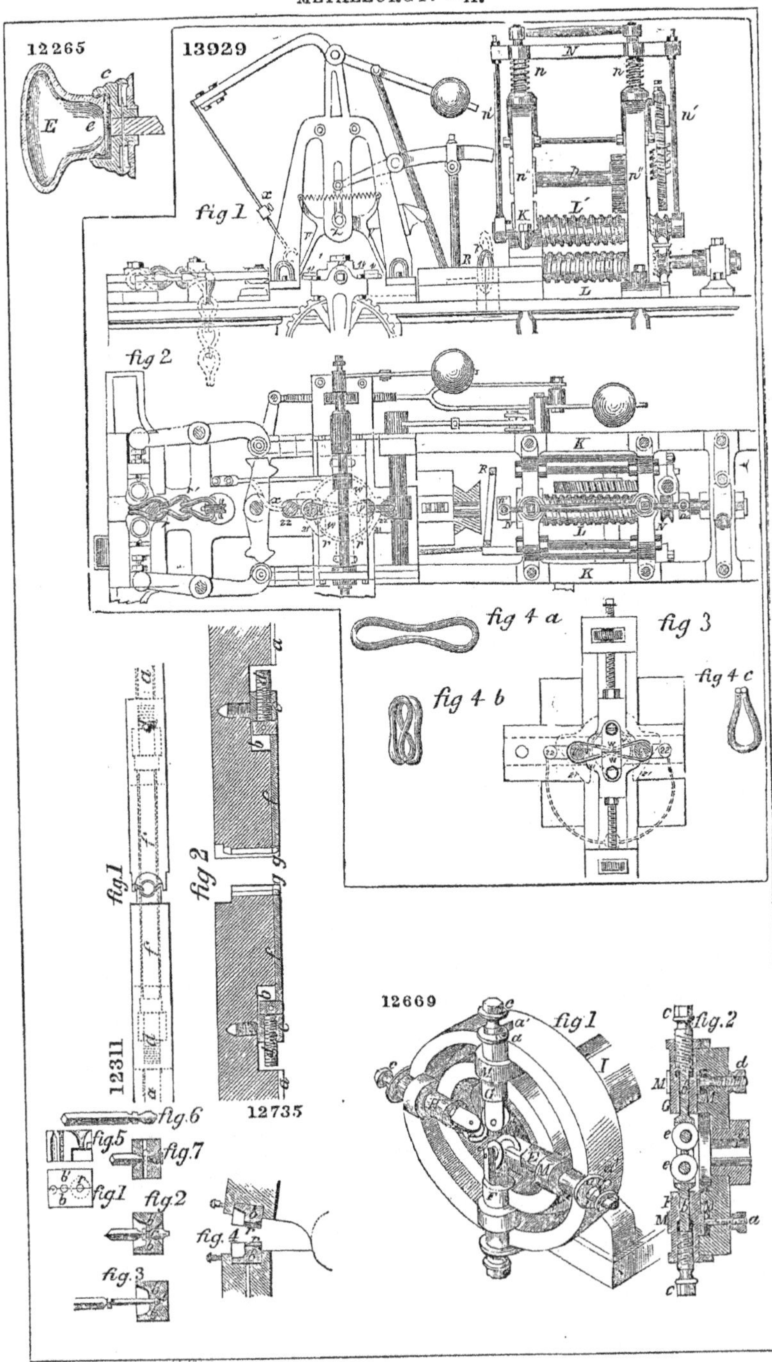
12265
13929
fig 1
fig 2
fig 3
fig 4 a
fig 4 b
fig 4 c
12311
12735
12669

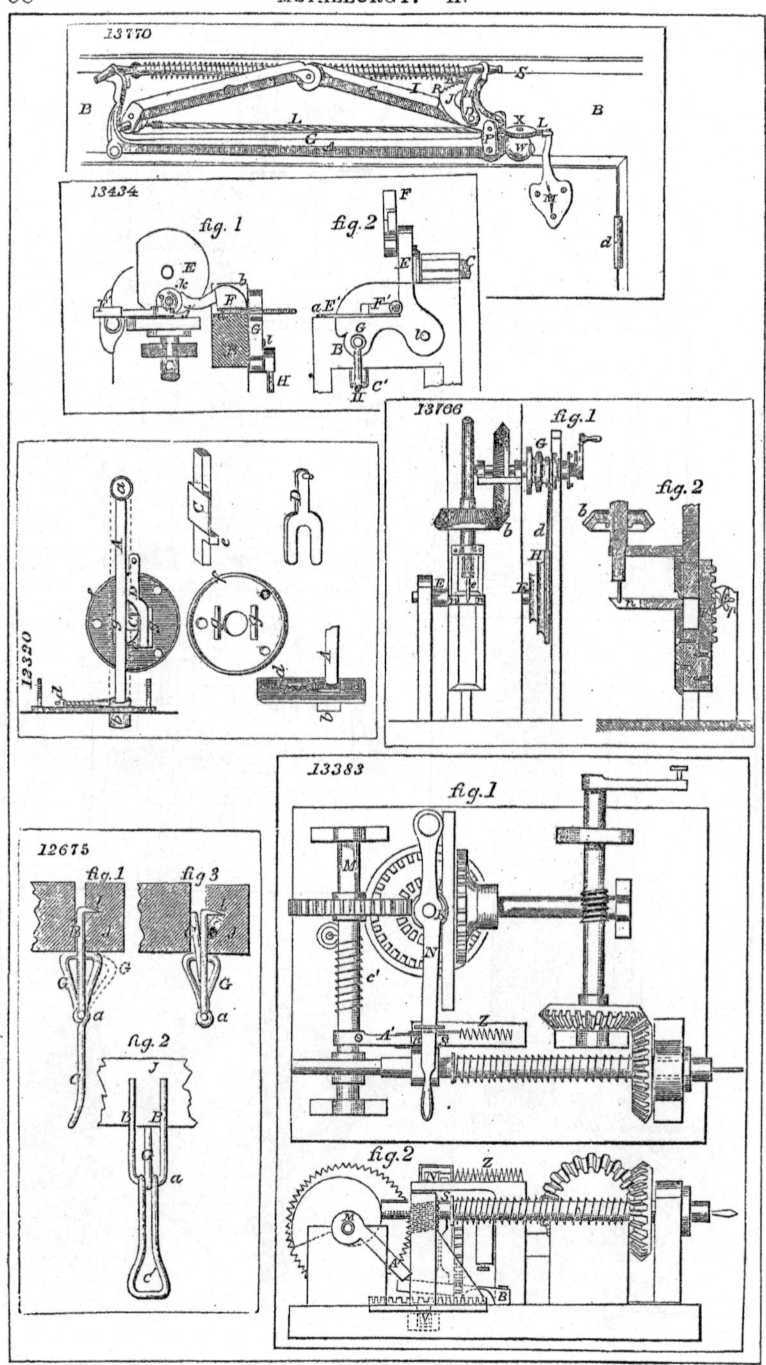
13770
13434
fig. 1
fig. 2
13766
fig. 1
fig. 2
12320
13383
fig. 1
fig. 2
12675
fig. 1
fig 3
fig. 2

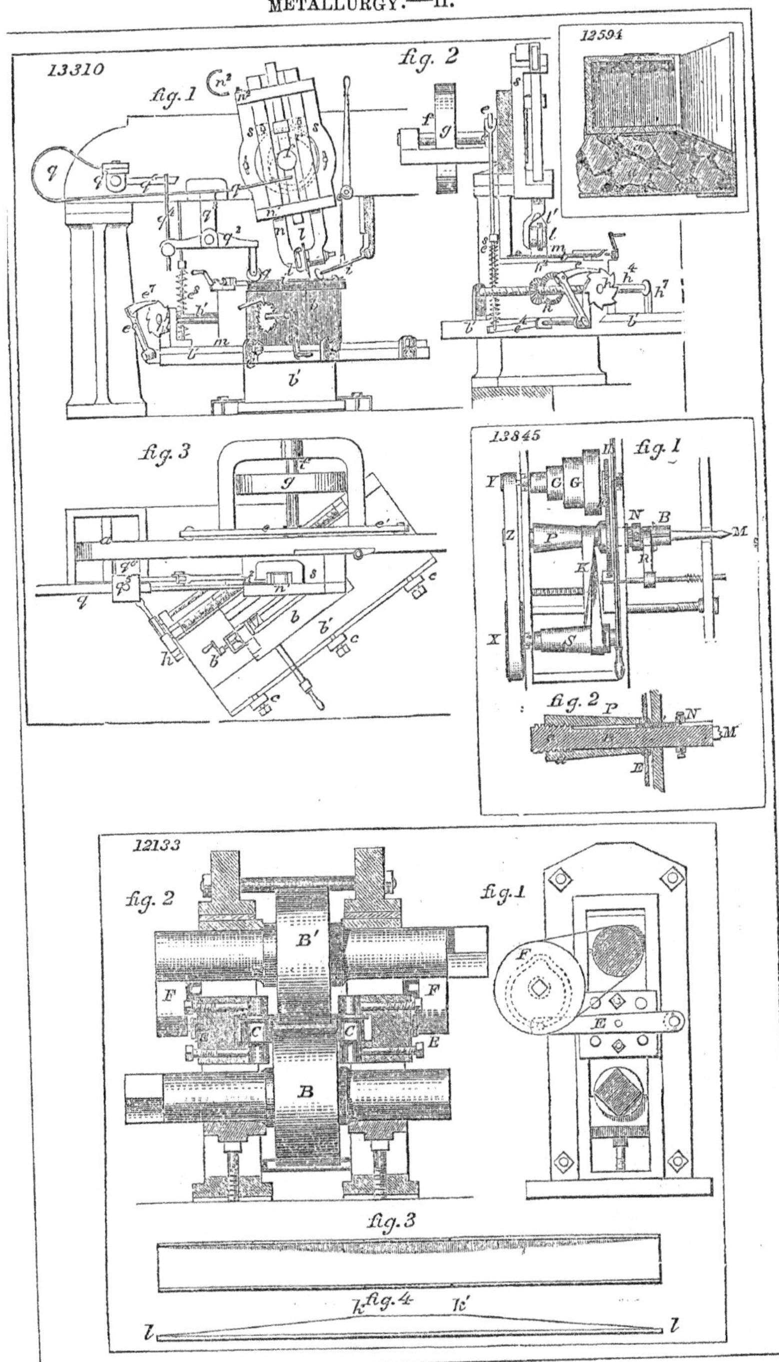
13310
fig. 1
fig. 2
fig. 3
12594
13845
fig. 1
fig. 2
12133
fig. 2
fig. 1
fig. 3
fig. 4

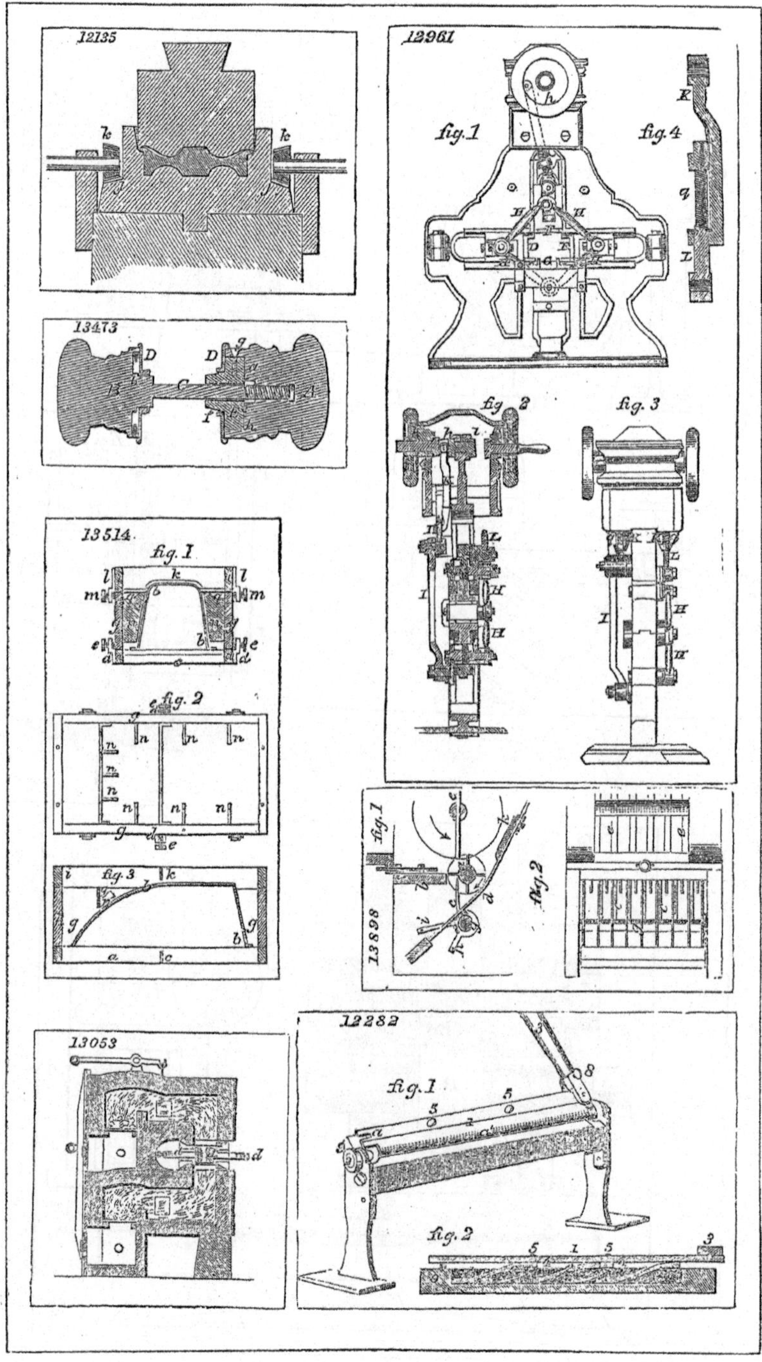
12135
k
k
12961
fig.1
fig.4
13473
D
D
C
fig.2
fig.3
13514.
fig.I
fig.2
fig.3
13898
fig.1
fig.2
13053
d
12282
fig.1
fig.2

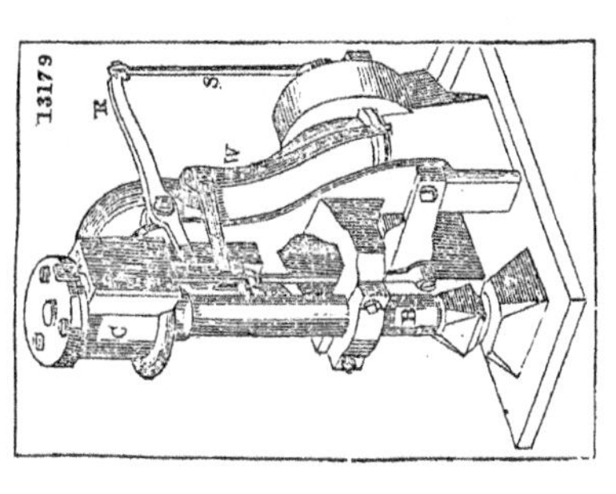
13179

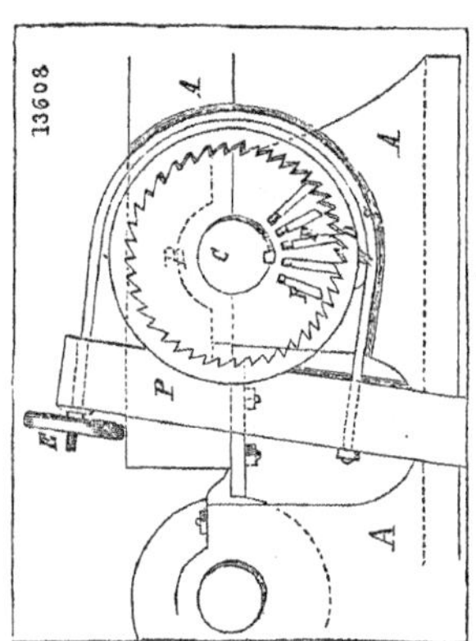
13608

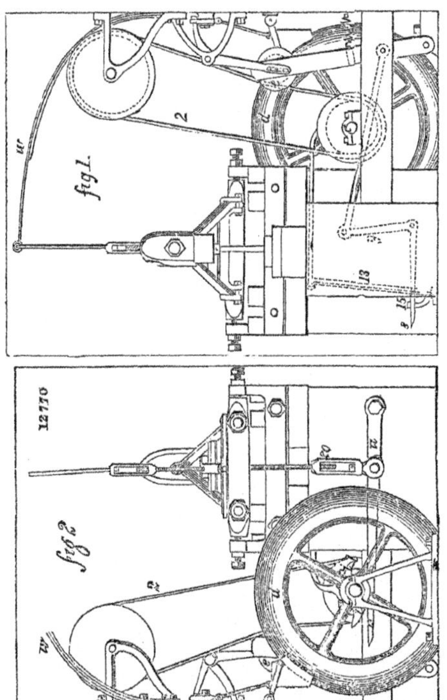
12770
fig 1.
fig 2

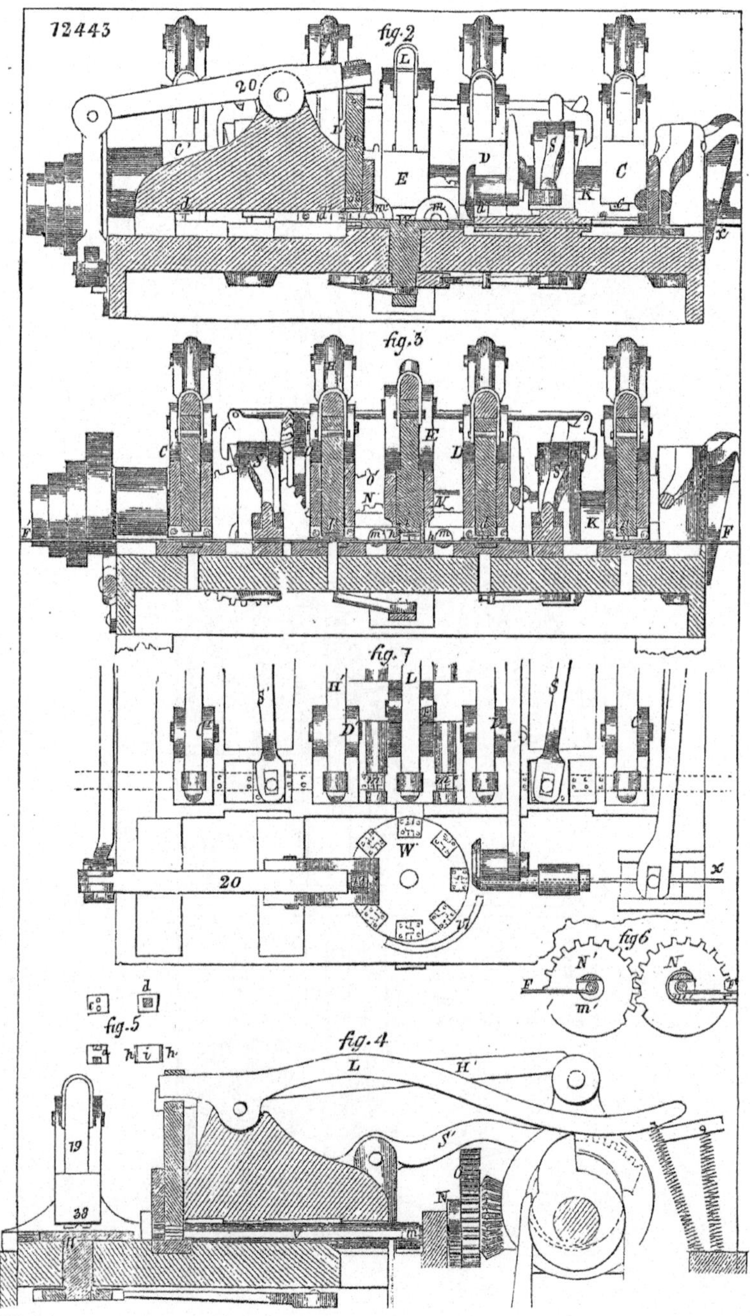
72443
fig. 2
fig. 3
fig. 7
fig. 6
fig. 5
fig. 4

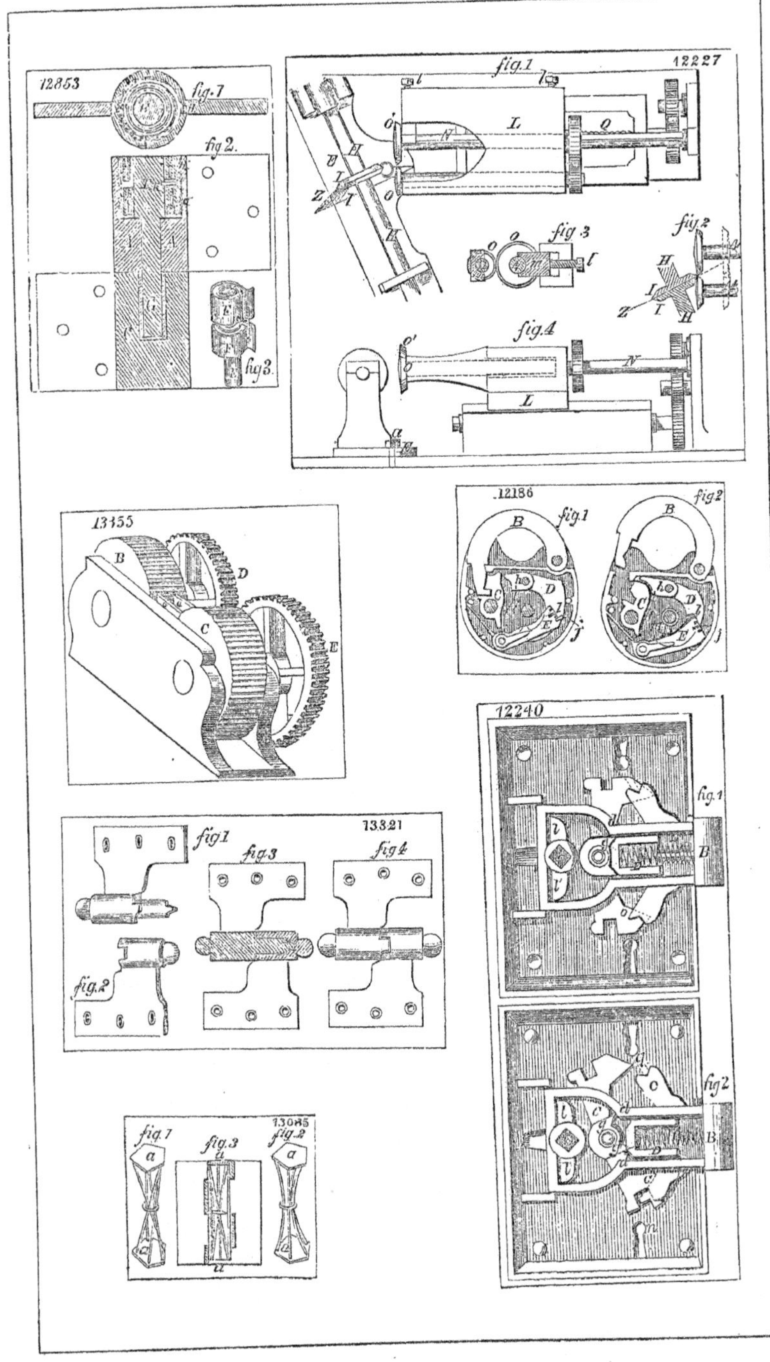
12853
fig.1
fig.2.
fig.3.
fig.1
12227
L
Q
N
fig.3
fig.2
fig.4
L
13355
B
D
C
E
12186
fig.1
fig.2
B
C
D
E
12240
fig.1
B
fig.2
B
13321
fig.1
fig.3
fig.4
fig.2
13085
fig.1
fig.3
fig.2
a

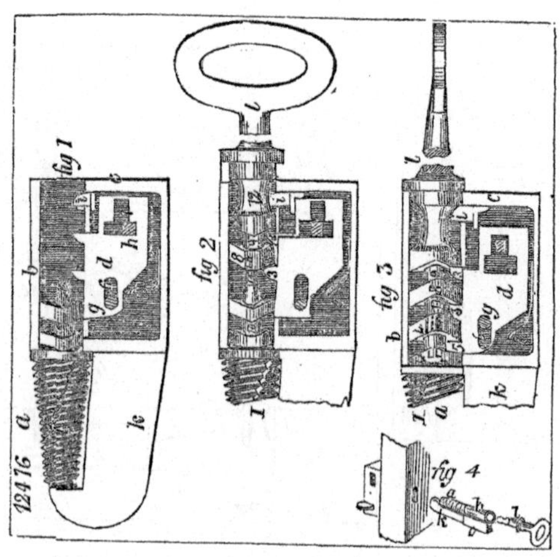
12476
fig 1
fig 2
fig 3
fig 4

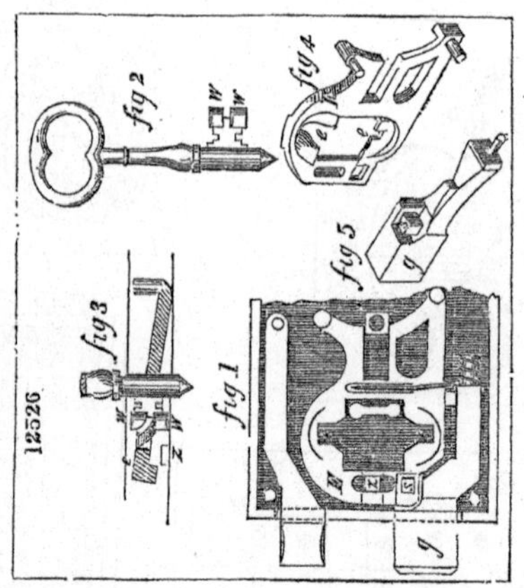
12526
fig 1
fig 2
fig 3
fig 4
fig 5

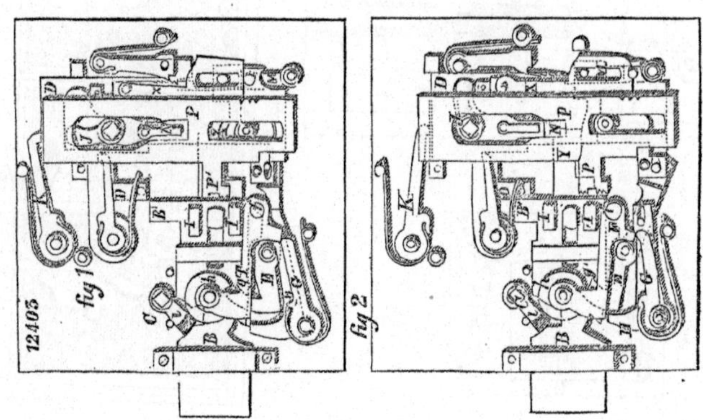
12405
fig 1
fig 2

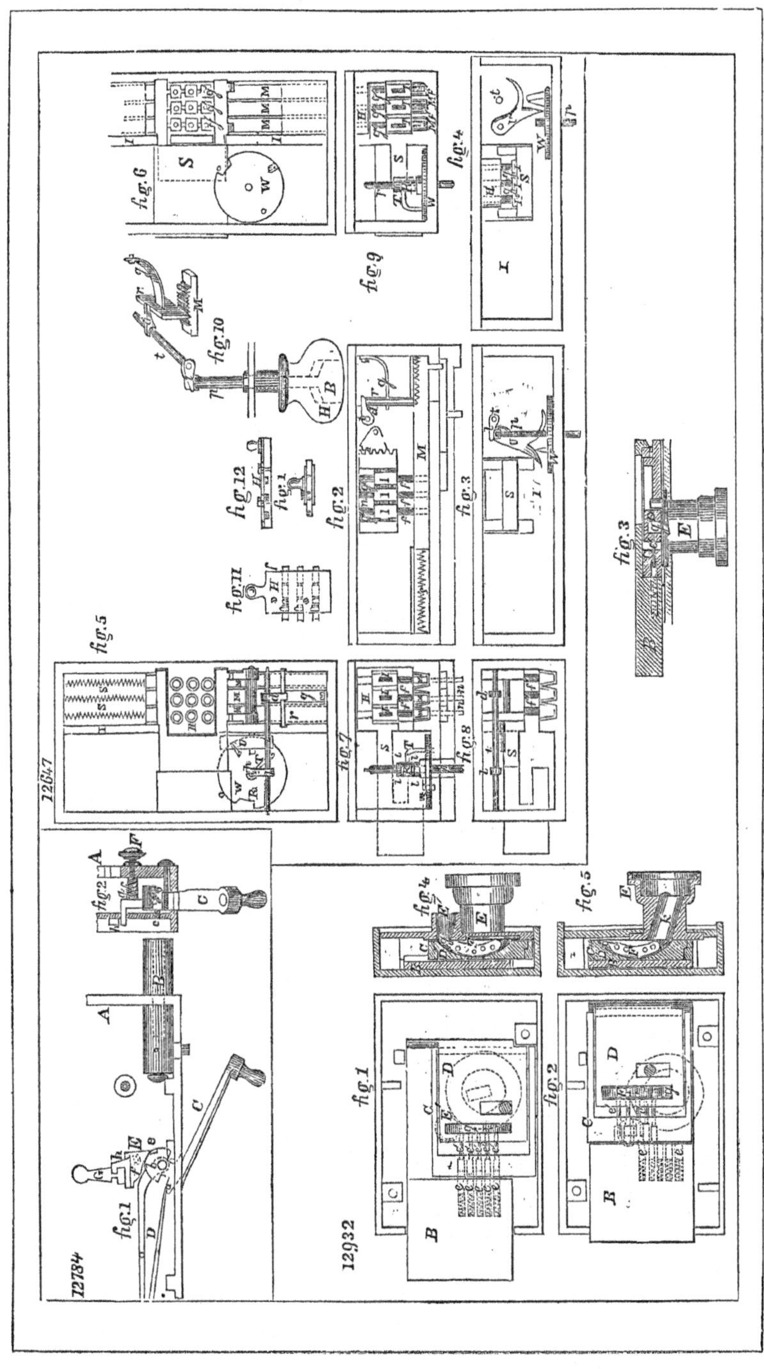
12734
fig.1
fig.2
12932
fig.1
fig.2
fig.3
fig.4
fig.5
12647
fig.5
fig.11
fig.12
fig.1
fig.2
fig.10
fig.6
fig.9
fig.4
fig.3
fig.7
fig.8
fig.3

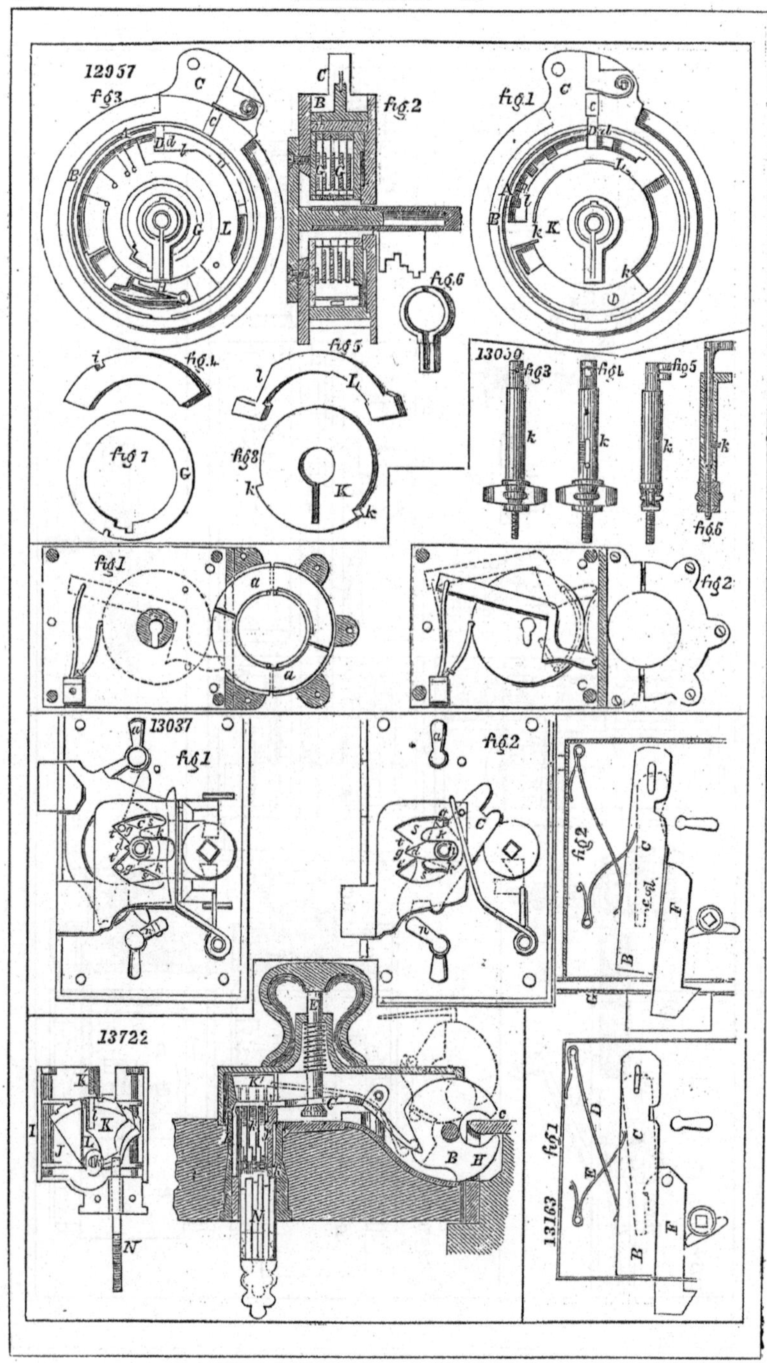
12957
13050
13037
13722
13163

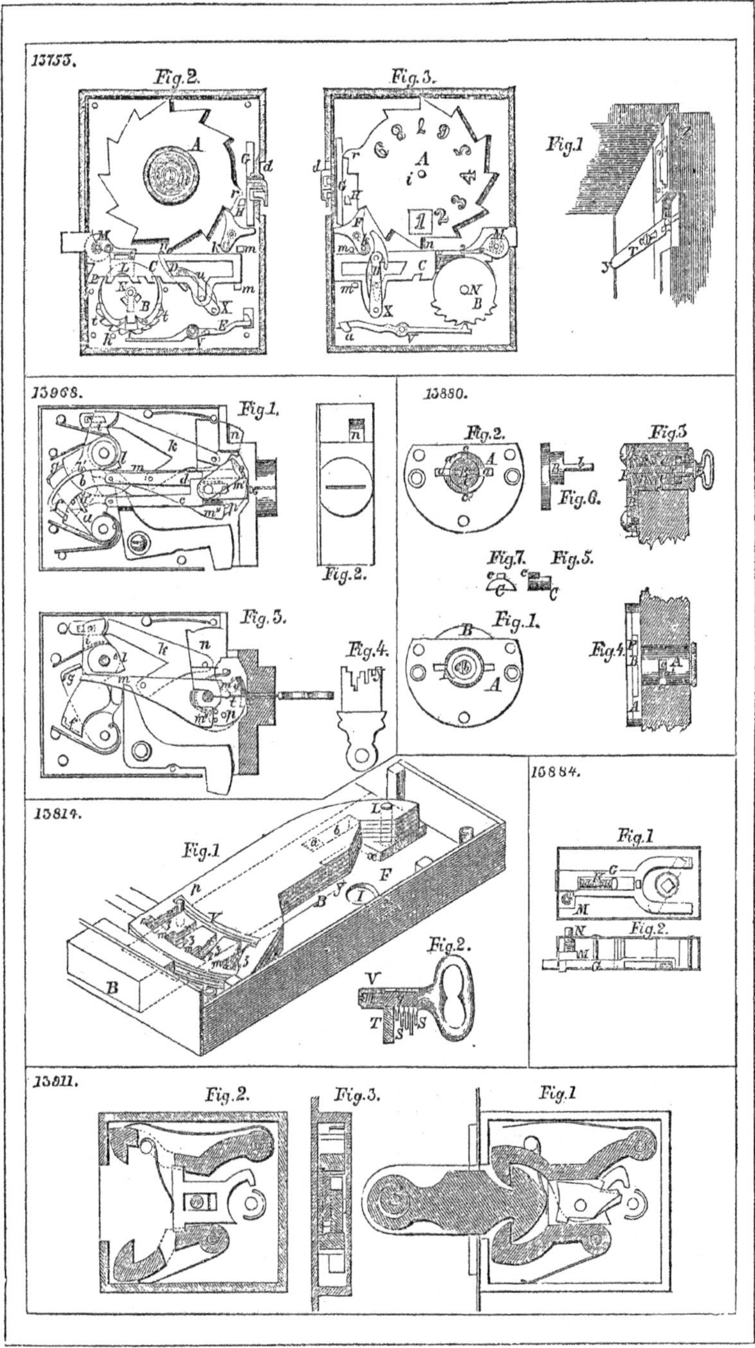
13753.
Fig.2.
Fig.3.
Fig.1
13968.
Fig.1.
Fig.2.
Fig.3.
Fig.4.
13880.
Fig.2.
Fig.3
Fig.6.
Fig.7.
Fig.5.
Fig.1.
Fig.4.
13884.
Fig.1
Fig.2.
13814.
Fig.1
Fig.2.
13811.
Fig.2.
Fig.3.
Fig.1

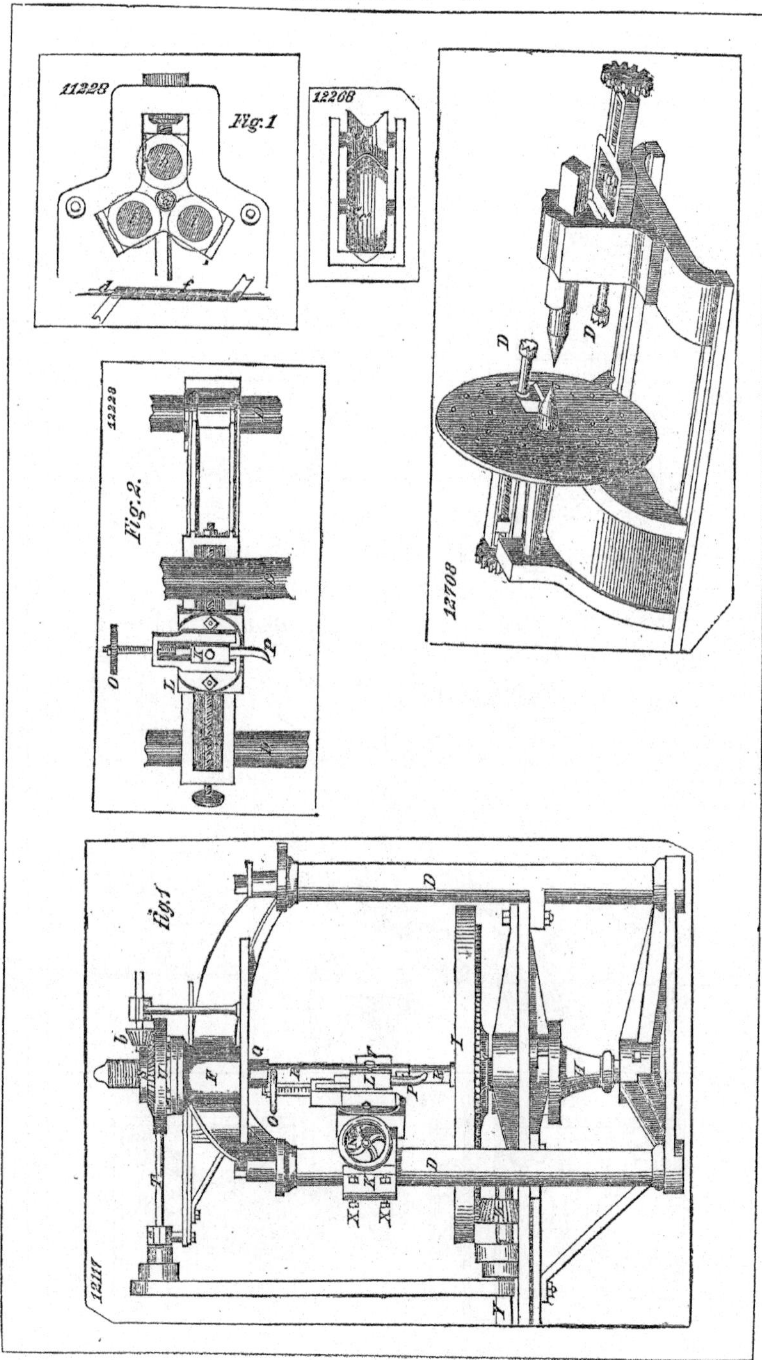

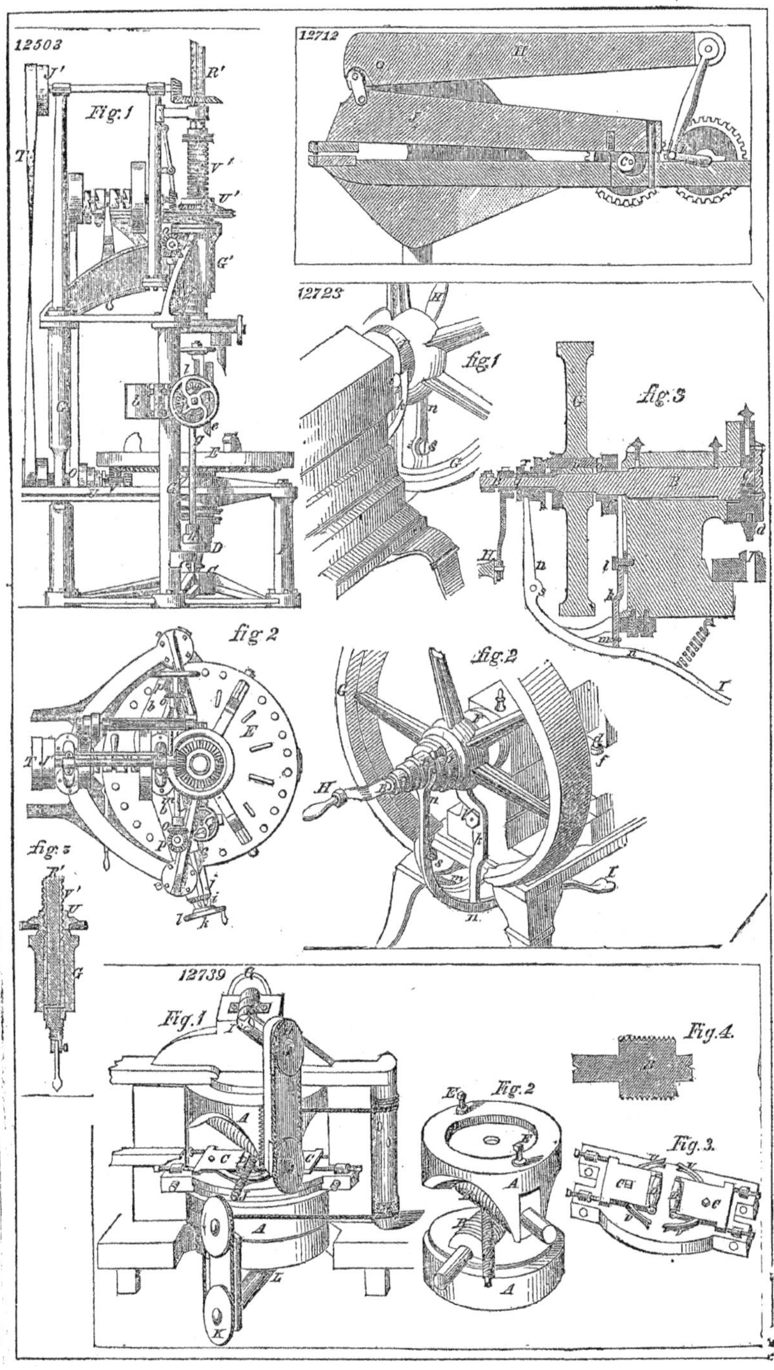
12503
Fig. 1
fig 2
fig. 3
12712
12723
fig 1
fig. 3
fig. 2
12739
Fig. 1
Fig. 2
Fig. 3.
Fig. 4.

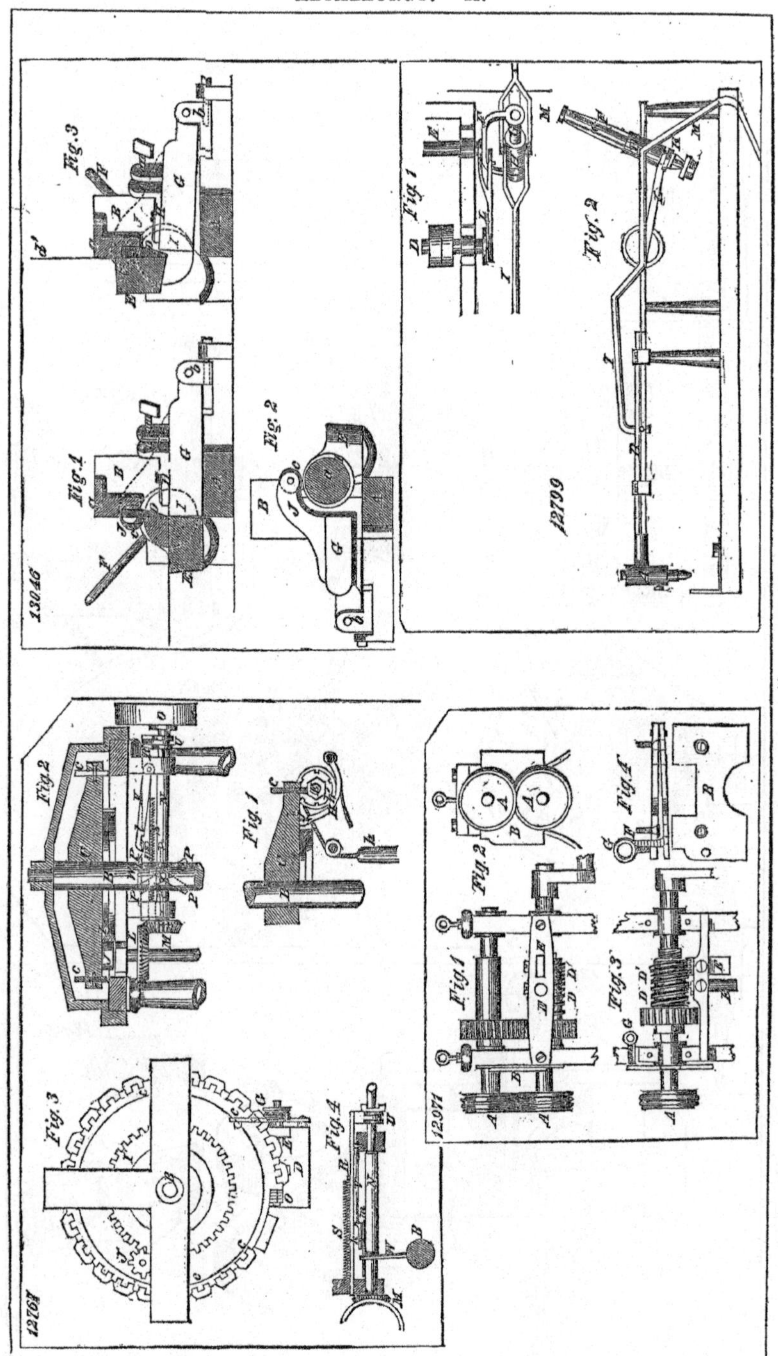
13046
12799
12767
12971

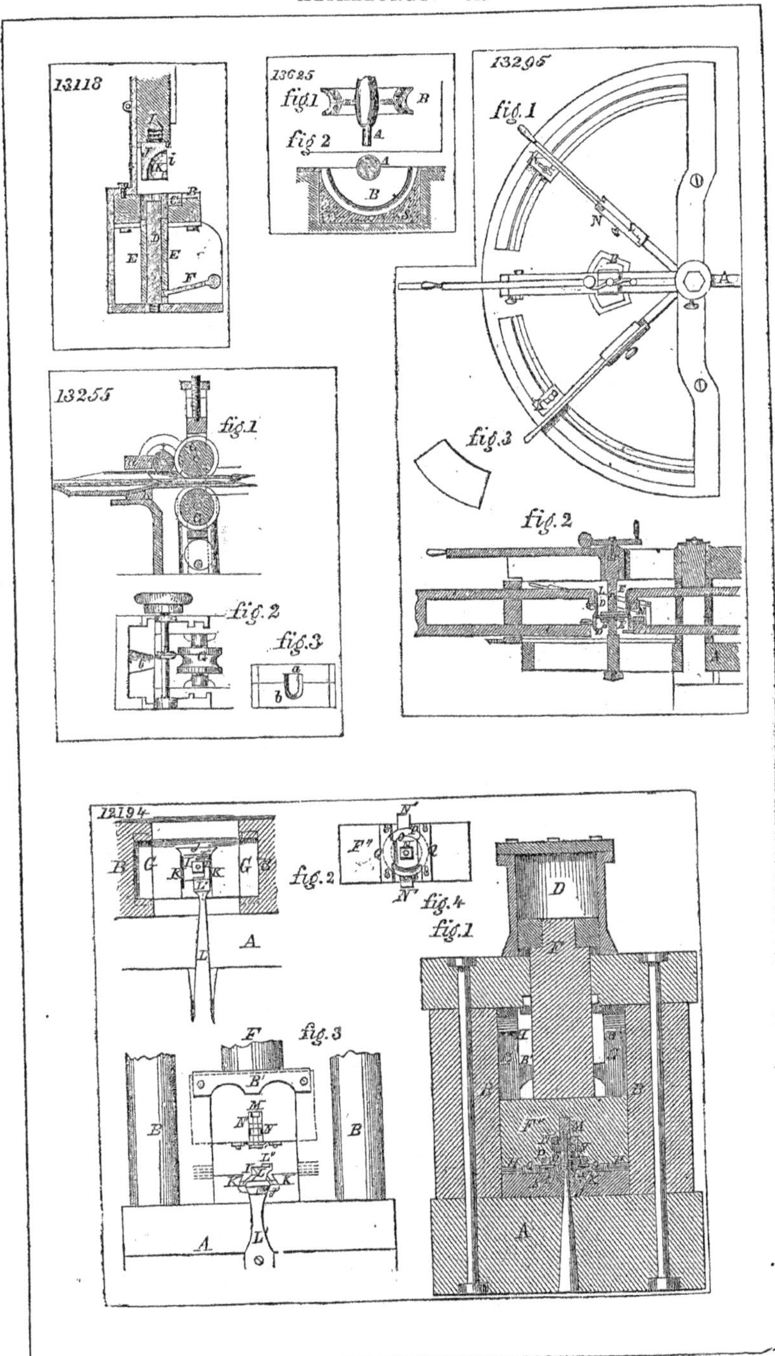
13118
13625
fig.1
fig 2
13295
fig.1
13255
fig.1
fig.2
fig.3
fig.3
fig.2
12194
fig.2
fig.4
fig.1
fig.3

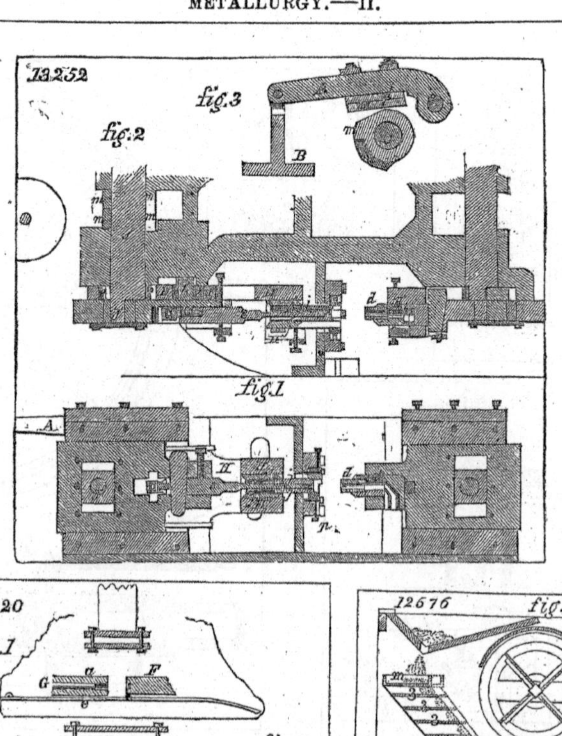
13252
fig.3
fig.2
fig.1

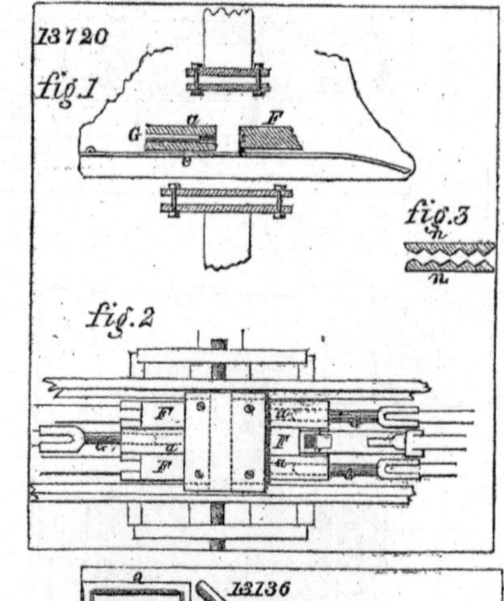
13720
fig.1
fig.2
fig.3

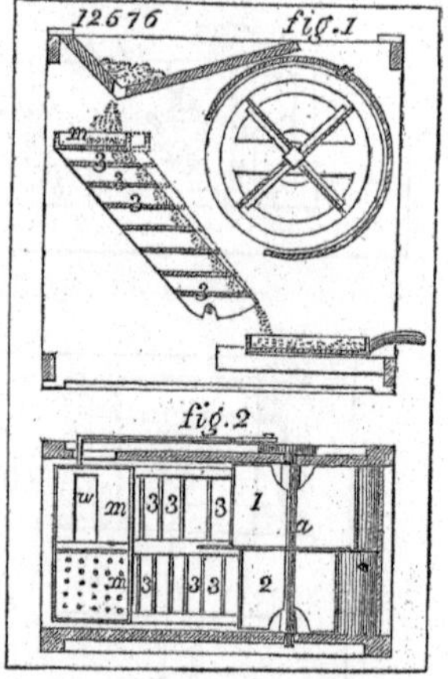
12676
fig.1
fig.2

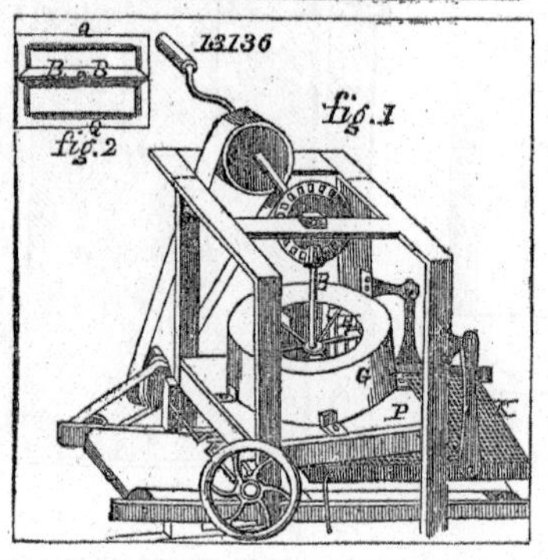
13136
fig.1
fig.2

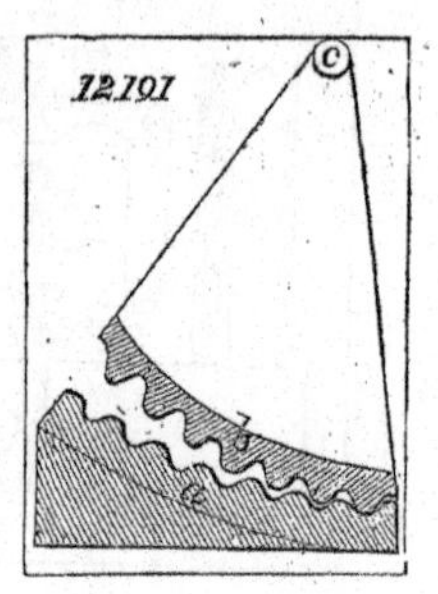
12191

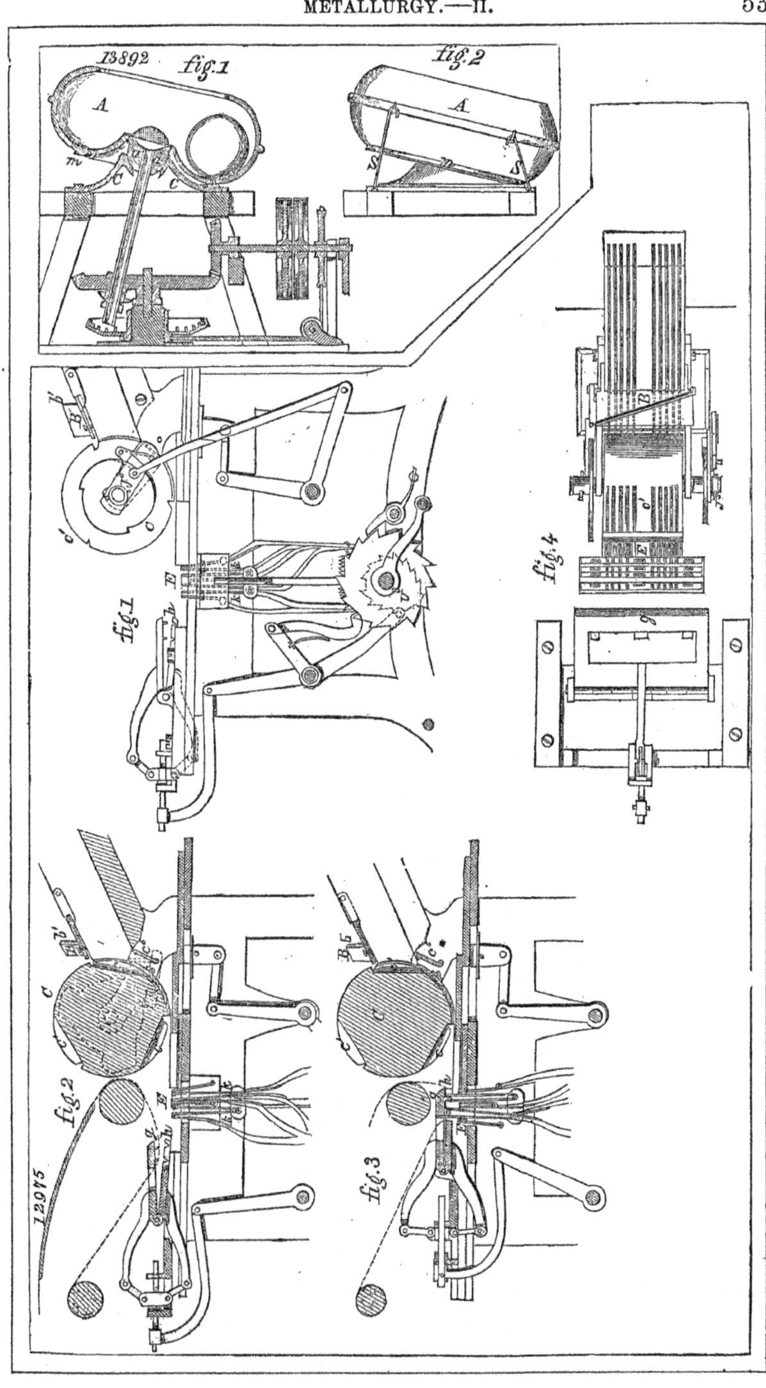

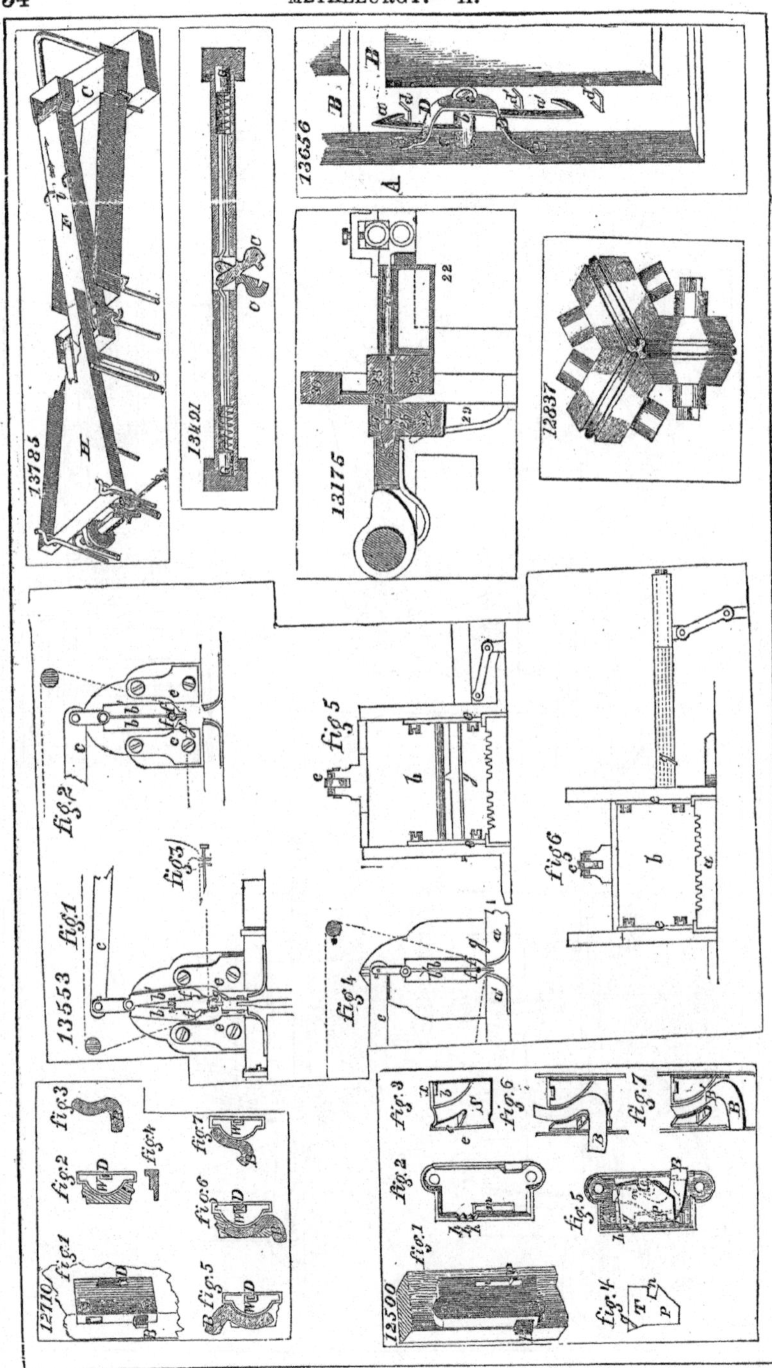
13785
13401
13656
13175
12837
13553
fig.1
fig.2
fig.3
fig.4
fig.5
fig.6
12710
12500

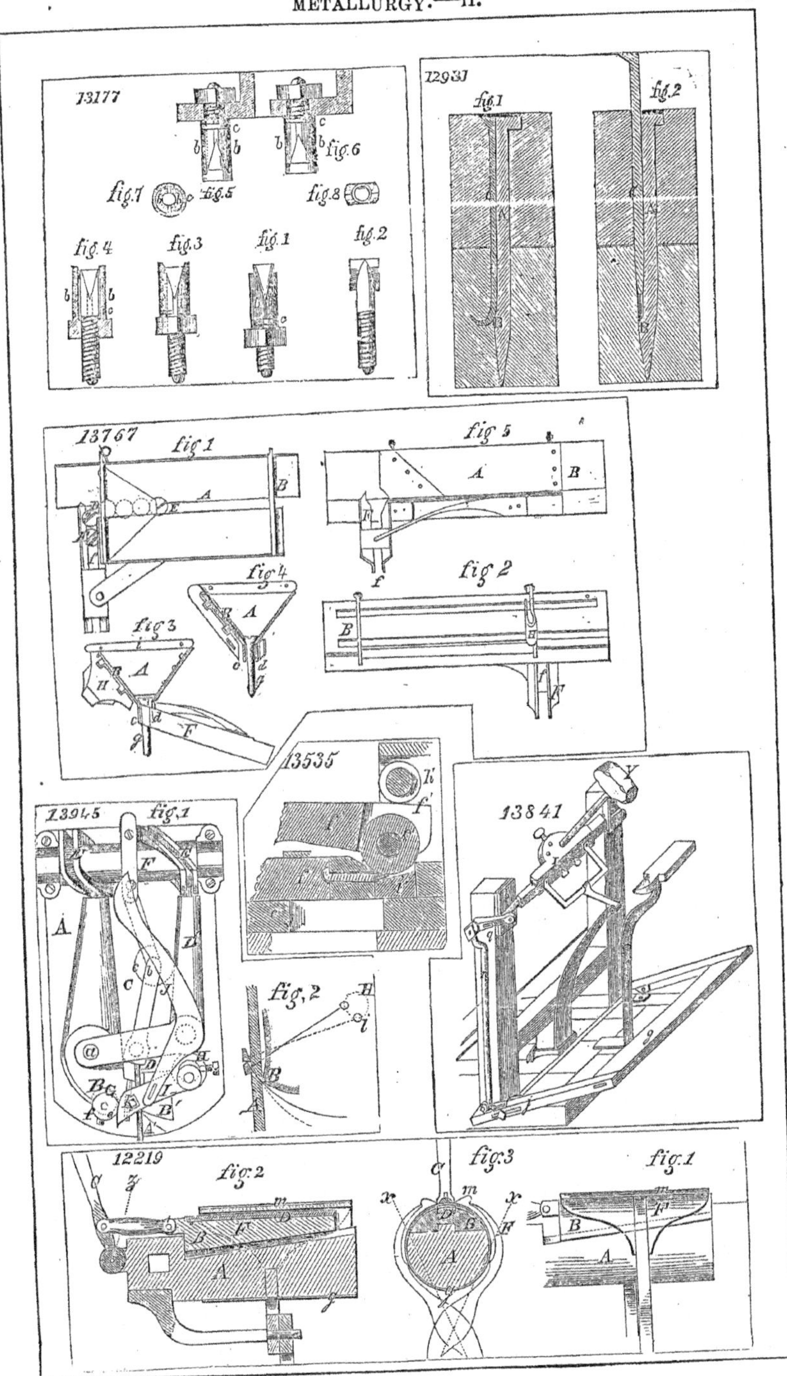
13177
fig.7
fig.5
fig.8
fig.6
fig.4
fig.3
fig.1
fig.2
12931
fig.1
fig.2
13767
fig 1
fig 5
fig 2
fig 3
fig 4
13535
13945
fig.1
fig.2
13841
12219
fig.2
fig.3
fig.1

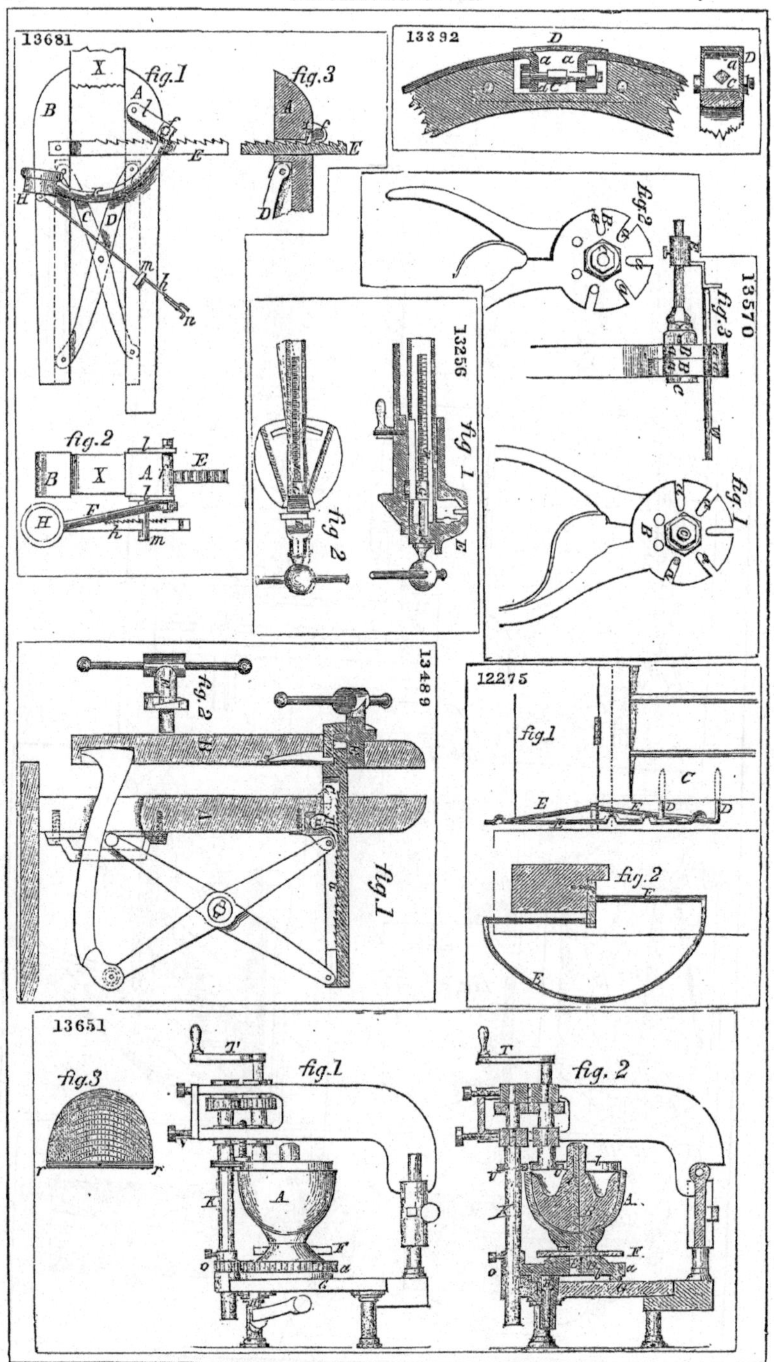
13681
fig.1
fig.3
fig.2
13392
13570
fig.2
fig.3
fig.1
13256
fig 1
fig 2
13489
fig.2
fig.1
12275
fig.1
fig.2
13651
fig.3
fig.1
fig. 2

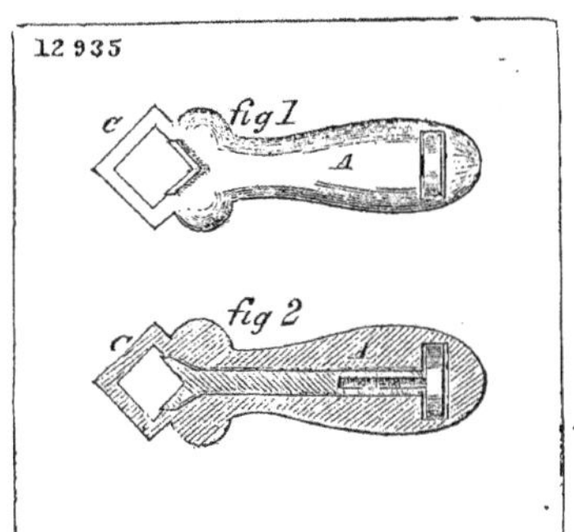
12935
fig 1
fig 2

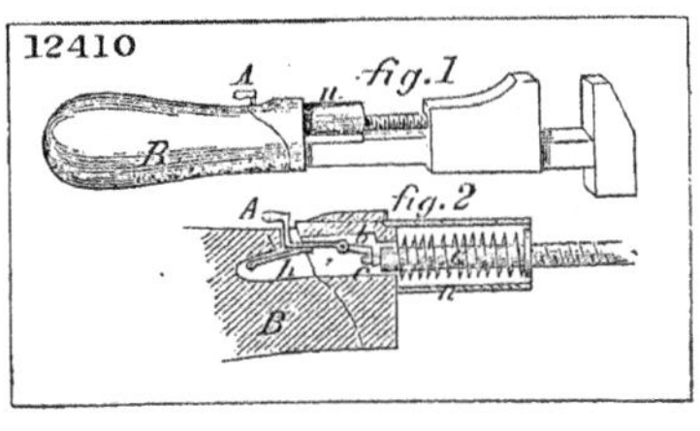
12410
fig. 1
fig. 2

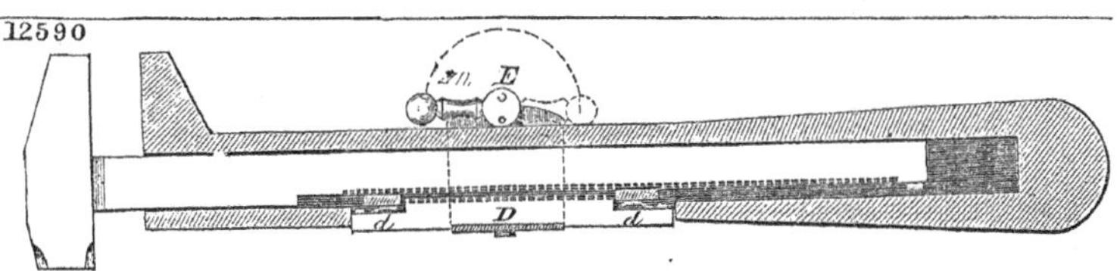
12590

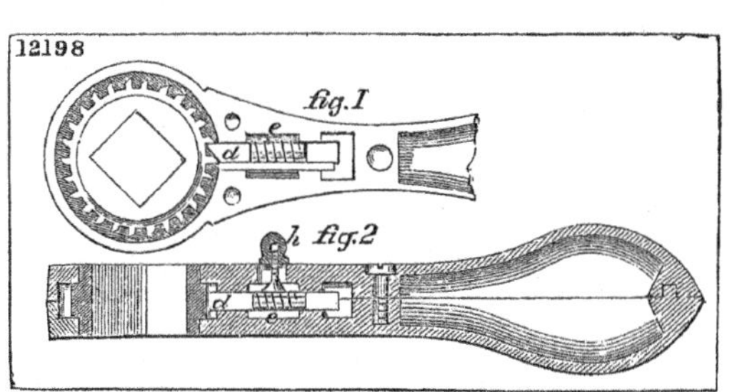
12198
fig. 1
fig. 2

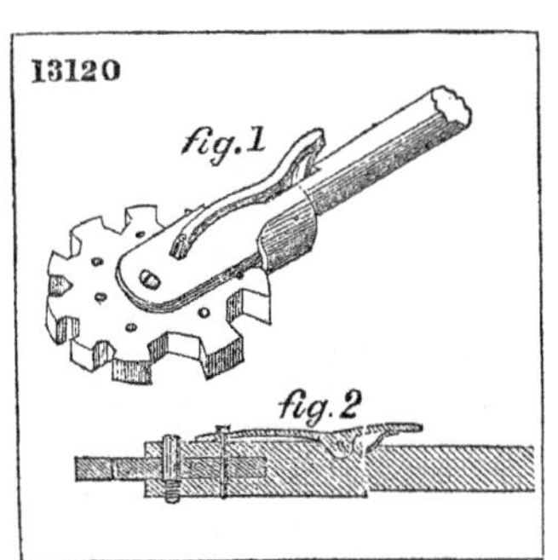
13120
fig. 1
fig. 2

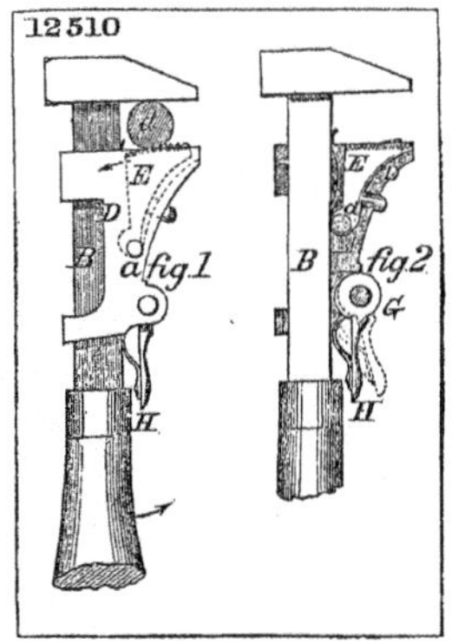
12510
fig 1
fig 2

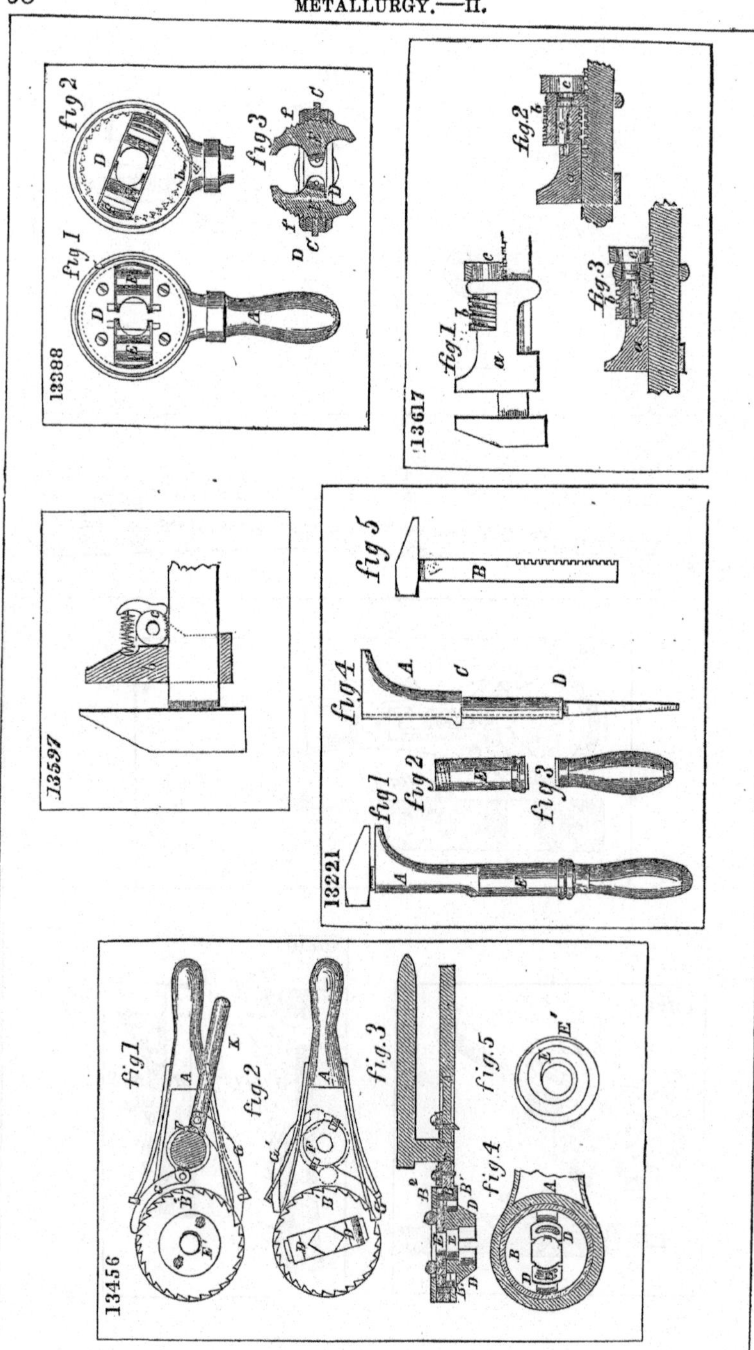
13388
fig 1
fig 2
fig 3
13617
fig.1
fig.2
fig.3
13597
13221
fig 1
fig 2
fig 3
fig 4
fig 5
13456
fig.1
fig.2
fig.3
fig.4
fig.5

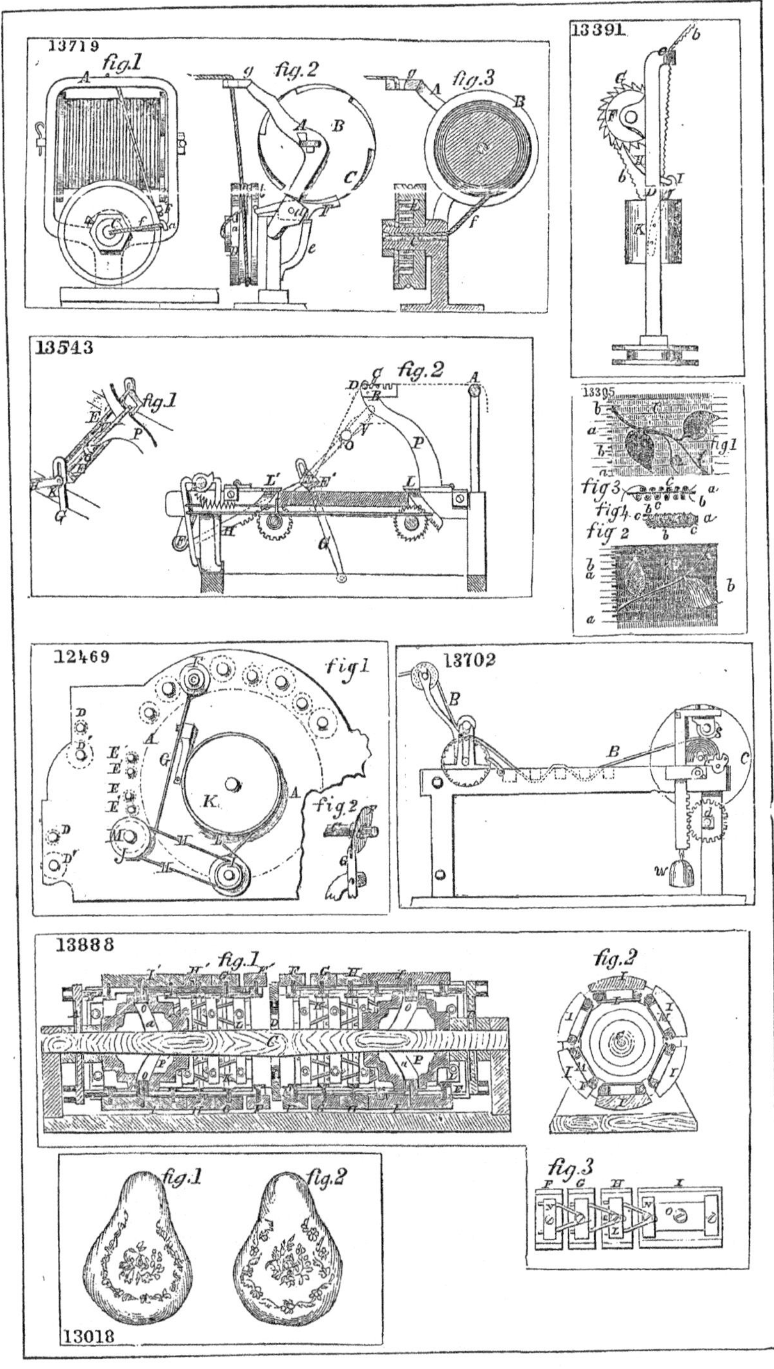
13719
fig.1
fig.2
fig.3
13391
13543
fig.1
fig.2
13305
fig1
fig3
fig4
fig 2
12469
fig1
fig.2
13702
13888
fig.1
fig.2
fig.3
fig.1
fig.2
13018

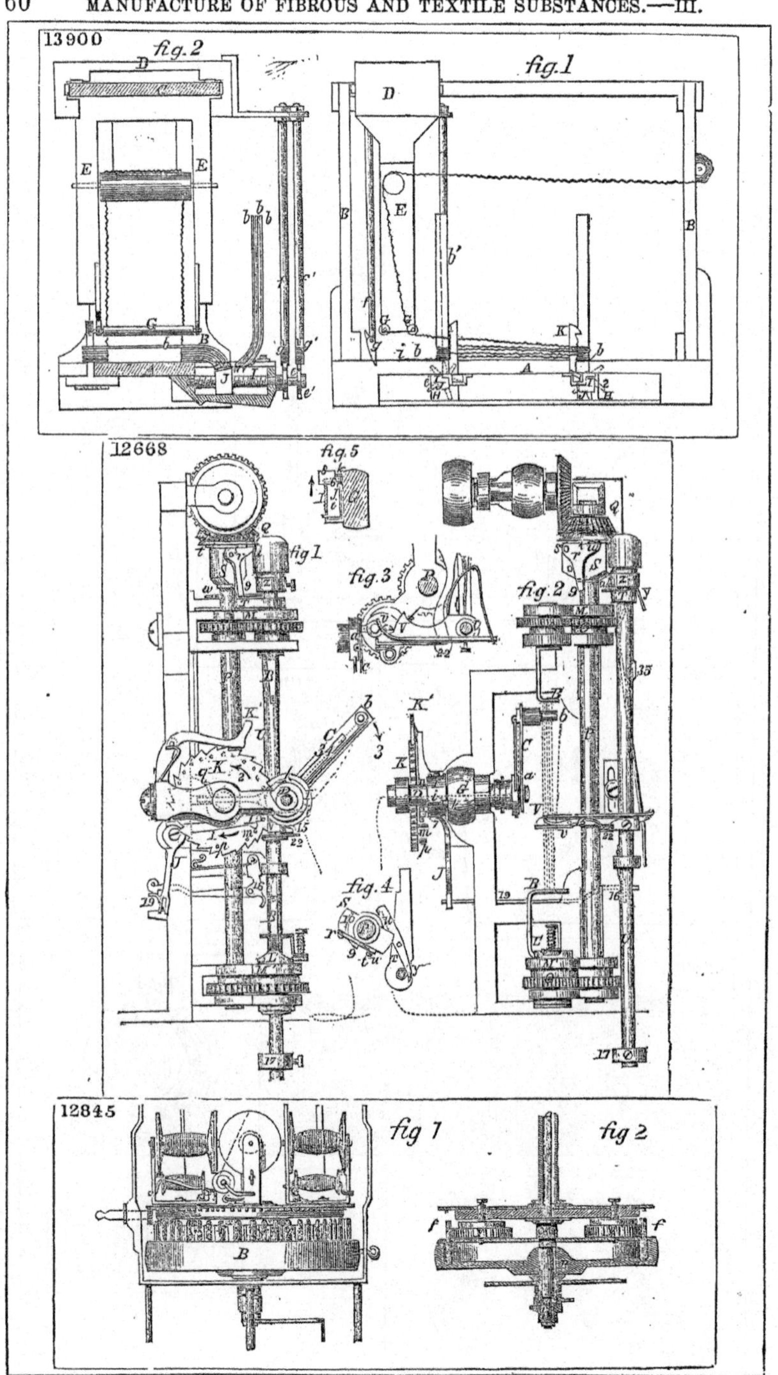
13900
fig. 2
fig. 1
12668
fig. 5
fig 1
fig. 3
fig. 2
fig. 4
12845
fig 1
fig 2

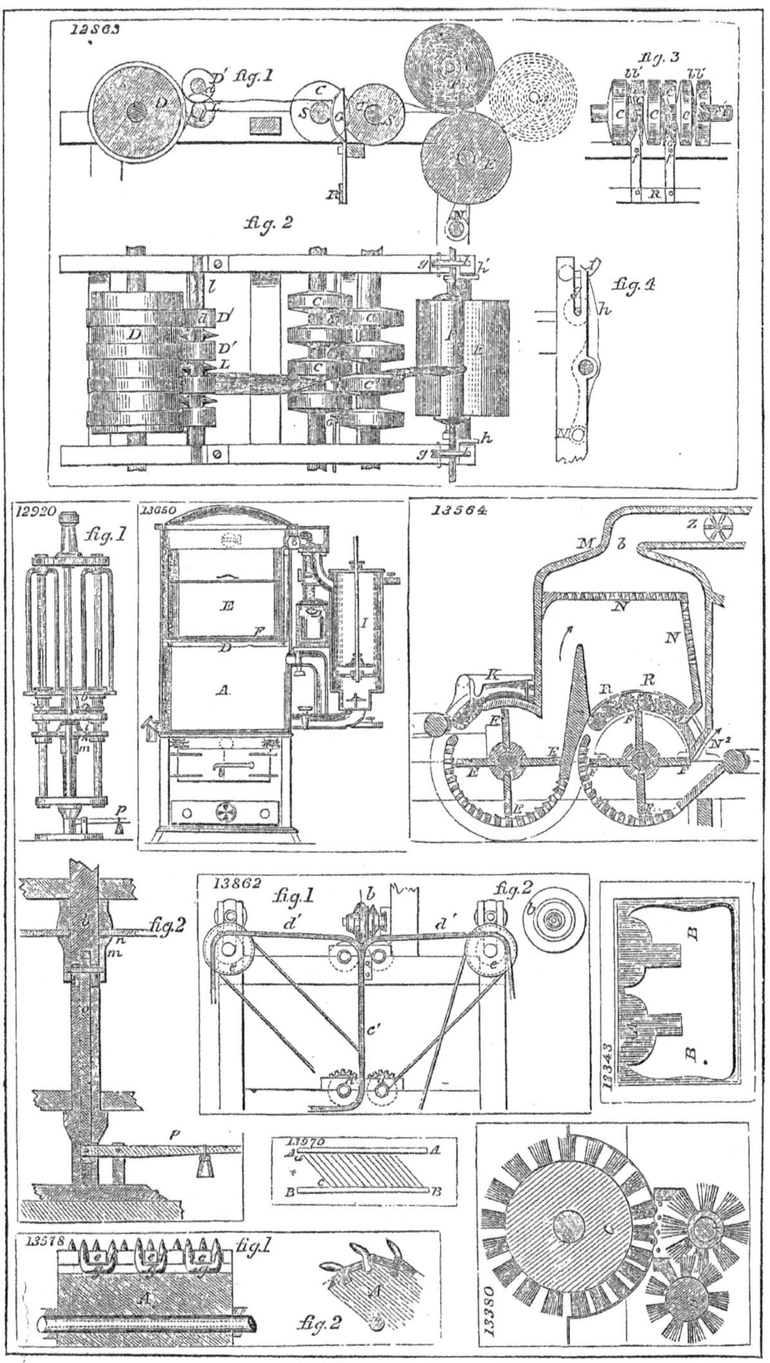

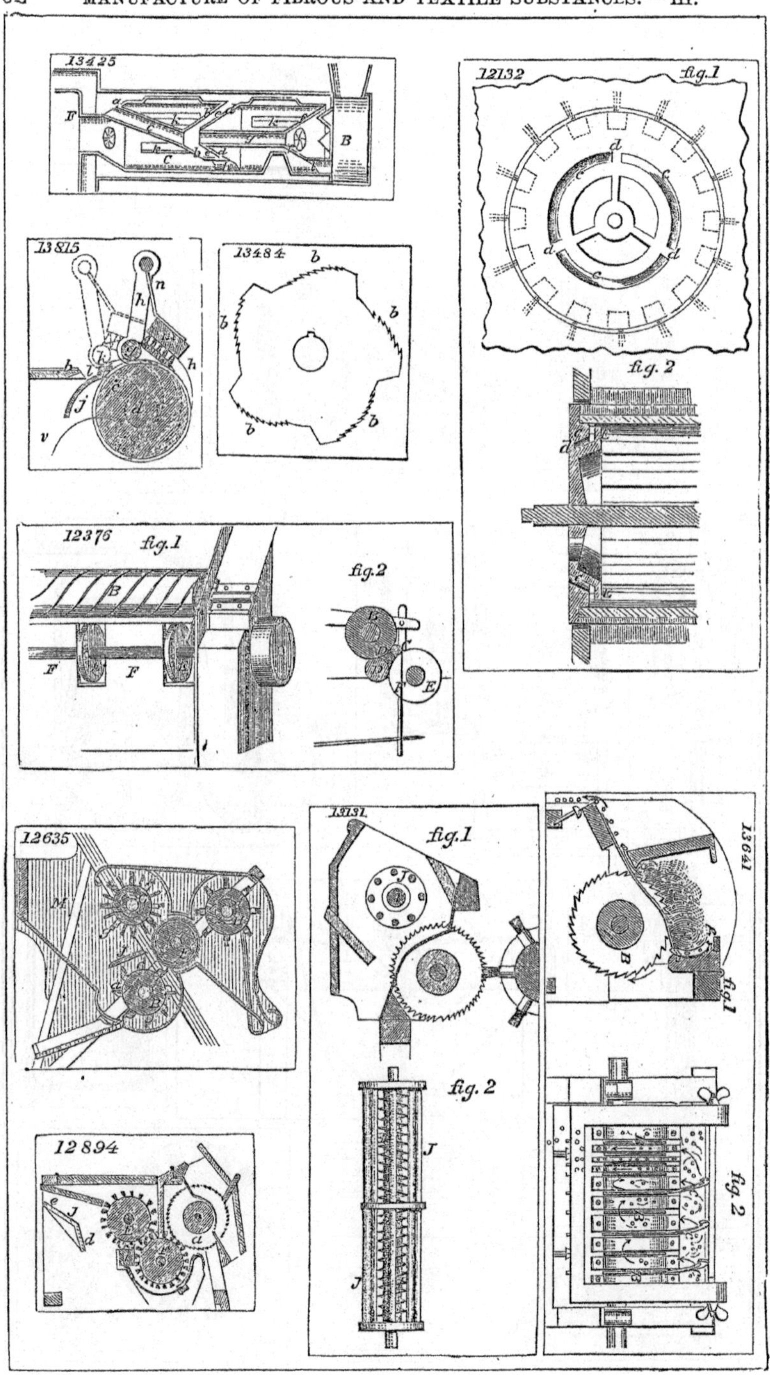
13425
12132
fig.1
13815
13484
fig. 2
12376
fig.1
fig.2
12635
13131
fig.1
13641
fig.1
fig. 2
12894
fig. 2

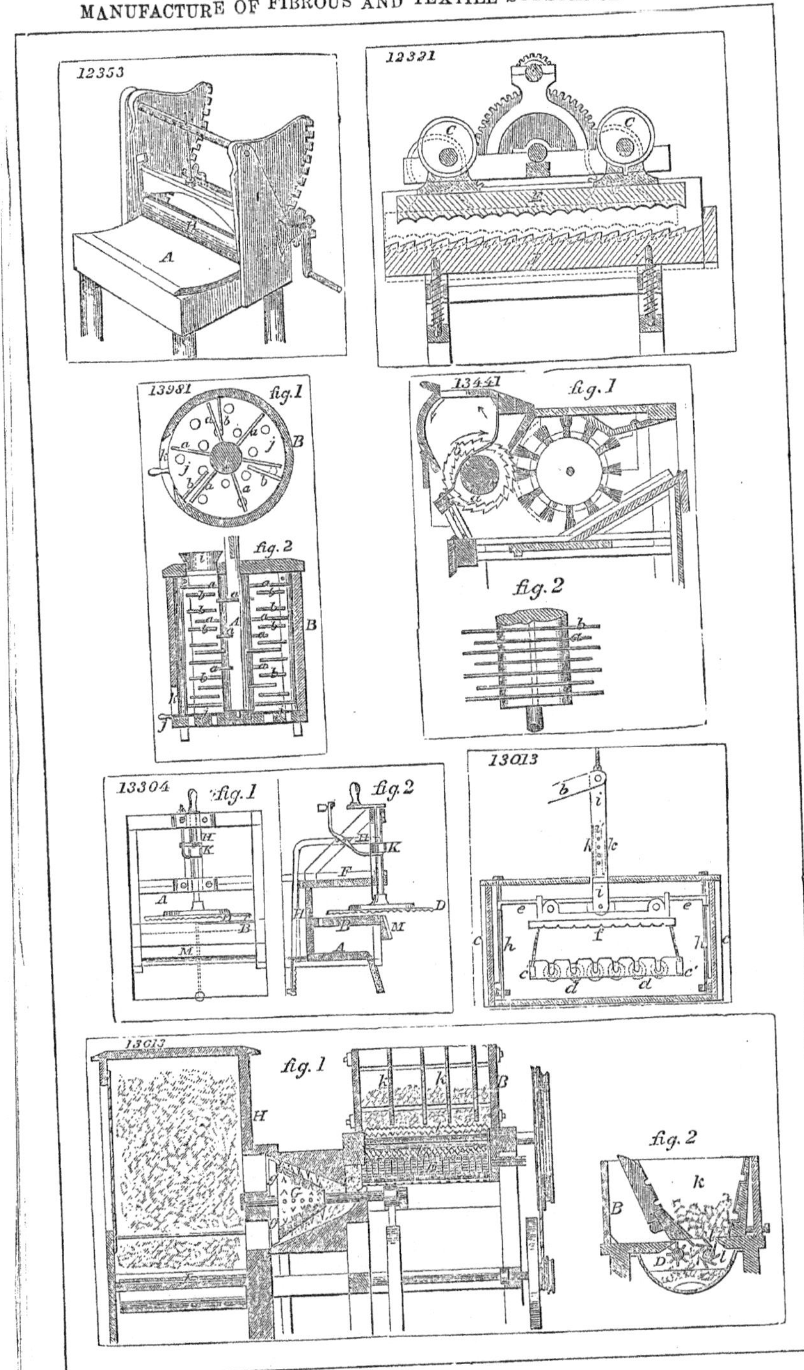
12353
12321
13981
fig. 1
fig. 2
13441
fig. 1
fig. 2
13304
fig. 1
fig. 2
13013
13613
fig. 1
fig. 2

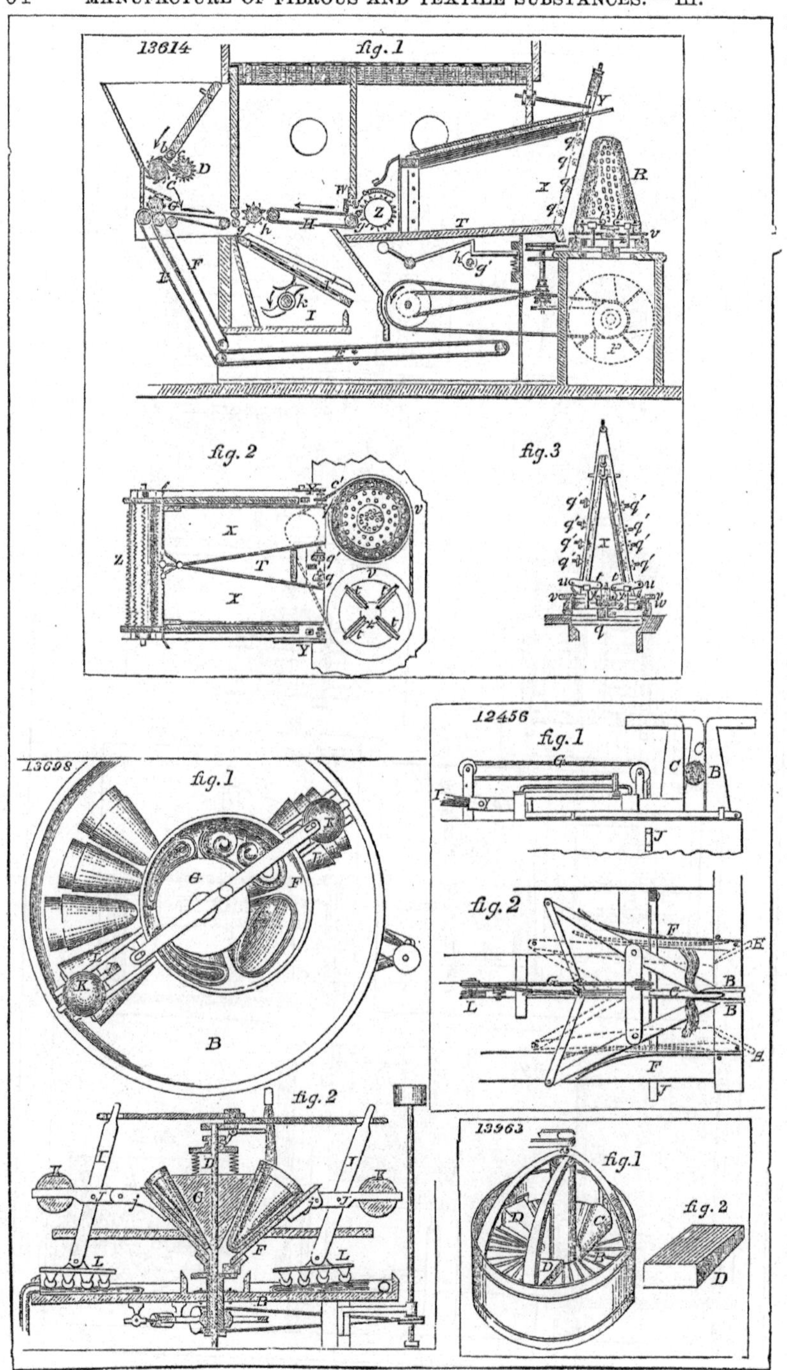
13614
fig. 1
fig. 2
fig. 3
12456
fig. 1
fig. 2
13698
fig. 1
fig. 2
13963
fig. 1
fig. 2

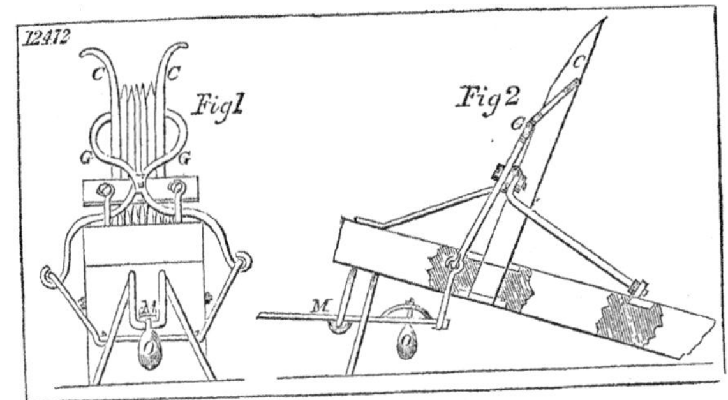
12412
Fig1
Fig2
C
G
M
O

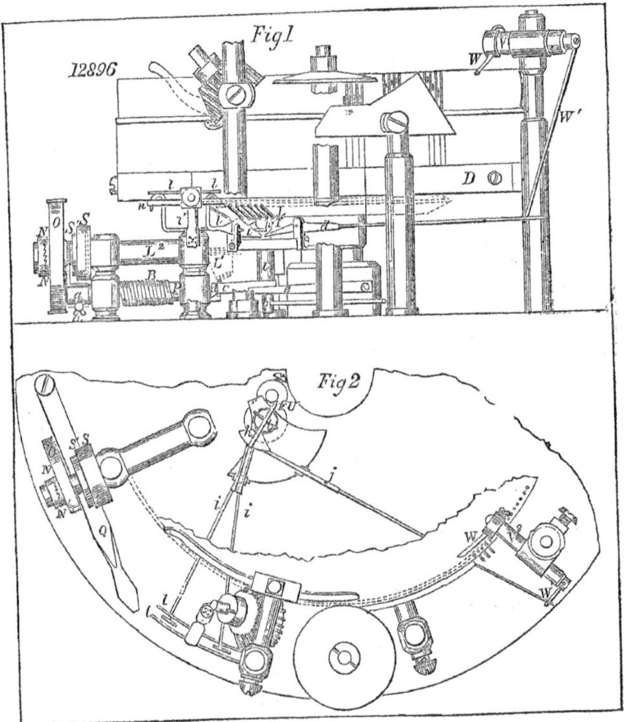
Fig1
12896
W
W'
D
O
S
N
B
P
Fig2
i
j
Q
l

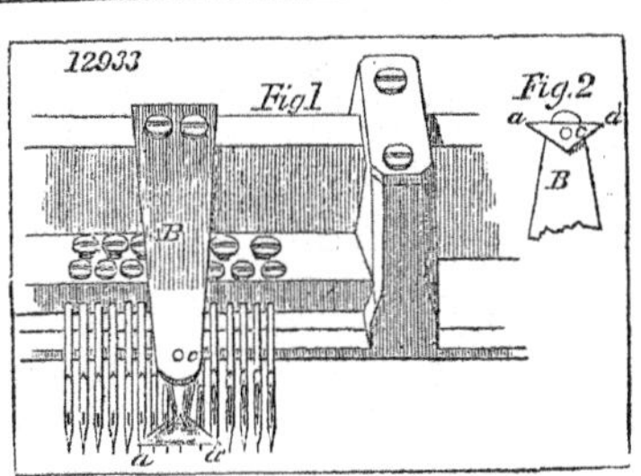
12933
Fig1
Fig.2
a
d
B

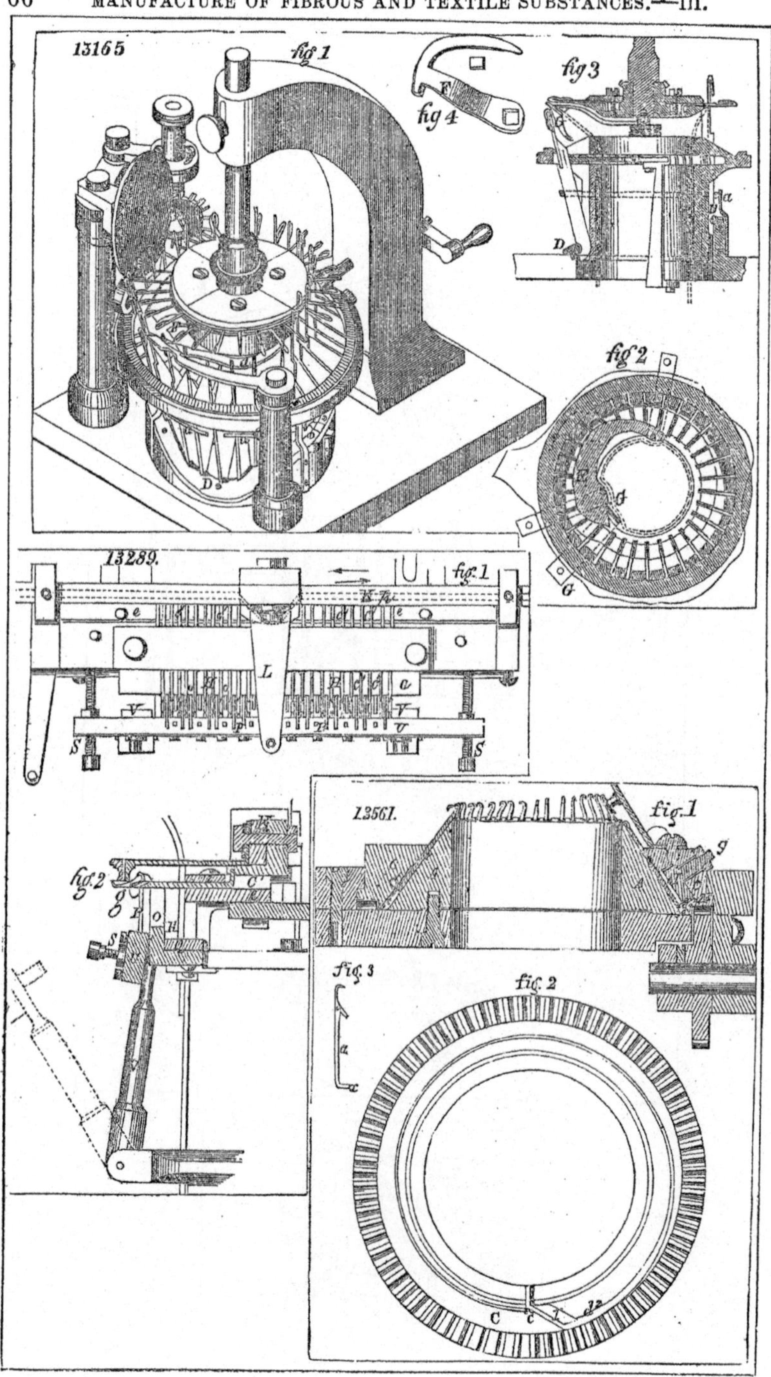
13165
fig 1
fig 4
fig 3
fig 2
13289.
fig. 1
13561.
fig. 1
fig. 2
fig. 3
fig. 2

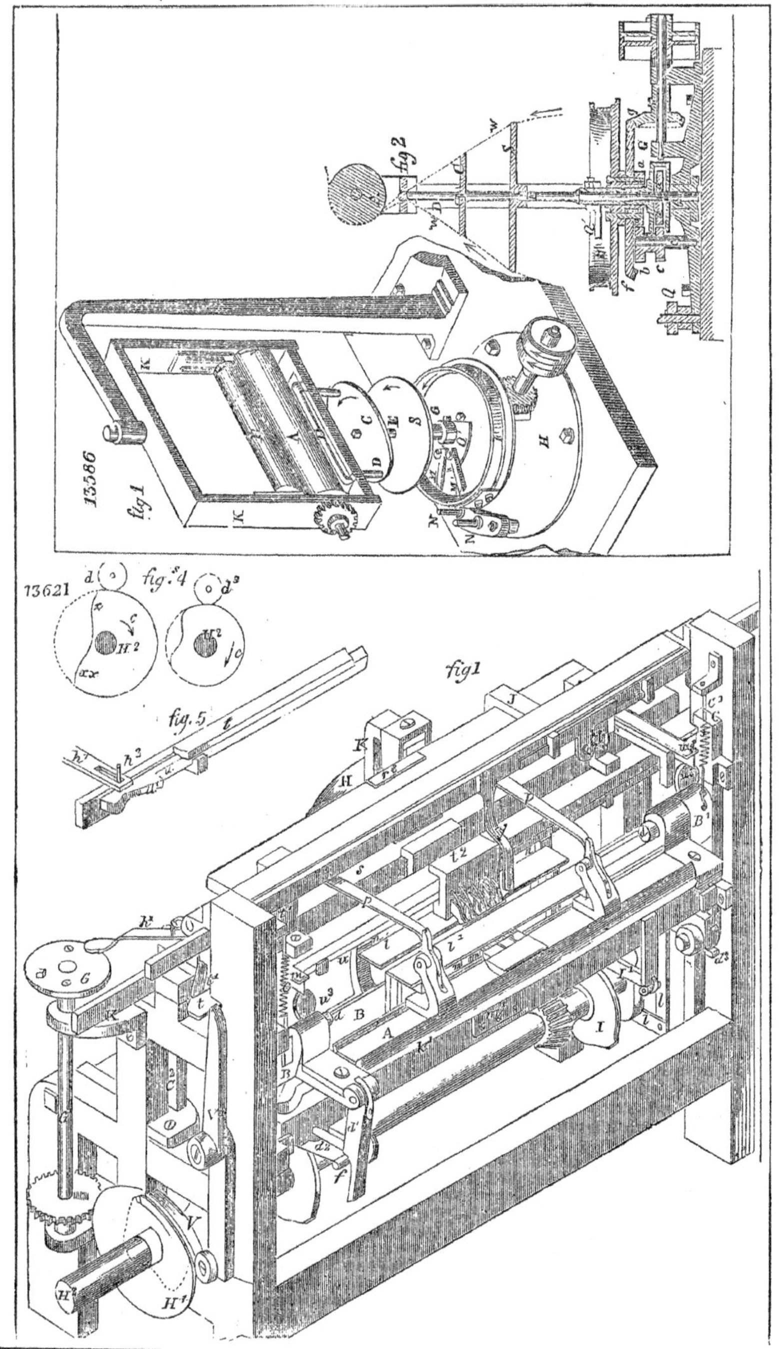
13586
fig 1
fig 2
13621
fig.s 4
fig. 5
fig1

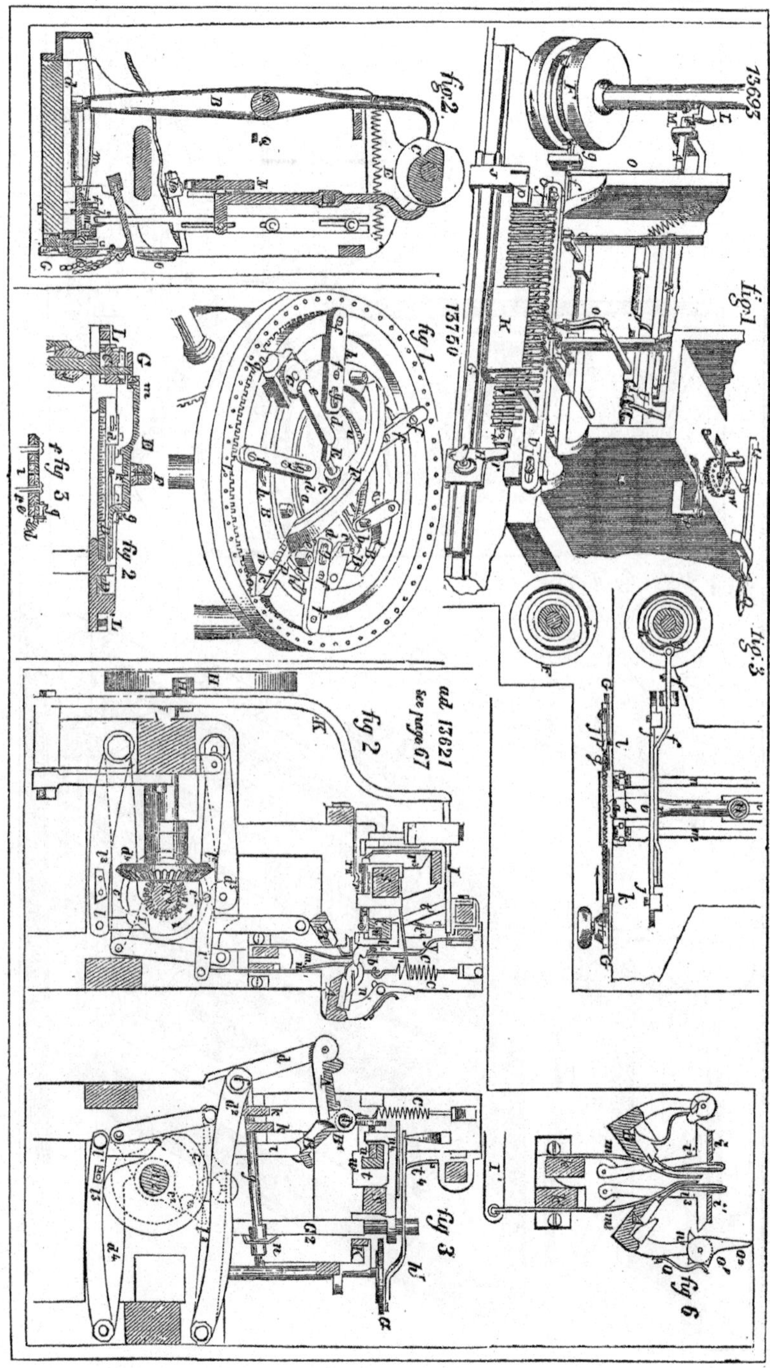
13693
fig. 1
fig. 2.
fig. 3
13750
fig 1
fig 2
fig 3
ad 13621
see page 67
fig 2
fig 3
fig 6

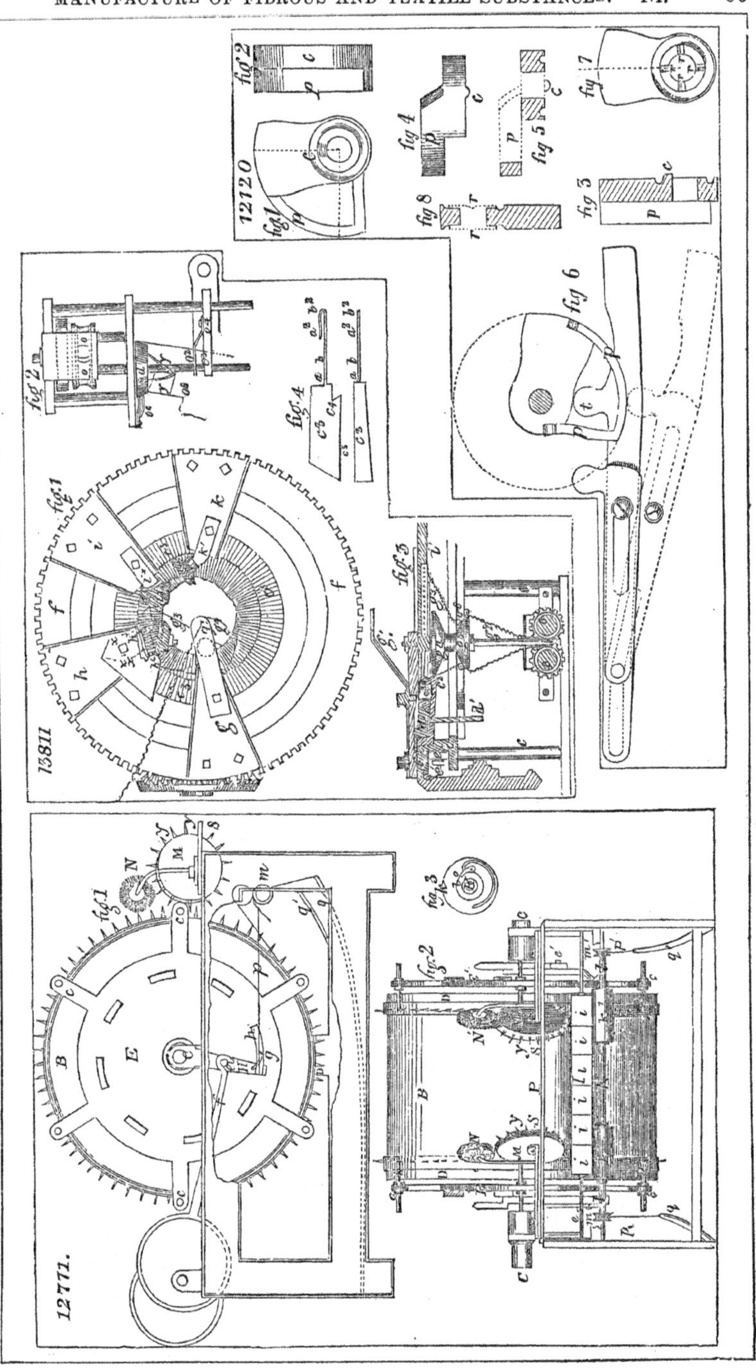
12771.
13811
12120

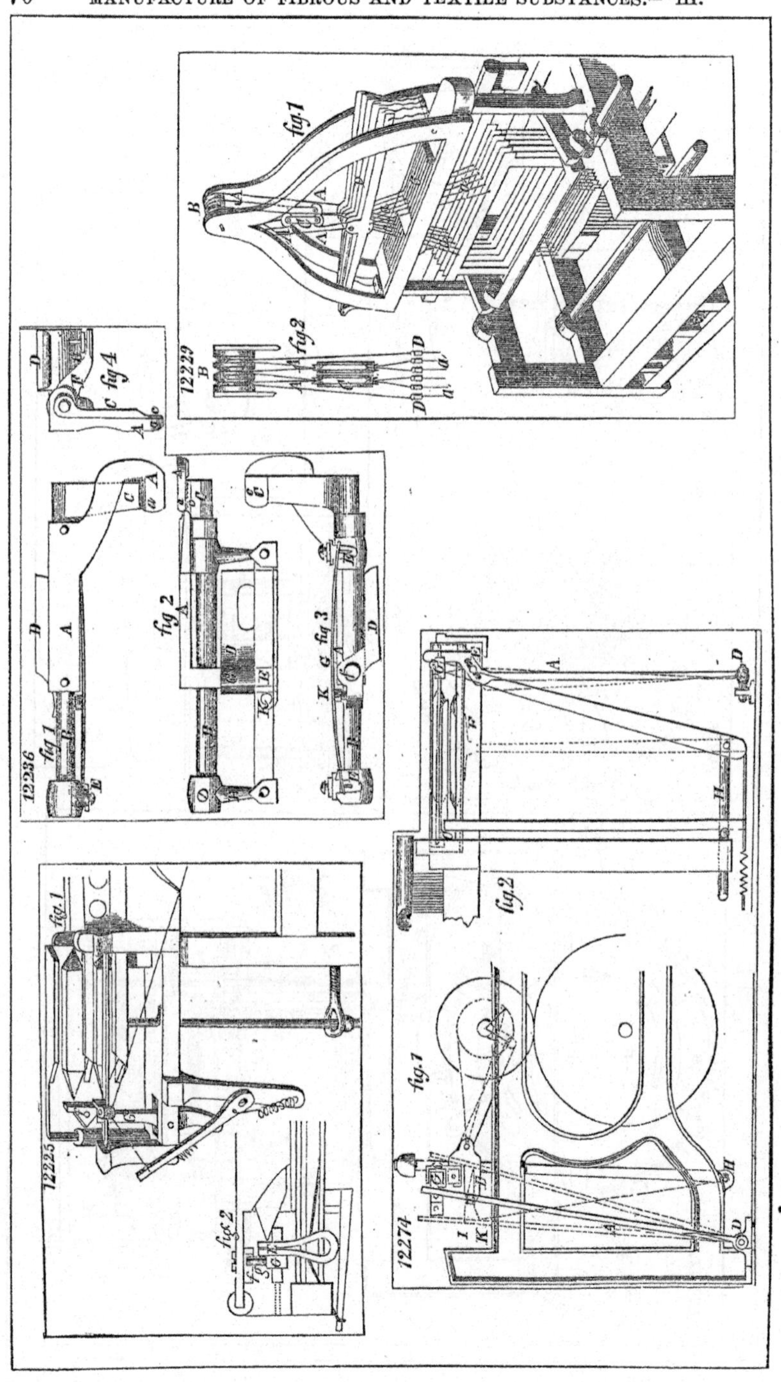
12229
fig.1
fig.2
12236
fig 1
fig 2
fig 3
fig 4
12225
fig.1
fig.2
12274
fig.1
fig.2

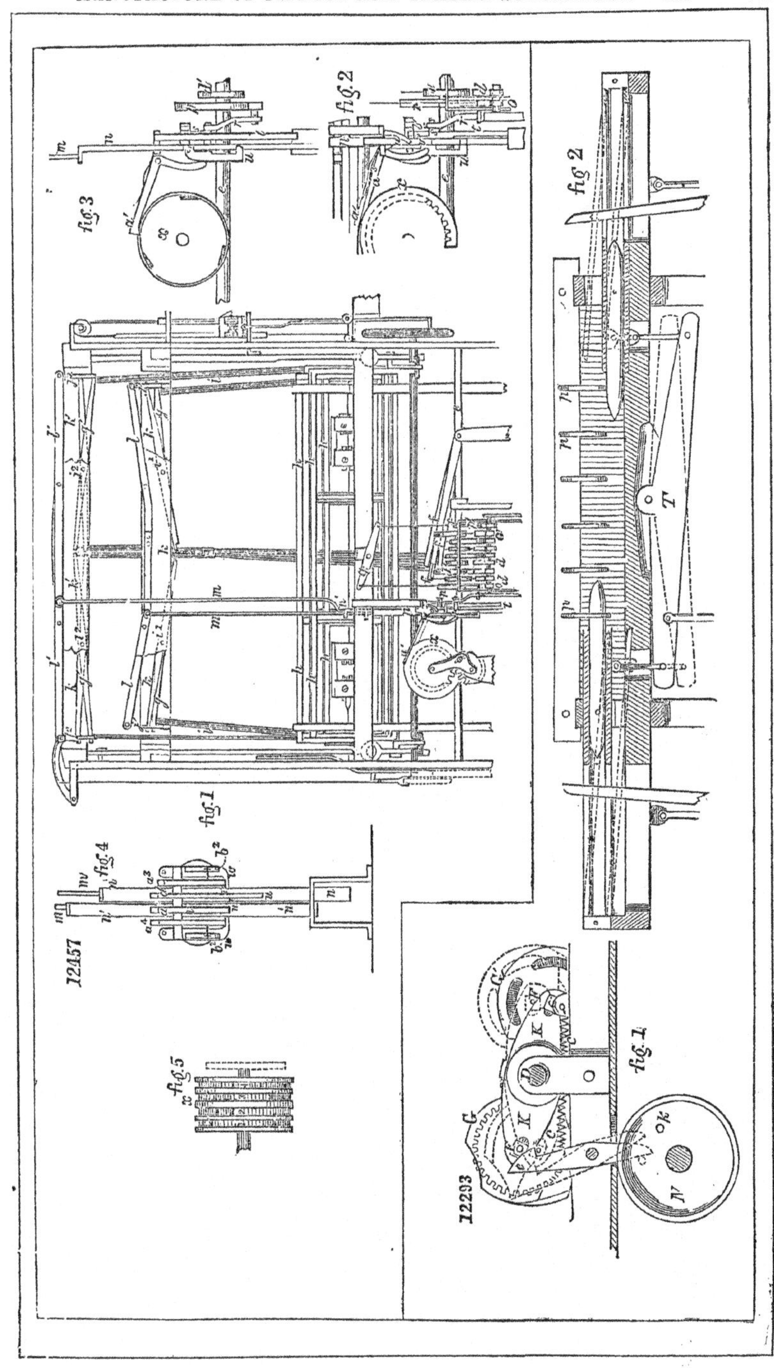
12457
fig. 1
fig. 2
fig. 3
fig. 4
fig. 5
12293
fig. 1
fig. 2

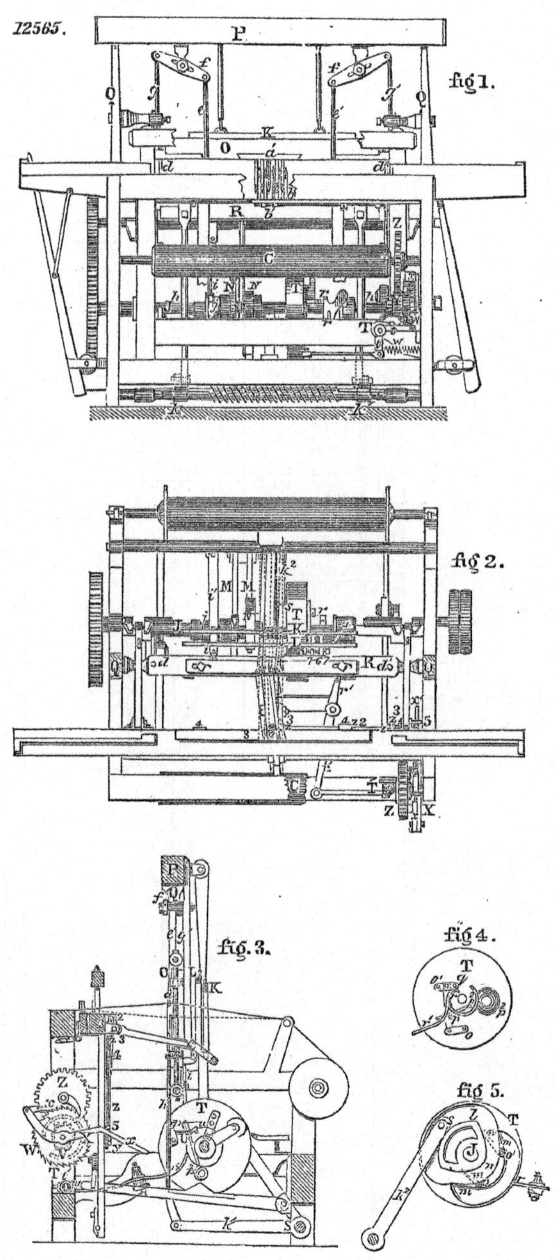
12565.
fig 1.
fig 2.
fig. 3.
fig 4.
fig 5.

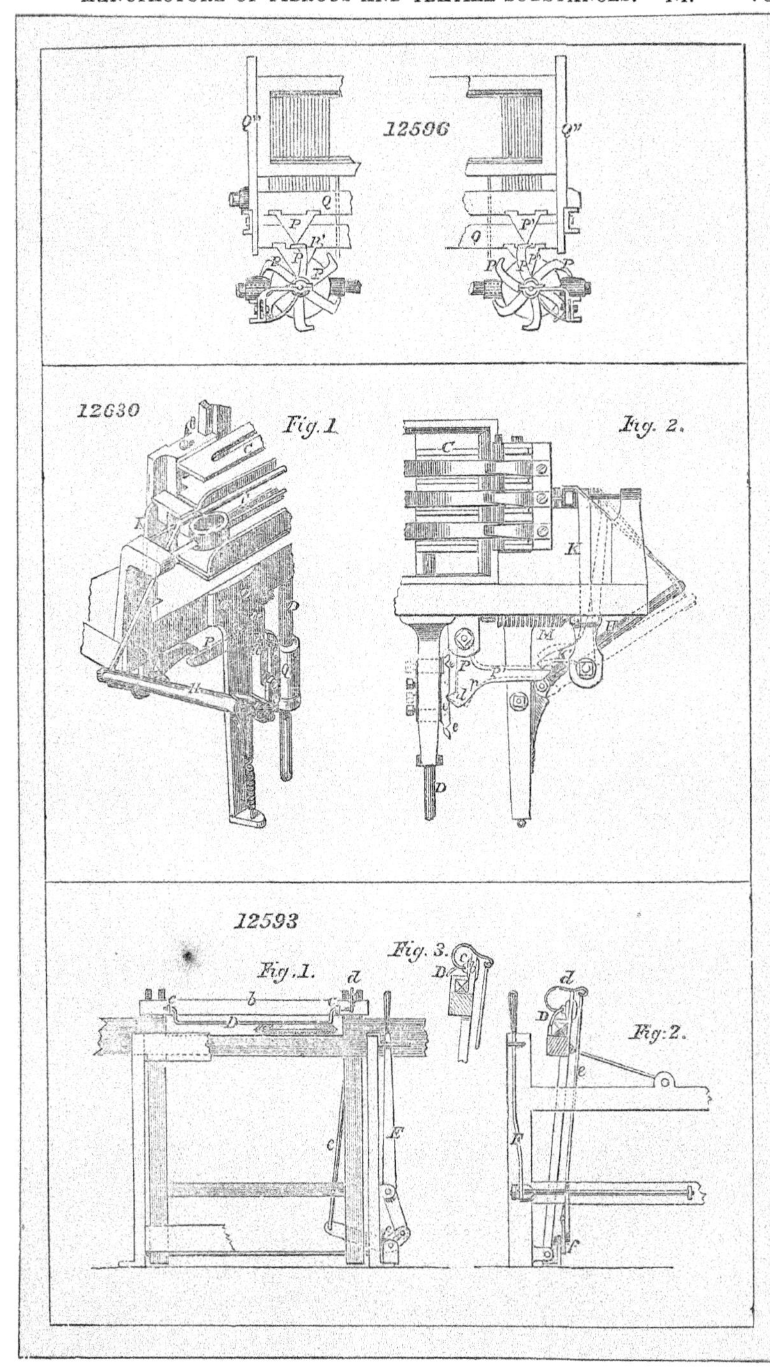
12596
12630
Fig. 1
Fig. 2.
12593
Fig. 1.
Fig. 2.
Fig. 3.

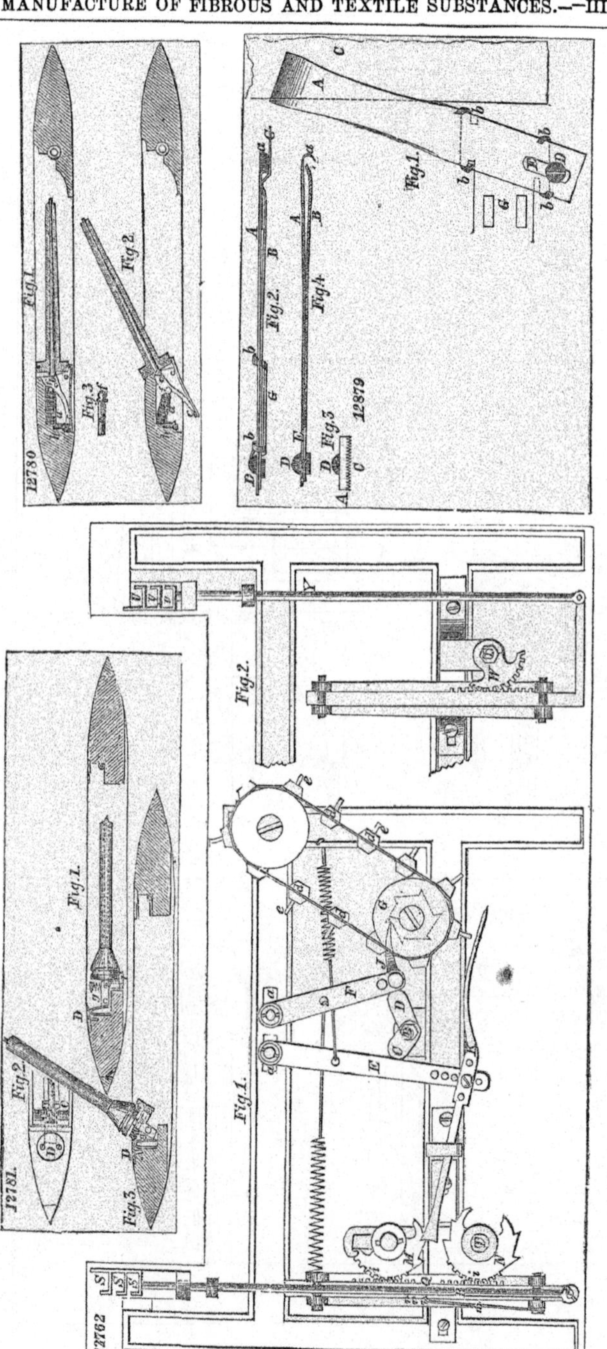

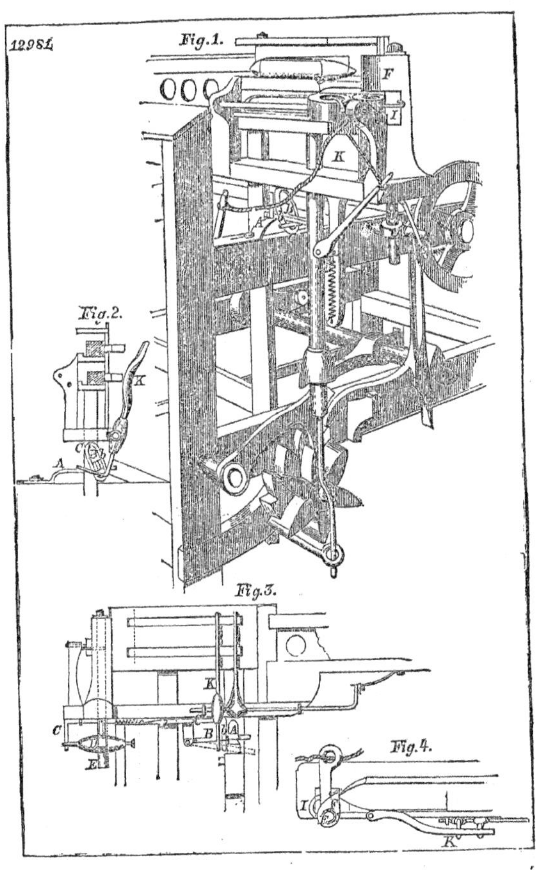
12984
Fig.1.
F
I
K
A
Fig.2.
K
C
A
Fig.3.
K
C
B
A
E
Fig.4.
I
F
E
K

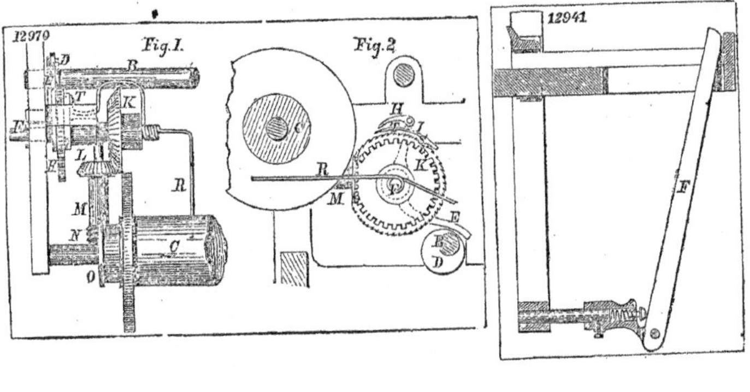
12970
Fig.1.
D
B
T
K
F
L
R
M
N
C
O
Fig.2
H
I
K
R
M
E
D
12941
F

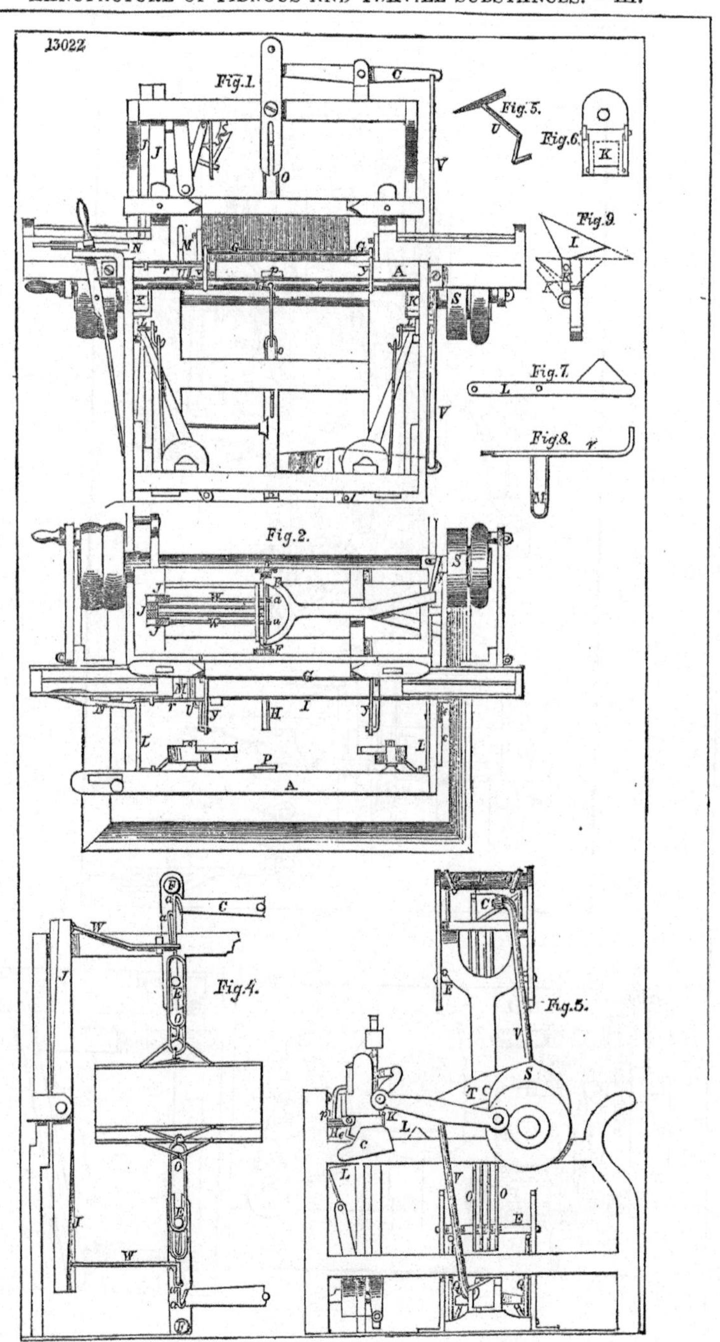
13022
Fig. 1.
Fig. 5.
Fig. 6.
Fig. 9.
Fig. 7.
Fig. 8.
Fig. 2.
Fig. 4.
Fig. 5.

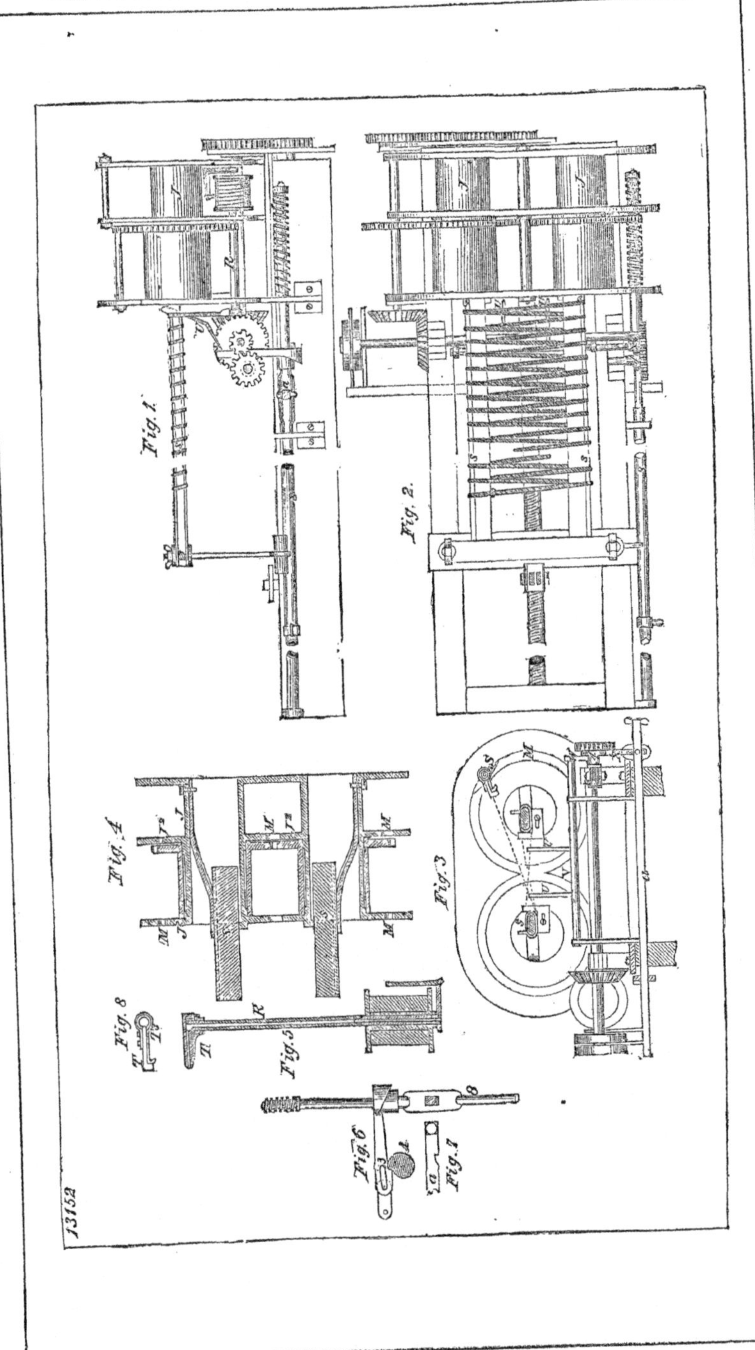
Fig. 1
Fig. 2
Fig. 3
Fig. 4
Fig. 5
Fig. 6
Fig. 7
Fig. 8
13152

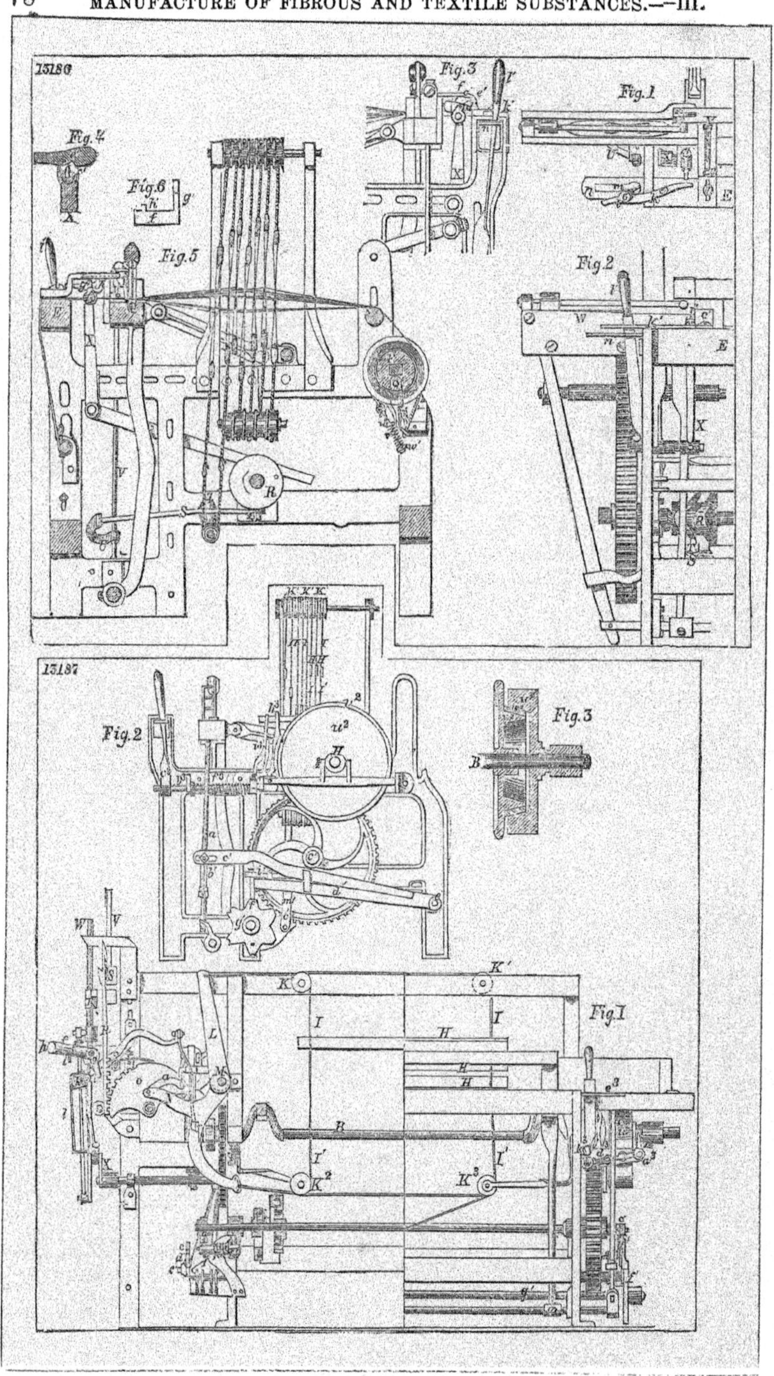
13186
Fig. 4
Fig. 6
Fig. 5
Fig. 3
Fig. 1
Fig. 2
13187
Fig. 2
Fig. 3
Fig. 1

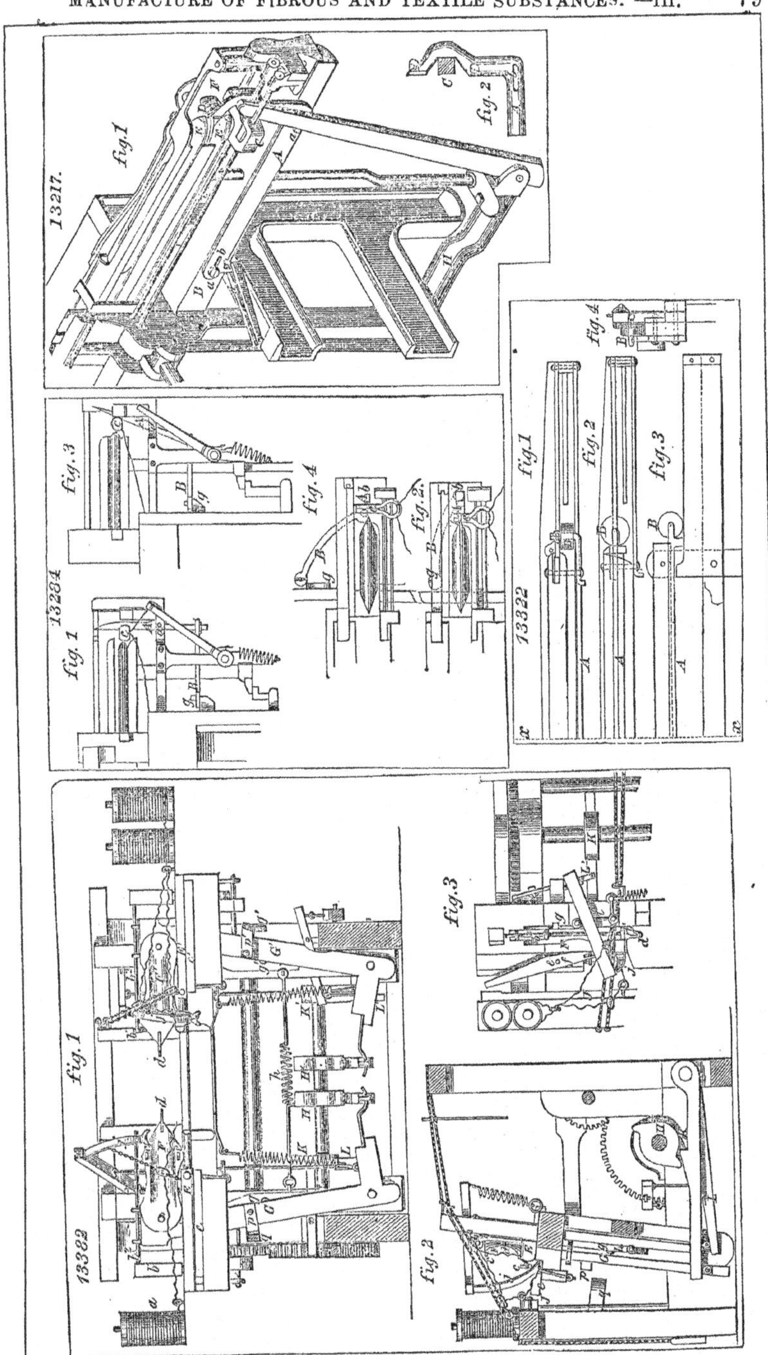

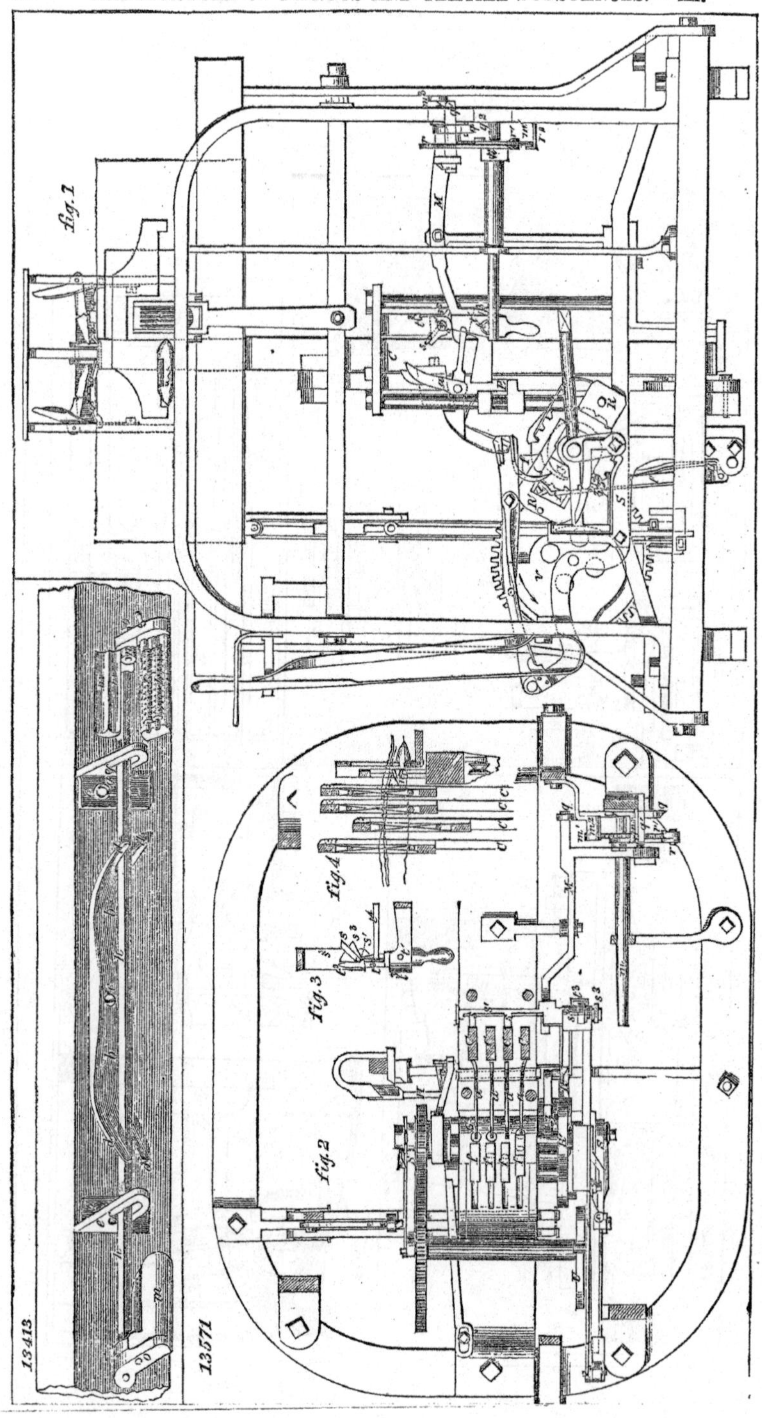
Fig. 1
13413
13571
Fig. 2
Fig. 3
Fig. 4

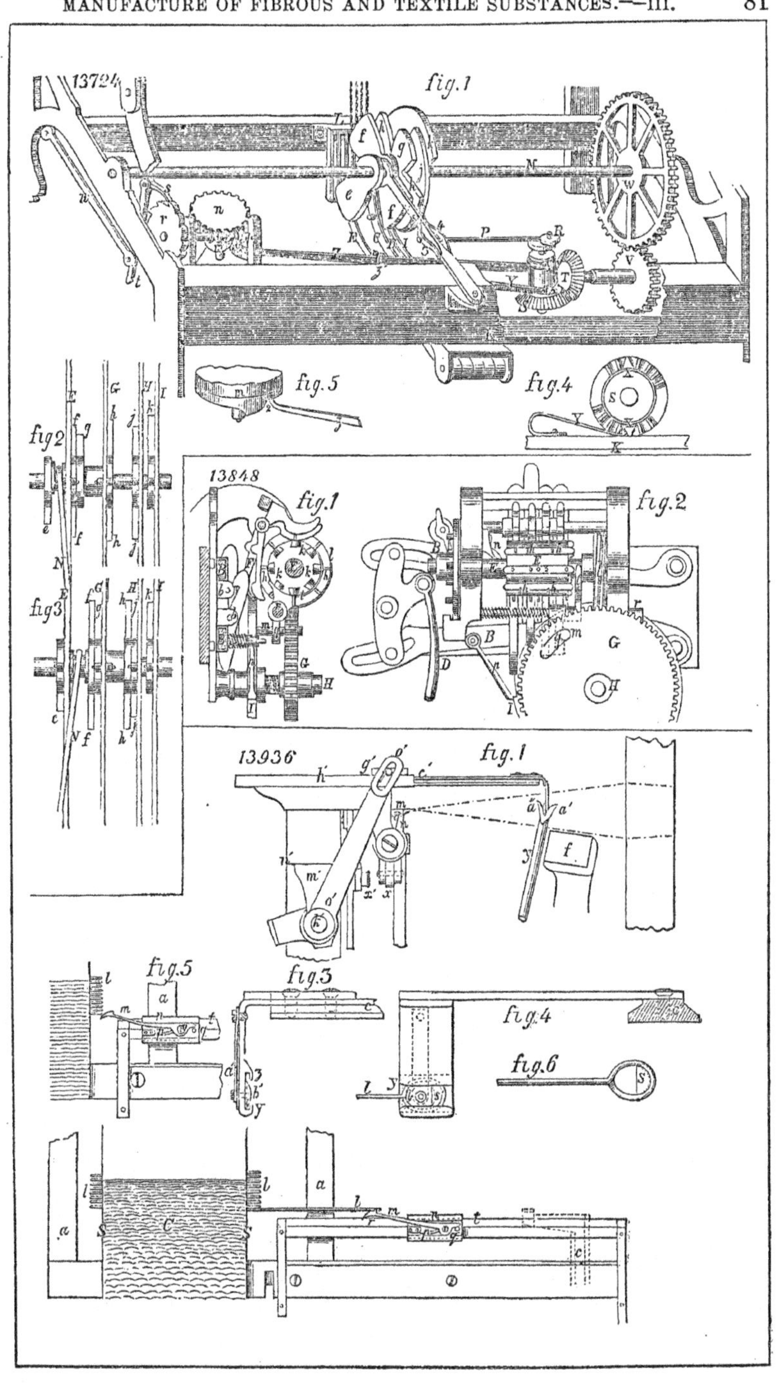

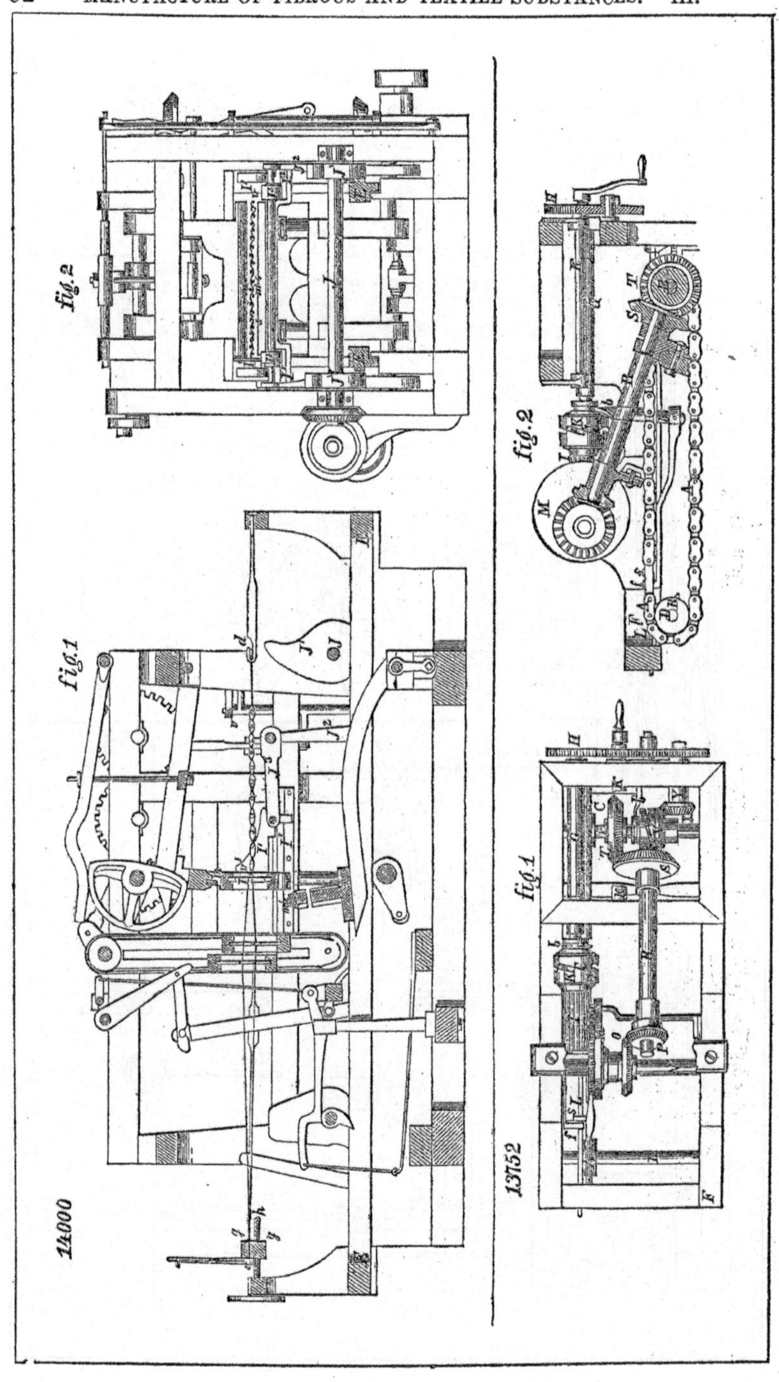
fig.2
fig.1
14000
fig.2
fig.1
13752

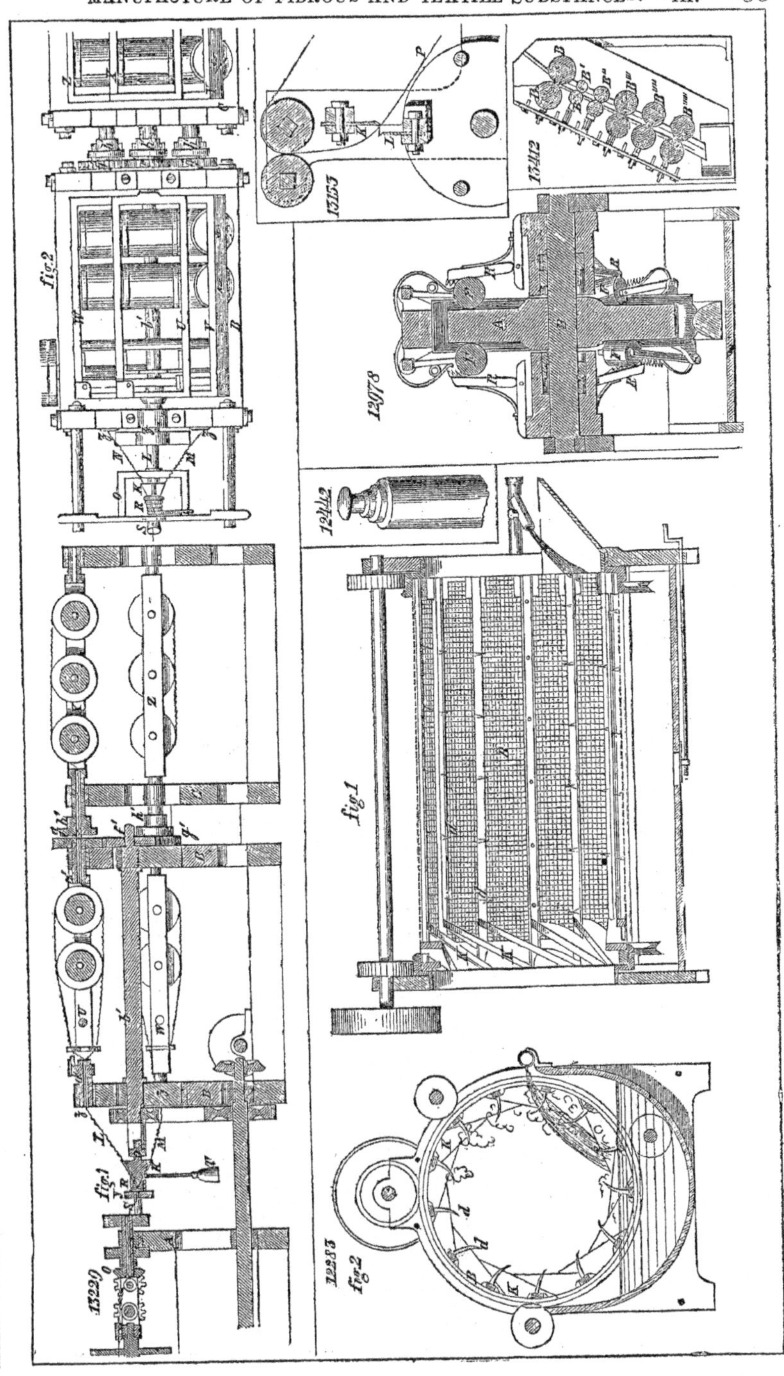
13229
fig.1
fig.2
13153
13412
12078
12412
fig.1
12283
fig.2

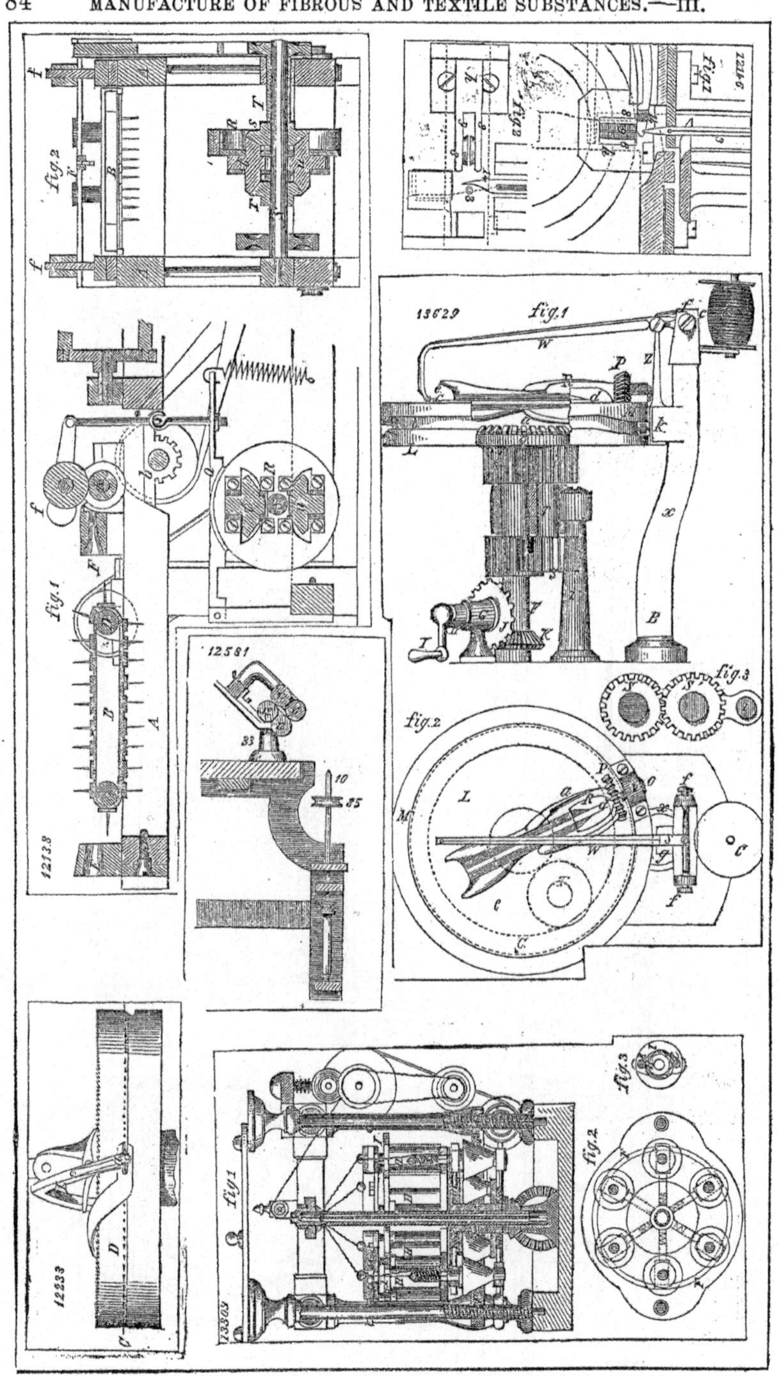
12138
fig.1
fig.2
12581
13629
fig.1
fig.2
fig.3
12146
12233
13303
fig.1
fig.2
fig.3

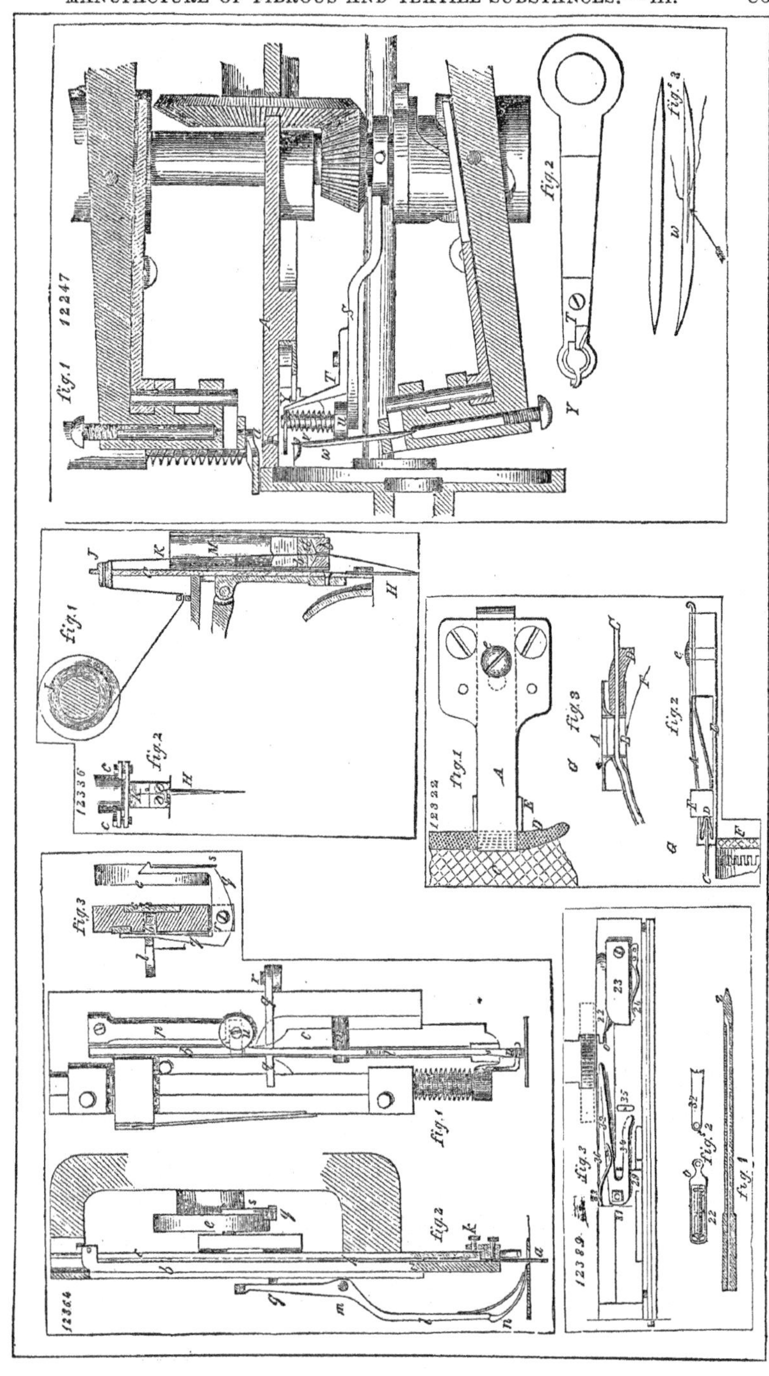

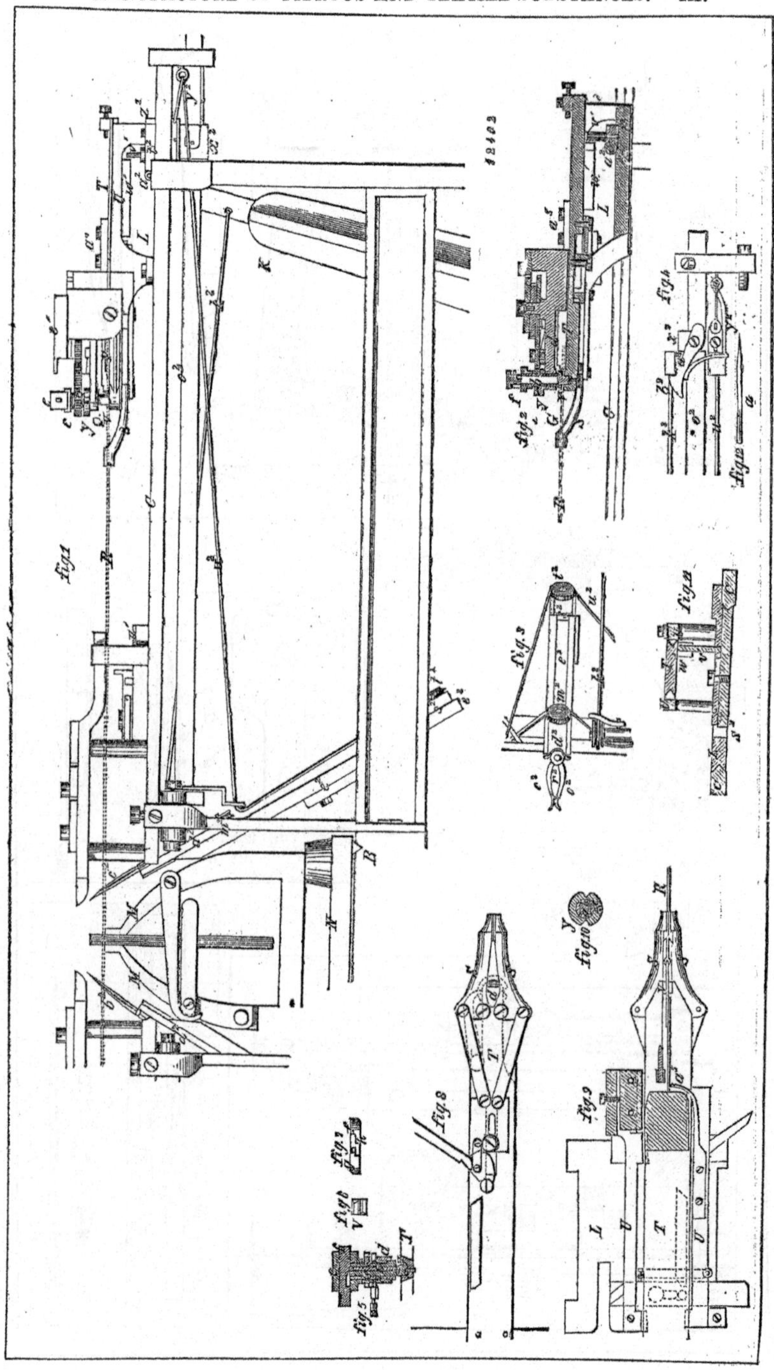

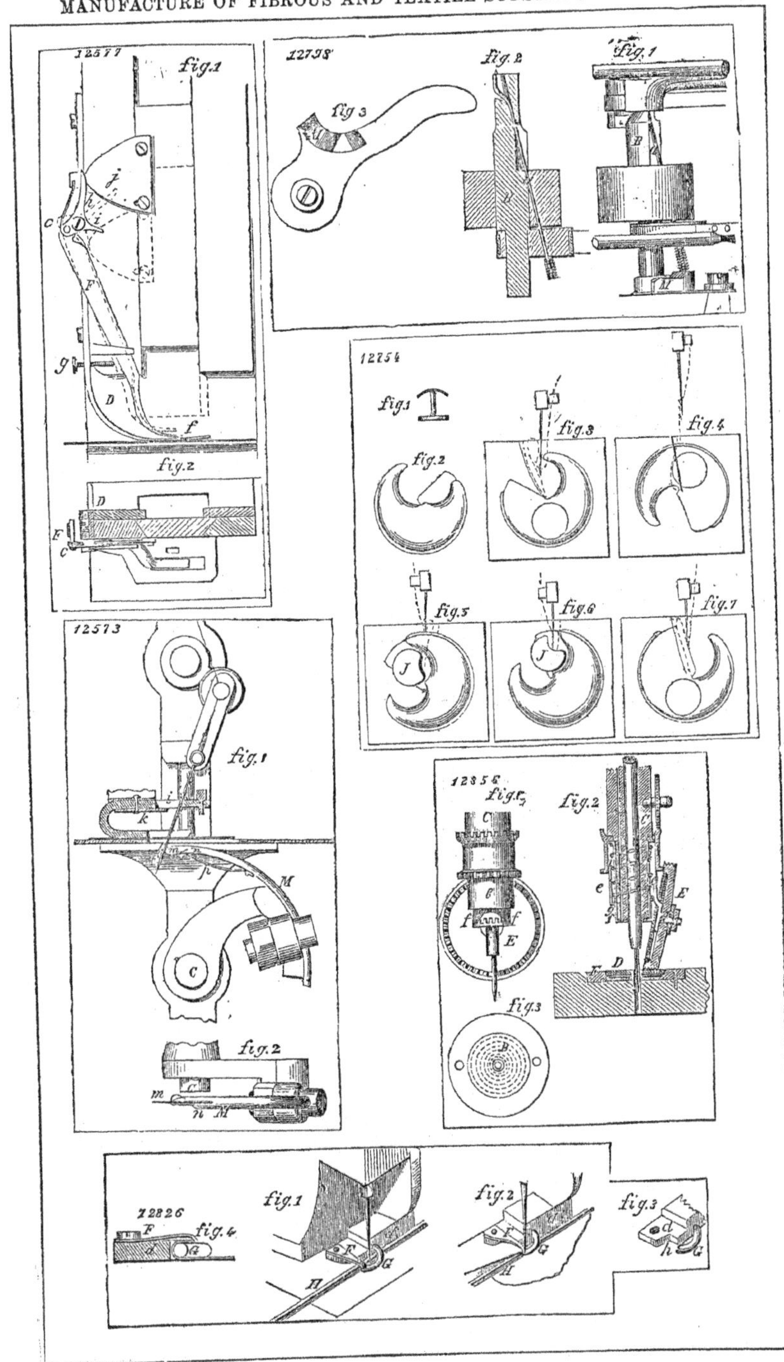
12577
fig.1
fig.2
12798
fig 3
fig. 2
fig. 1
12754
fig.1
fig.2
fig.3
fig.4
fig.5
fig.6
fig.7
12573
fig.1
fig.2
12856
fig.2
fig.3
12826
fig.4
fig.1
fig.2
fig.3

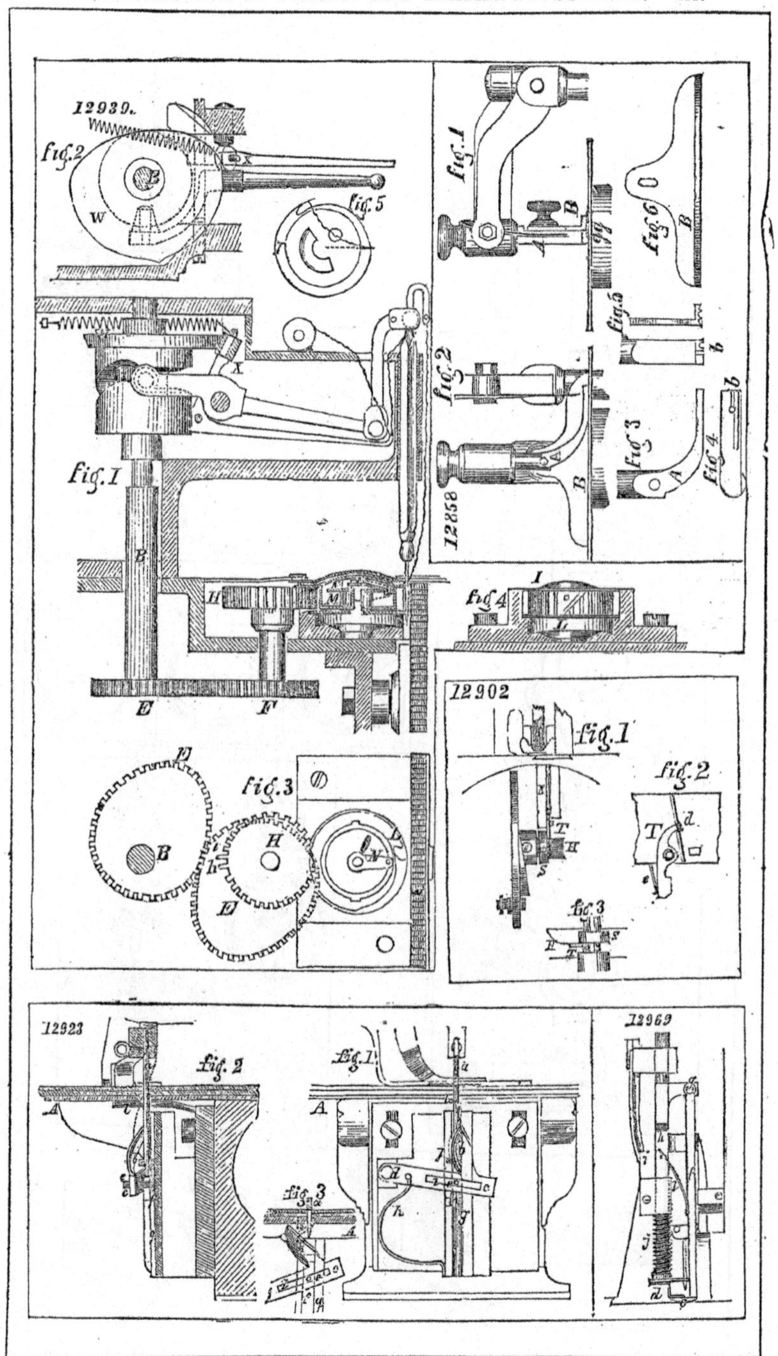
12939.
fig. 2
fig. 5
fig. 1
fig. 3
fig. 4
12858
fig. 1
fig. 2
fig. 3
fig. 4
fig. 5
fig. 6
12902
fig. 1
fig. 2
fig. 3
12923
fig. 2
fig. 1
fig. 3
12969

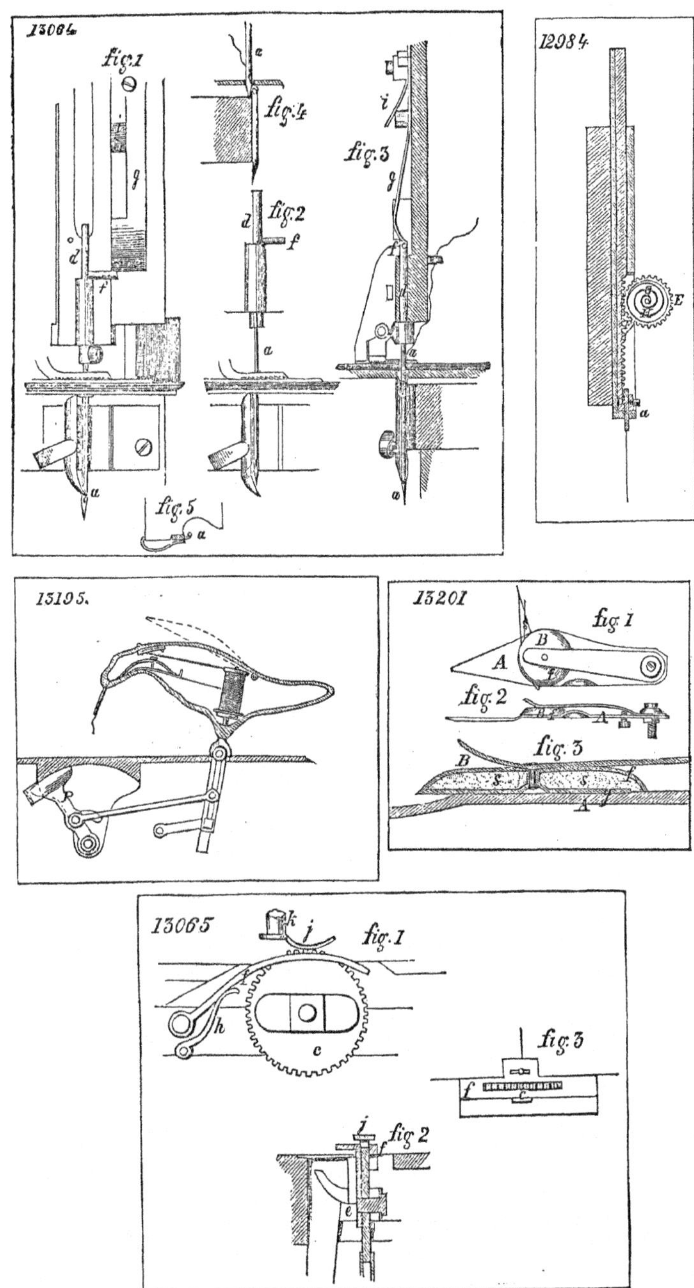
13064
fig.1
fig.4
fig.3
fig.2
fig.5
12984
13195.
13201
fig.1
fig.2
fig.3
13065
fig.1
fig.3
fig 2

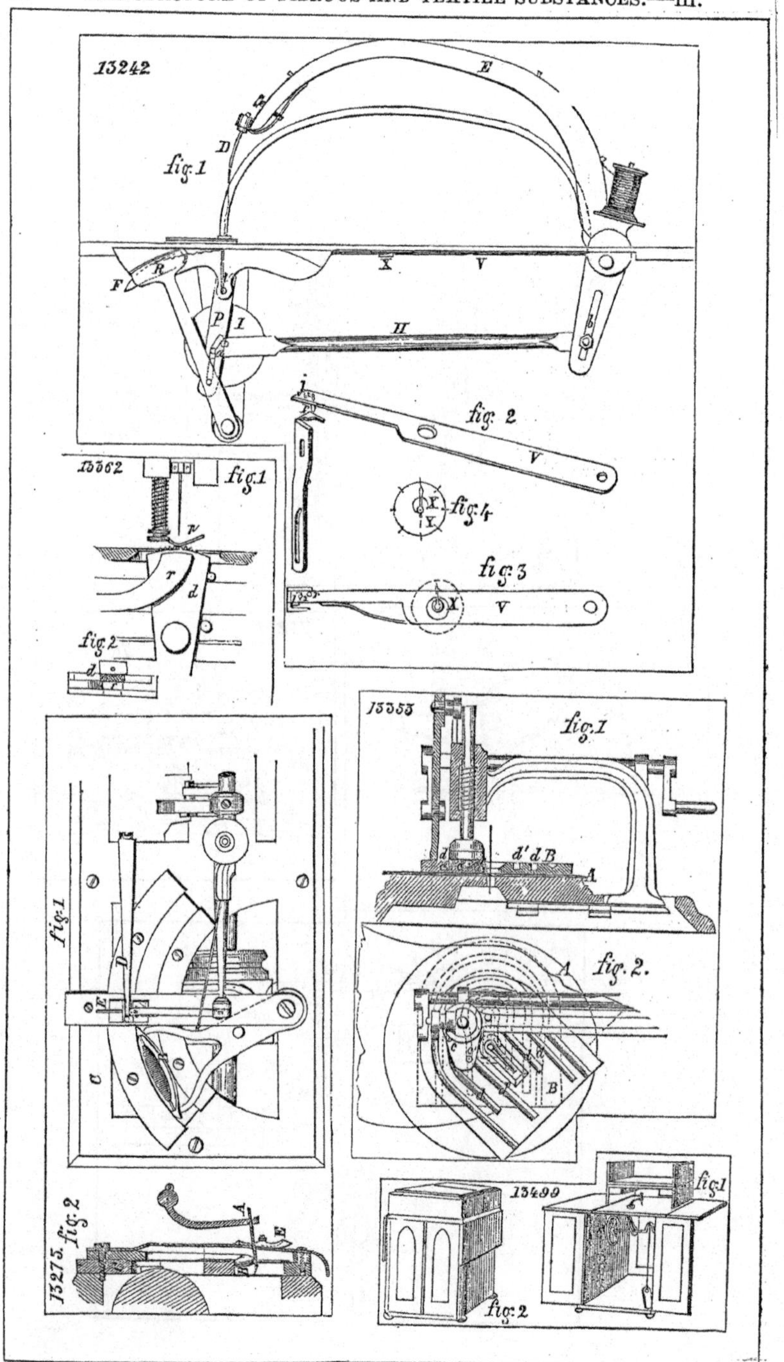
13242
fig. 1
fig. 2
fig. 3
fig. 4
13362
fig. 1
fig. 2
13353
fig. 1
fig. 2.
13275. fig. 2
fig. 1
13499
fig. 1
fig. 2

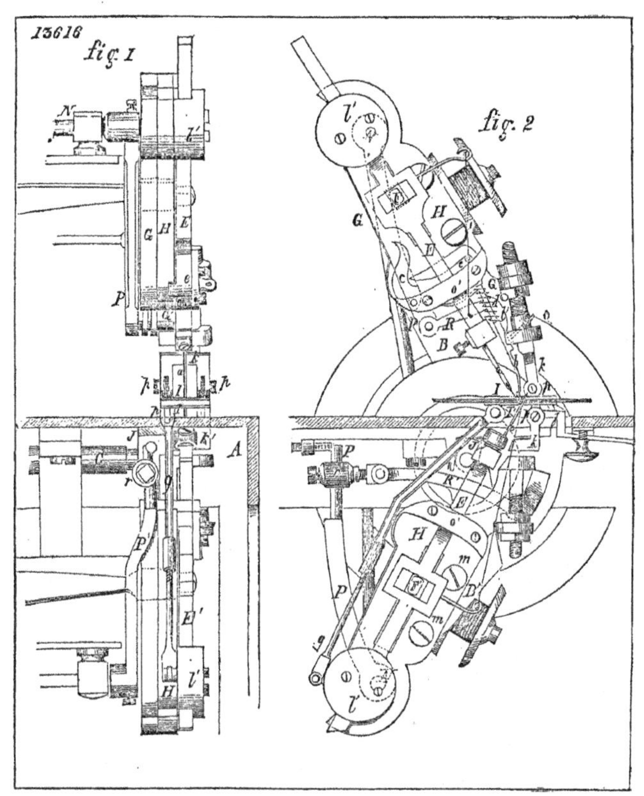
13616
fig. 1
fig. 2

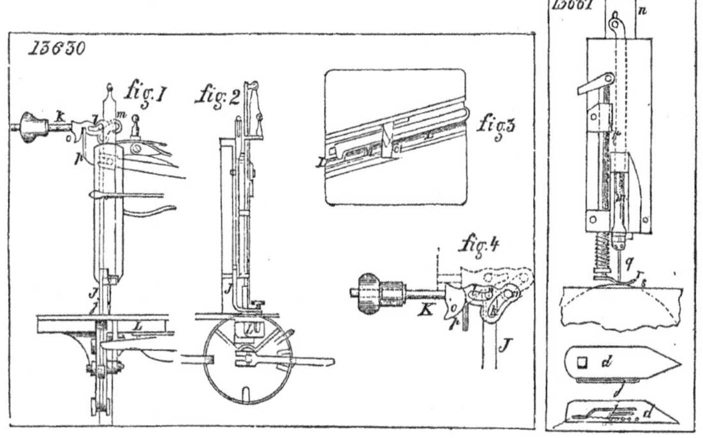
13630
fig. 1
fig. 2
fig. 3
fig. 4
13661

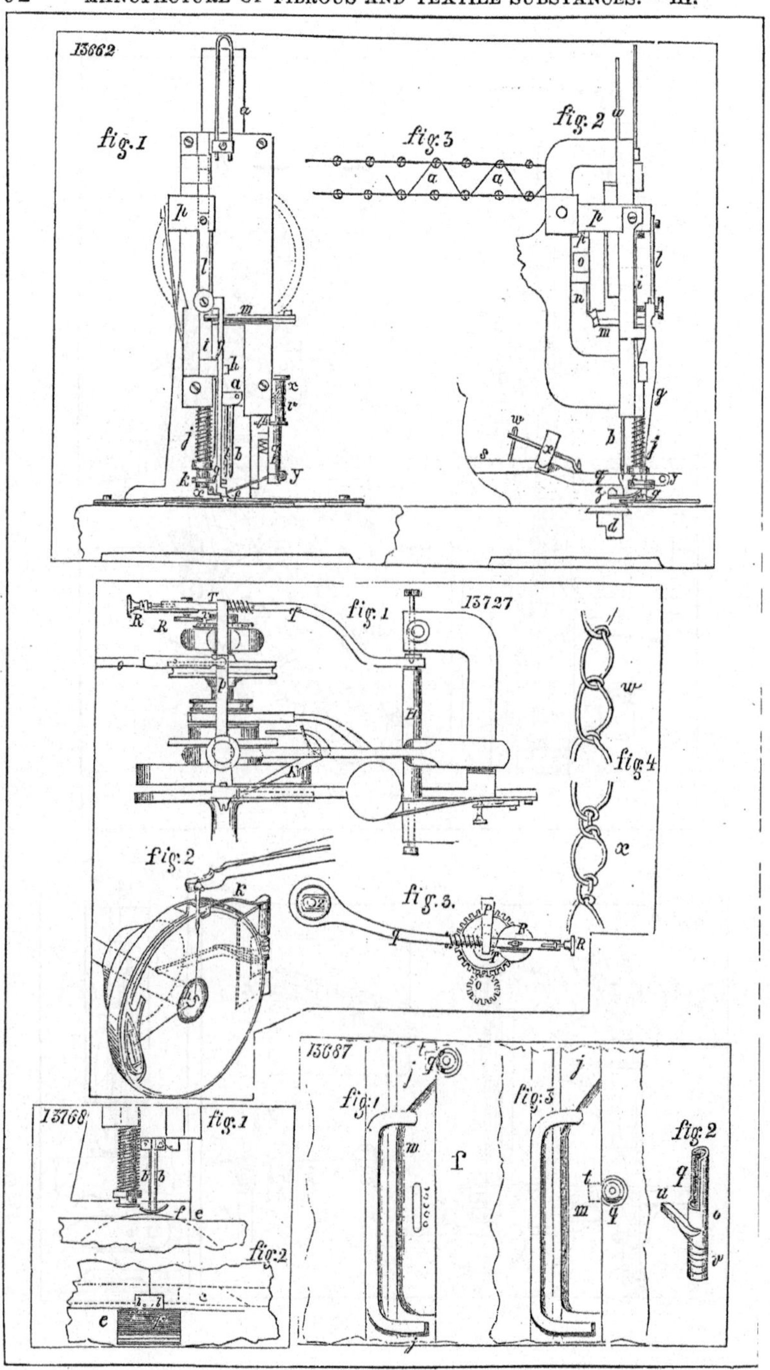
13662
fig. 1
fig. 3
fig. 2
13727
fig. 1
fig. 4
fig. 2
fig. 3.
13768
fig. 1
fig. 2
13687
fig. 1
fig. 3
fig. 2

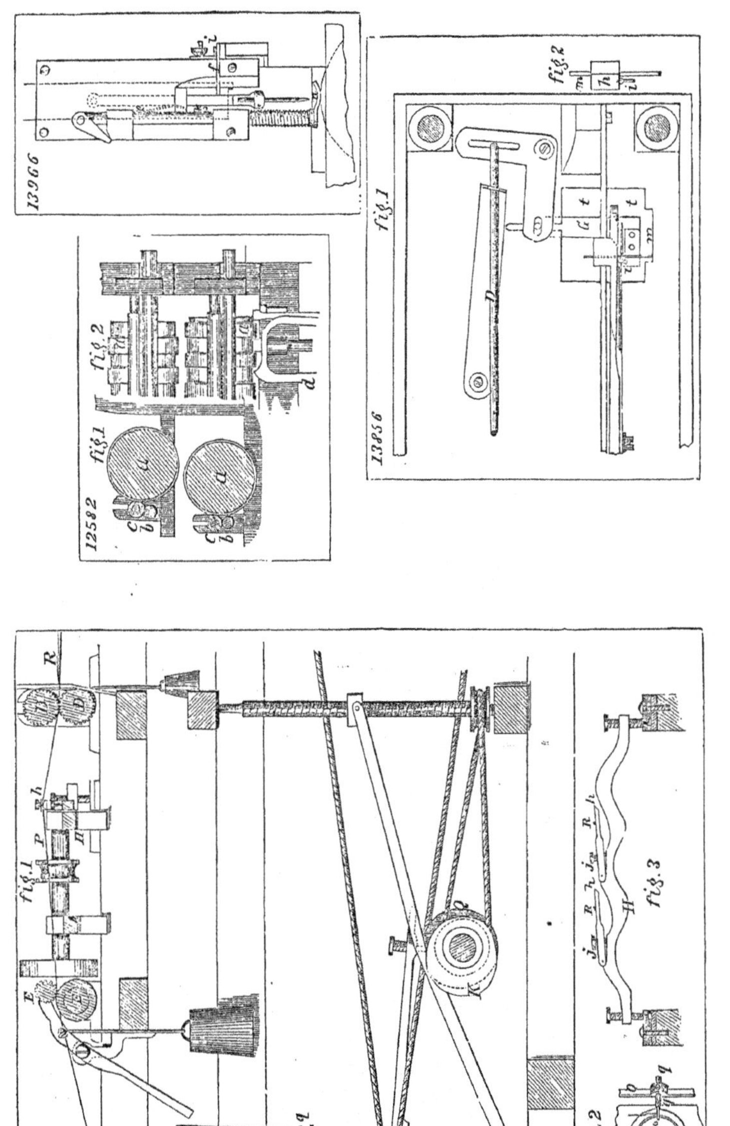
13966
12582
fig.1
fig.2
13856
fig.1
fig.2
12532
fig.1
fig.2
fig.3

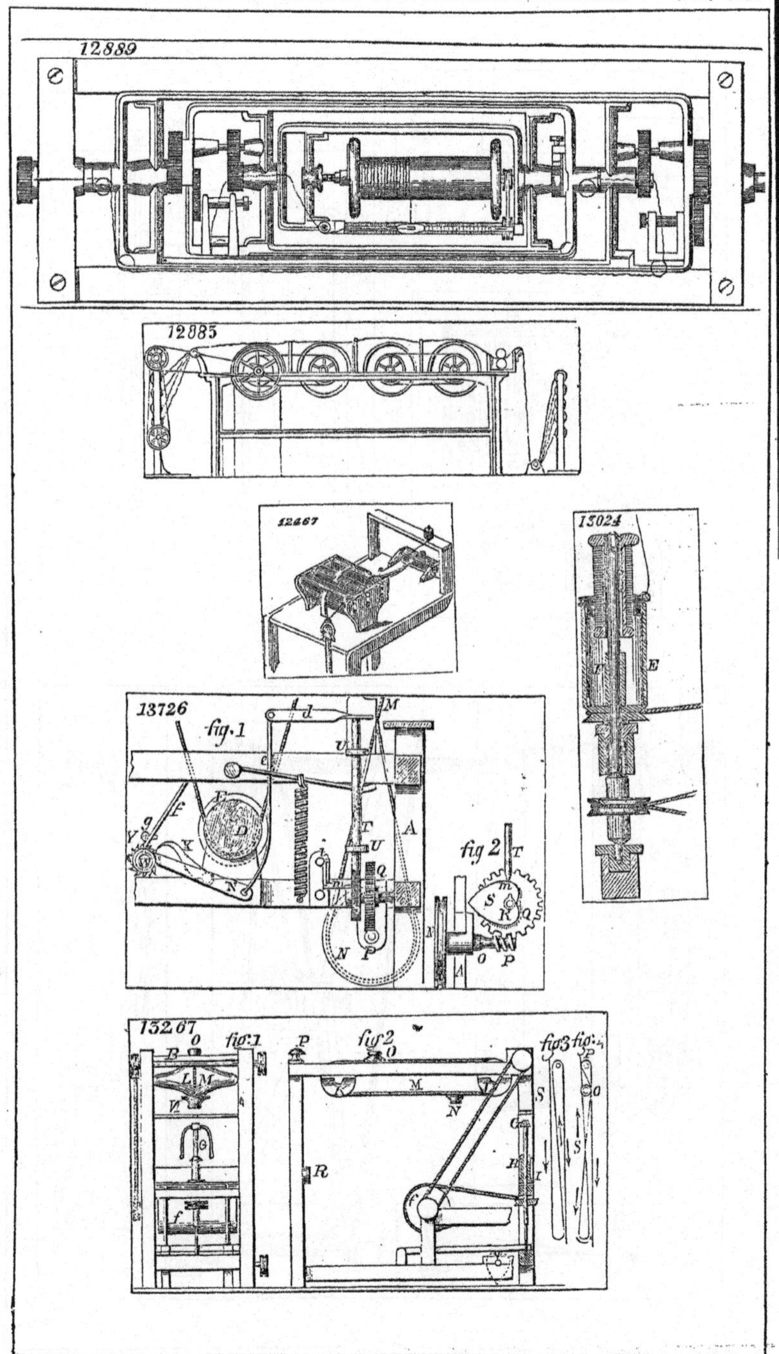
12889
12885
12467
13024
13726
fig. 1
fig 2
13267
fig. 1
fig 2
fig 3
fig. 4

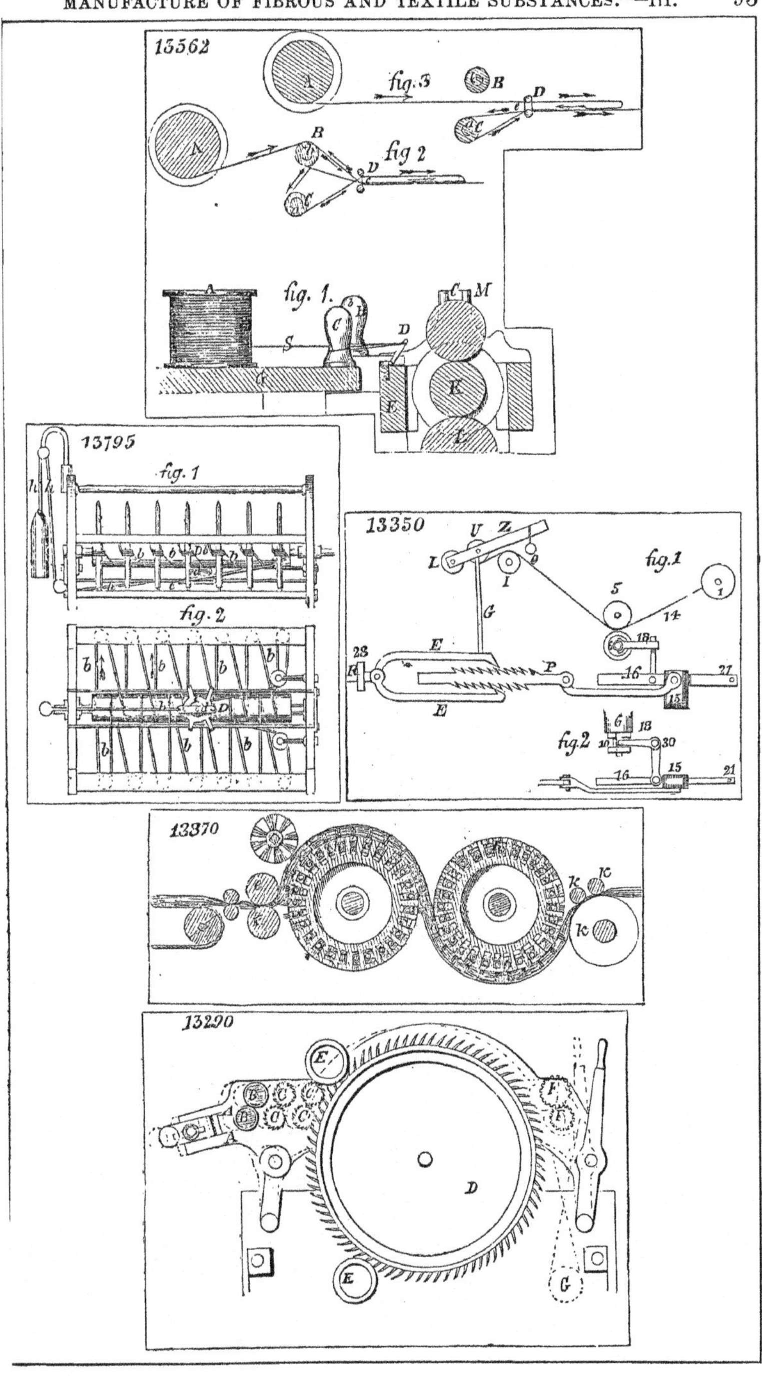
13562
fig. 3
fig 2
fig. 1.
13795
fig. 1
fig. 2
13350
fig.1
fig.2
13370
13290

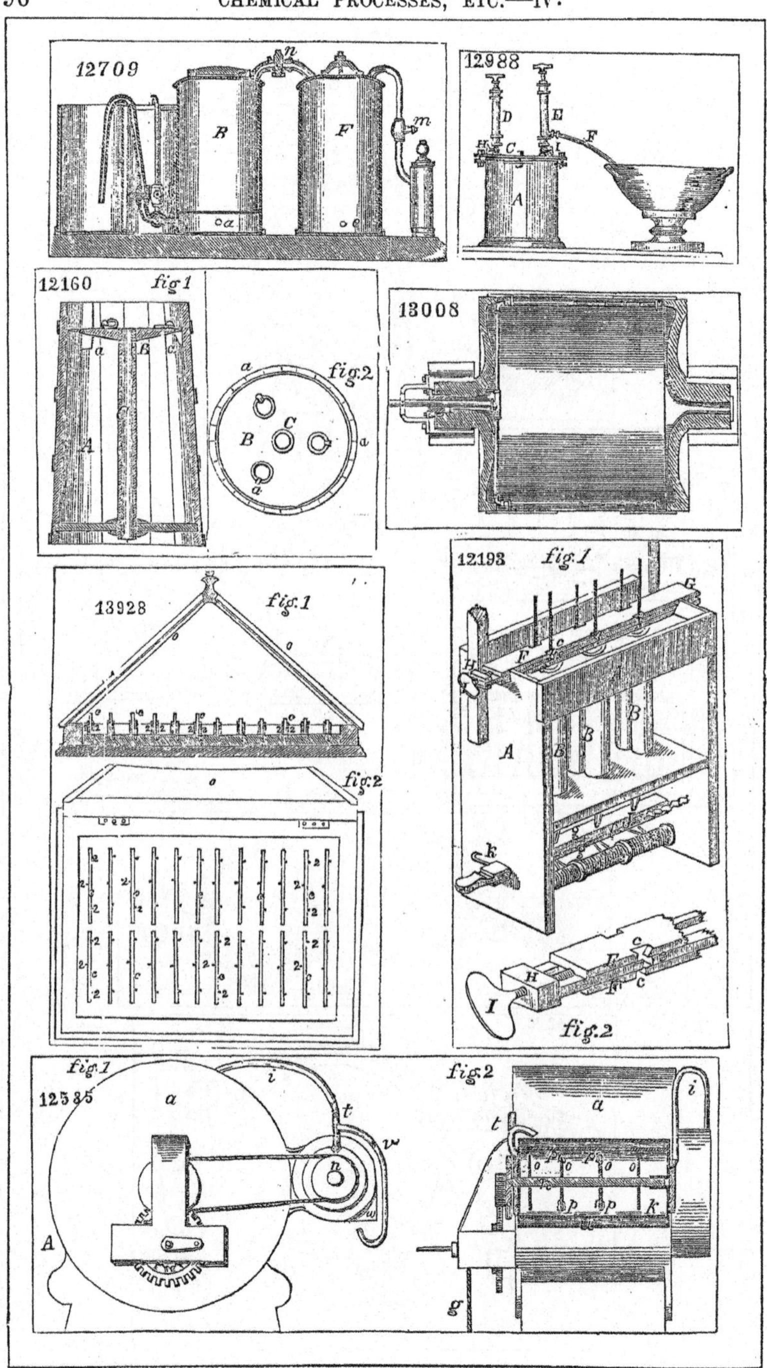
12709
12988
12160
fig 1
fig.2
13008
13928
fig.1
fig.2
12193
fig.1
fig.2
fig.1
12585
fig.2

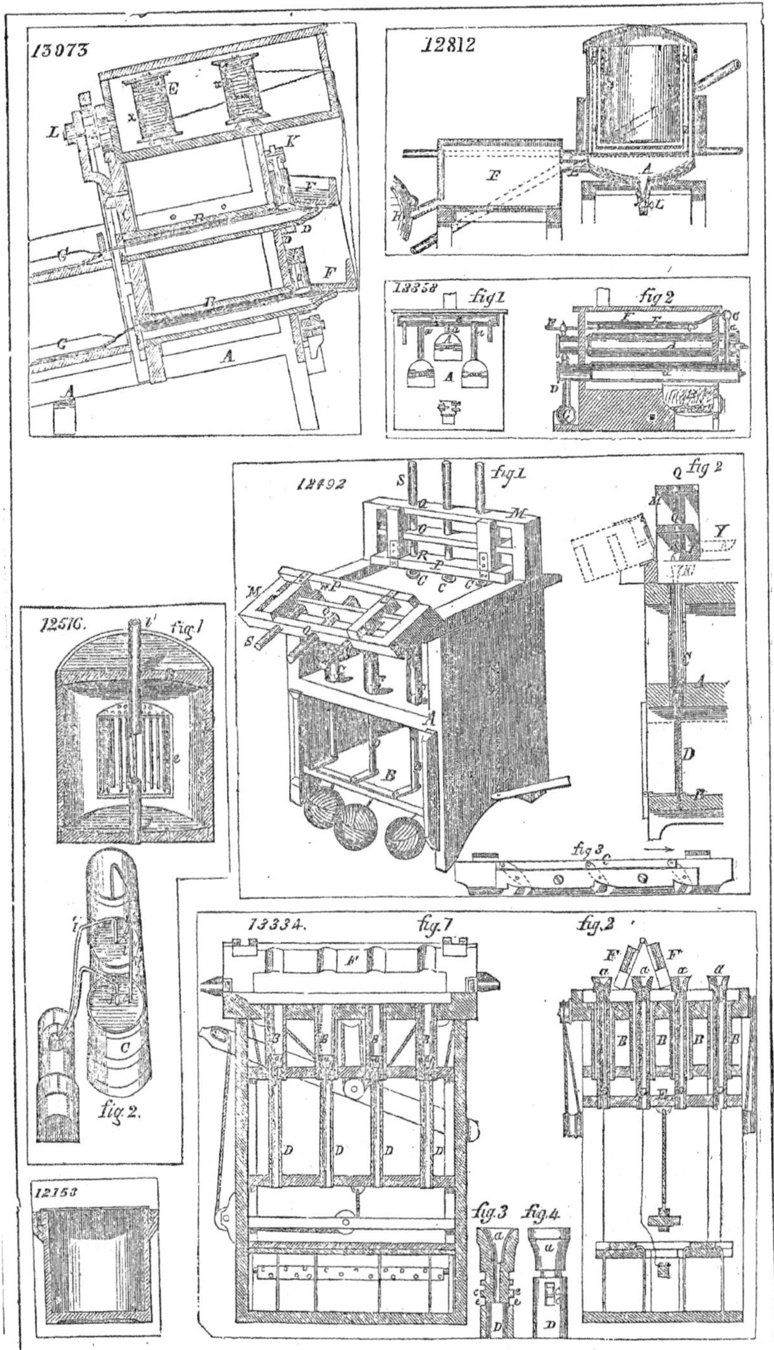
13973
12812
13358 fig 1
fig 2
12492 fig 1
fig 2
fig 3
12516. fig. 1
fig. 2.
12153
13334. fig. 1
fig. 2
fig. 3
fig. 4

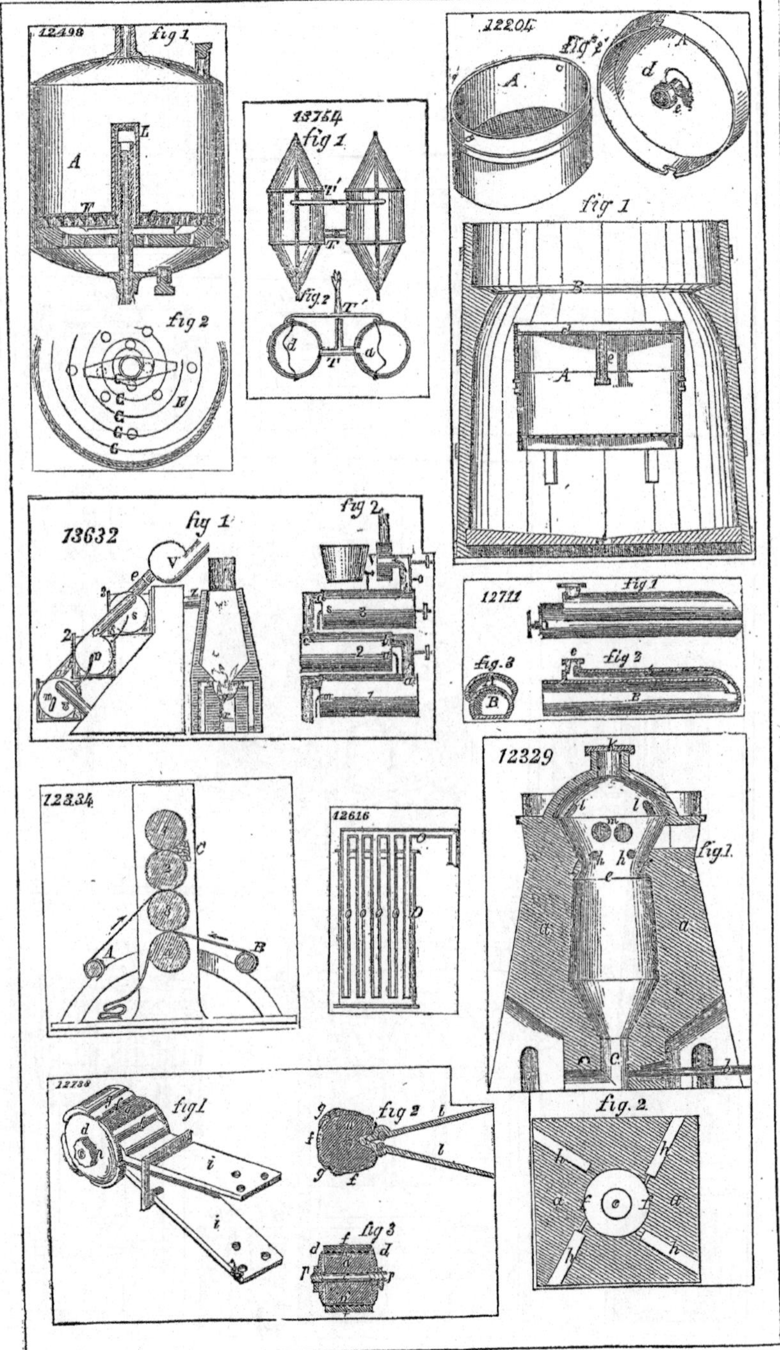
12498
fig 1
fig 2
13754
fig 1
fig 2
12204
fig 1
13632
fig 1
fig 2
12711
fig 1
fig 2
fig 3
12329
fig.1.
fig. 2.
12334
12616
12738
fig1
fig 2
fig 3

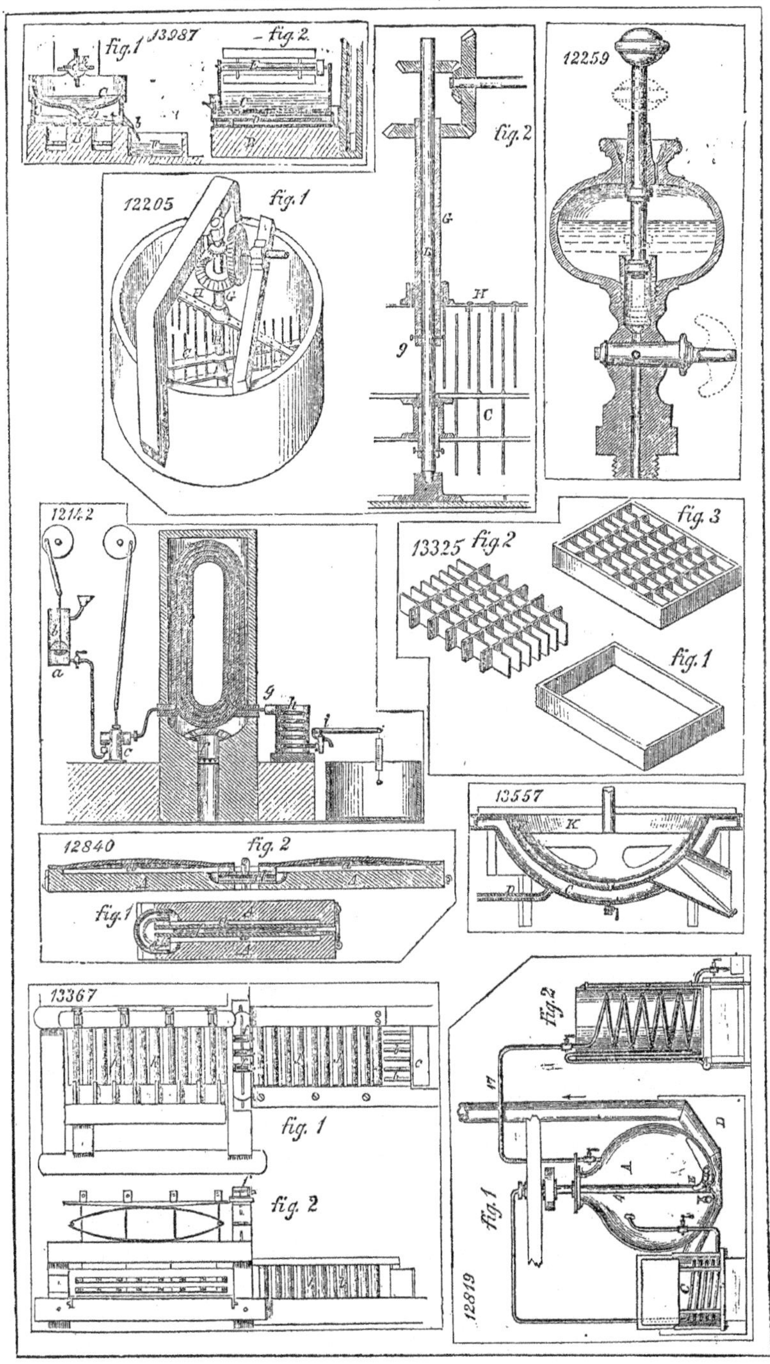
fig.1 13987
fig.2
12259
12205
fig.1
fig.2
12142
13325 fig.2
fig.3
fig.1
13557
12840
fig. 2
fig.1
13367
fig. 1
fig. 2
fig.2
fig.1
12819

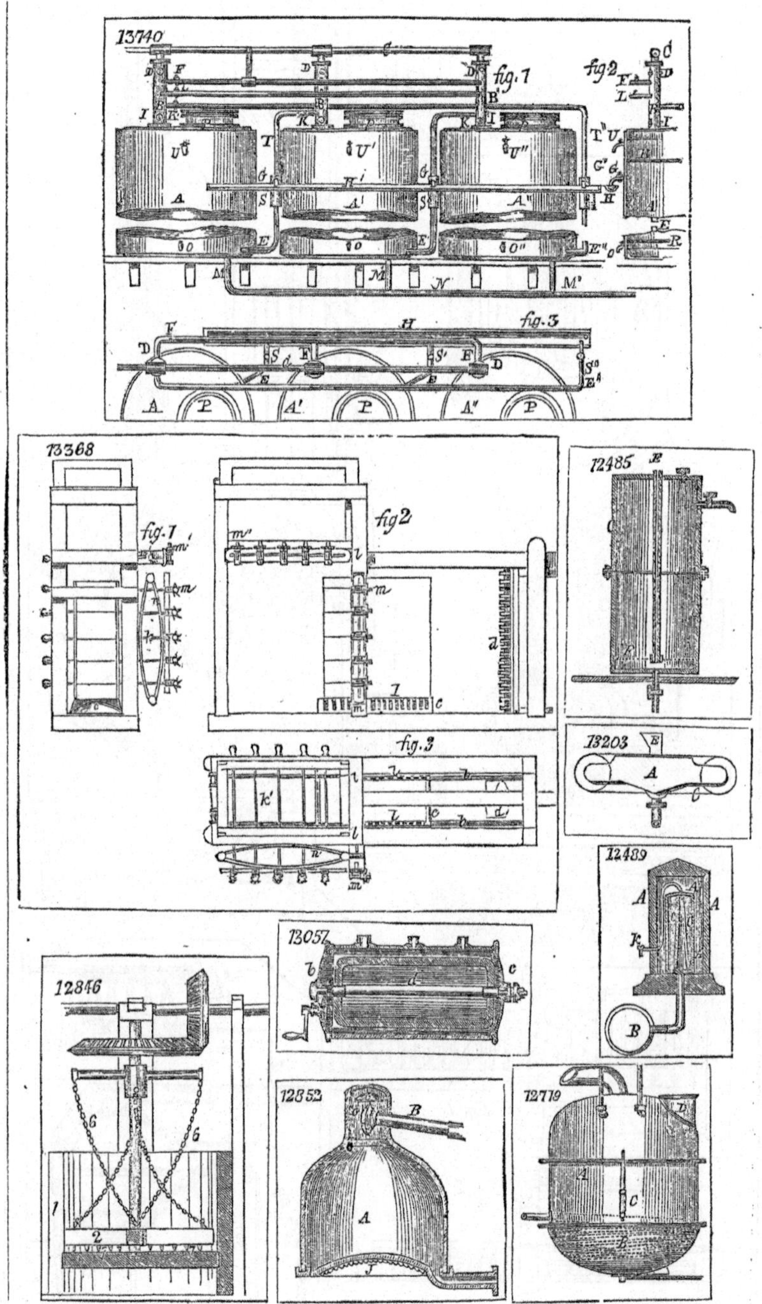
13740
fig. 1
fig. 2
fig. 3
13368
fig. 1
fig. 2
fig. 3
12485
13203
12489
13057
12846
12852
12719

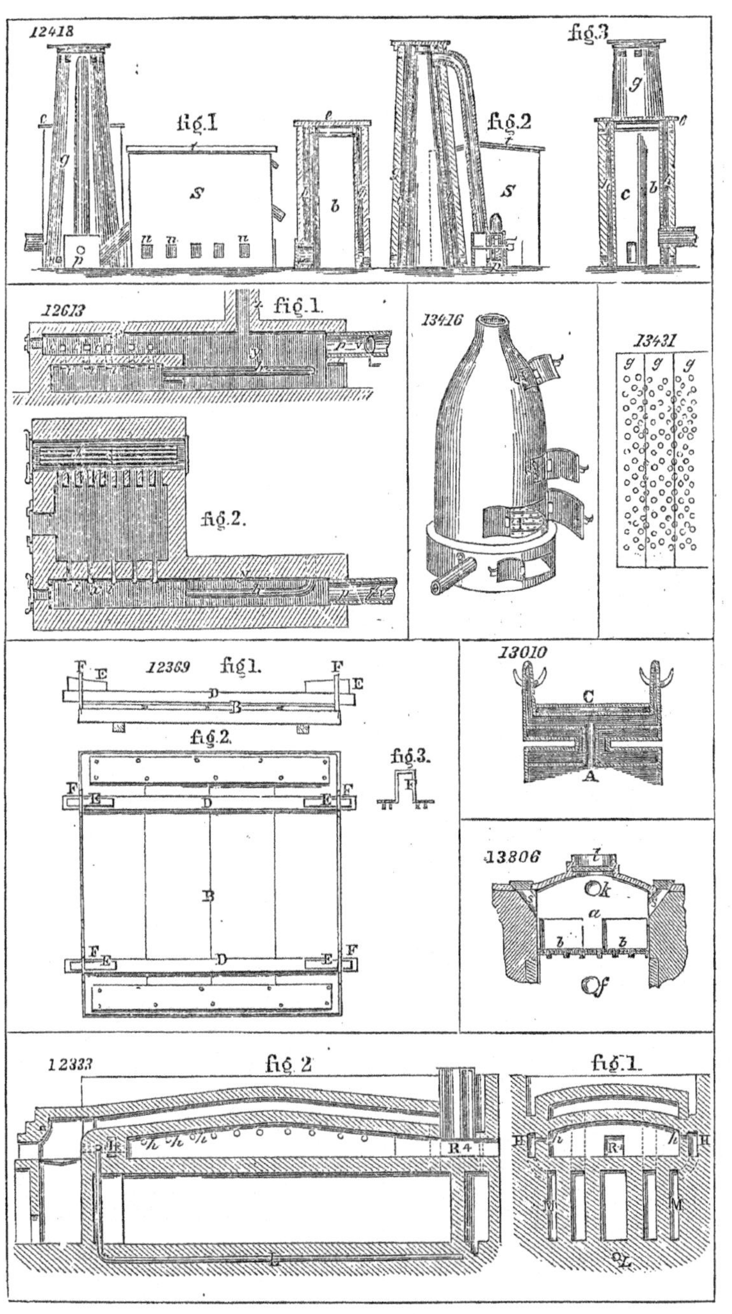

12418
fig.1
fig.2
fig.3
12613
fig.1.
fig.2.
13416
13431
12369
fig.1.
fig.2.
fig.3.
13010
13806
12333
fig.2
fig.1.

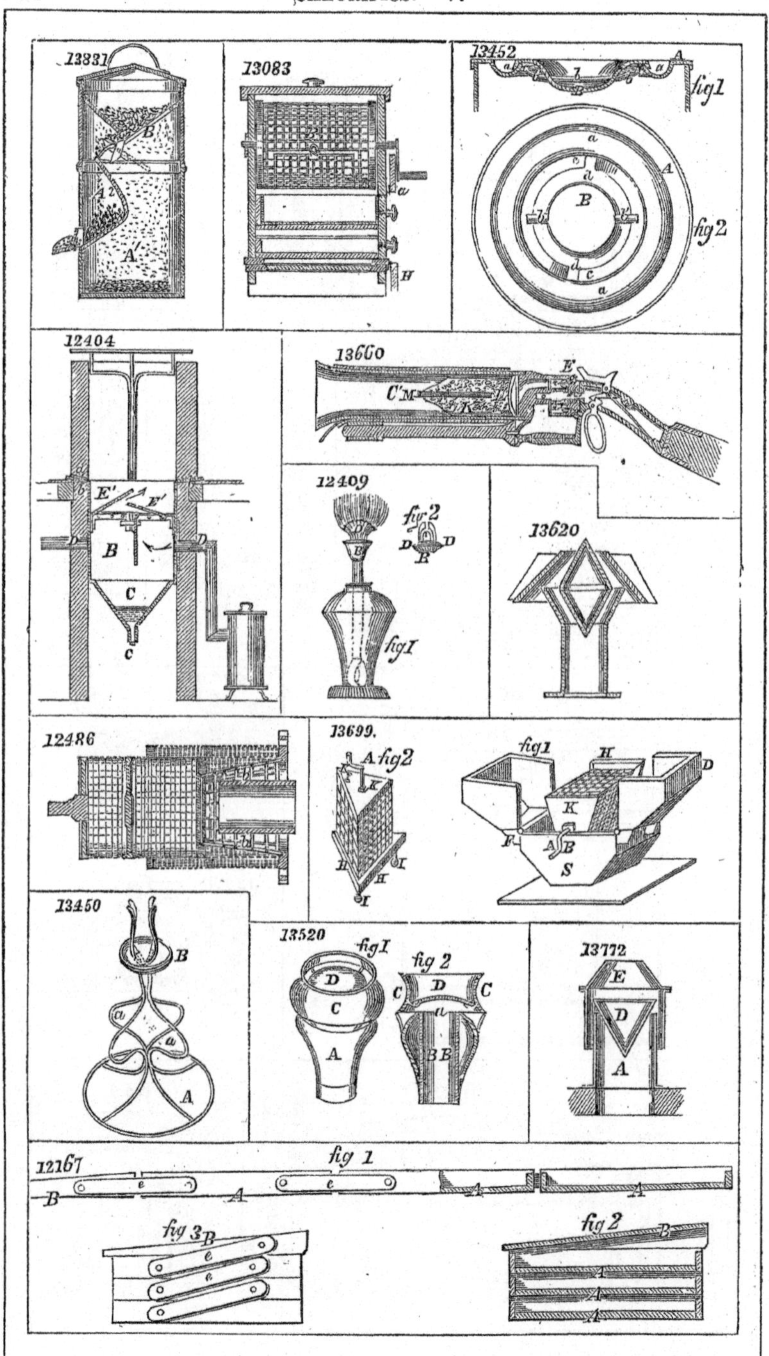

13331
13083
13452
fig1
fig2
12404
13660
12409
fig 2
fig1
13620
12486
13699.
fig2
fig1
13450
13520
fig1
fig 2
13772
12167
fig 1
fig 3
fig 2

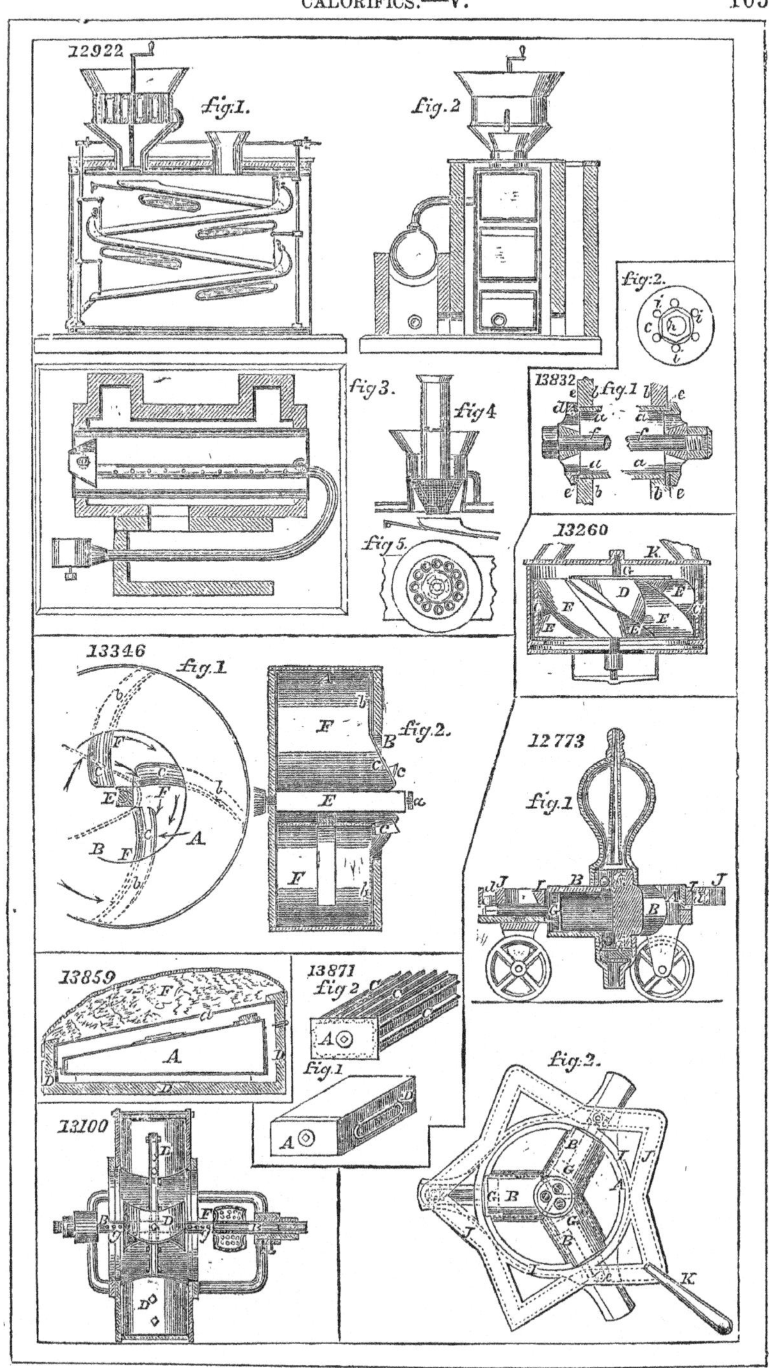
12922
fig.1.
fig.2
fig:2.
13832
fig.1
fig 3.
fig 4
fig 5.
13260
13346
fig.1
fig.2.
12773
fig.1
13859
13871
fig 2
fig.1
fig.2.
13100

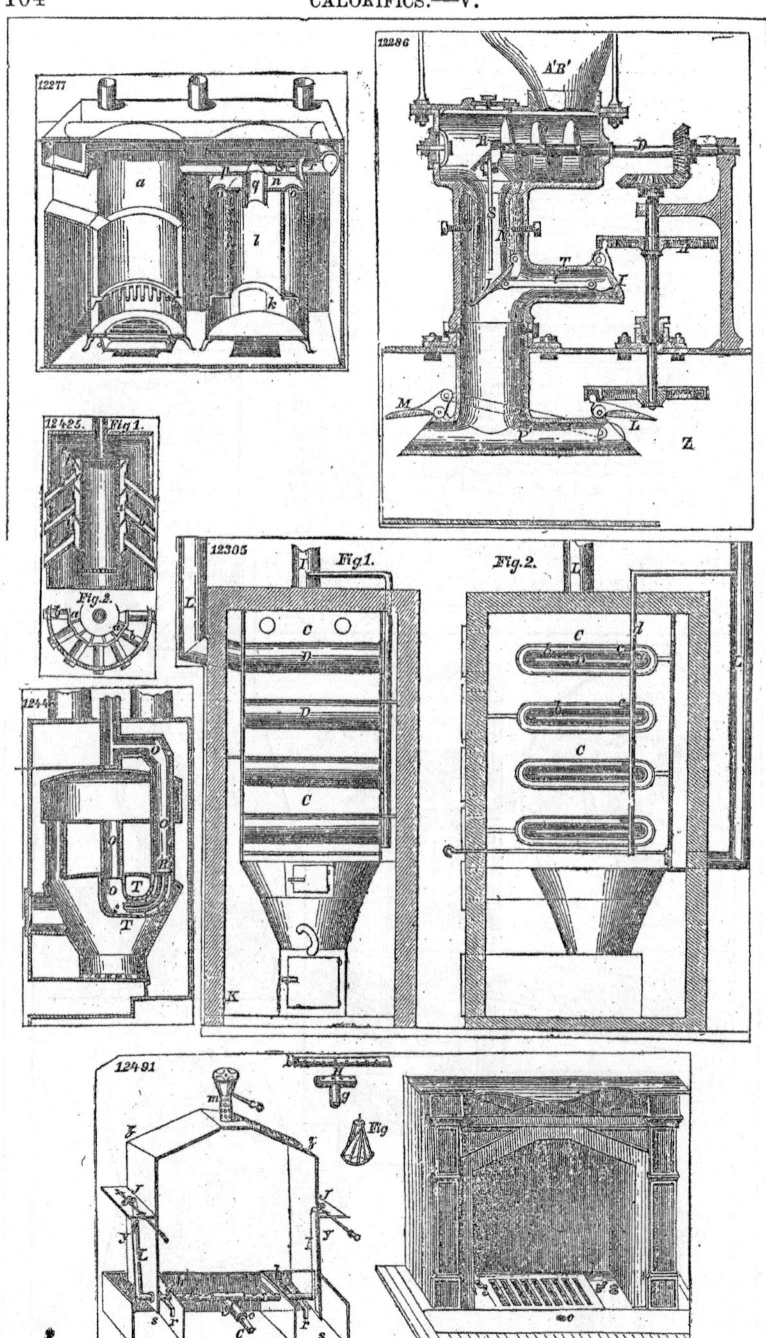
12277
12286
12425. Fig 1.
Fig. 2.
12305
Fig 1.
Fig. 2.
12491

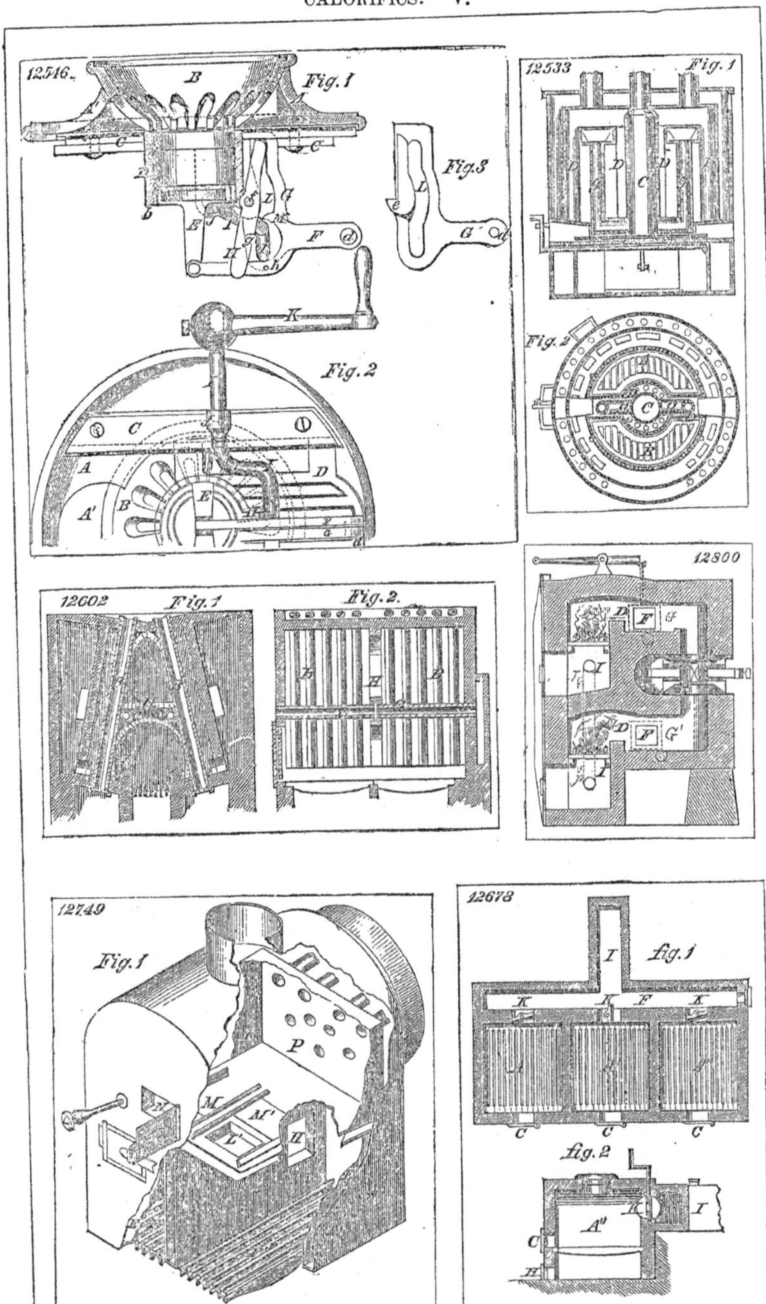
12546
Fig. 1
Fig. 2
Fig. 3
12533
Fig. 1
Fig. 2
12602
Fig. 1
Fig. 2.
12800
12749
Fig. 1
12678
fig. 1
fig. 2

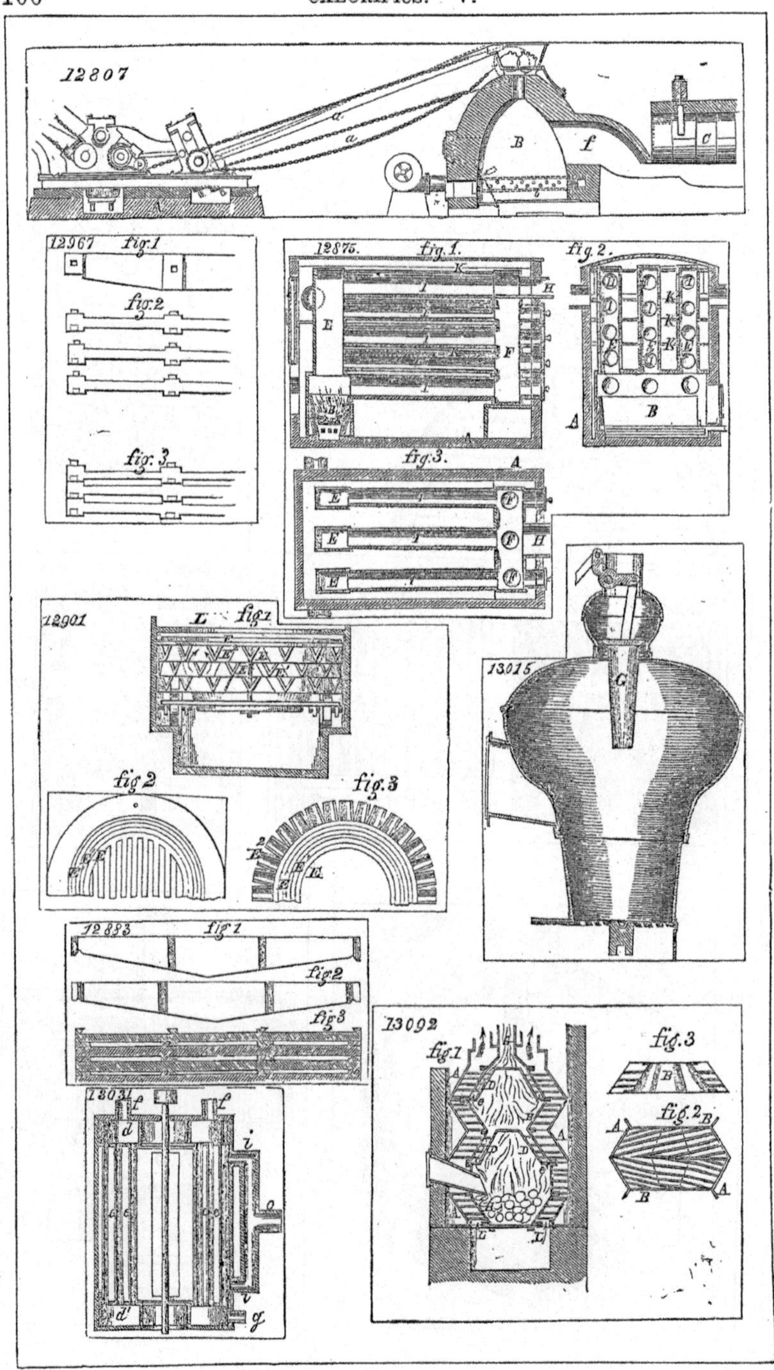
12807
a
a
B
f
C
12967 fig.1
fig.2
fig.3
12875. fig.1.
K
I
H
E
F
A
fig.2.
B
fig.3.
A
E
F
H
12901
L fig.1
fig.2
fig.3
13015
G
12883 fig.1
fig.2
fig.3
13031
f
d
i
o
d'
g
13092
fig.1
A
D
E
fig.3
B
fig.2
A
B

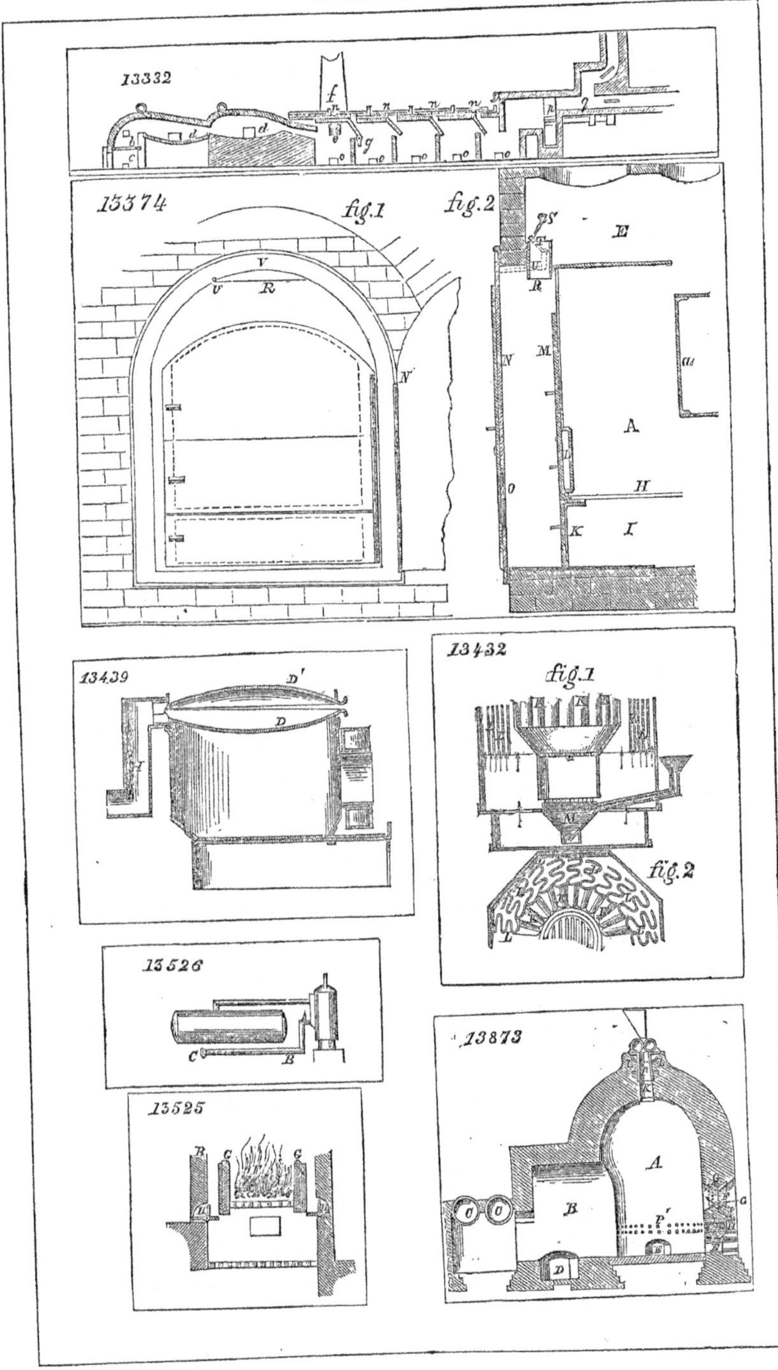
13332
13374
fig.1
fig.2
13439
13432
fig.1
fig.2
13526
13873
13525

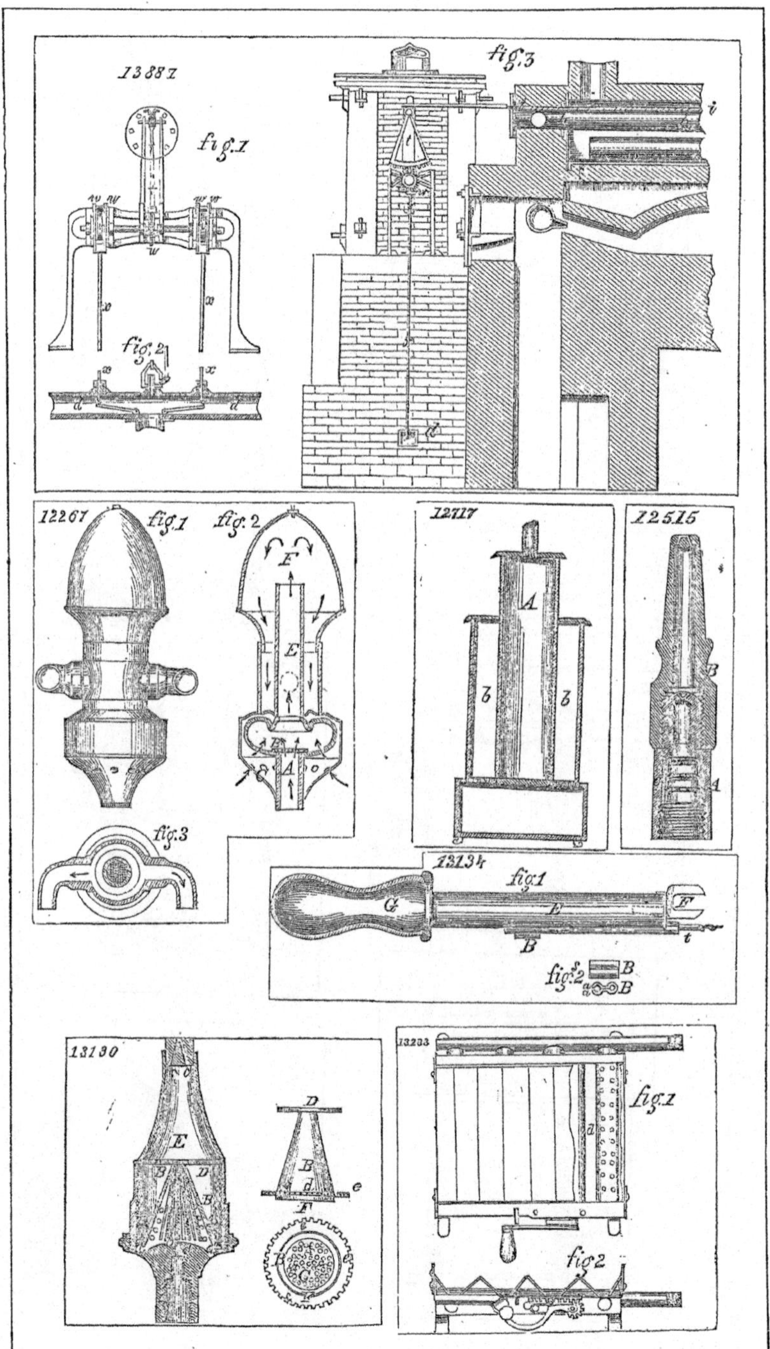
13887
fig.1
fig.2
fig.3
12267
fig.1
fig.2
fig.3
12717
12515
13134
fig.1
fig.2
13130
13233
fig.1
fig.2

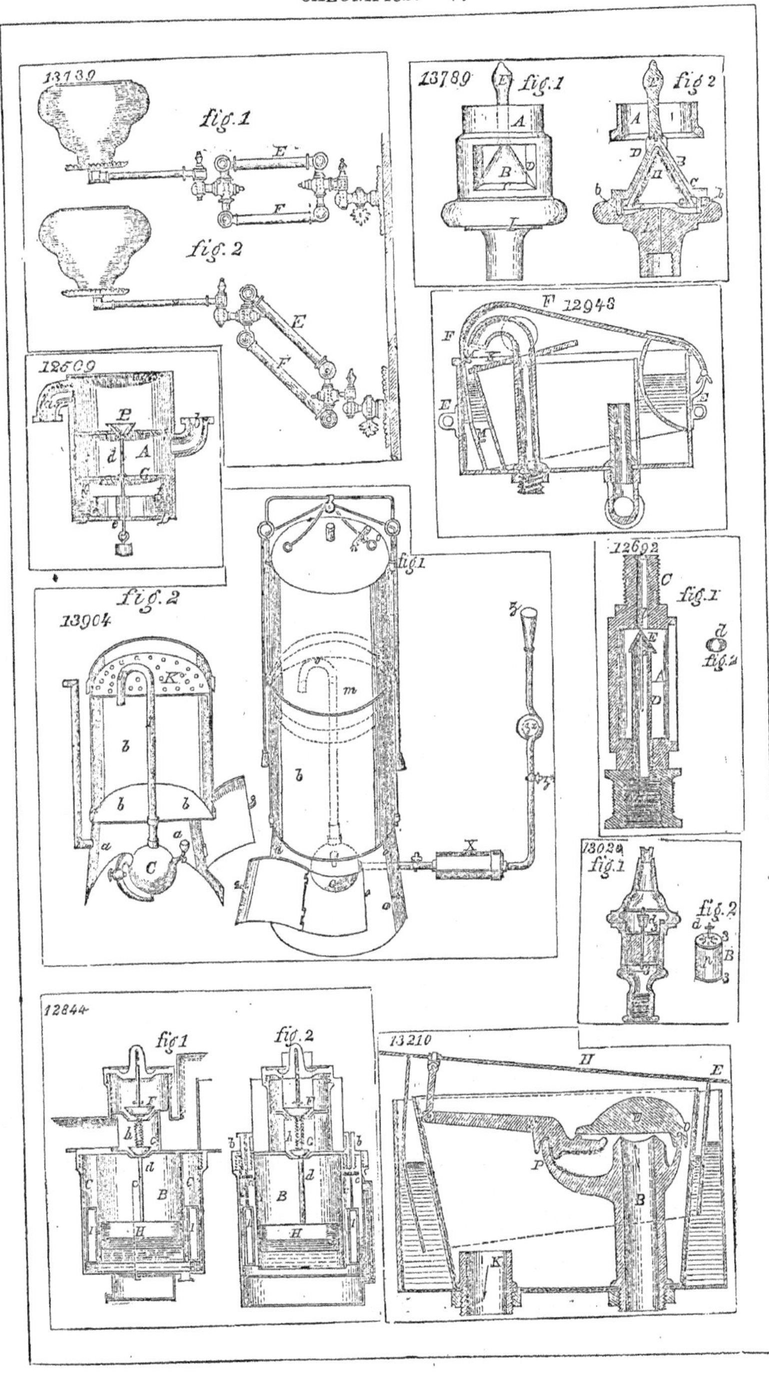
13139
fig.1
fig.2
13789
fig.1
fig 2
12948
12609
fig.2
13904
fig1
12692
fig.1
fig.2
13020
fig.1
fig.2
12844
fig1
fig.2
13210

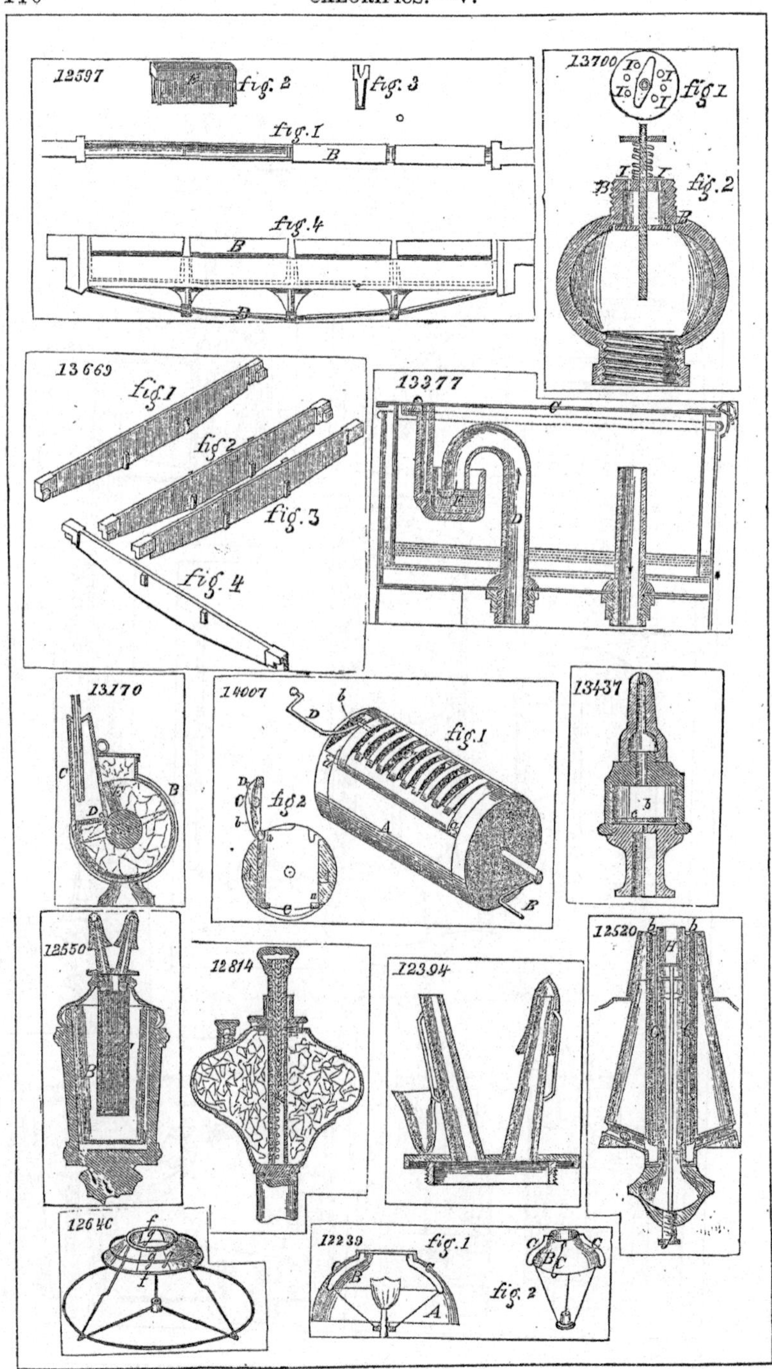
12597
fig. 2
fig. 3
fig. 1
fig. 4
13700
fig. 1
fig. 2
13669
fig. 1
fig. 2
fig. 3
fig. 4
13377
13170
14007
fig. 1
fig. 2
13437
12550
12814
12394
12520
12640
12239
fig. 1
fig. 2

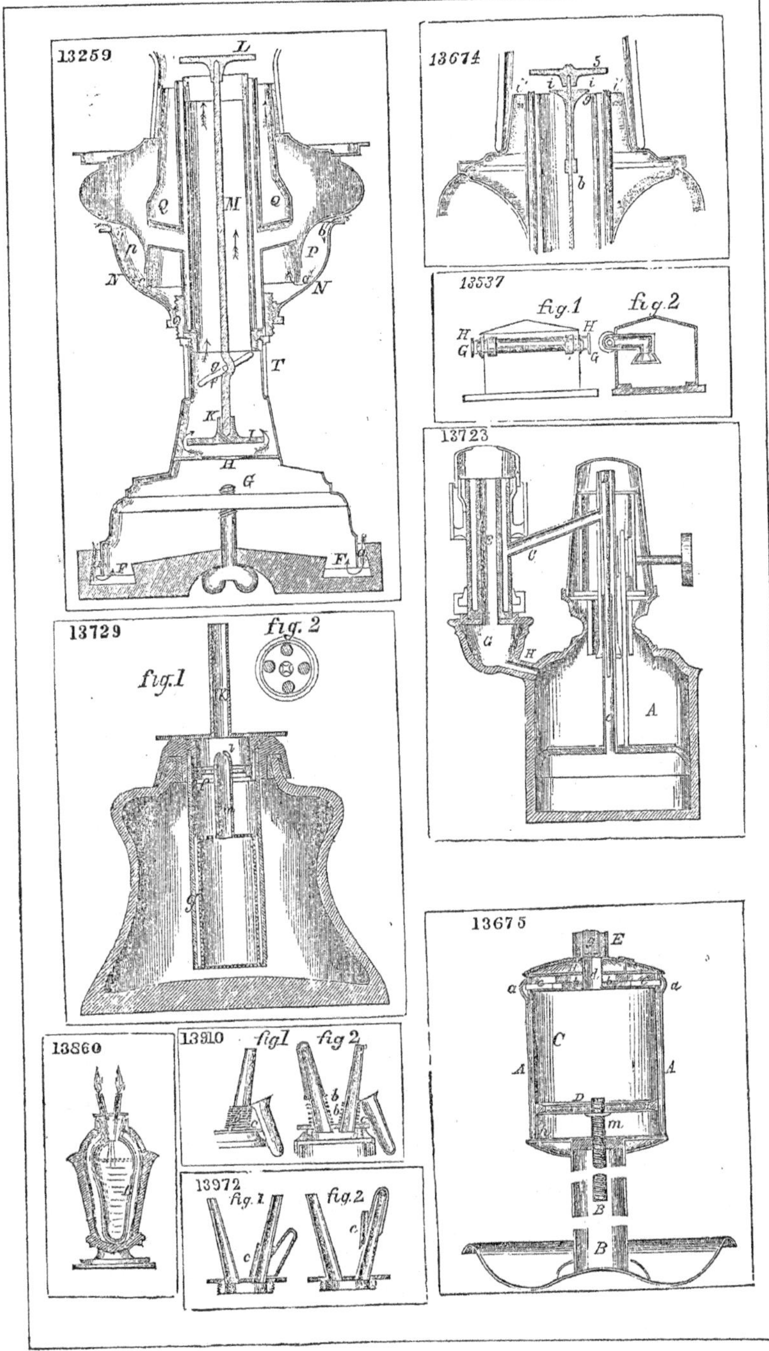
13259
L
M
Q
Q
P
p
N
N
T
K
H
G
F
F
13674
5
i
i
b
13537
fig.1
fig.2
H
H
G
G
13723
C
G
H
A
13729
fig. 2
fig.1
K
13675
E
a
a
C
A
A
D
m
B
B
13860
13910
fig1
fig2
13972
fig.1
fig.2
c
c

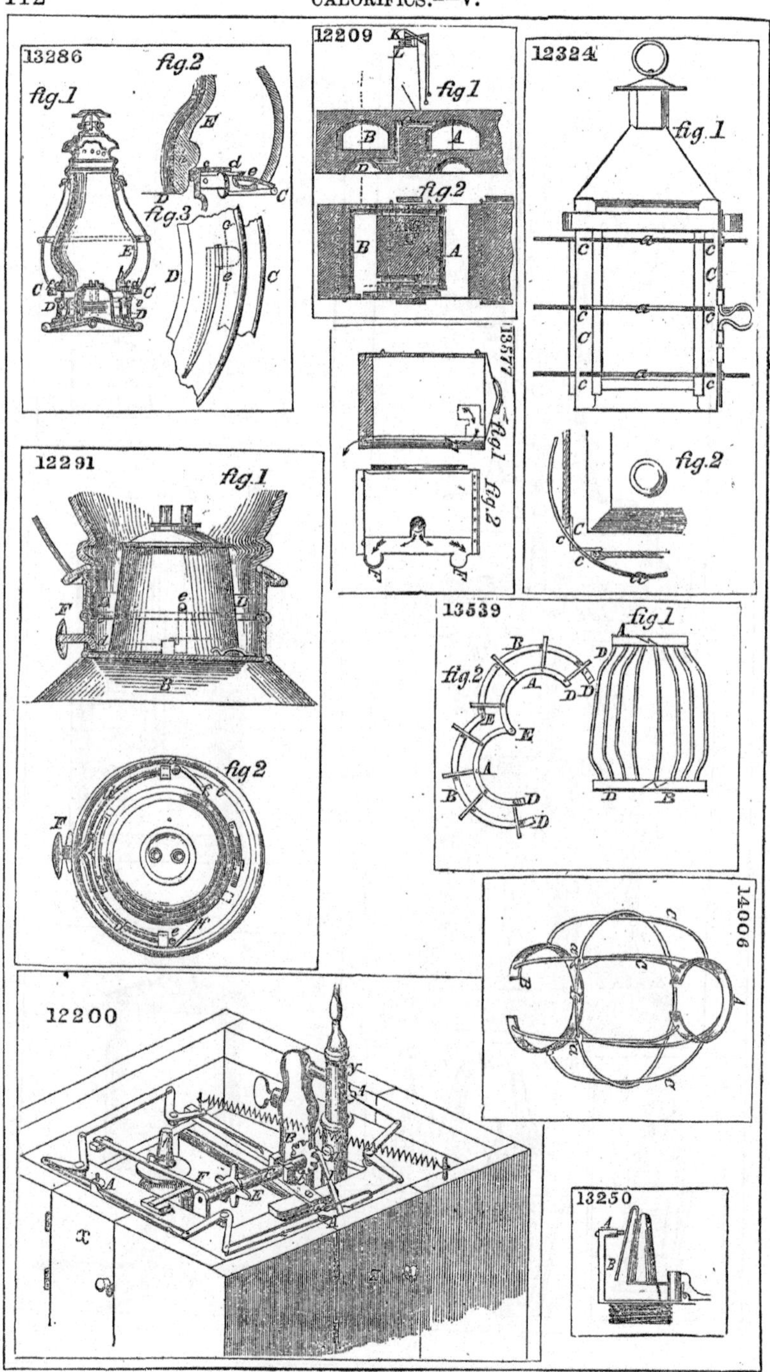
13286
fig.1
fig.2
fig.3
12209
fig1
fig.2
12324
fig.1
fig.2
13577
fig.1
fig.2
12291
fig.1
fig2
13539
fig1
fig.2
14006
12200
13250

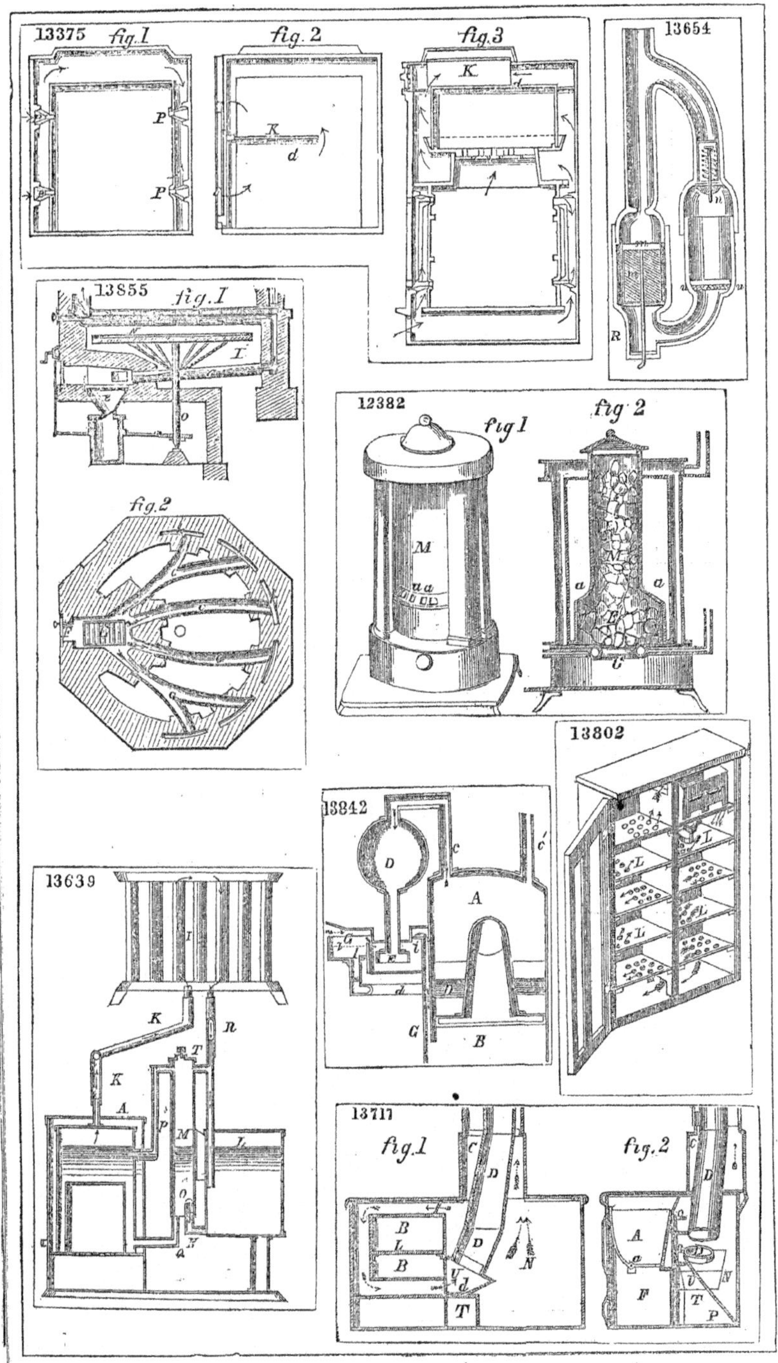
13375
fig.1
fig. 2
fig.3
K
P
P
K
d
13654
R
13855
fig. I
fig.2
12382
fig 1
fig 2
M
a a
M
a
a
13802
L
13842
D
A
G
B
13639
K
R
T
K
A
M
L
13717
fig.1
fig. 2
B
L
B
D
D
N
T
A
F
T
P

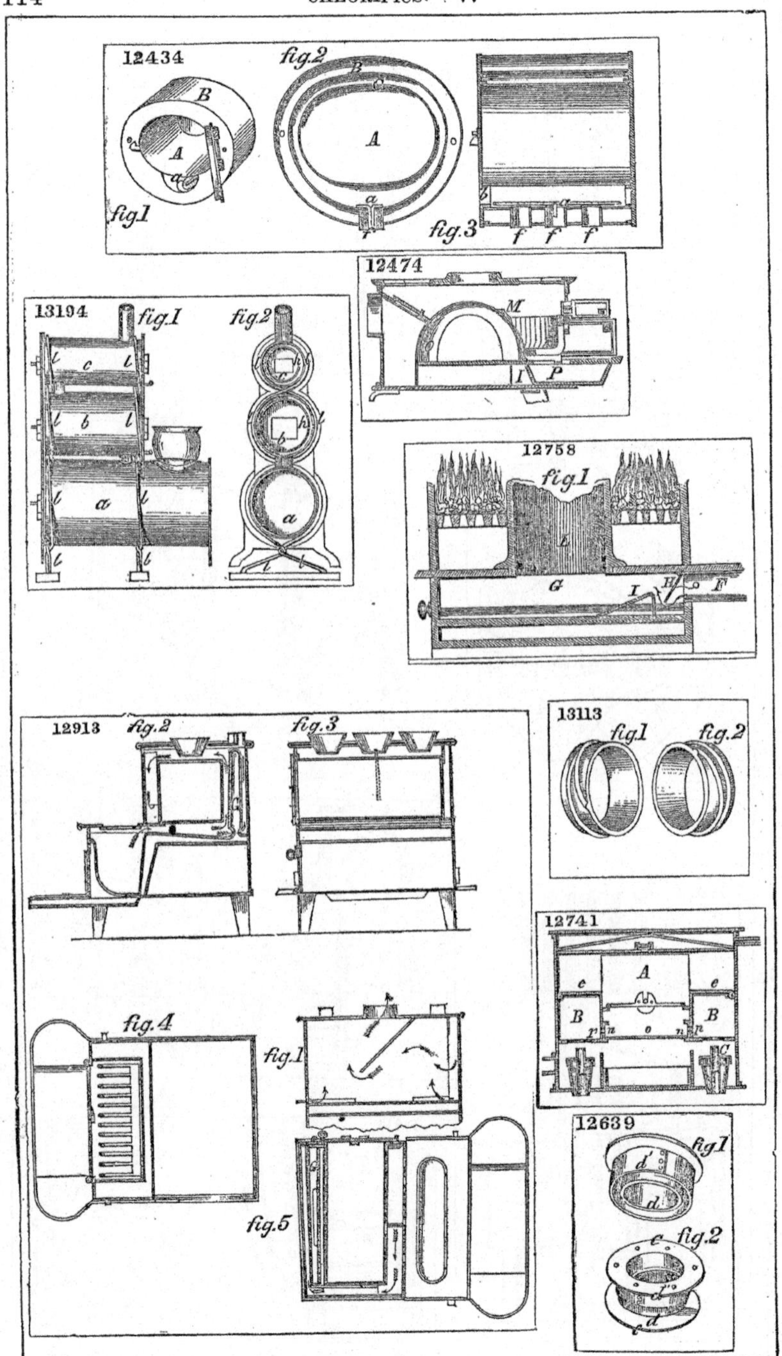
12434
fig.1
fig.2
fig.3
12474
13194
fig.1
fig.2
12758
fig.1
12913
fig.2
fig.3
fig.4
fig.1
fig.5
13113
fig.1
fig.2
12741
12639
fig.1
fig.2

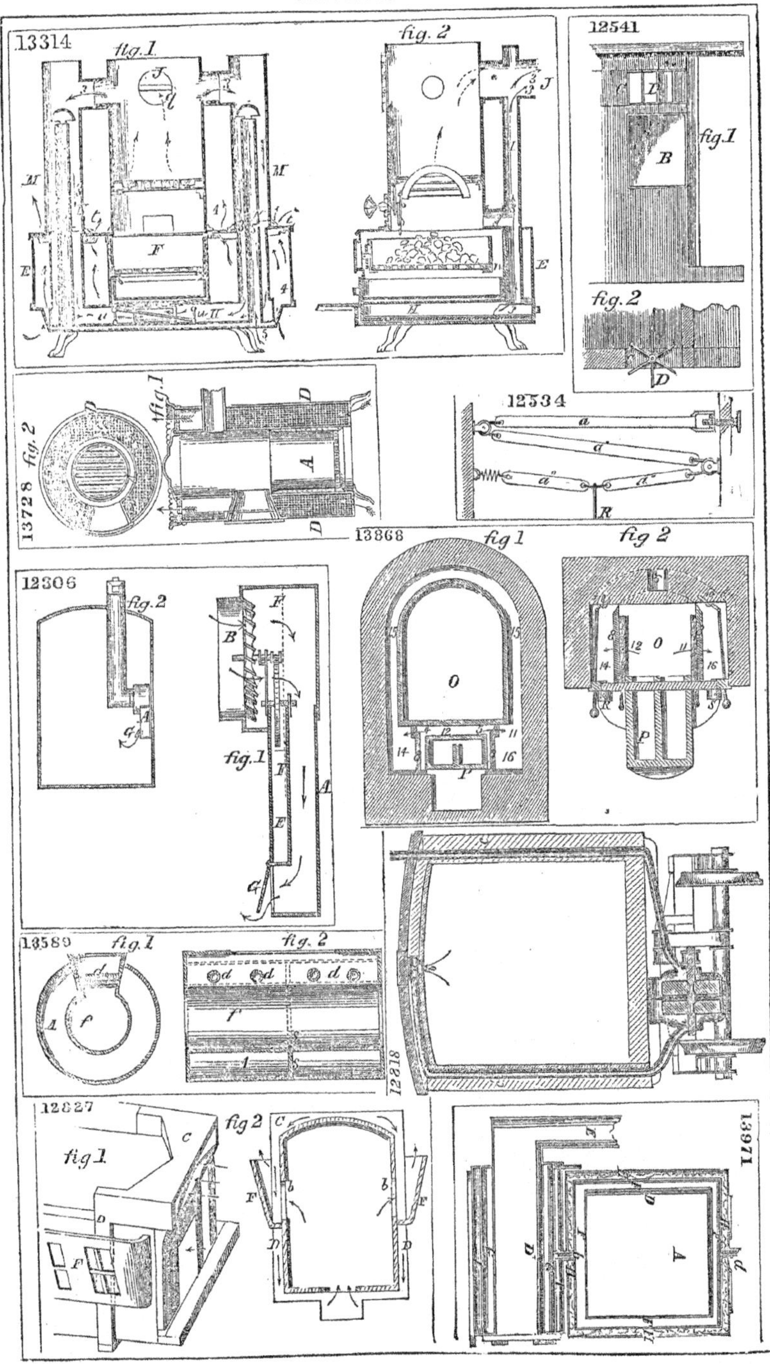
13314
fig. 1
fig. 2
12541
fig. 1
fig. 2
13728
fig. 2
fig. 1
12534
13868
fig 1
fig 2
12306
fig. 2
fig. 1
13589
fig. 1
fig. 2
12818
12827
fig 1
fig 2
13971

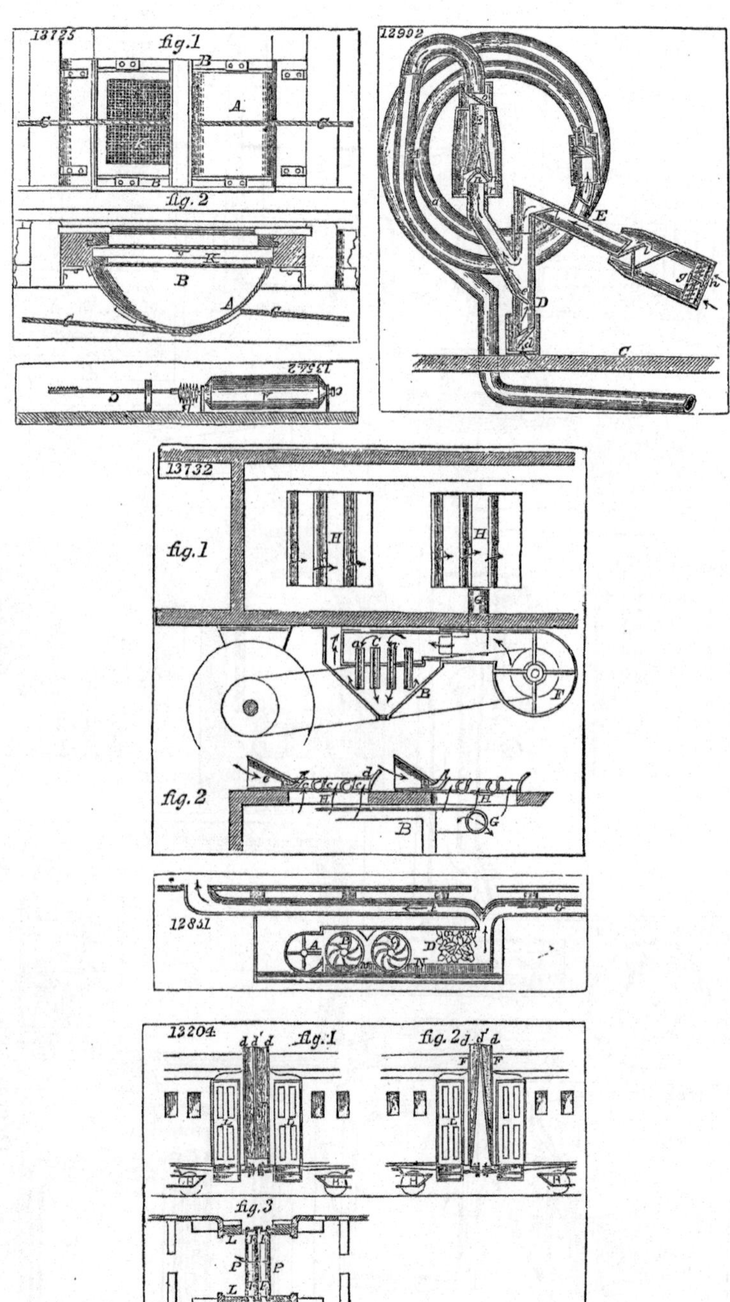
13725
fig.1
fig.2
12992
13732
fig.1
fig.2
12851
13204
fig.1
fig.2
fig.3

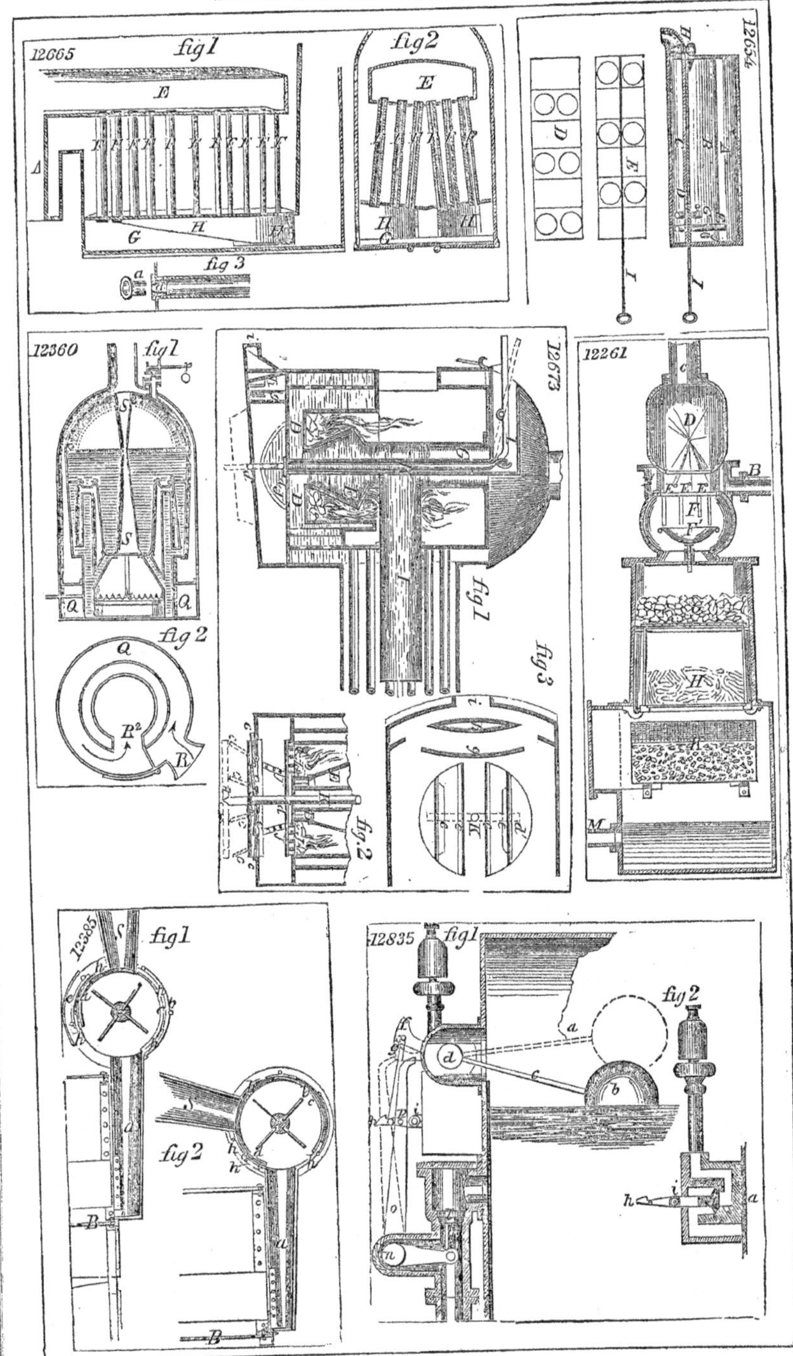
12665
fig 1
fig 2
fig 3
12654
12360
12673
12261
12385
12835

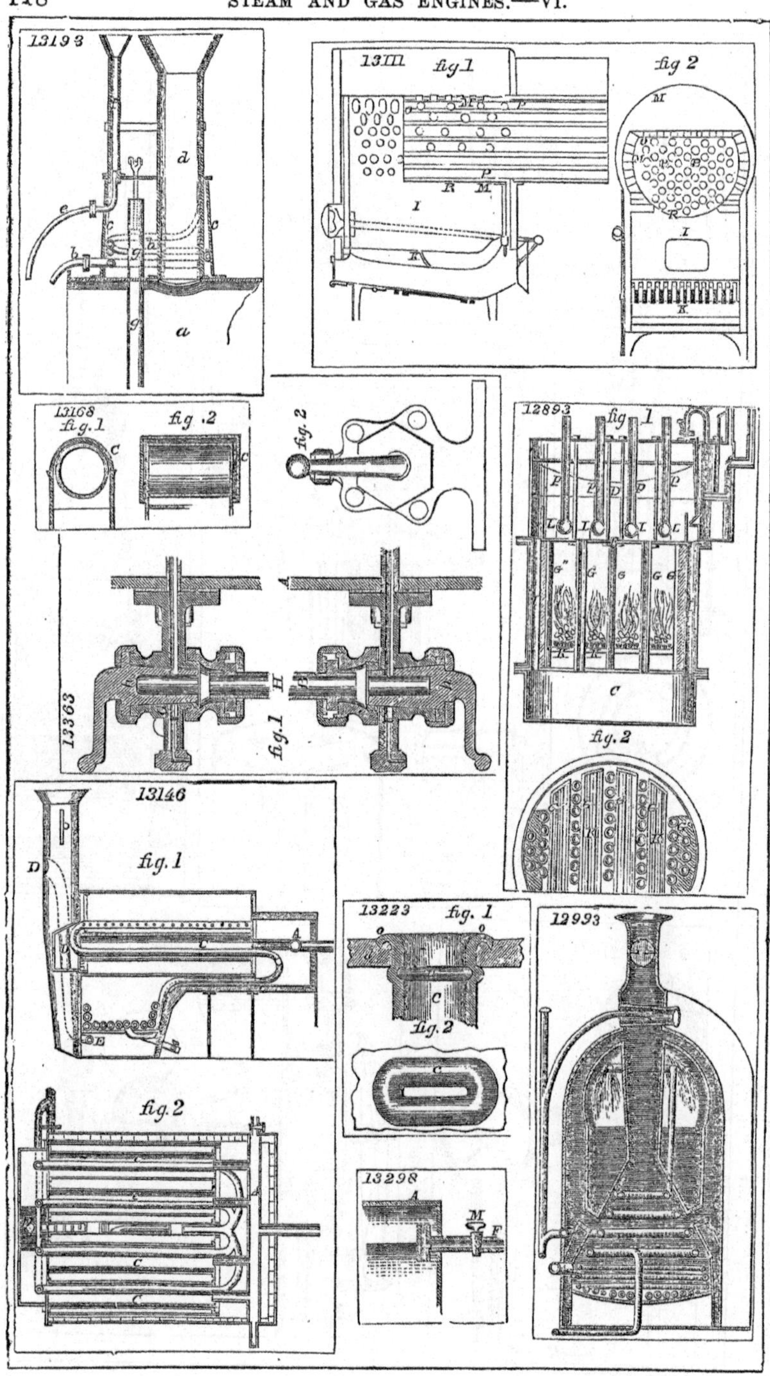
13193
13111 fig 1
fig 2
13168 fig. 1
fig. 2
fig. 2
12893 fig 1
fig. 2
13363
fig. 1
13146
fig. 1
fig. 2
13223 fig. 1
fig. 2
12993
13298

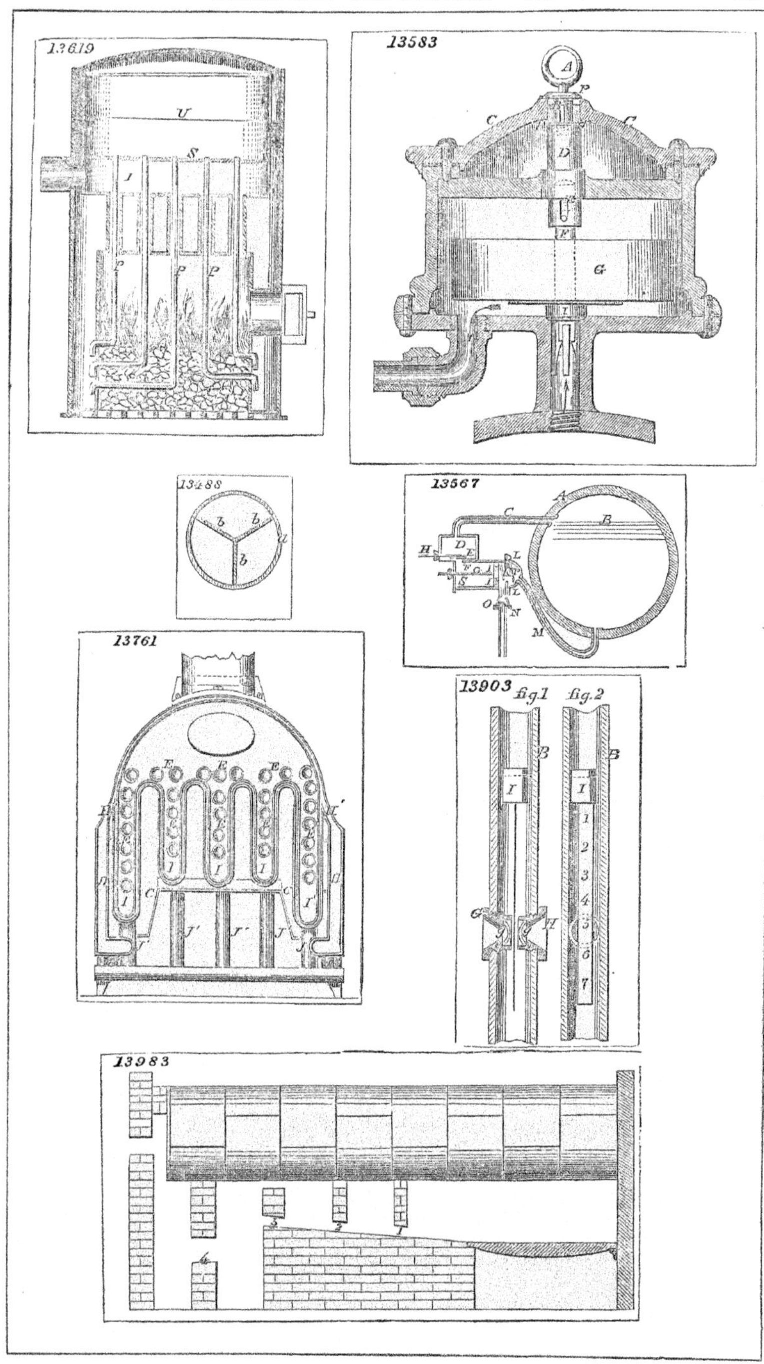
13619
13583
13488
13567
13761
13903
fig.1
fig.2
13983

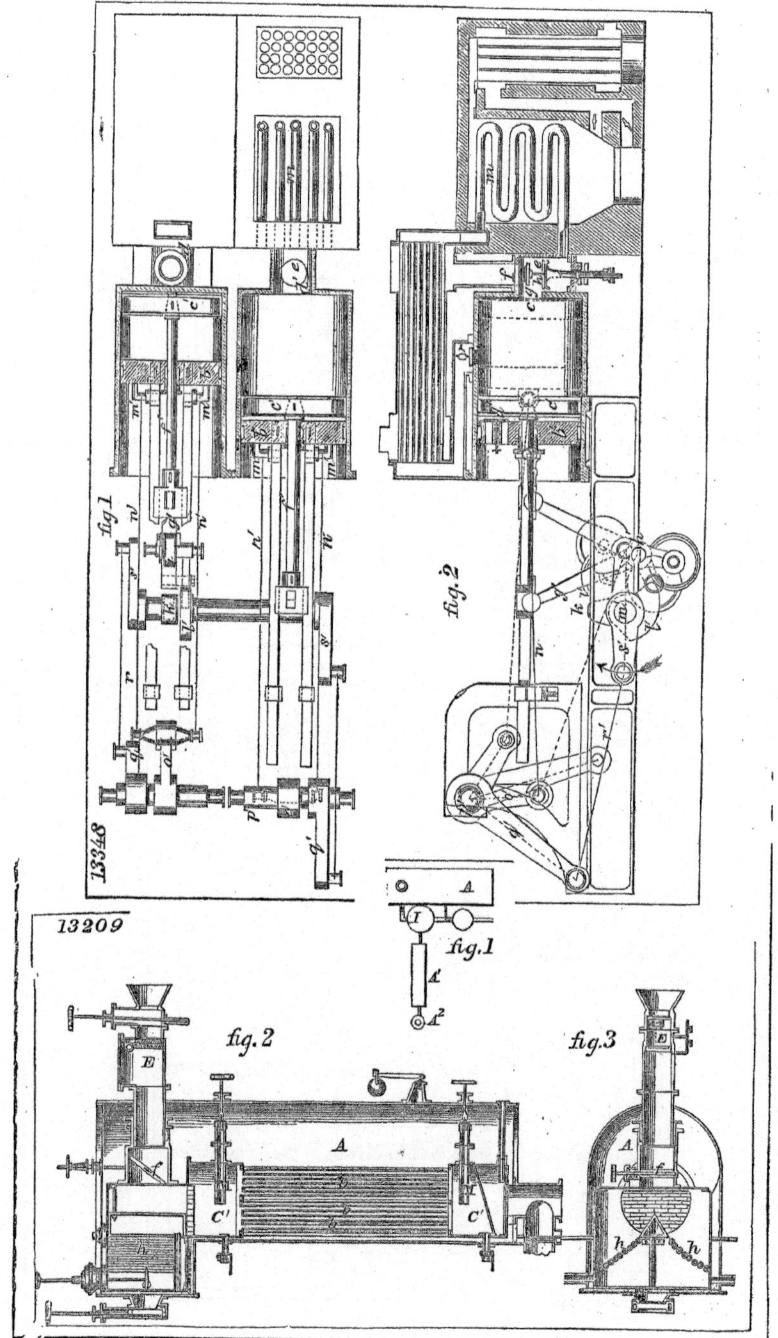
13348
fig.1
fig.2
13209
fig.1
fig.2
fig.3

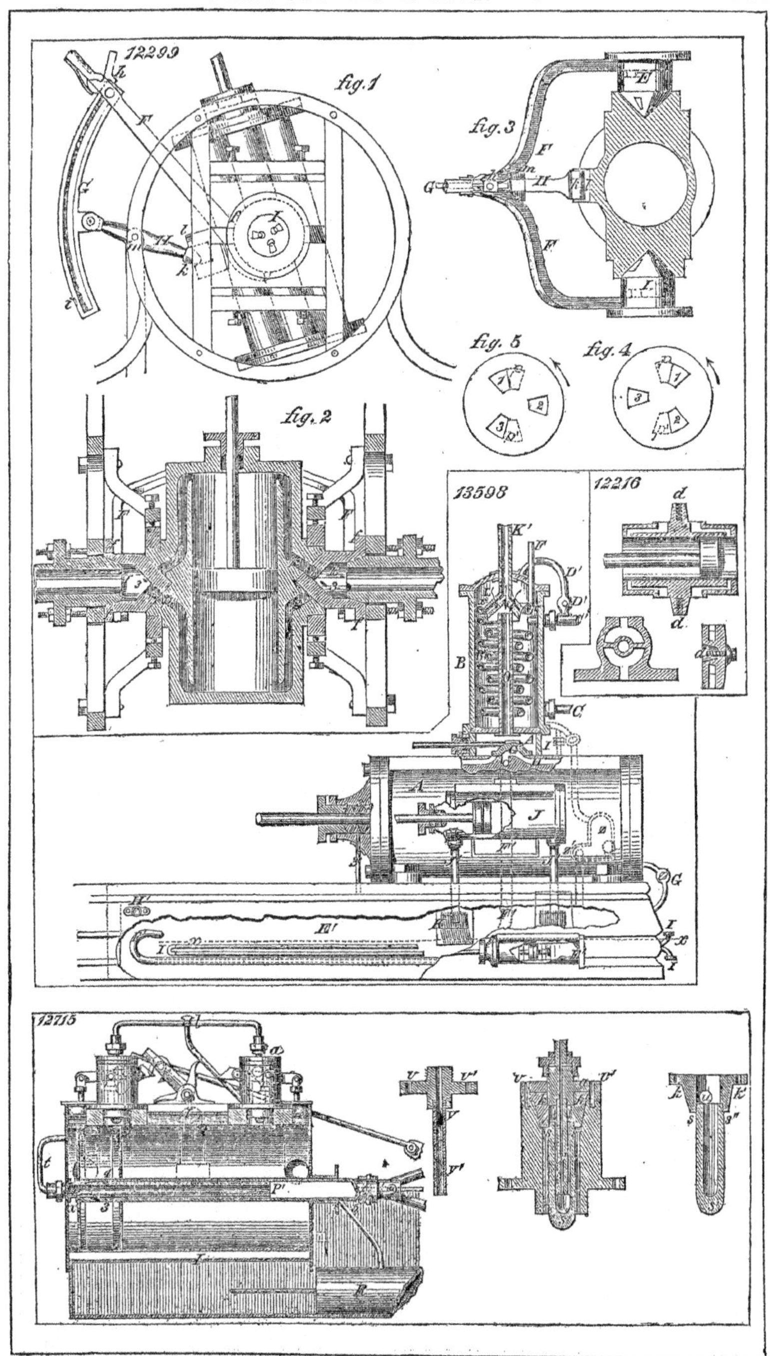
12299
fig.1
fig.2
fig.3
fig. 5
fig. 4
13598
12216
12715

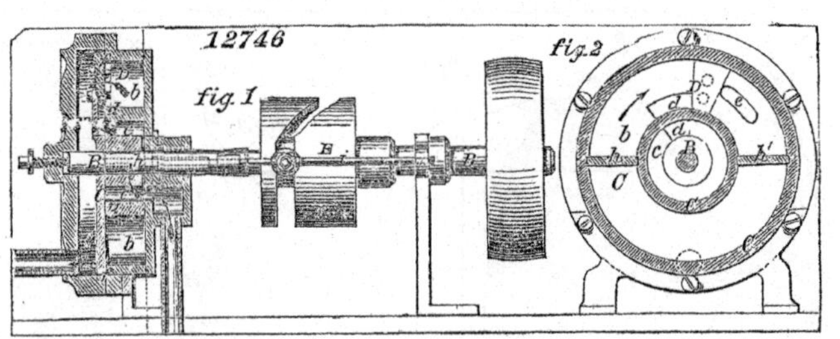
12746
fig. 1
fig. 2

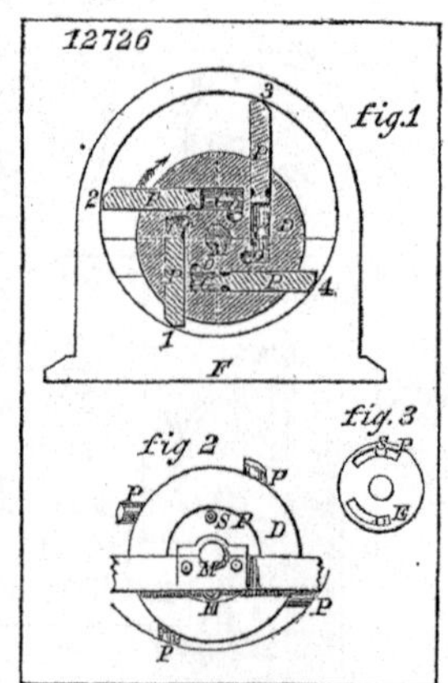
12726
fig. 1
fig 2
fig. 3

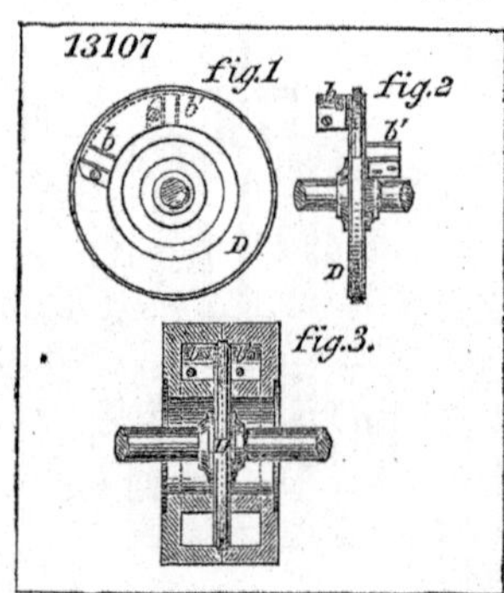
13107
fig. 1
fig. 2
fig. 3.

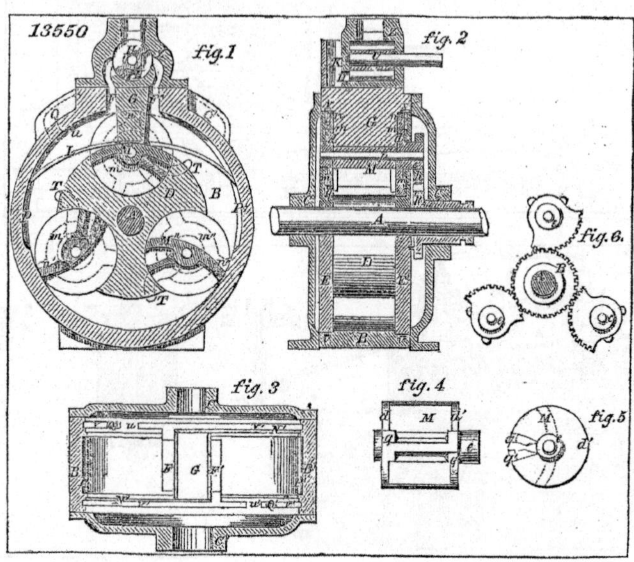
13550
fig. 1
fig. 2
fig. 3
fig. 4
fig. 5
fig. 6.

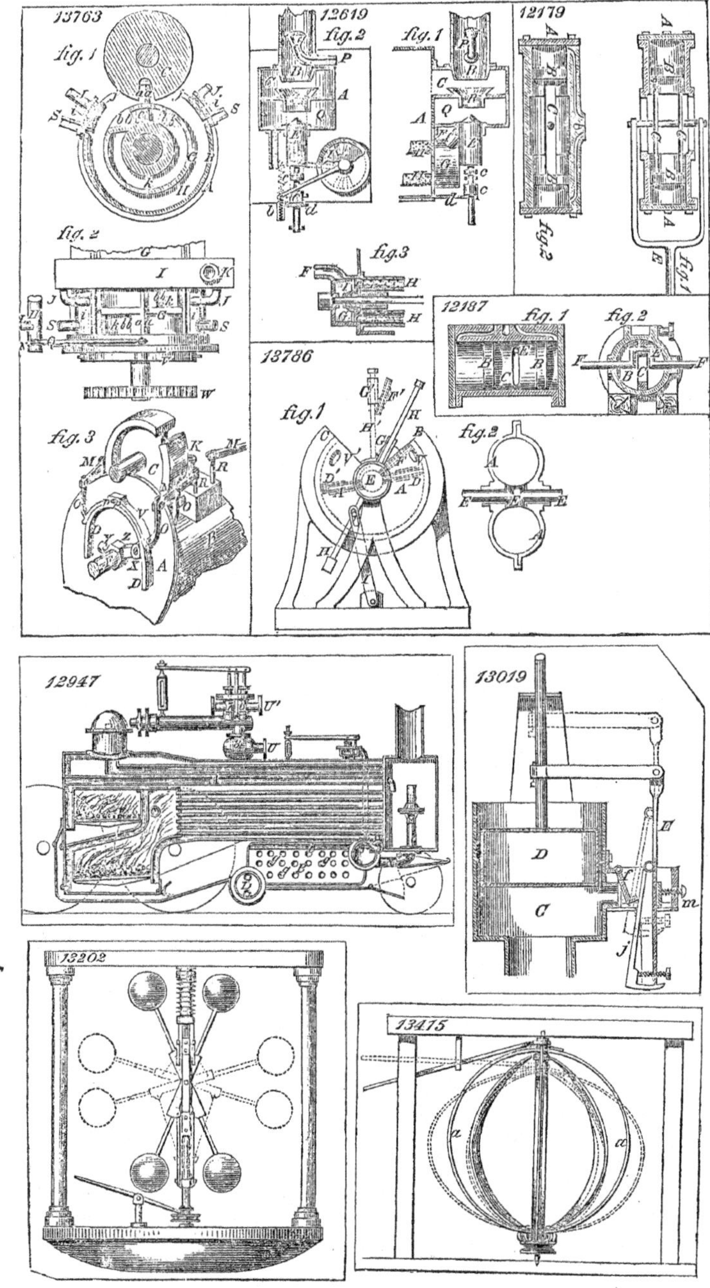
13763
fig. 1
fig. 2
fig. 3
12619
fig. 2
fig. 1
fig. 3
12179
fig. 2
fig. 1
12187
fig. 1
fig. 2
13786
fig. 1
fig. 2
12947
13019
13202
13475

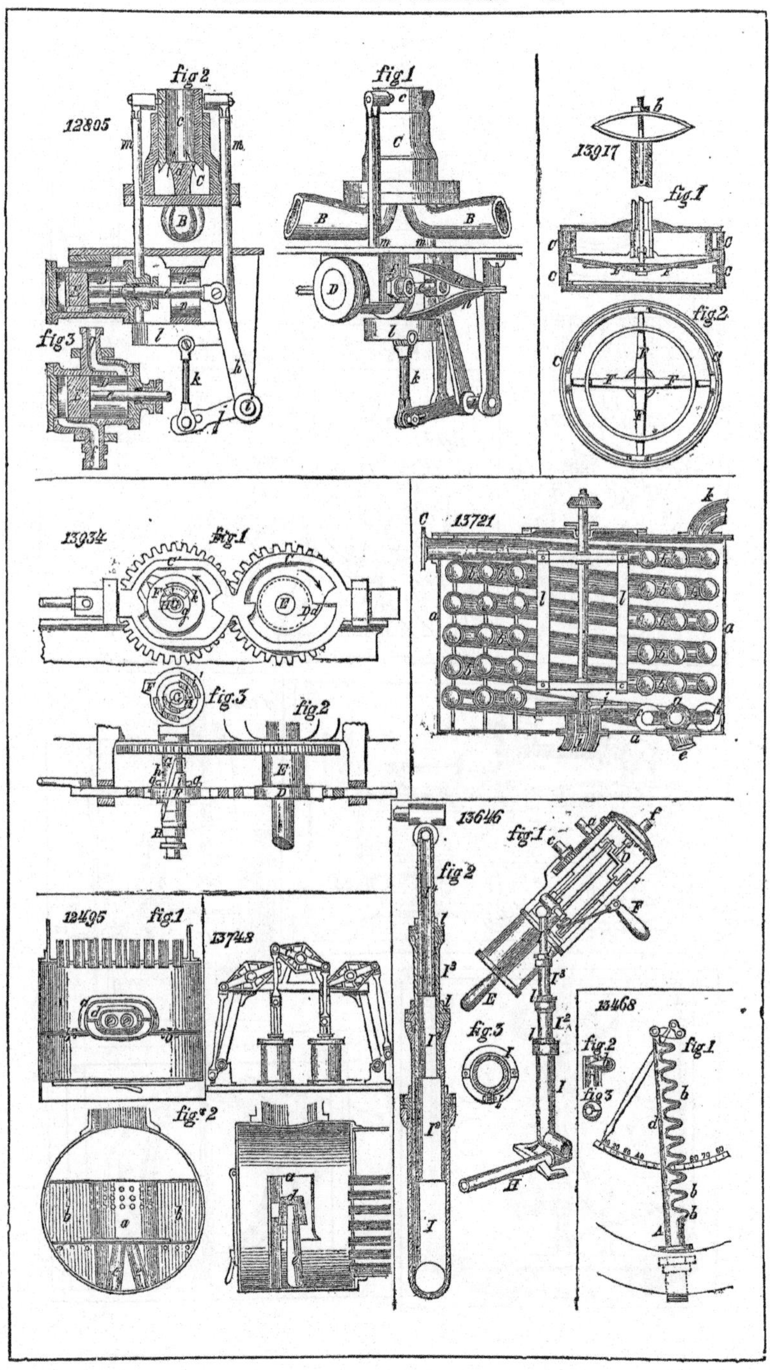
12805
fig 2
fig 1
fig 3
13917
fig.1
fig.2
13934
fig.1
fig.3
fig.2
13721
13646
fig.1
fig 2
fig.3
12495
fig.1
fig. 2
13748
13468
fig.2
fig.1
fig 3

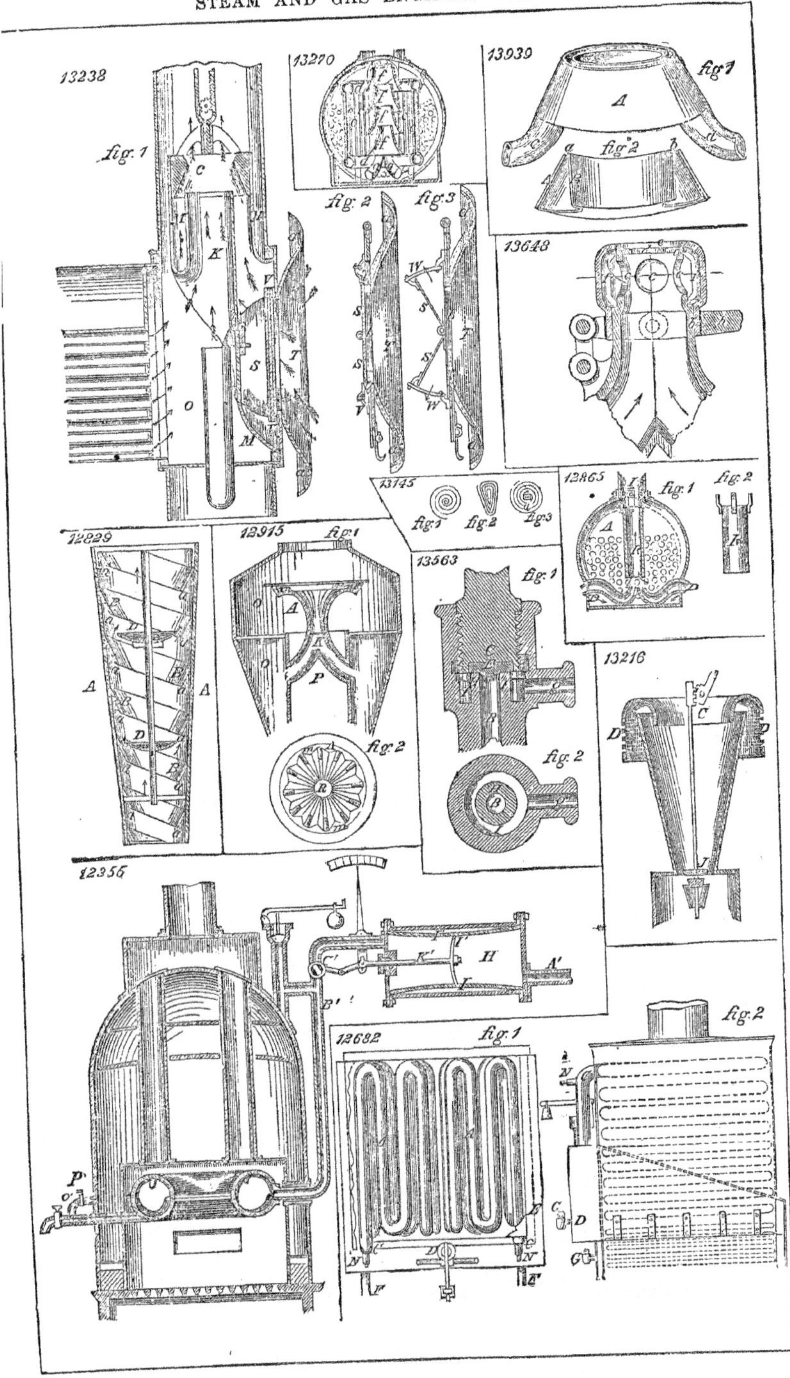

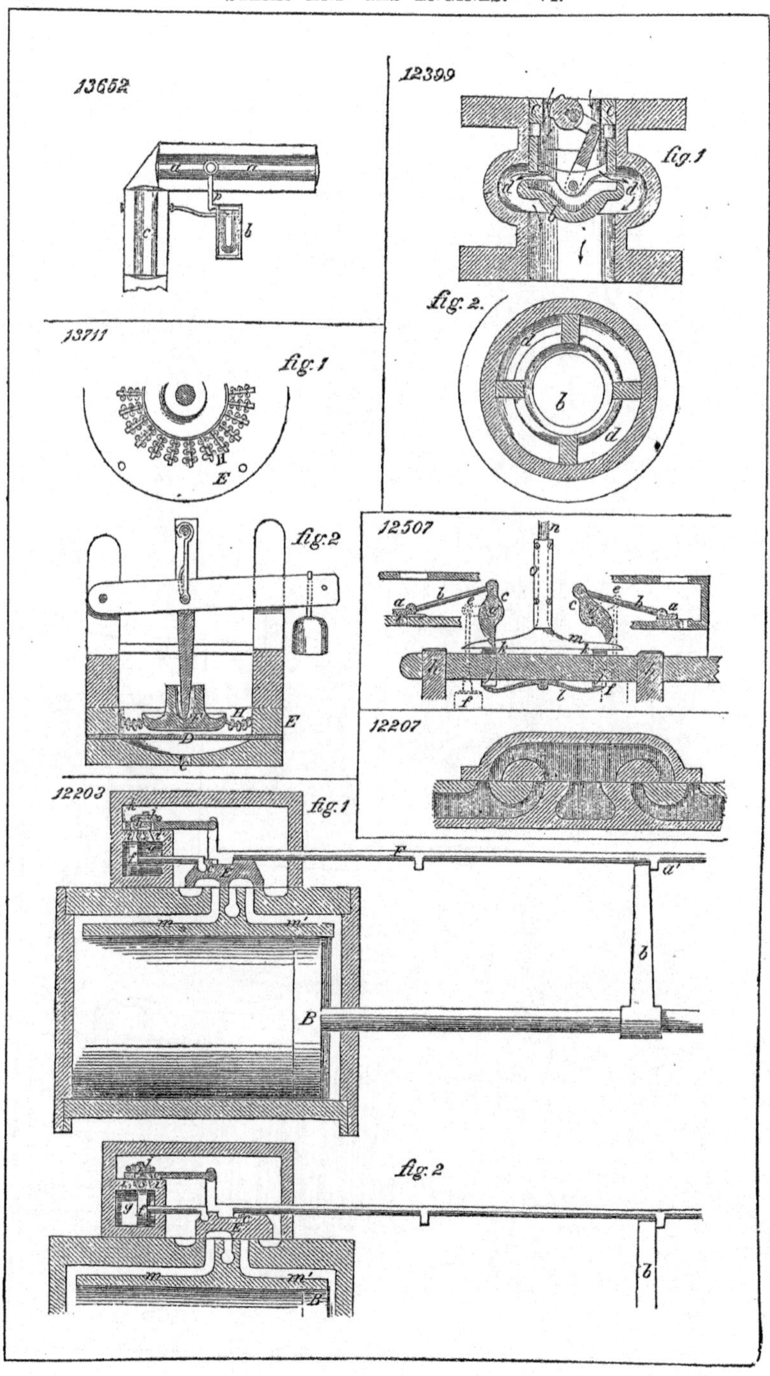
13652
12399
fig.1
fig. 2.
13711
fig.1
fig.2
12507
12207
12203
fig.1
fig. 2

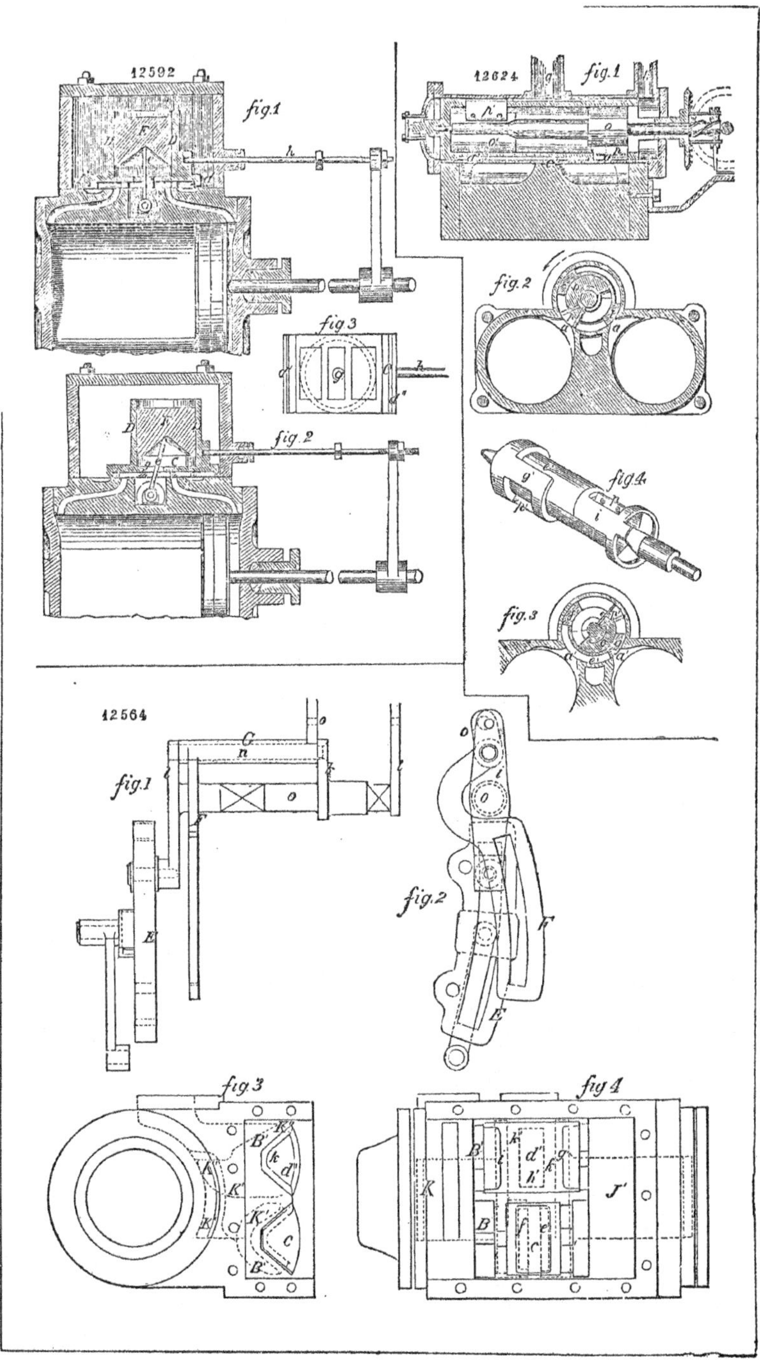
12592
fig.1
fig.3
fig.2
12624
fig.1
fig.2
fig.4
fig.3
12564
fig.1
fig.2
fig.3
fig.4

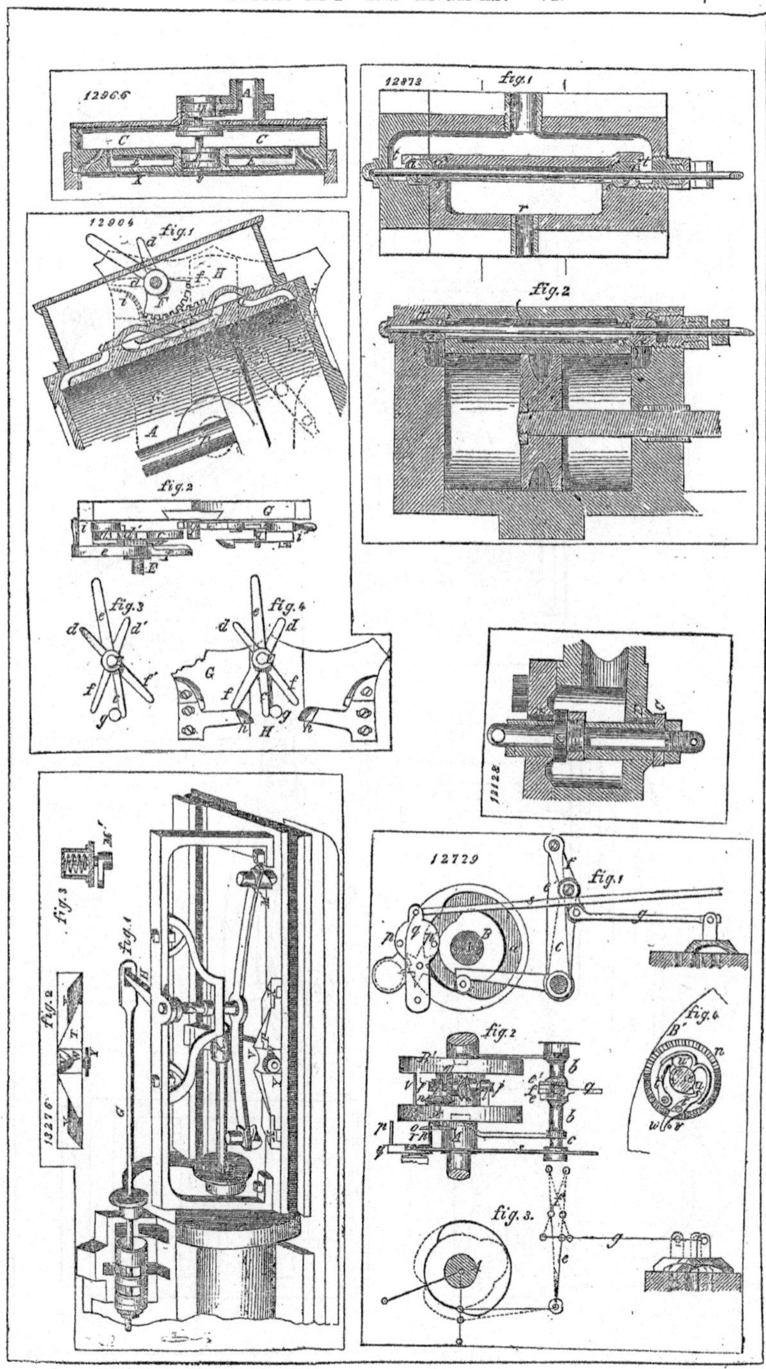

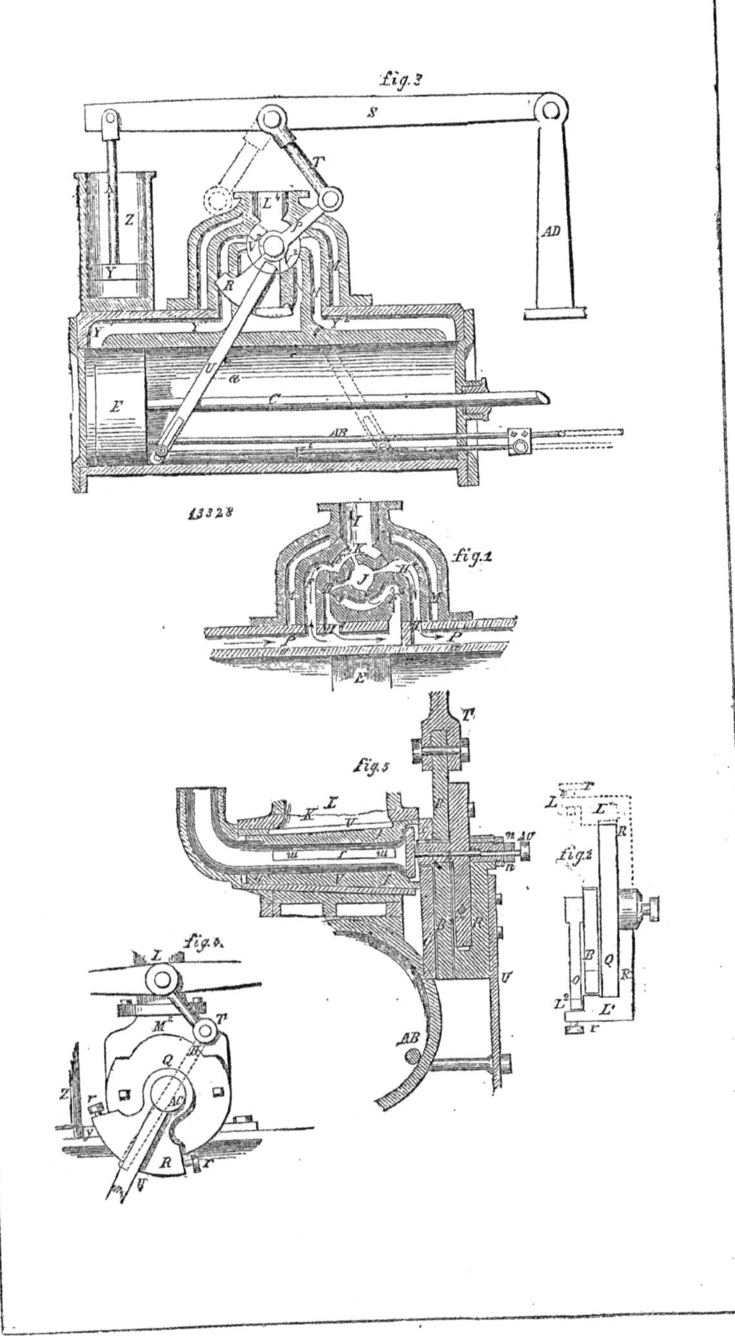
fig.3
S
T
AD
Z
Y
L
R
U
E
C
AB
13328
fig.1
I
K
J
H
P
E
fig.5
L
K
V
u
AC
T
U
AB
fig.2
B
Q
R
O
fig.4
L
M
T
Q
AC
Z
R
U

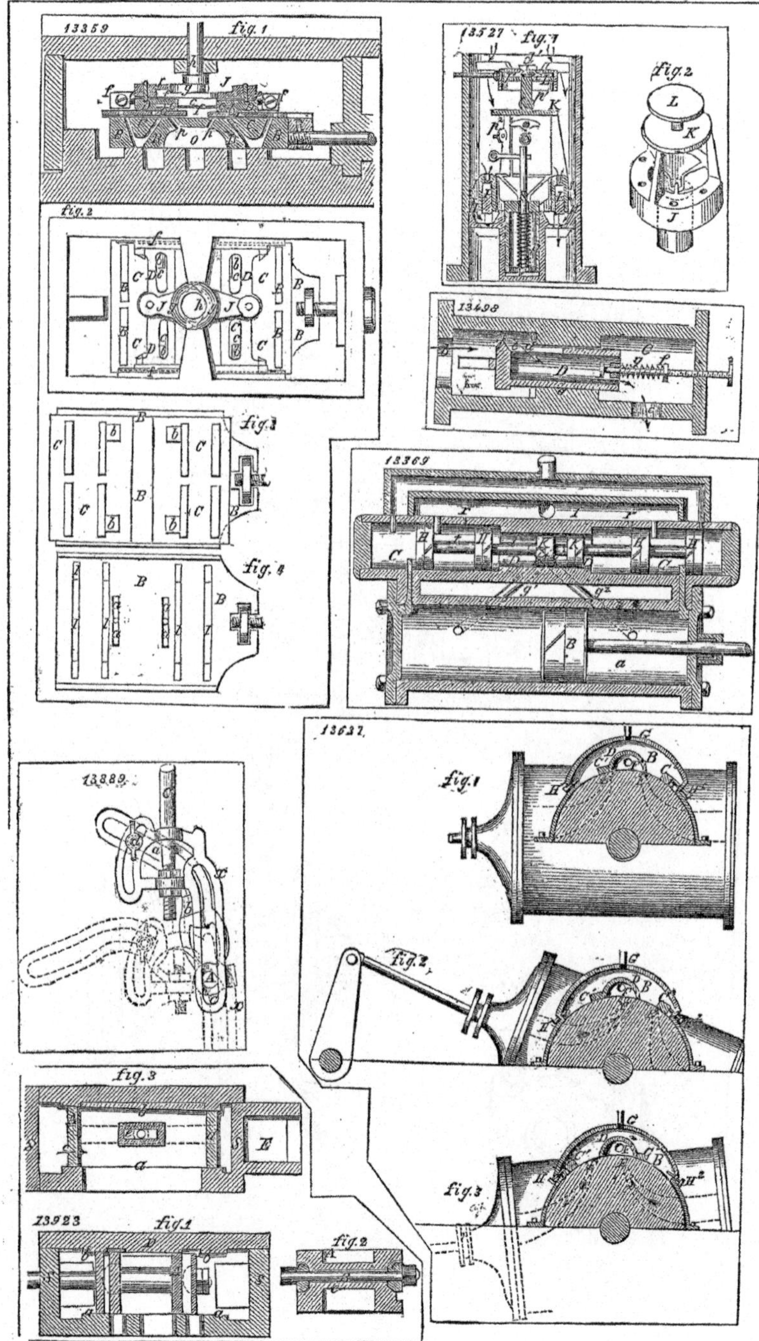

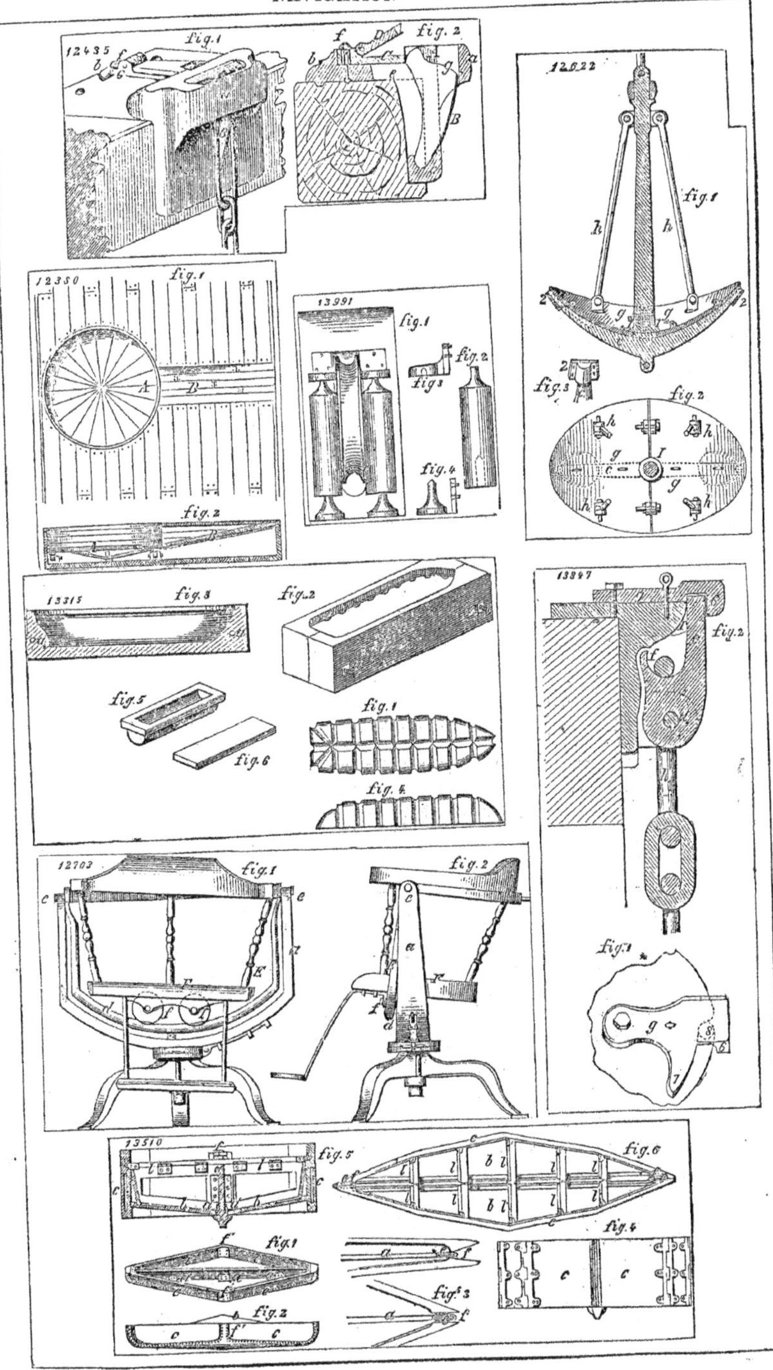

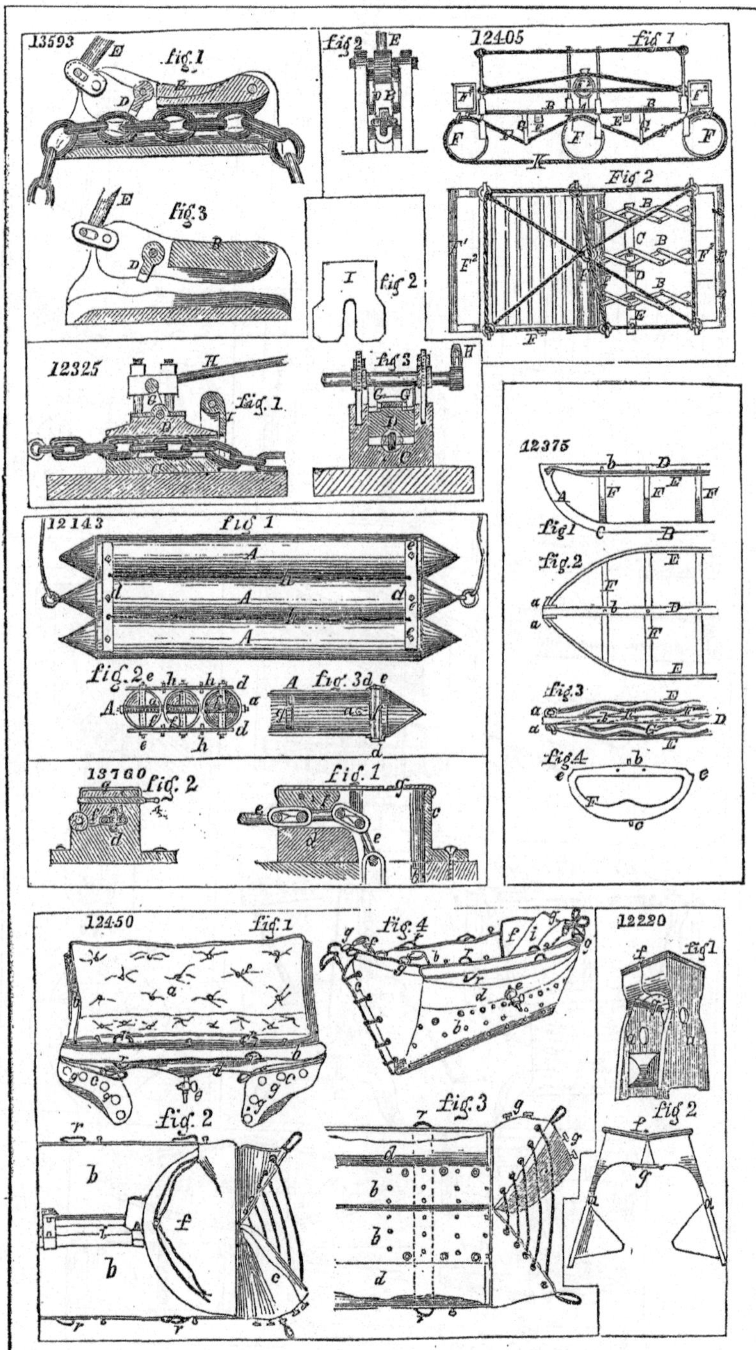
13593
fig. 1
fig. 3
fig 2
12405
fig 1
Fig 2
fig 2
12325
fig. 1
fig 3
12375
fig 1
fig. 2
fig. 3
fig 4
12143
fig 1
fig. 2
fig. 3
13760
fig. 2
fig. 1
12450
fig. 1
fig. 4
fig. 2
fig. 3
12220
fig 1
fig 2

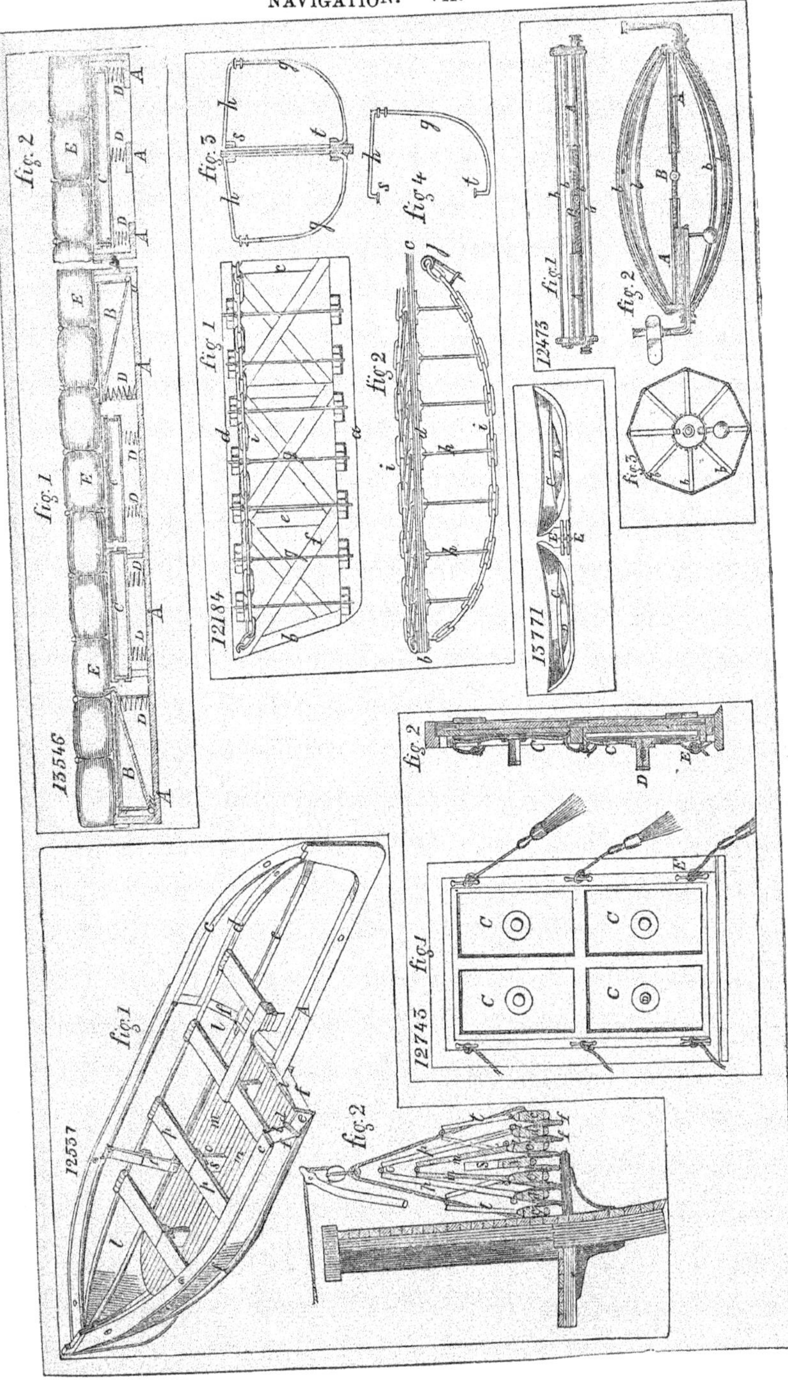
13546
fig. 1
fig. 2
12184
fig. 1
fig. 2
fig. 3
fig. 4
13771
12473
fig. 1
fig. 2
fig. 3
12557
fig. 1
fig. 2
12745
fig. 1
fig. 2

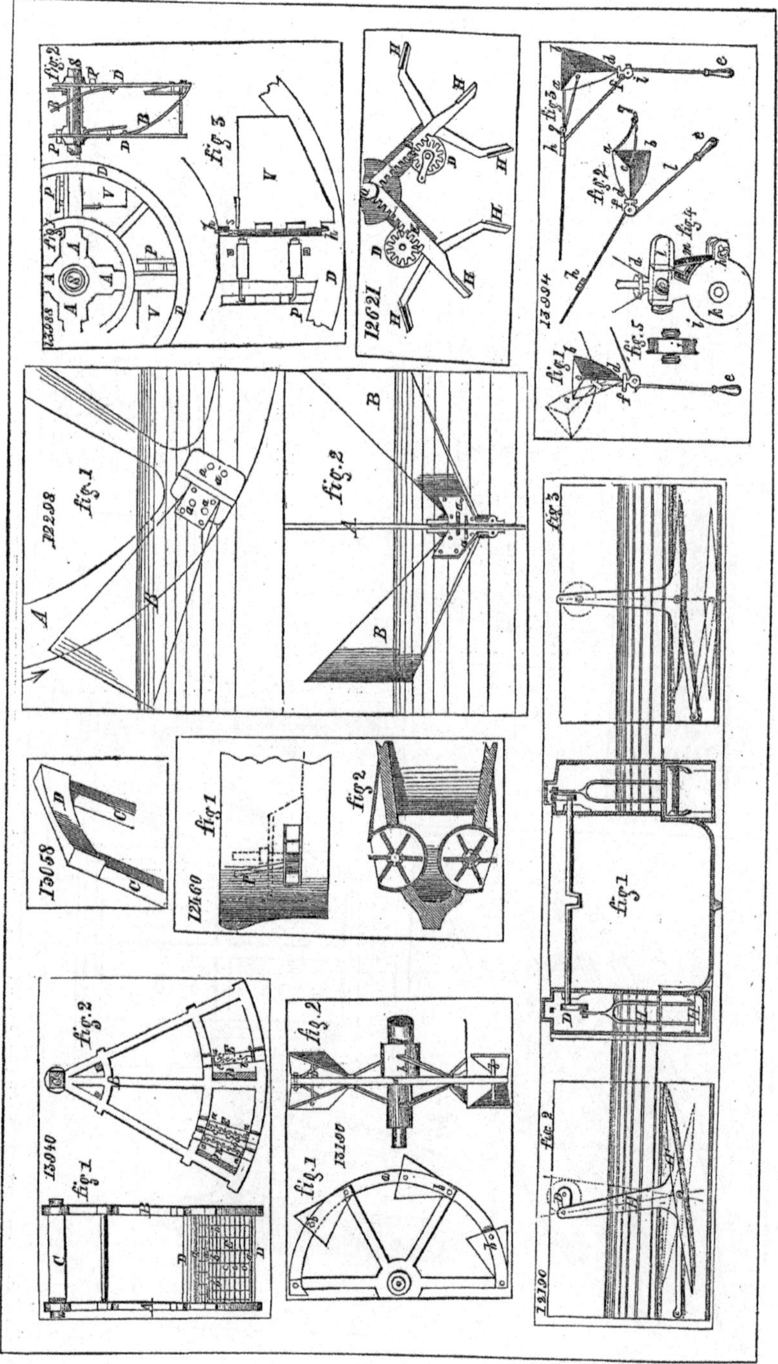

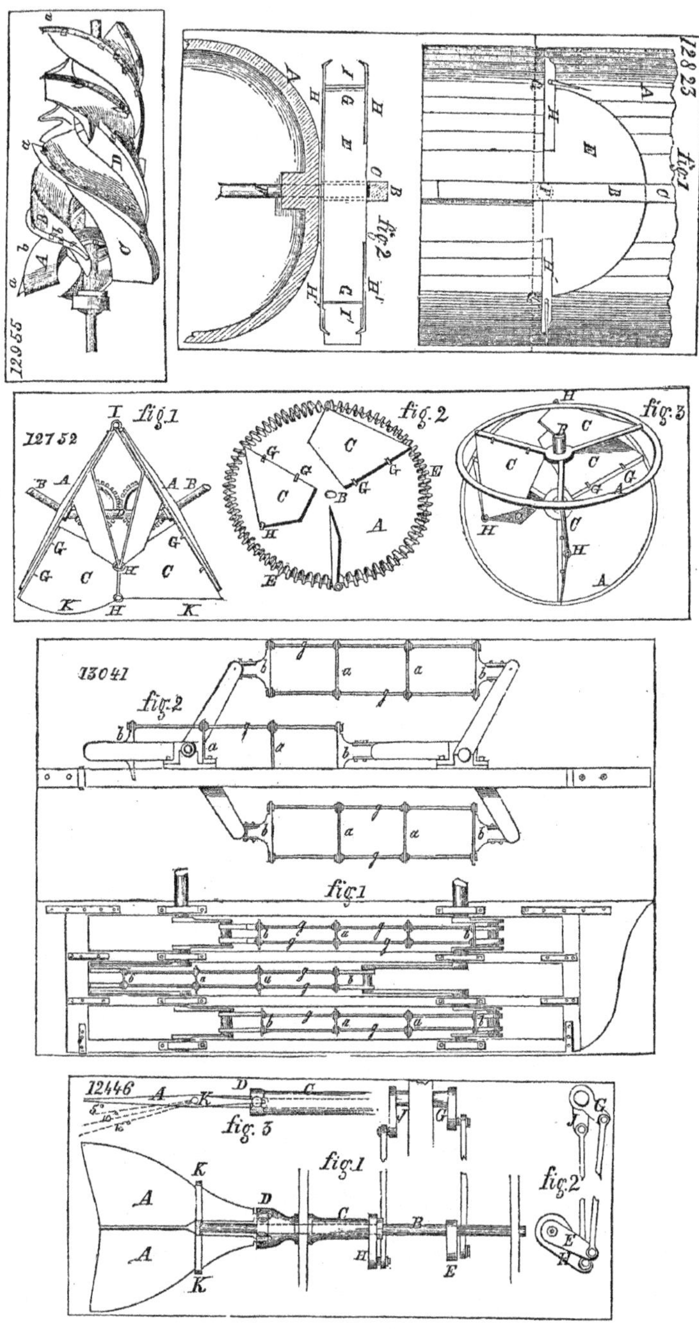
12955
12823
fig.1
fig.2
12752
fig.1
fig.2
fig.3
13041
fig.2
fig.1
12446
fig.3
fig.1
fig.2

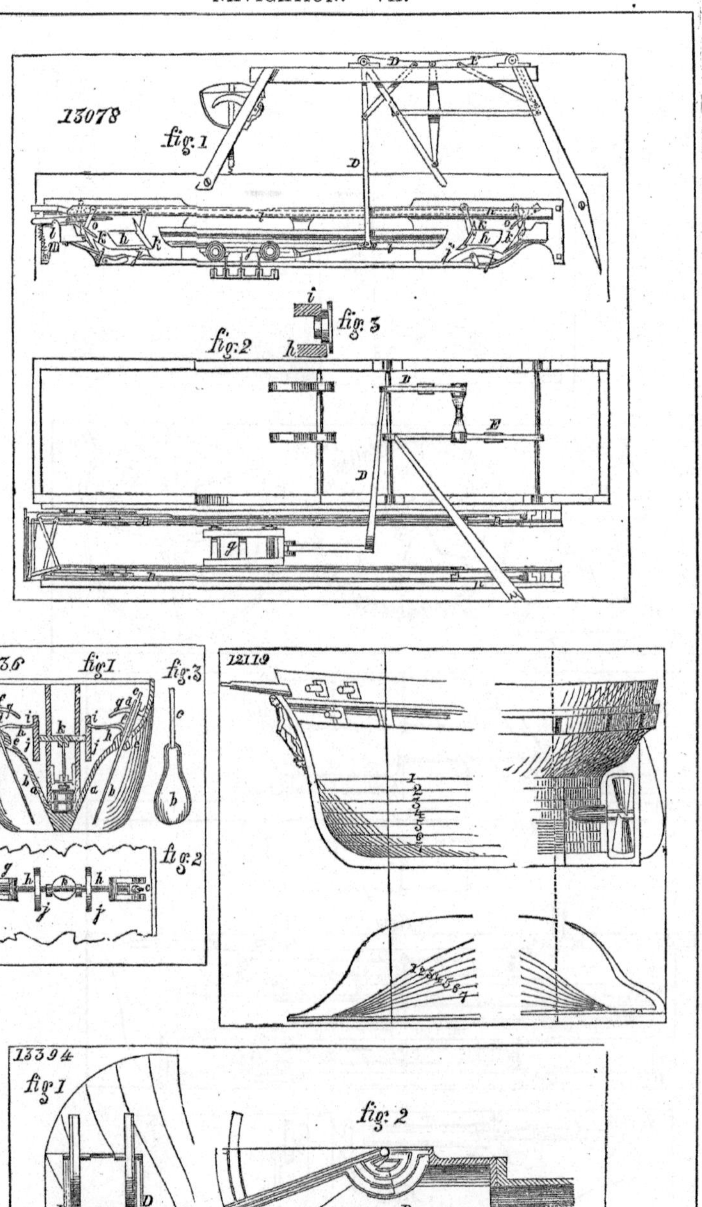
13078
fig. 1
fig. 2
fig. 3
13336
fig 1
fig. 3
fig. 2
12119
13394
fig 1
fig. 2

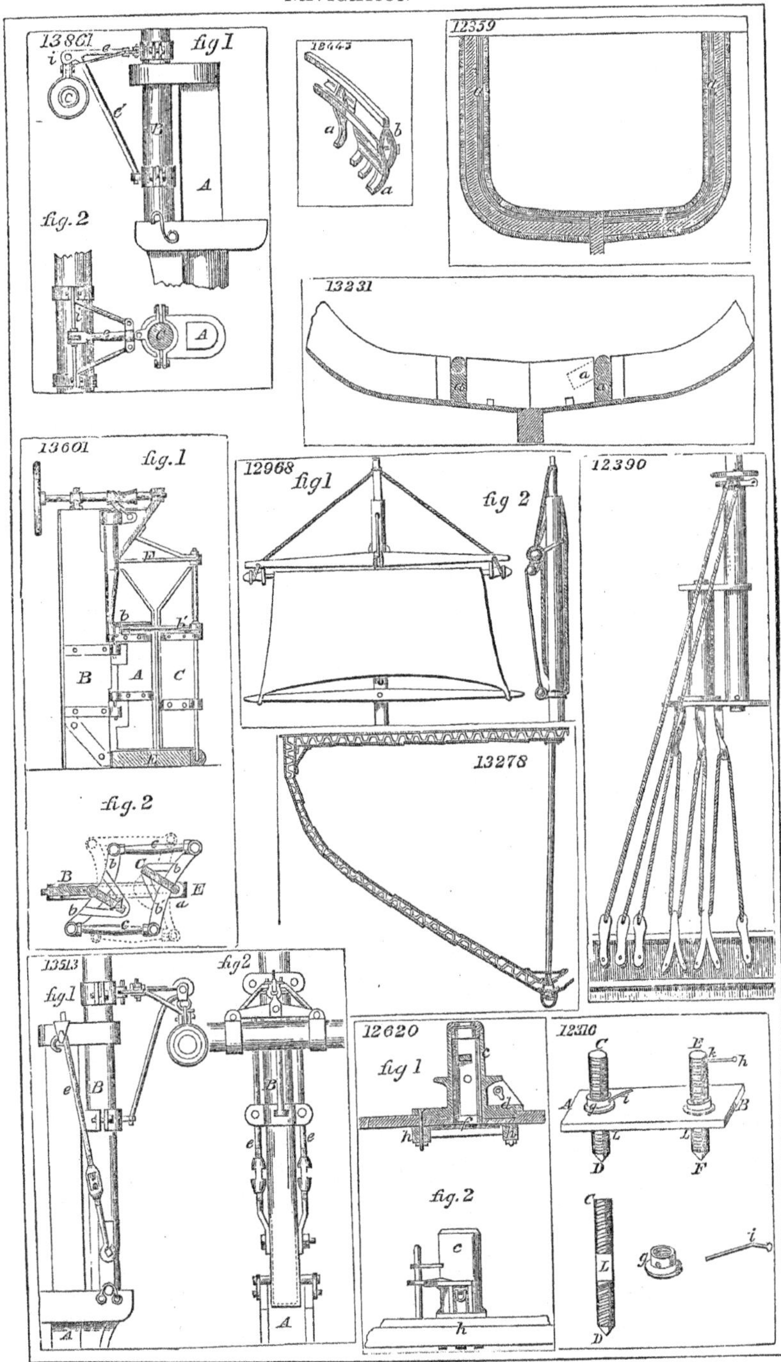
13801
fig 1
fig. 2
18443
12359
13231
13601
fig. 1
fig. 2
12968
fig 1
fig 2
12390
13278
13513
fig. 1
fig 2
12620
fig 1
fig. 2
12310

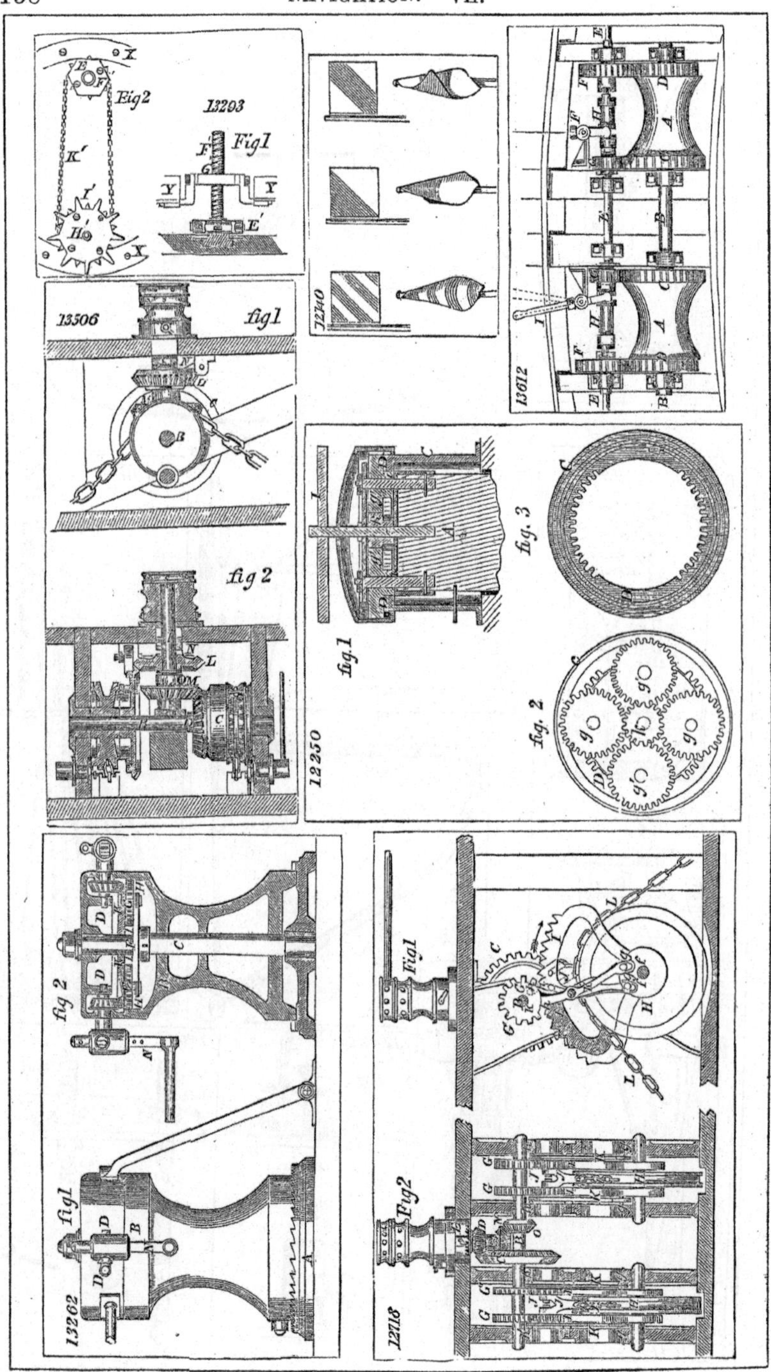
13203
Fig1
Fig2
13506
fig1
fig2
12740
13612
12250
fig. 1
fig. 2
fig. 3
13262
fig1
fig 2
12718
Fig1
Fig2

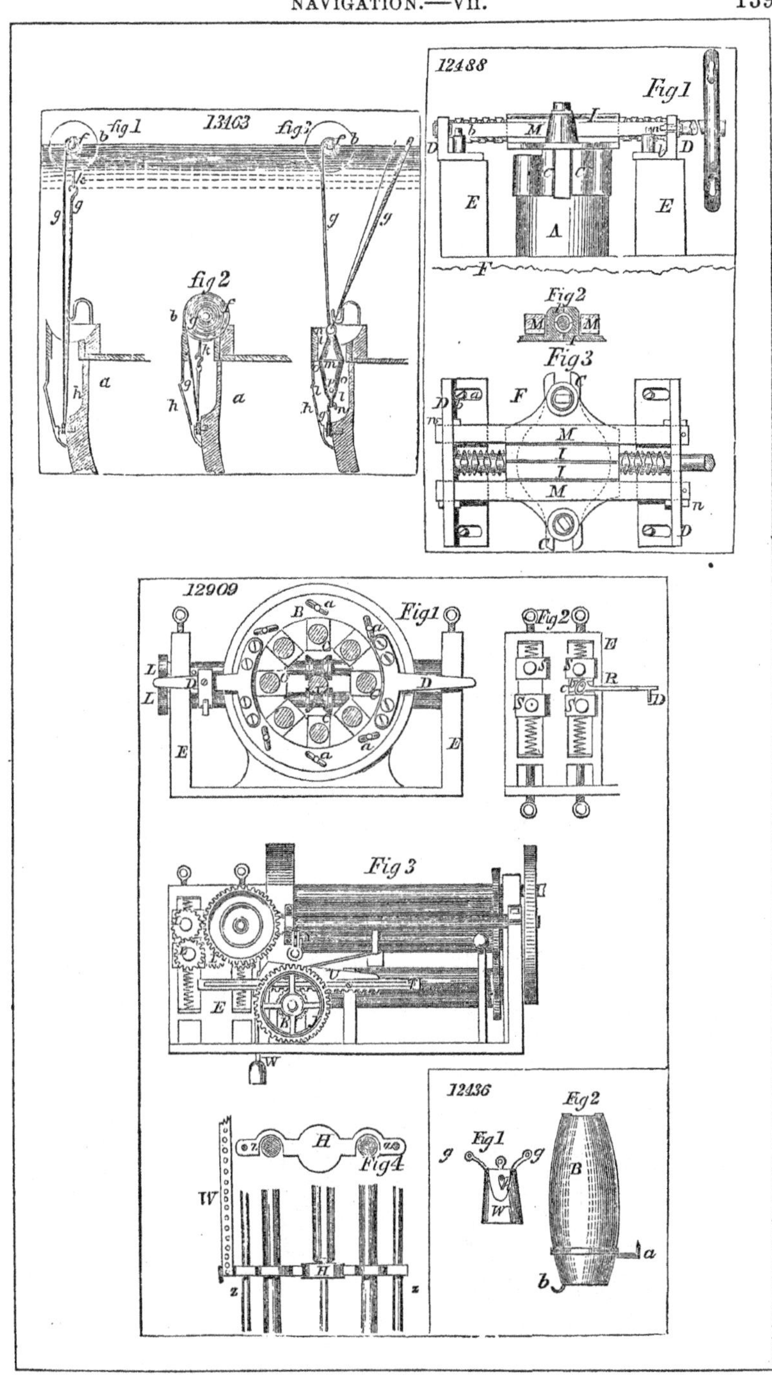
13463
fig 1
fig 2
12488
Fig 1
Fig 2
Fig 3
12909
Fig 1
Fig 2
Fig 3
Fig 4
12436
Fig 1
Fig 2

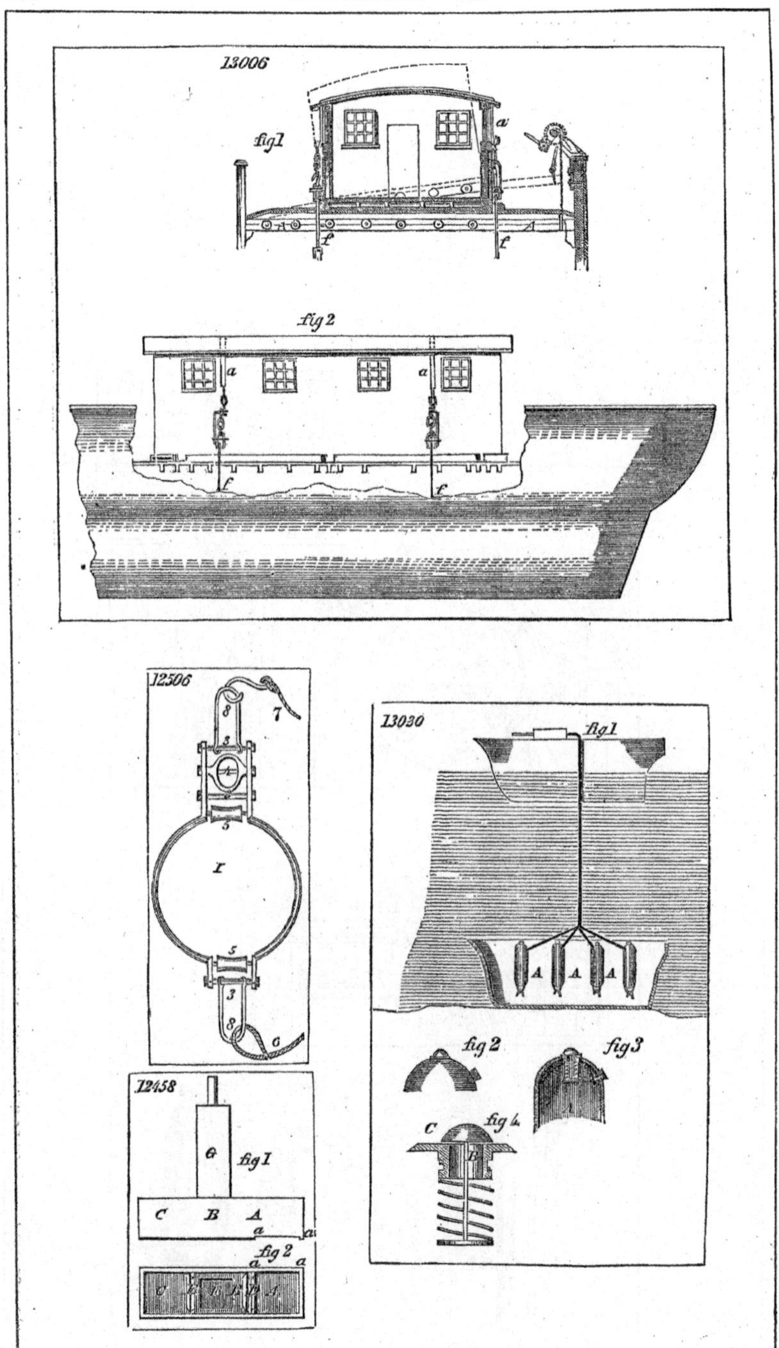
13006
fig1
a
f
f
fig 2
a
a
f
f
12506
7
8
3
5
I
5
3
8
6
13030
fig1
A
A
A
fig 2
fig3
C
fig 4
B
12458
G
fig 1
C
B
A
a
a
fig 2
a
a

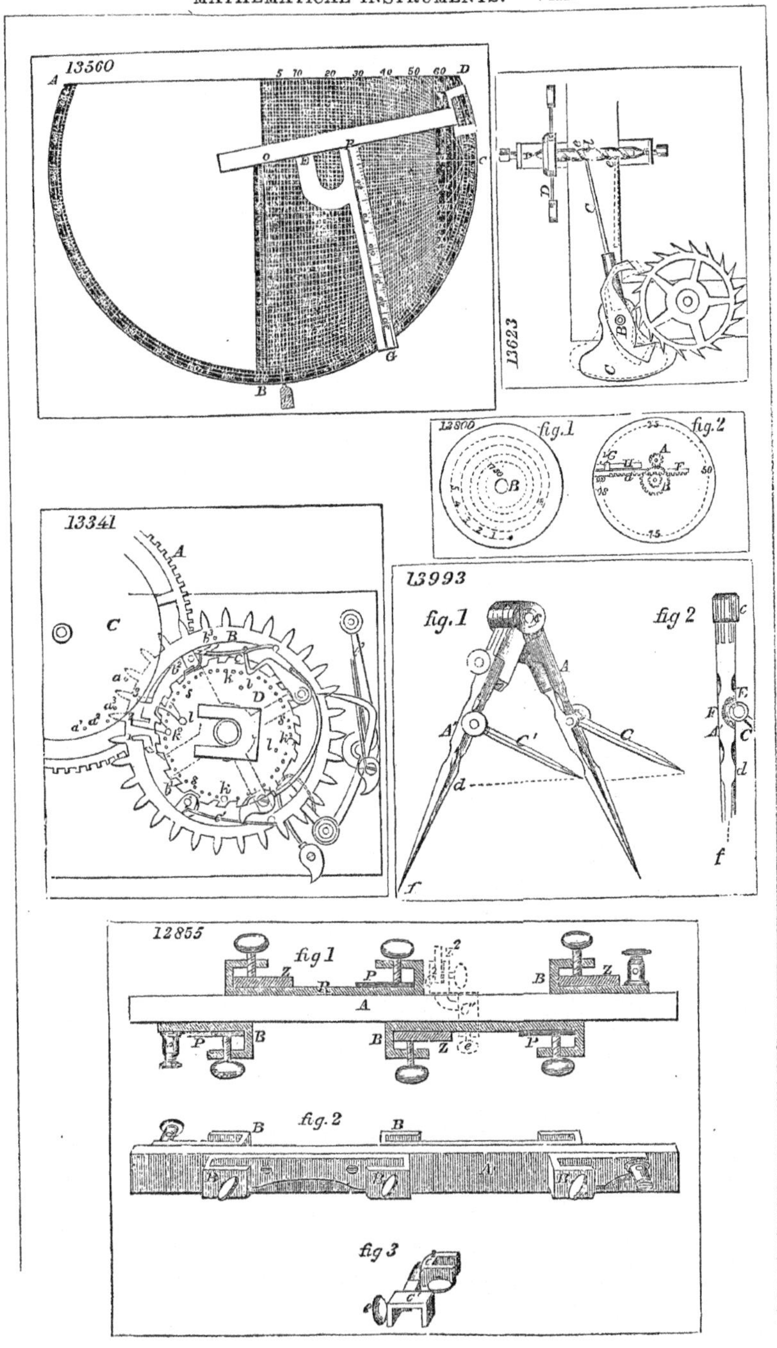
13560
13623
12800
fig.1
fig.2
13341
13993
fig.1
fig 2
12855
fig 1
fig.2
fig 3

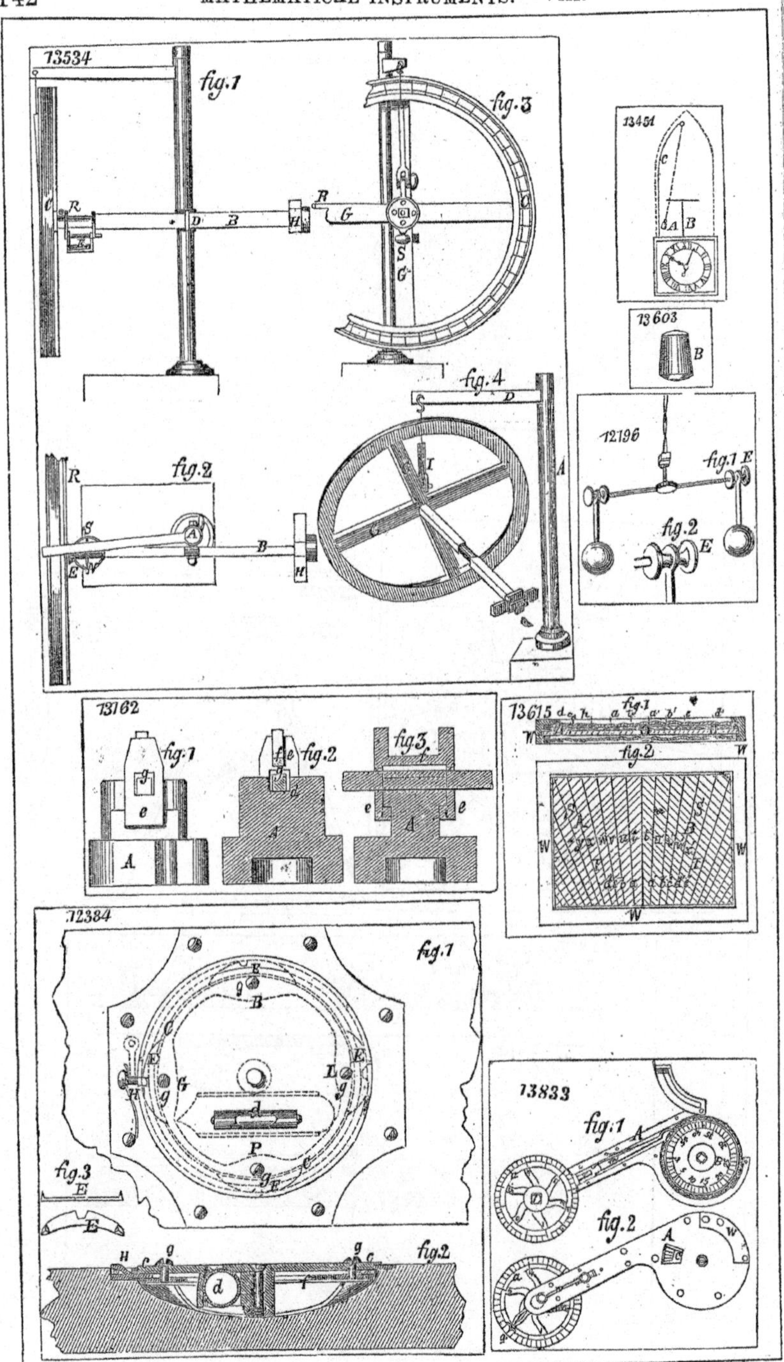
13534
fig.1
fig.3
fig.2
fig.4
13451
13603
12196
fig.1
fig.2
13162
fig.1
fig.2
fig.3
13615
fig.1
fig.2
12384
fig.1
fig.3
fig.2
13833
fig.1
fig.2

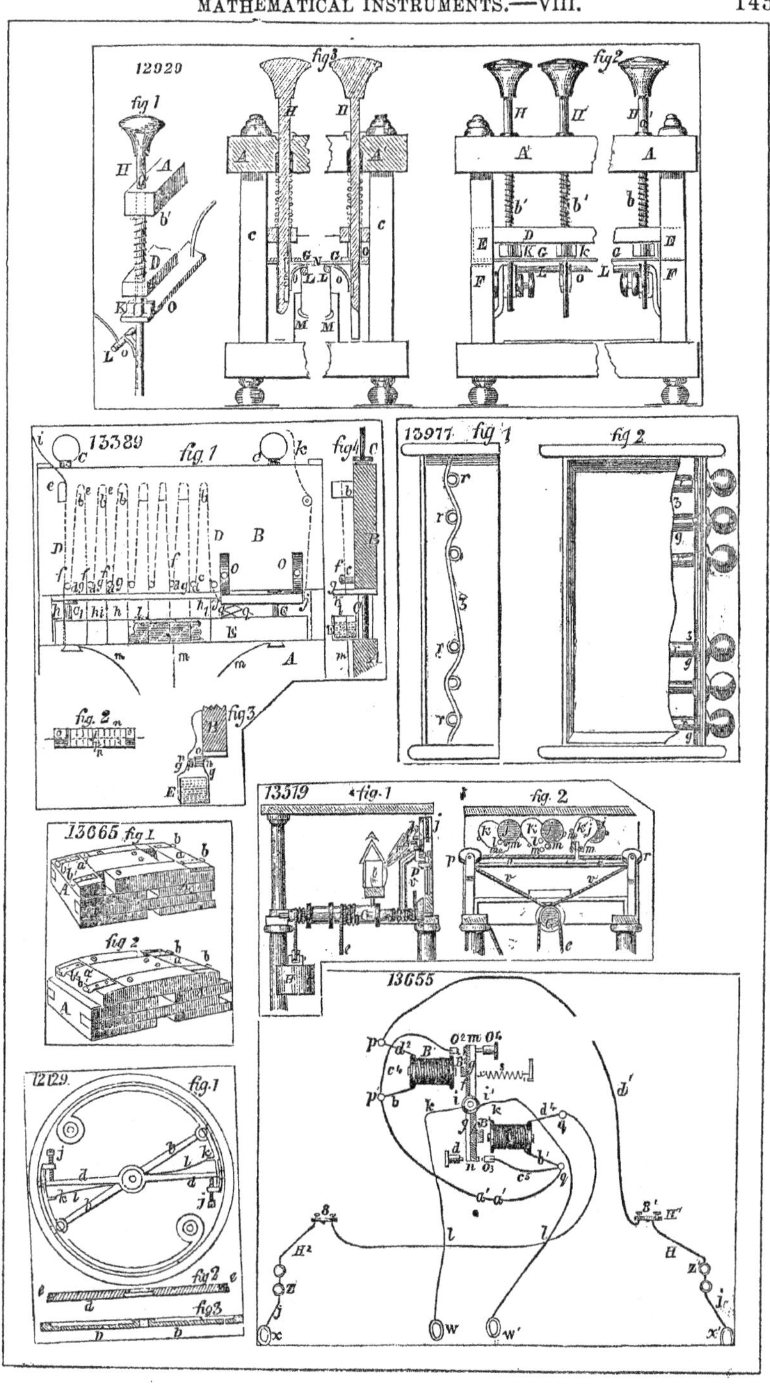
12920
fig 1
fig 3
fig 2
13389
fig 1
fig 4
fig. 2
fig 3
13977
fig 1
fig 2
13519
fig. 1
fig. 2
13665
fig 1
fig 2
13655
12129
fig. 1
fig 2
fig 3

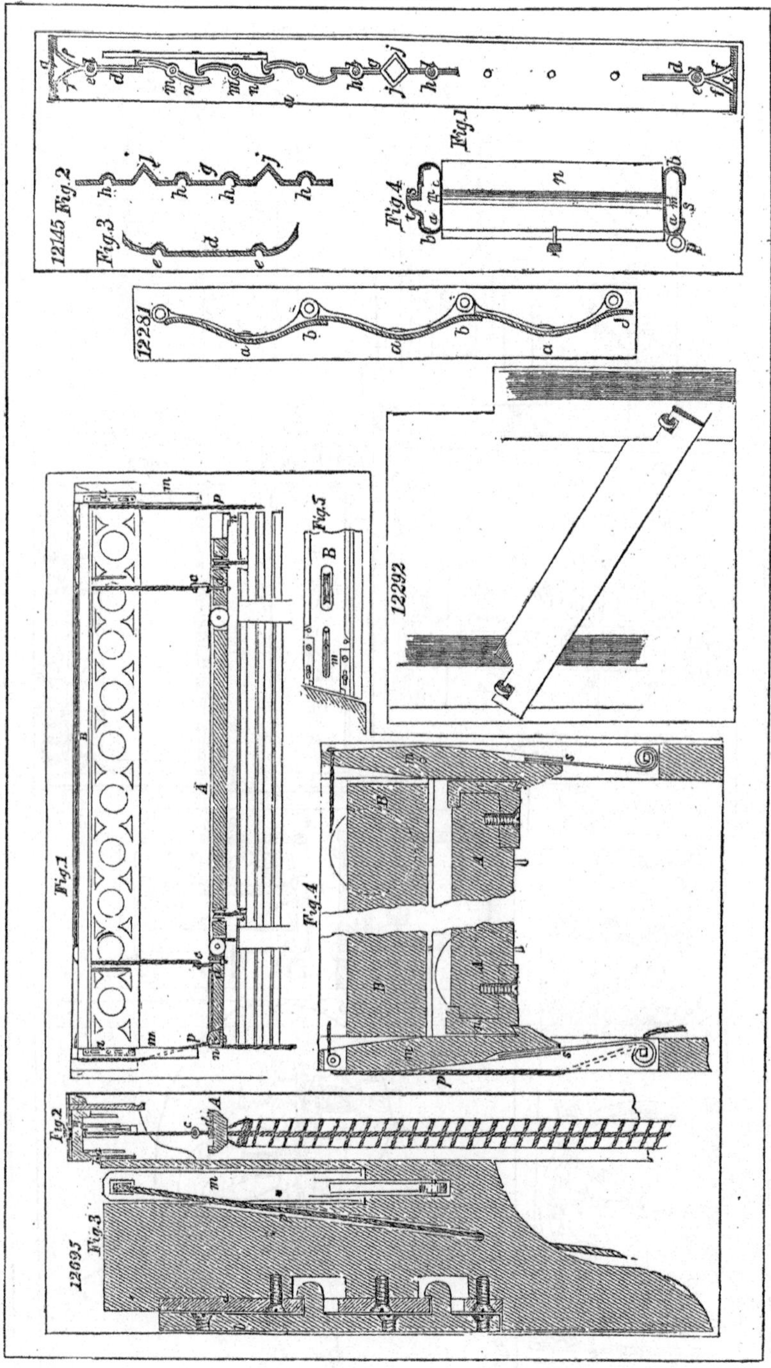

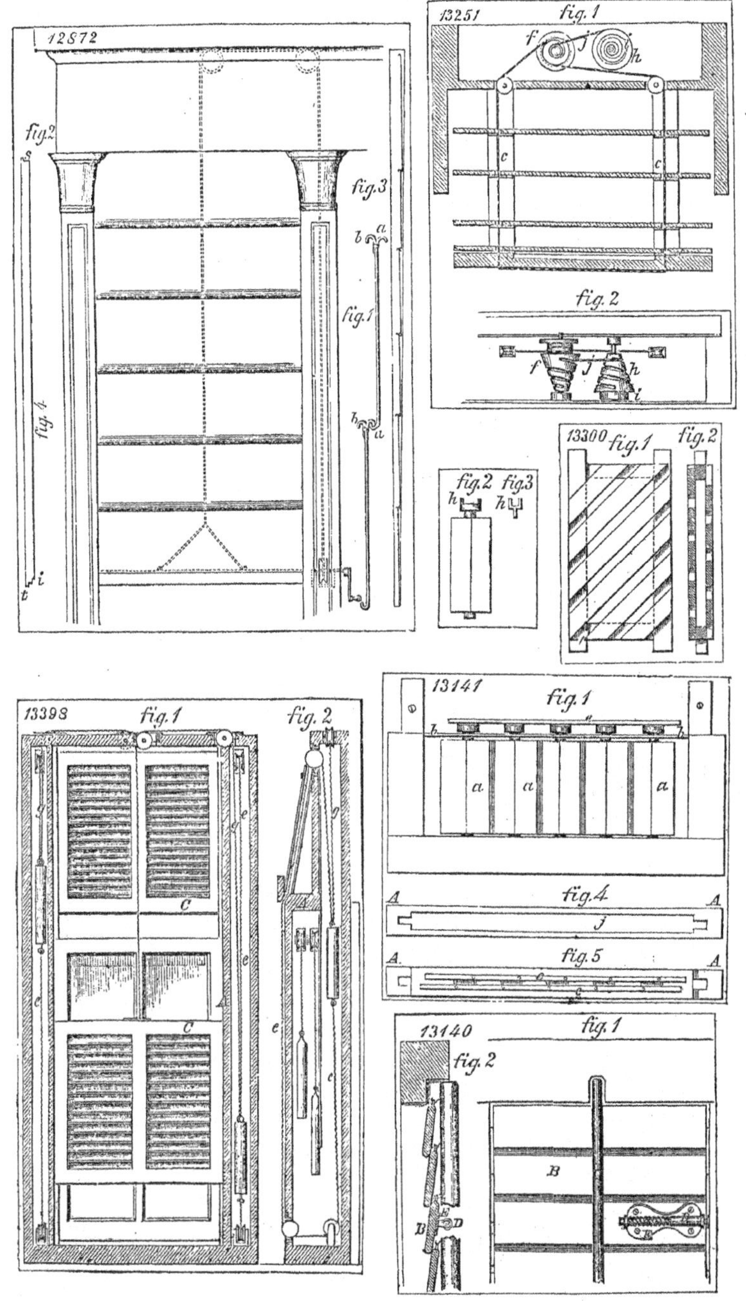
12872
fig.2
fig.3
fig.1
fig.4
13251
fig.1
fig.2
13300
fig.1
fig.2
fig.2
fig.3
13398
fig.1
fig.2
13141
fig.1
fig.4
fig.5
13140
fig.1
fig.2

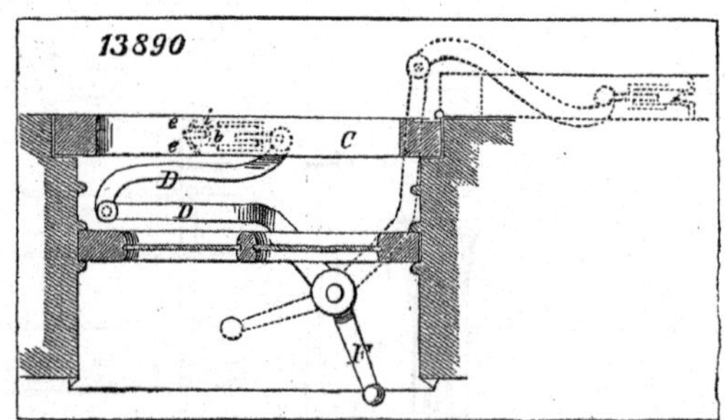
13890
C
D
D
F

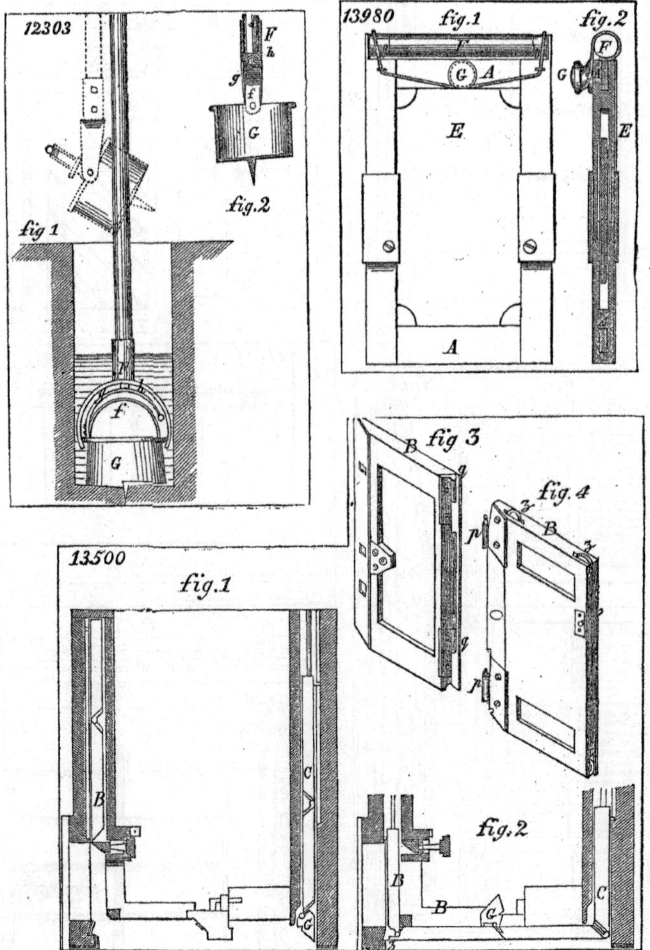
12303
fig 1
fig.2
G
F
13980
fig.1
fig.2
E
A
fig 3
B
fig. 4
B
13500
fig.1
B
C
fig.2
B
B
C

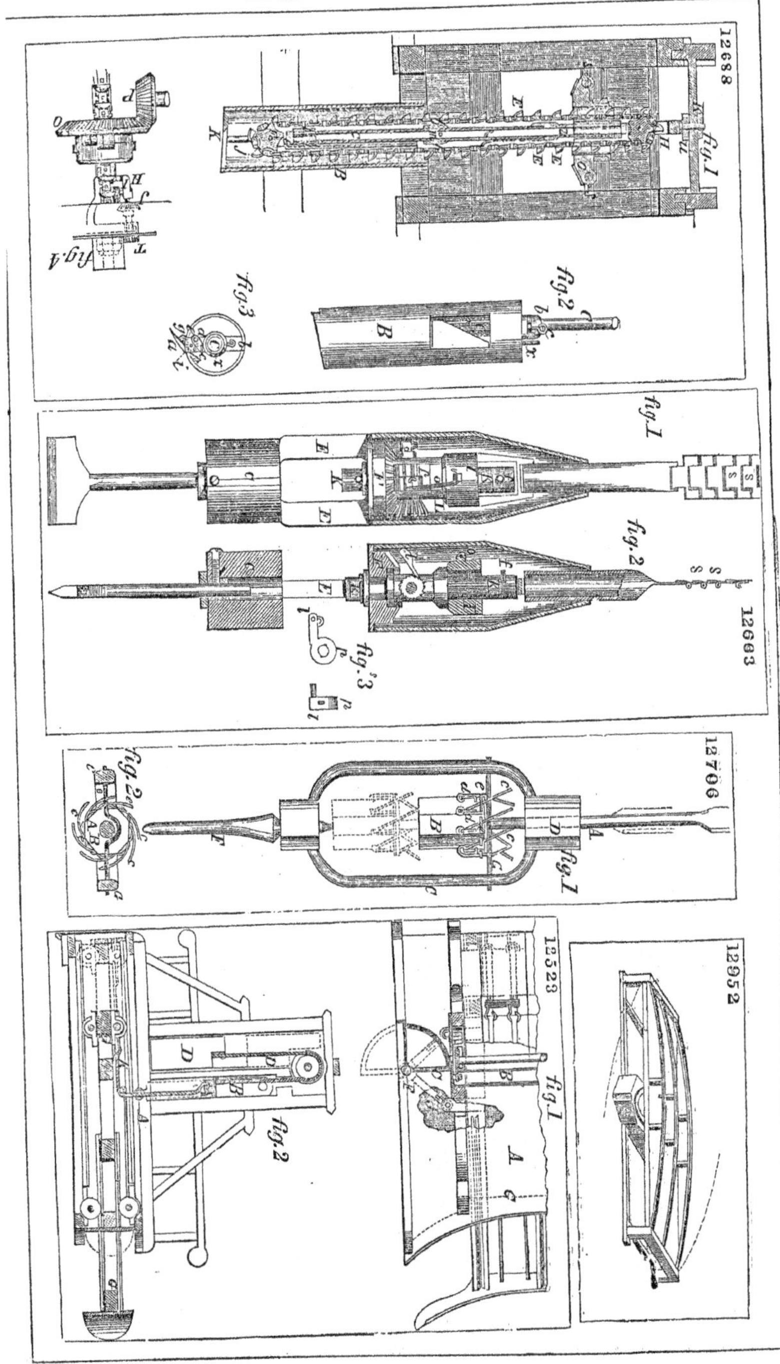
12688
12663
12706
12523
12952

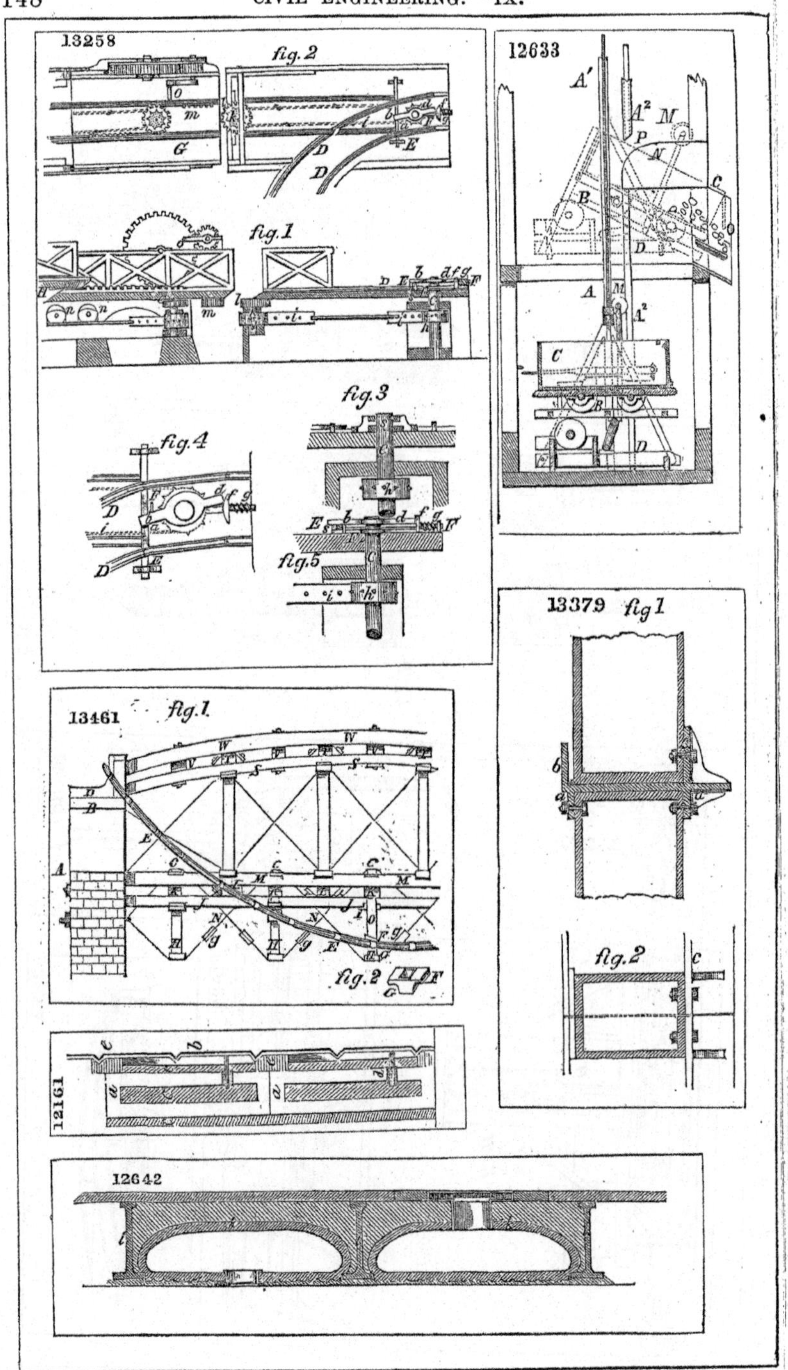
13258
fig. 2
fig. 1
fig. 3
fig. 4
fig. 5
12633
13379 fig 1
fig. 2
13461 fig. 1
fig. 2
12161
12642

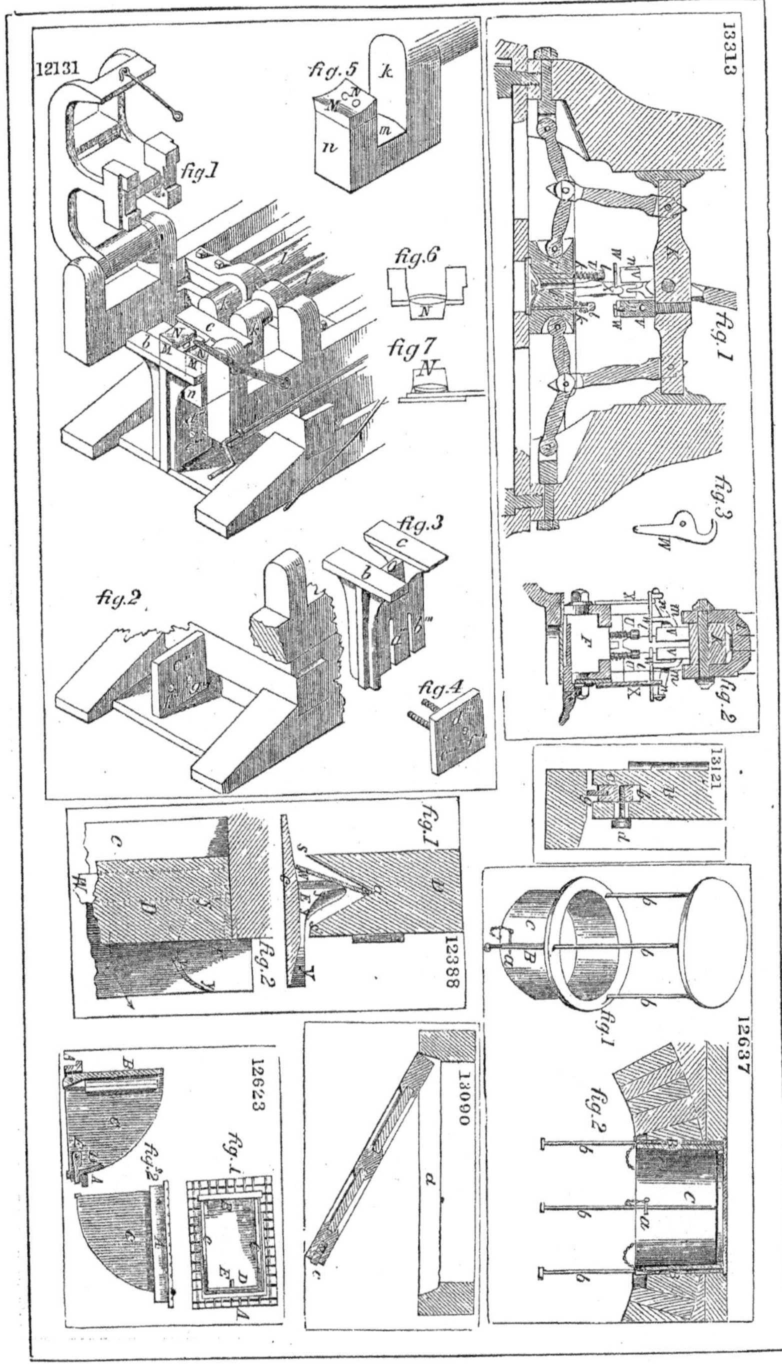

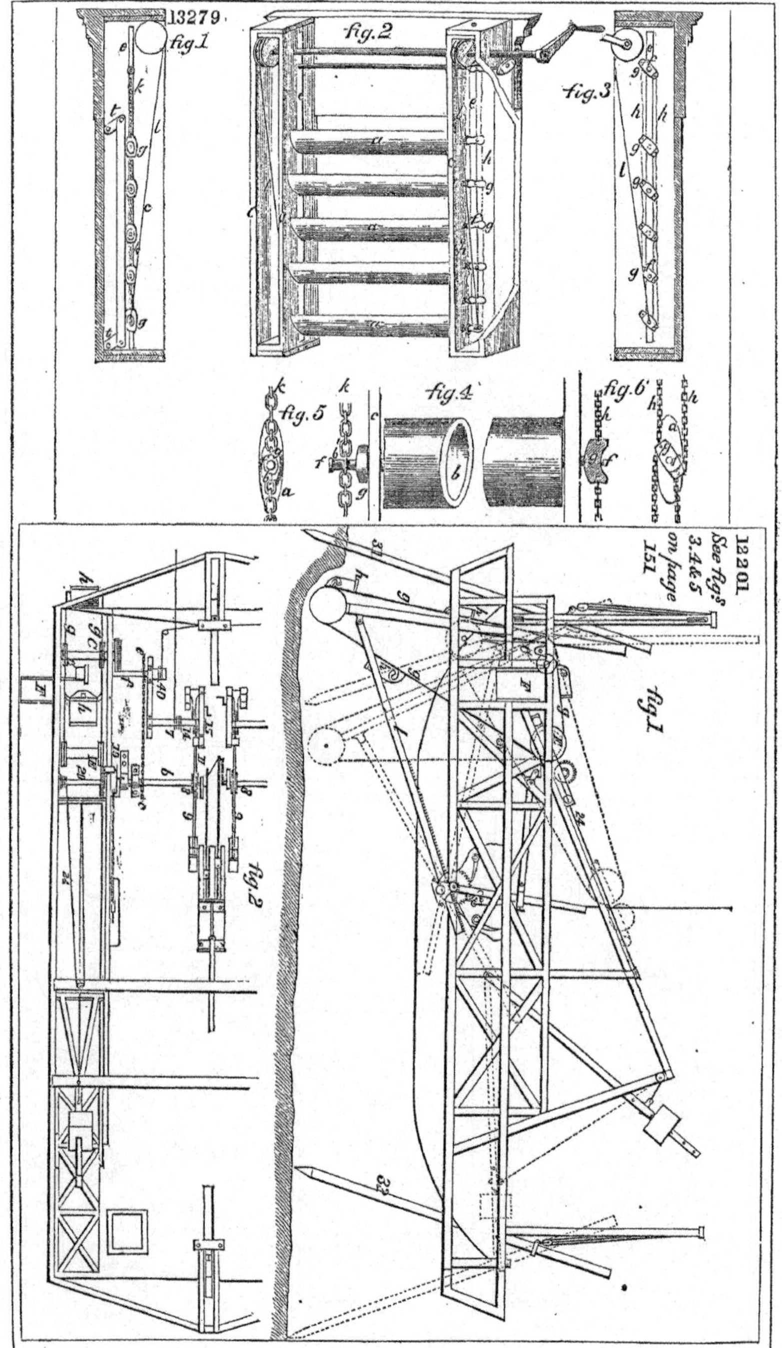
13279
fig.1
fig.2
fig.3
fig.4
fig.5
fig.6
12201
See figs 3.4 & 5 on page 151
fig.1
fig.2

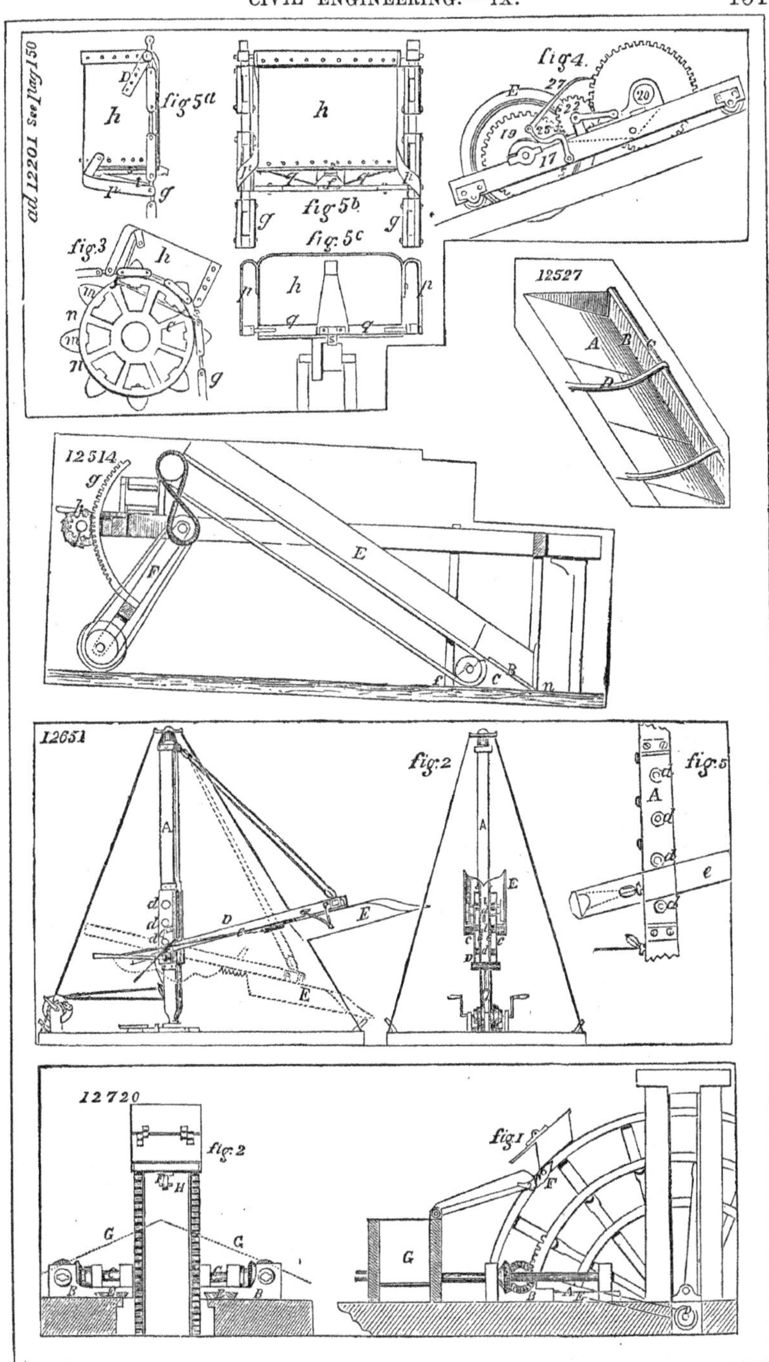
ad 12201 See Pag 150
fig 5a
fig 5b
fig 5c
fig 3
fig 4
12527
12514
12651
fig 2
fig 5
12720
fig 2
fig 1

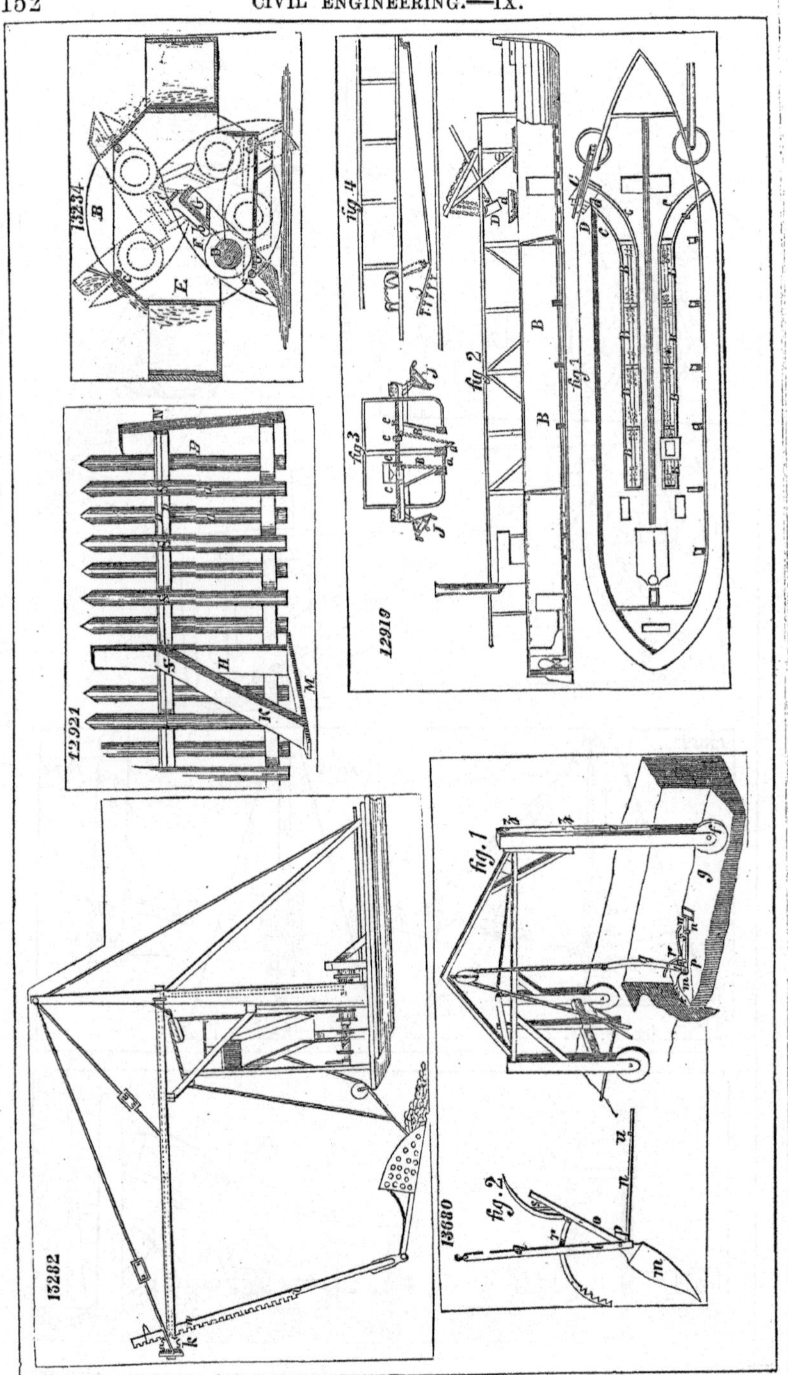
13234
12921
12919
fig. 1
fig. 2
fig. 3
fig. 4
15282
13680
fig. 1
fig. 2

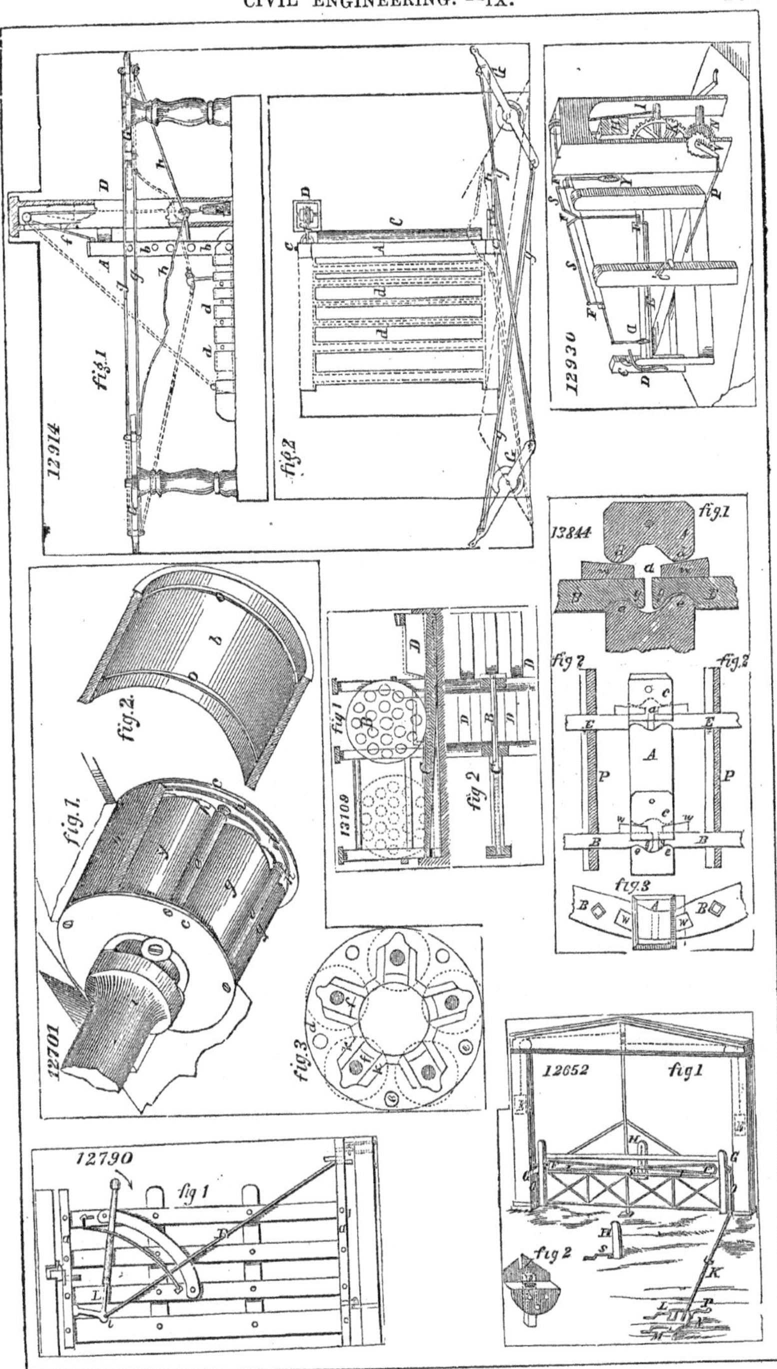

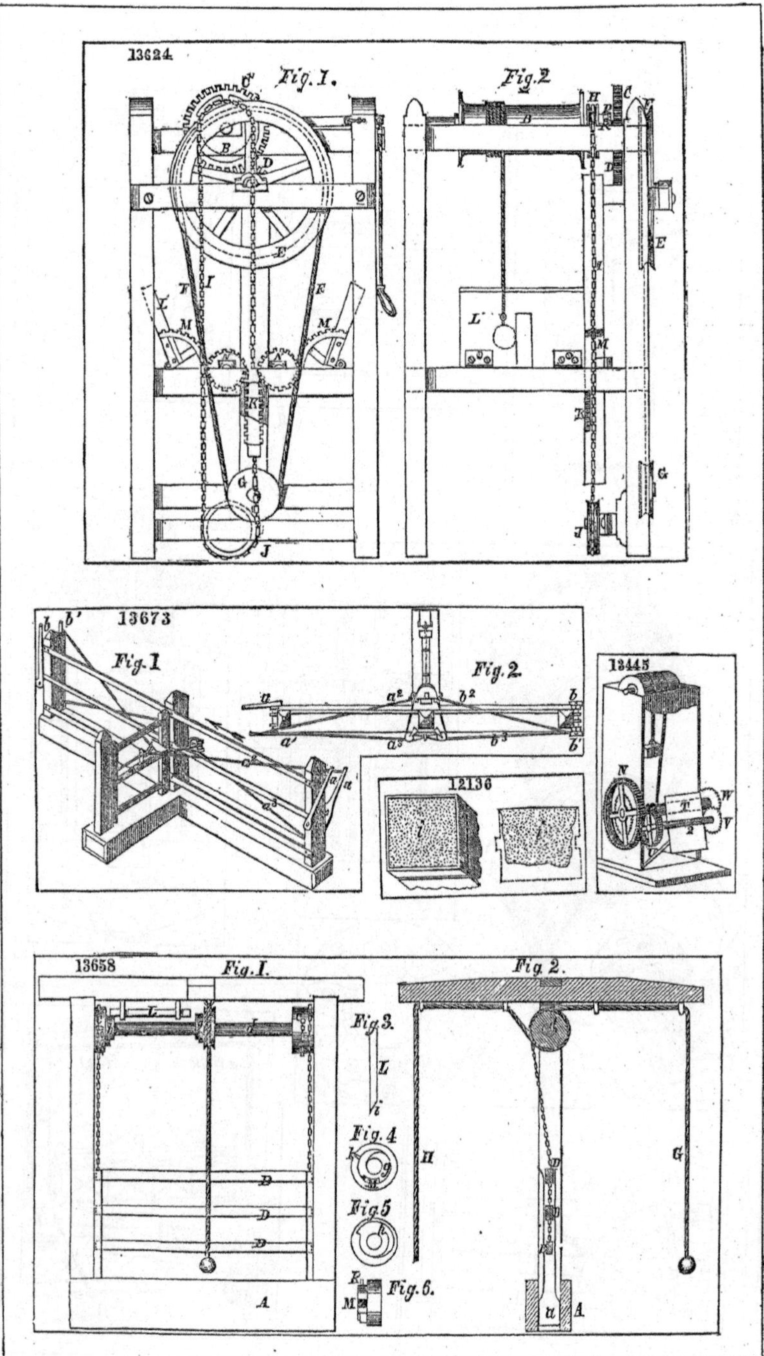
13624
Fig. 1.
Fig. 2
13673
Fig. 1
Fig. 2
13445
12136
13658
Fig. I.
Fig. 2.
Fig. 3.
Fig. 4
Fig. 5
Fig. 6.

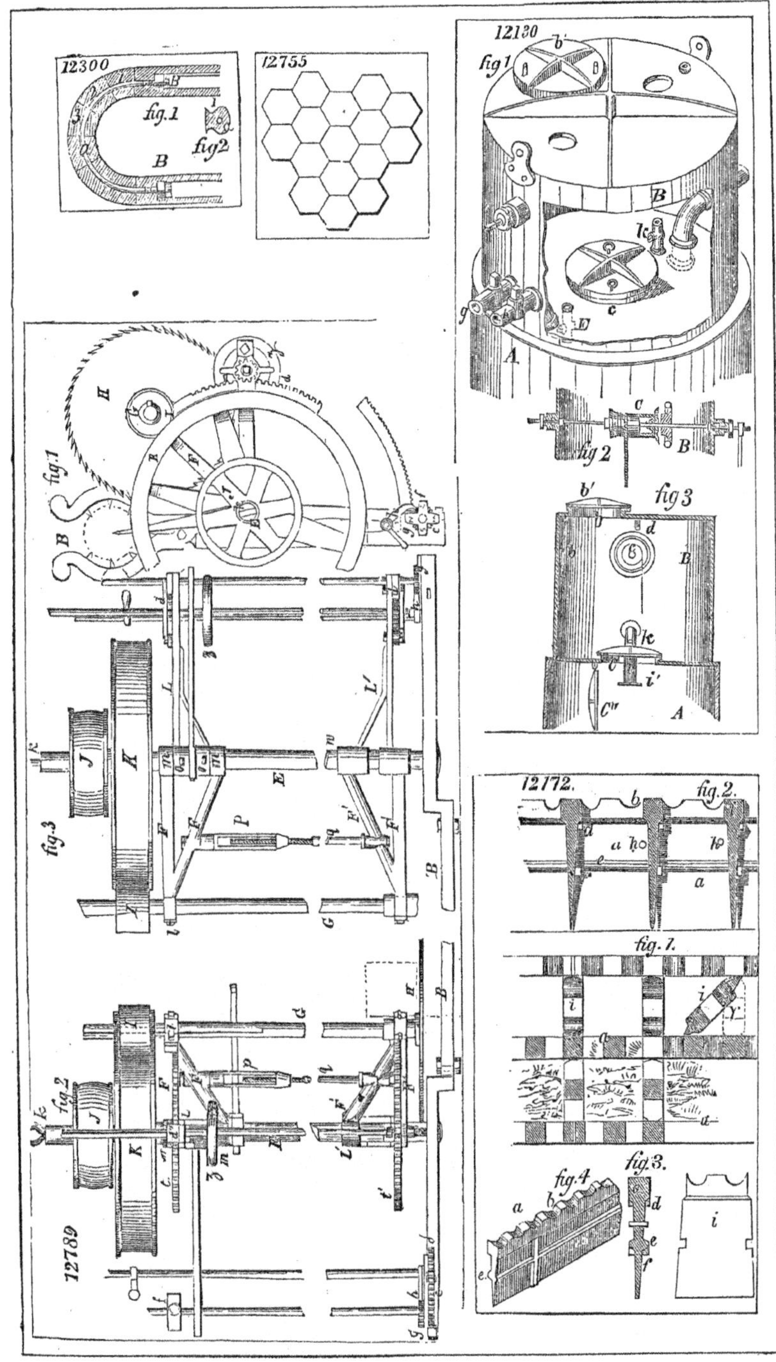
12300
fig.1
fig2
B
12755
12130
fig1
fig 2
fig 3
12789
fig.2
fig.3
fig.1
12172.
fig.2.
fig. 1.
fig.4
fig.3.

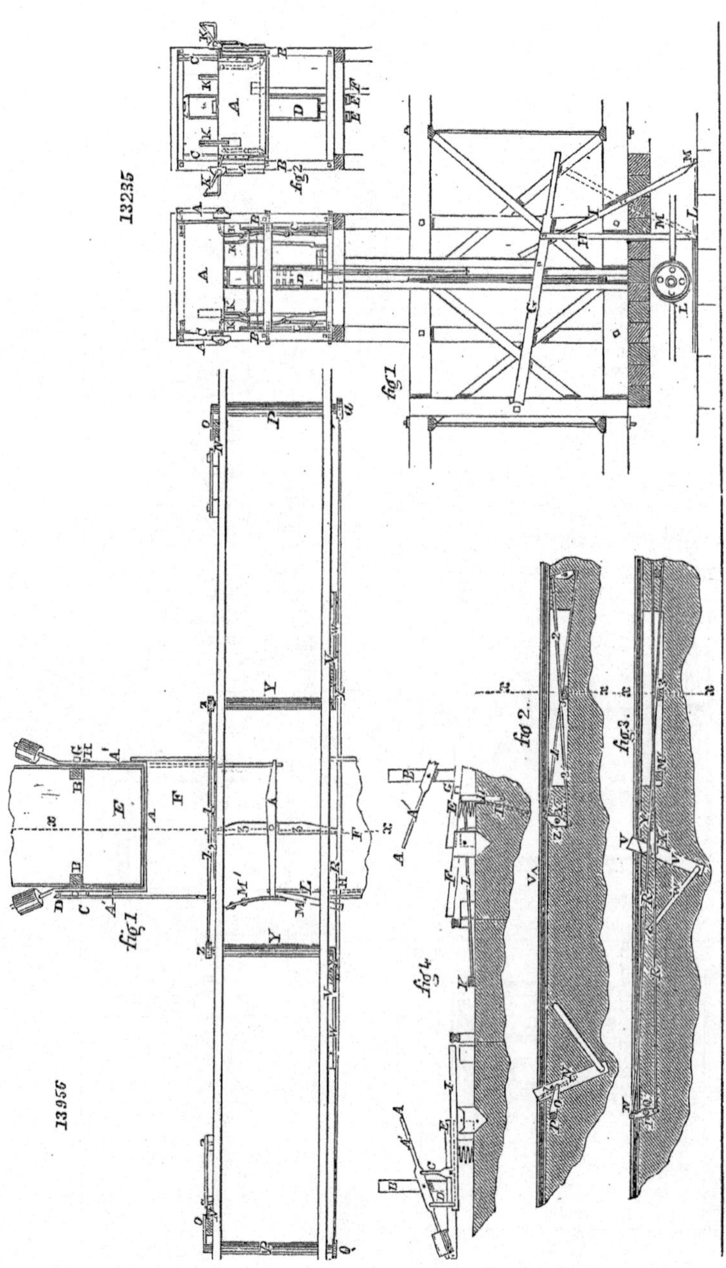
13235
13956

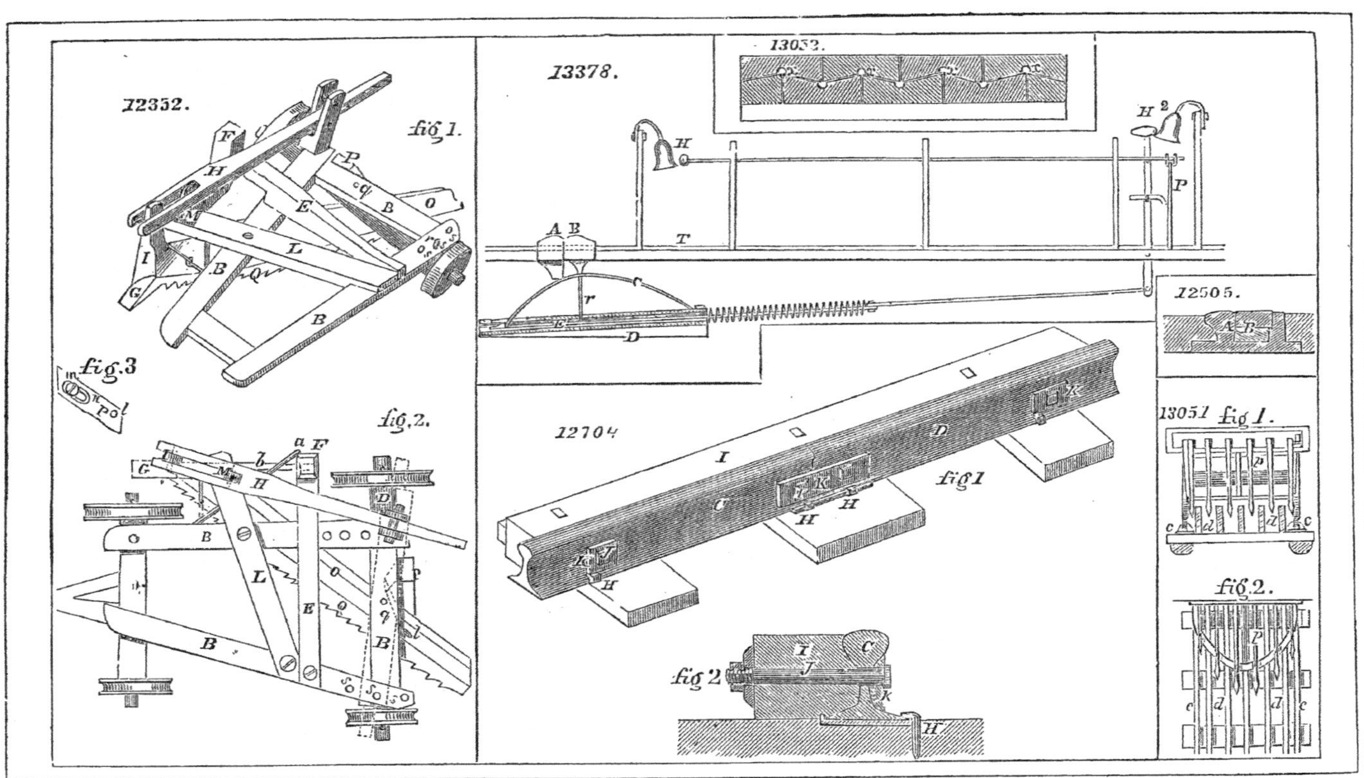
12352.
fig 1.
fig 2.
fig 3
13378.
13032.
12505.
12704
fig 1
fig 2
13051 fig 1.
fig 2.

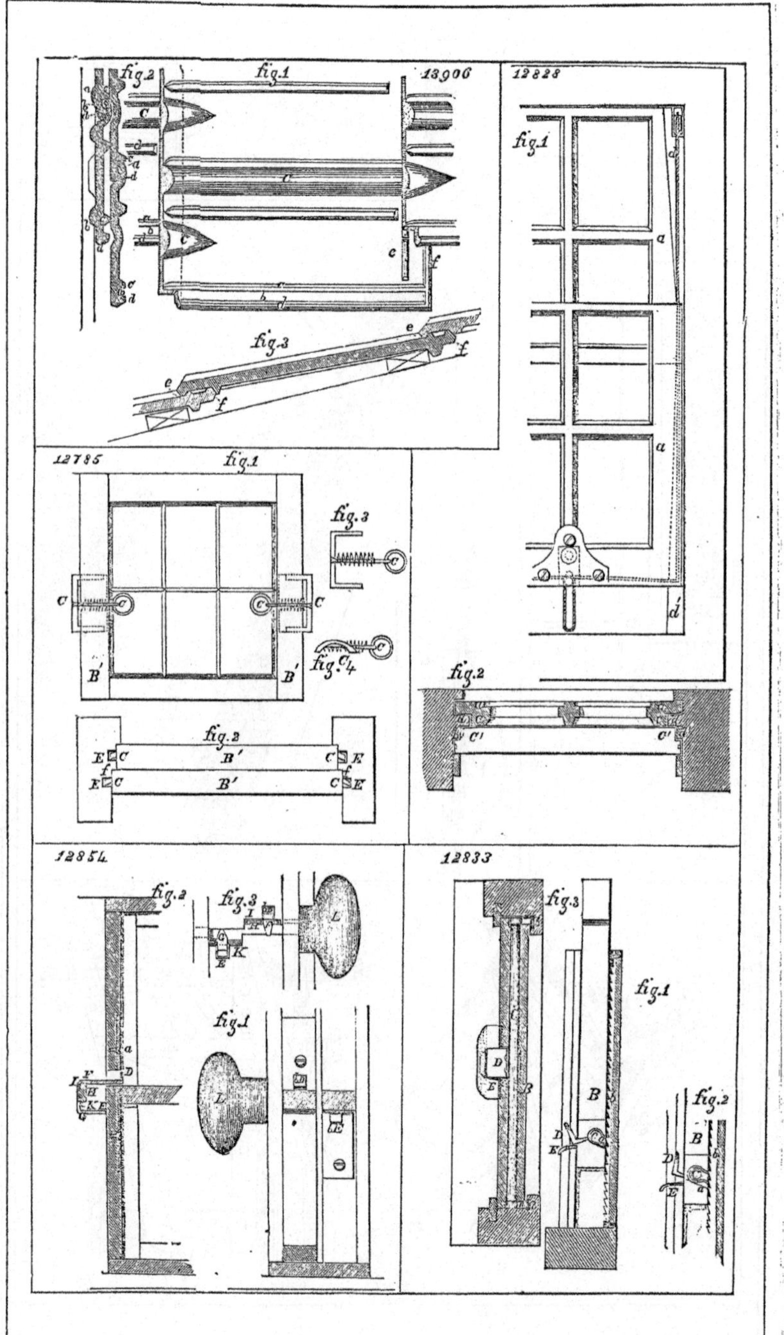
fig.2
fig.1
13906
12828
fig.1
fig.3
12785
fig.1
fig.3
fig.4
fig.2
fig.2
12854
fig.2
fig.3
fig.1
12833
fig.3
fig.1
fig.2

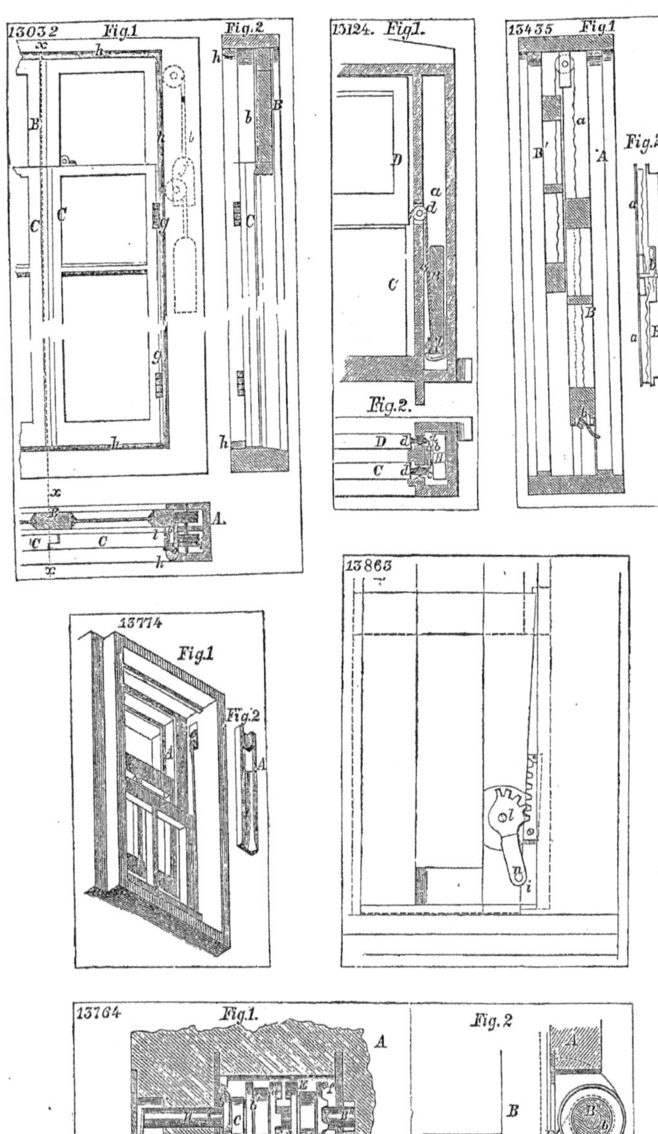
13032
Fig.1
Fig.2
13124.
Fig.1.
Fig.2.
13435
Fig.1
Fig.2
13774
Fig.1
Fig.2
13863
13764
Fig.1.
Fig.2

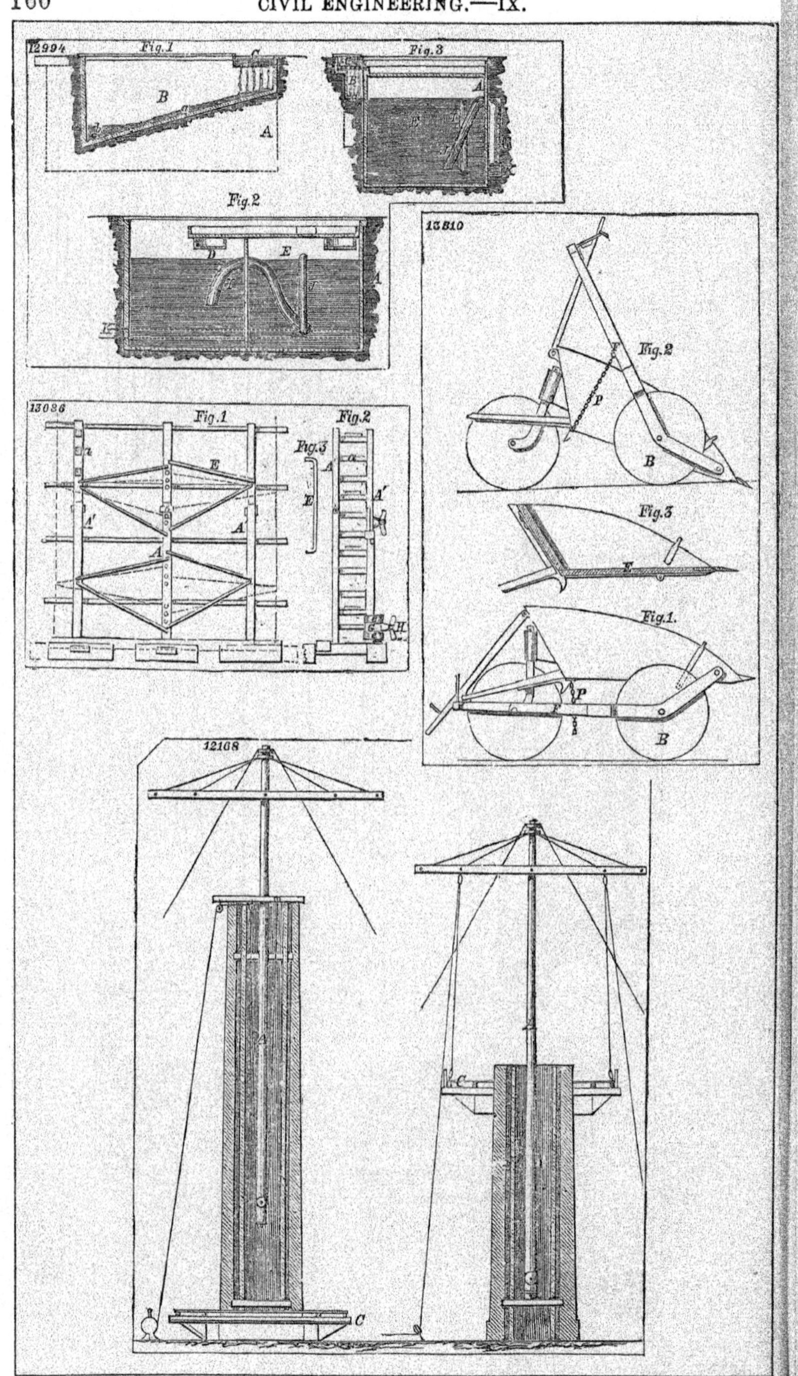
12994
Fig.1
Fig.3
Fig.2
13810
Fig.2
Fig.3
Fig.1.
13086
Fig.1
Fig.2
Fig.3
12168

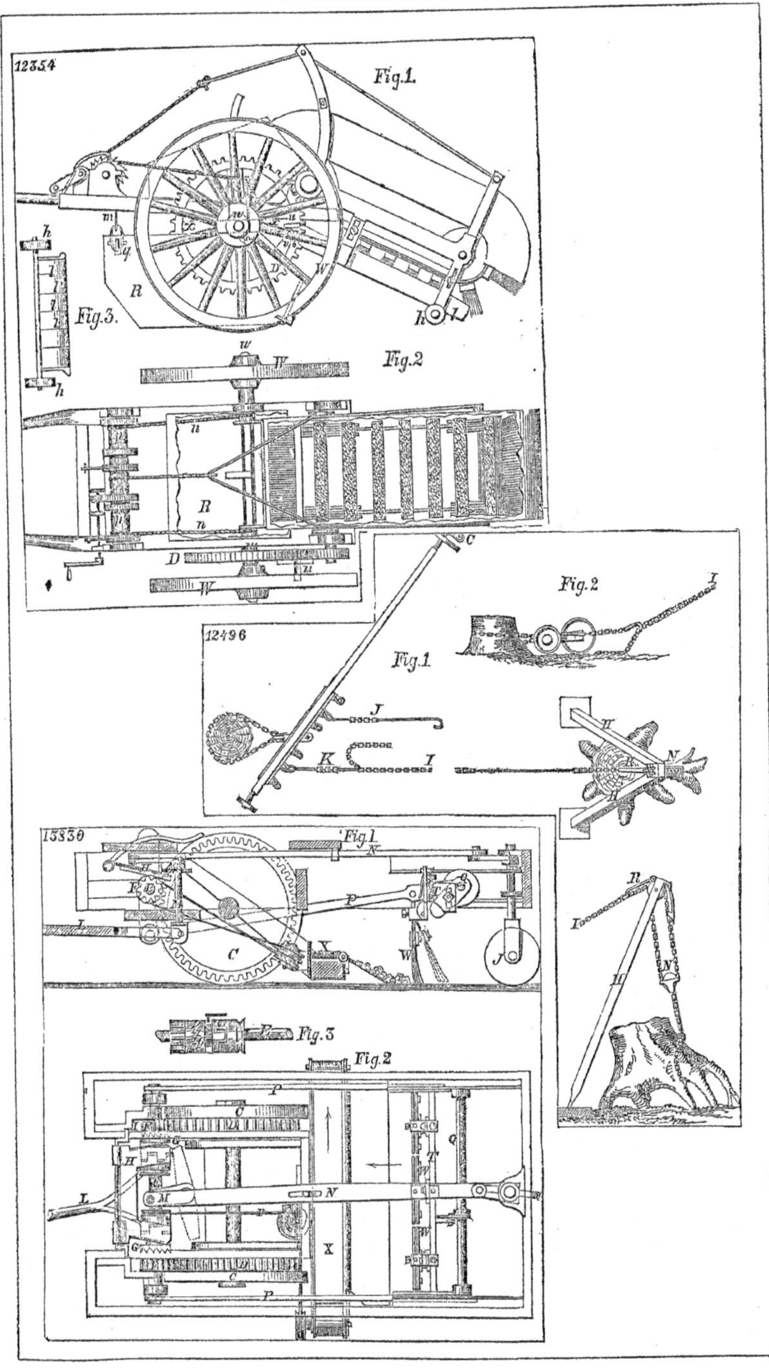
12354
Fig.1.
Fig.2
Fig.3.
12496
Fig.1
Fig.2
15830
Fig.1
Fig.2
Fig.3

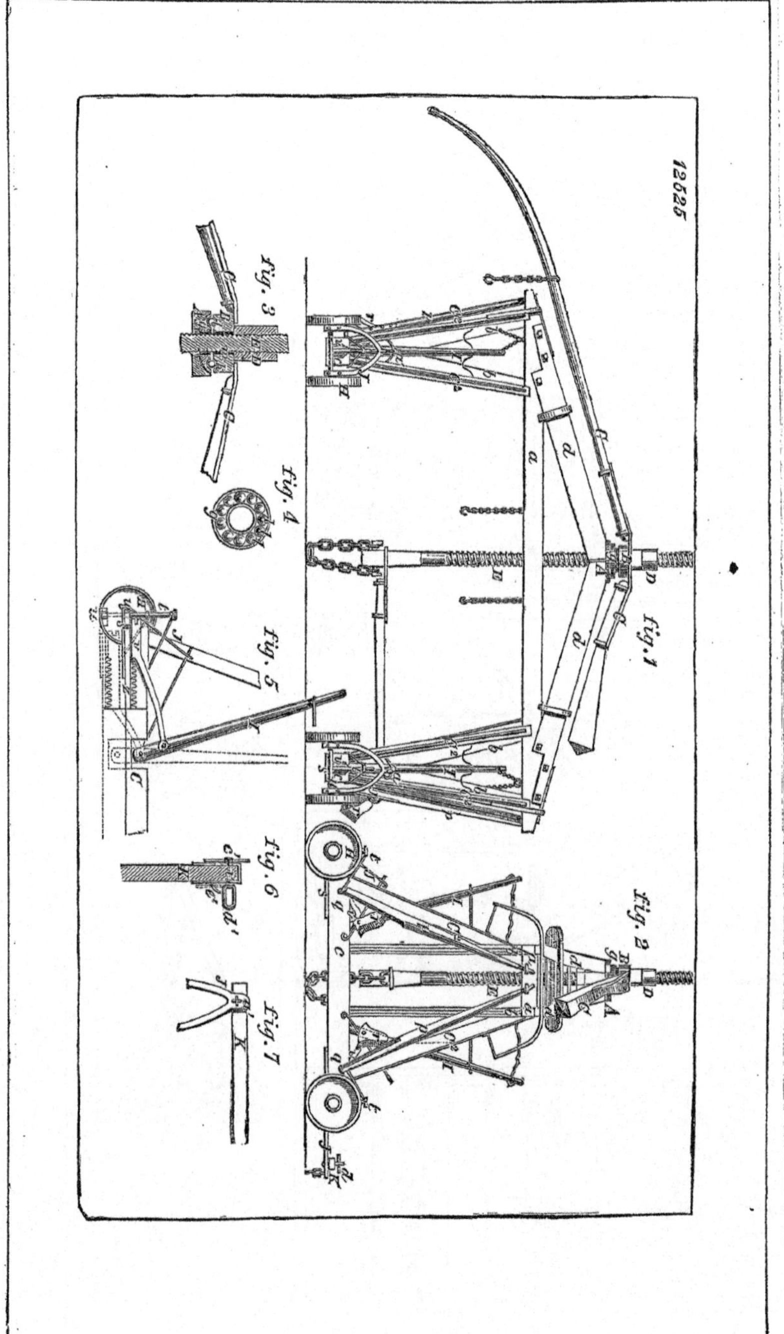
12525
fig. 1
fig. 2
fig. 3
fig. 4
fig. 5
fig. 6
fig. 7

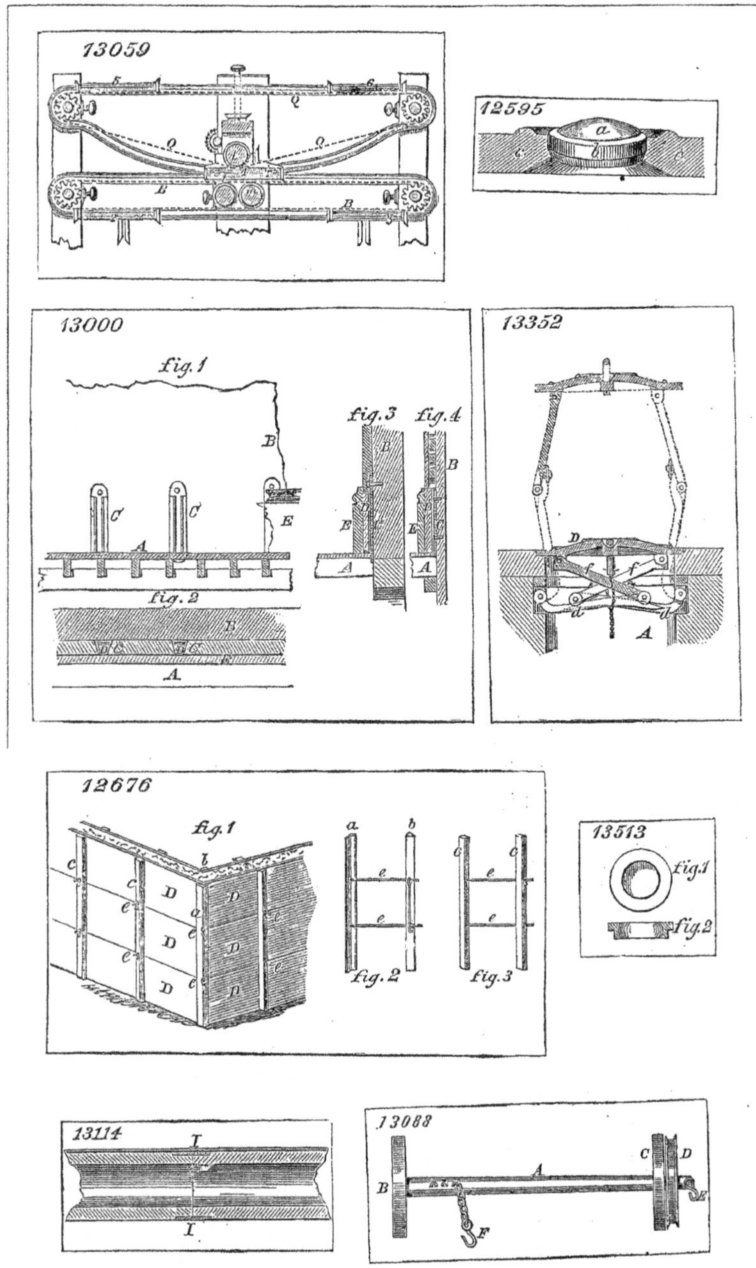
13059
12595
13000
fig. 1
fig. 2
fig. 3
fig. 4
13352
12676
fig. 1
fig. 2
fig. 3
13513
fig. 1
fig. 2
13114
13088

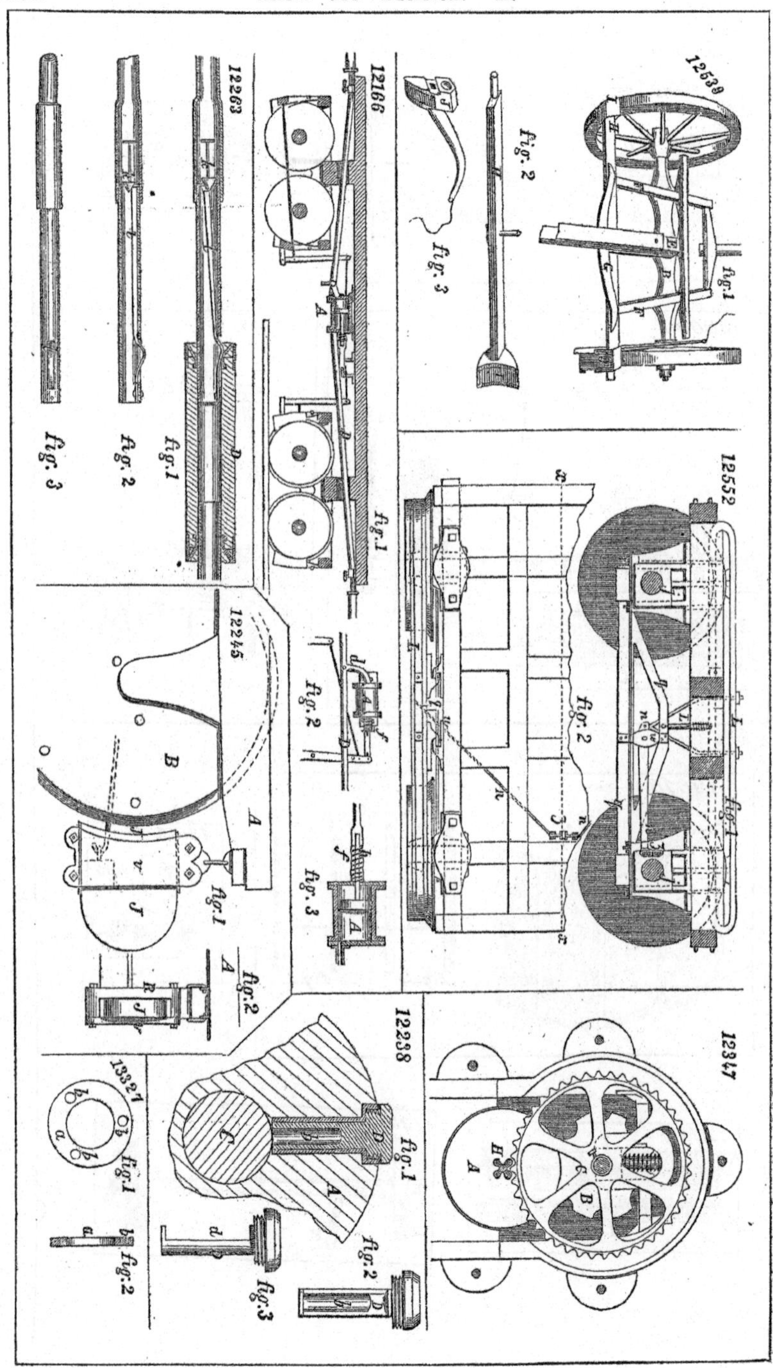
12539
12552
12347
12166
12263
12245
12238
13327

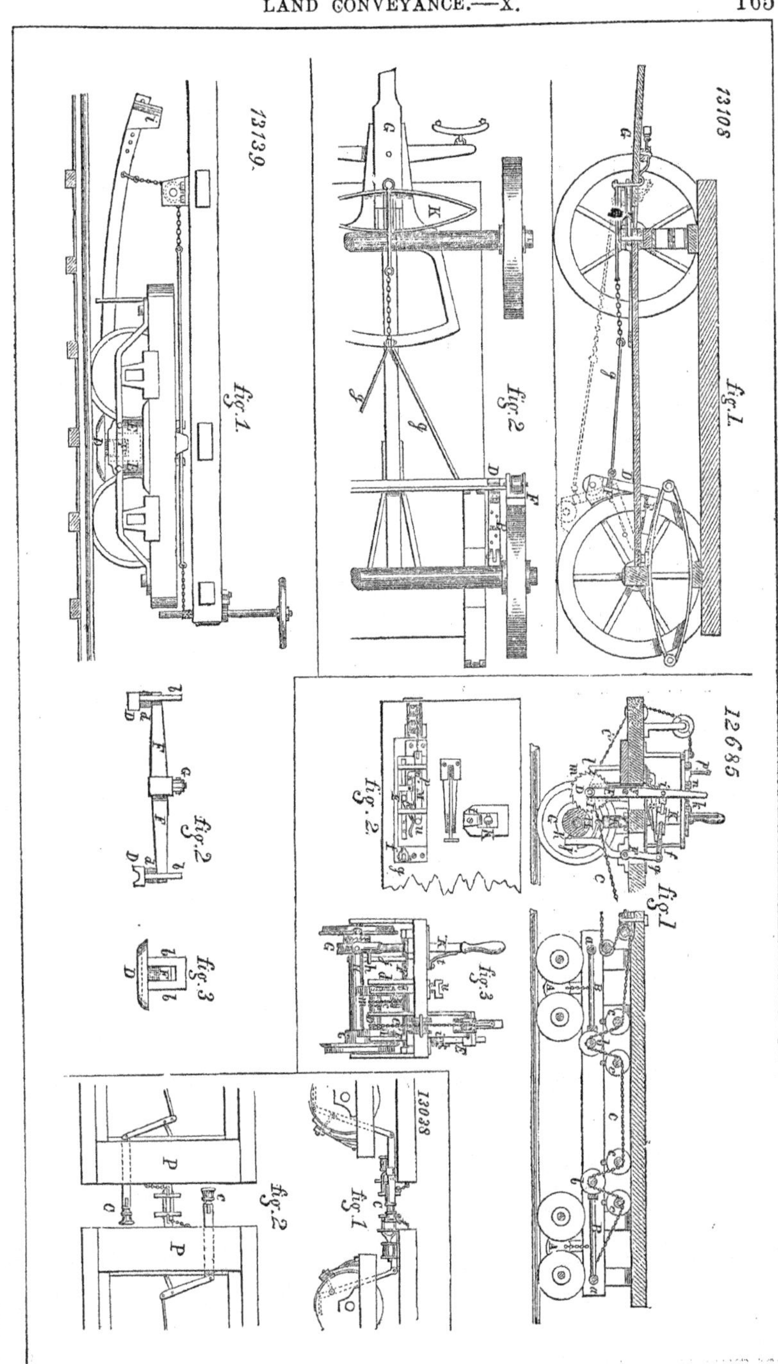
13108
fig.1.
fig.2
13139.
fig.1.
fig.2
fig.3
12685
fig.1
fig.2.
fig.3
13038
fig.1
fig.2

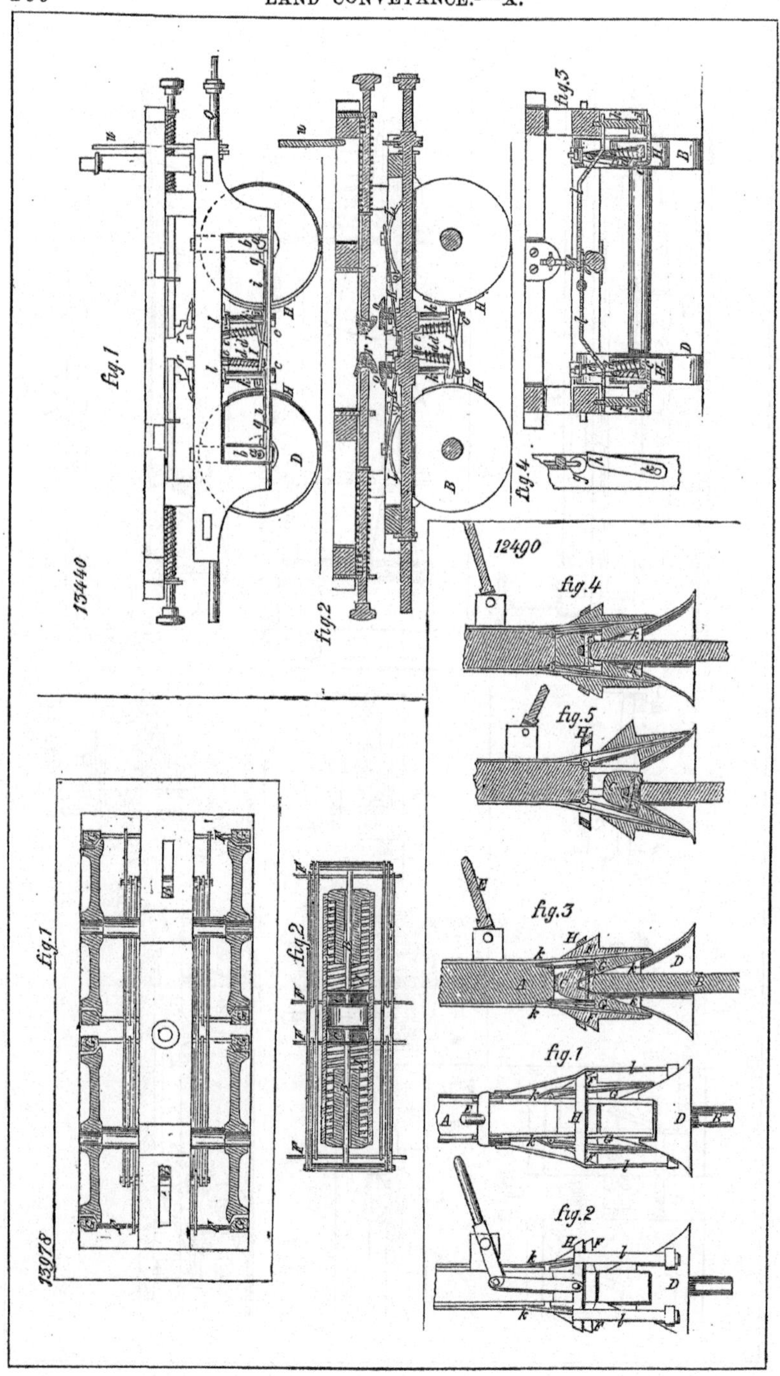
13440
fig.1
fig.2
fig.3
fig.4
12490
fig.4
fig.5
fig.3
fig.1
fig.2
13978
fig.1
fig.2

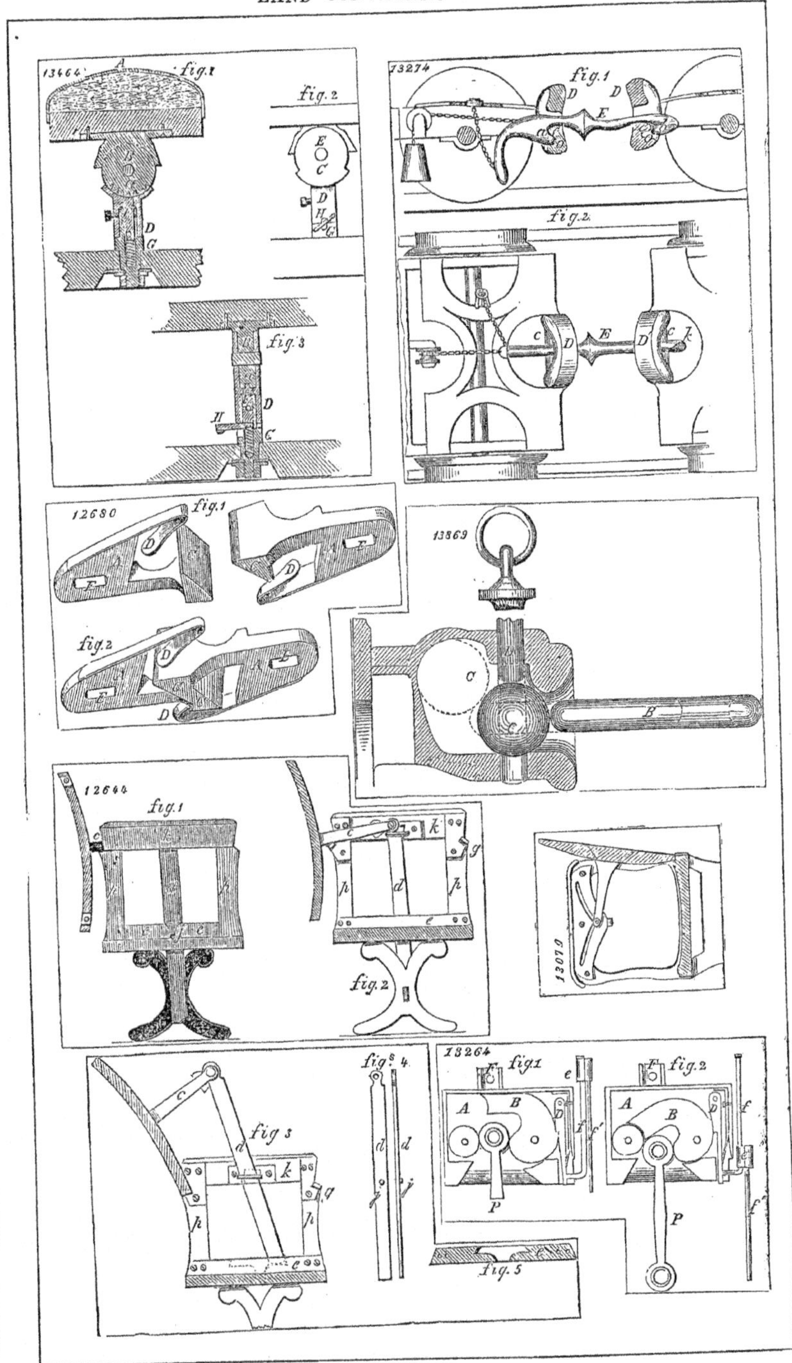
13464
fig. 1
fig. 2
fig. 3
13274
fig. 1
fig. 2
12680
fig. 1
fig. 2
13869
12644
fig. 1
fig. 2
13070
fig. 3
figs 4
fig. 5
13264
fig. 1
fig. 2

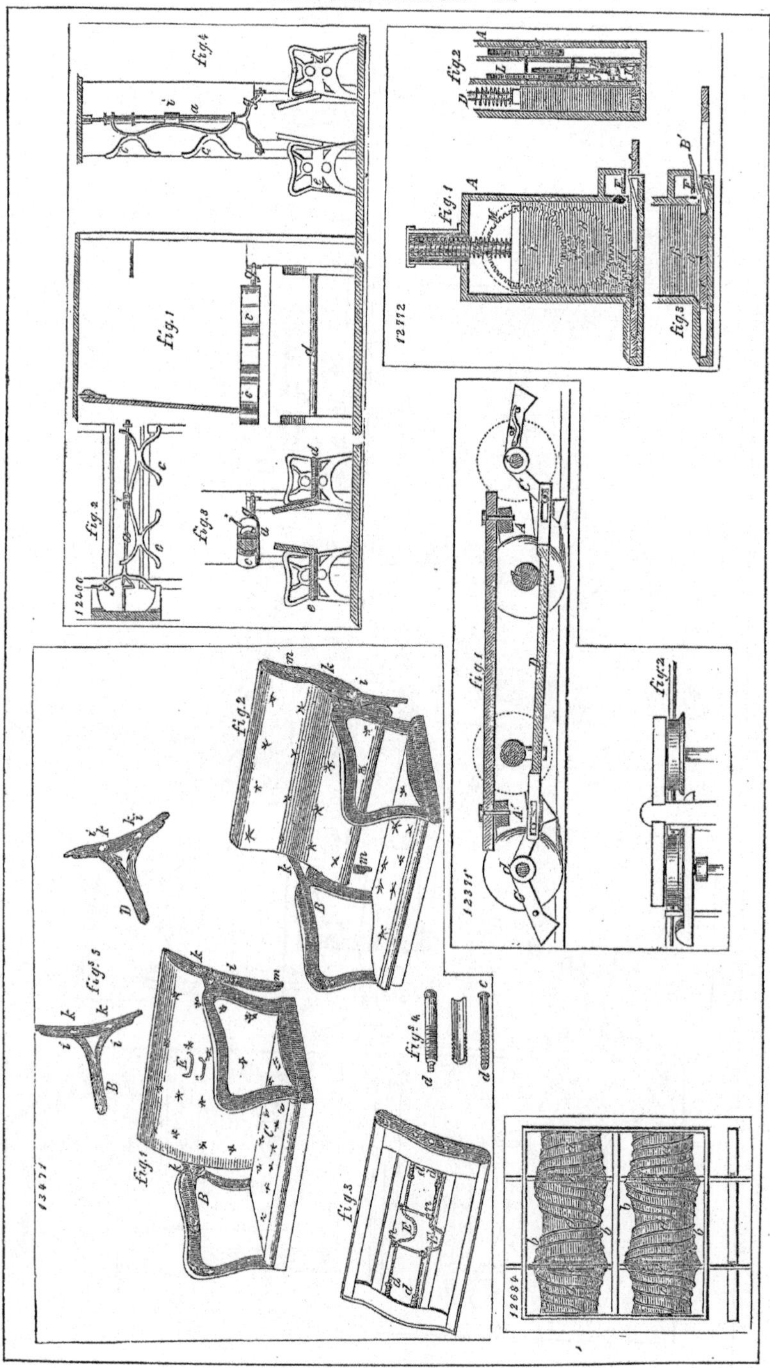

12400
12772
12371
13471
12684

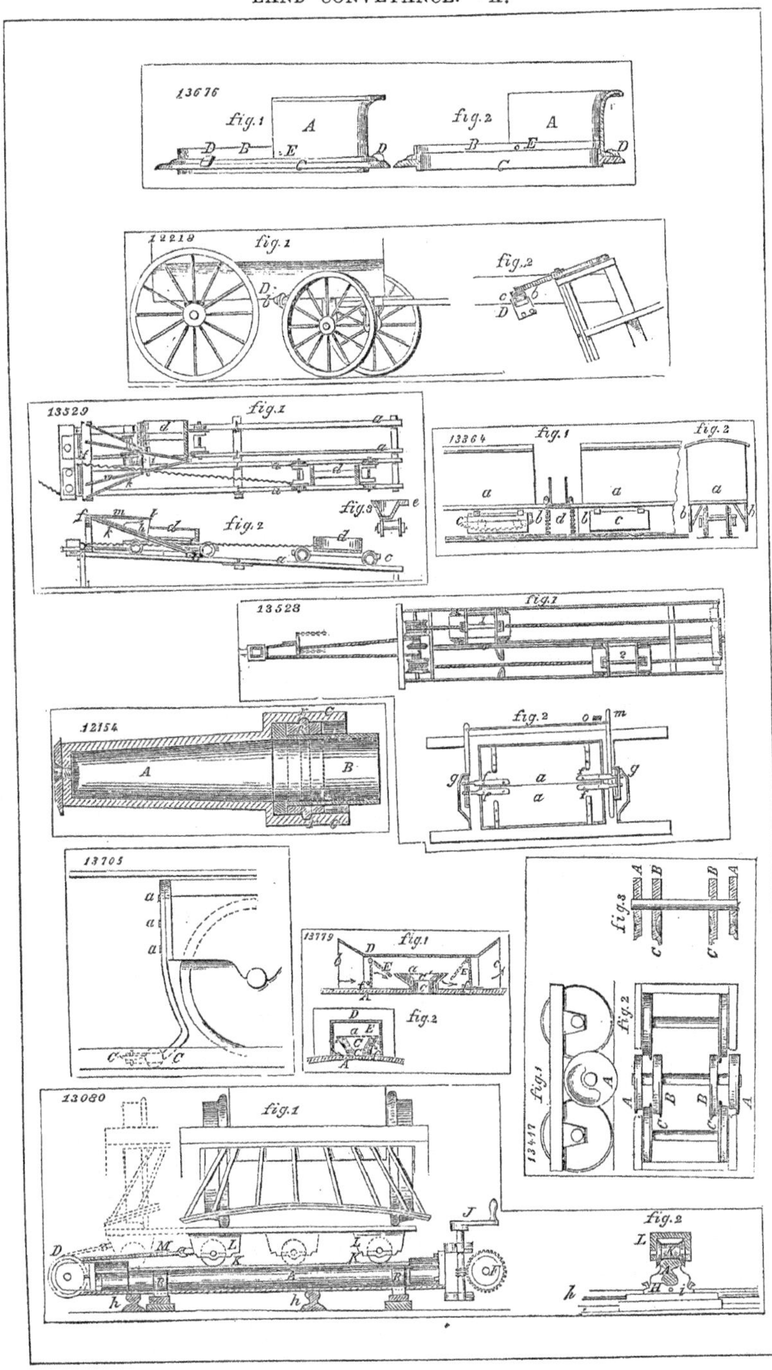

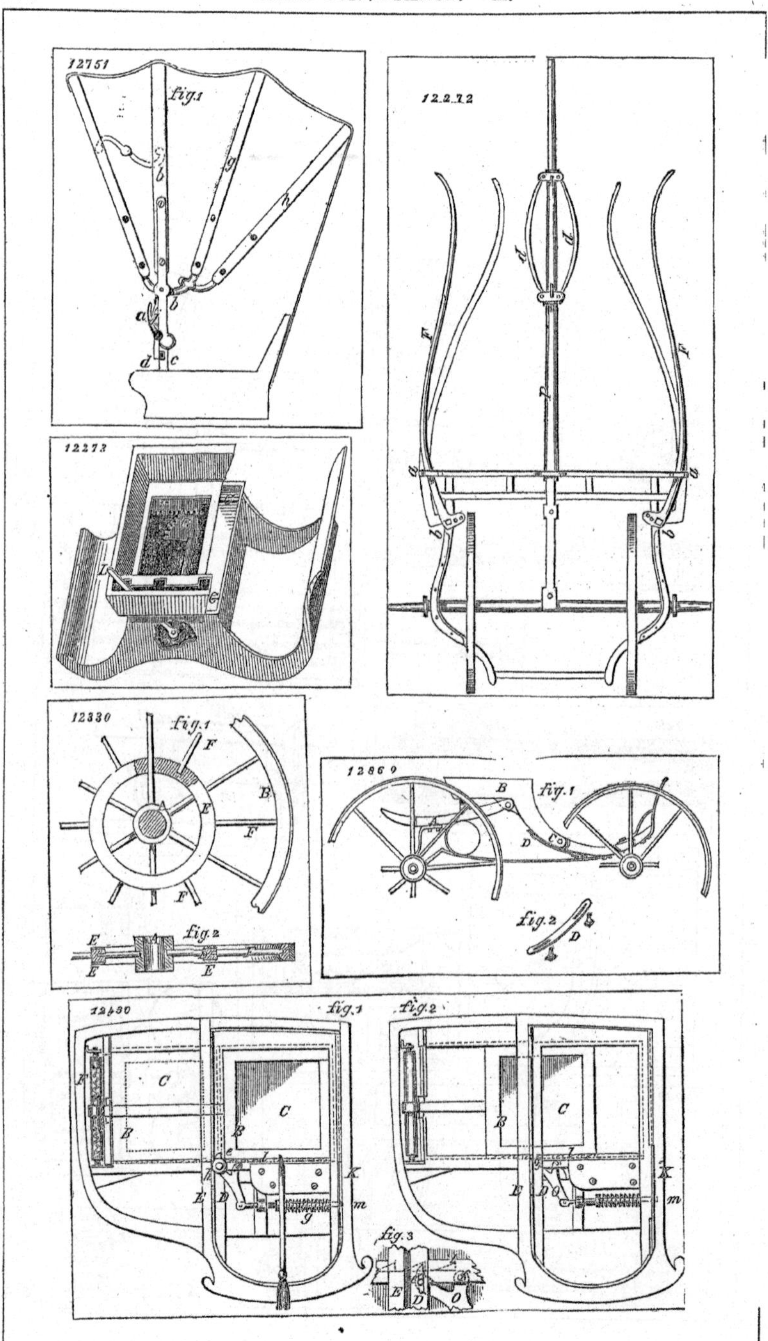
12751
fig.1
12272
12273
12330
fig.1
fig.2
12869
fig.1
fig.2
12430
fig.1
fig.2
fig.3

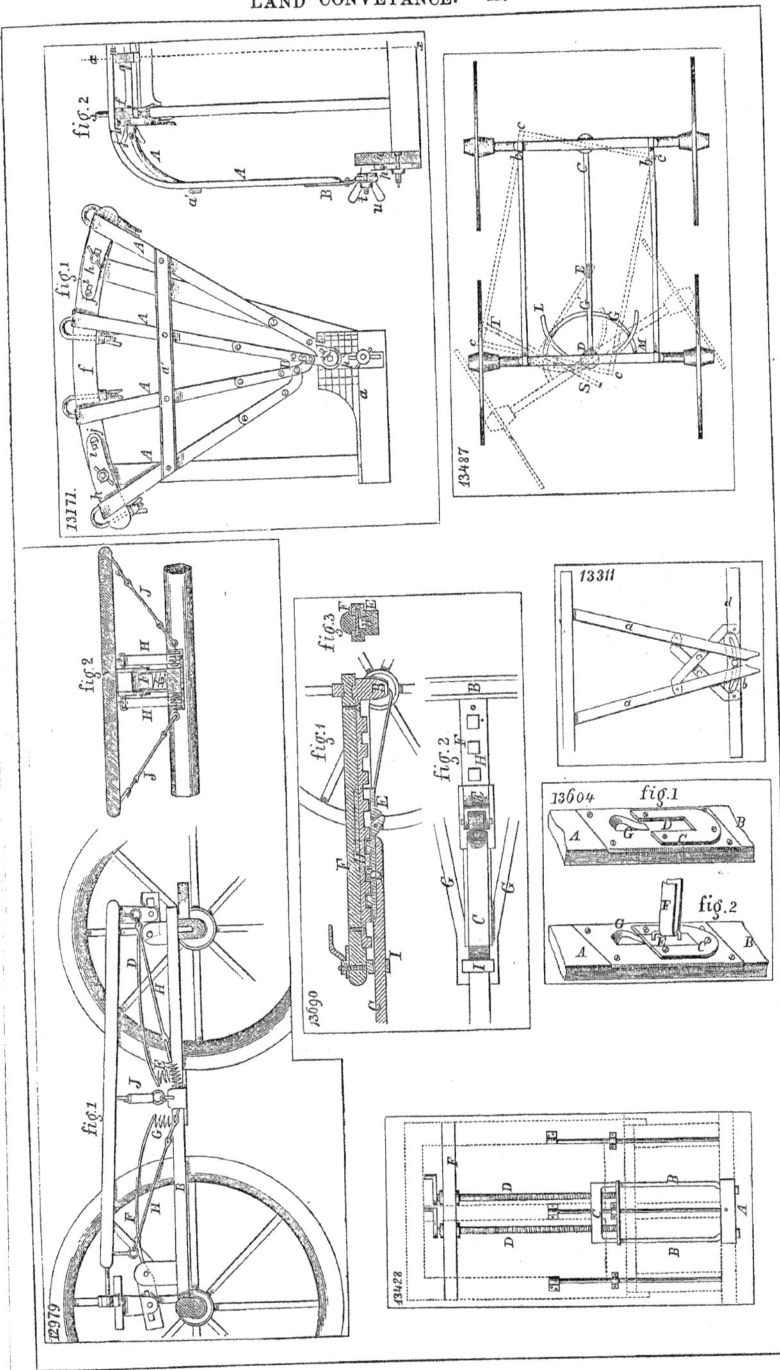

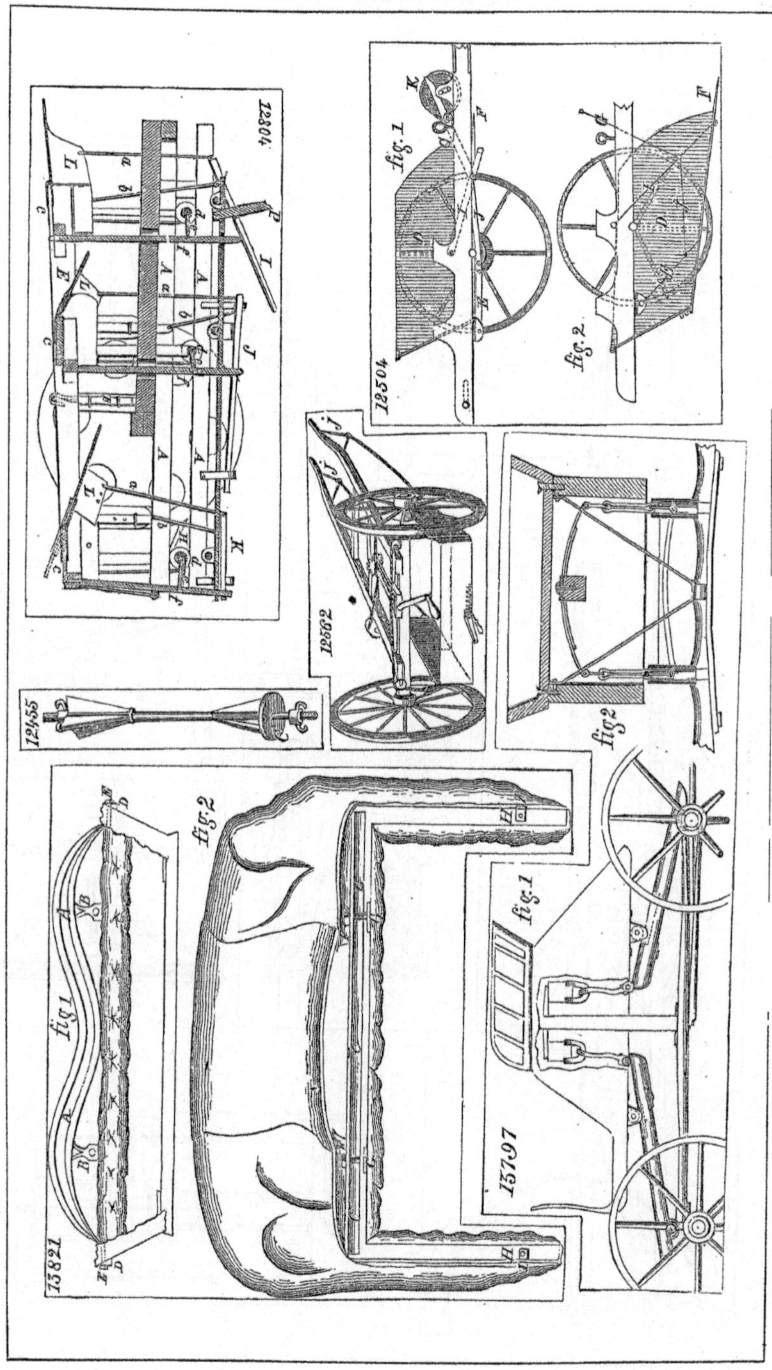
12804
12504
fig. 1
fig. 2
12562
fig 2
12455
13821
fig 1
fig. 2
13797
fig. 1

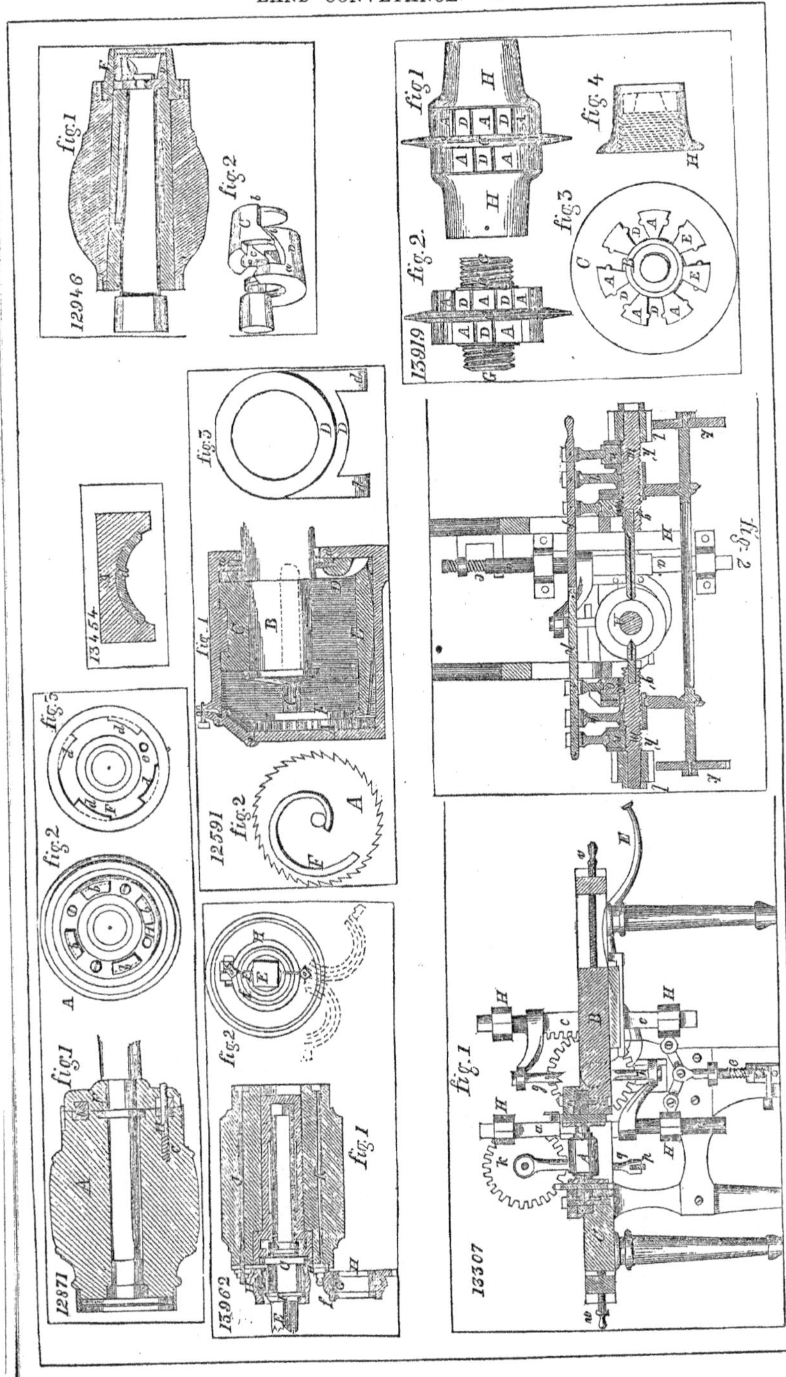
12946
fig.1
fig.2
13919
fig.1
fig.2.
fig.3
fig.4
13454
12591
fig.1
fig.2
fig.3
fig. 2
12871
fig.1
fig.2
fig.3
13962
fig.1
fig.2
13307
fig.1

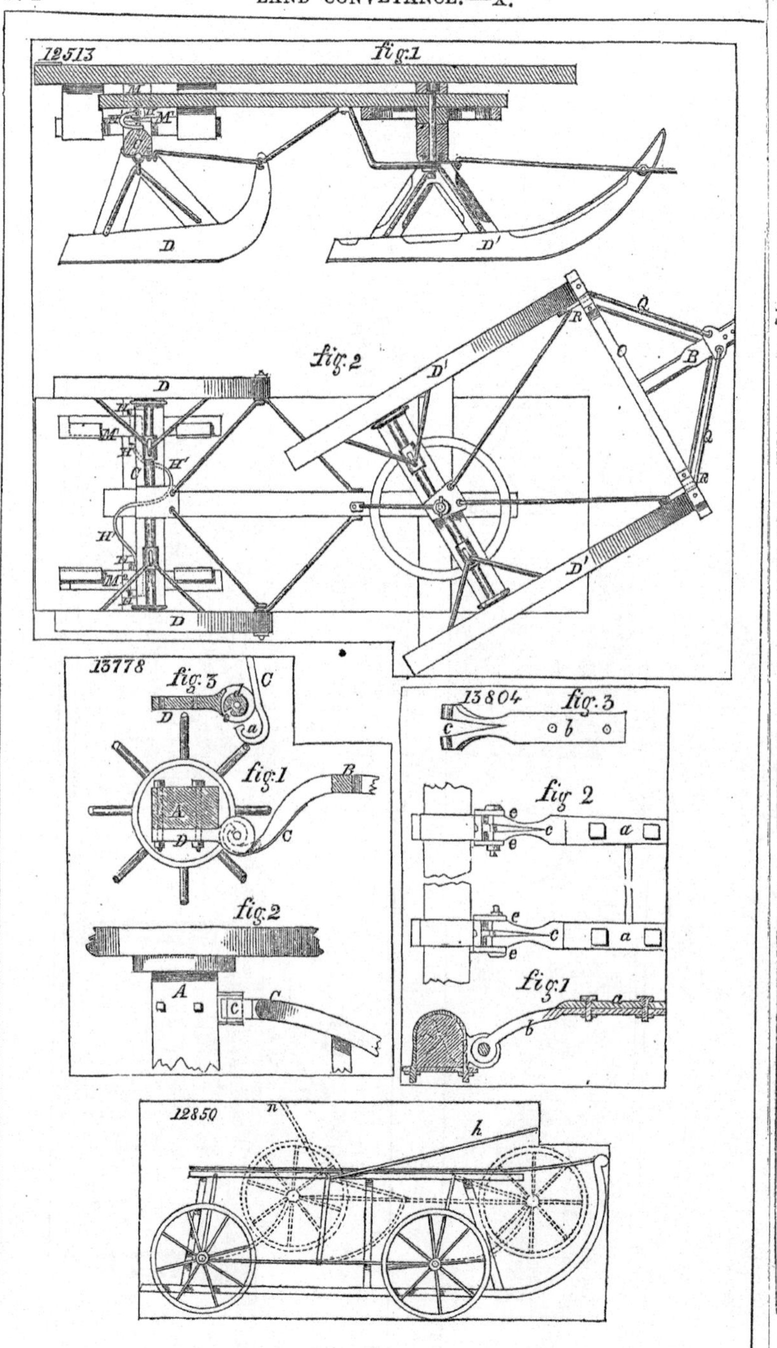
12513
fig.1
fig.2
13778
fig.3
fig.1
fig.2
13804
fig.3
fig 2
fig.1
12850

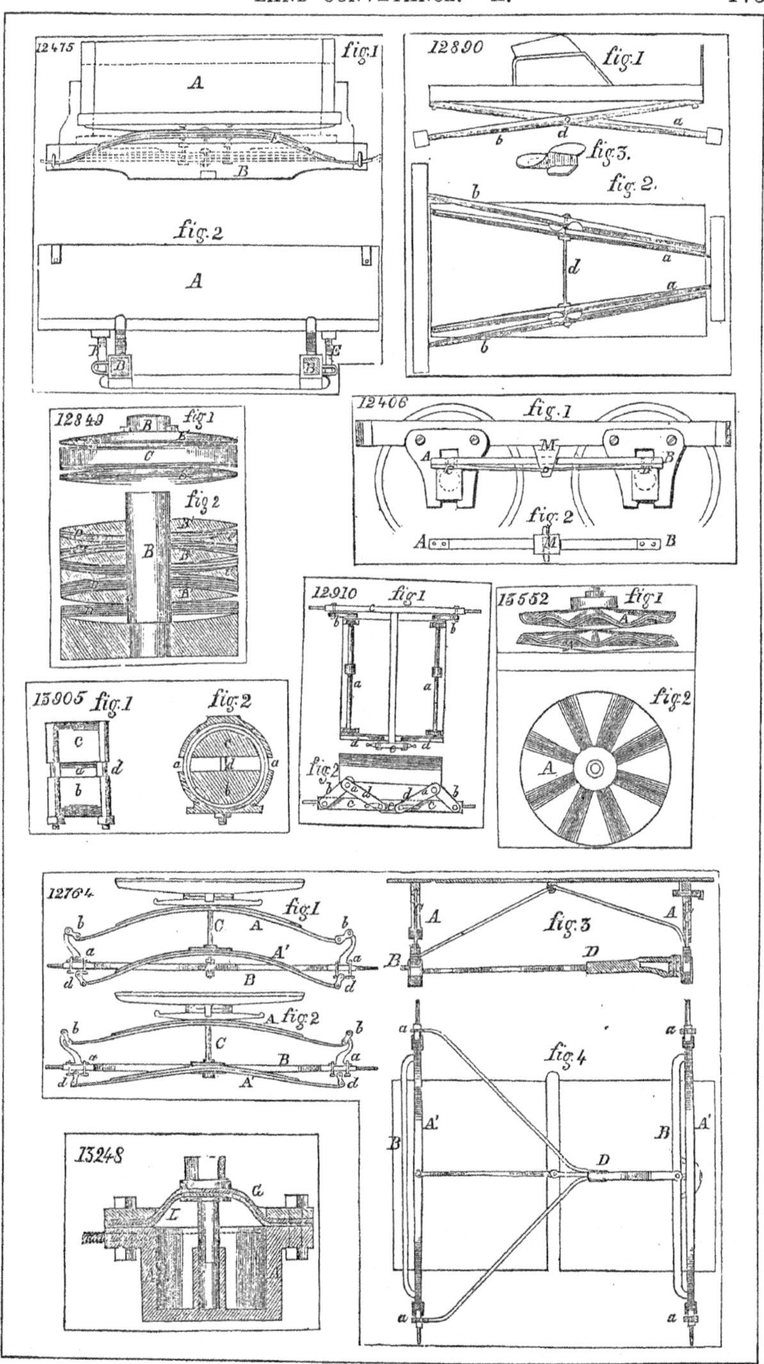
12475
fig.1
A
B
fig.2
A
E
B
B
E
12890
fig.1
a
b
d
fig.3
fig.2
b
a
d
a
b
12849
fig 1
B
E
C
fig 2
B
B
B
B
12406
fig.1
M
A
B
C
D
fig.2
A
M
B
12910
fig 1
b
b
a
a
d
d
fig 2
b
c
c
b
13552
fig 1
A
fig 2
A
13905
fig.1
c
a
b
d
fig 2
c
a
a
b
12764
fig.1
C
A
b
b
a
A'
a
B
d
d
A fig.2
b
b
C
a
B
a
d
A'
d
fig.3
A
A
B
D
fig.4
a
a
A'
B
B
A'
D
a
a
13248
a
E
A
A

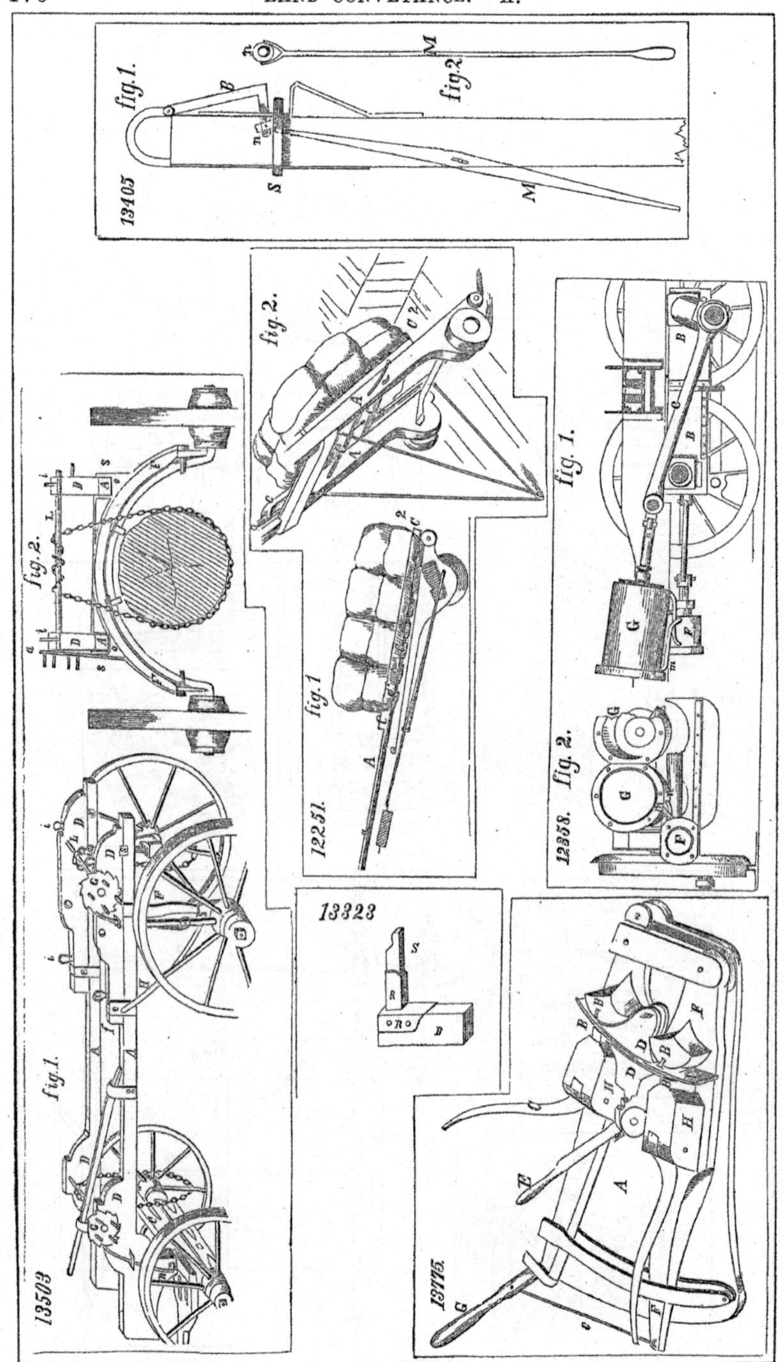
13405
fig.1.
fig.2
12251.
fig.1
fig.2.
12358.
fig. 1.
fig. 2.
13509
fig.1.
fig.2.
13323
13775

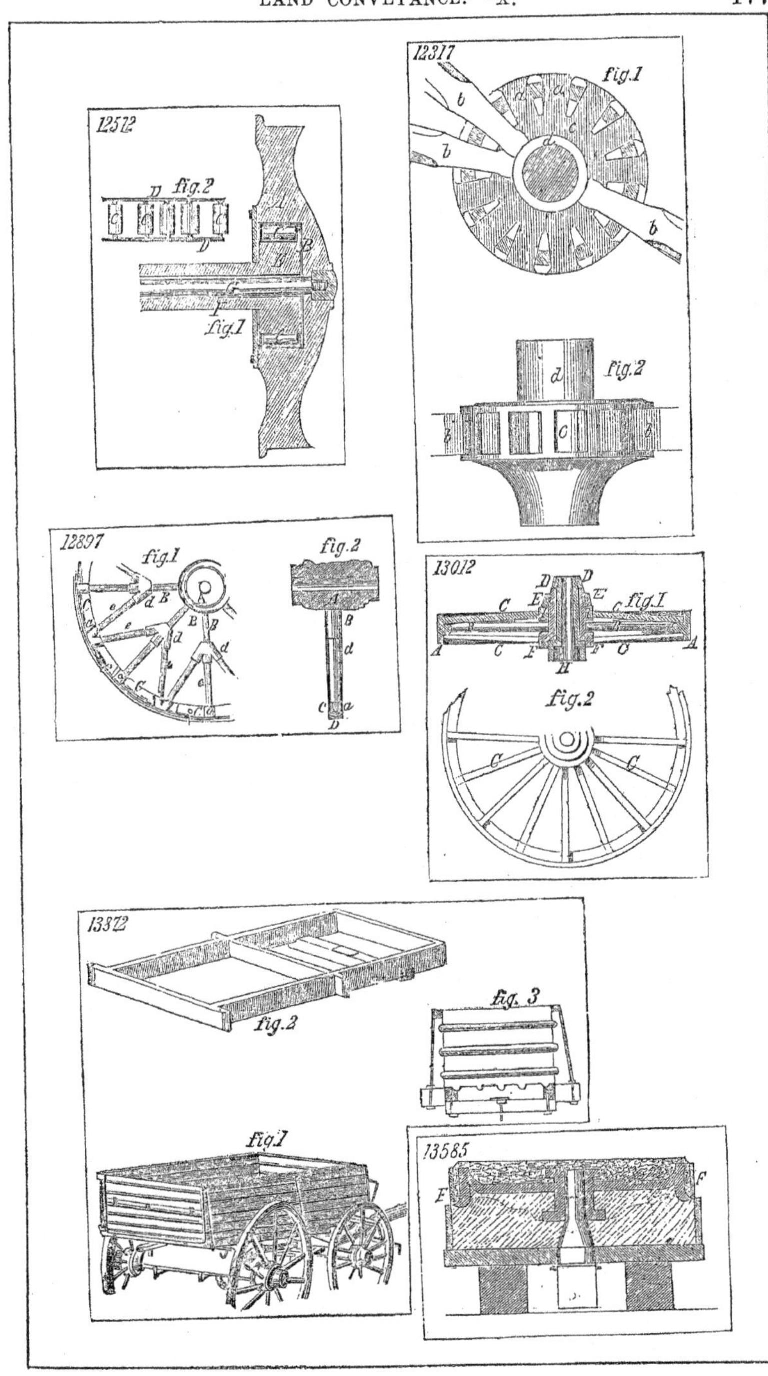
12512
fig.2
fig.1
12317
fig.1
fig.2
12897
fig.1
fig.2
13012
fig.1
fig.2
13372
fig.2
fig. 3
fig.1
13585

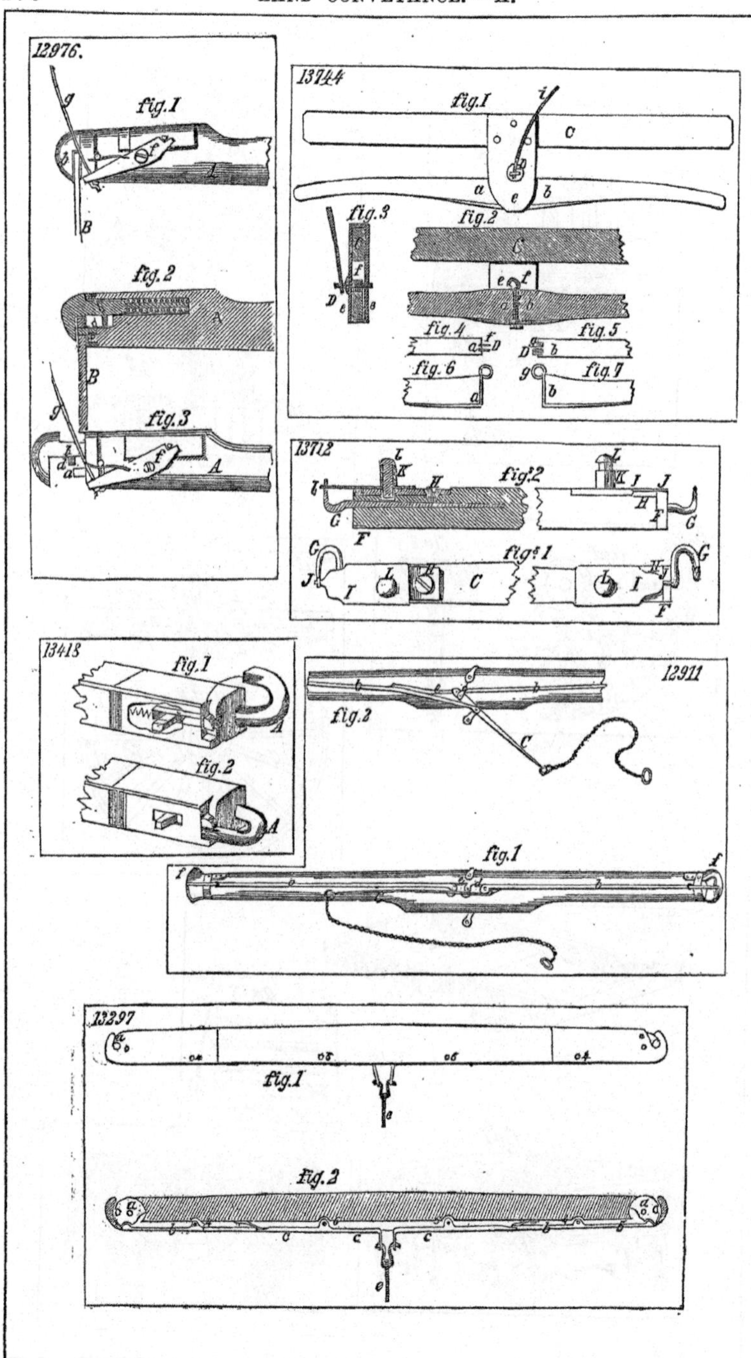
12976.
fig.1
fig.2
fig.3
13744
fig.1
fig.2
fig.3
fig.4
fig.5
fig.6
fig.7
13712
fig.2
fig.1
13418
fig.1
fig.2
12911
fig.2
fig.1
13297
fig.1
fig.2

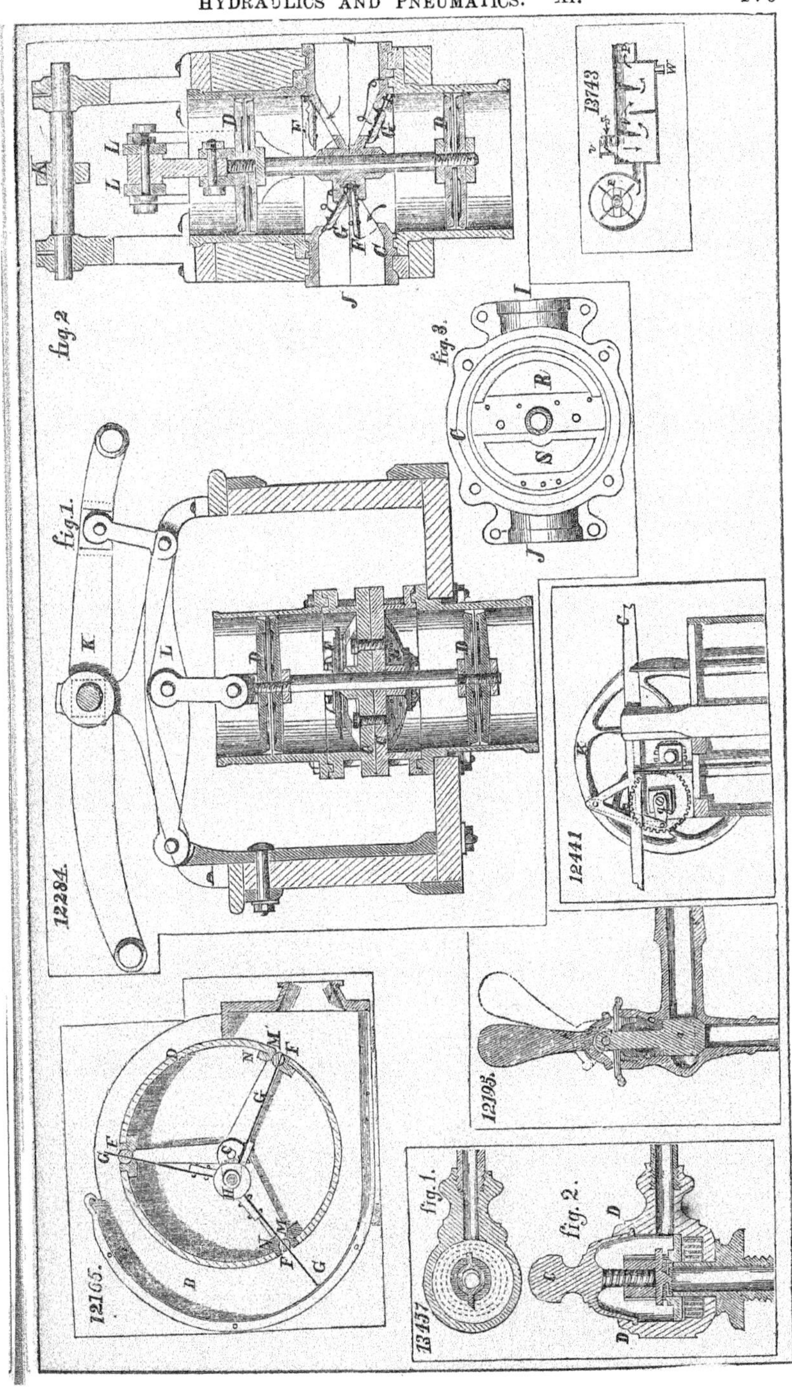
12284.
fig.1.
fig.2
fig.3.
13743
12441
12165.
12195.
13457
fig.1.
fig.2.

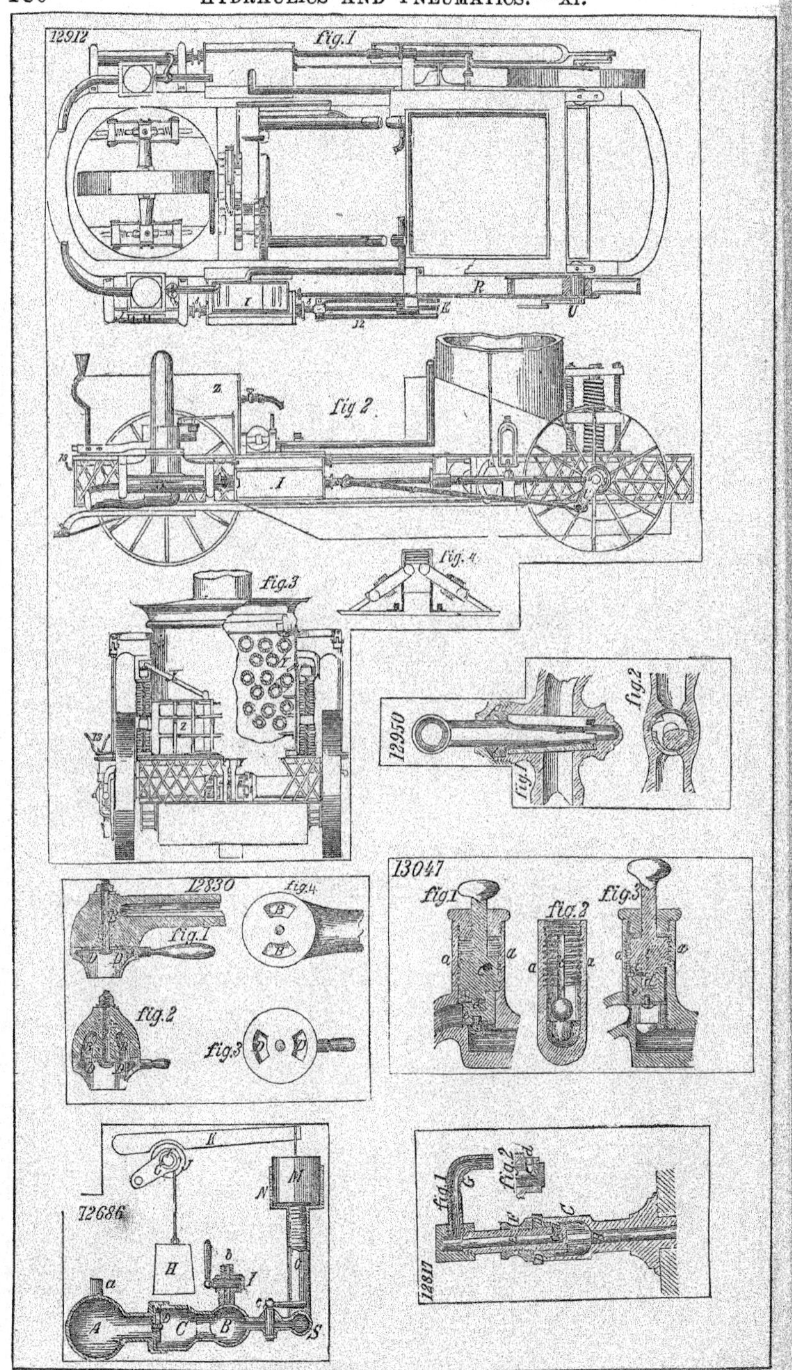
12912
fig.1
fig 2
fig.3
fig.4
12950
fig.1
fig.2
12830
fig.1
fig.2
fig.3
fig.4
13047
fig.1
fig.2
fig.3
72686
12817
fig.1
fig.2

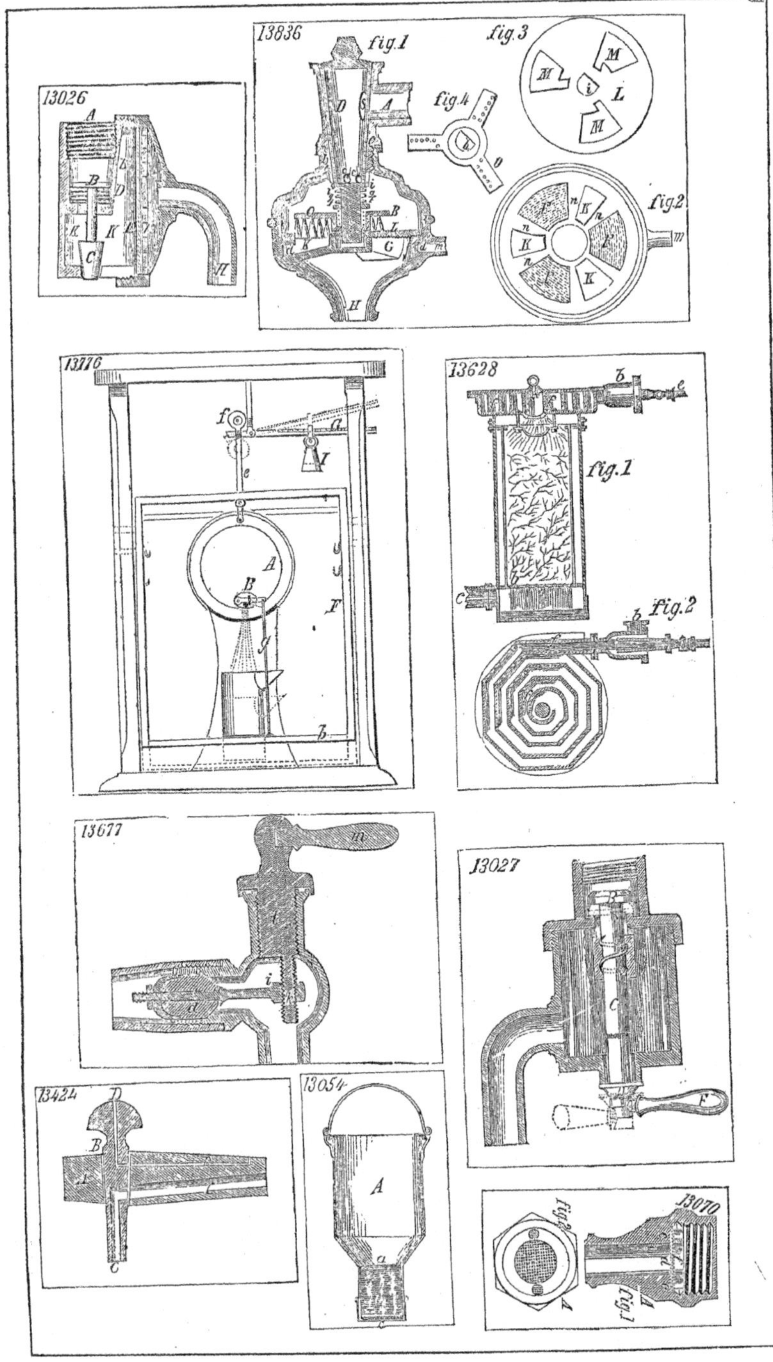
13026
13836
fig.1
fig.3
fig.4
fig.2
13776
13628
fig.1
fig.2
13677
13027
13424
13054
13070
fig.2
fig.1

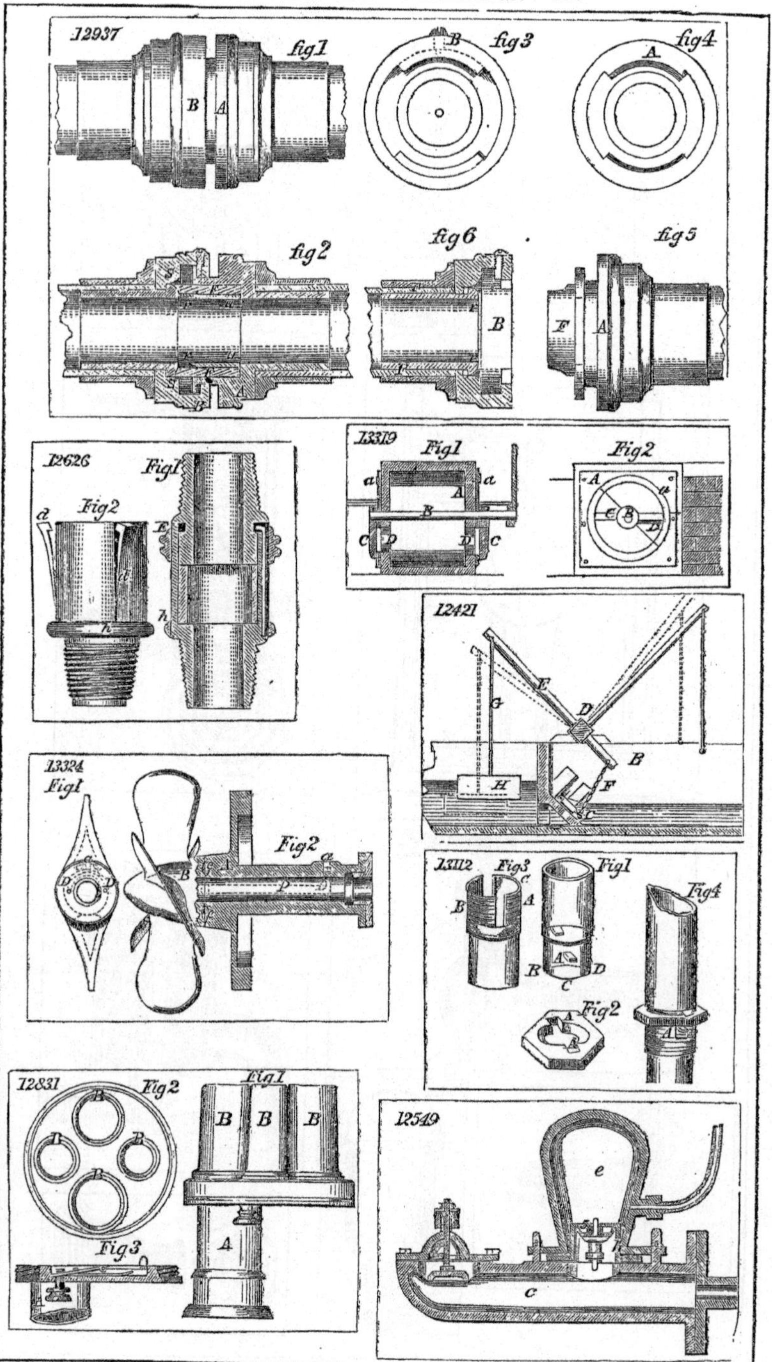
12937
fig1
fig3
fig4
fig2
fig6
fig5
12626
Fig1
Fig2
13319
Fig1
Fig2
12421
13324
Fig1
Fig2
13112
Fig3
Fig1
Fig4
Fig2
12831
Fig2
Fig1
Fig3
12549

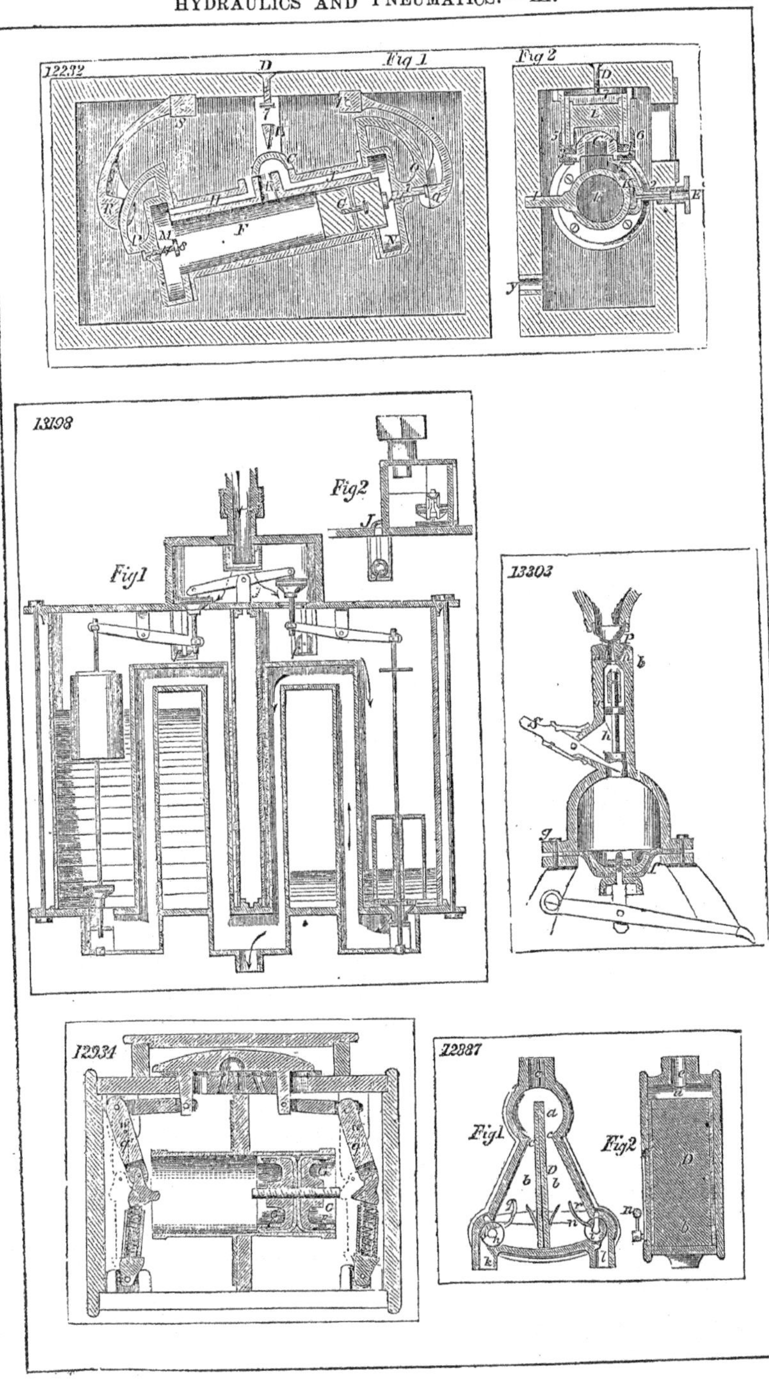
12232
Fig 1
Fig 2
13198
Fig1
Fig2
13303
12934
12887
Fig1
Fig2

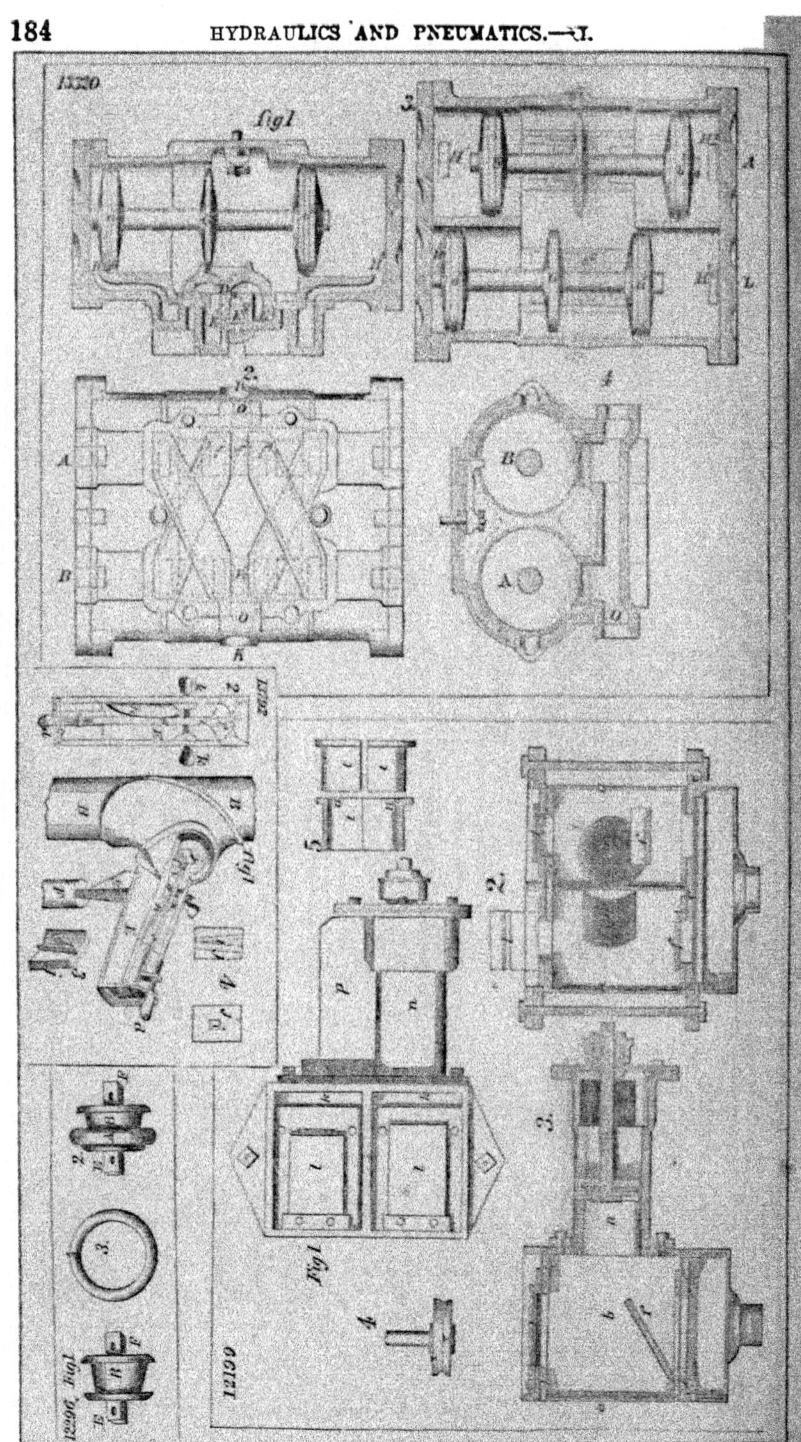

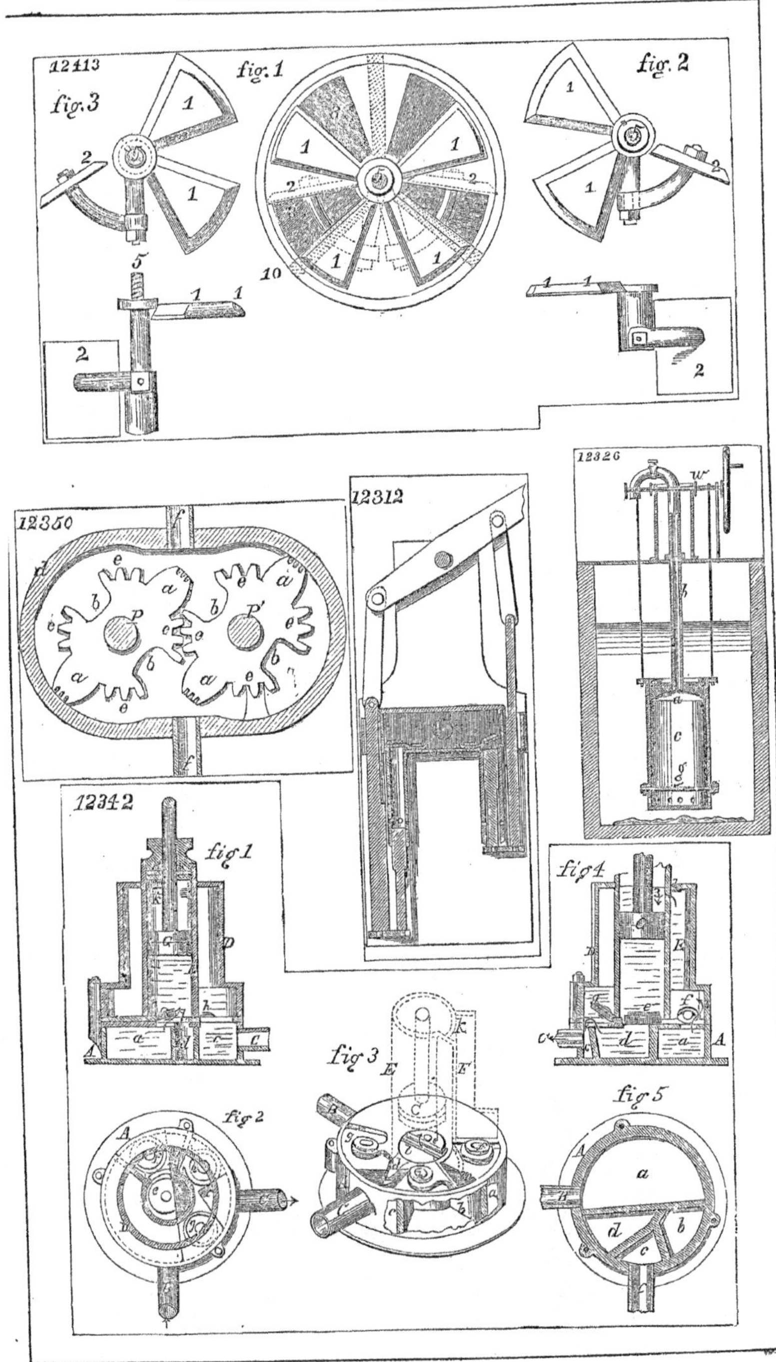
12413
fig. 1
fig. 2
fig. 3
12350
12312
12326
12342
fig 1
fig 2
fig 3
fig 4
fig 5

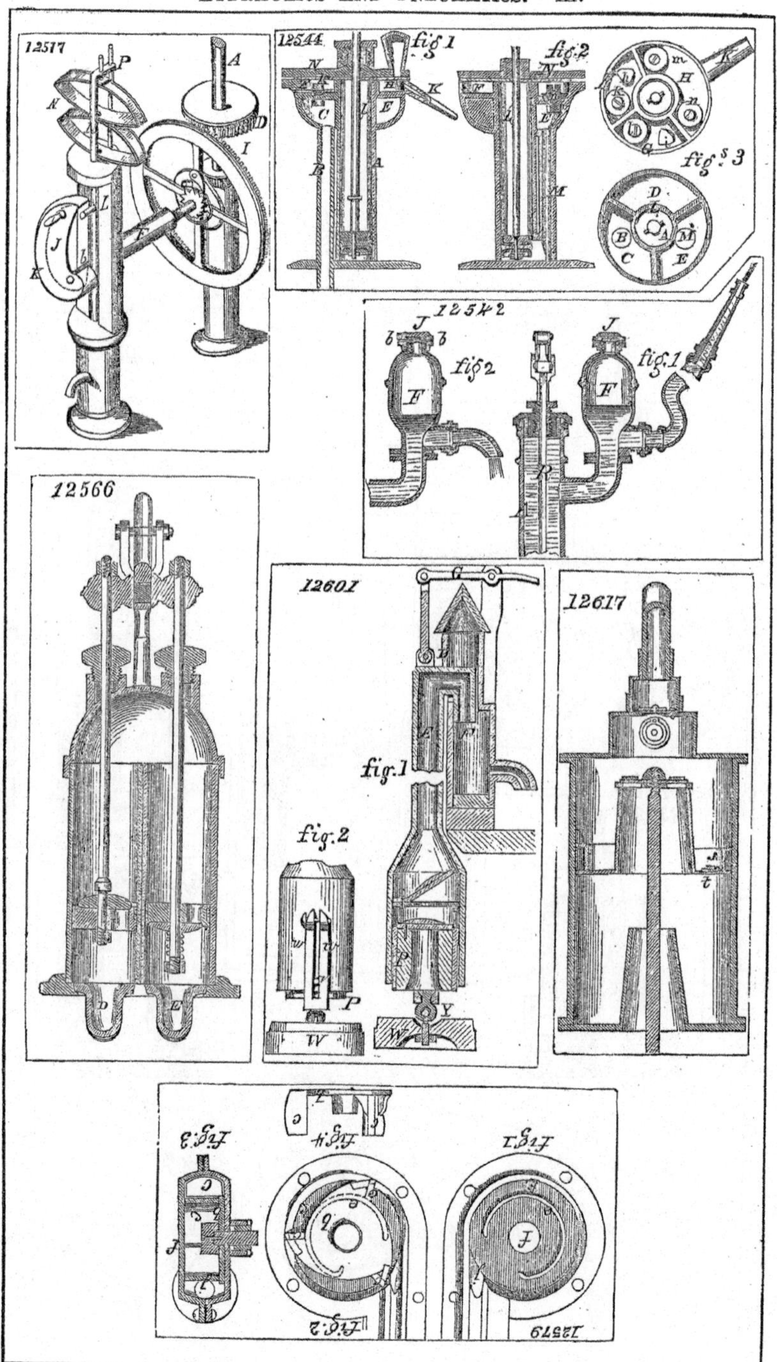
12517
12544
fig 1
fig 2
figs 3
12542
fig 2
fig 1
12566
12601
fig. 1
fig. 2
12617
12579
fig. 1
fig. 2
fig. 3
fig. 4

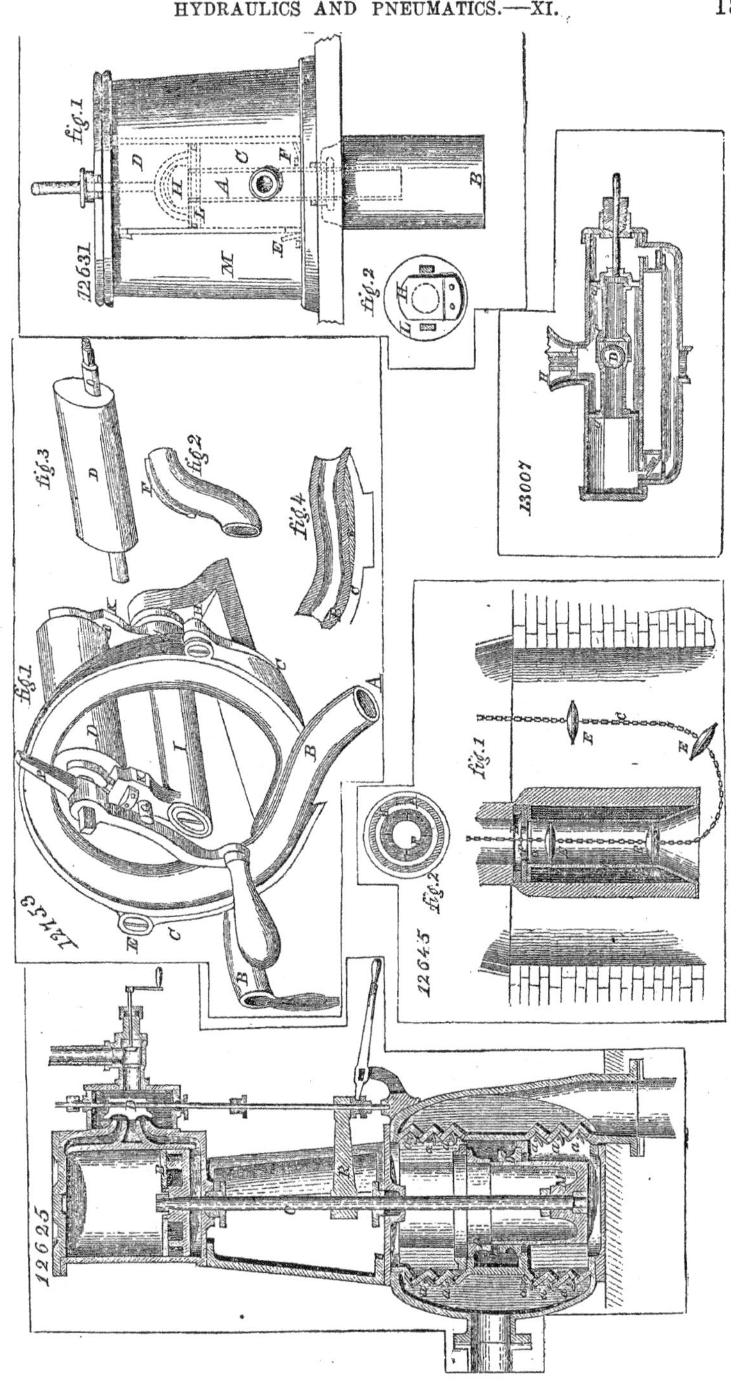
12631
fig.1
fig.2
fig.3
fig.4
13001
12645
12625

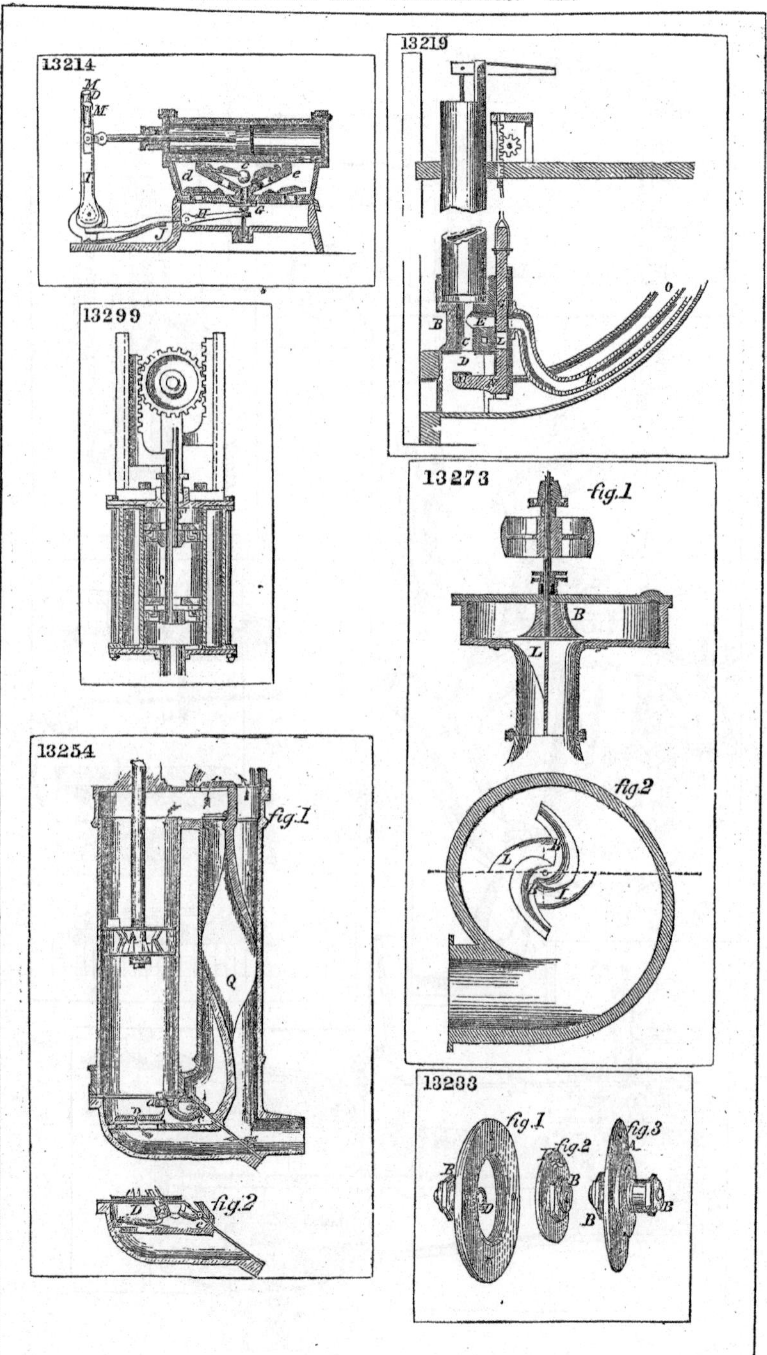
13214
13219
13299
13273
fig.1
fig.2
13254
fig.1
fig.2
13283
fig.1
fig.2
fig.3

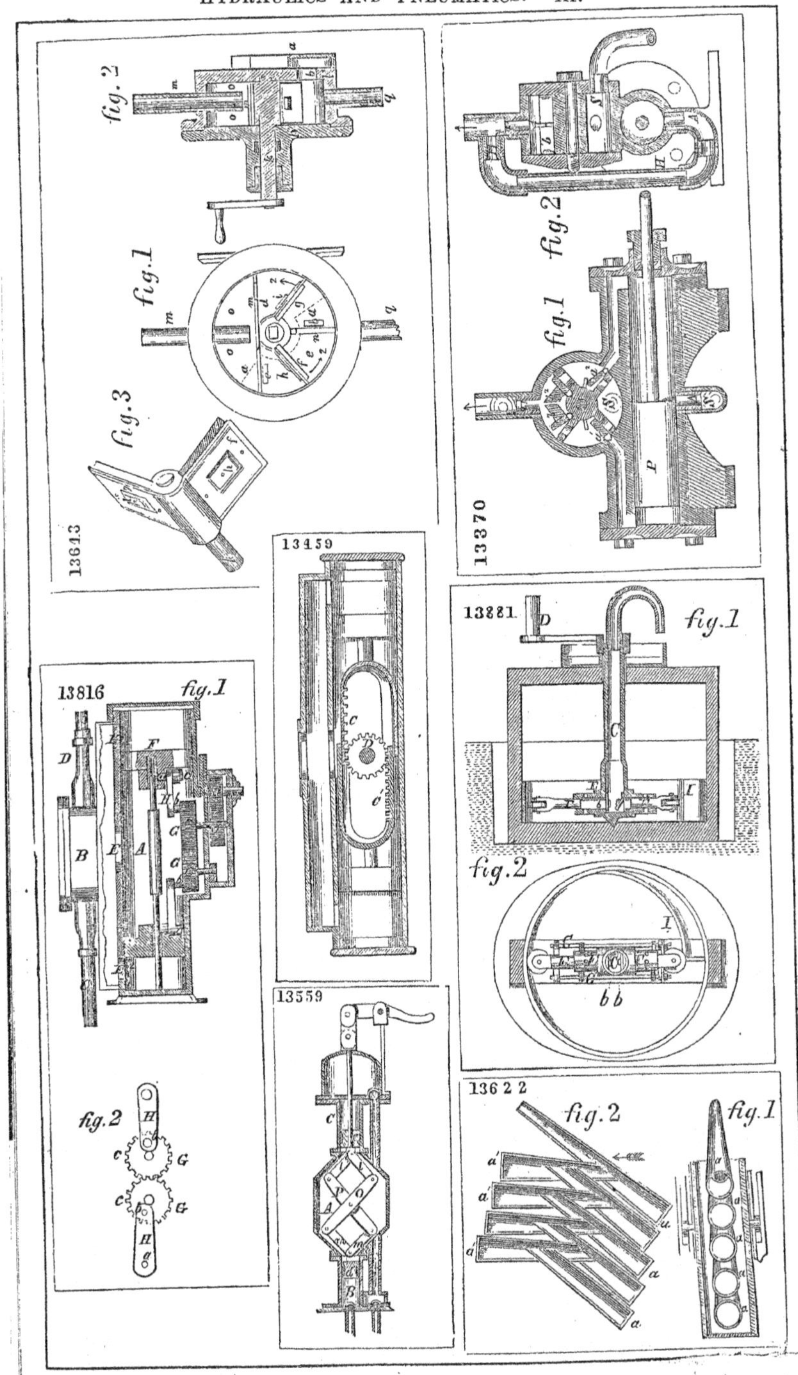
13643
fig. 1
fig. 2
fig. 3
13370
fig. 1
fig. 2
13459
13816
fig. 1
fig. 2
13881
fig. 1
fig. 2
13559
136 2 2
fig. 2
fig. 1

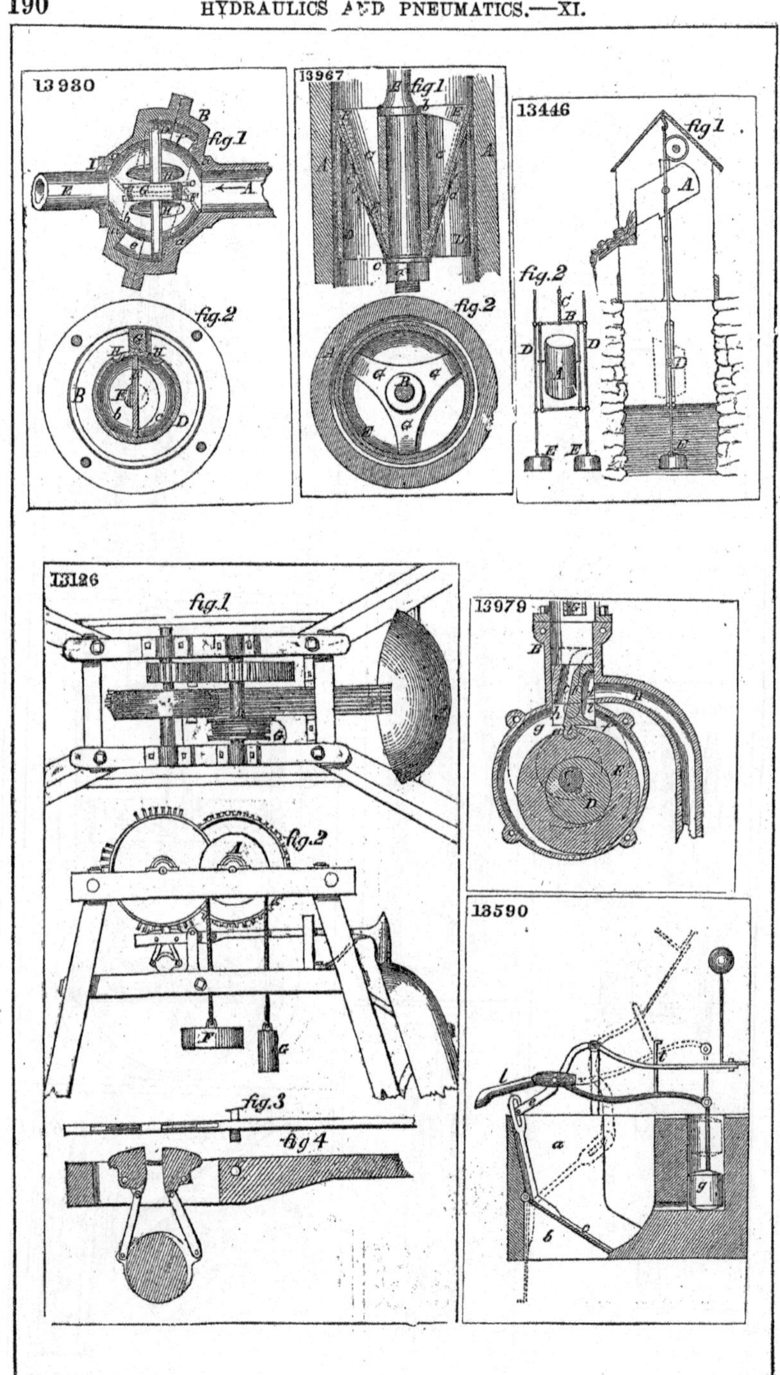
13930
fig.1
fig.2
13967
fig.1
fig.2
13446
fig.1
fig.2
13126
fig.1
fig.2
fig.3
fig.4
13979
13590

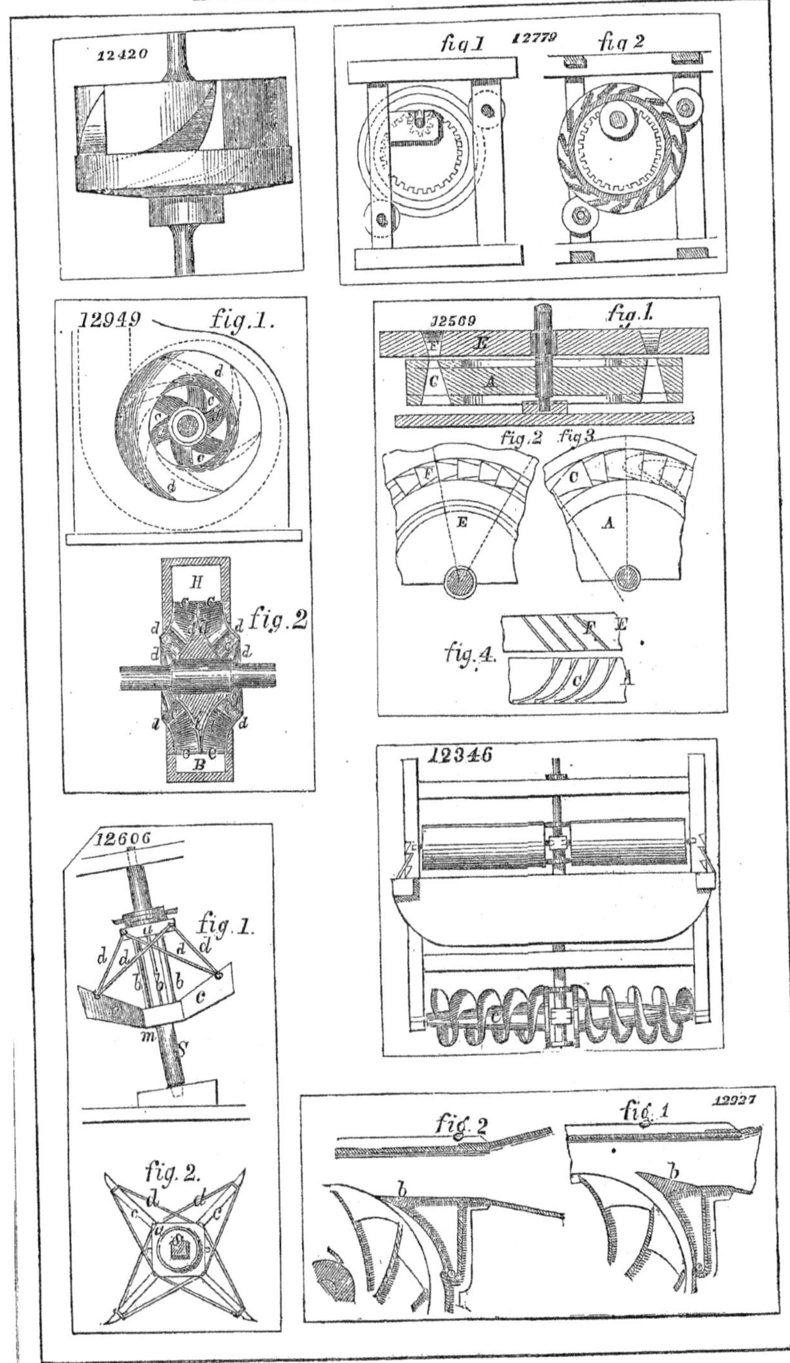
12420
fig 1
12779
fig 2
12949
fig. 1.
fig. 2
12569
fig. 1.
fig. 2
fig 3
fig. 4.
12346
12606
fig. 1.
fig. 2.
12927
fig. 2
fig. 1

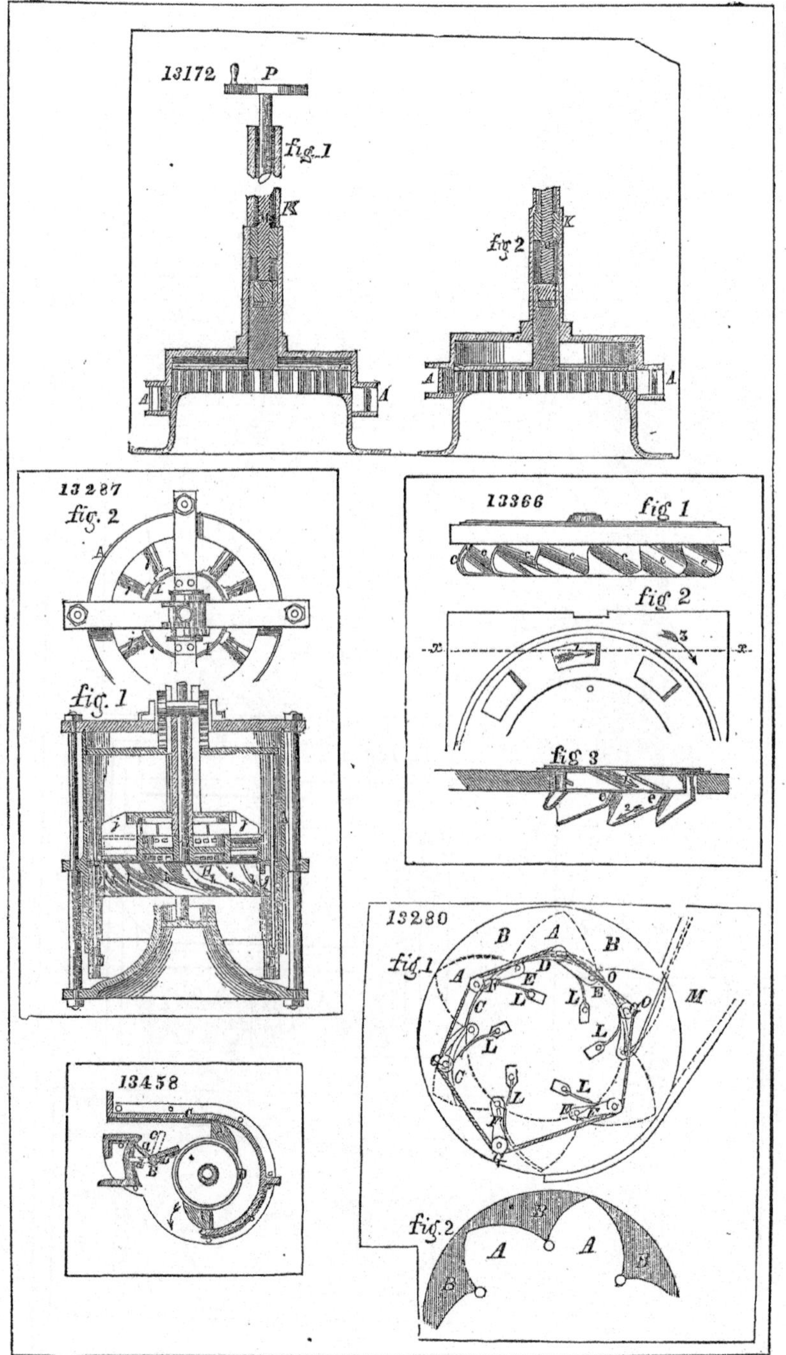
13172
P
fig. 1
K
fig 2
K
A
A
A
A
13287
fig. 2
fig. 1
13366
fig 1
fig 2
fig 3
13280
fig. 1
B
A
B
D
E
L
M
fig. 2
13458

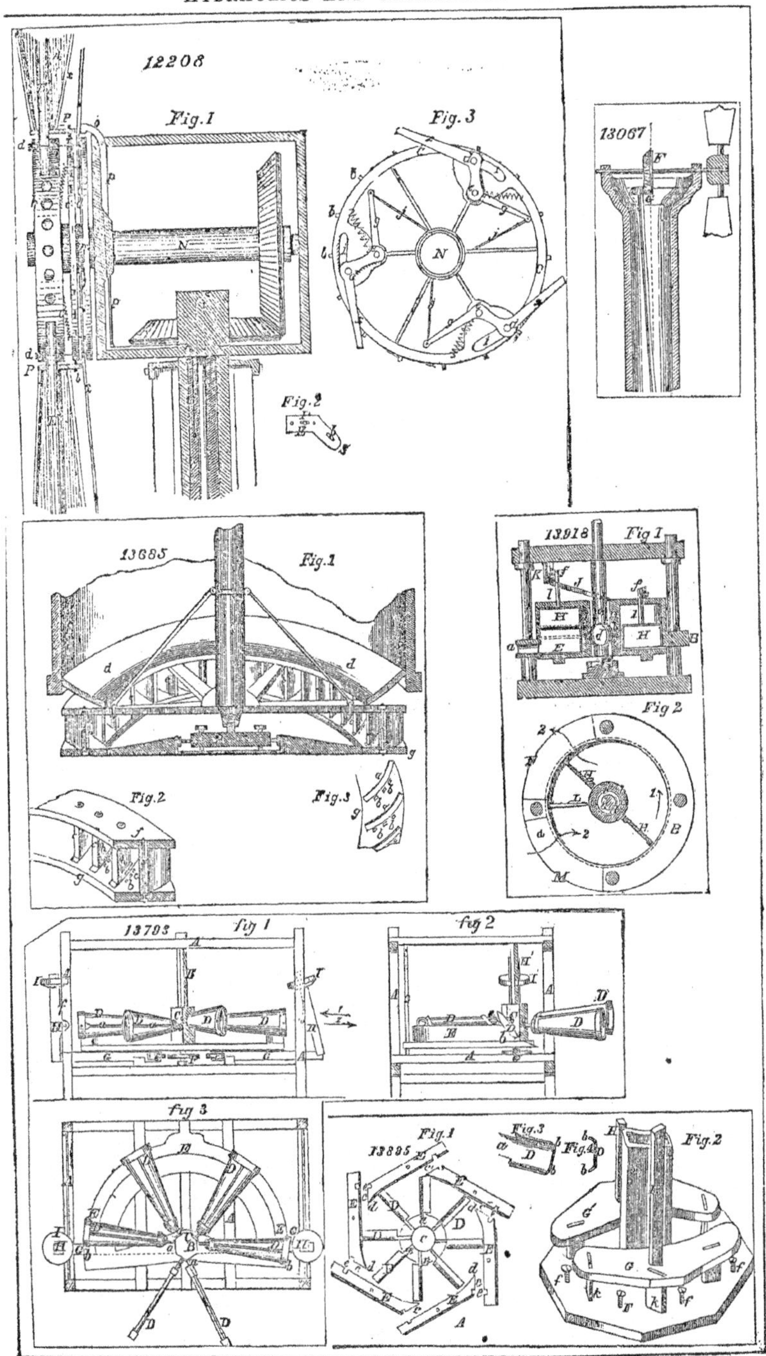
12208
Fig.1
Fig.3
Fig.2
13067
13685
Fig.1
Fig.2
Fig.3
13918
Fig 1
Fig 2
13793
fig 1
fig 2
fig 3
13885
Fig.1
Fig.3
Fig.4
Fig.2

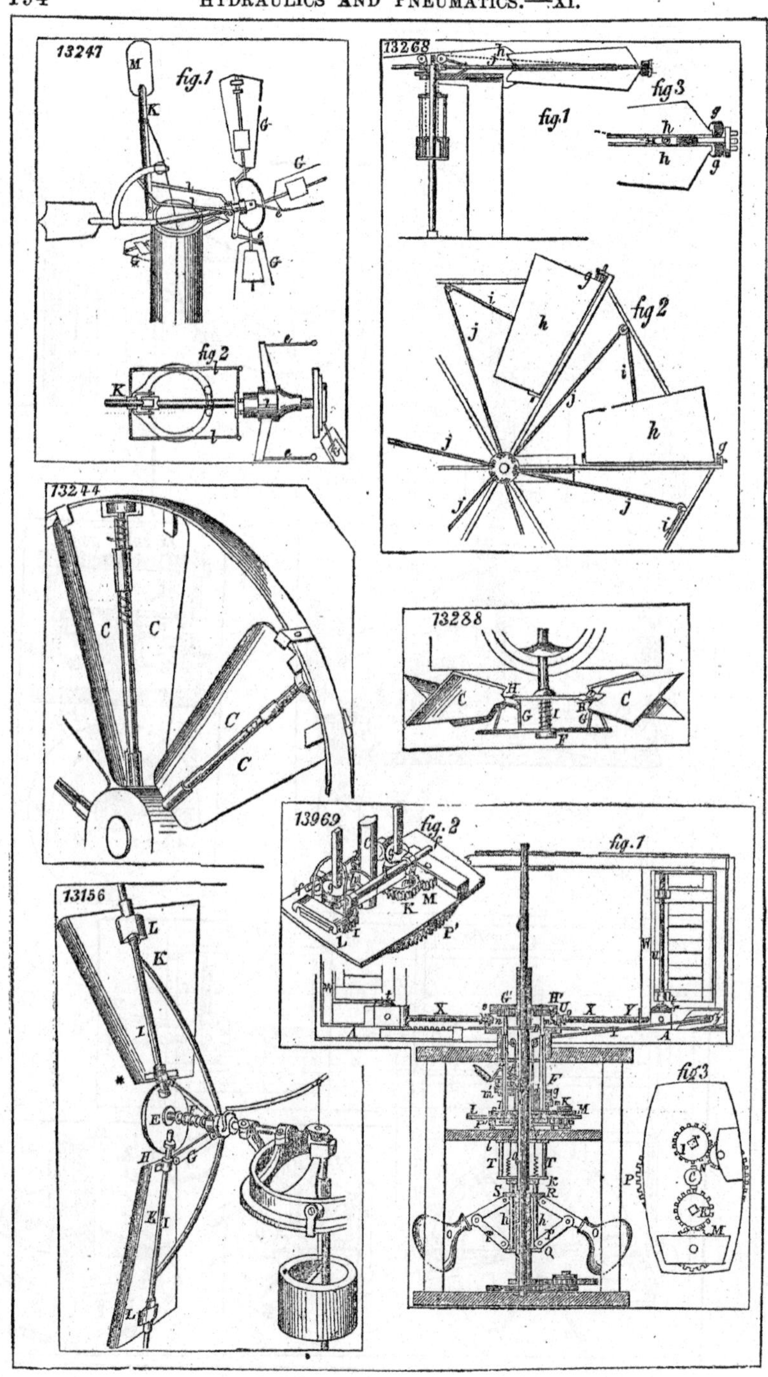
13247
fig.1
fig 2
13268
fig.1
fig 3
fig 2
13244
13288
13969
fig. 2
fig.1
fig 3
13156

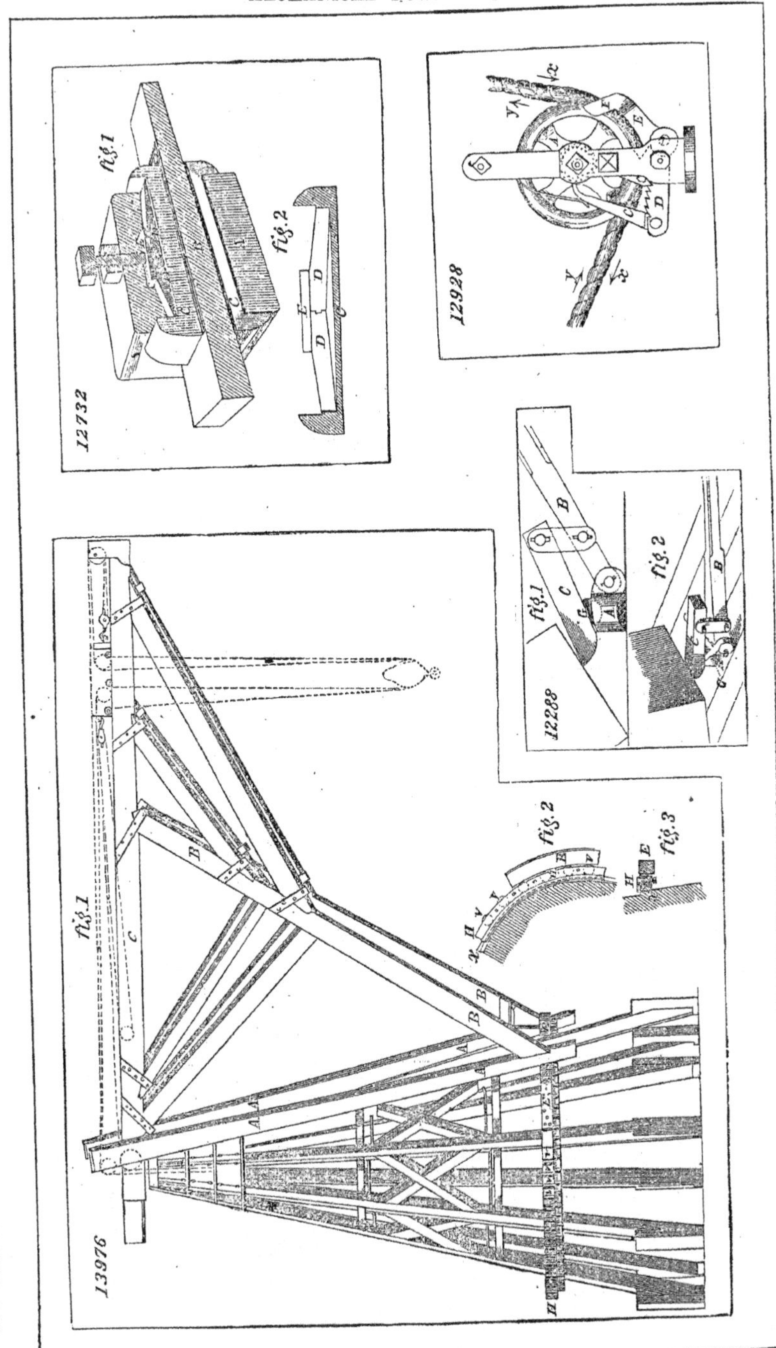
12732
fig.1
fig.2
12928
12288
fig.1
fig.2
13976
fig.1
fig.2
fig.3

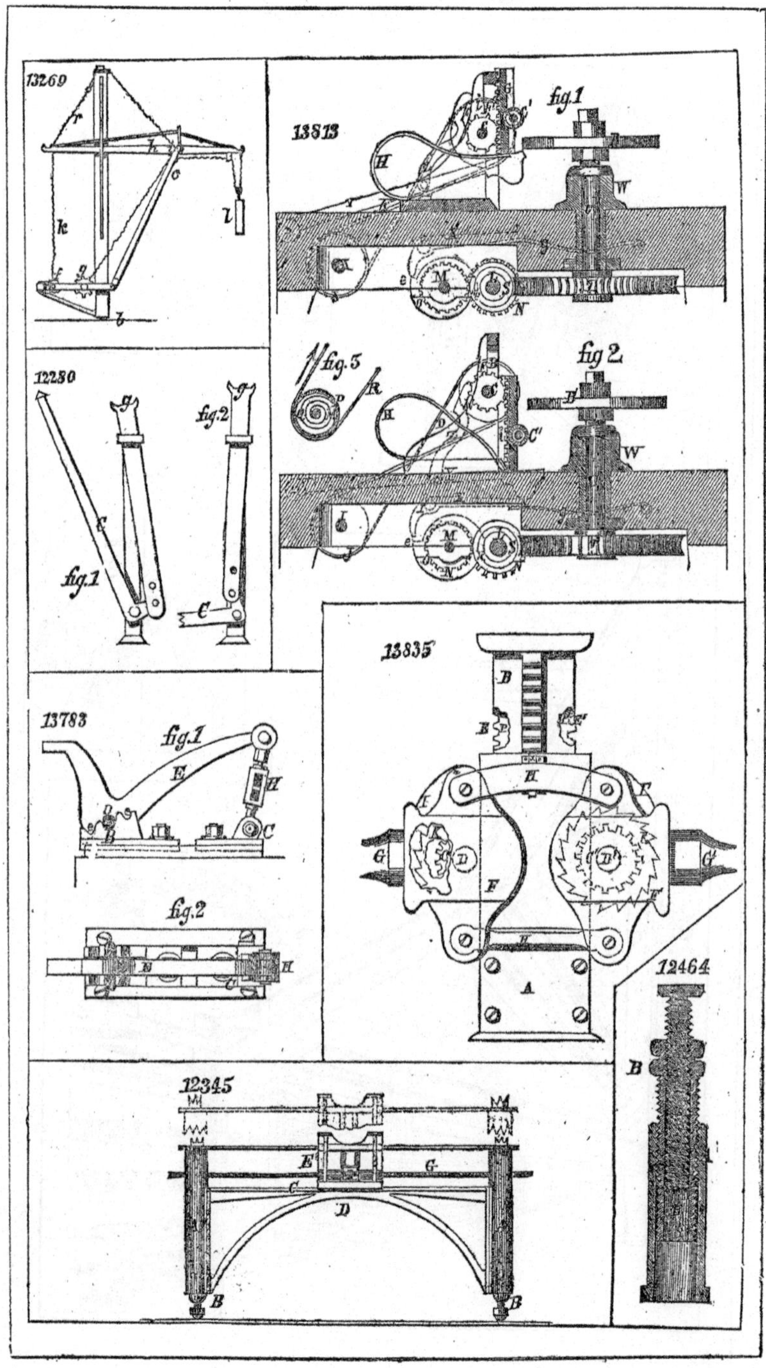
13269
13813
fig.1
fig 2.
fig. 3
12280
fig.1
fig.2
13783
fig.1
fig.2
13835
12464
12345

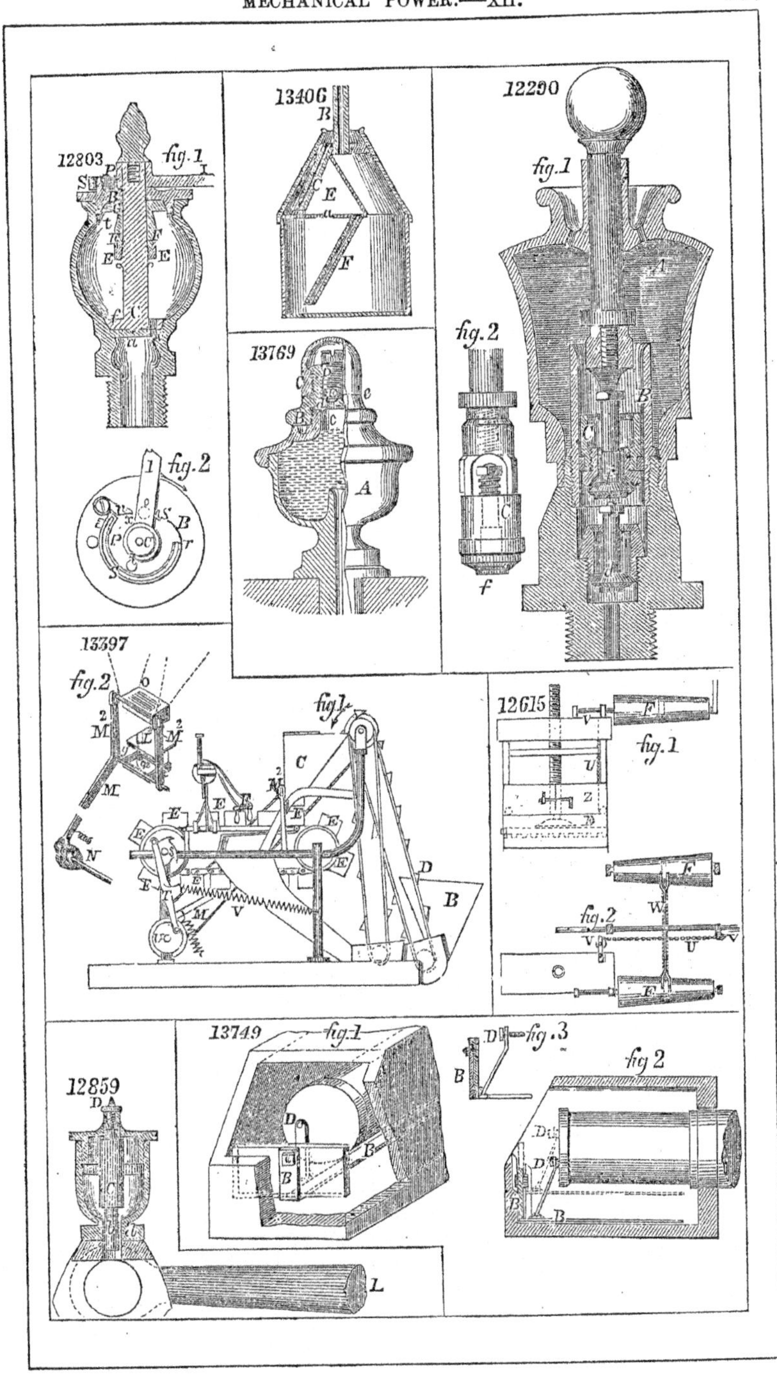
12803
fig. 1
fig. 2
13406
13769
12290
fig. 1
fig. 2
13397
fig. 2
fig. 1
12615
fig. 1
fig. 2
12859
13749
fig. 1
fig. 3
fig 2

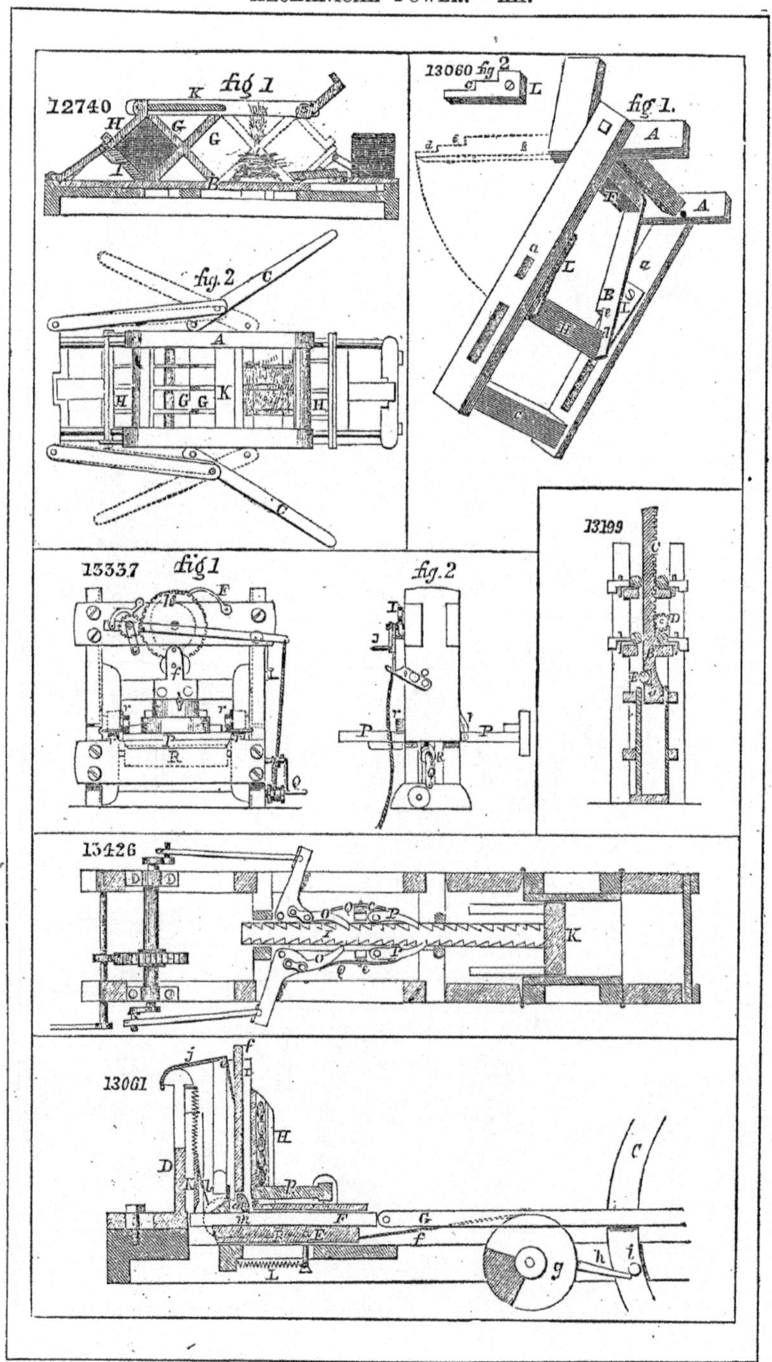
12740
fig 1
fig. 2
13060 fig 2
fig 1.
13337
fig 1
fig. 2
13199
13426
13061

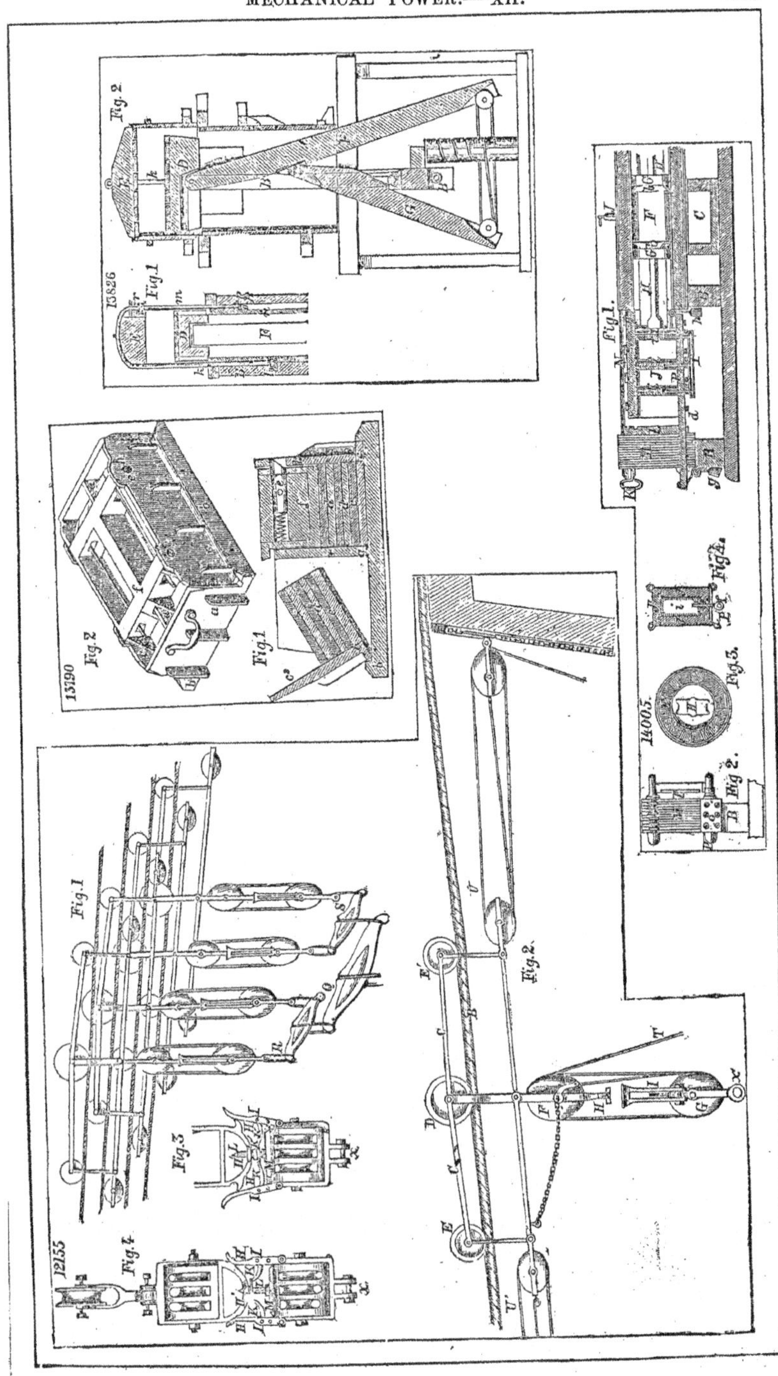
13826
Fig. 1
Fig. 2
13190
Fig. 2
Fig. 1
12155
Fig. 1
Fig. 2
Fig. 3
Fig. 4
Fig. 1
14005
Fig. 2
Fig. 3
Fig. 4

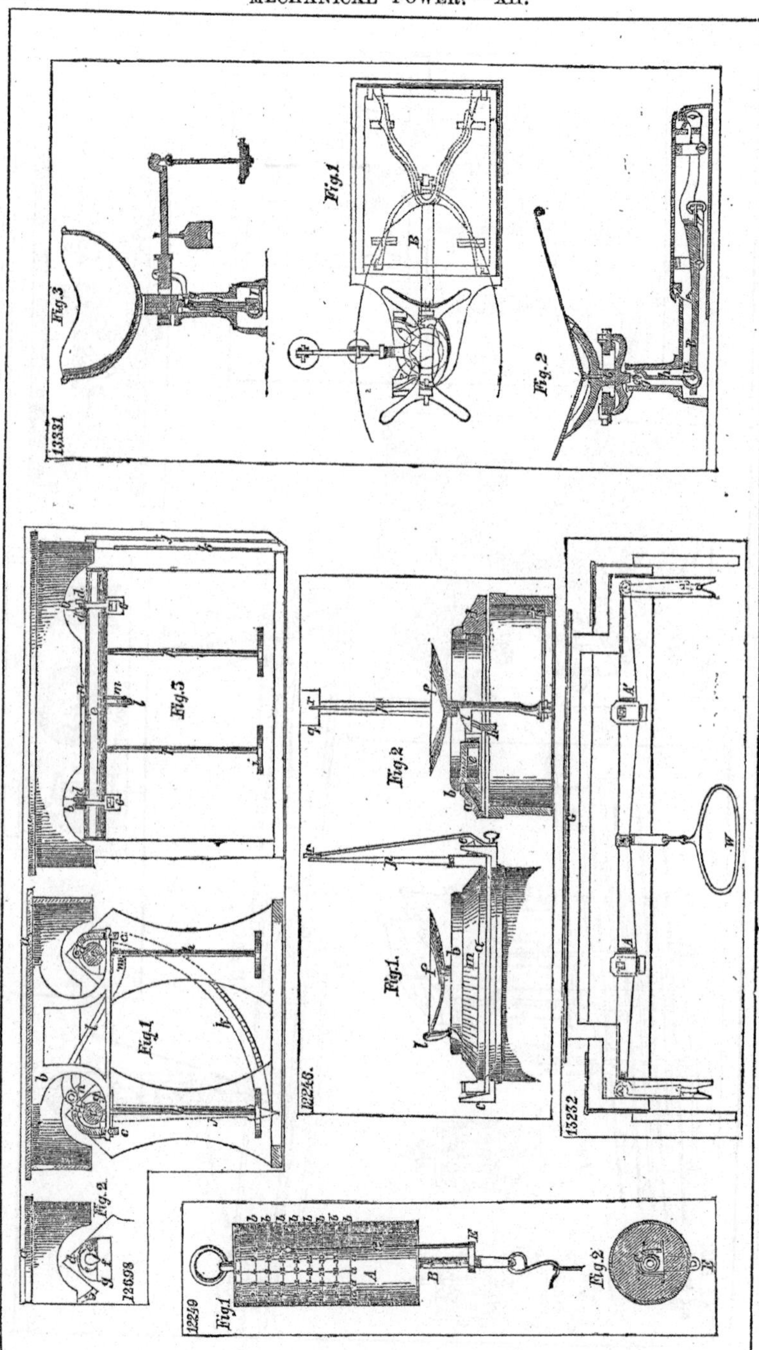

13331
Fig.1
Fig.2
Fig.3
12698
Fig.1
Fig.2
Fig.3
12248.
Fig.1.
Fig.2
13232
12259
Fig.1
Fig.2

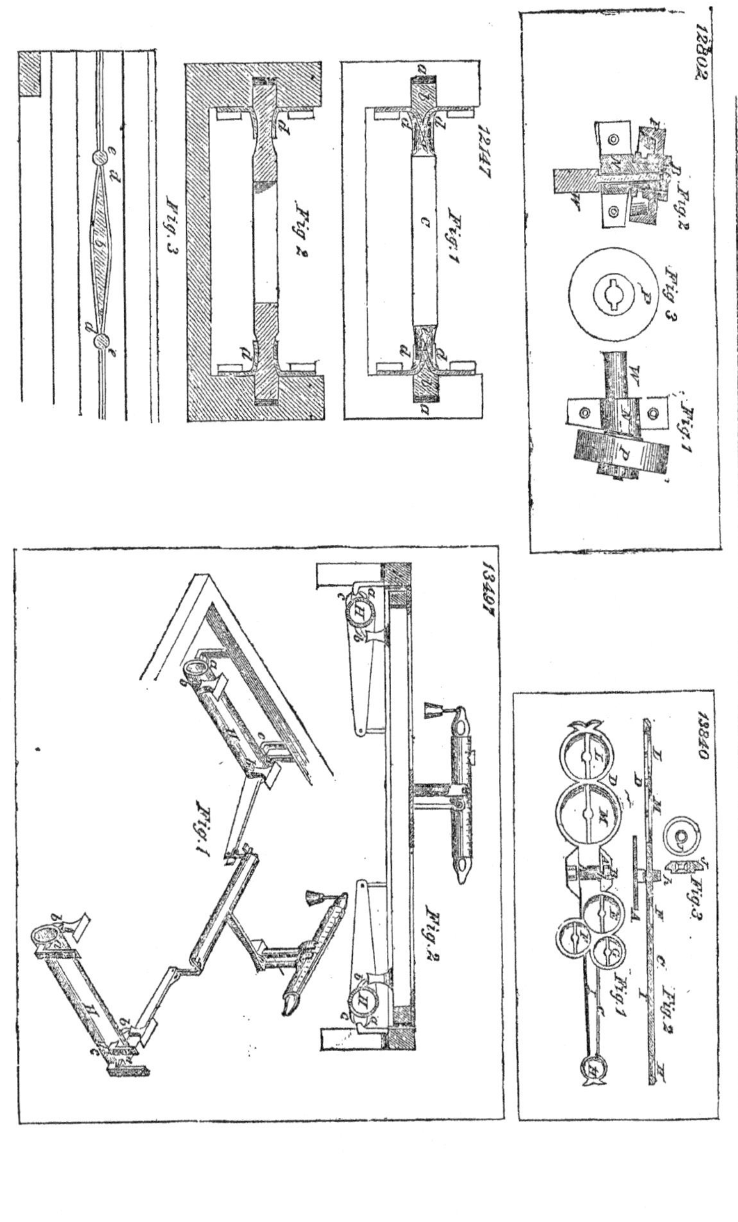
12802
12147
13497
13840

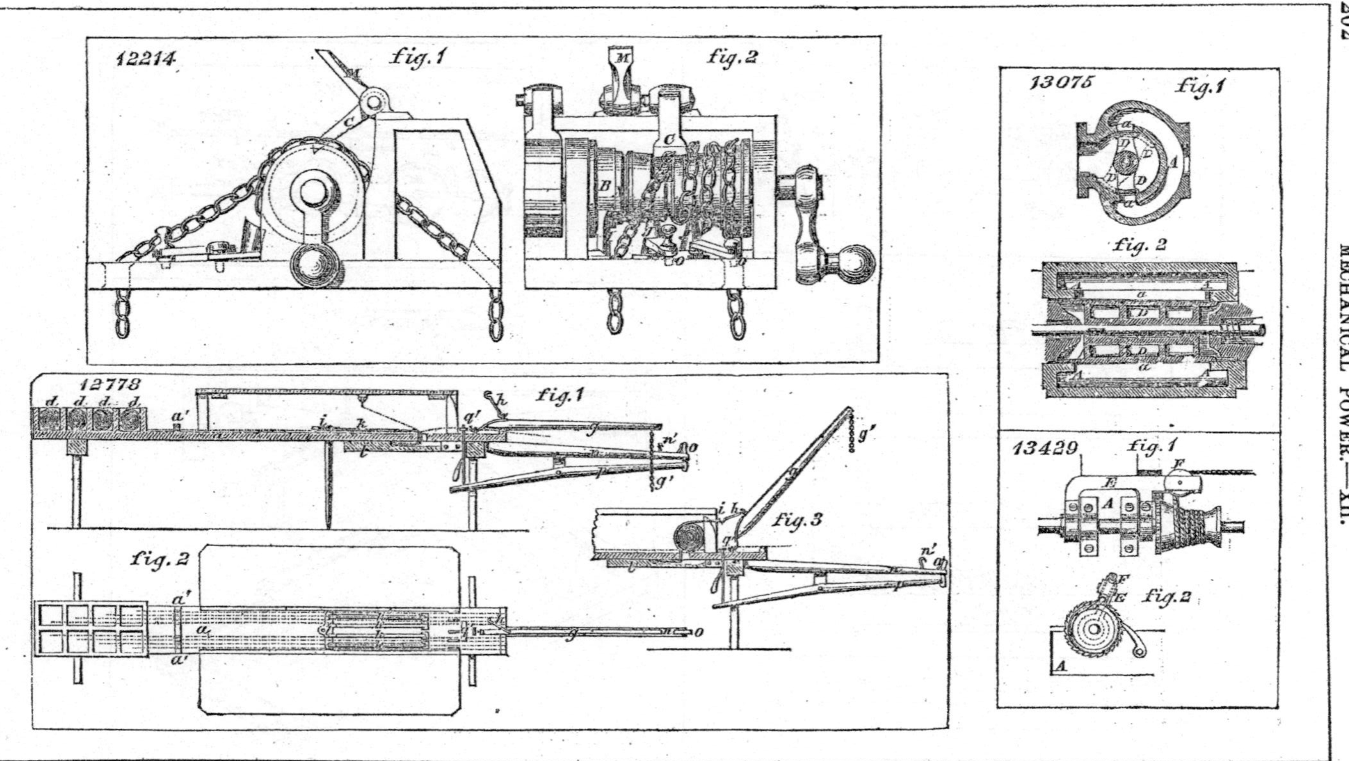
12214
fig.1
fig.2
12778
fig.1
fig.2
fig.3
13075
fig.1
fig. 2
13429
fig.1
fig.2

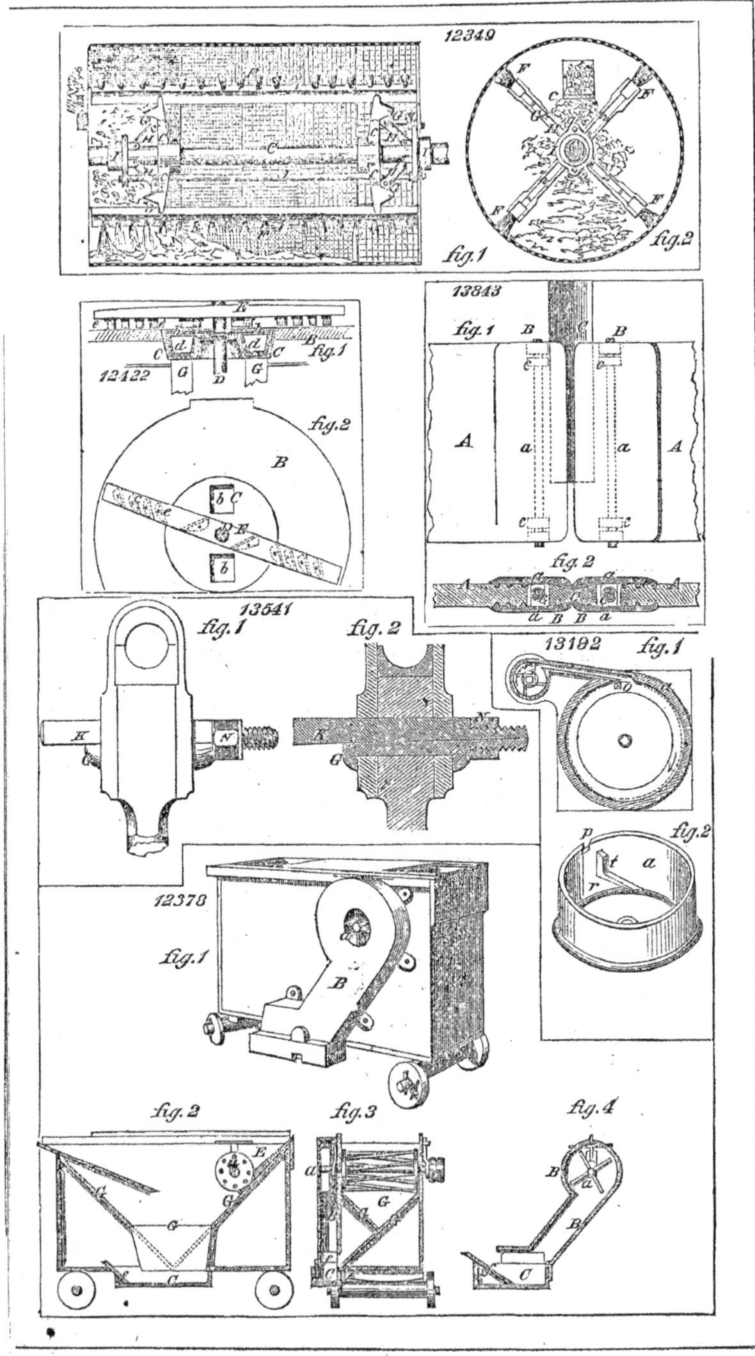
12349
fig.1
fig.2
12422
fig.1
fig.2
13843
fig.1
fig.2
13541
fig.1
fig.2
13192
fig.1
fig.2
12378
fig.1
fig.2
fig.3
fig.4

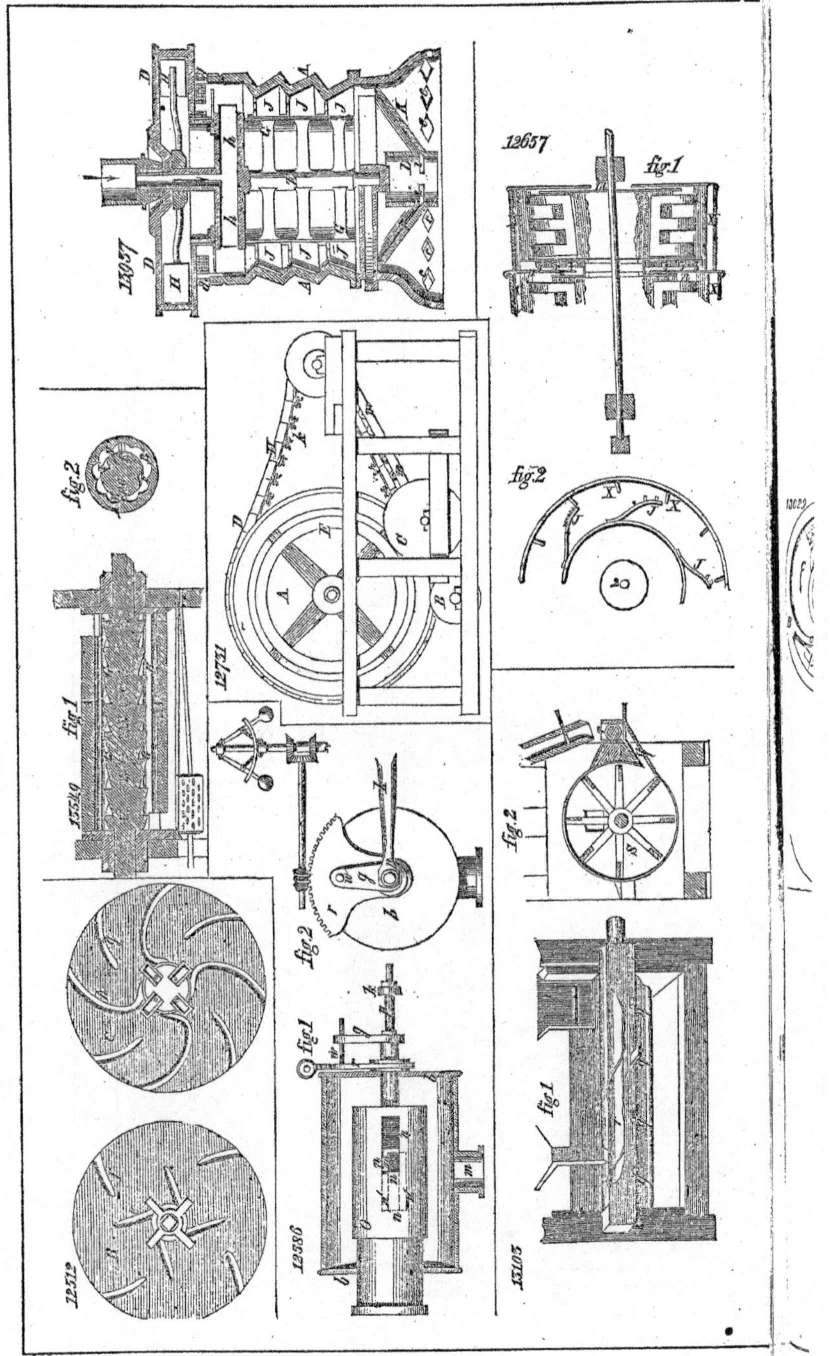

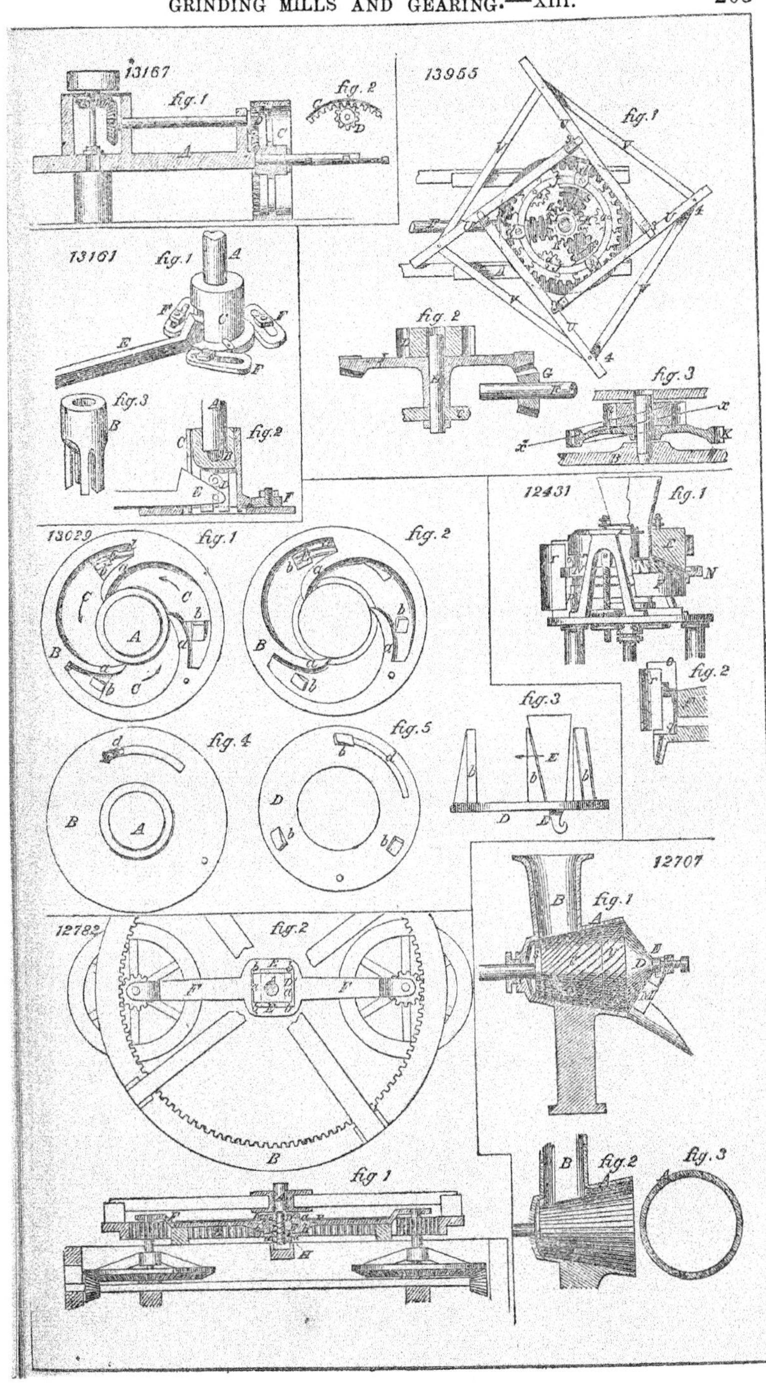
13167
fig. 1
fig. 2
13955
fig. 1
fig. 2
fig. 3
13161
fig. 1
fig. 3
fig. 2
12431
fig. 1
fig. 2
13029
fig. 1
fig. 2
fig. 4
fig. 5
fig. 3
12707
fig. 1
12782
fig. 2
fig. 1
fig. 2
fig. 3

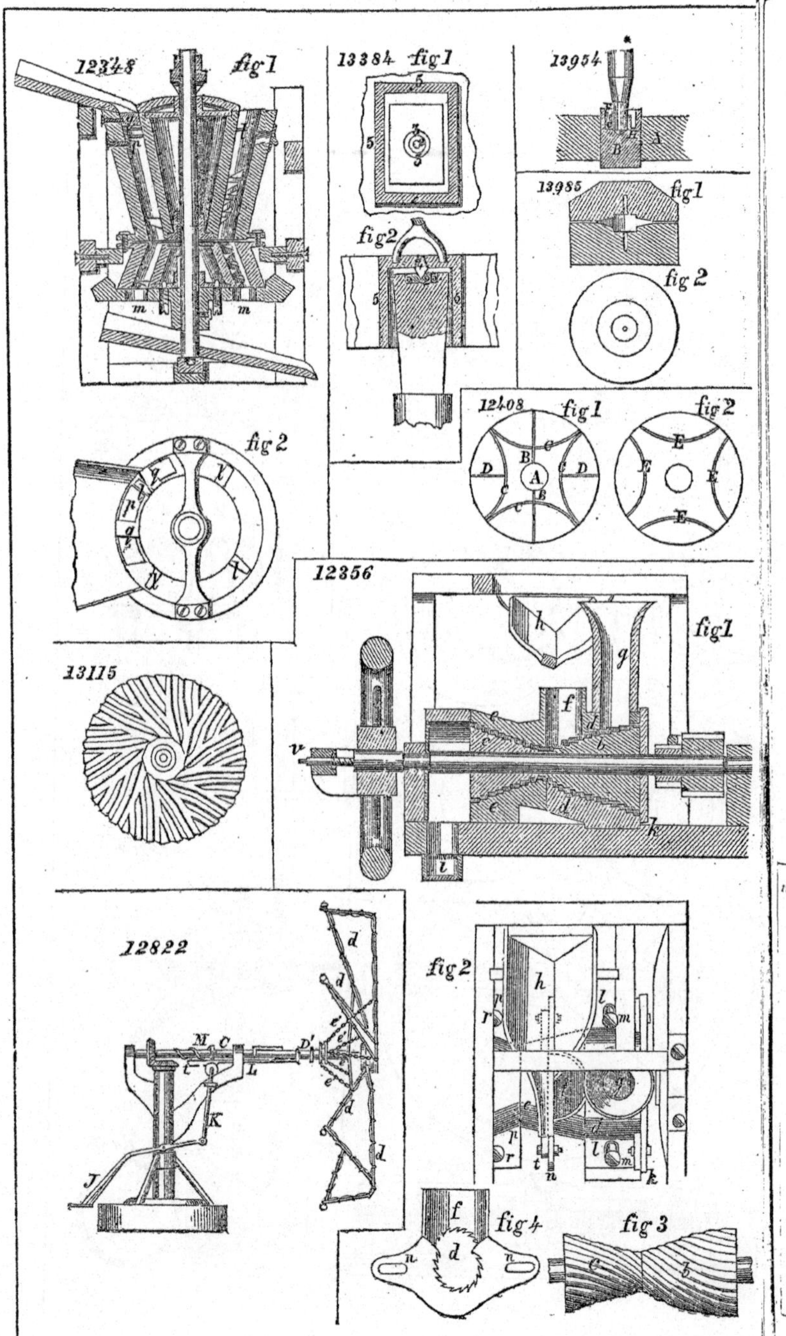
12348
fig 1
fig 2
13384 fig 1
fig 2
13954
13985
fig 1
fig 2
12408
fig 1
fig 2
12356
fig 1
fig 2
fig 3
fig 4
13115
12822

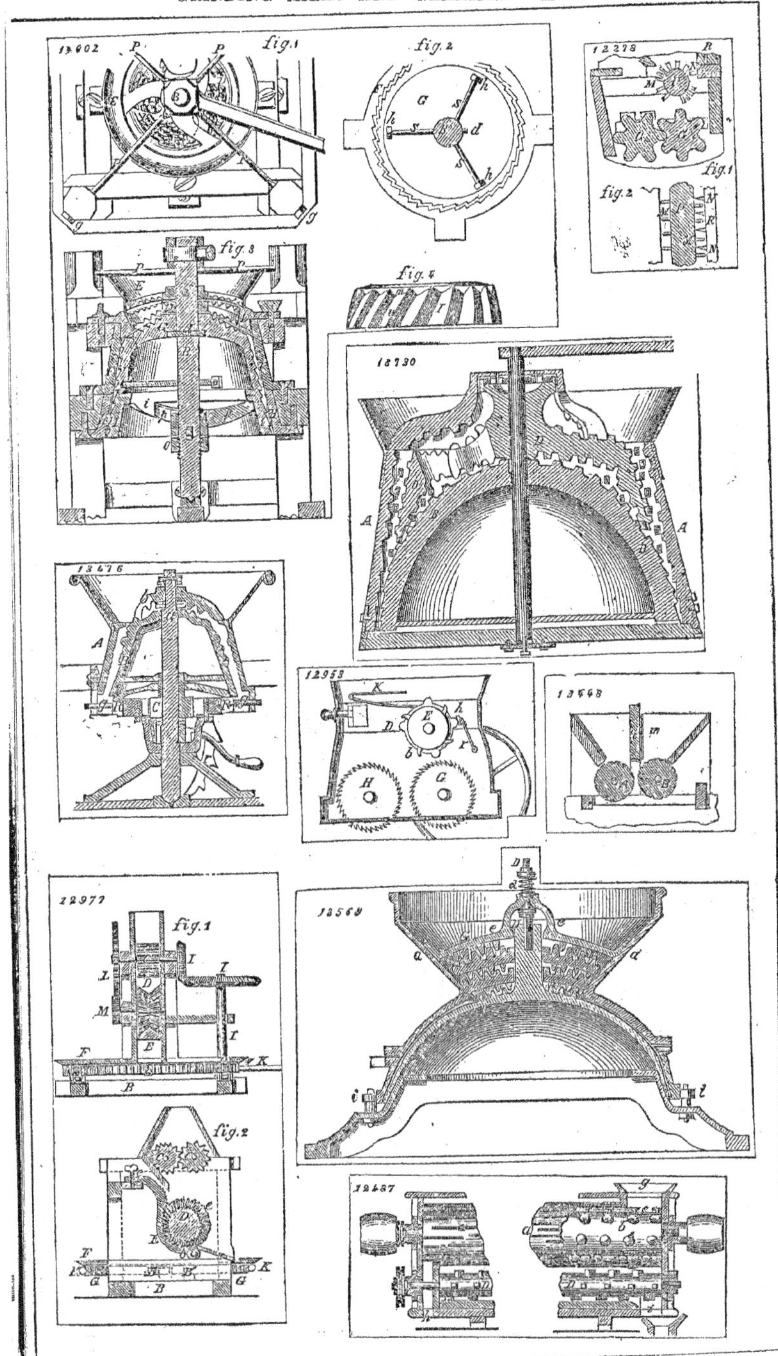

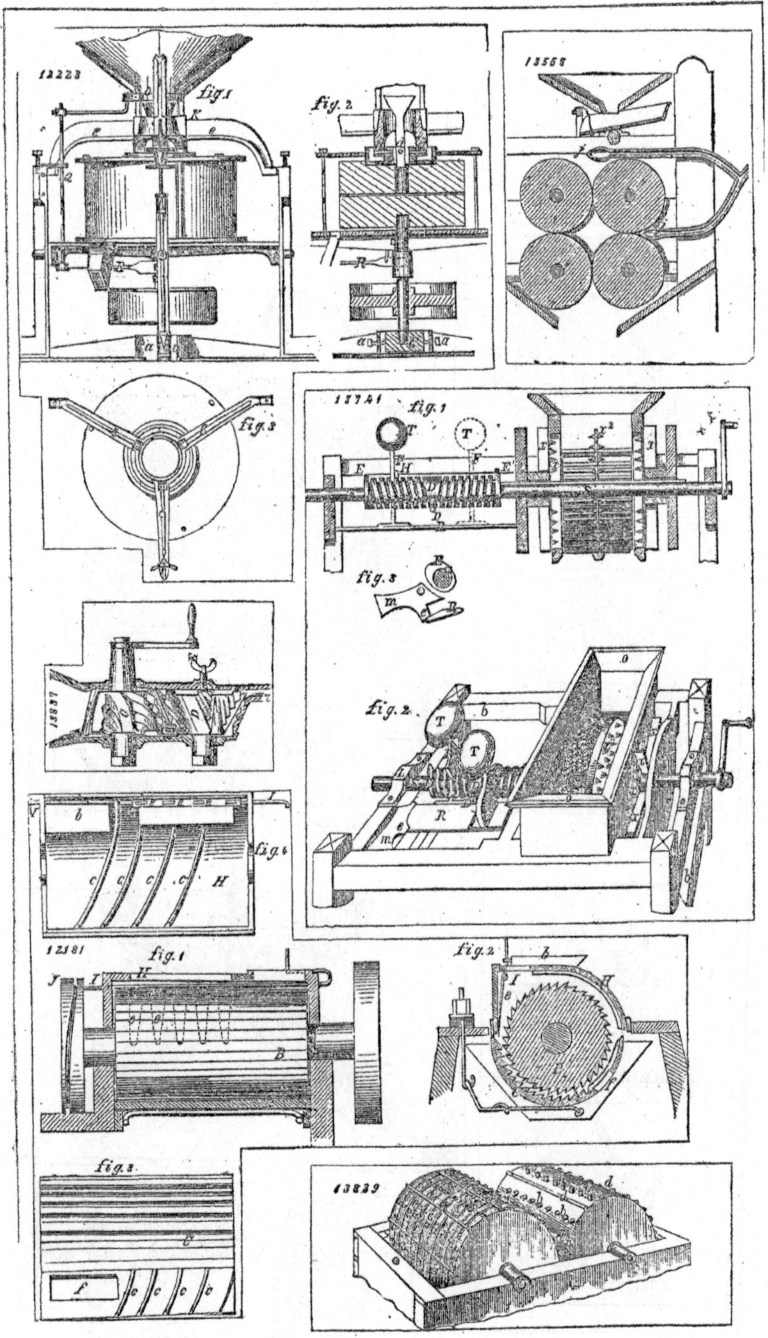

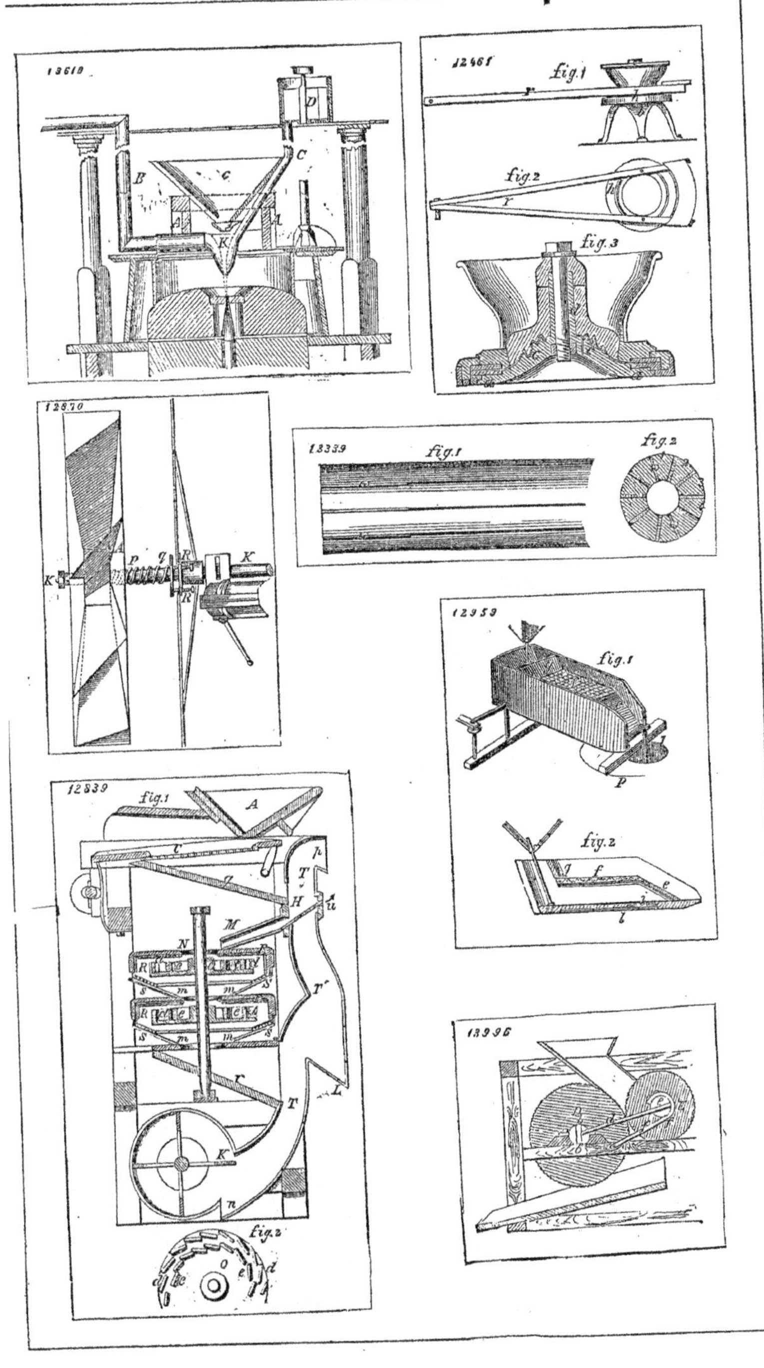
13618
12461
fig.1
fig.2
fig.3
12870
13339
fig.1
fig.2
12959
fig.1
fig.2
12839
fig.1
fig.2
13996

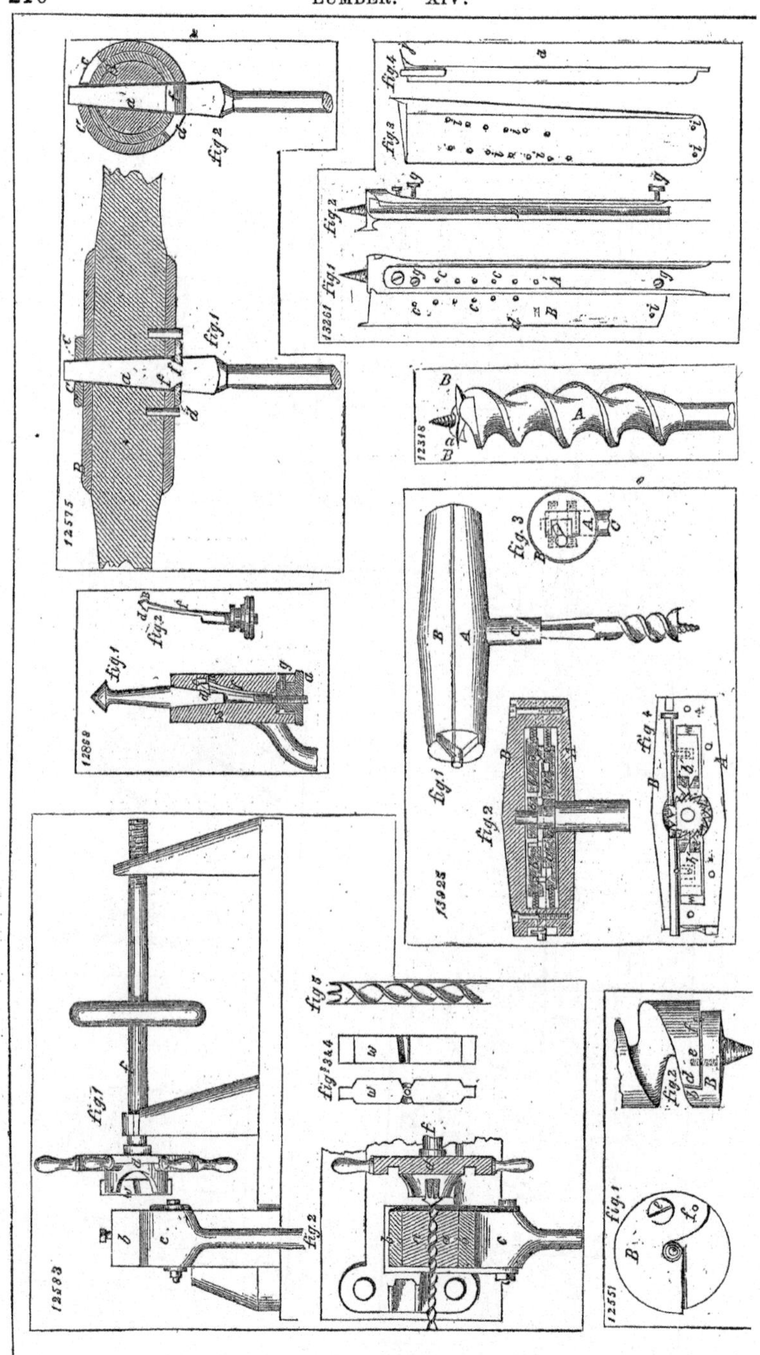
12575
13261
12318
12869
13923
12583
12551

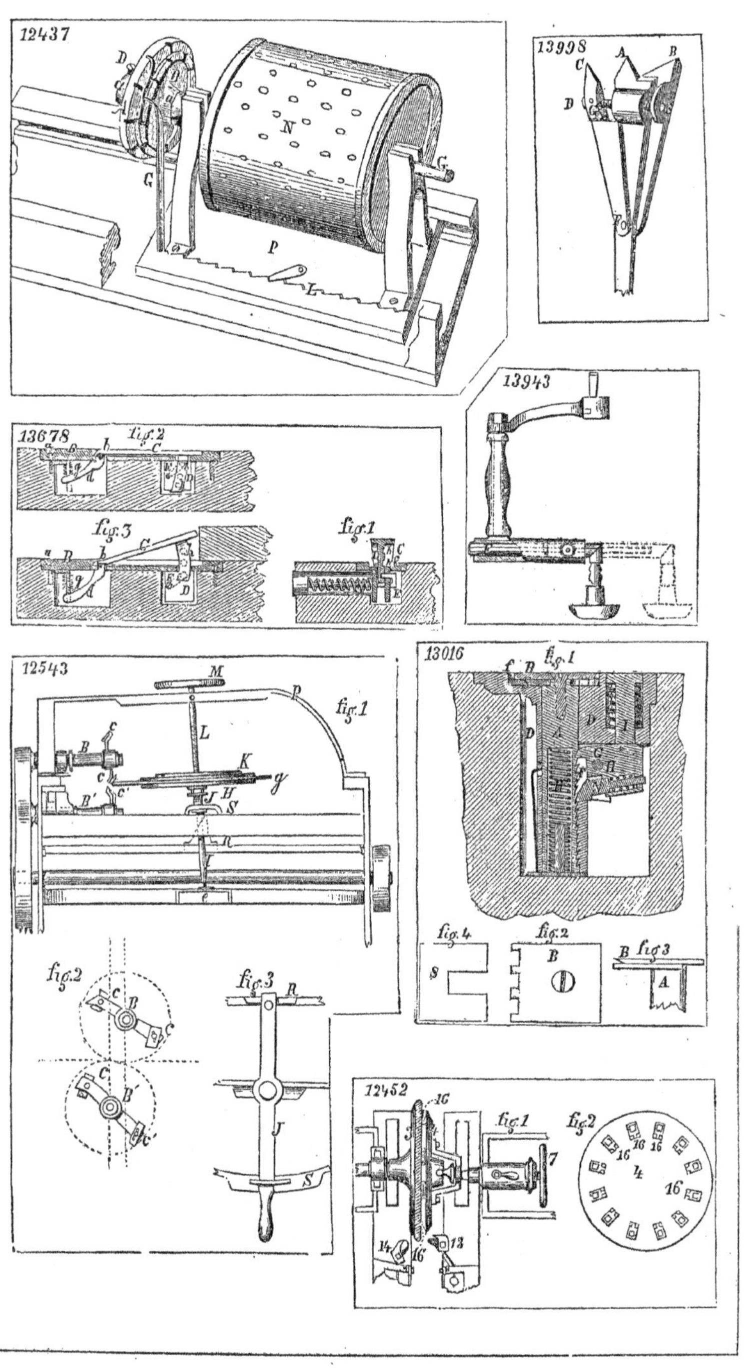
12437
13998
13943
13678
fig.2
fig.3
fig.1
12543
13016
fig.4
12452

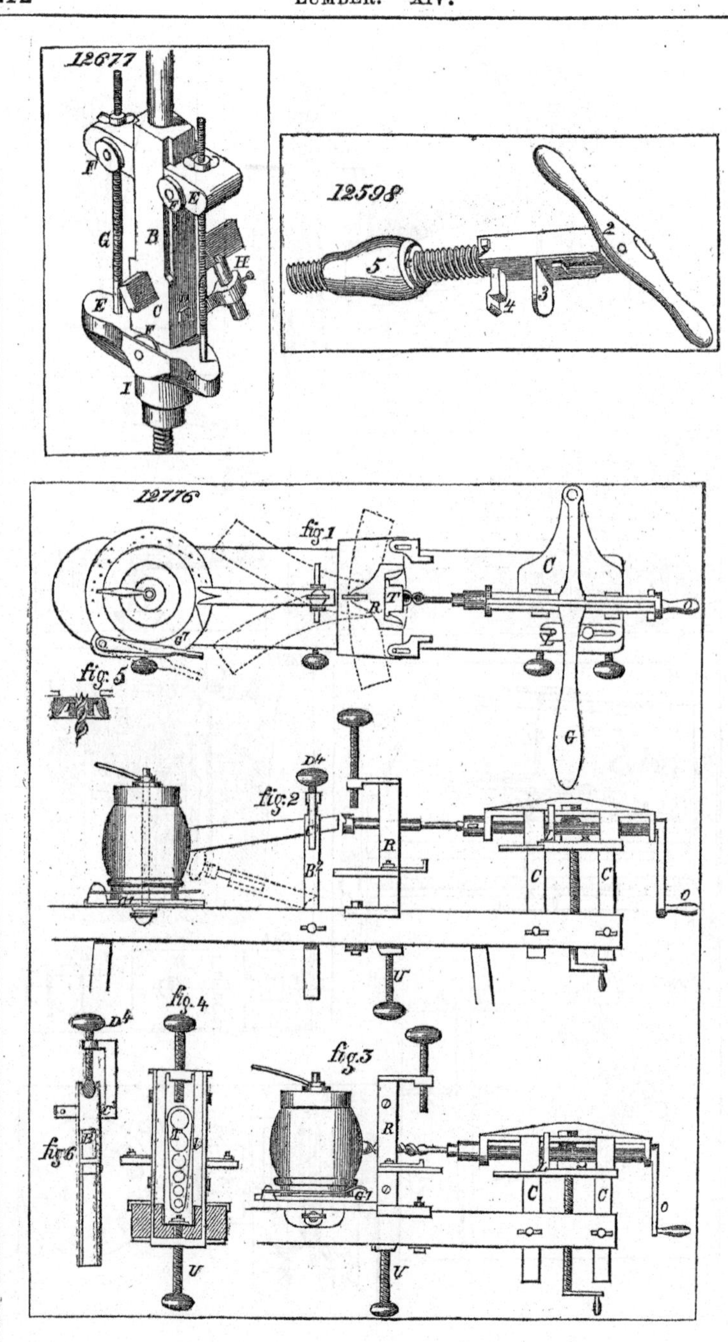
12677
12598
12776
fig.1
fig.2
fig.3
fig.4
fig.5
fig.6

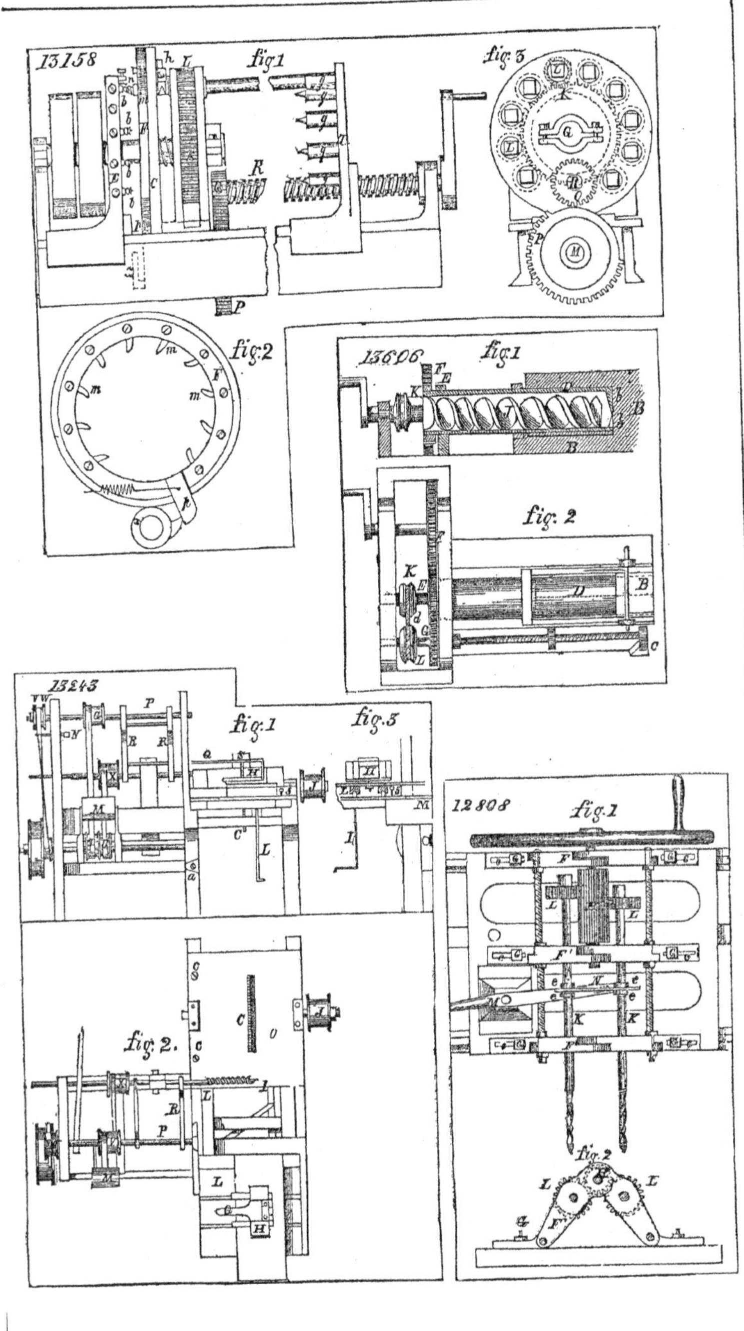
13158
fig 1
fig 3
fig 2
13606
fig 1
fig 2
13243
fig 1
fig 3
fig 2.
12808
fig 1
fig 2

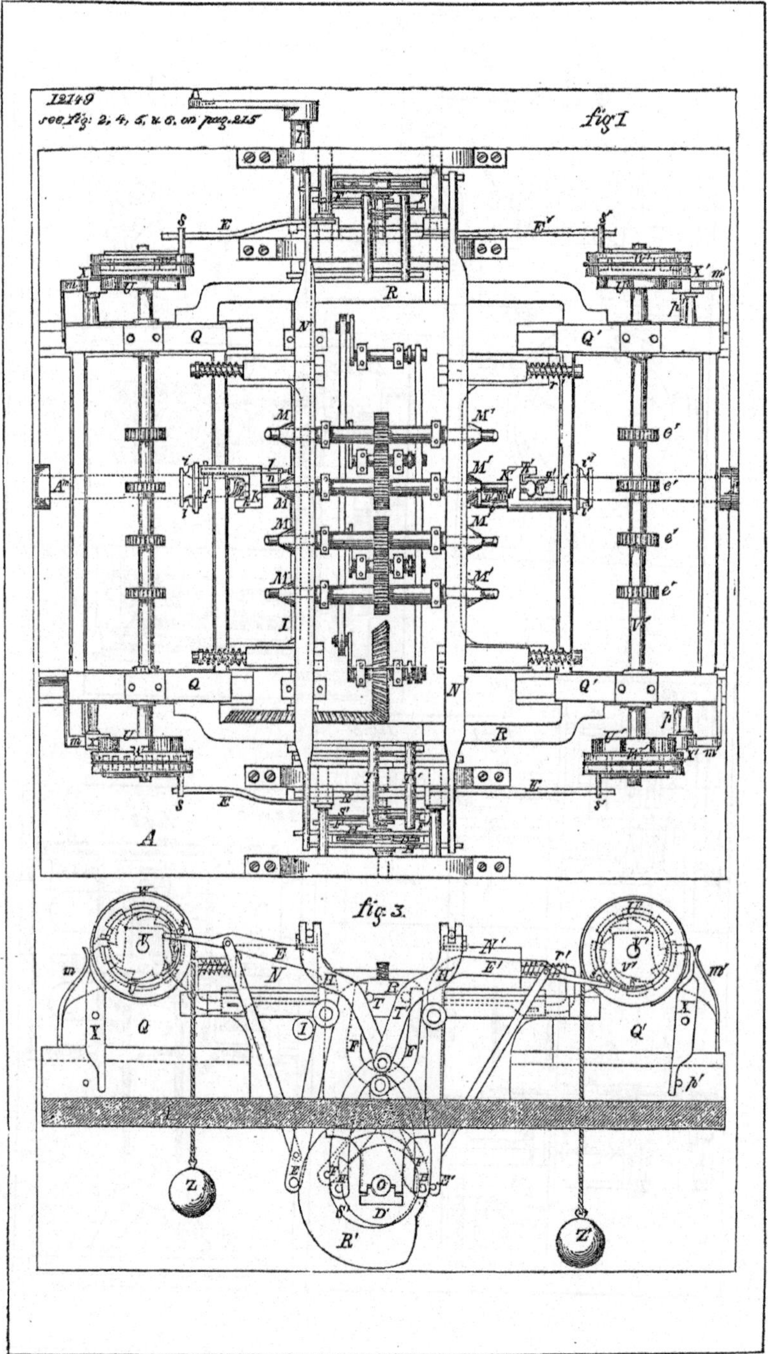
12749
see fig: 2, 4, 5, & 6. on pag. 215
fig 1
fig. 3.

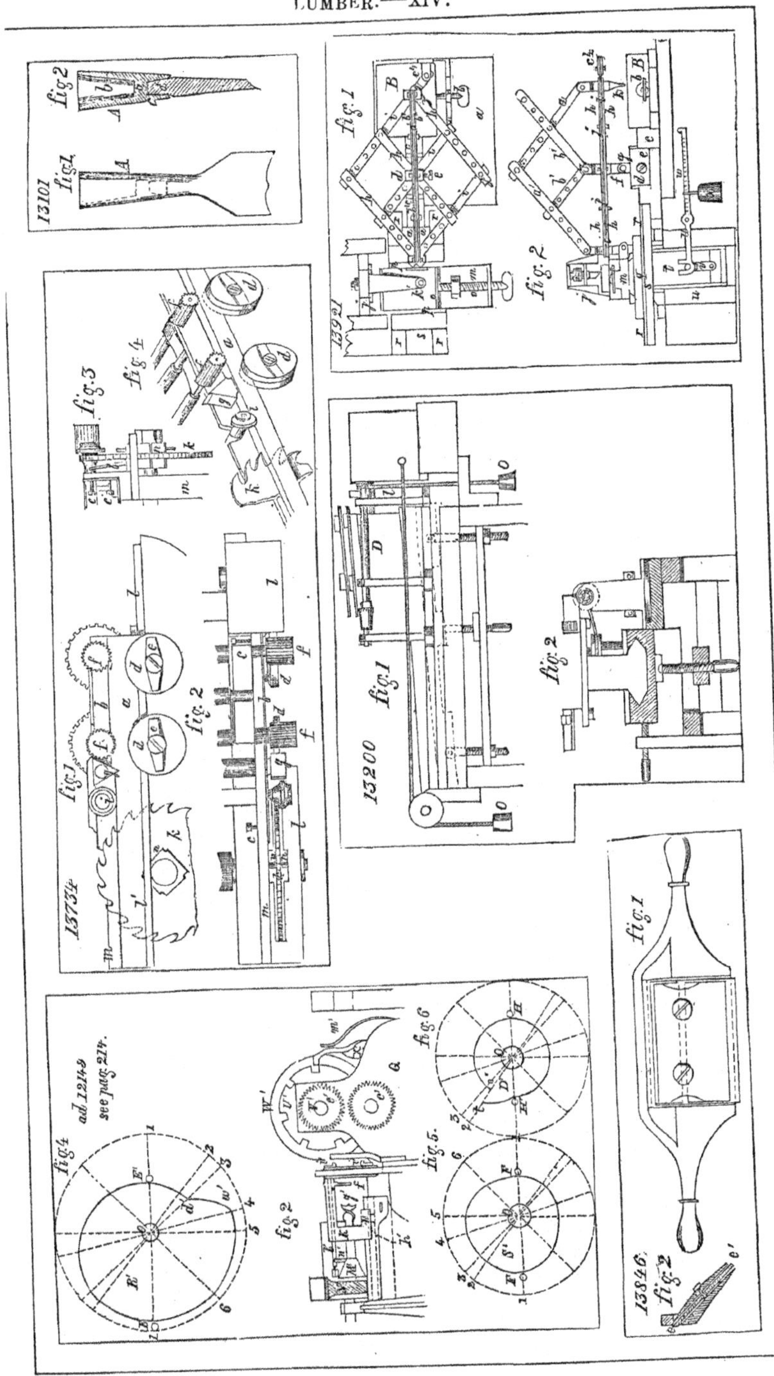
13101
13734
13200
13921
ad 12148
see pag. 214.
13846

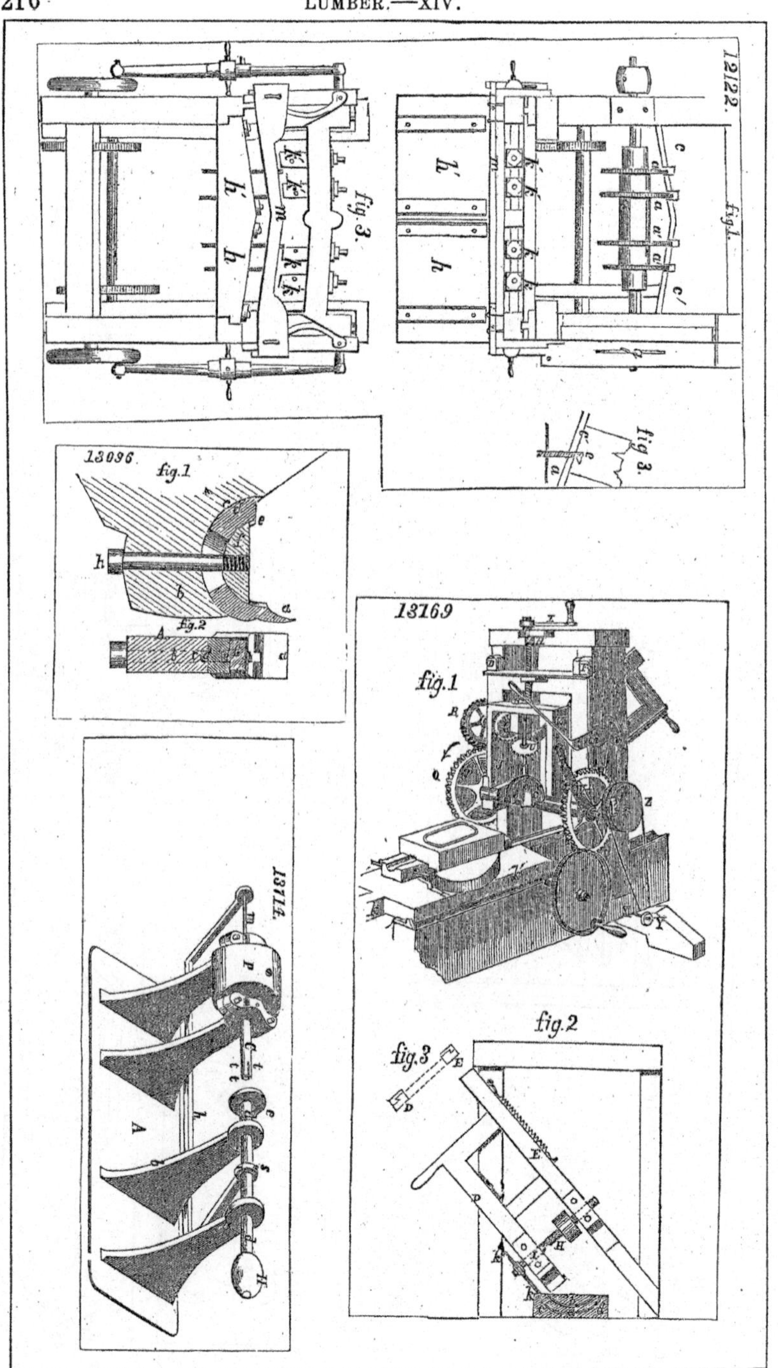
12122.
fig. 1.
fig. 3.
13096
fig. 1
fig. 2
13169
fig. 1
fig. 2
fig. 3
13714.

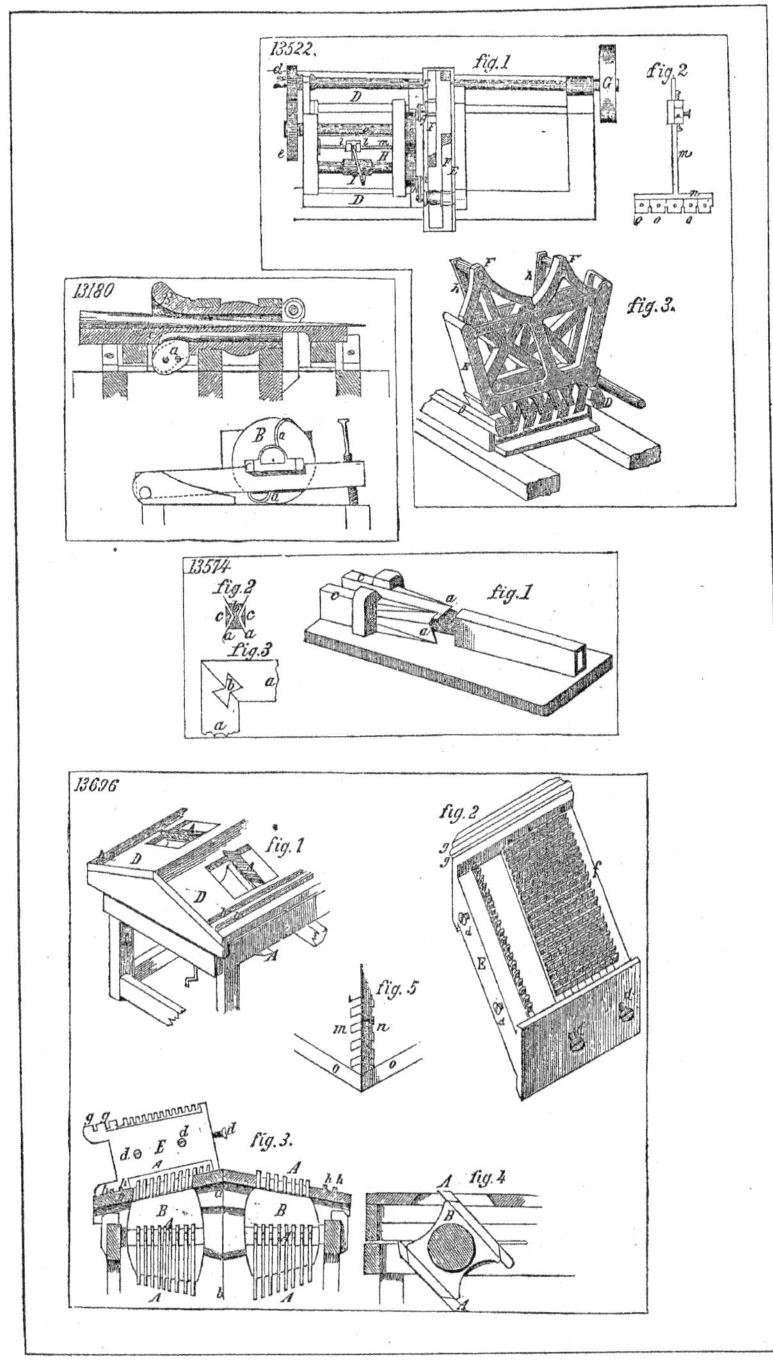
13522.
fig.1
fig.2
fig.3.
13180
13574
fig.2
fig.3
fig.1
13696
fig.1
fig.2
fig.5
fig.3.
fig.4

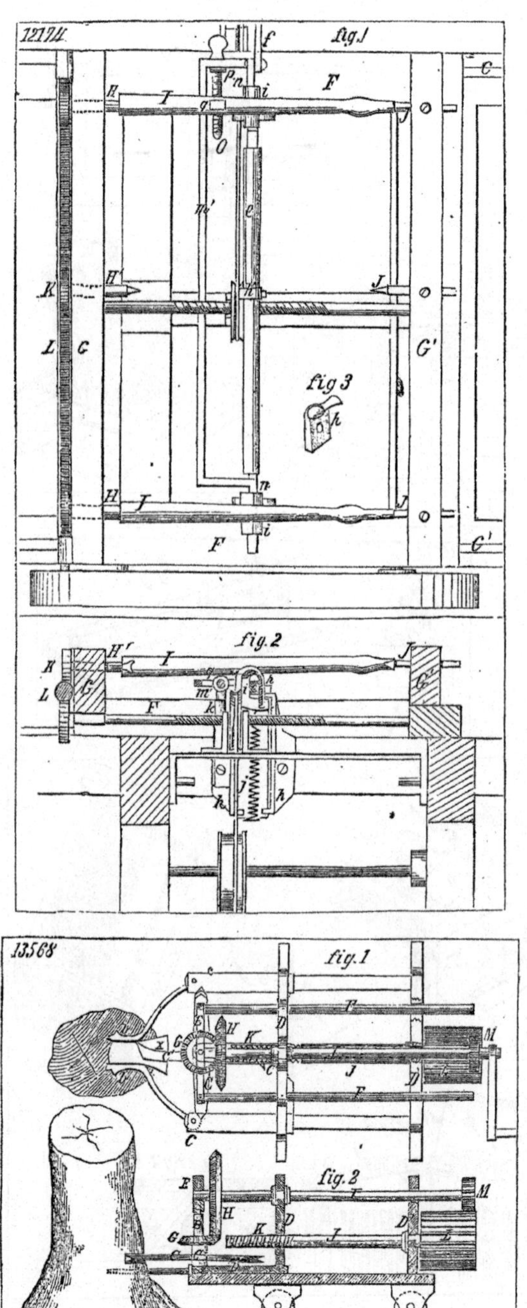
12174
fig.1
fig 3
fig.2
13568
fig.1
fig.2

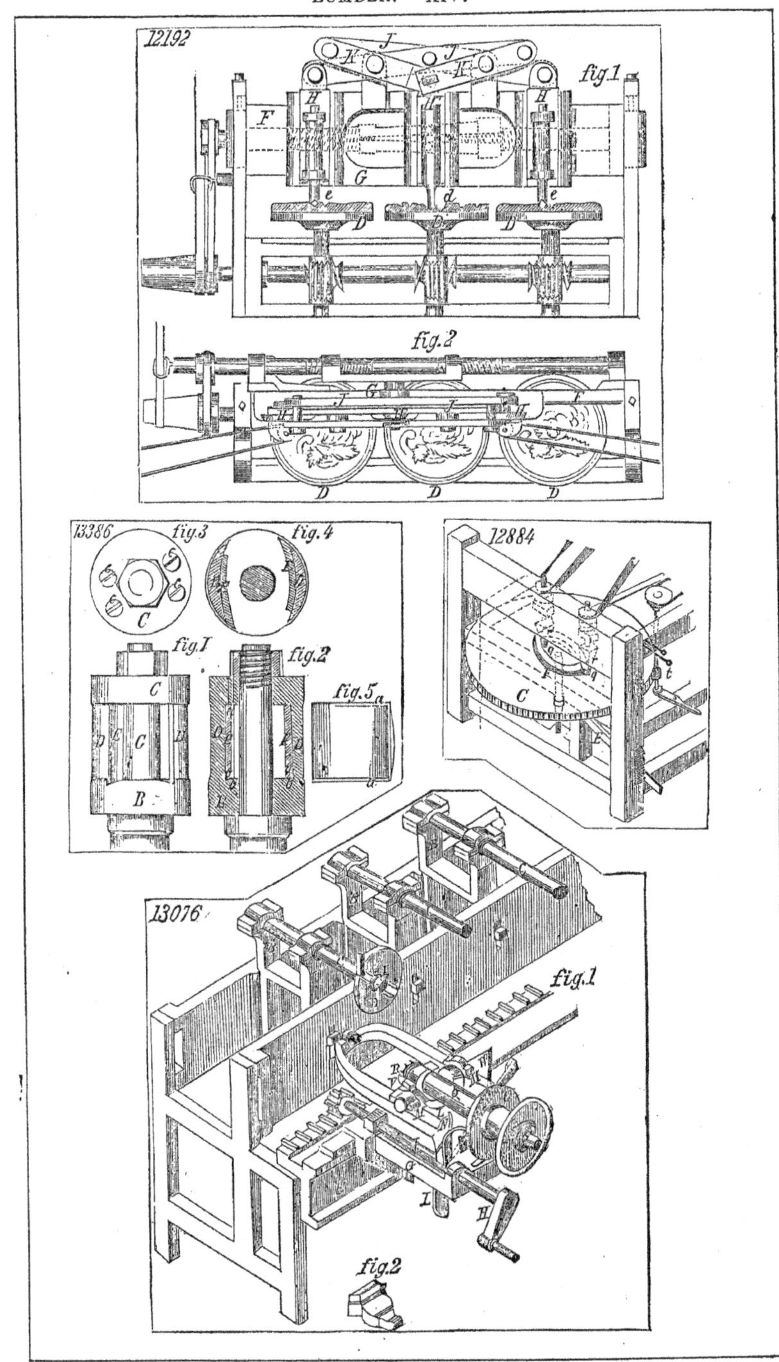
12192
fig.1
fig.2
13386
fig.3
fig.4
fig.1
fig.2
fig.5
12884
13076
fig.1
fig.2

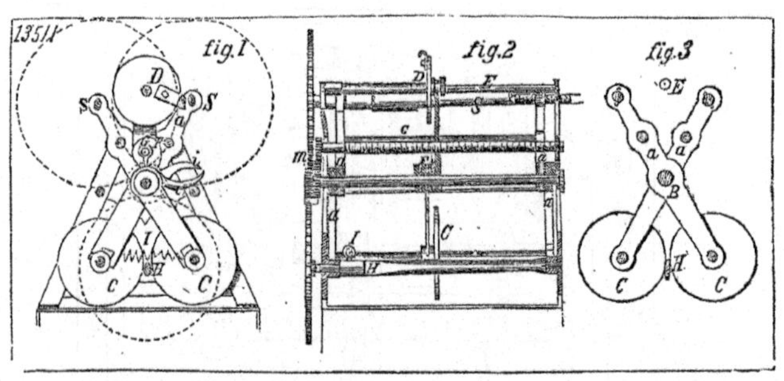
13511
fig.1
fig.2
fig.3

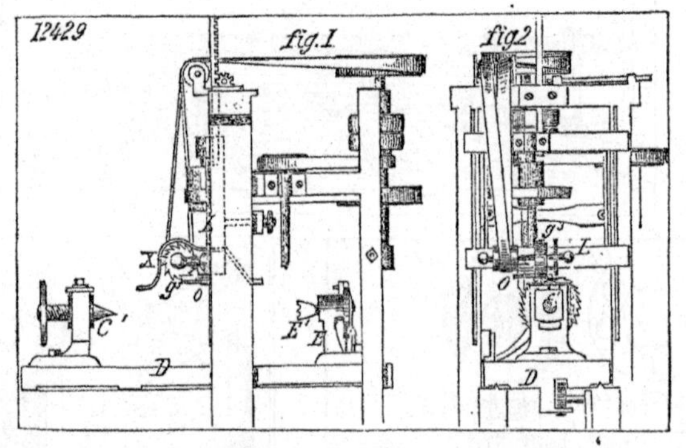
12429
fig.1
fig.2

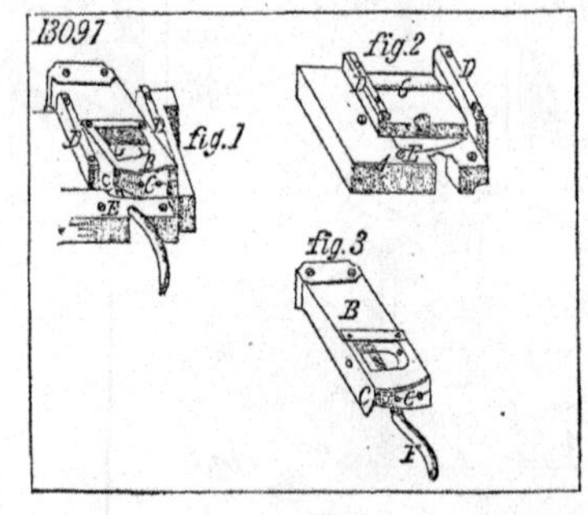
13097
fig.1
fig.2
fig.3

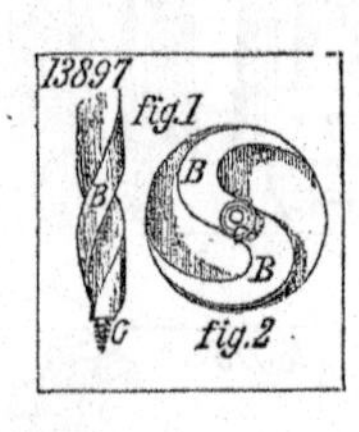
13897
fig.1
fig.2

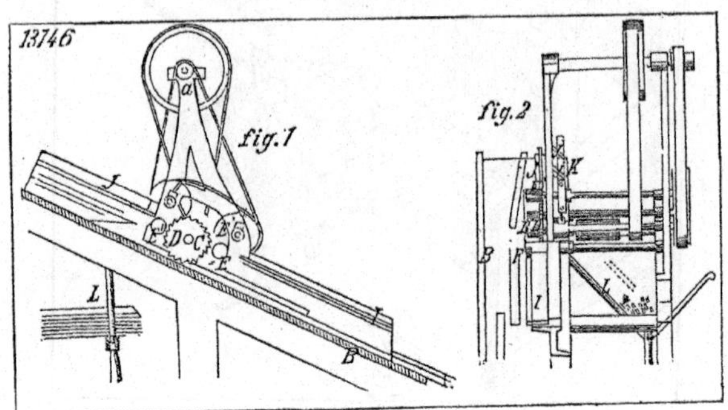
13146
fig.1
fig.2

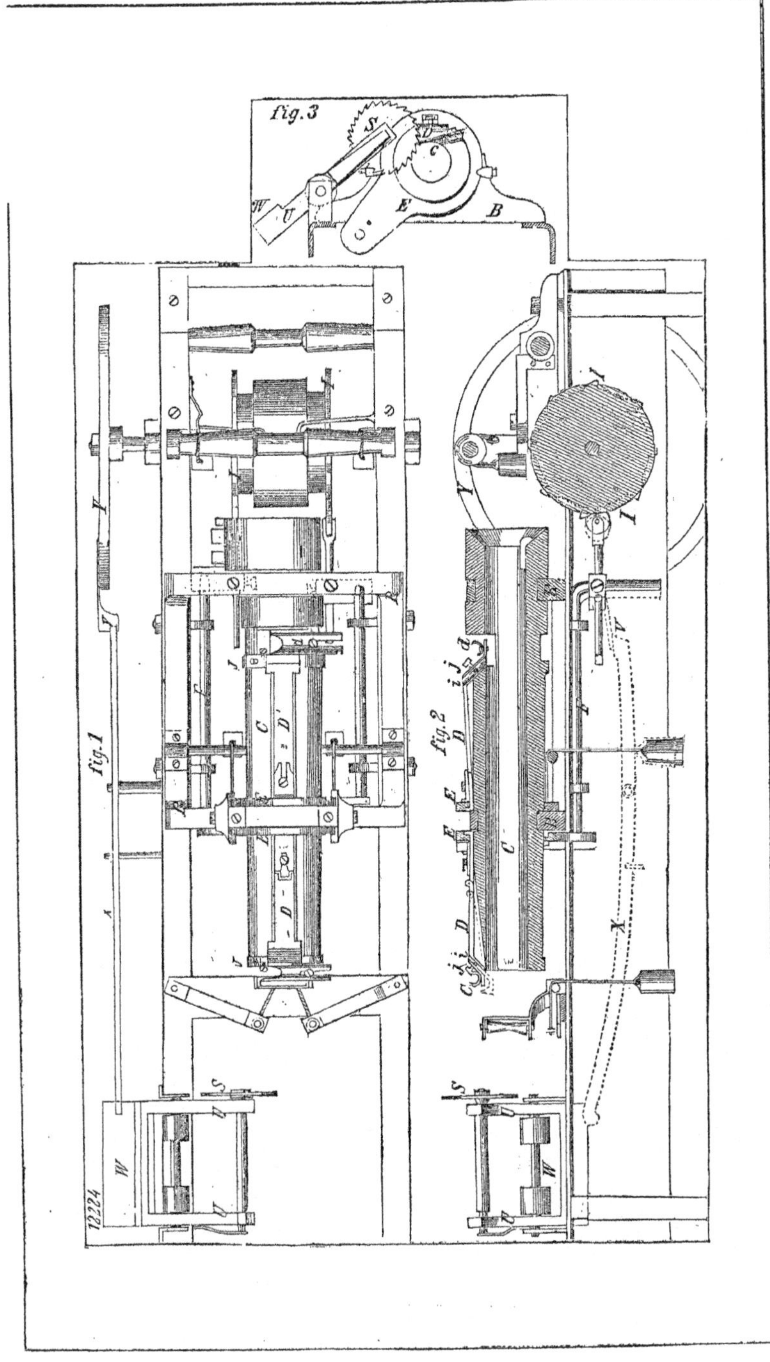
fig. 3
fig. 1
fig. 2
12224

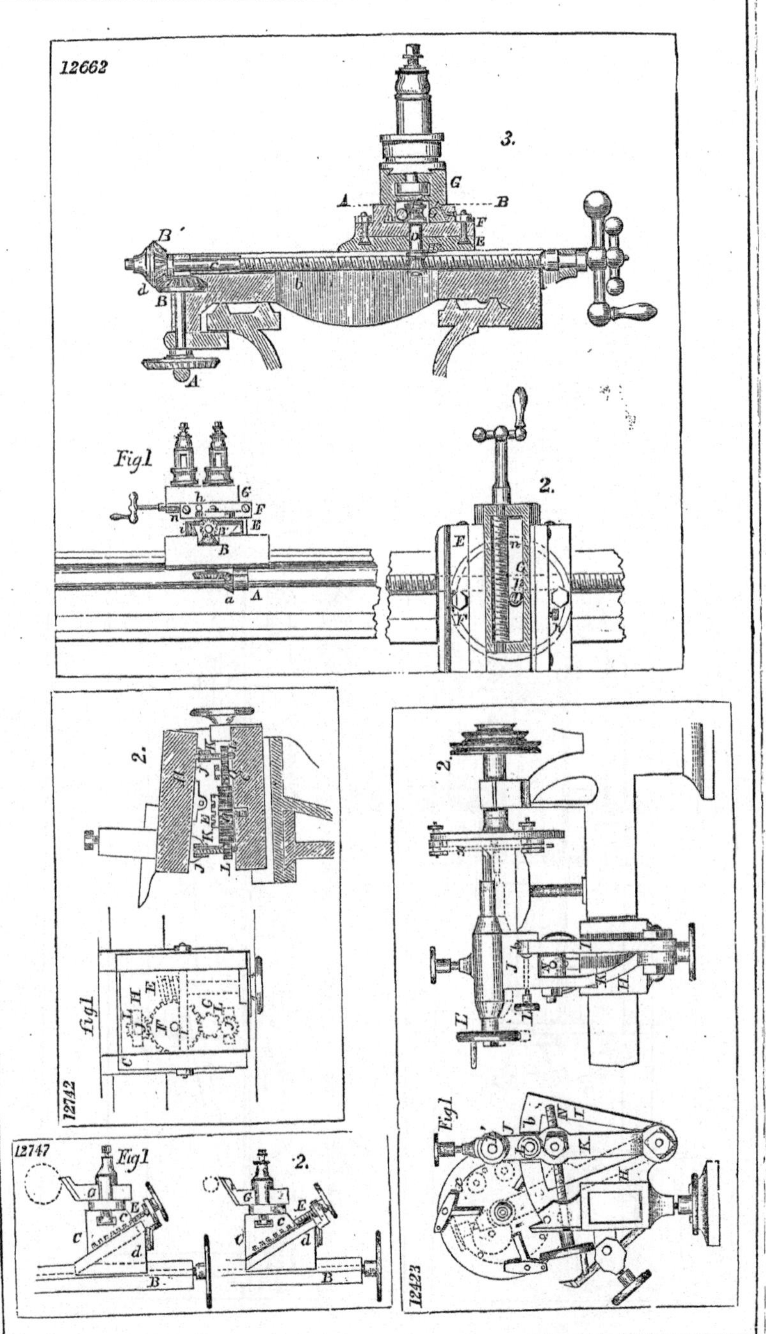

12662
3.
Fig1
2.
12742
2.
fig1
12423
2.
Fig1
12747
Fig1
2.

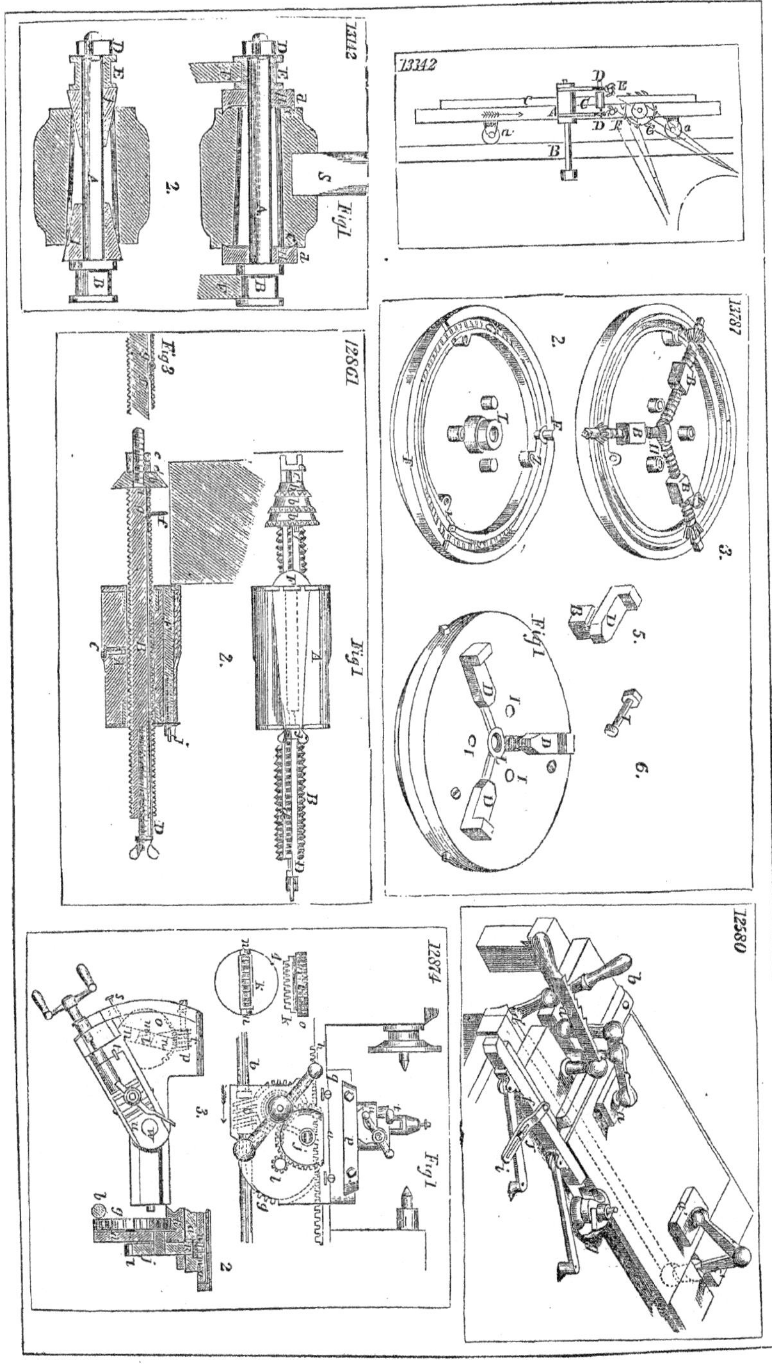

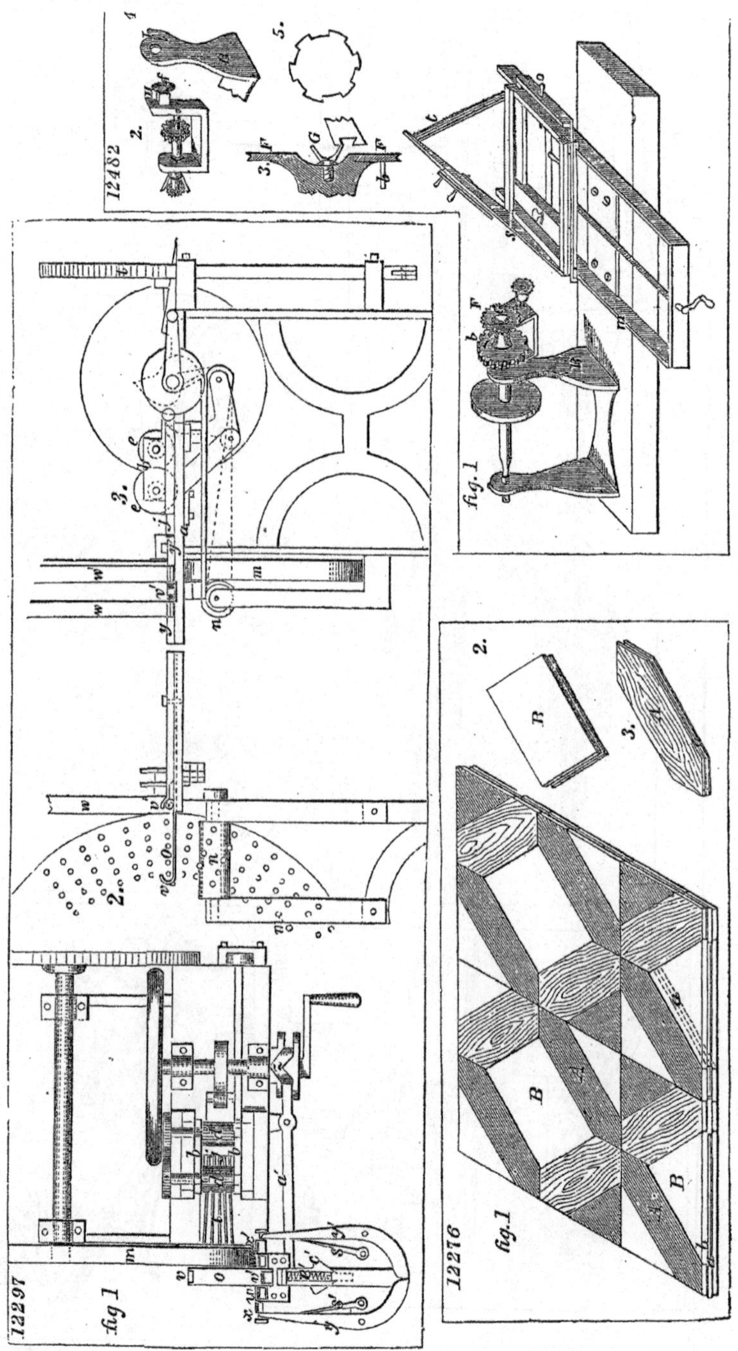
12297
fig 1
2.
3.
12482
fig. 1
12276
fig. 1
2.
3.

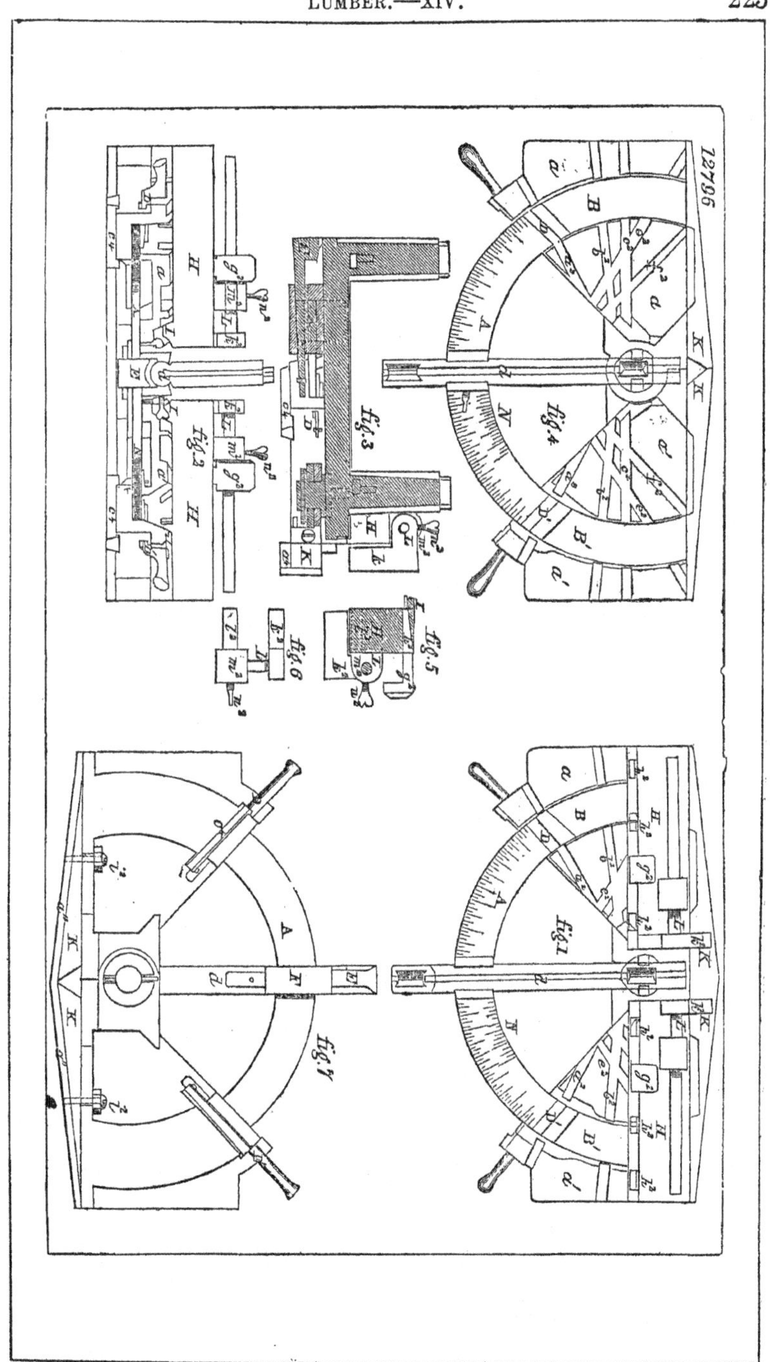
12796
fig.1
fig.2
fig.3
fig.4
fig.5
fig.6
fig.7

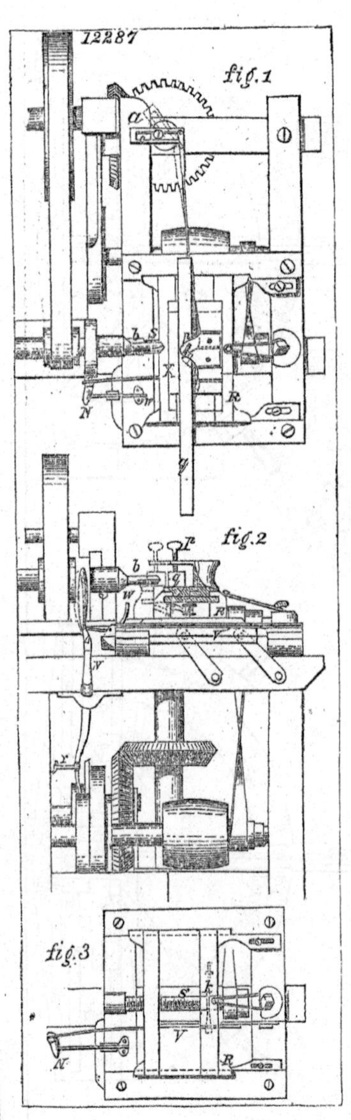
12287
fig.1
fig.2
fig.3

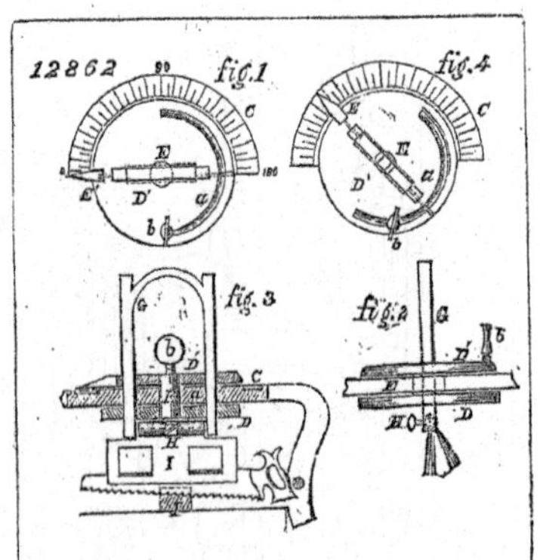
12862
fig.1
fig.4
fig.3
fig.2

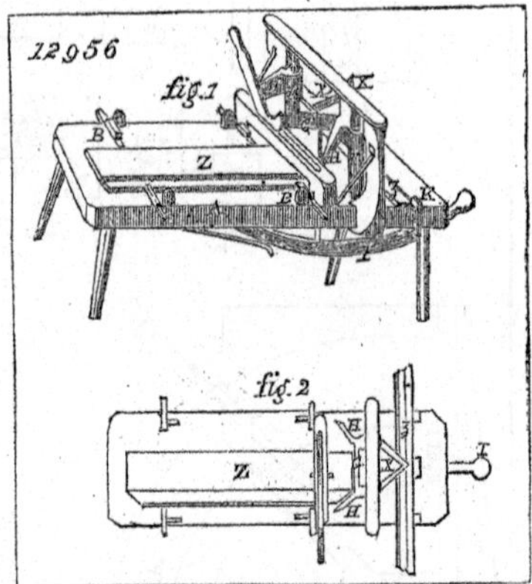
12956
fig.1
fig.2

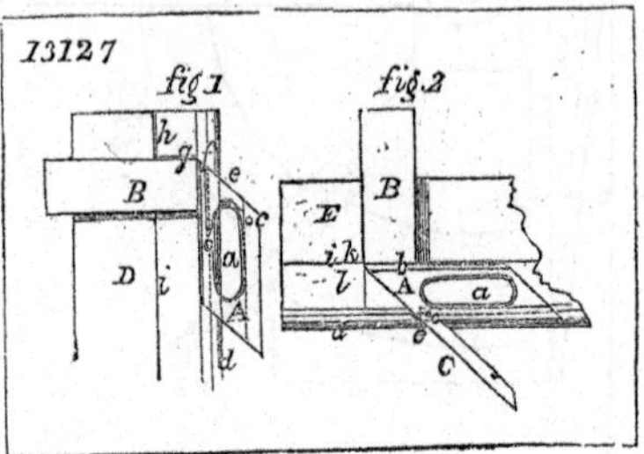
13127
fig.1
fig.2

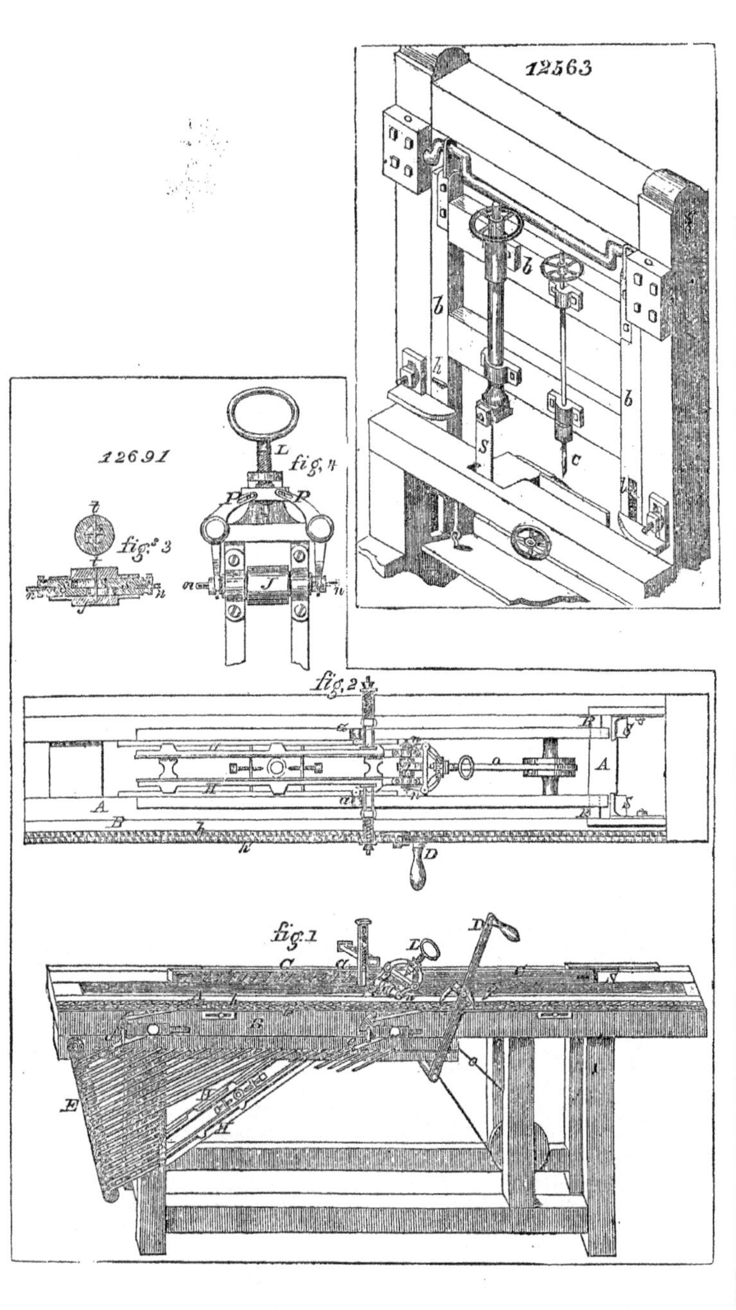
12563
12691
fig. 4
fig. 3
fig. 2
fig. 1

13184

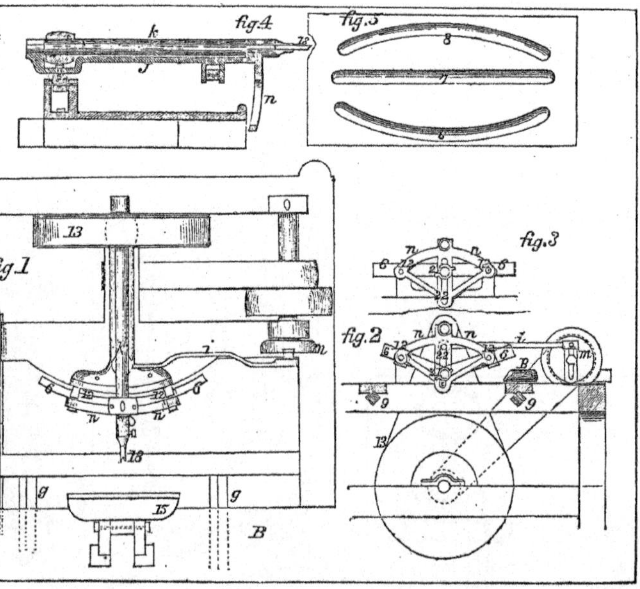

13271

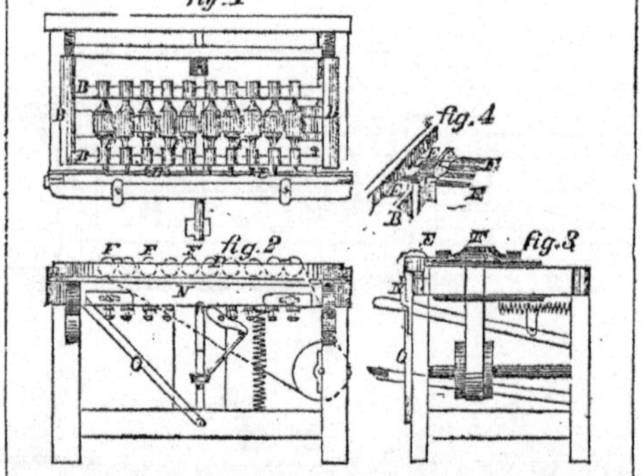

13594

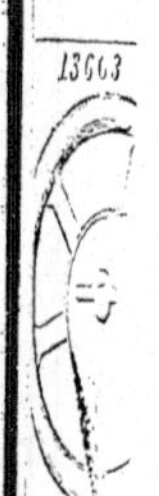

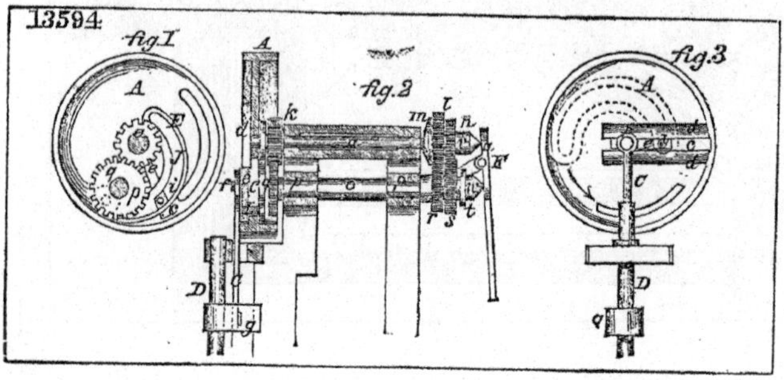

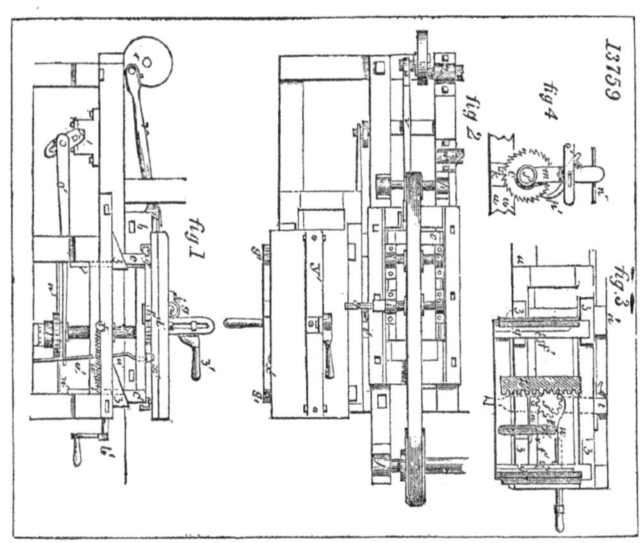
13759
fig 1
fig 2
fig 3
fig 4

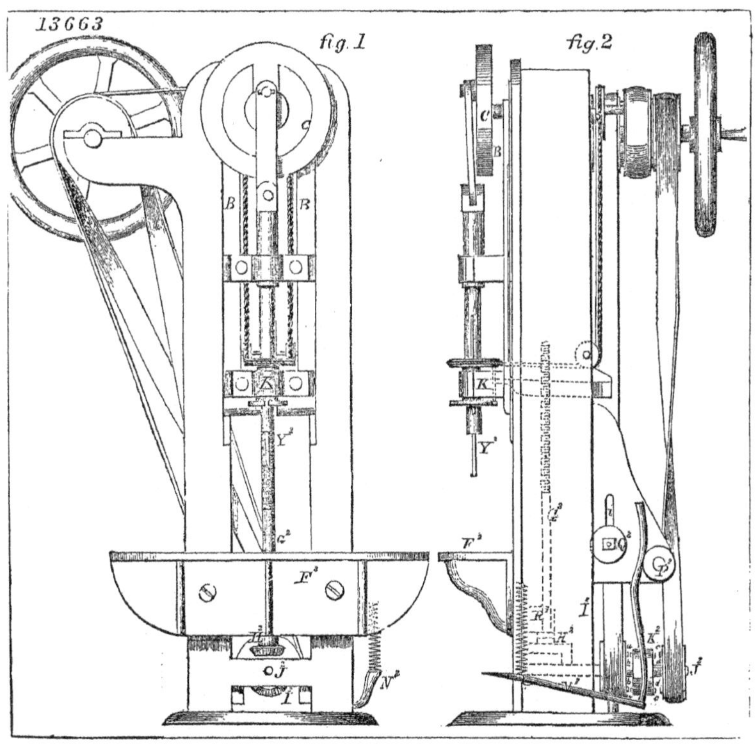
13663
fig. 1
fig. 2

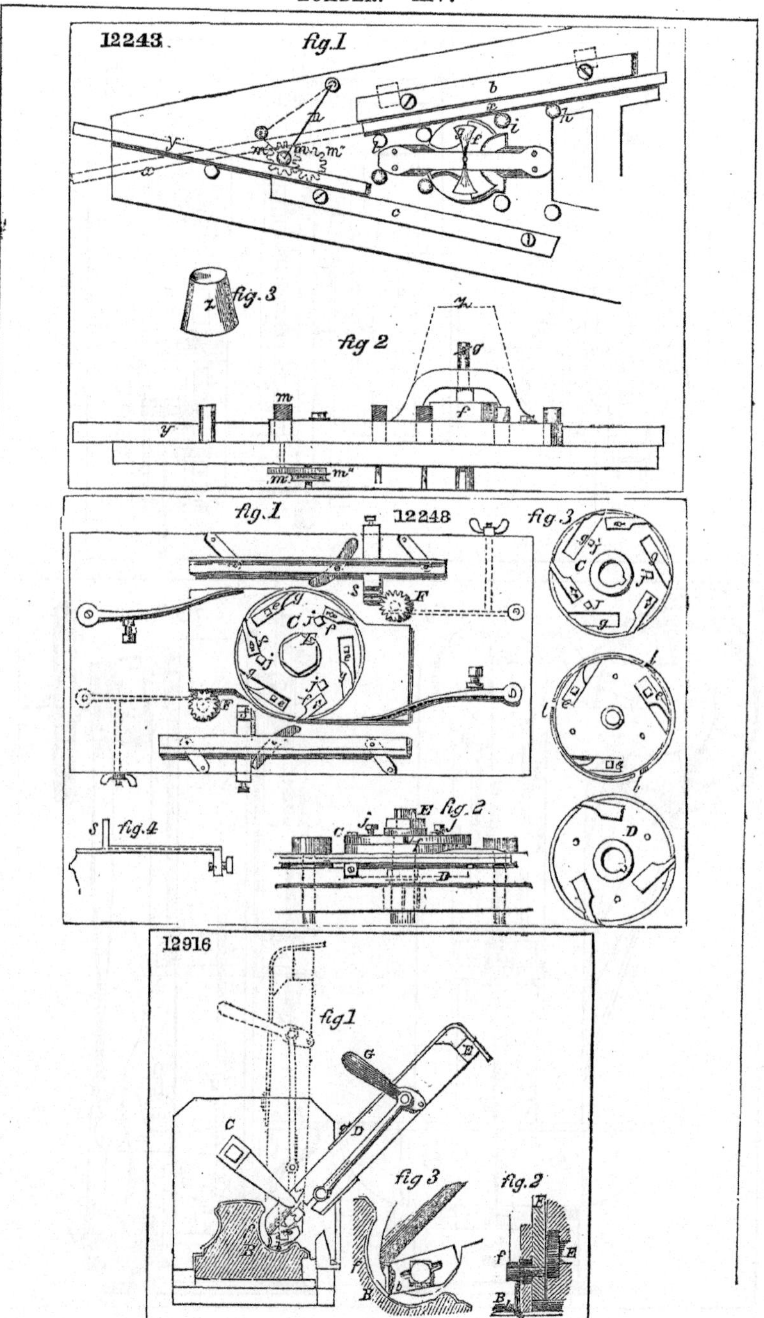
12243
fig.1
fig.3
fig 2
fig.1
12248
fig.3
fig.4
fig.2
12916
fig1
fig 3
fig.2

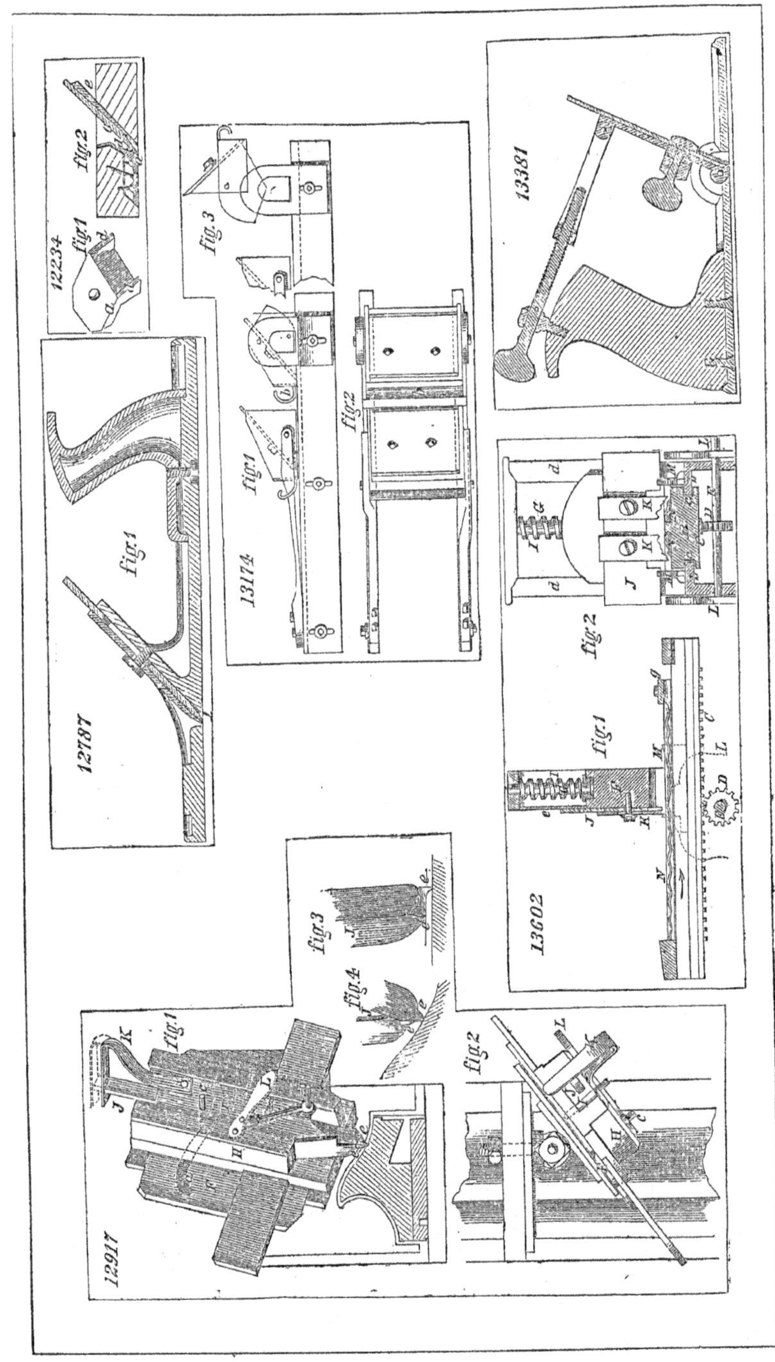
12234
fig.1
fig.2
13174
fig.1
fig.2
fig.3
12787
fig.1
13381
13602
fig.1
fig.2
12917
fig.1
fig.2
fig.3
fig.4

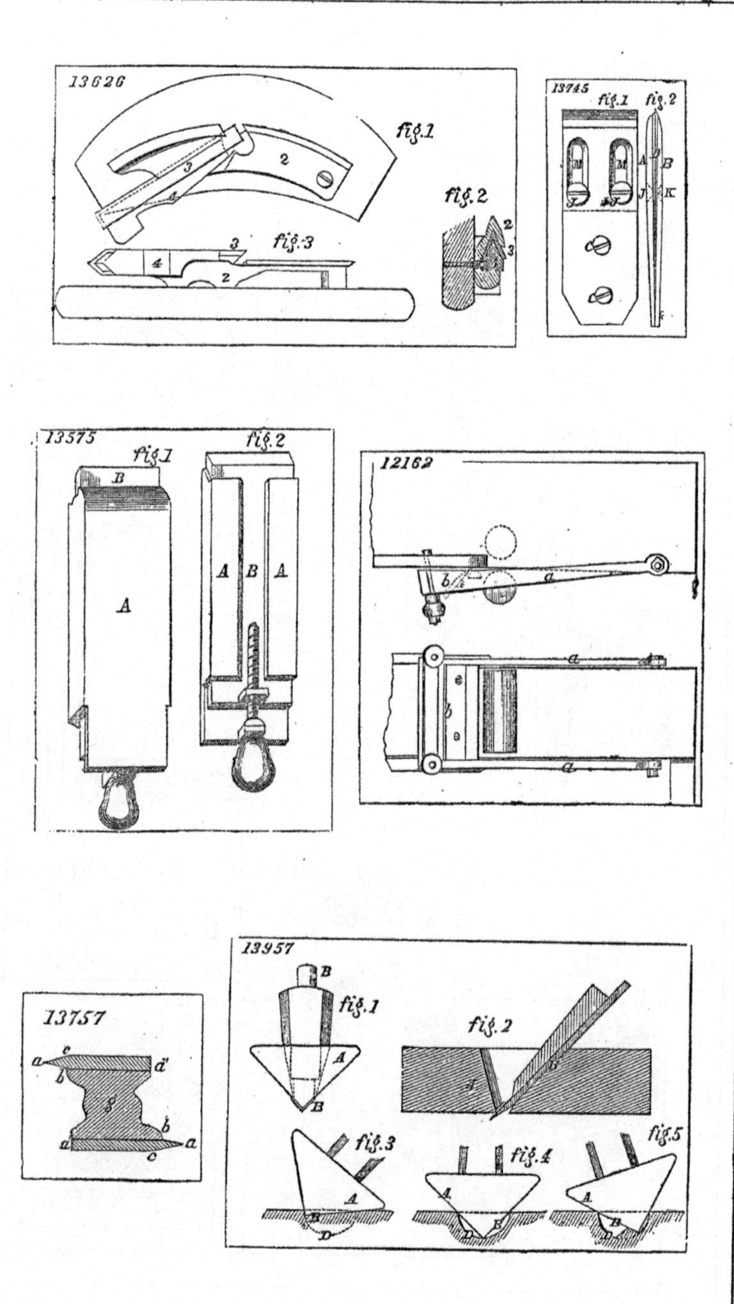
13626
fig.1
fig.2
fig.3
13745
fig.1
fig.2
13575
fig.1
fig.2
12162
13757
13957
fig.1
fig.2
fig.3
fig.4
fig.5

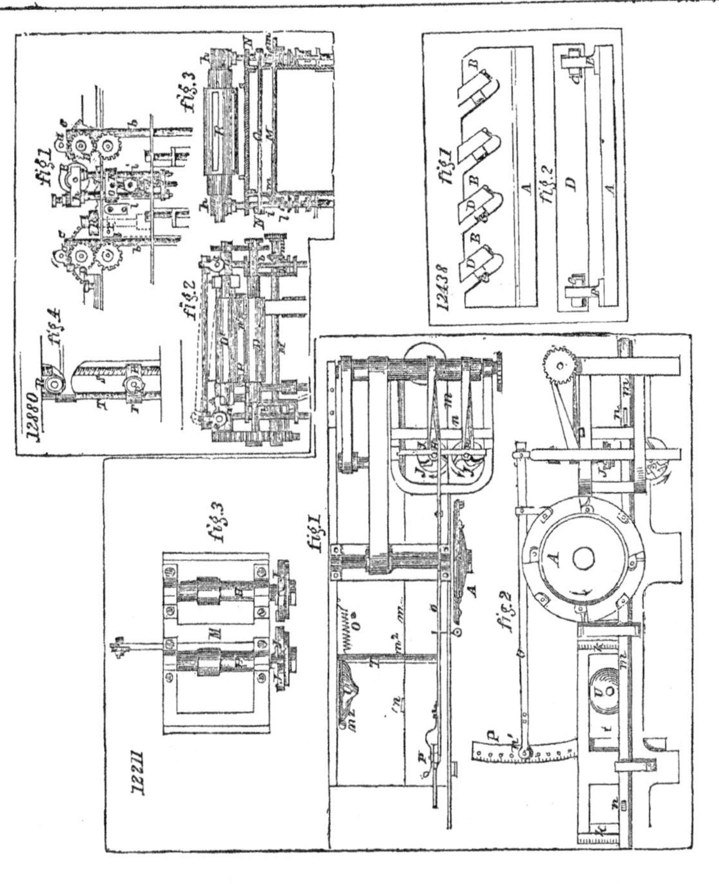
12880
12438
12211

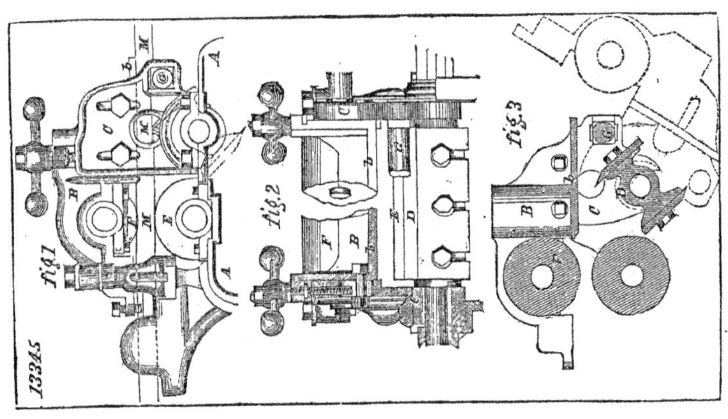
13345

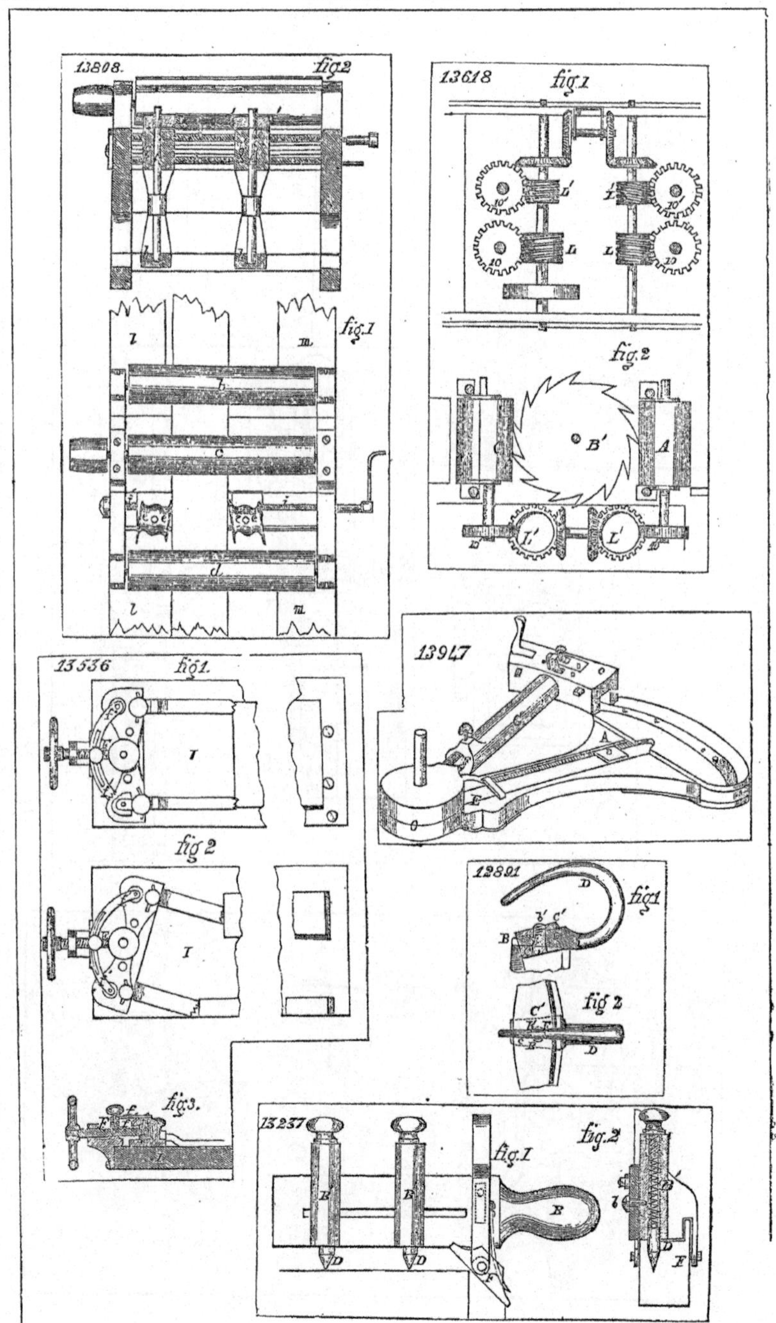
13808.
fig 2
fig.1
13618
fig.1
fig.2
13536
fig.1.
fig 2
fig.3.
13947
12891
fig.1
fig 2
13237
fig.1
fig.2

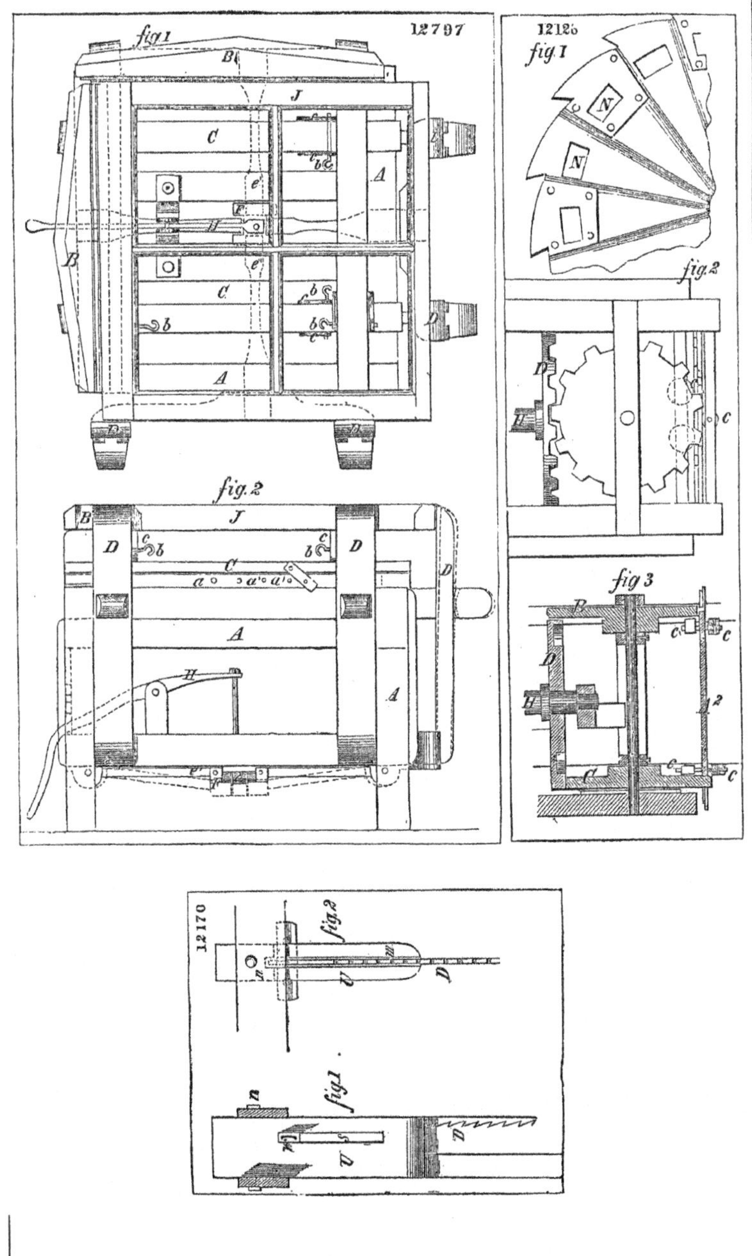
12797
fig 1
fig. 2
12126
fig. 1
fig. 2
fig 3
12170
fig. 2
fig. 1

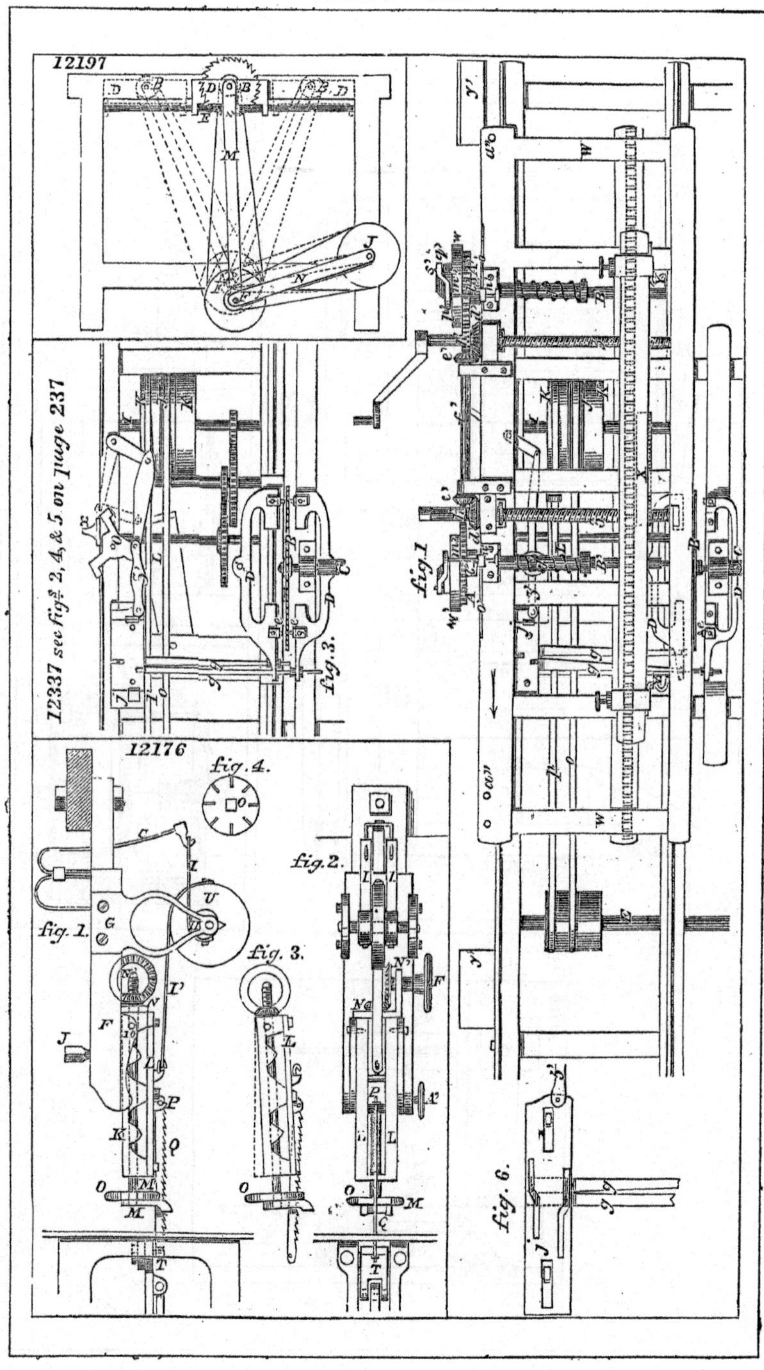
12197
12337 see figs 2, 4, & 5 on page 237
fig. 3.
fig. 1
12176
fig. 4.
fig. 1.
fig. 3.
fig. 2.
fig. 6.

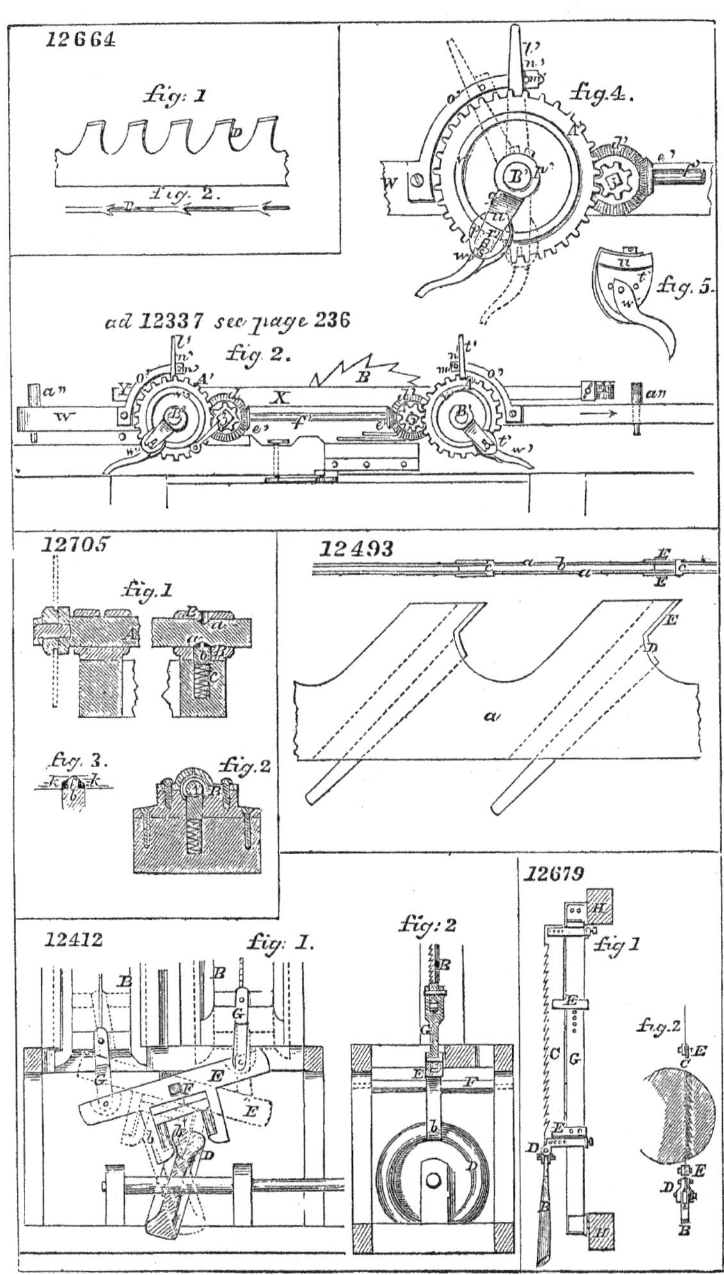
12664
fig: 1
fig. 2.
fig.4.
fig. 5.
ad 12337 see page 236
fig. 2.
12705
fig.1
fig. 3.
fig.2
12493
12679
fig 1
fig.2
12412
fig: 1.
fig: 2

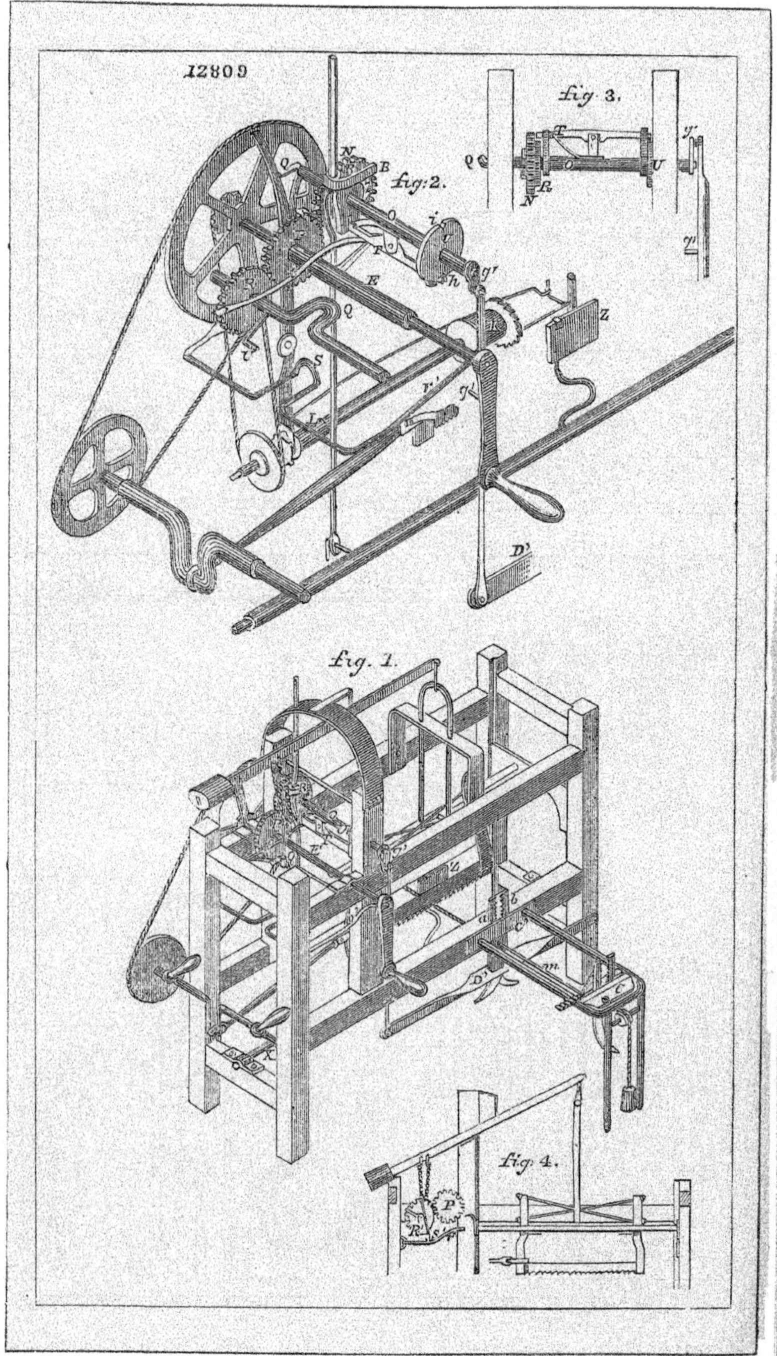
12809
Fig. 1.
Fig. 2.
Fig. 3.
Fig. 4.

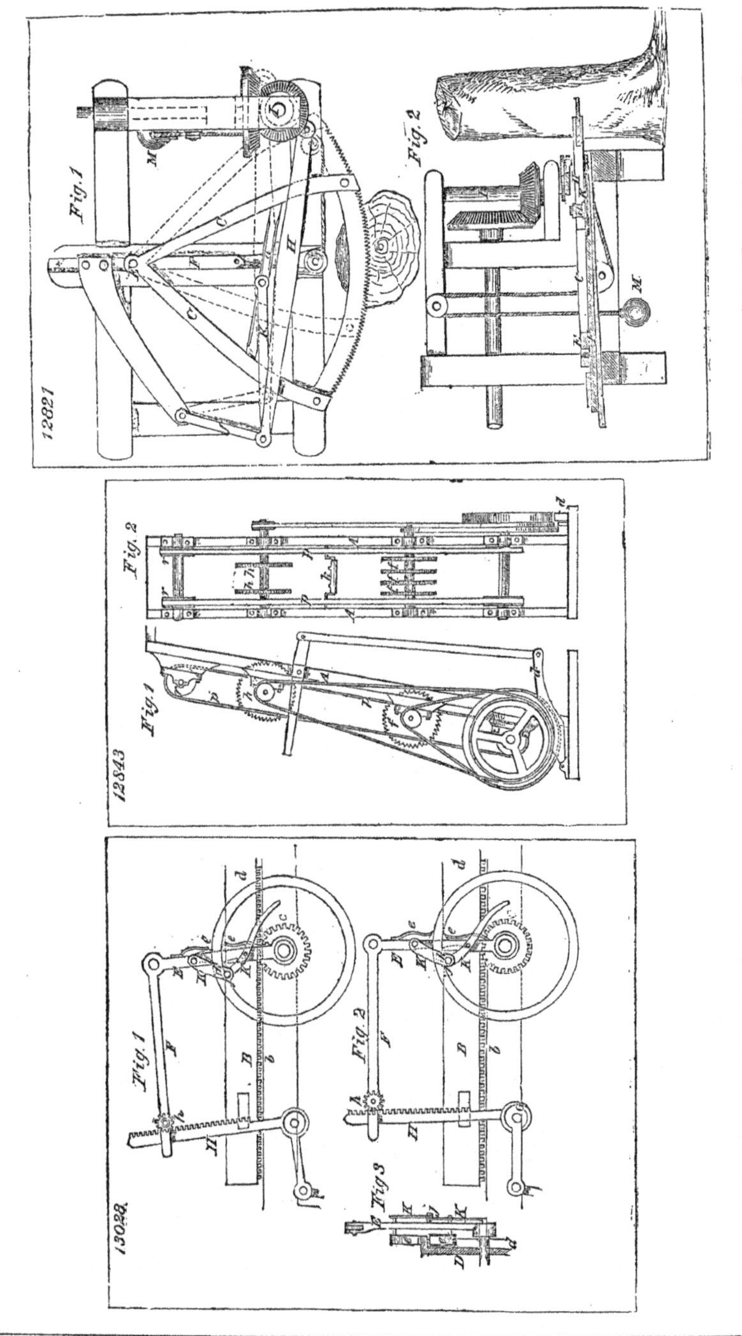
12821
Fig. 1
Fig. 2
12843
Fig. 1
Fig. 2
13022
Fig. 1
Fig. 2
Fig. 3

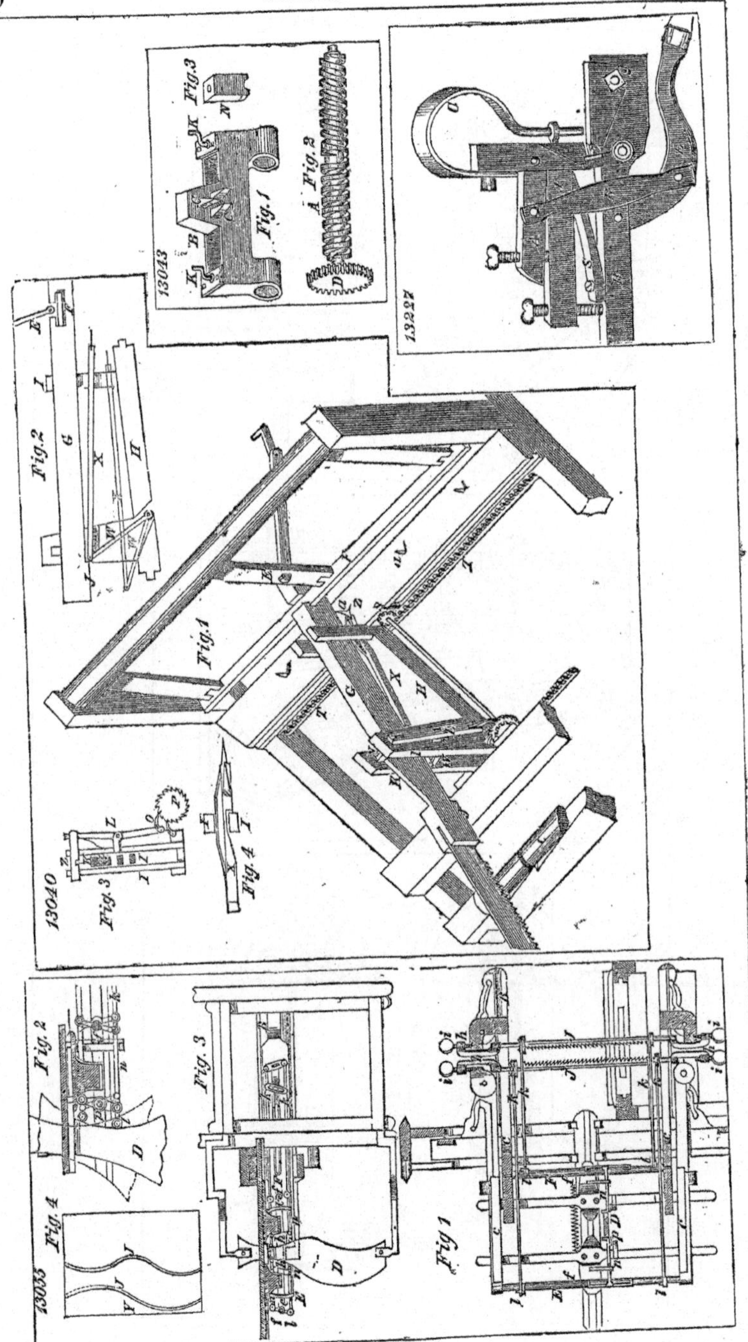
13043
Fig.1
Fig.2
Fig.3
13227
13040
Fig.1
Fig.2
Fig.3
Fig.4
13035
Fig 1
Fig. 2
Fig. 3
Fig. 4

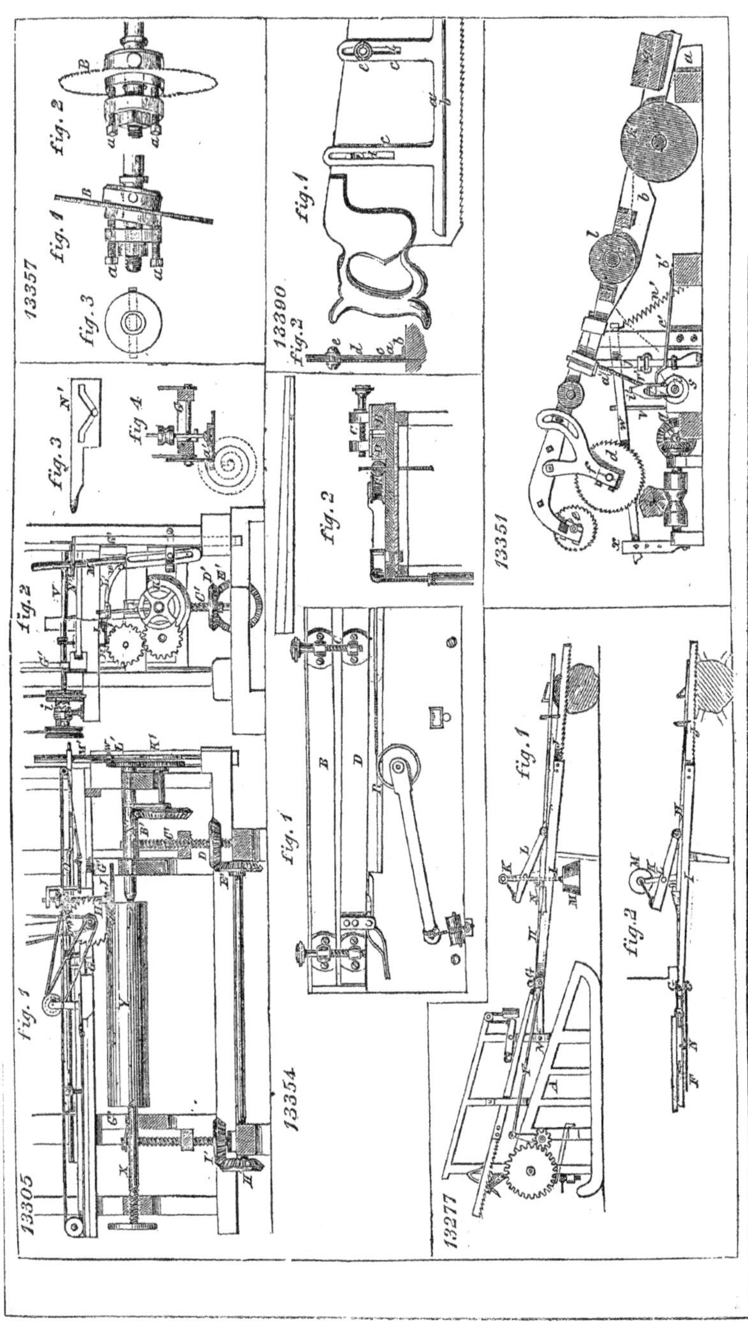
13357
fig. 1
fig. 2
fig. 3
13390
fig. 1
fig. 2
13305
fig. 1
fig. 2
fig. 3
fig. 4
13354
fig. 1
fig. 2
13351
13277
fig. 1
fig. 2

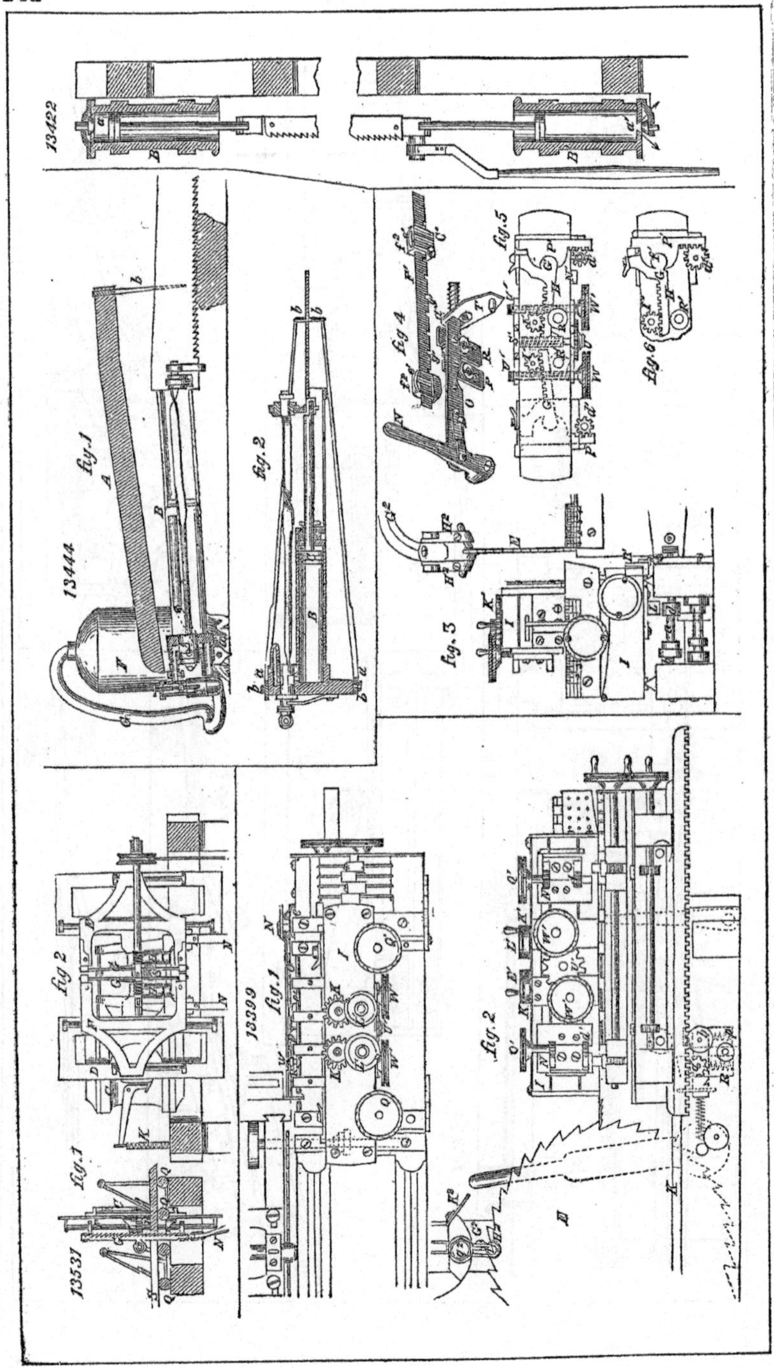
13422
13444
fig.1
fig.2
fig 4
fig.5
fig.6
fig. 3
13537
fig.1
fig 2
13399
fig.1
fig.2

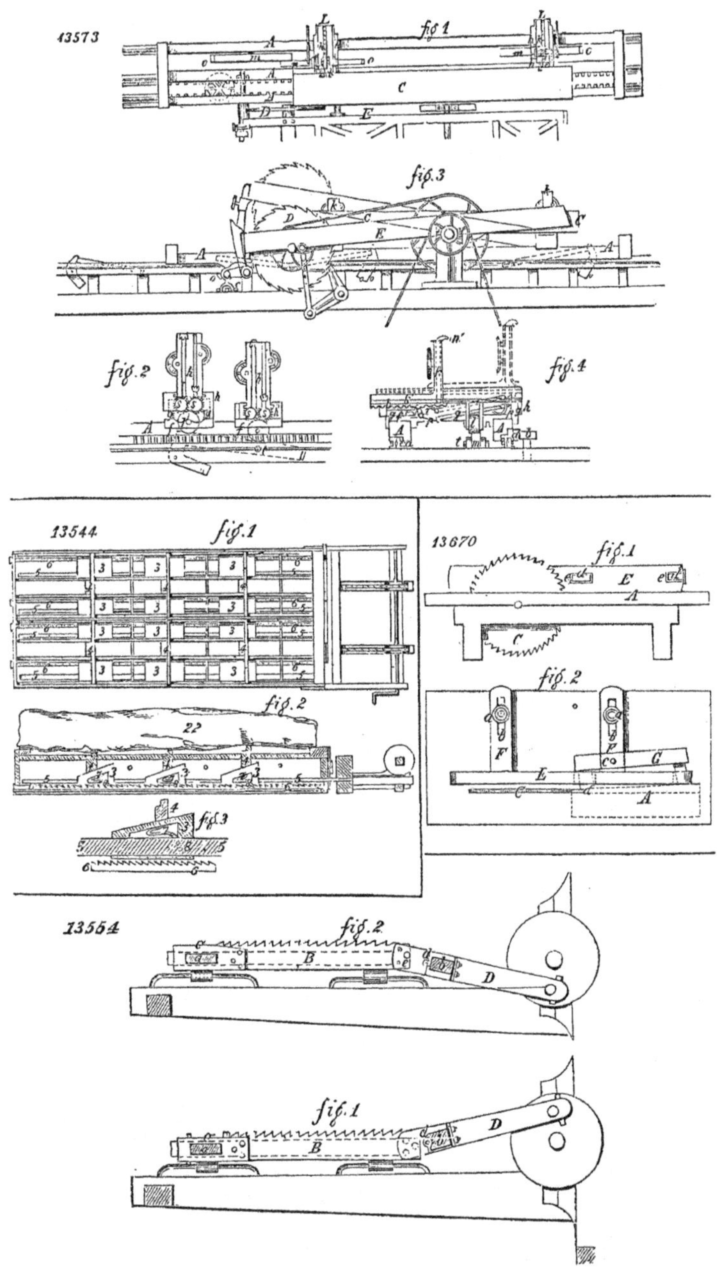
13573
fig 1
fig.3
fig.2
fig.4
13544
fig.1
fig.2
fig.3
13670
fig.1
fig.2
13554
fig.2
fig.1

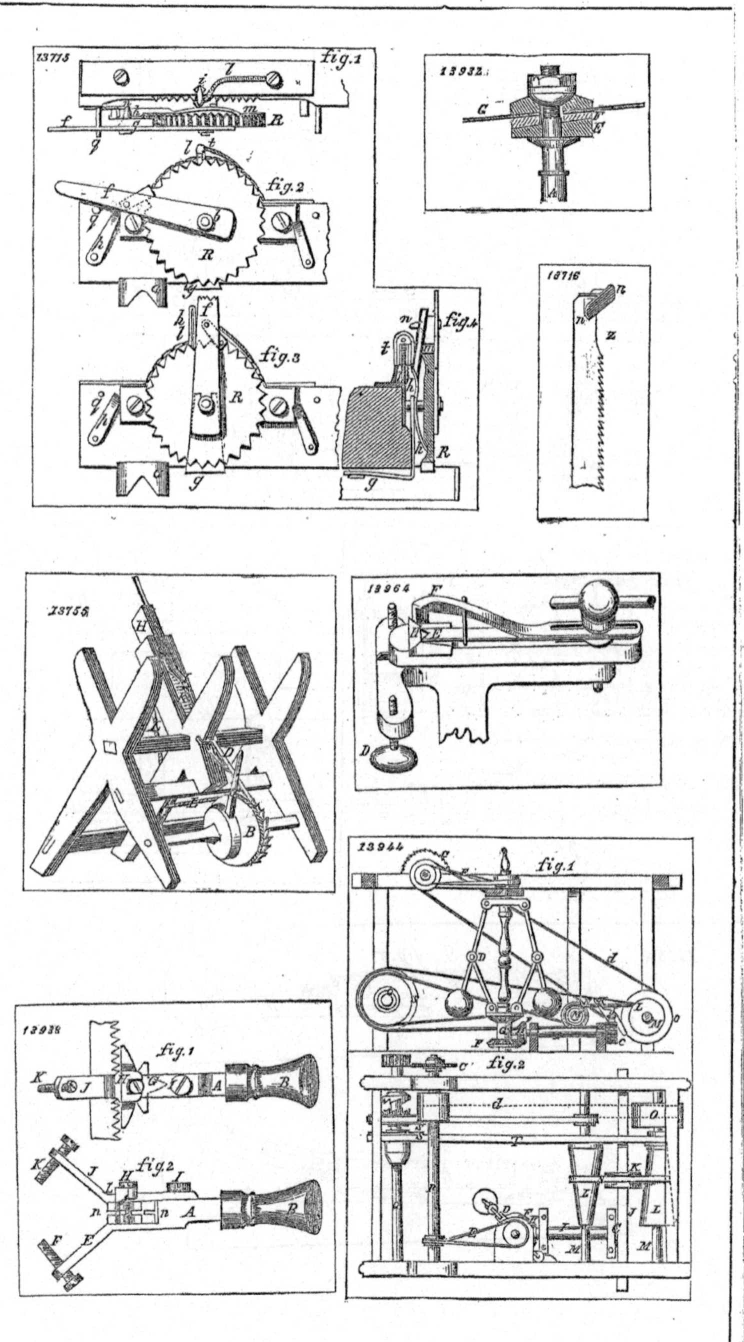
13715
fig.1
fig.2
fig.3
fig.4
13932
13716
13755
13964
13944
13938

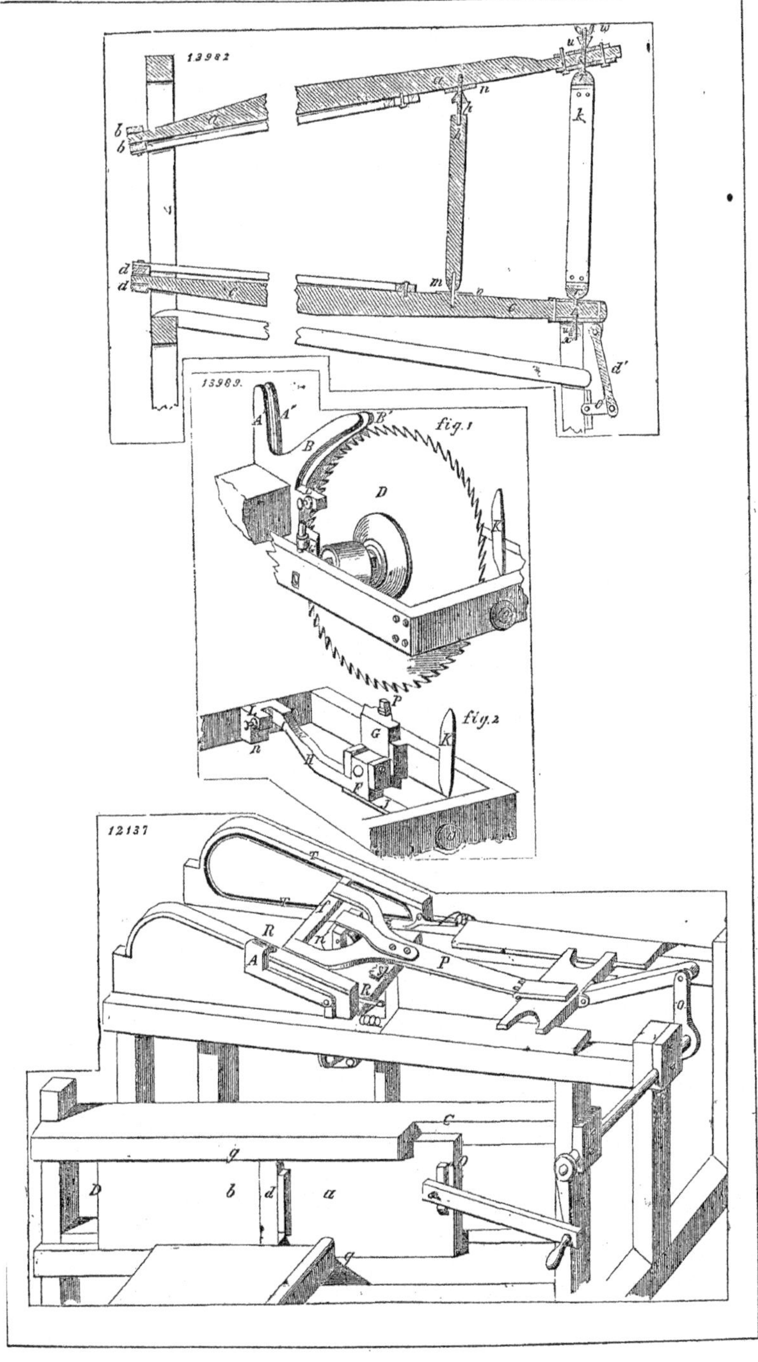
13982
13989.
fig.1
fig.2
12137

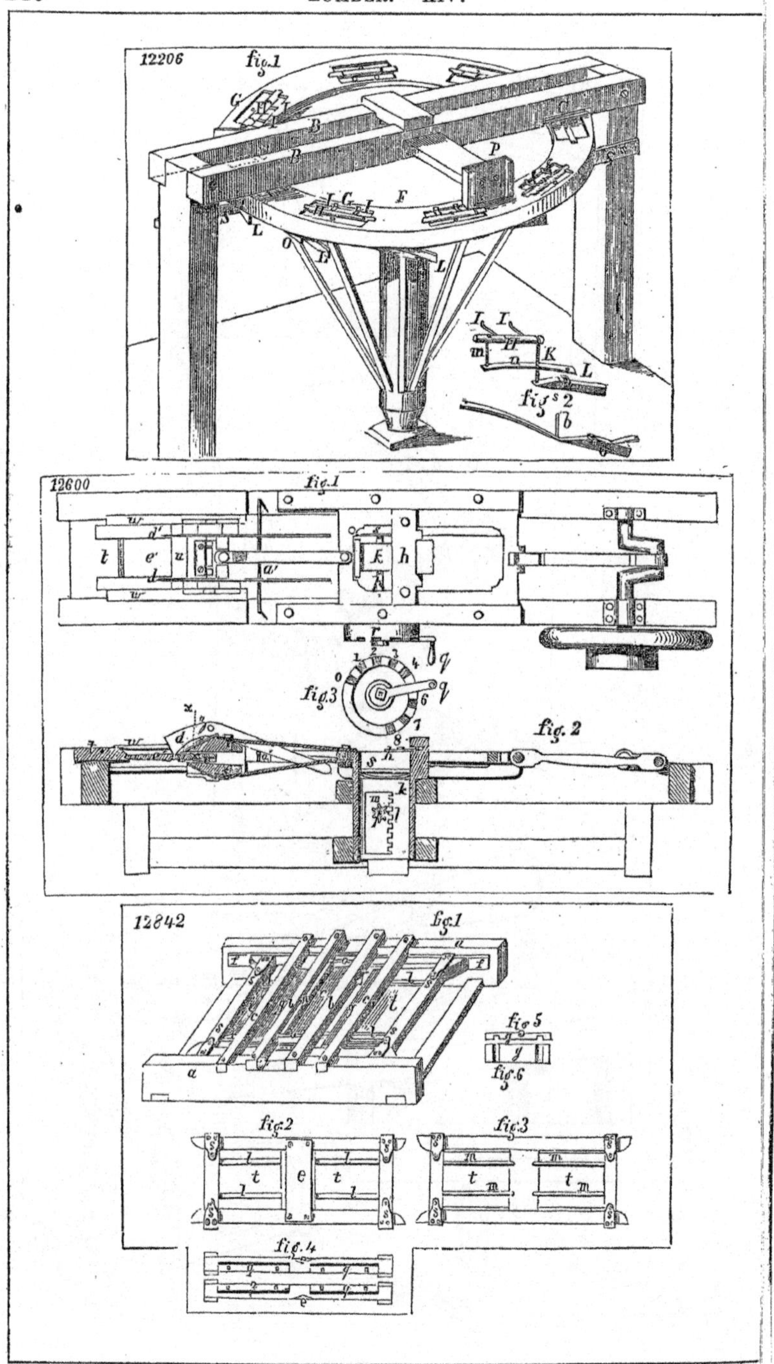
12206
fig. 1
figs 2
12600
fig. 1
fig. 2
fig. 3
12842
fig. 1
fig. 2
fig. 3
fig. 4
fig. 5
fig. 6

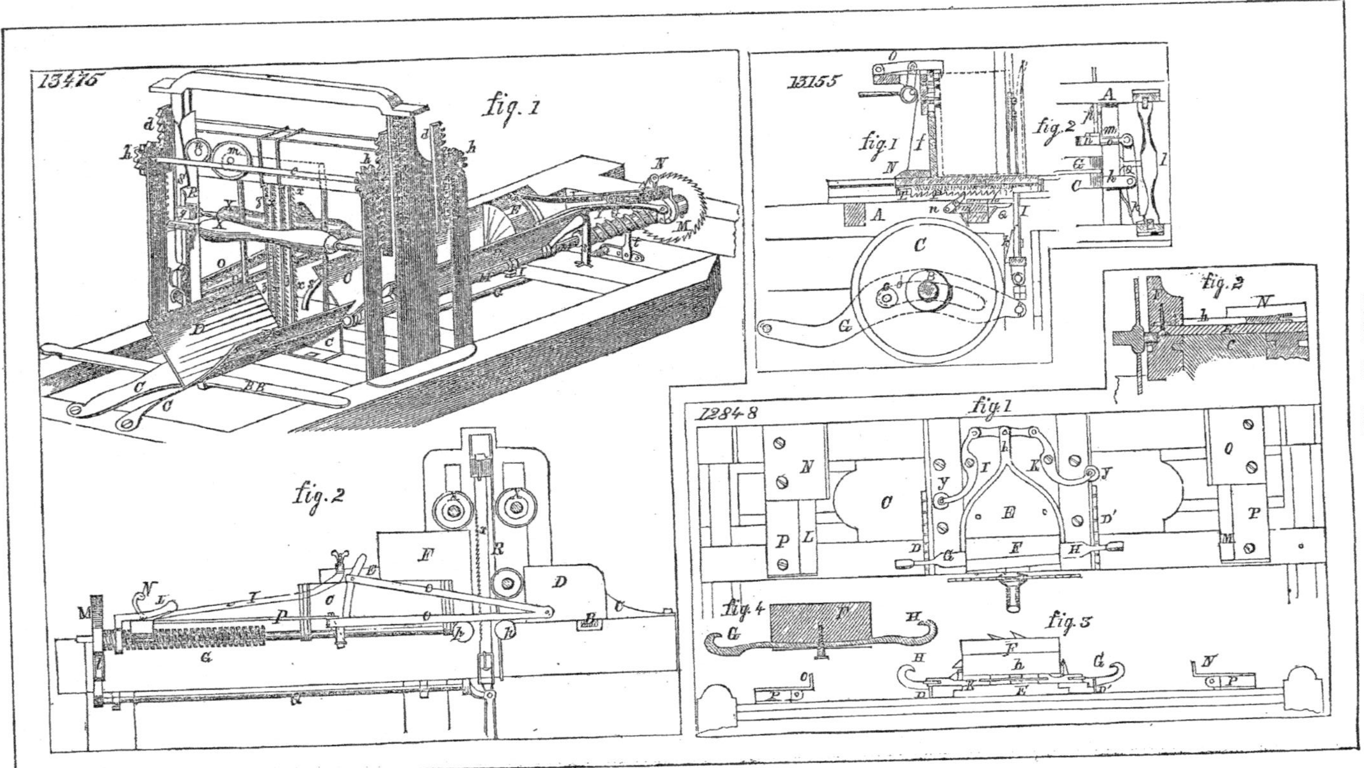
13475
fig. 1
fig. 2
13155
fig. 1
fig. 2
fig. 2
12848
fig. 1
fig. 4
fig. 3

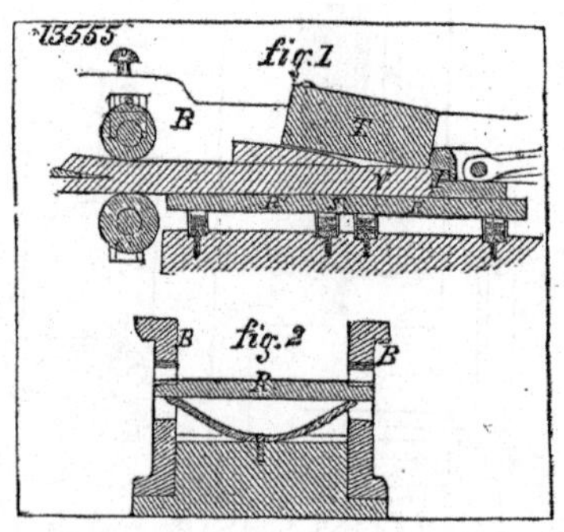
13555
fig. 1
B
T
V
fig. 2
B
B
R

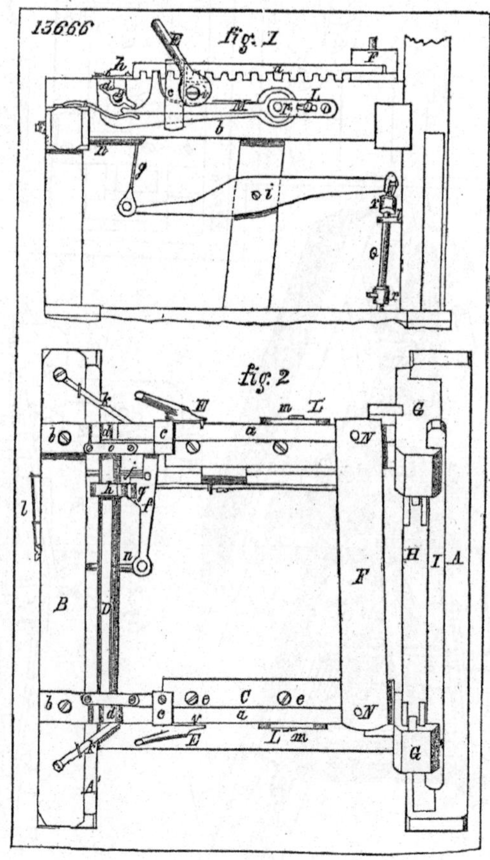
13666
fig. 1
E
F
a
h
d
c
M
L
b
n
g
i
G
fig. 2
E
m
L
G
b
d
c
a
N
h
f
l
n
H
I
A
F
B
D
C
e
E
A'

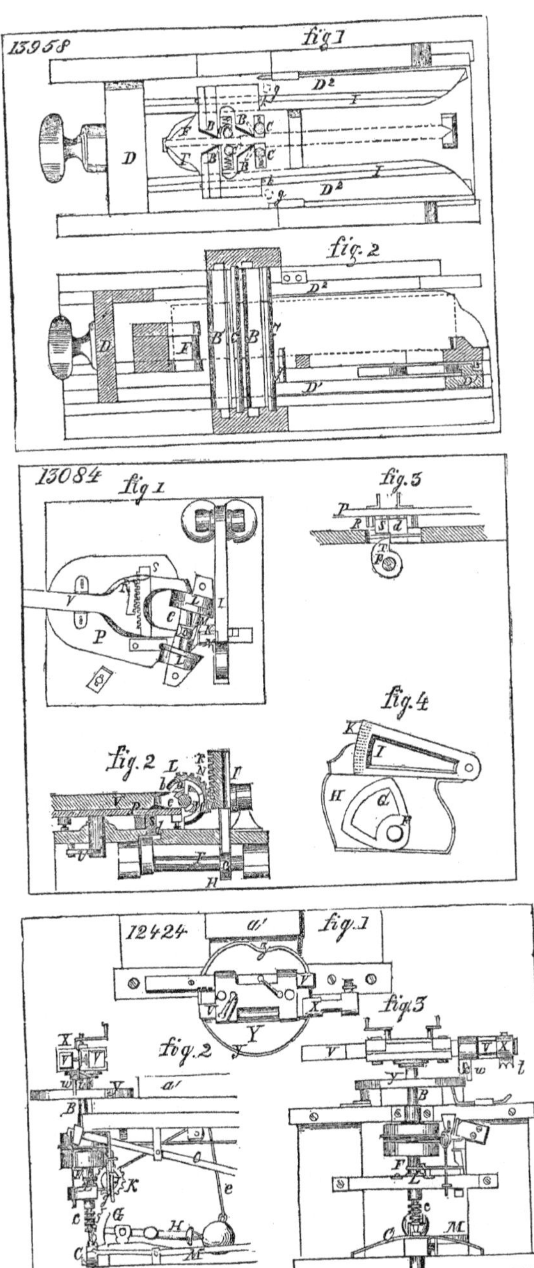
13958
fig 1
fig. 2
13084
fig 1
fig. 3
fig. 2
fig. 4
12424
fig. 1
fig. 2
fig. 3

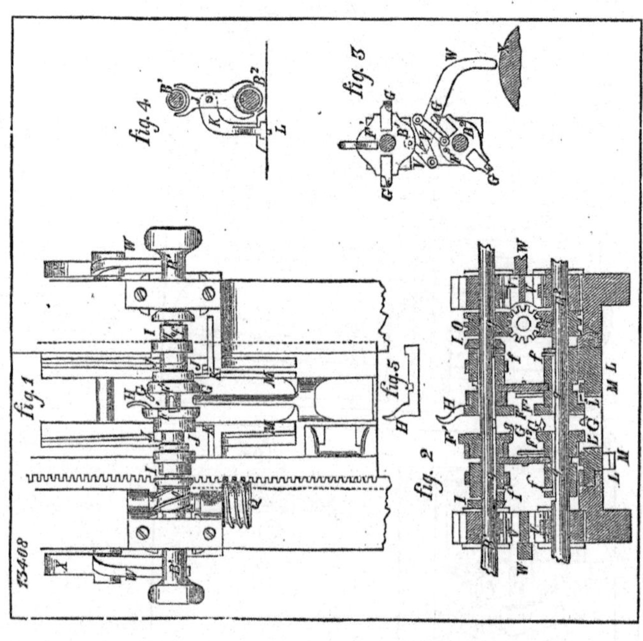
13408
fig.1
fig. 2
fig.3
fig.4
fig.5

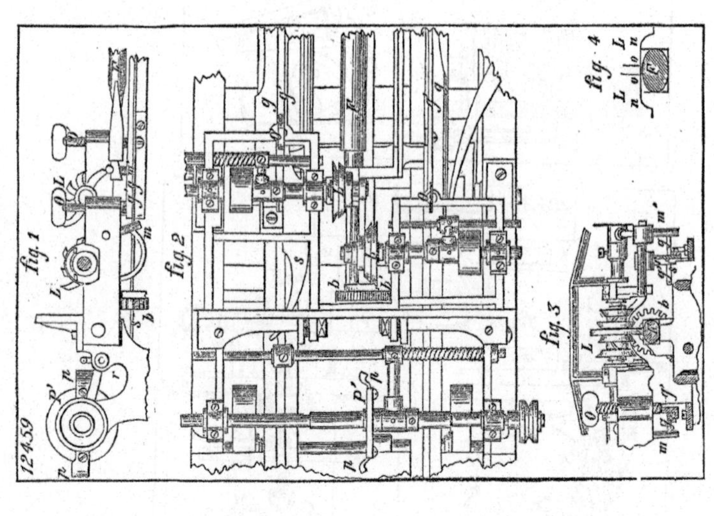
12459
fig.1
fig.2
fig.3
fig. 4

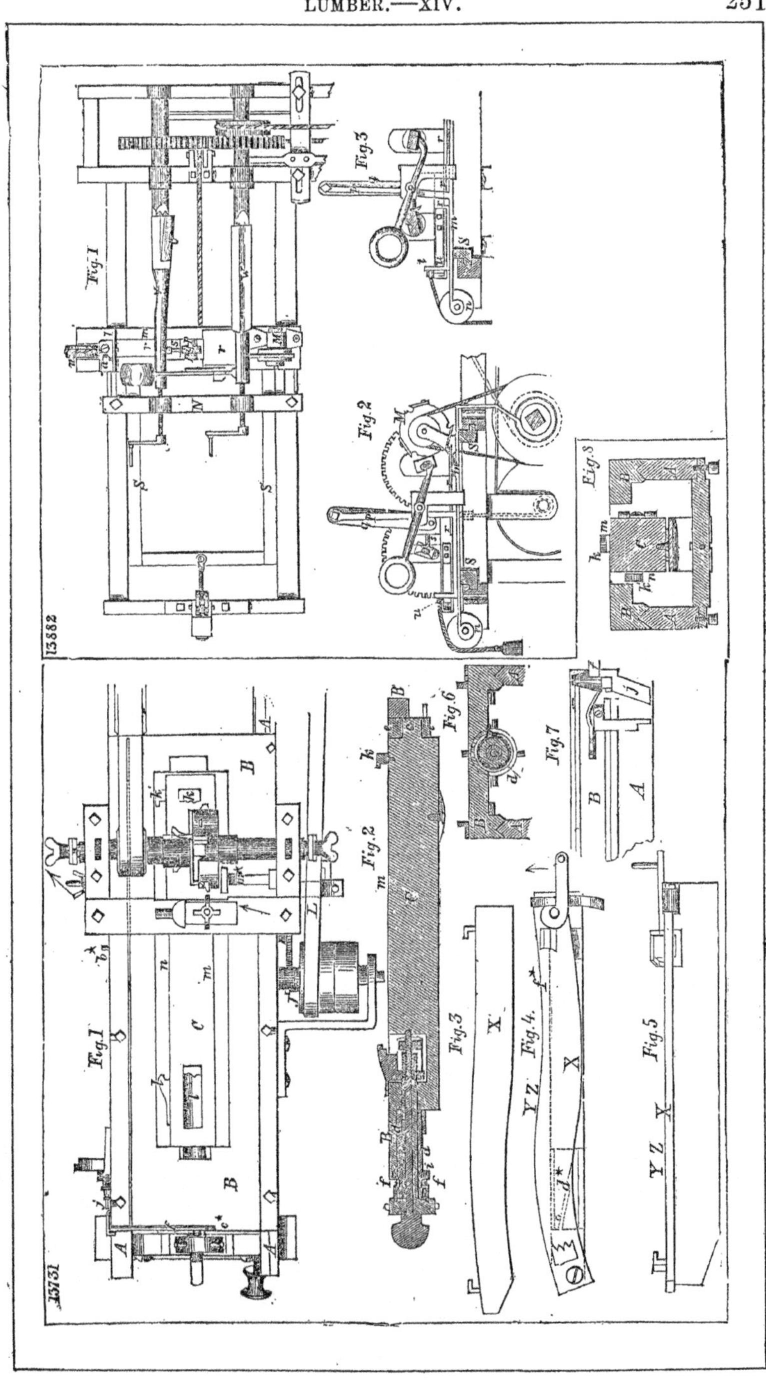
13882
Fig. 1
Fig. 2
Fig. 3
Fig. 8
15731
Fig. 1
Fig. 2
Fig. 3
Fig. 4
Fig. 5
Fig. 6
Fig. 7

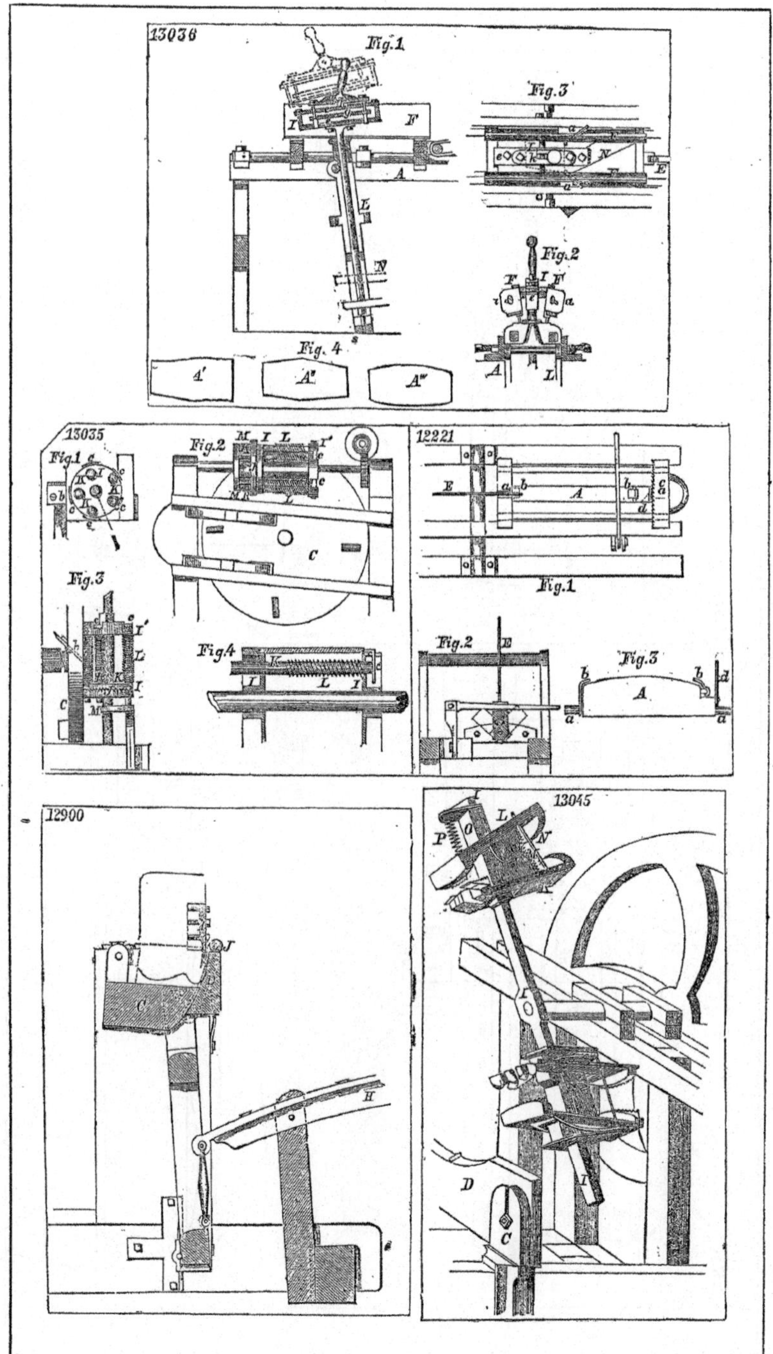
13036
Fig. 1
Fig. 2
Fig. 3
Fig. 4
13035
Fig. 1
Fig. 2
Fig. 3
Fig. 4
12221
Fig. 1
Fig. 2
Fig. 3
12900
13045

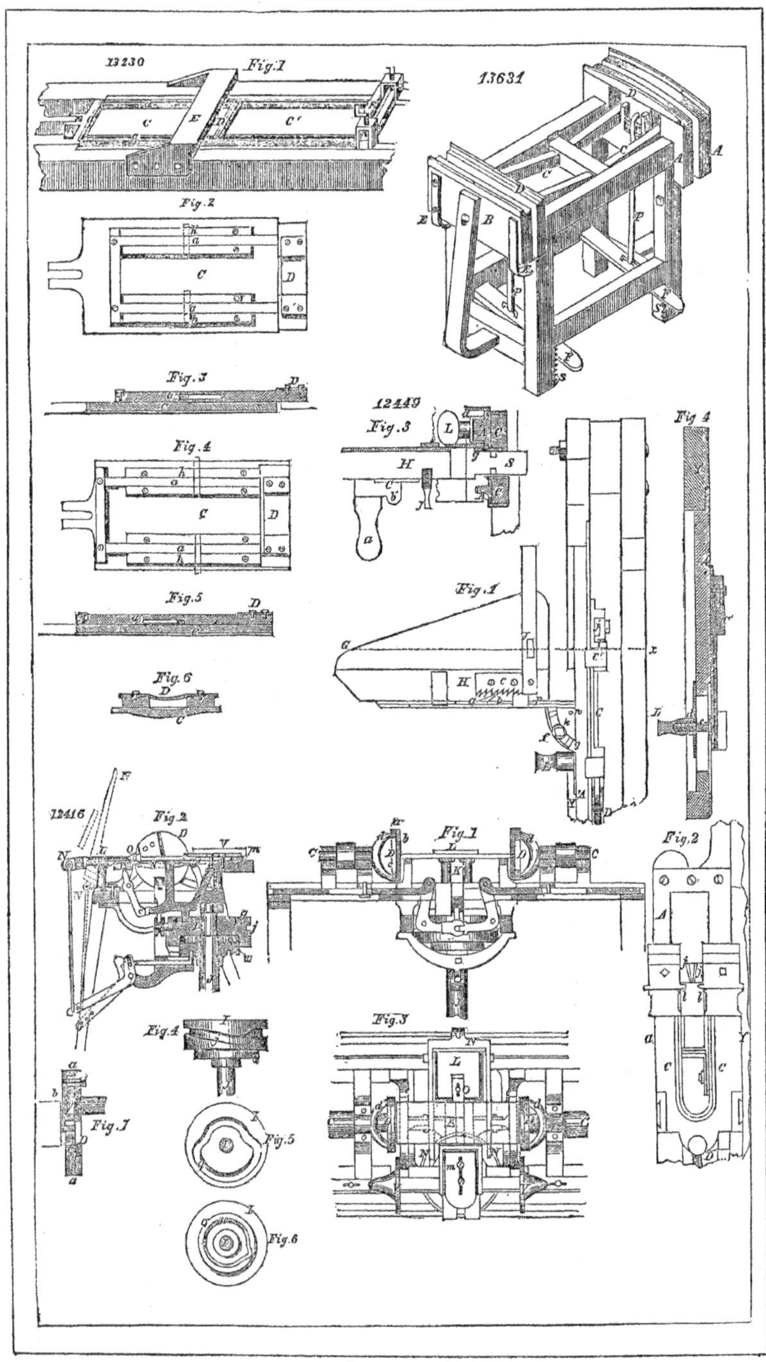
13230
Fig.1
Fig.2
Fig.3
Fig.4
Fig.5
Fig.6
13631
12449
Fig.3
Fig.1
Fig 4
12416
Fig.2
Fig.1
Fig.2
Fig.4
Fig.3
Fig.7
Fig.5
Fig.6

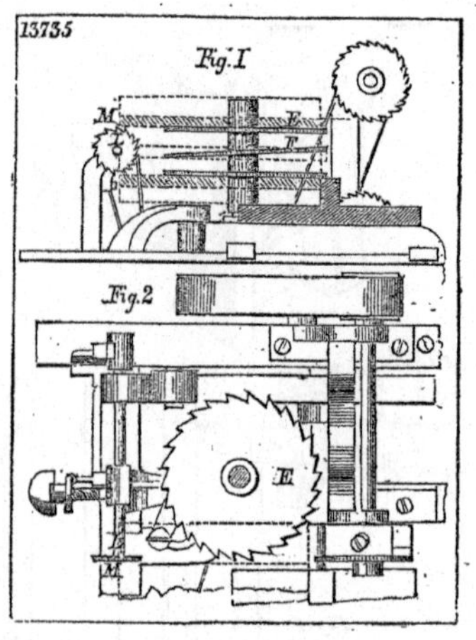
13735
Fig. I
Fig. 2
E

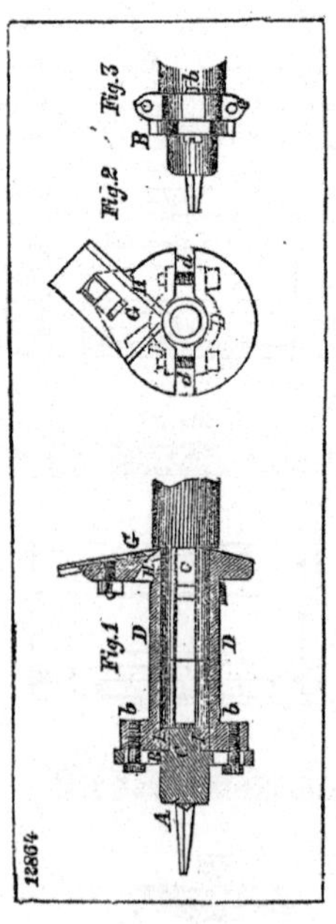
12864
Fig. 1
Fig. 2
Fig. 3

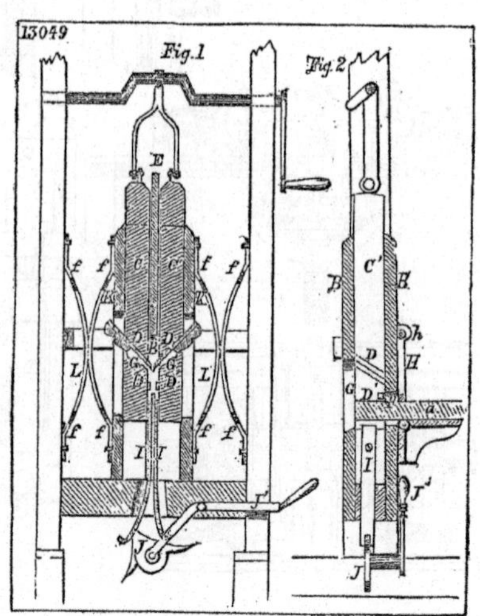
13049
Fig. 1
Fig. 2

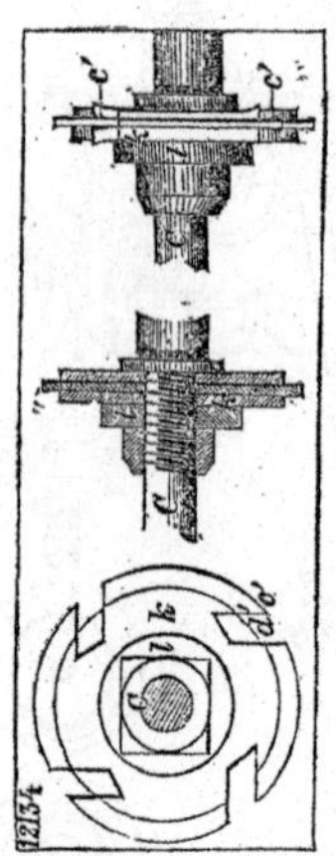
12134

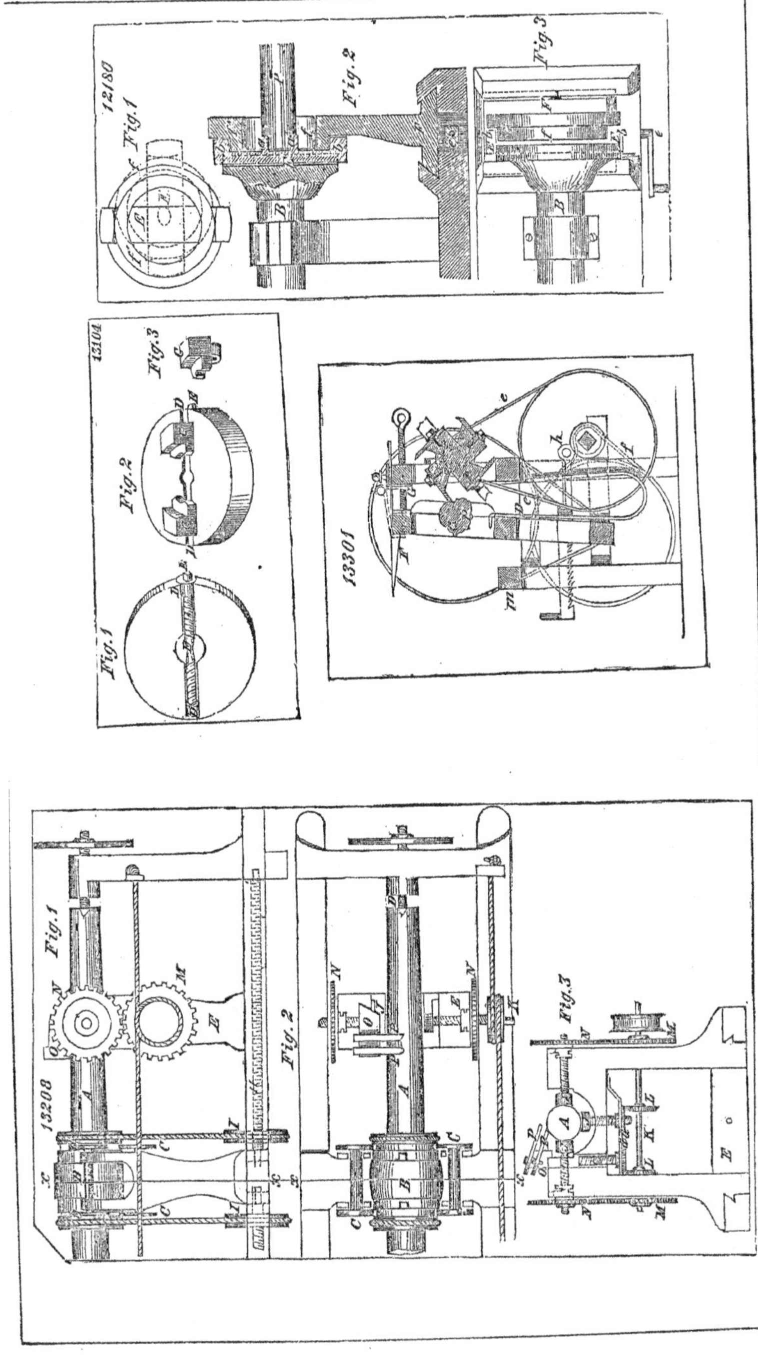
12180
Fig.1
Fig.2
Fig.3
13104
Fig.1
Fig.2
Fig.3
13301
13208
Fig.1
Fig.2
Fig.3

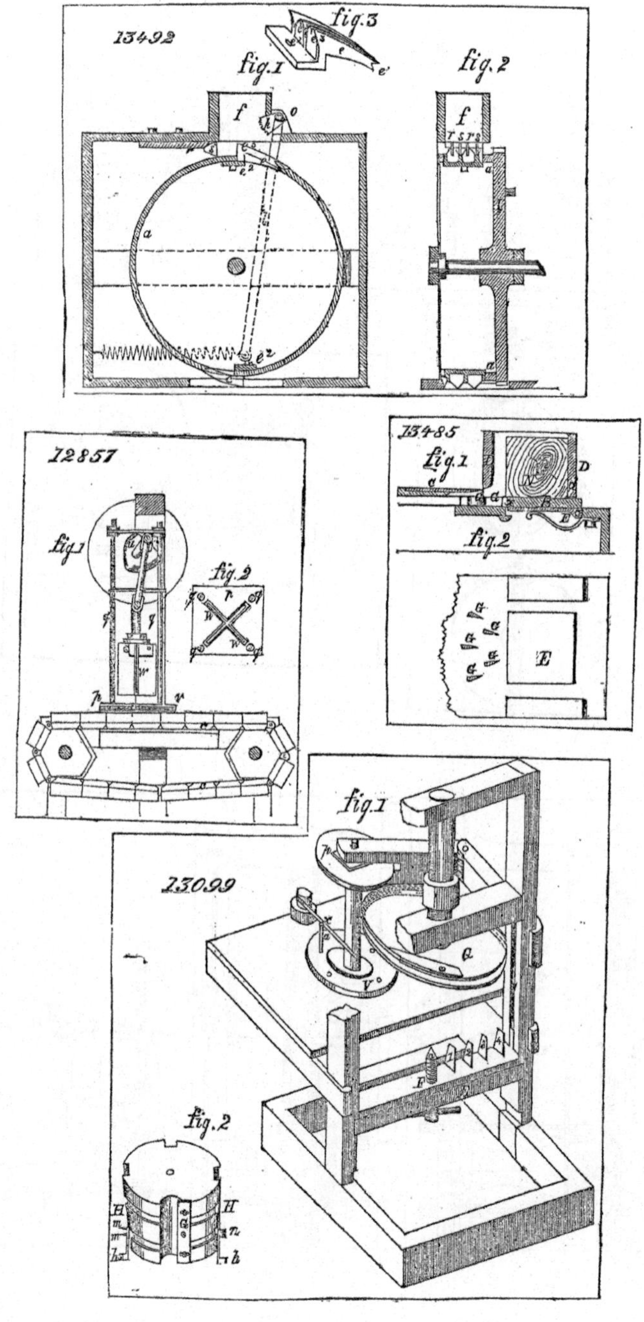
13492
fig.3
fig.1
fig.2
12857
fig.1
fig.2
13485
fig.1
fig.2
13099
fig.1
fig.2

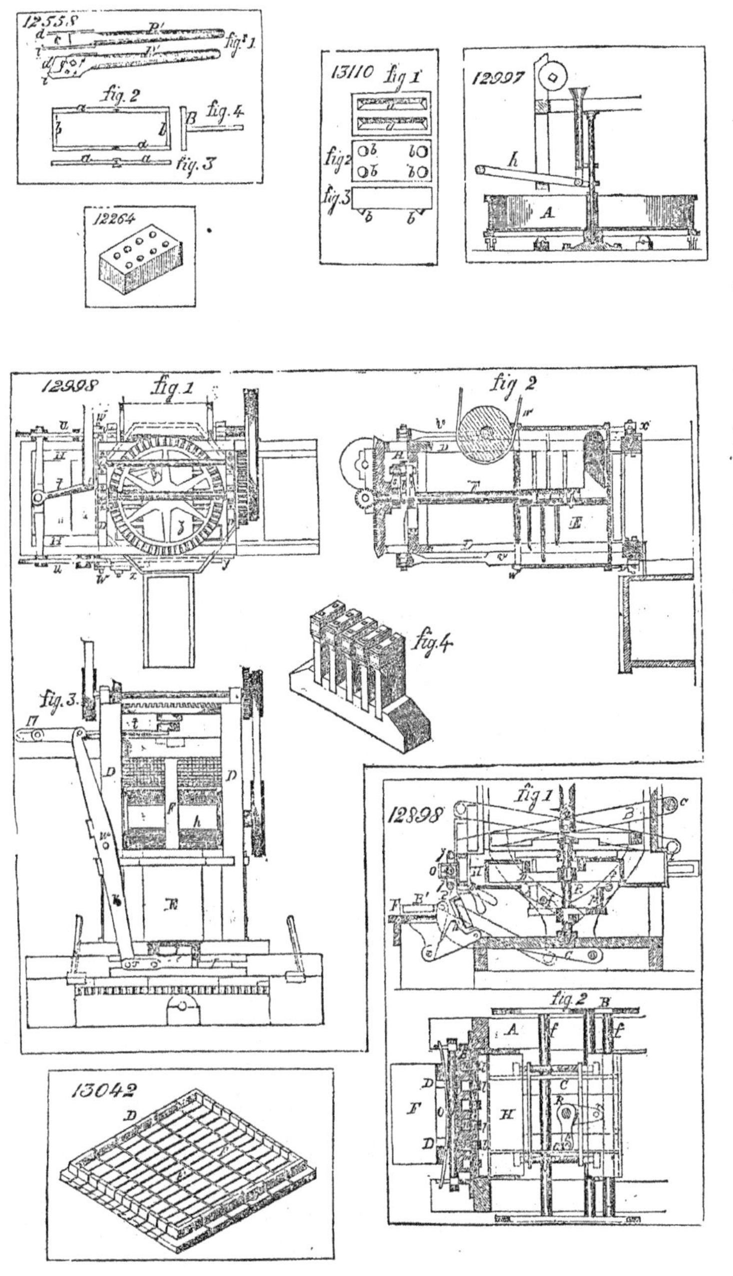
12558
fig. 1
fig. 2
fig. 4
fig. 3
12264
13110
fig 1
fig 2
fig 3
12997
12998
fig. 1
fig 2
fig. 4
fig. 3.
12898
fig. 1
fig. 2
13042

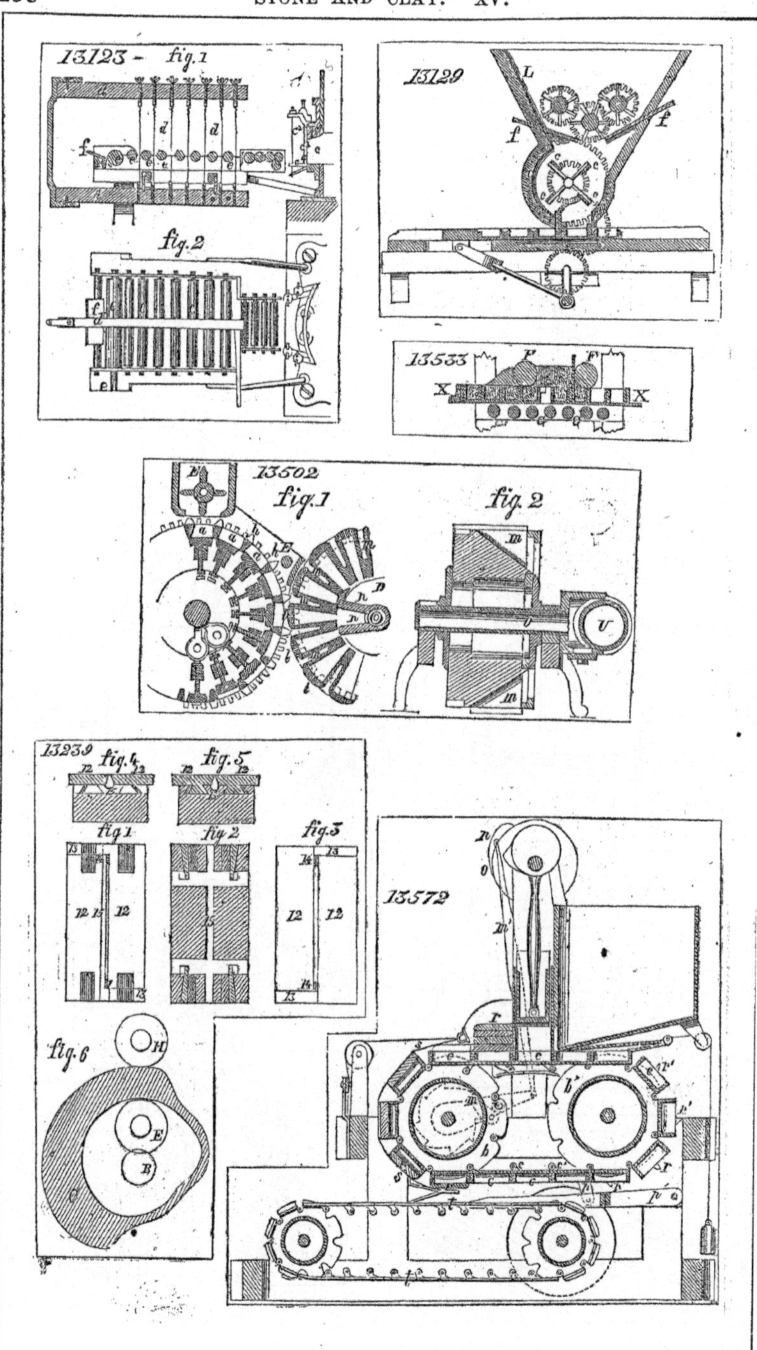
13123 - fig. 1
fig. 2
13129
13533
13502
fig. 1
fig. 2
13239
fig. 4
fig. 5
fig 1
fig 2
fig. 3
fig. 6
13572

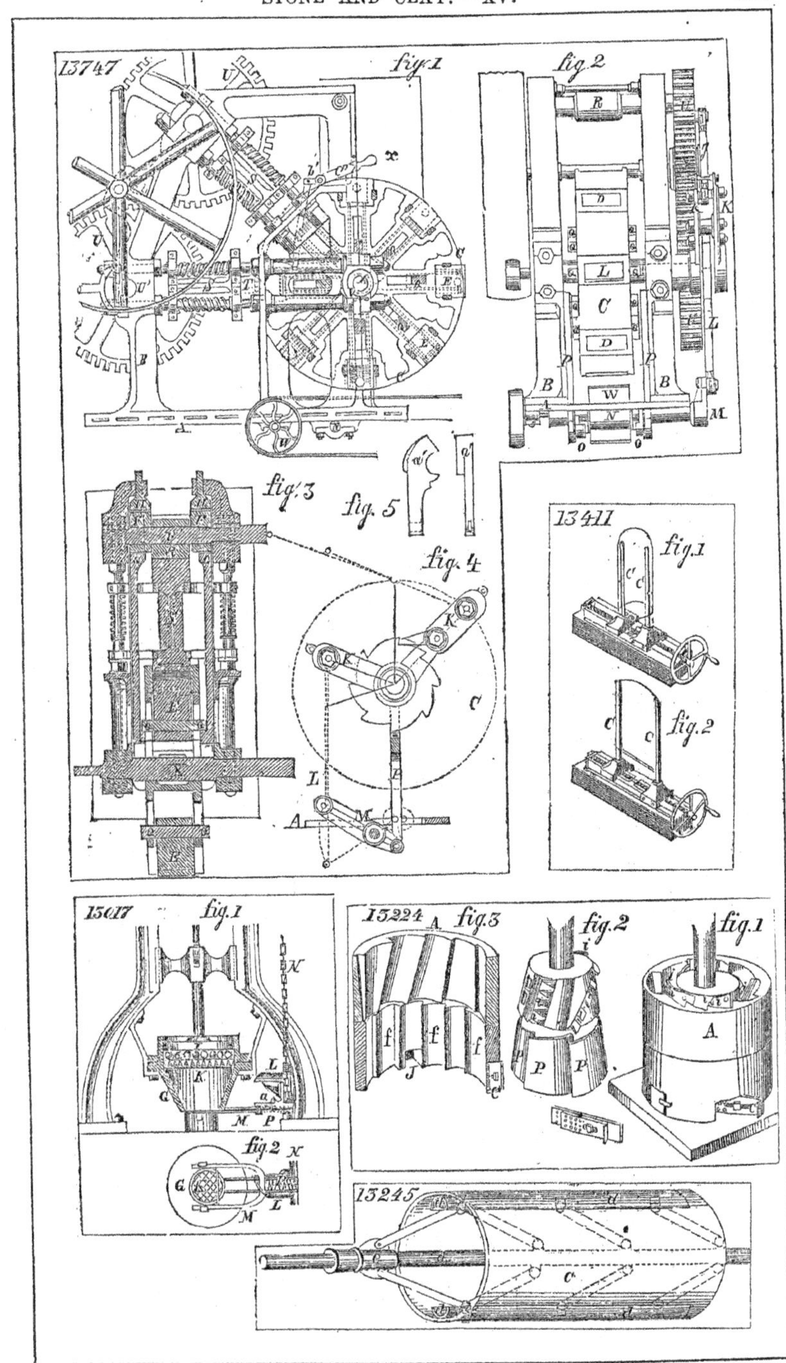
13747
fig. 1
fig. 2
fig. 3
fig. 4
fig. 5
13411
fig. 1
fig. 2
13017
fig. 1
fig. 2
13224
fig. 3
fig. 2
fig. 1
13245

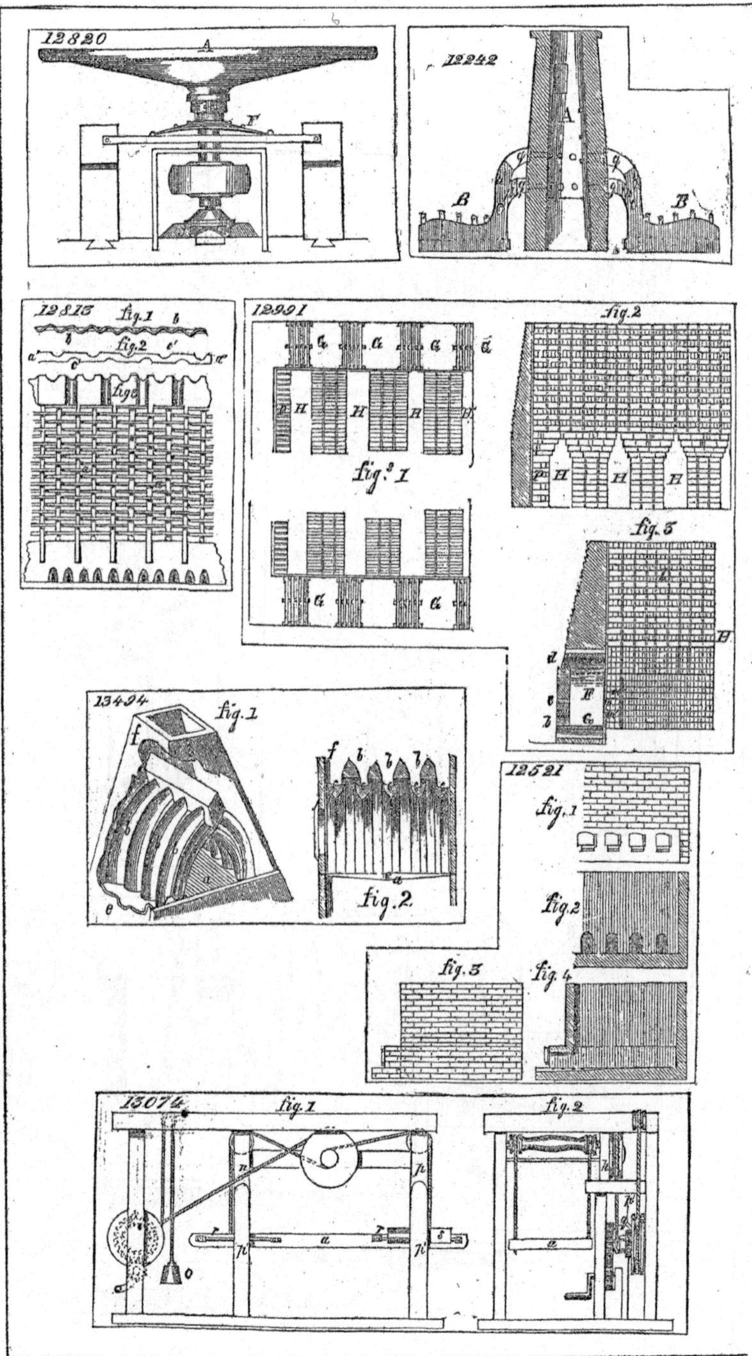
12820
12242
12813
12991
fig.2
fig.3
13494
fig.1
fig.2
12521
fig.1
fig.2
fig.3
fig.4
13074
fig.1
fig.2

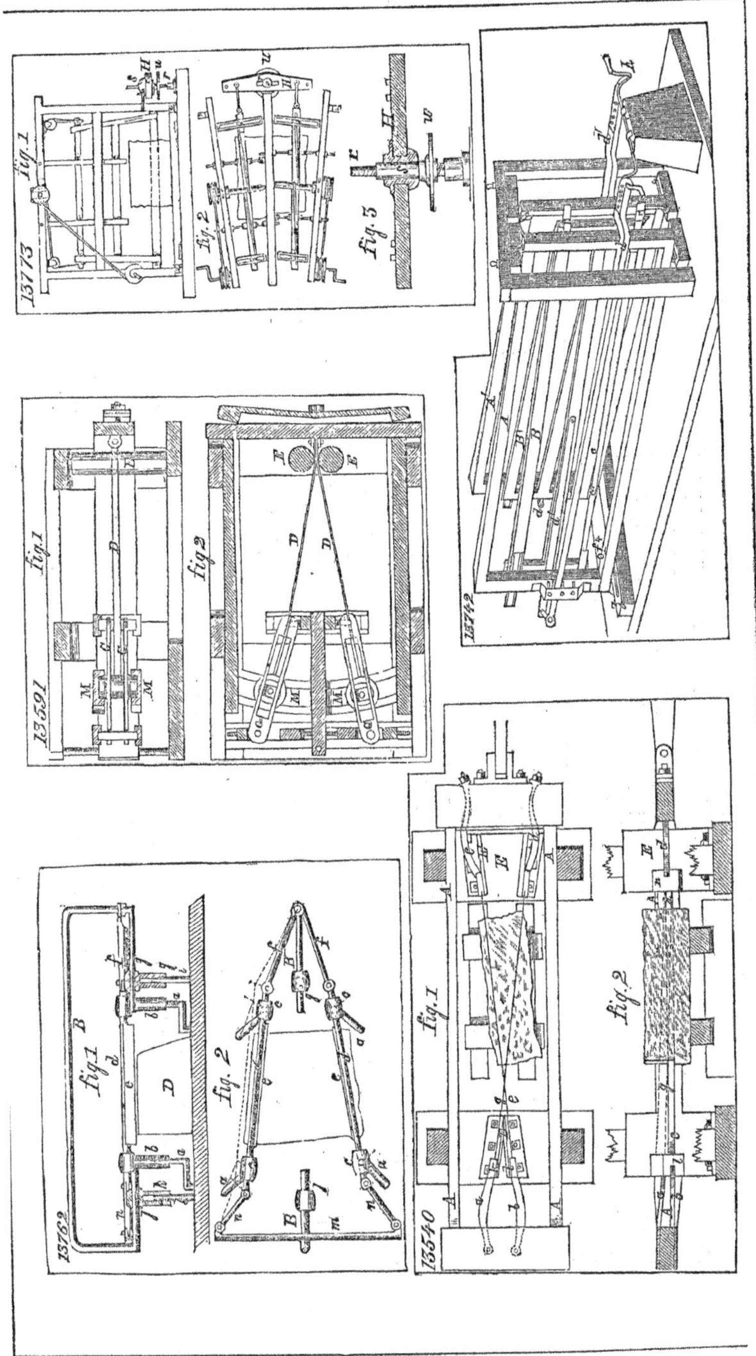

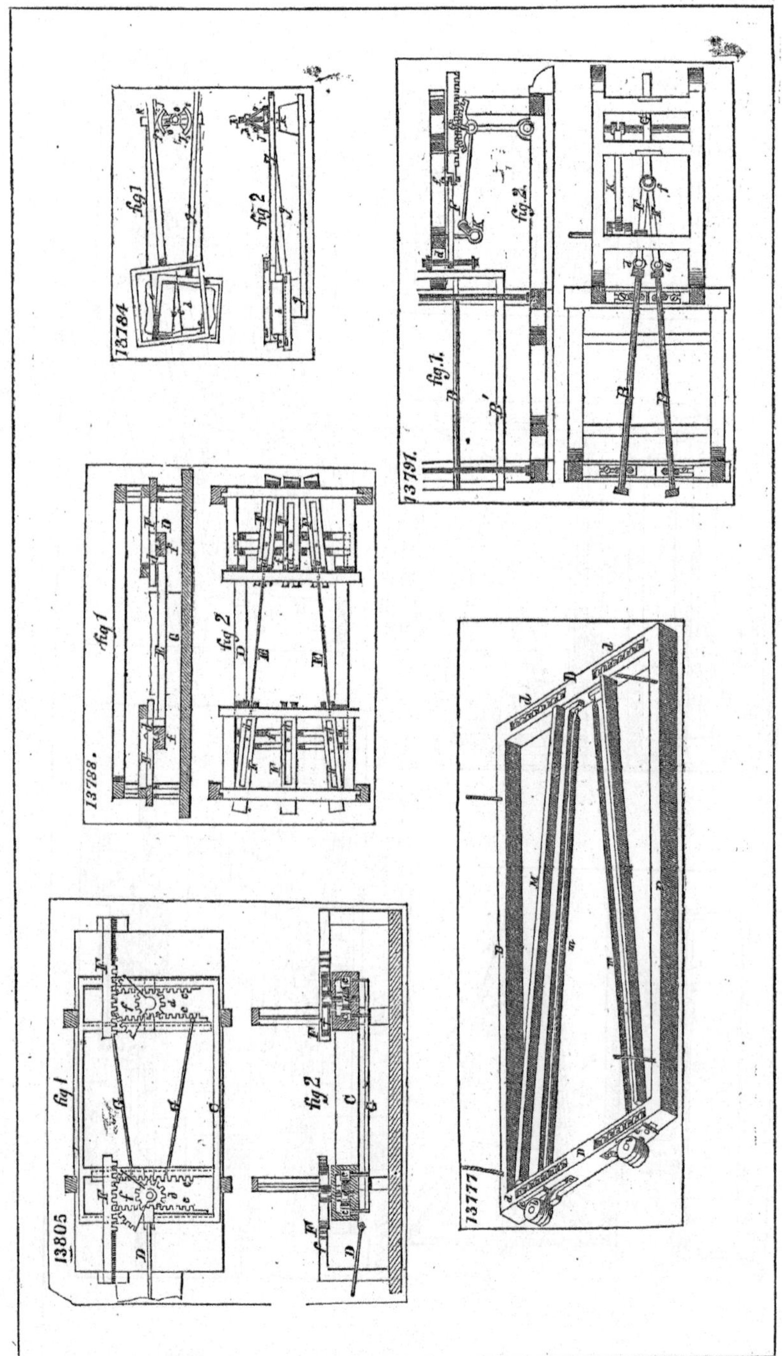
13784
fig 1
fig 2
13791.
fig.1.
fig.2.
13788.
fig 1
fig 2
13805
fig 1
fig 2
13777

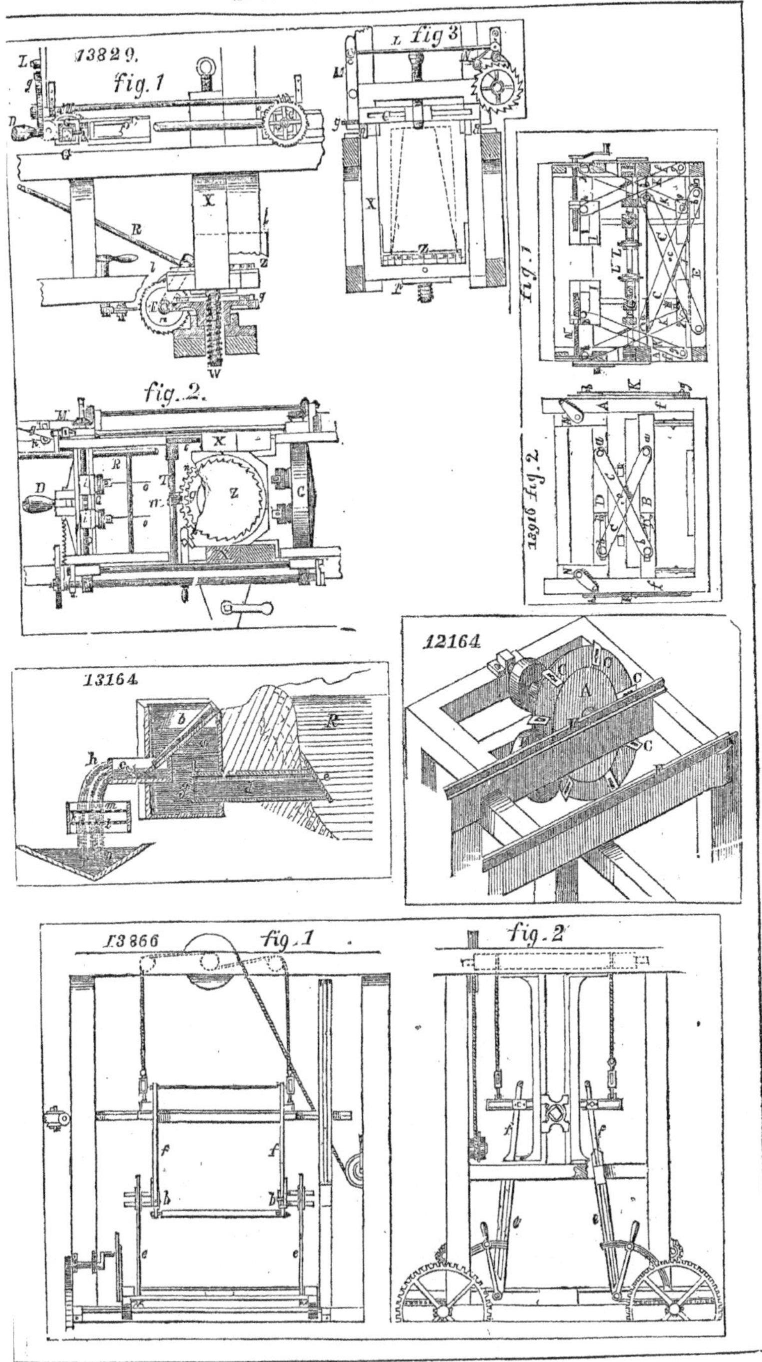
13829.
fig. 1
fig 3
fig. 2.
13916 fig. 2
fig. 1
13164
12164
13866
fig. 1
fig. 2

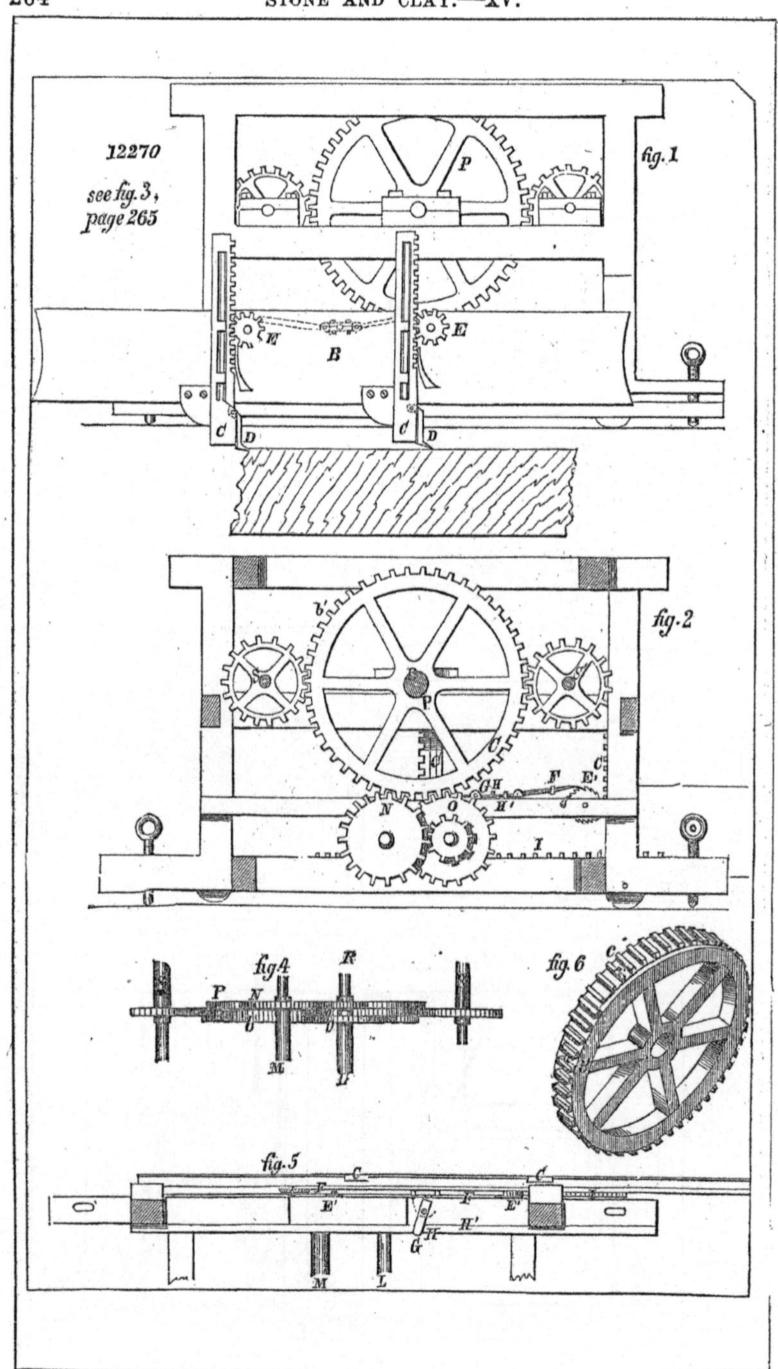

12270
see fig. 3, page 265
fig. 1
P
E
B
E
C
D
C
D
fig. 2
b'
P
C
G H
F
E
C
H'
N
O
I
fig 4
R
P
N
b
b
M
L
fig. 6
c
fig. 5
C
C
E
E'
I
E'
H'
H
G
M
L

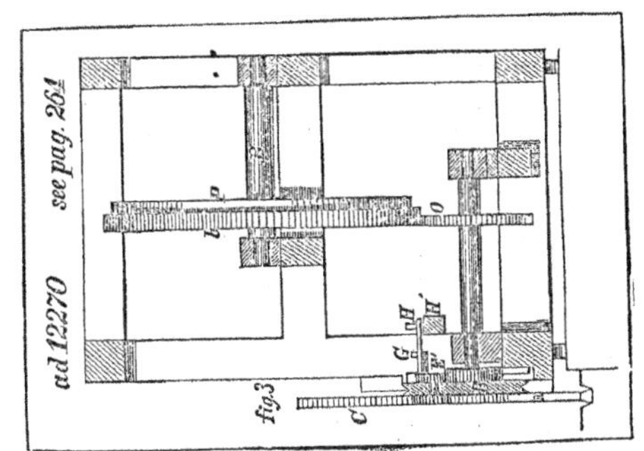
ad 12270
see pag. 261
fig.3

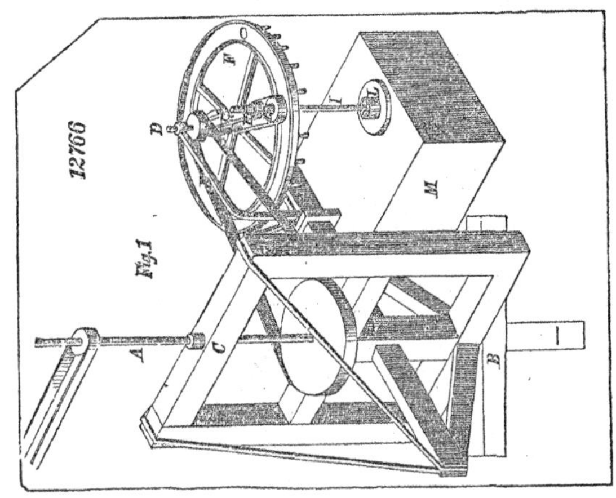
12766
Fig.1

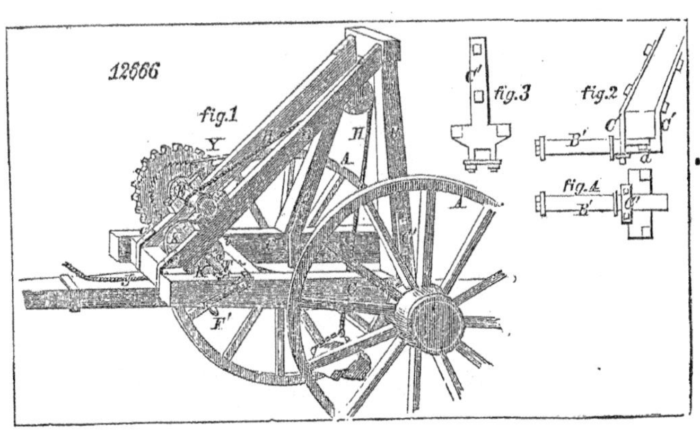
12666
fig.1
fig.3
fig.2
fig.4

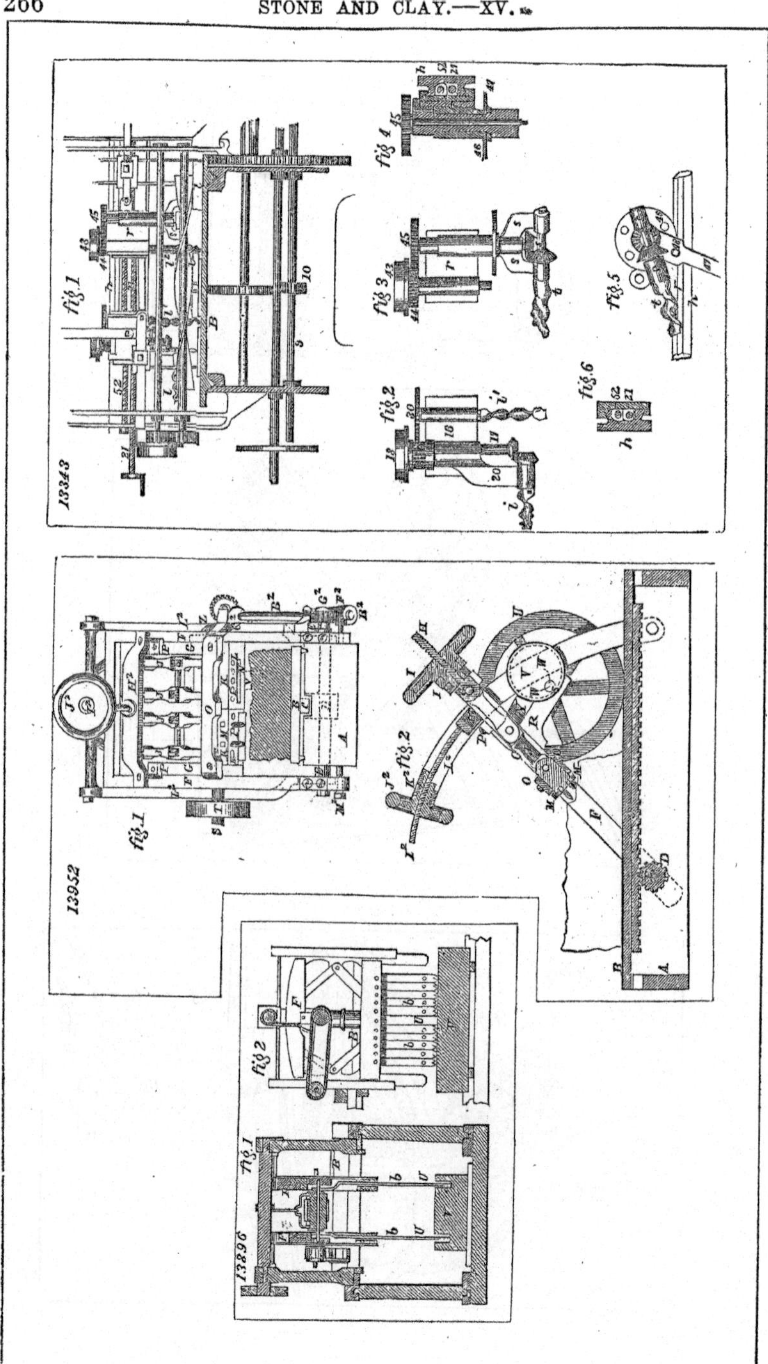

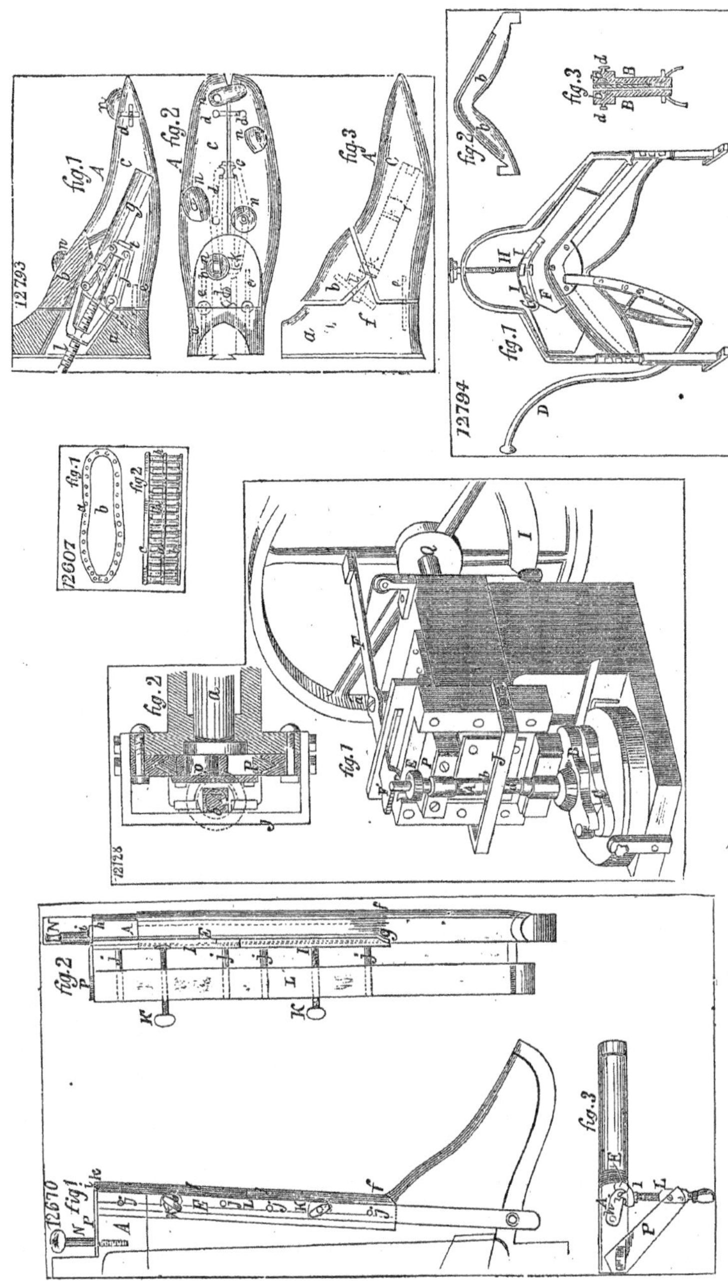
72793
72794
72607
72728
72670

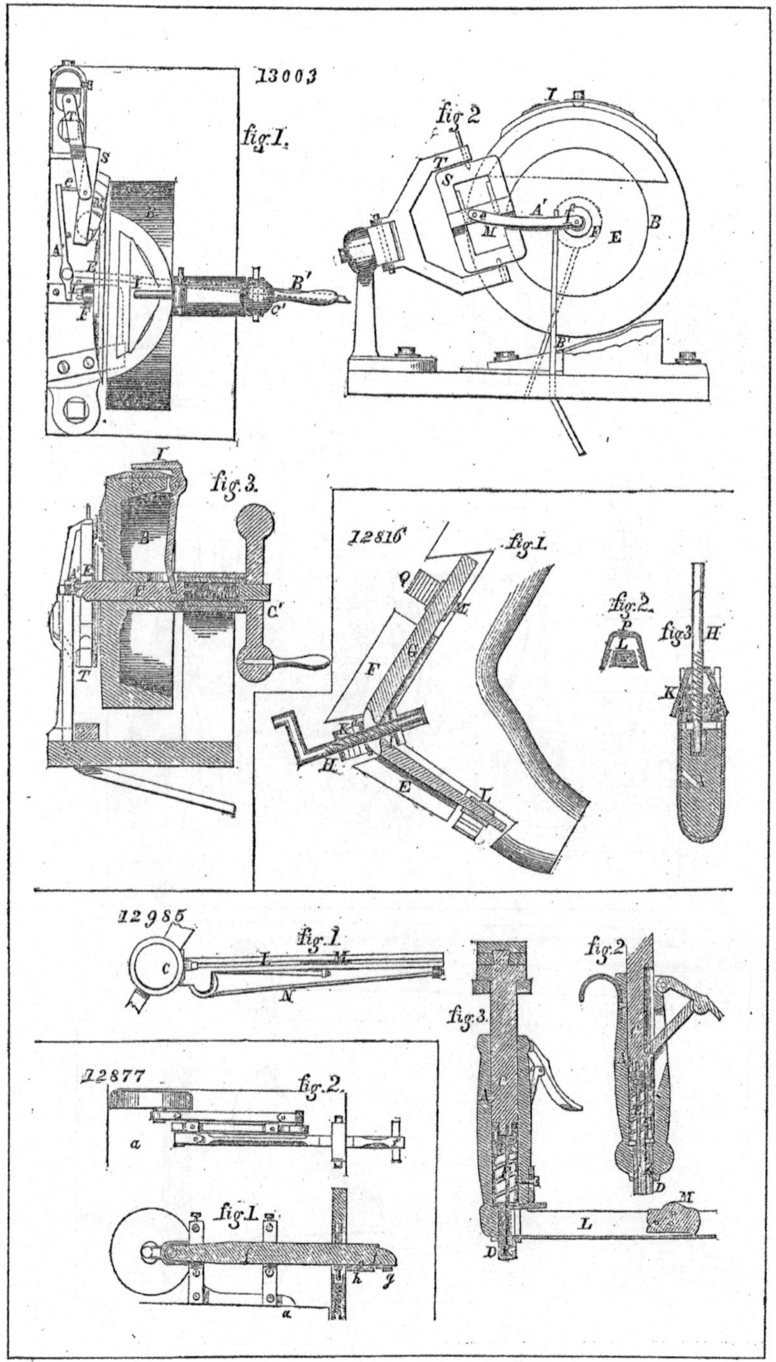
13003
fig.1.
fig.2
fig.3.
12816
fig.1.
fig.2.
fig.3
12985
fig.1.
fig.2
fig.3.
12877
fig.2.
fig.1.

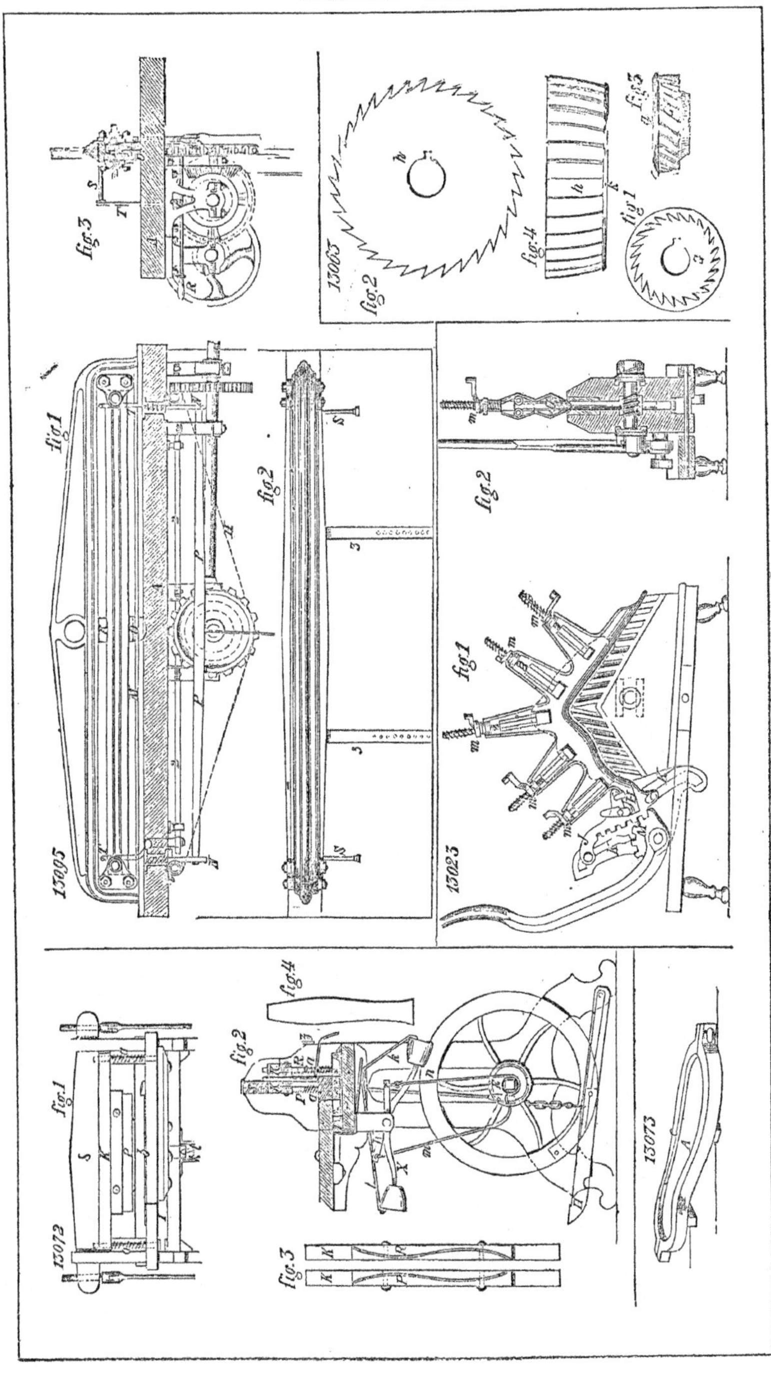

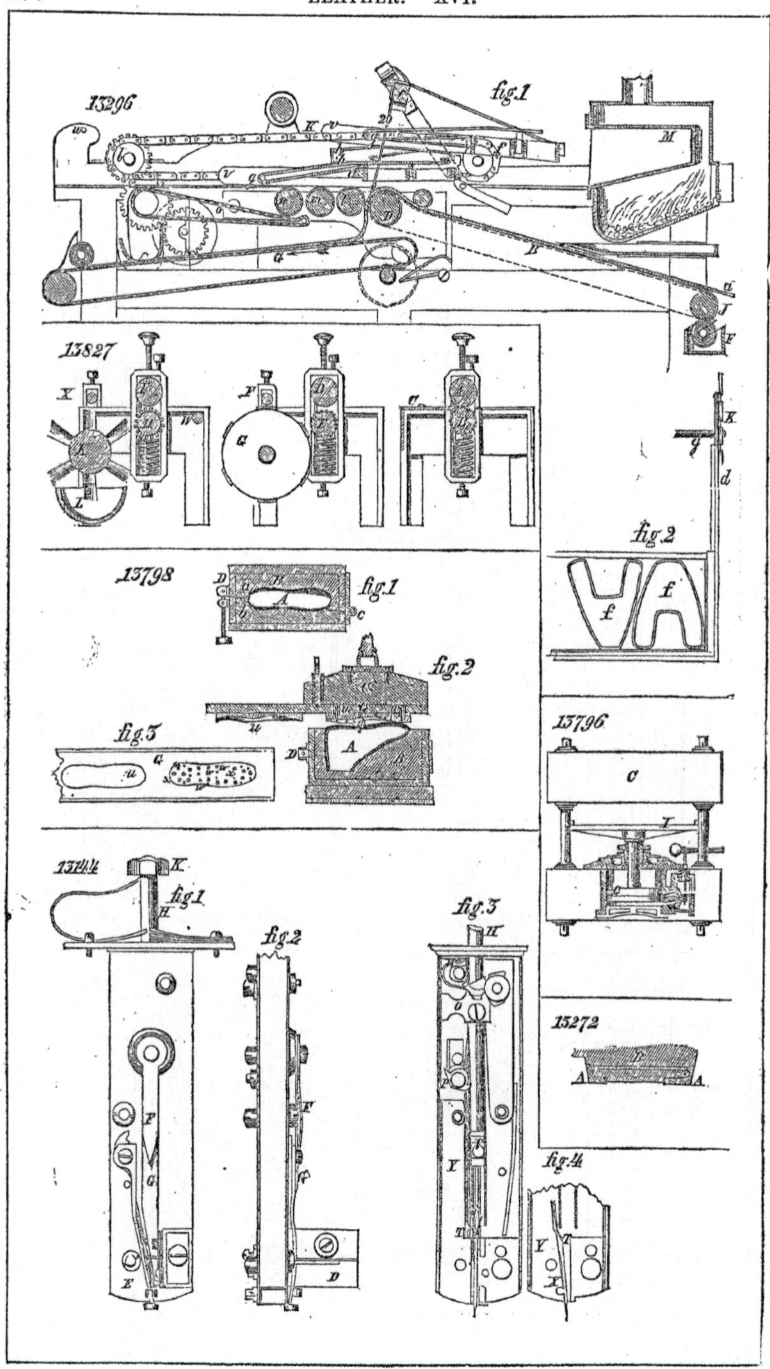
13296
fig.1
13827
13798
fig.1
fig.2
fig.3
fig.2
13796
13272
13144
fig.1
fig.2
fig.3
fig.4

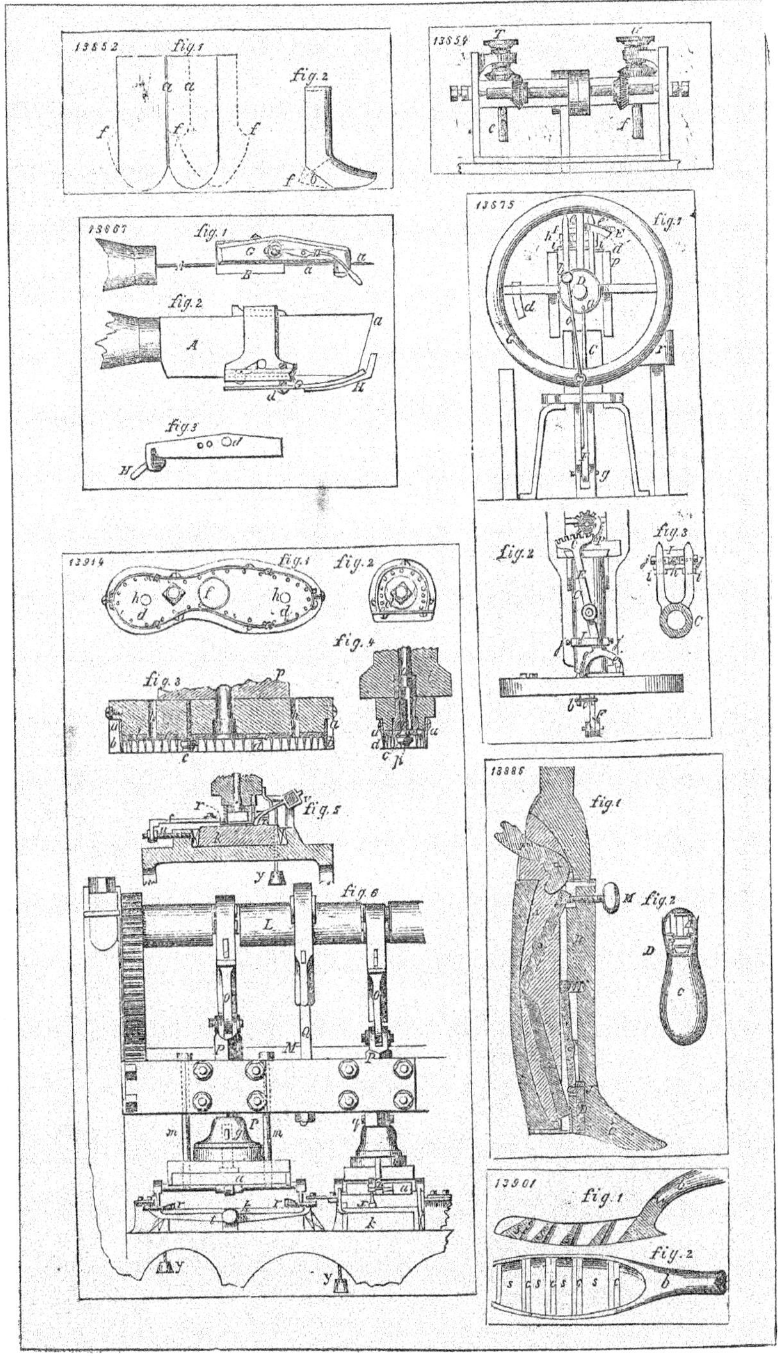
13852
fig. 1
fig. 2
13854
13867
fig. 1
fig. 2
fig. 3
13875
fig. 1
fig. 2
fig. 3
13914
fig. 1
fig. 2
fig. 3
fig. 4
fig. 5
fig. 6
13886
fig. 1
fig. 2
13901
fig. 1
fig. 2

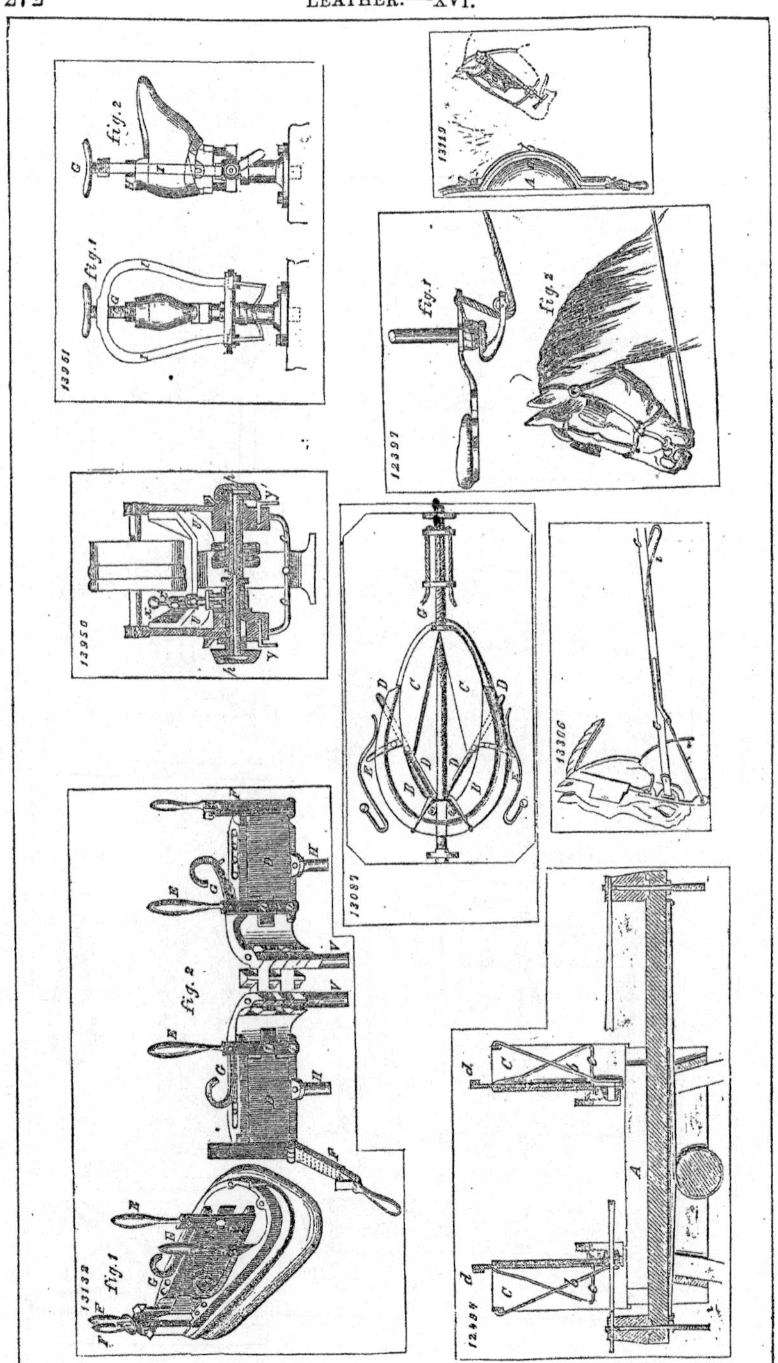
13951
fig.1
fig.2
13950
13132
fig.1
fig.2
13087
12397
fig.1
fig.2
13306
12494

13949
13858
13965
13189

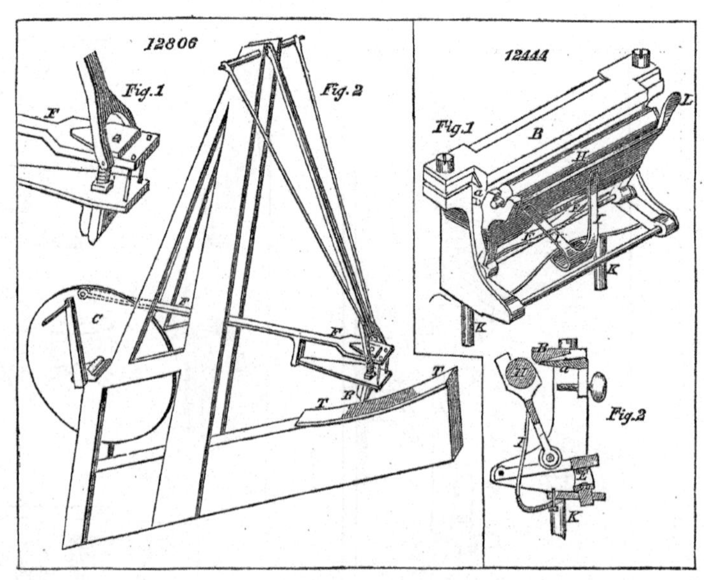
12806
Fig.1
Fig.2
12444
Fig.1
Fig.2

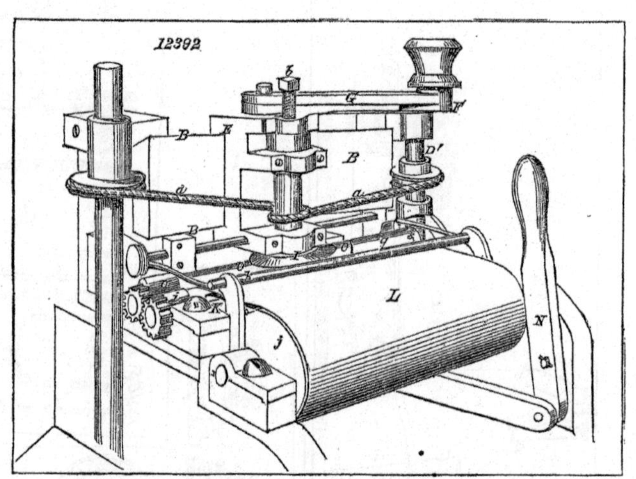
12392

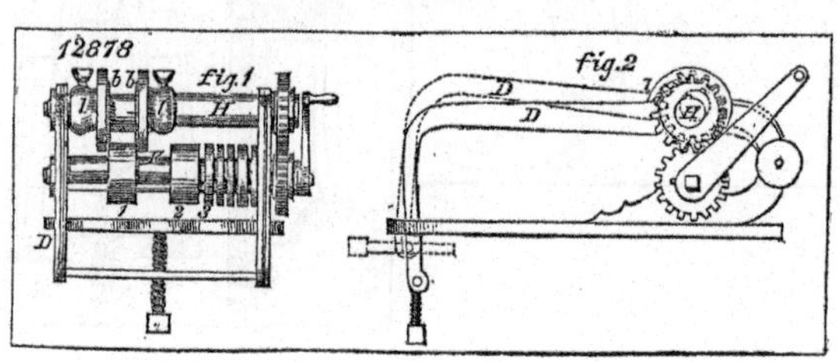
12878
fig.1
fig.2

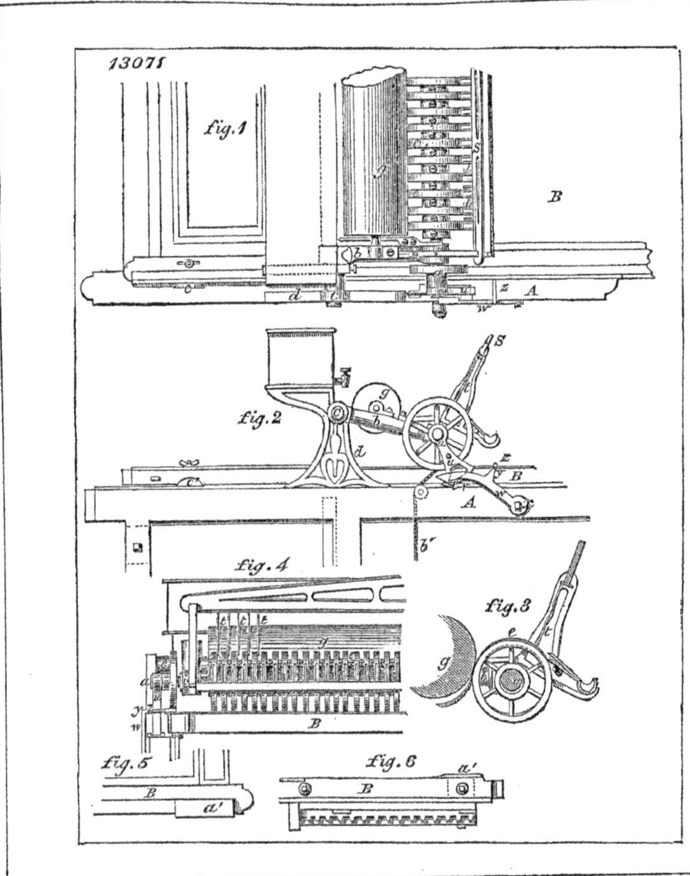
13071
fig. 1
fig. 2
fig. 3
fig. 4
fig. 5
fig. 6

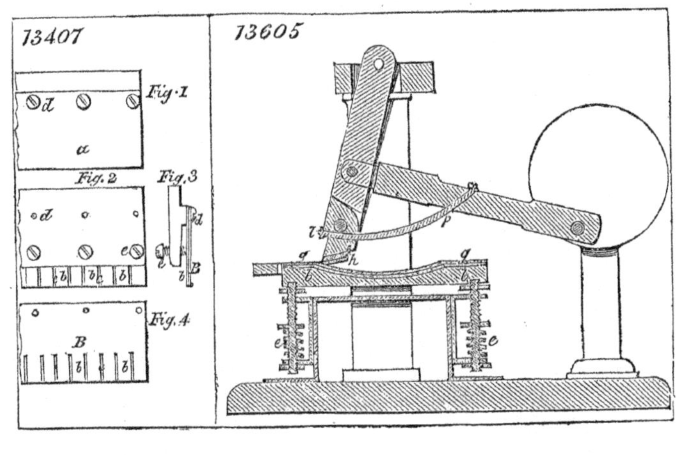
13407
Fig. 1
Fig. 2
Fig. 3
Fig. 4
13605

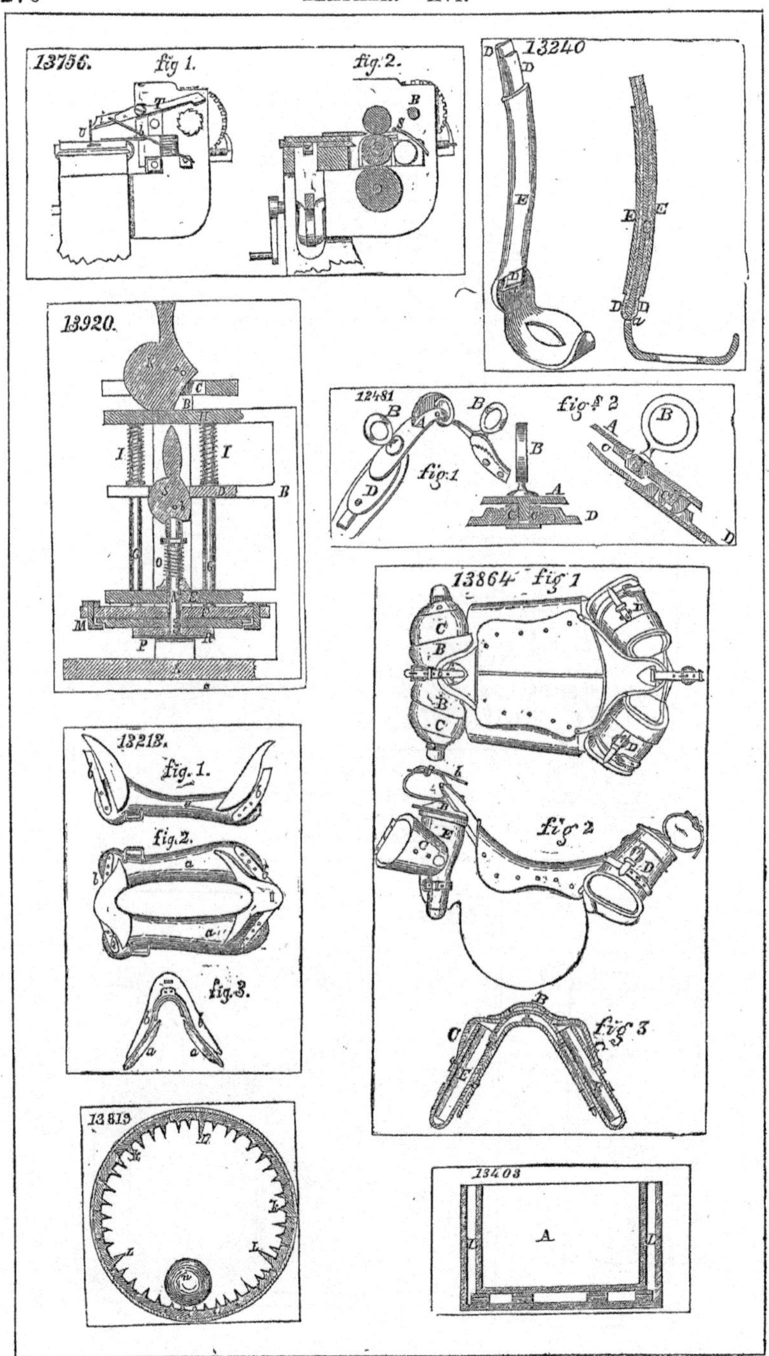
13756. fig 1.
fig. 2.
13240
13920.
12481
fig. 1
figs 2
13864 fig 1
fig 2
fig 3
13218.
fig. 1.
fig. 2.
fig. 3.
13813
13403
A

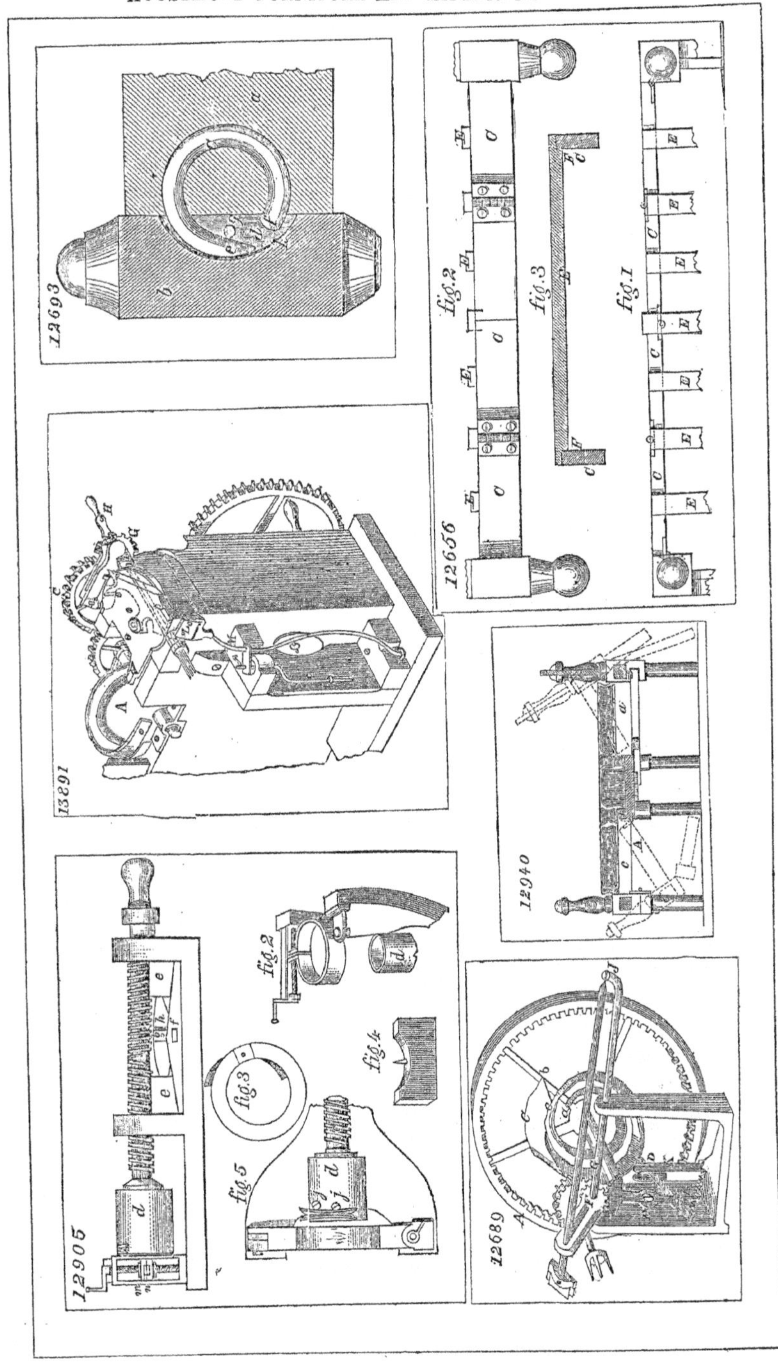
12693
13291
12905
12656
12940
12689

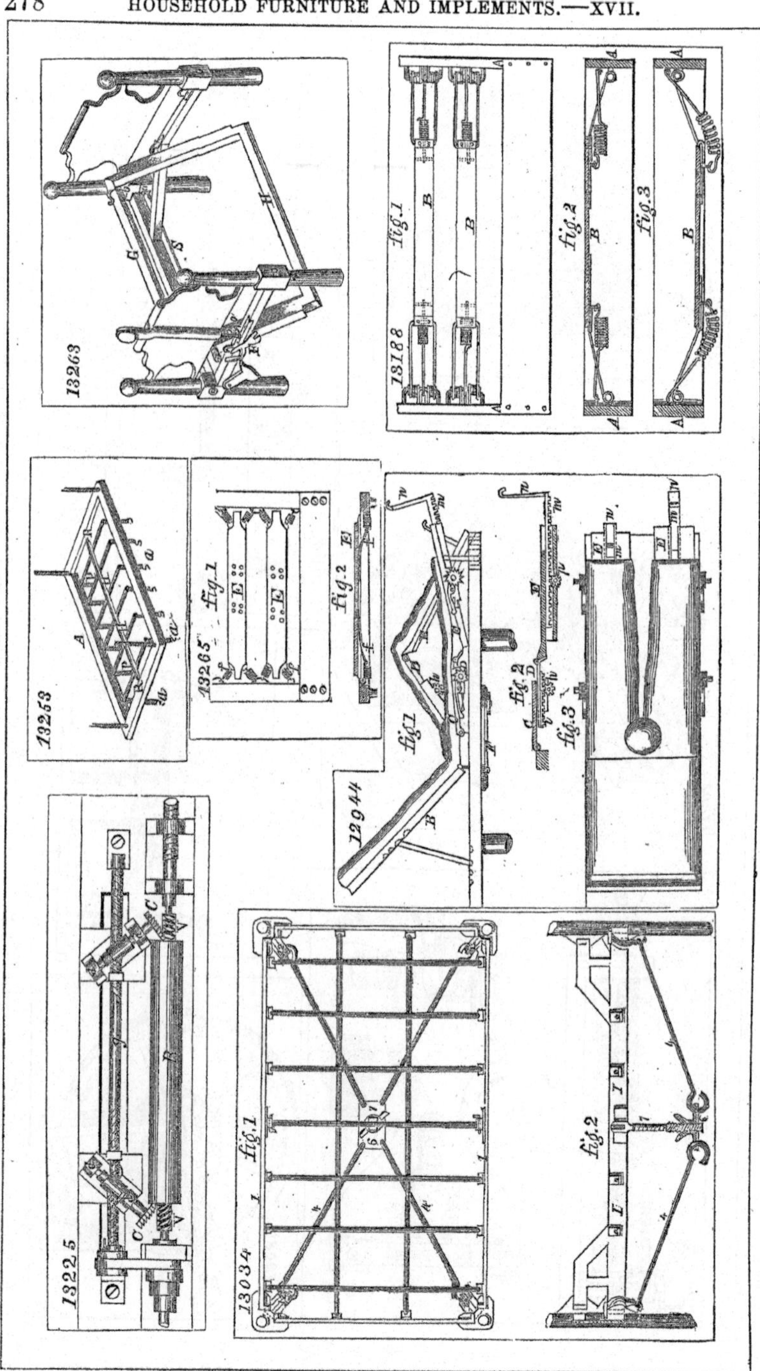
13263
13188
13253
13265
12944
13225
13034

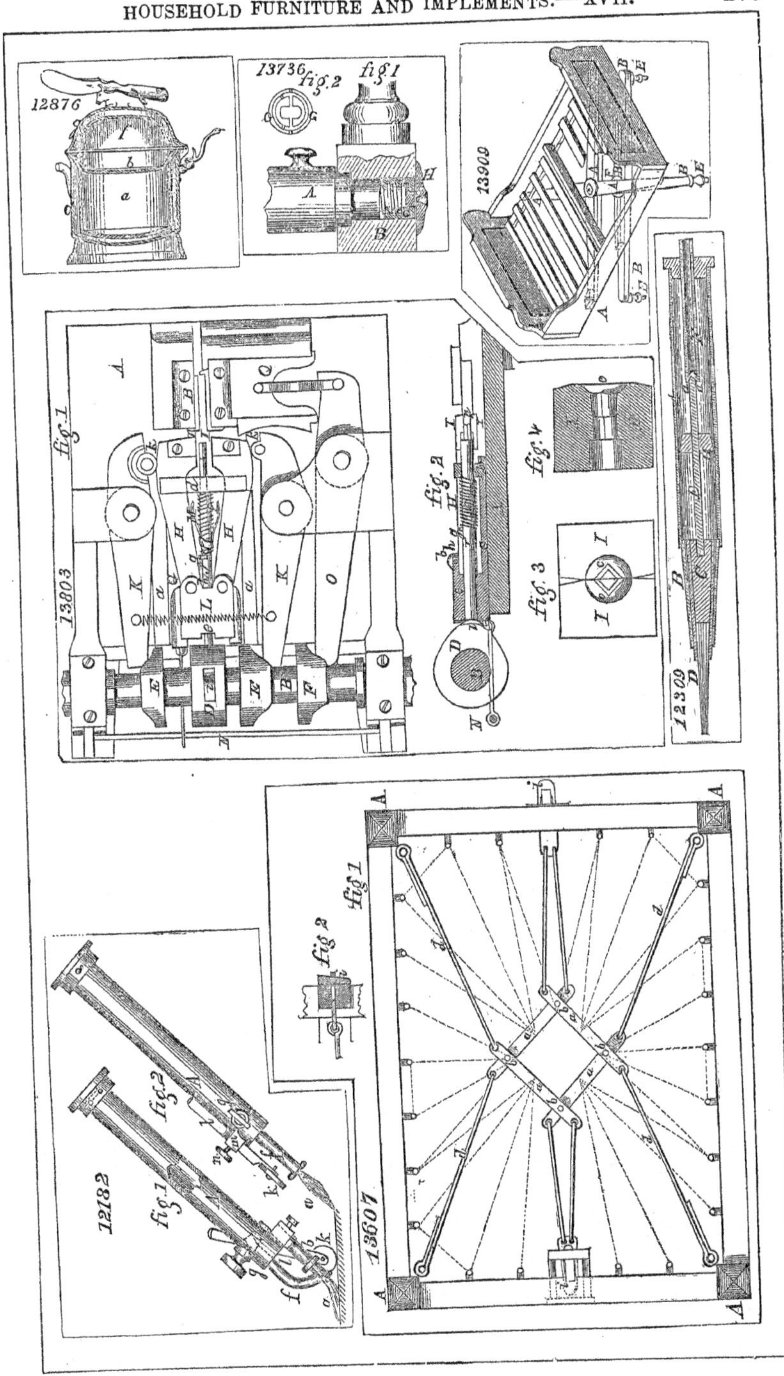
12876
13736
13803
12309
12182
13607

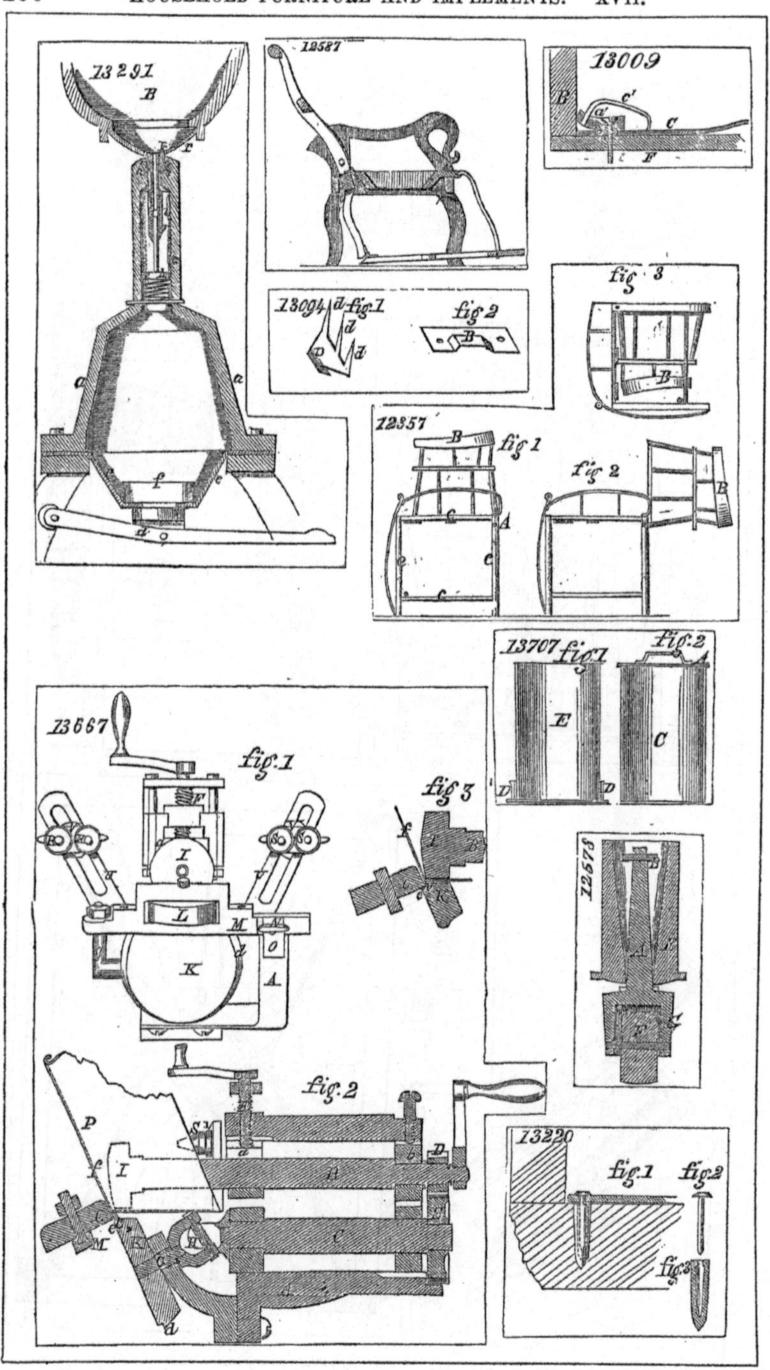
13291
12587
13009
13094 fig.1
fig.2
fig 3
12357 fig 1
fig 2
13707 fig.1
fig.2
13667
fig.1
fig 3
12578
fig.2
13220
fig.1
fig.2
fig.3

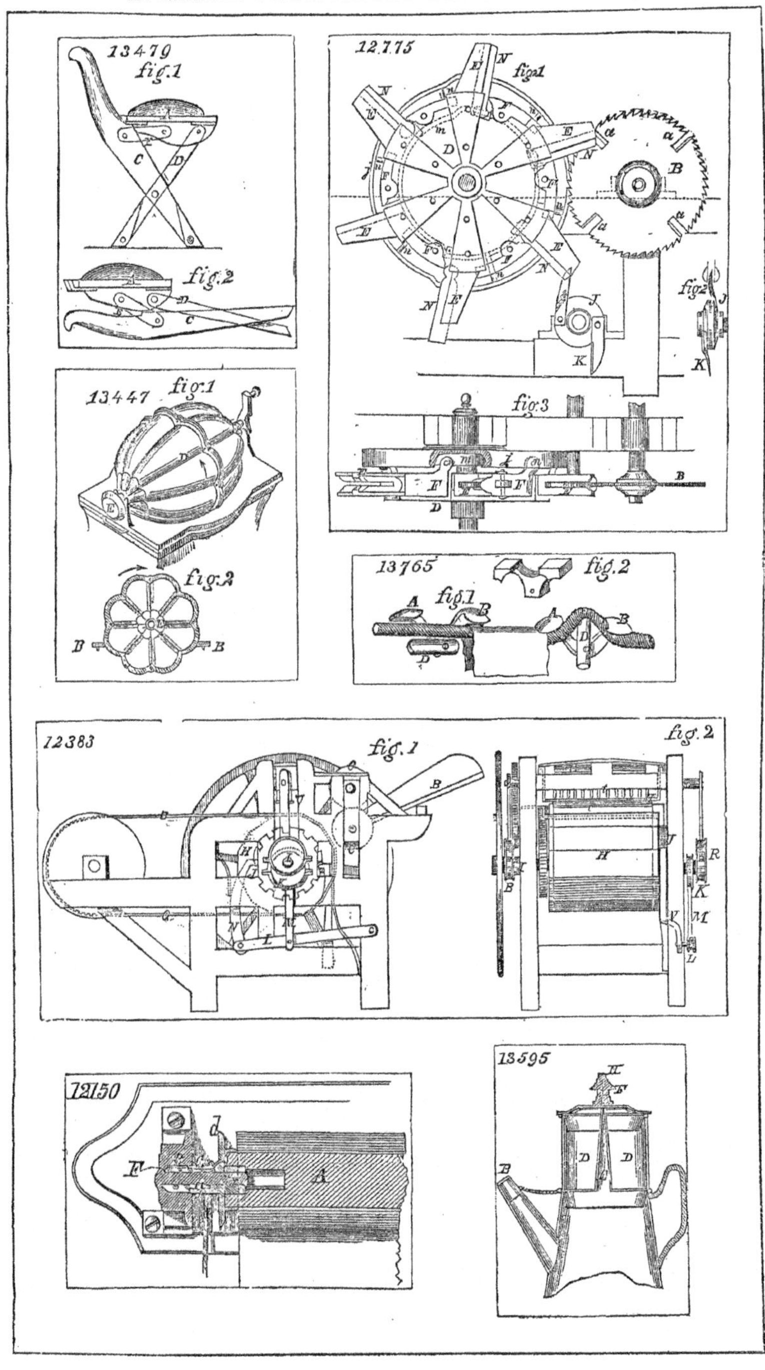
13479
fig. 1
fig. 2
12775
fig. 1
fig. 2
fig. 3
13447
fig. 1
fig. 2
13765
fig. 1
fig. 2
12383
fig. 1
fig. 2
12150
13595

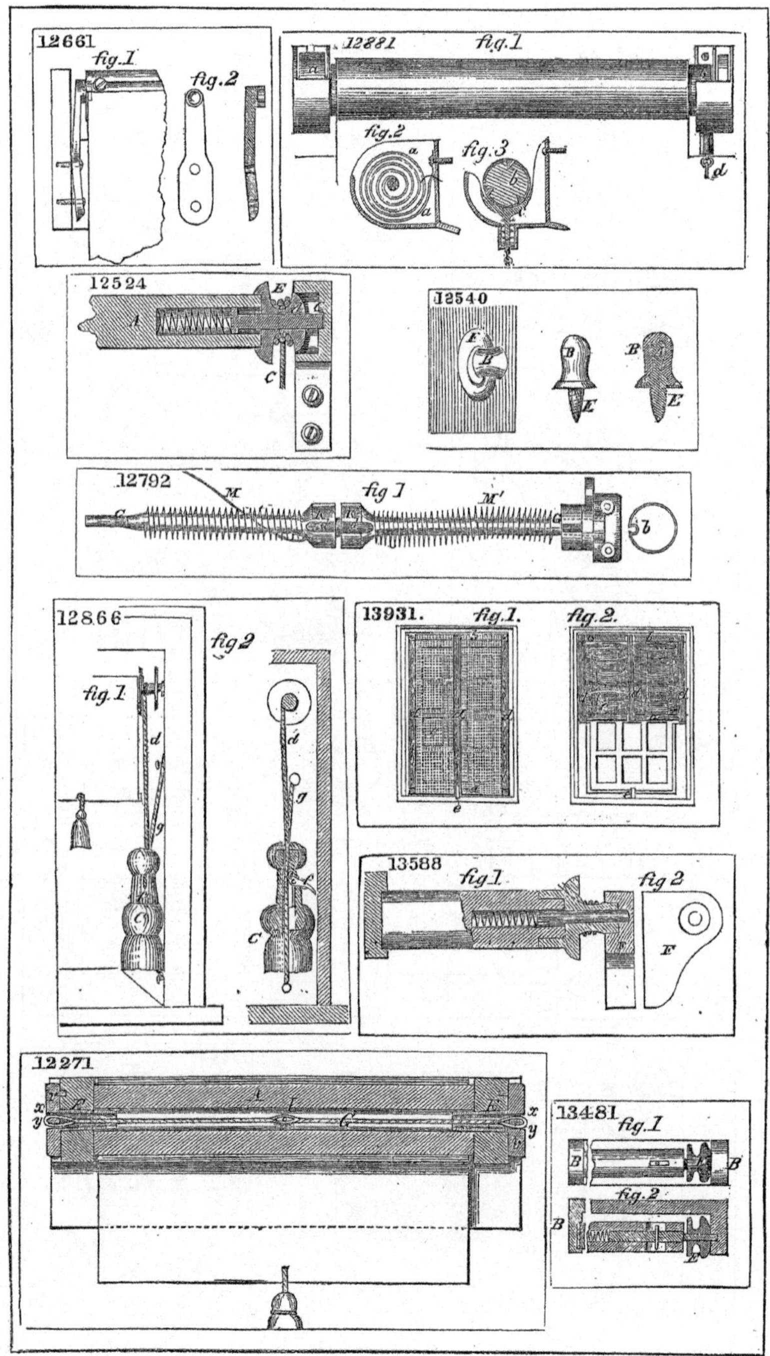
12661
fig. 1
fig. 2
12881
fig. 1
fig. 2
fig. 3
12524
12540
12792
fig 1
12866
fig 1
fig 2
13931.
fig. 1.
fig. 2.
13588
fig 1
fig 2
12271
13481
fig. 1
fig. 2

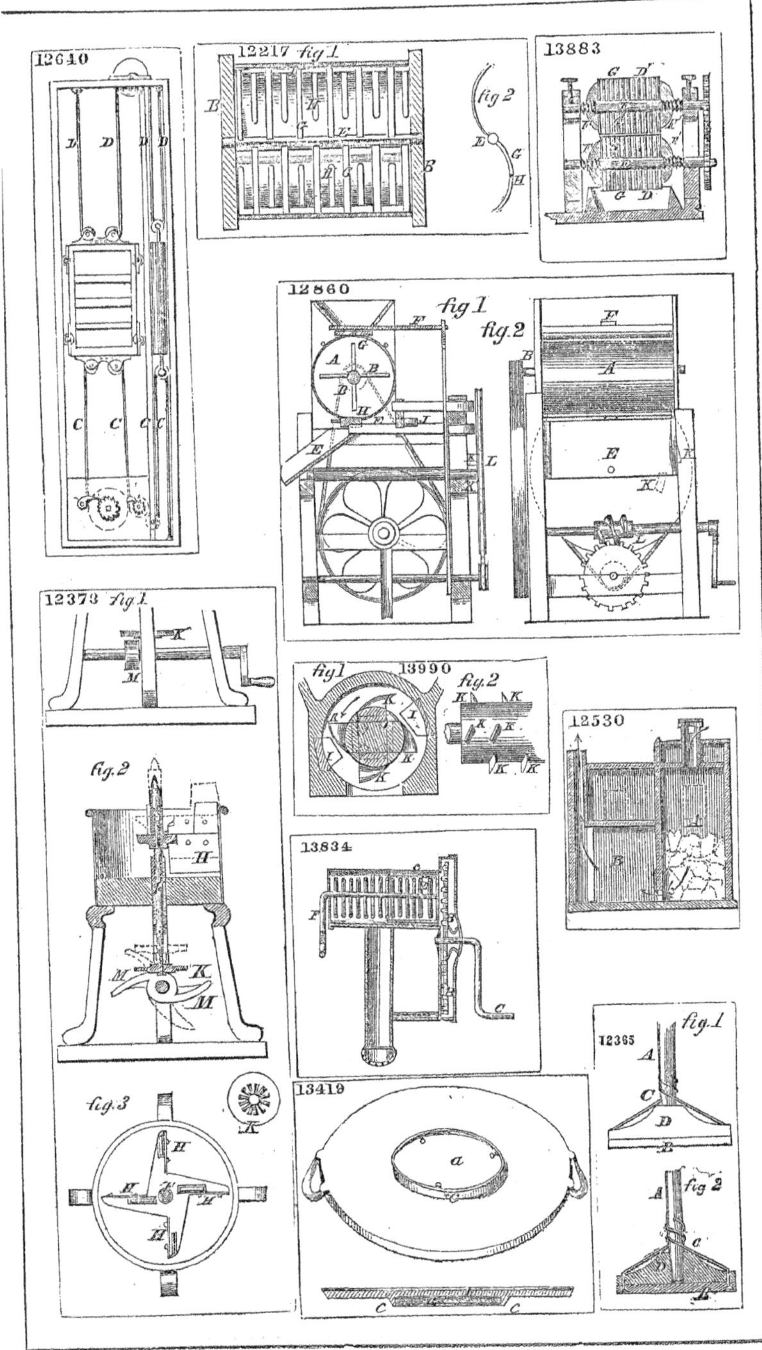
12640
12217 fig 1
fig 2
13883
12860
fig 1
fig 2
12373 fig 1
fig 2
fig 3
13990
12530
13834
13419
12365
fig 1
fig 2

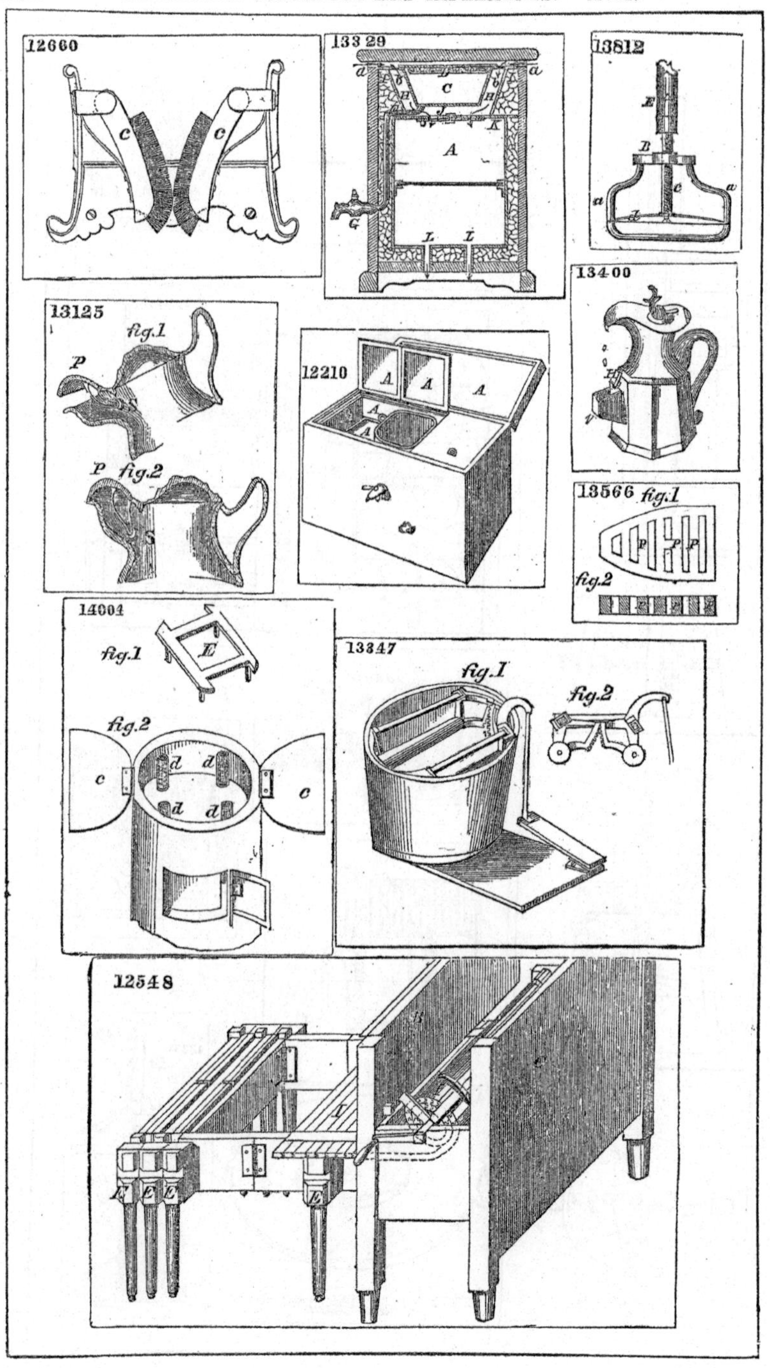
12660
13329
13812
13400
13125
fig.1
fig.2
12210
13566 fig.1
fig.2
14004
fig.1
fig.2
13347
fig.1
fig.2
12548

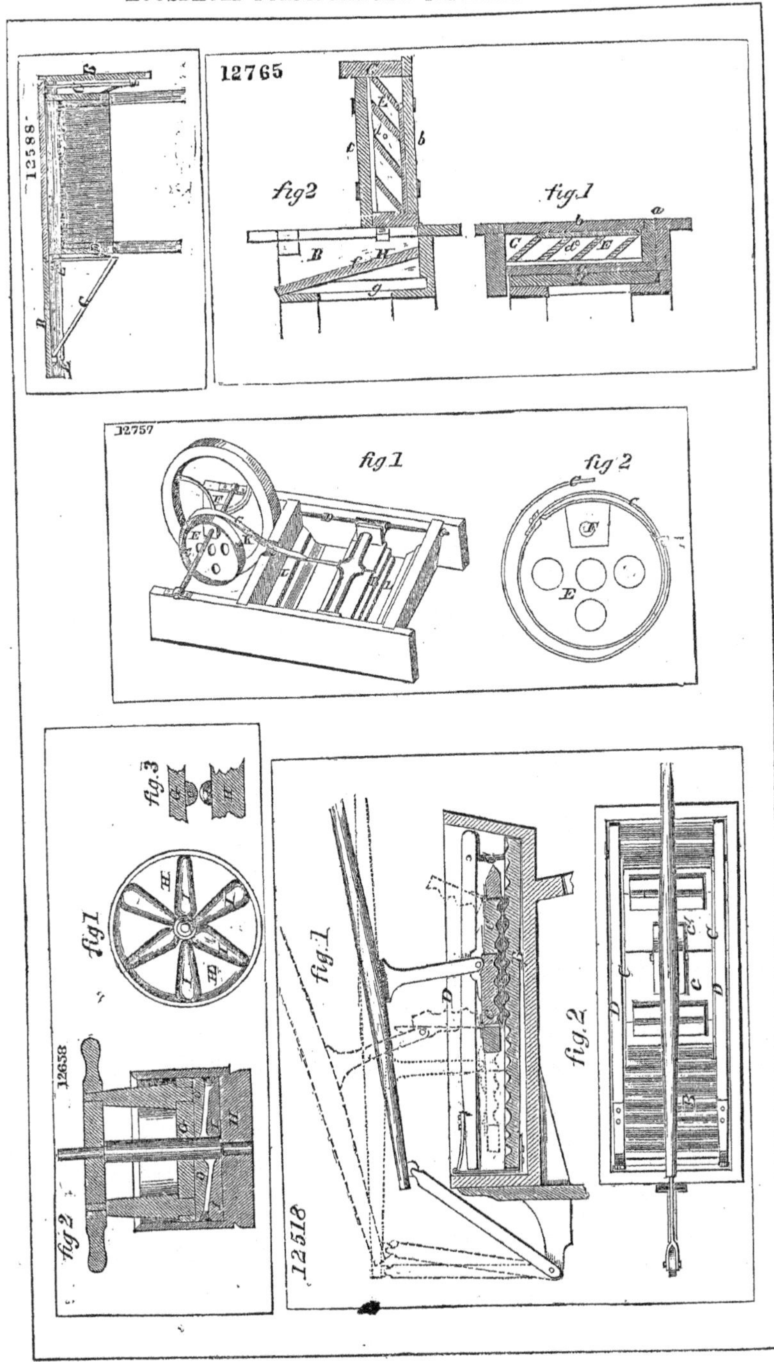
12588
12765
fig2
fig.1
12757
fig.1
fig.2
12658
fig.1
fig.2
fig.3
12518
fig.1
fig.2

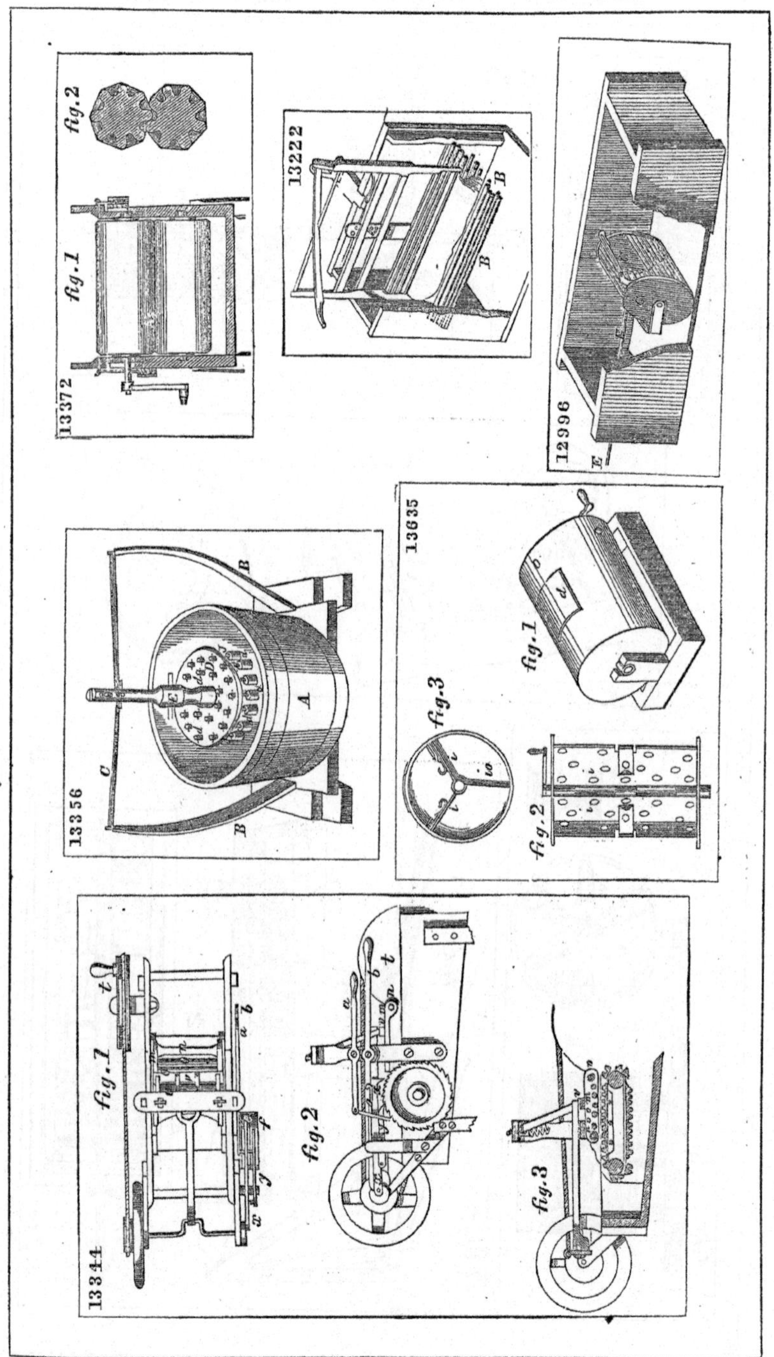
13372
fig.1
fig.2
13222
12996
13356
13635
fig.1
fig.2
fig.3
13344
fig.1
fig.2
fig.3

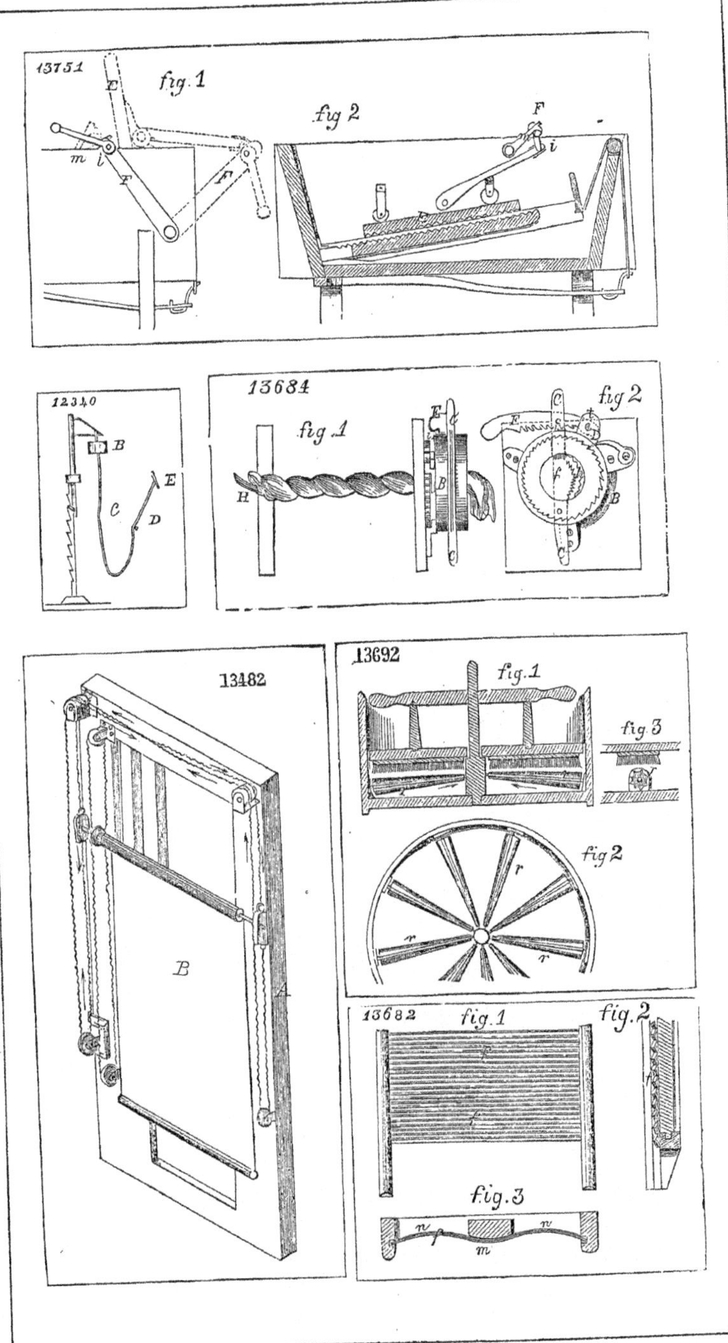
13751
fig. 1
fig 2
12340
13684
fig. 1
fig 2
13482
13692
fig. 1
fig. 3
fig 2
13682
fig. 1
fig. 2
fig. 3

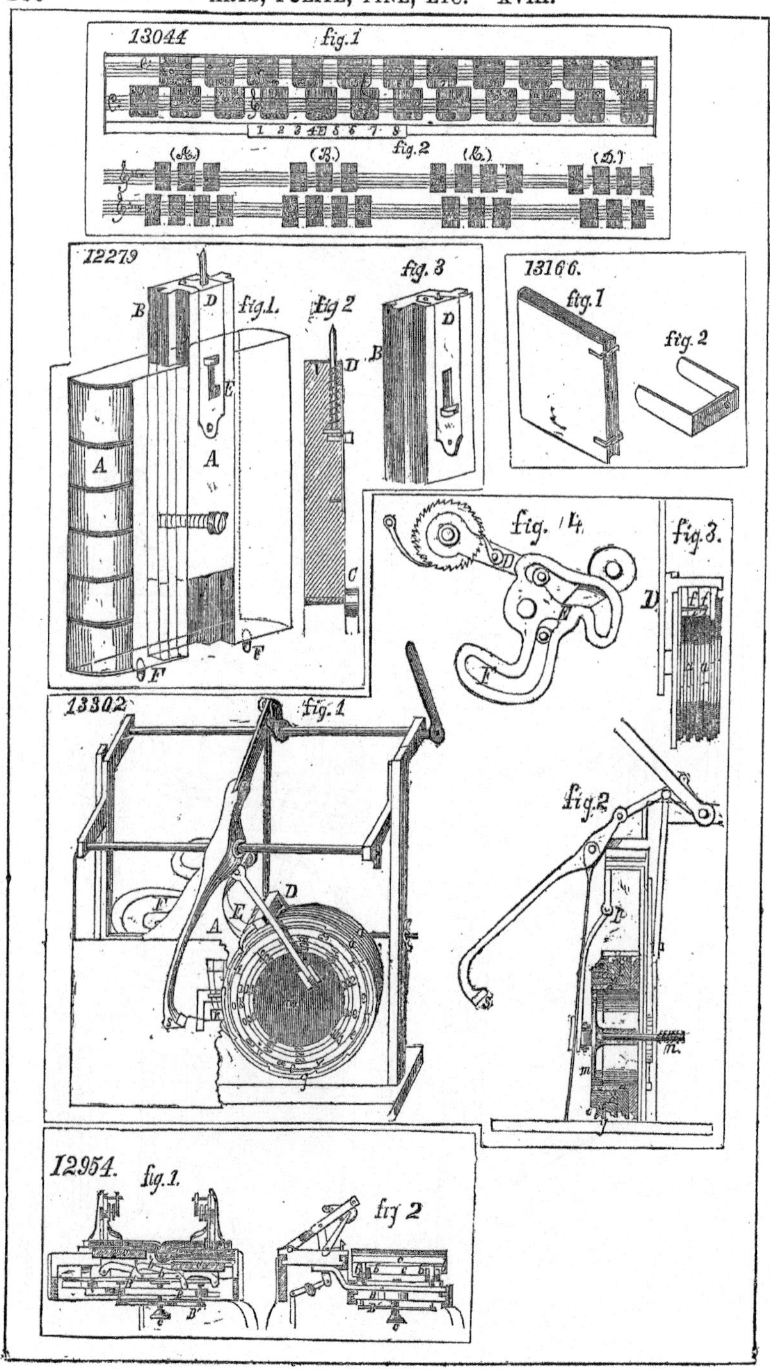
13044
fig.1
fig.2
12279
fig.1.
fig 2
fig.3
13166.
fig.1
fig.2
fig.4
fig.3.
13302
fig.1
fig.2
12954.
fig.1.
fig 2

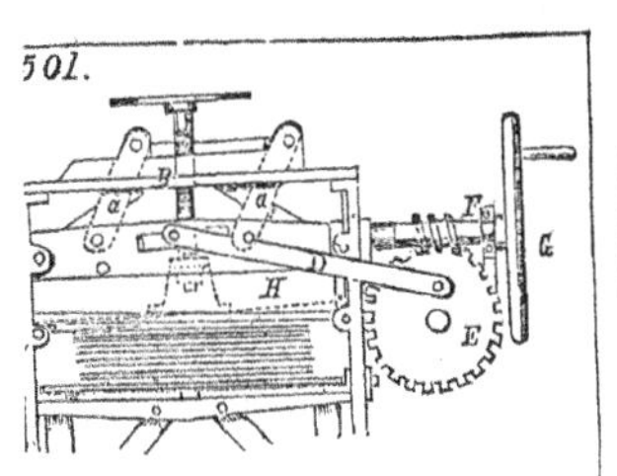
501.
E
G
H
D
F

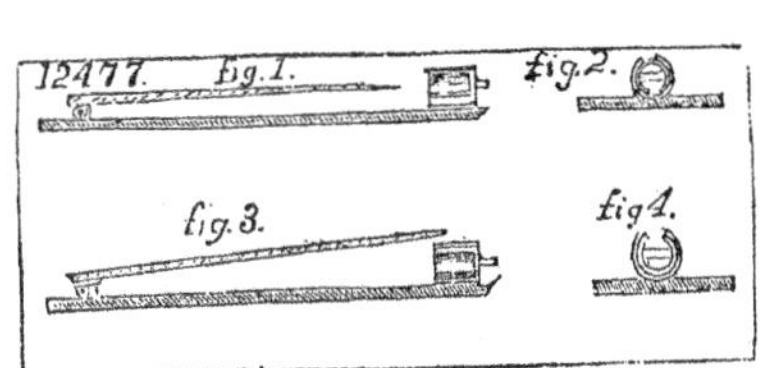
12477. fig.1.
fig.2.
fig.3.
fig.4.

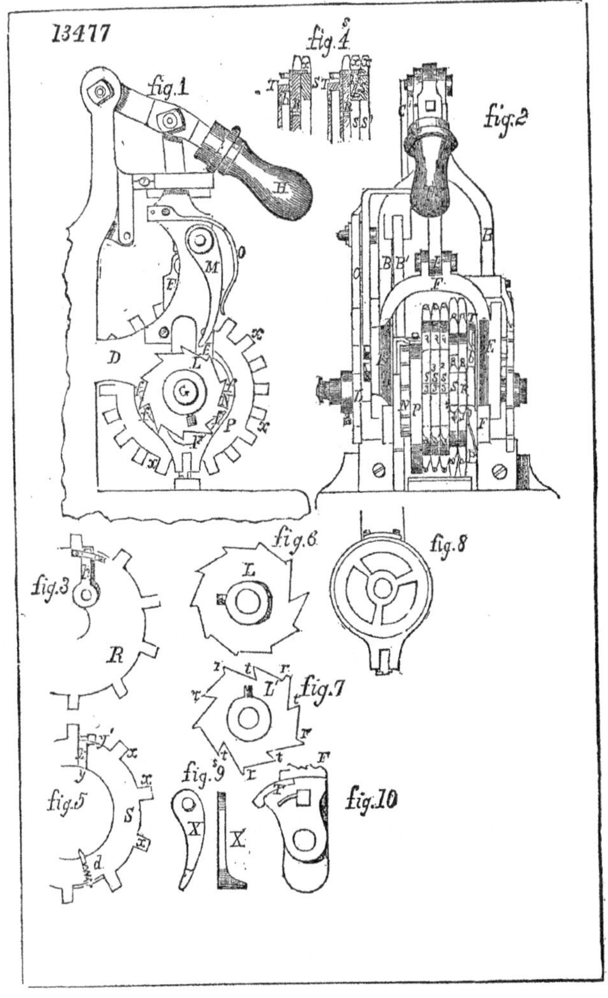
13477
fig.1
fig.2
fig.4
fig.3
fig.5
fig.6
fig.7
fig.8
fig.9
fig.10

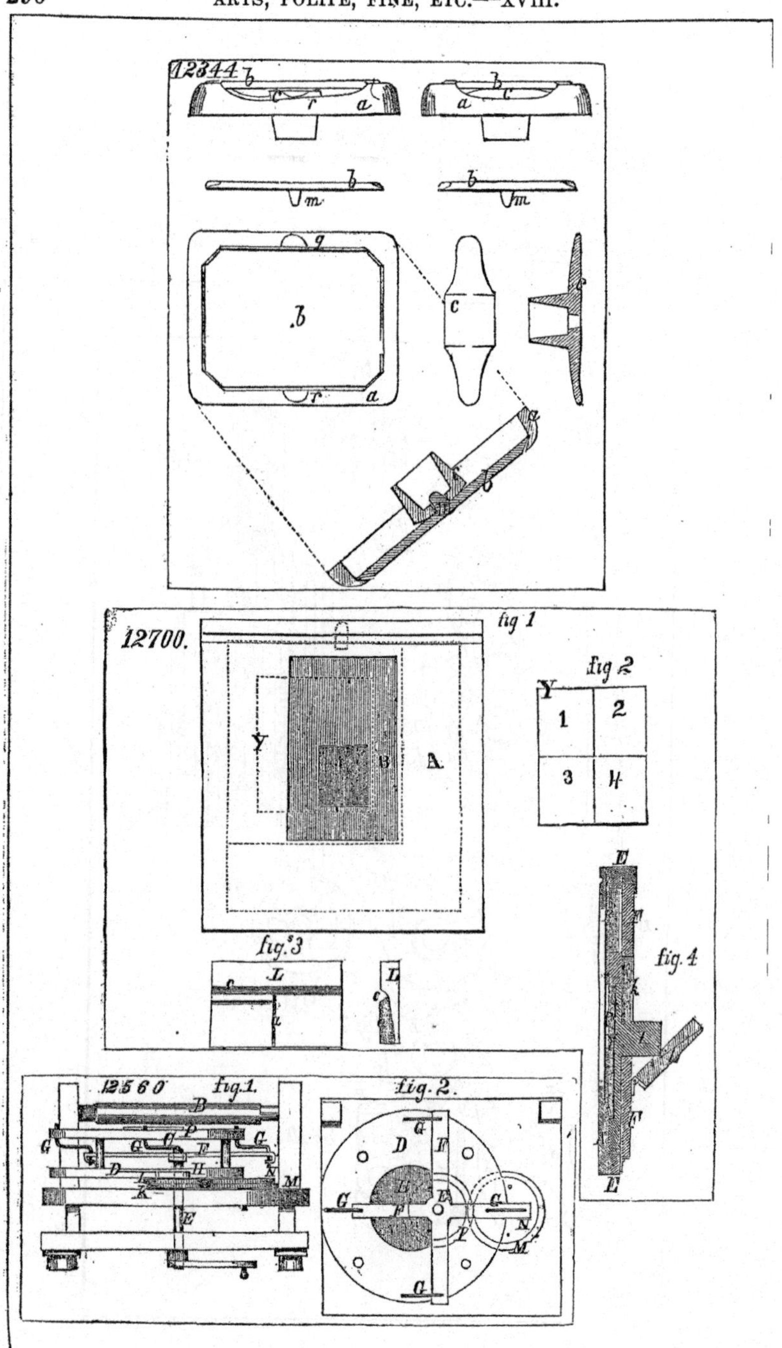
12344.
12700.
fig 1
fig 2
fig.s 3
fig. 4
12560
fig.1.
fig.2.

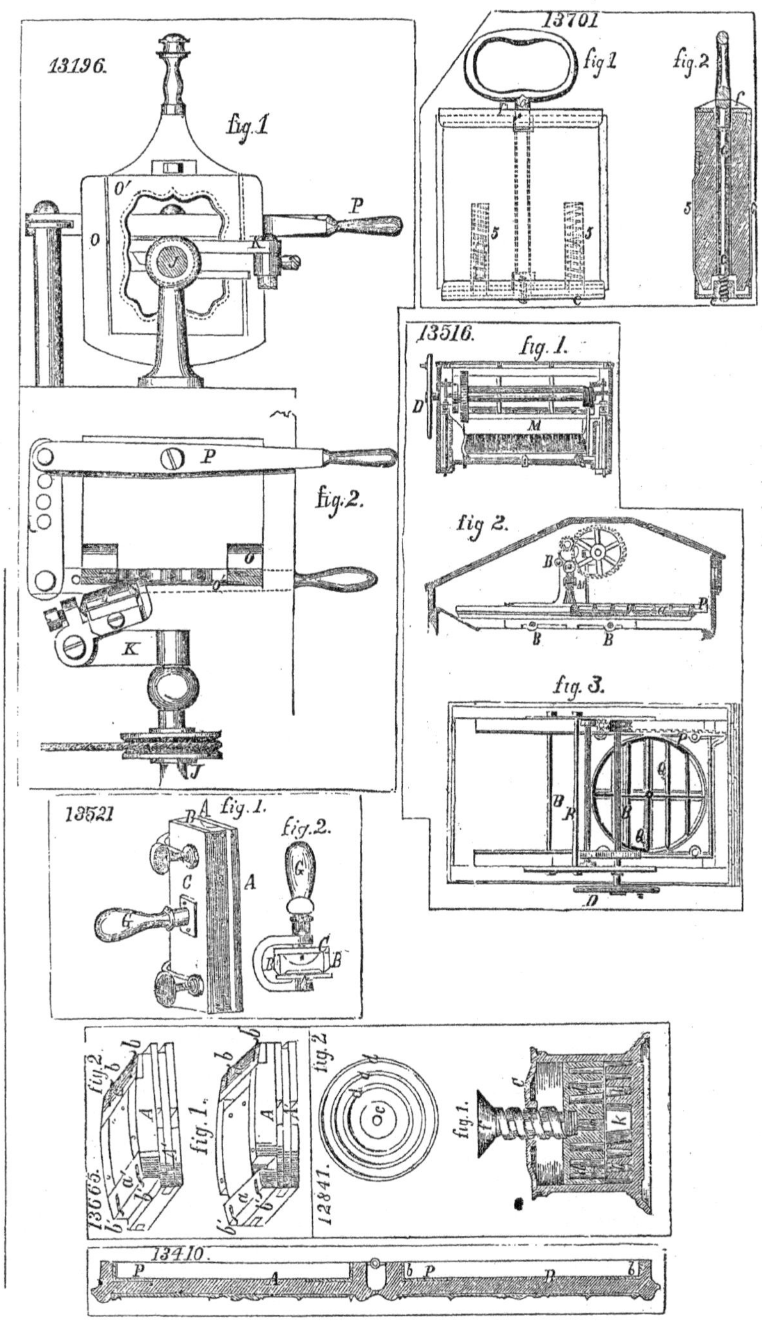
13196.
fig. 1
fig. 2.
13701
fig 1
fig. 2
13516.
fig. 1.
fig 2.
fig. 3.
13521
fig. 1.
fig. 2.
13665.
fig. 2
fig. 1.
12841.
fig. 2
fig. 1.
13410.

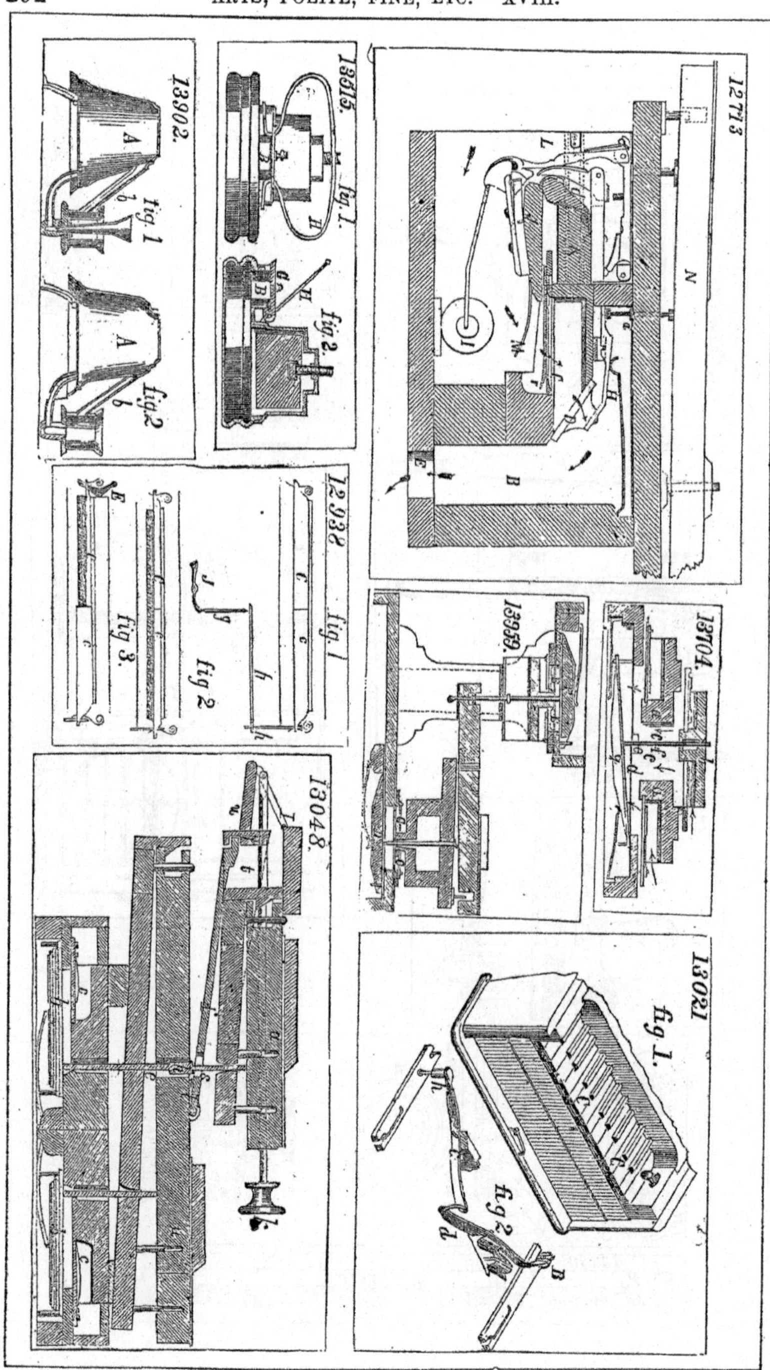
13902.
fig. 1
fig. 2
13615.
fig. 1.
fig. 2.
12773
12938
fig. 1
fig 2
fig 3.
13959.
13704.
13048.
13021
fig 1.
fig 2

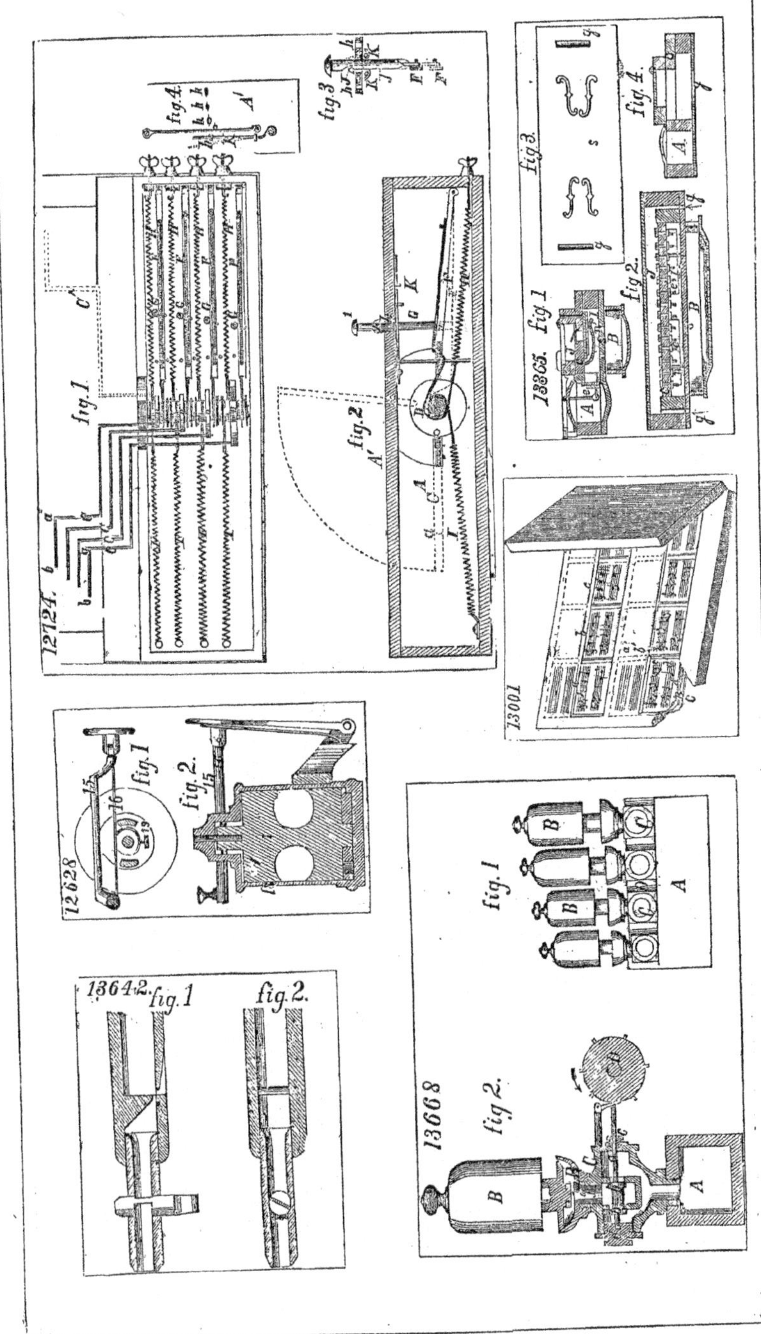
12724.
13365.
13001
12628
13642.
13668

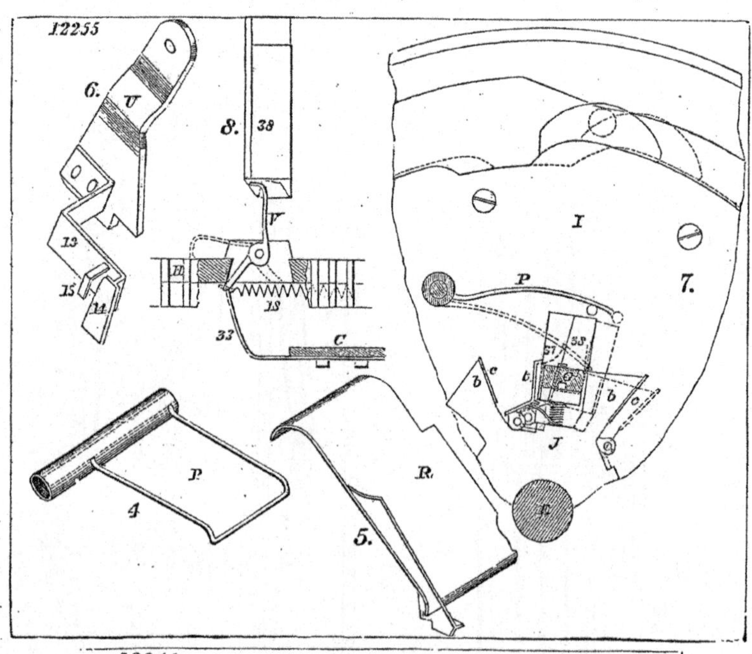
12255
6.
U
8.
38
V
13
15
14
18
33
C
I
P
7.
b
c
J
4
5.
R

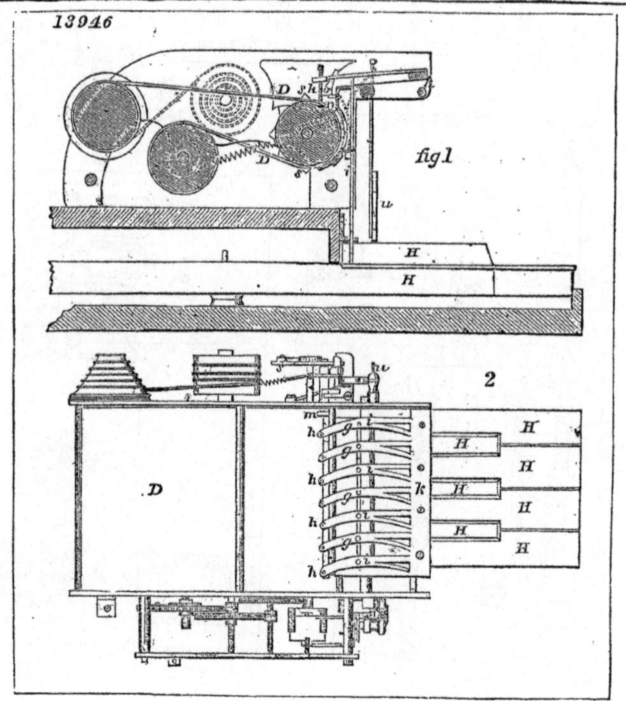
13946
fig1
D
H
2
h
g
k

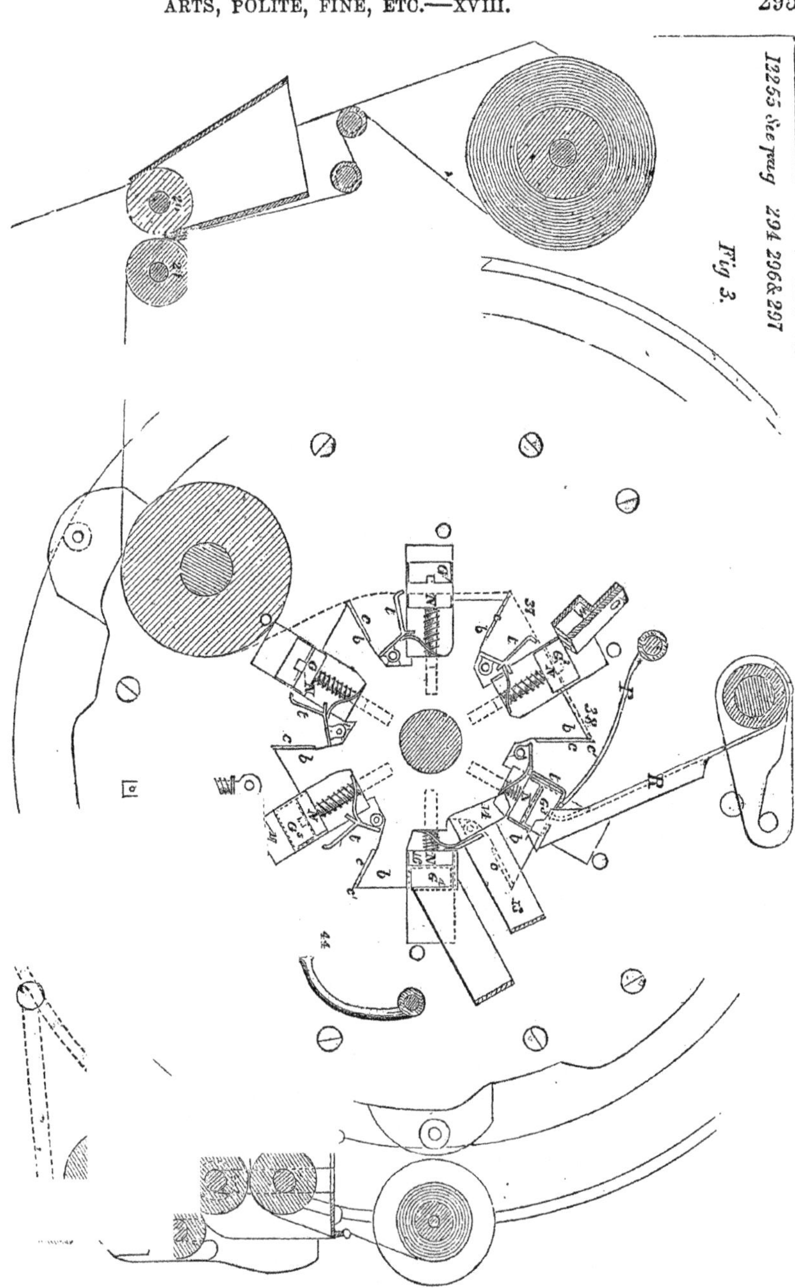

Fig 3.

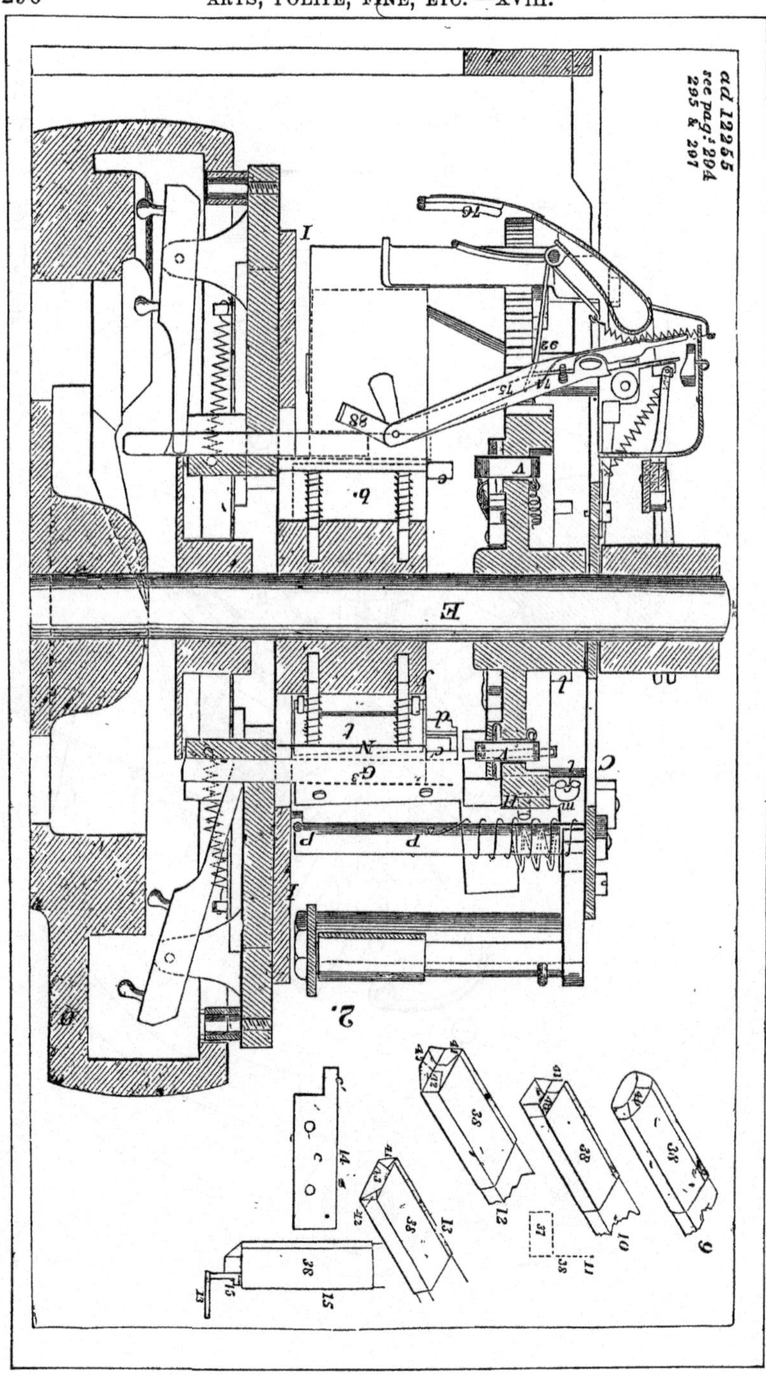
ad 12255
see pag.s 294
295 & 297
2.

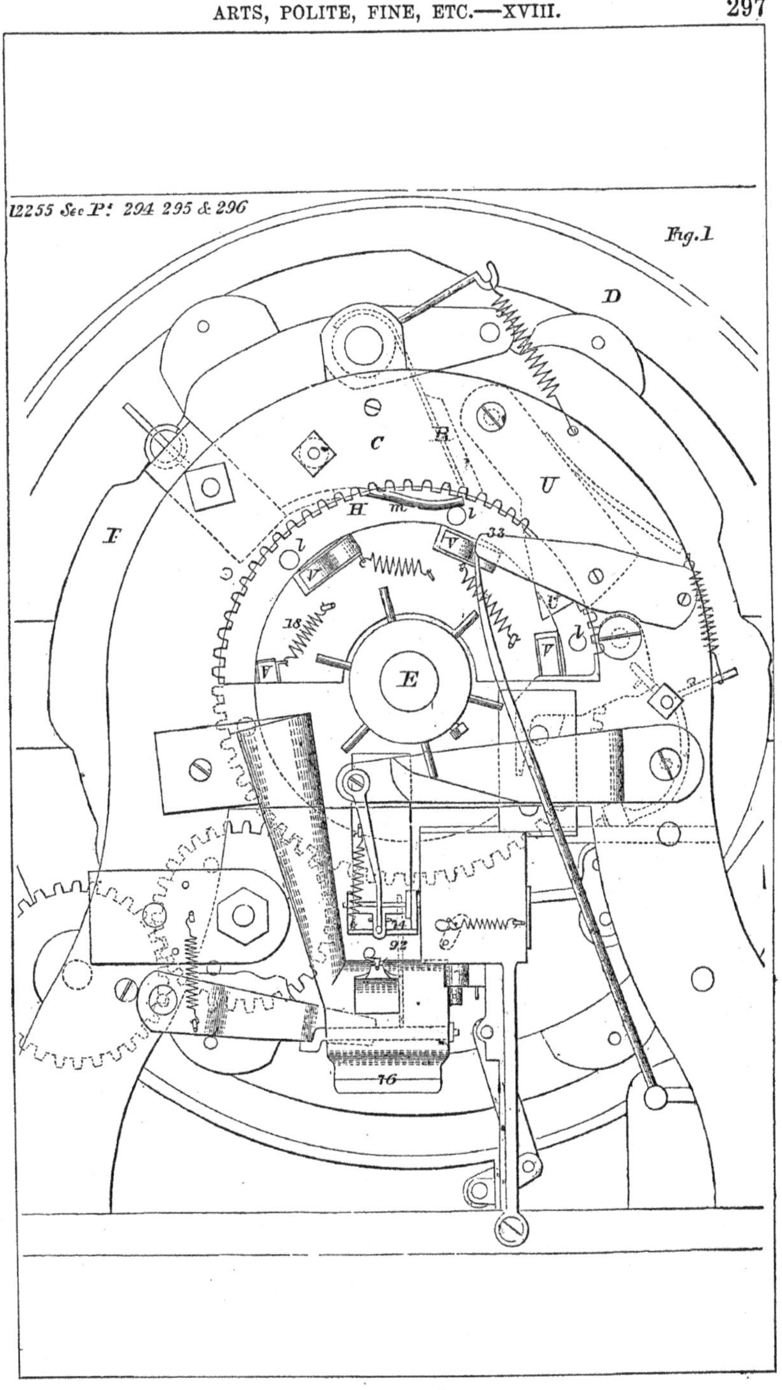
12255 See P: 294 295 & 296
Fig. 1
D
C
B
U
T
H
m
l
V
33
18
E
92
14
76

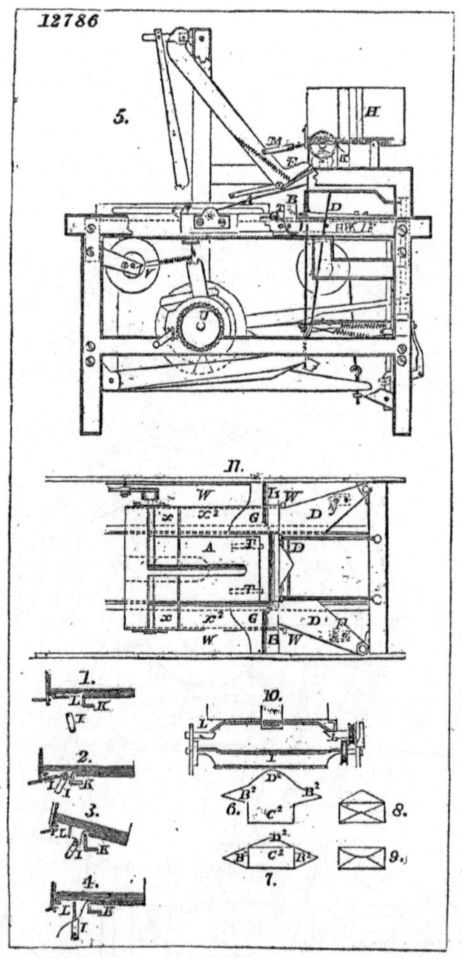
12786
5.
H
11.
1.
2.
3.
4.
10.
6.
7.
8.
9.

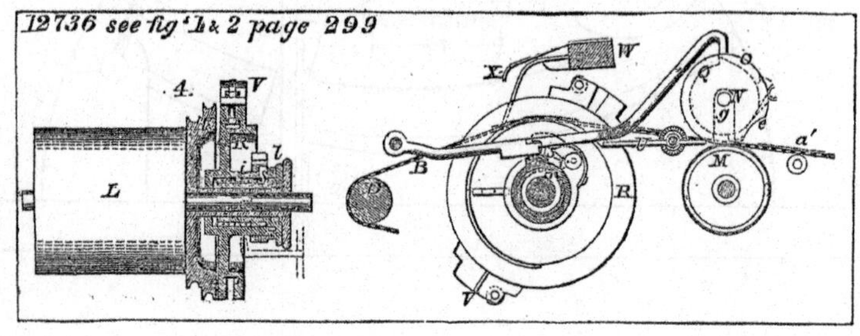
12736 see fig 1 & 2 page 299
4.
L
W
X
M
B
R

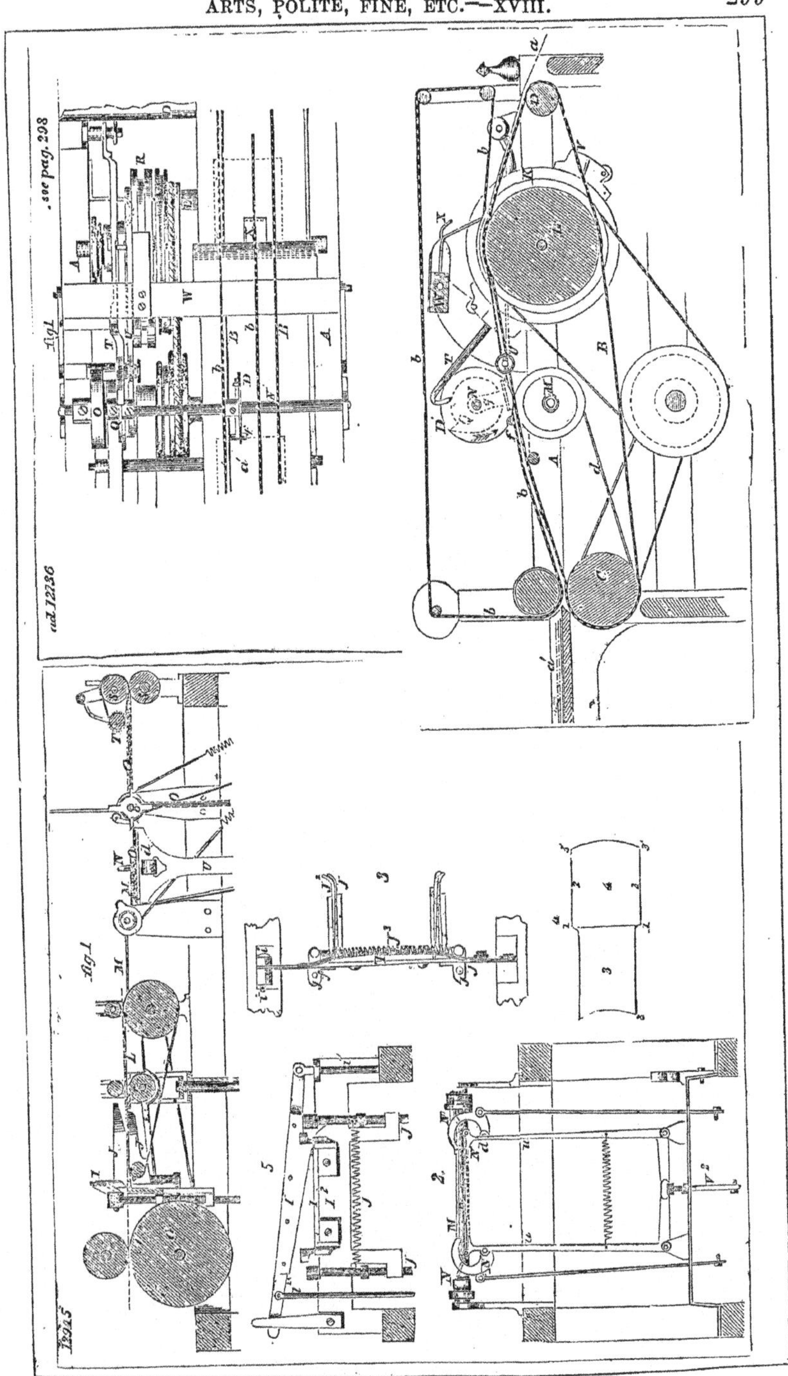

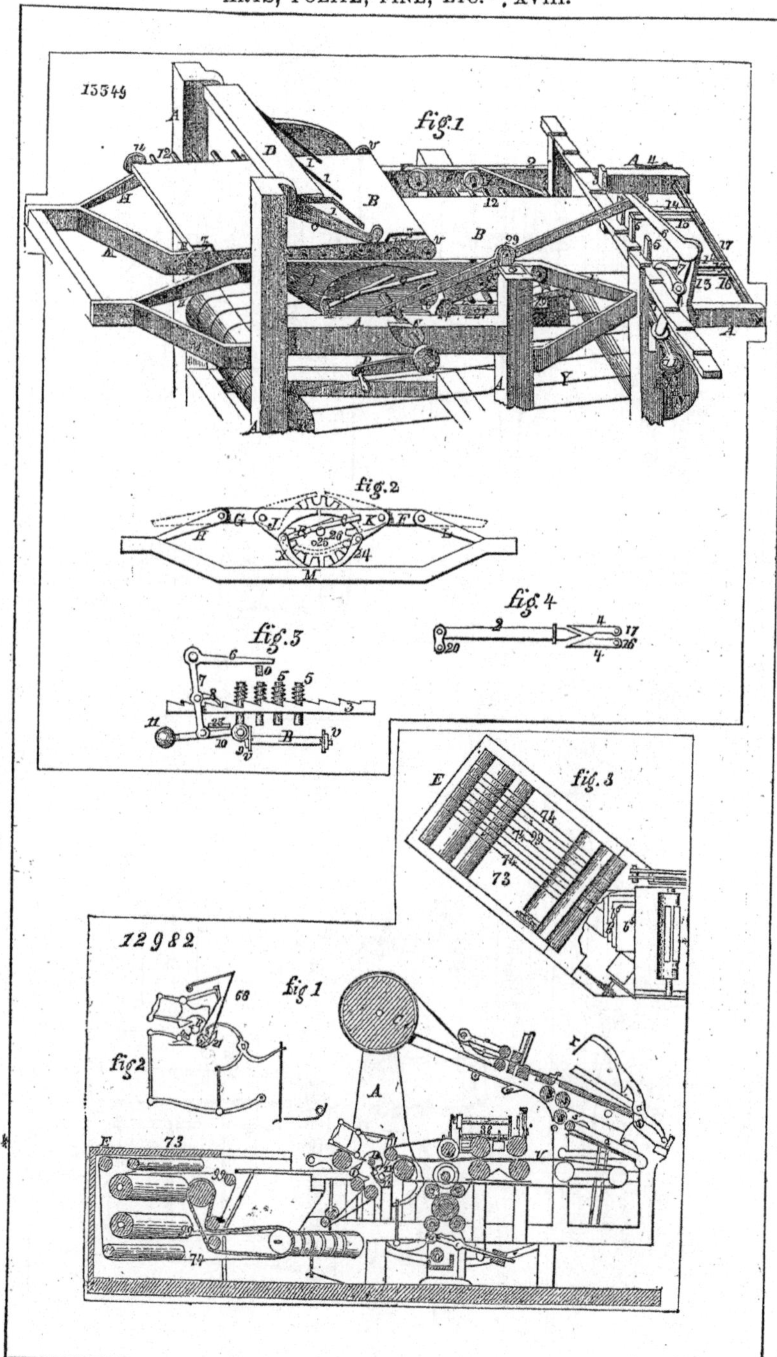
13349
fig. 1
fig. 2
fig. 3
fig. 4
12982
fig 1
fig 2
fig. 3

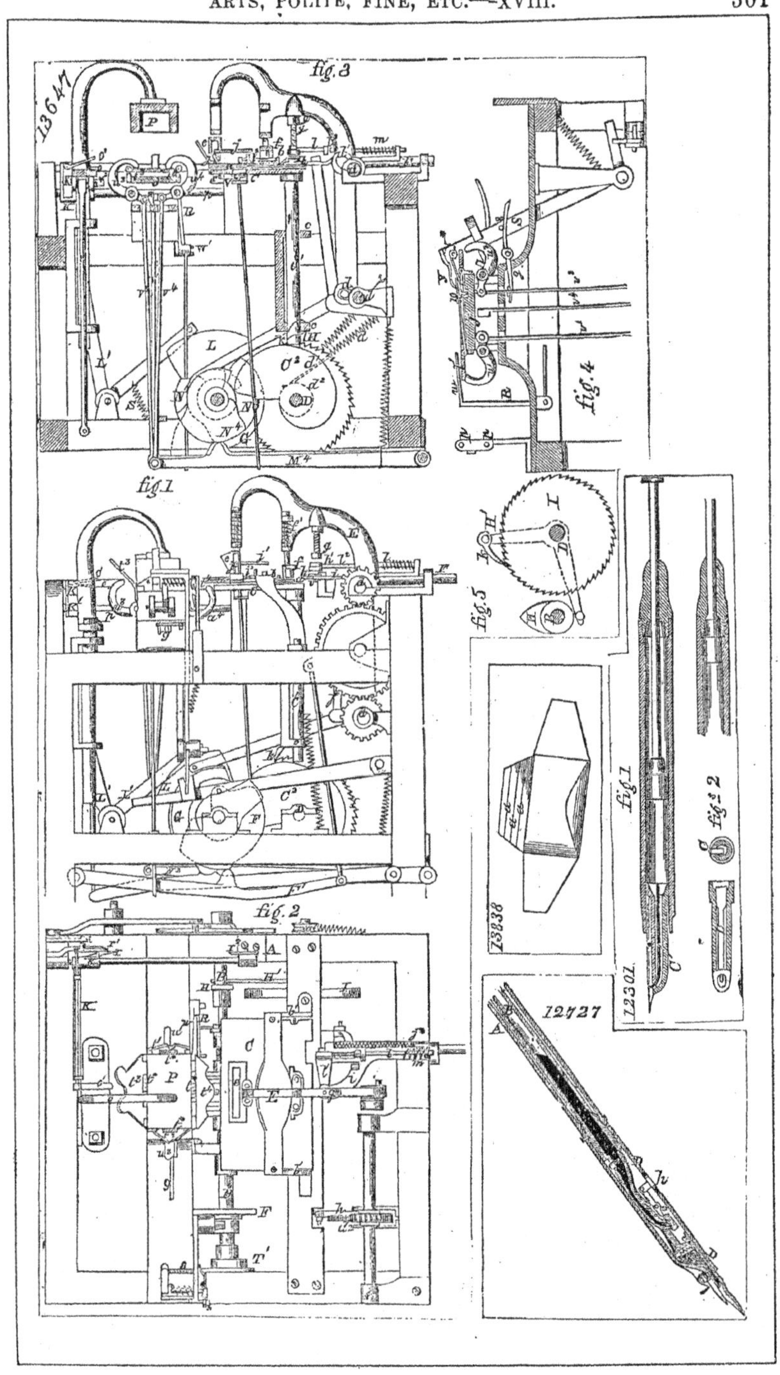
13647
fig. 3
fig. 1
fig. 2
fig. 4
fig. 5
13838
12301
fig. 1
fig. 2
12727

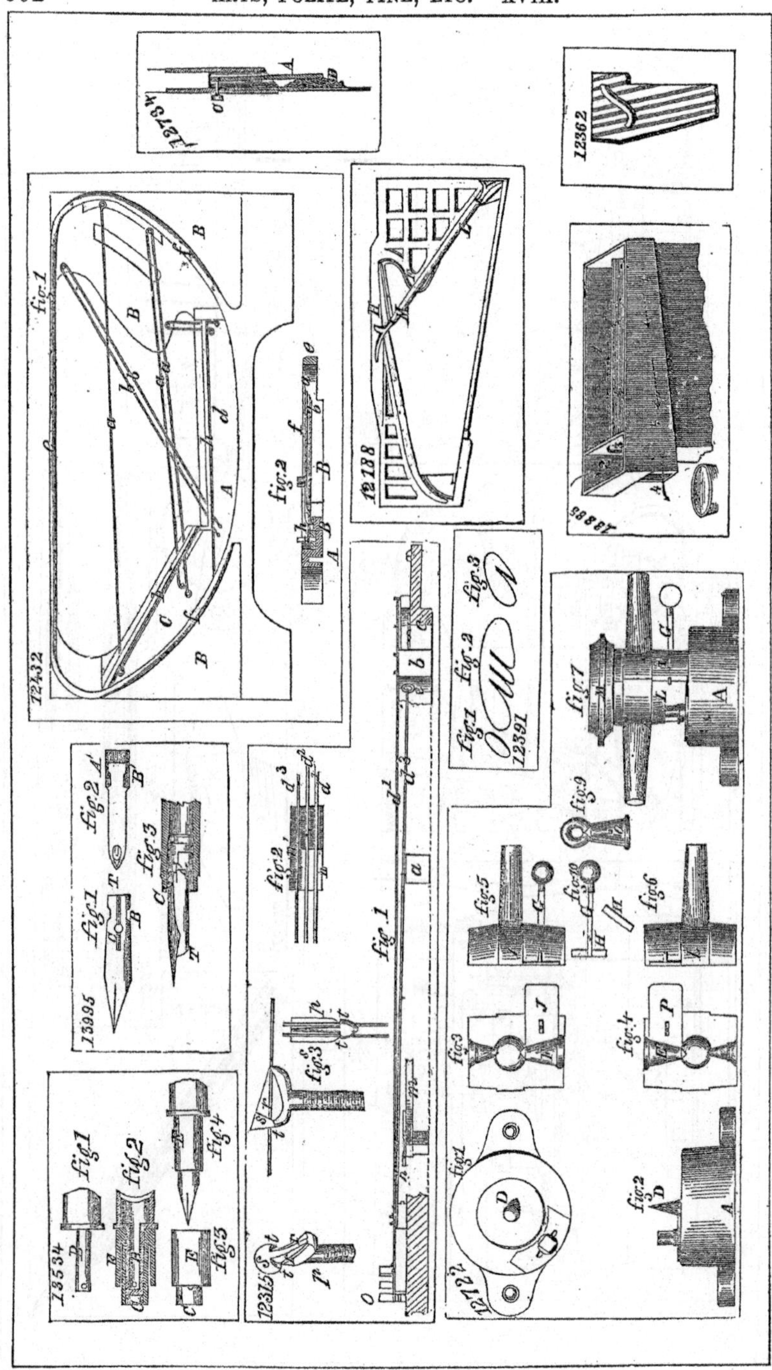

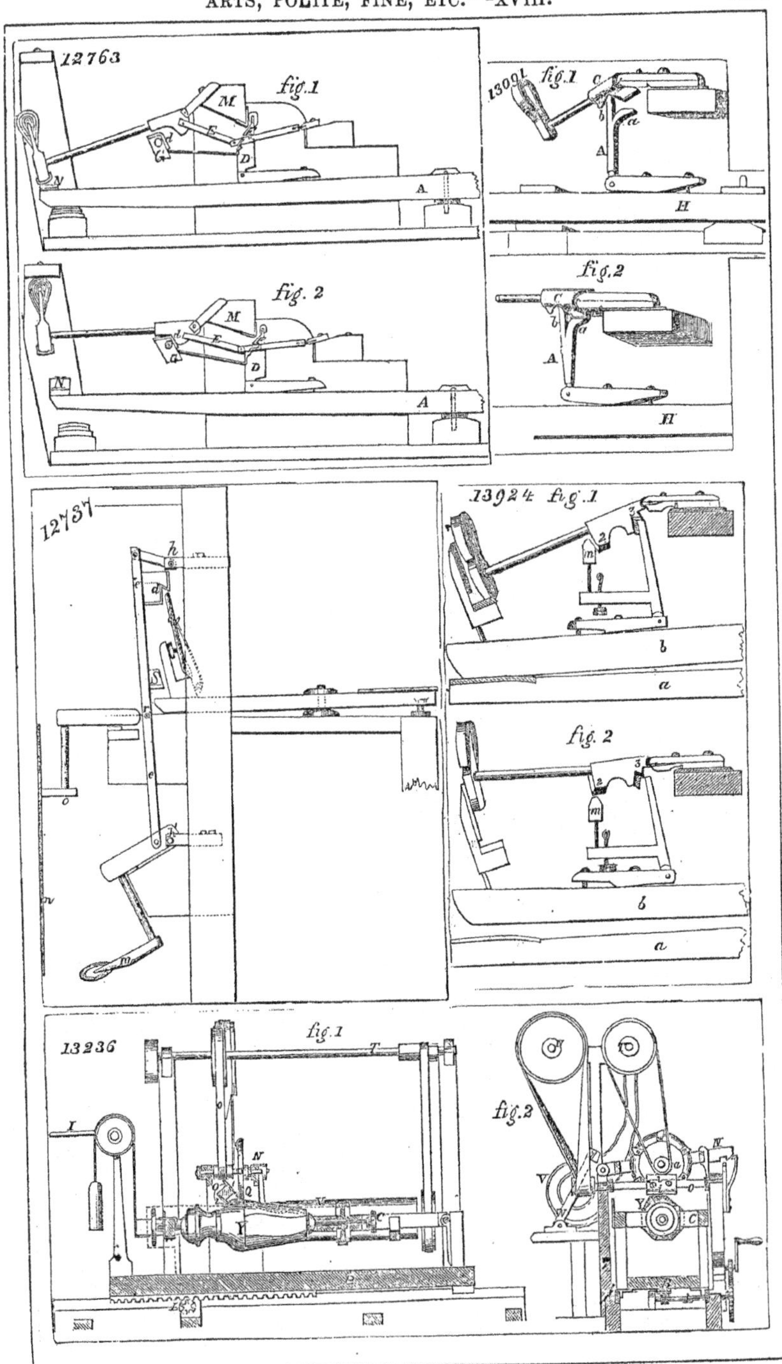
12763
fig.1
fig. 2
13001 fig.1
fig.2
12737
13924 fig.1
fig. 2
13236
fig.1
fig.2

13942
fig.1
fig.2
fig.3
fig.4

13580
fig.1
fig.2

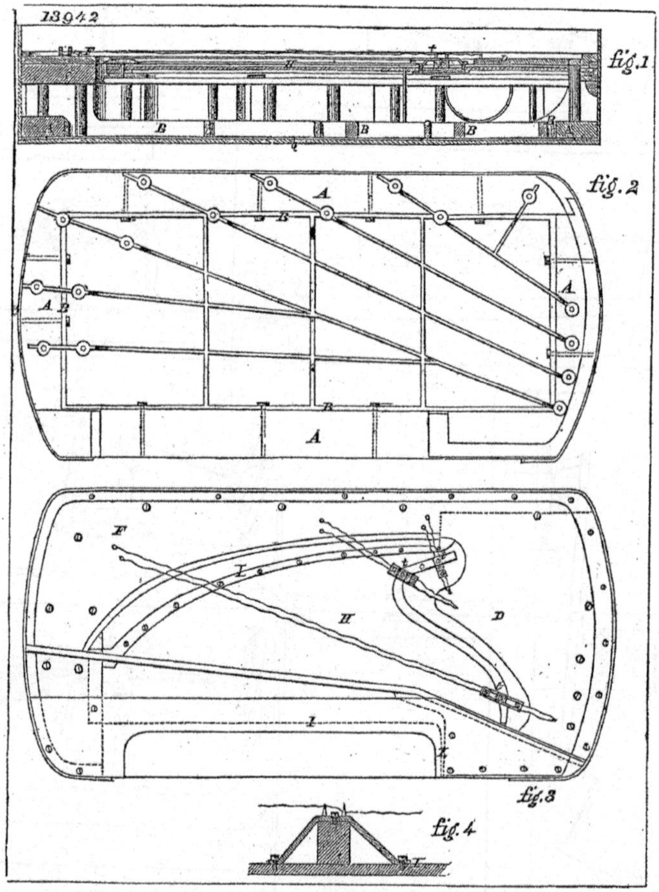

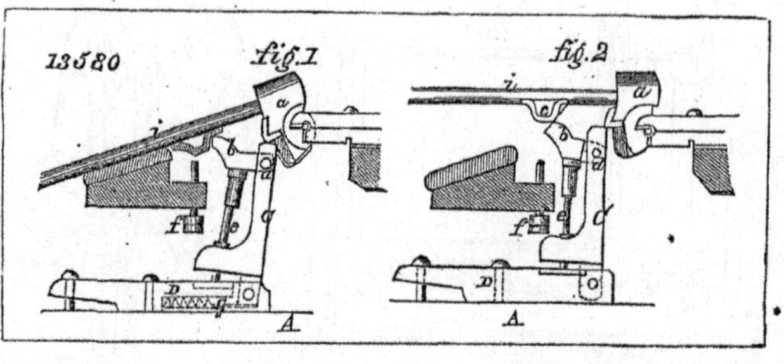

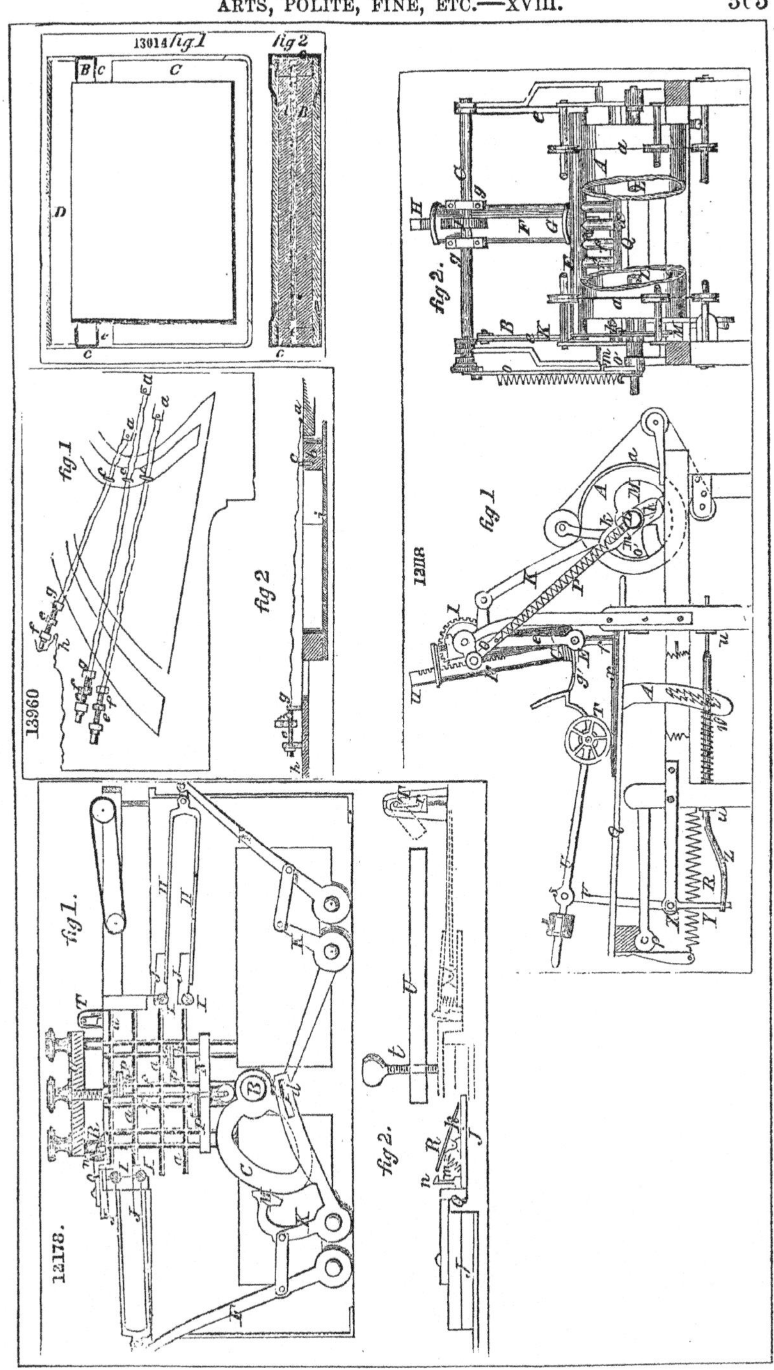
13014 fig 1
fig 2
13960
fig 1
fig 2
12178.
fig 1.
fig 2.
12118
fig 1
fig 2.

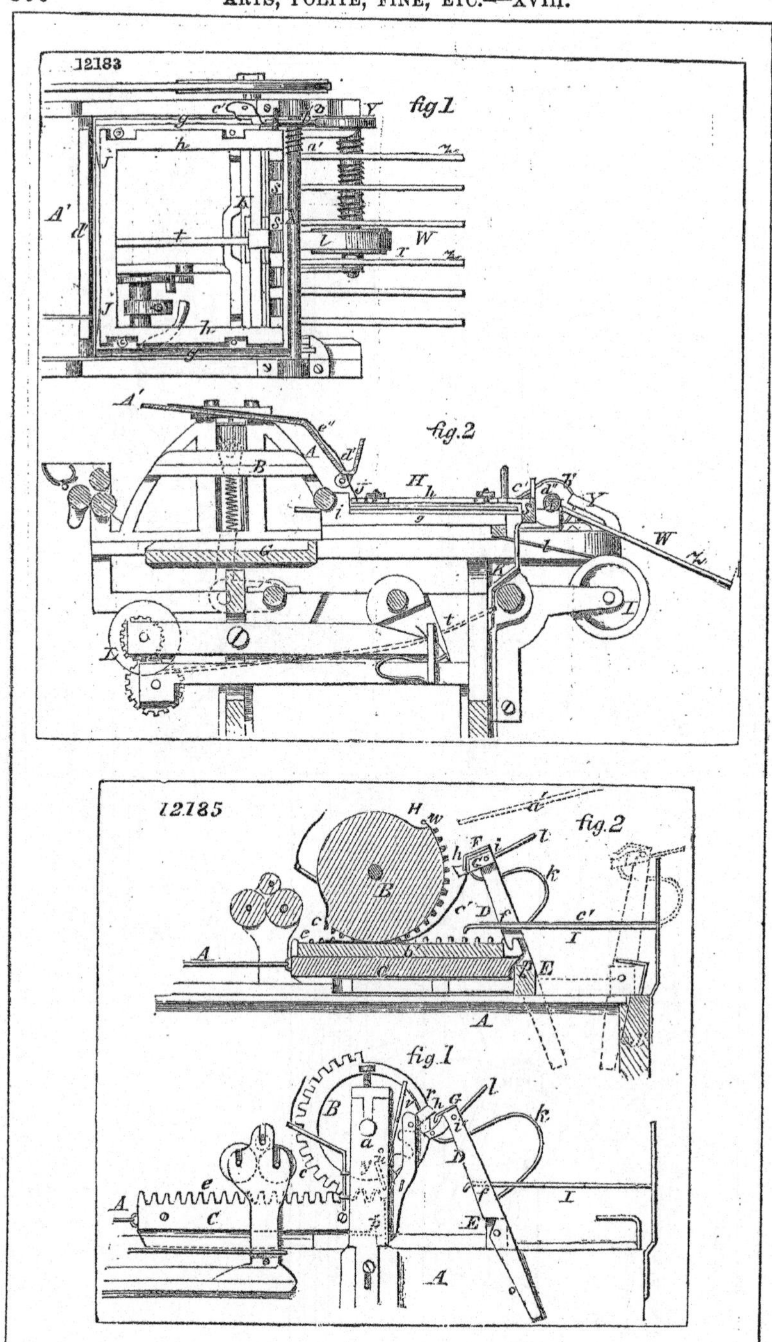
12183
fig.1
fig.2
12185
fig.2
fig.1

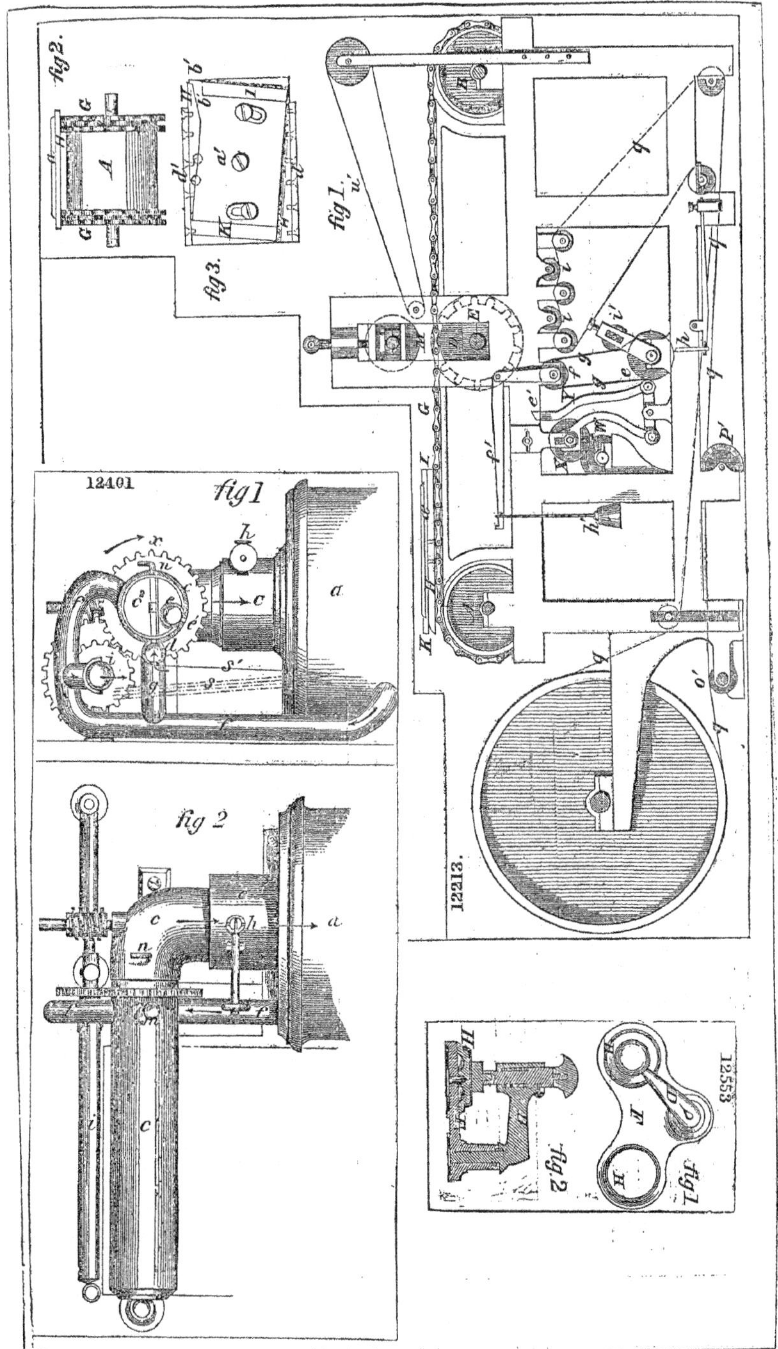
12401
fig1
fig 2
12213.
12553

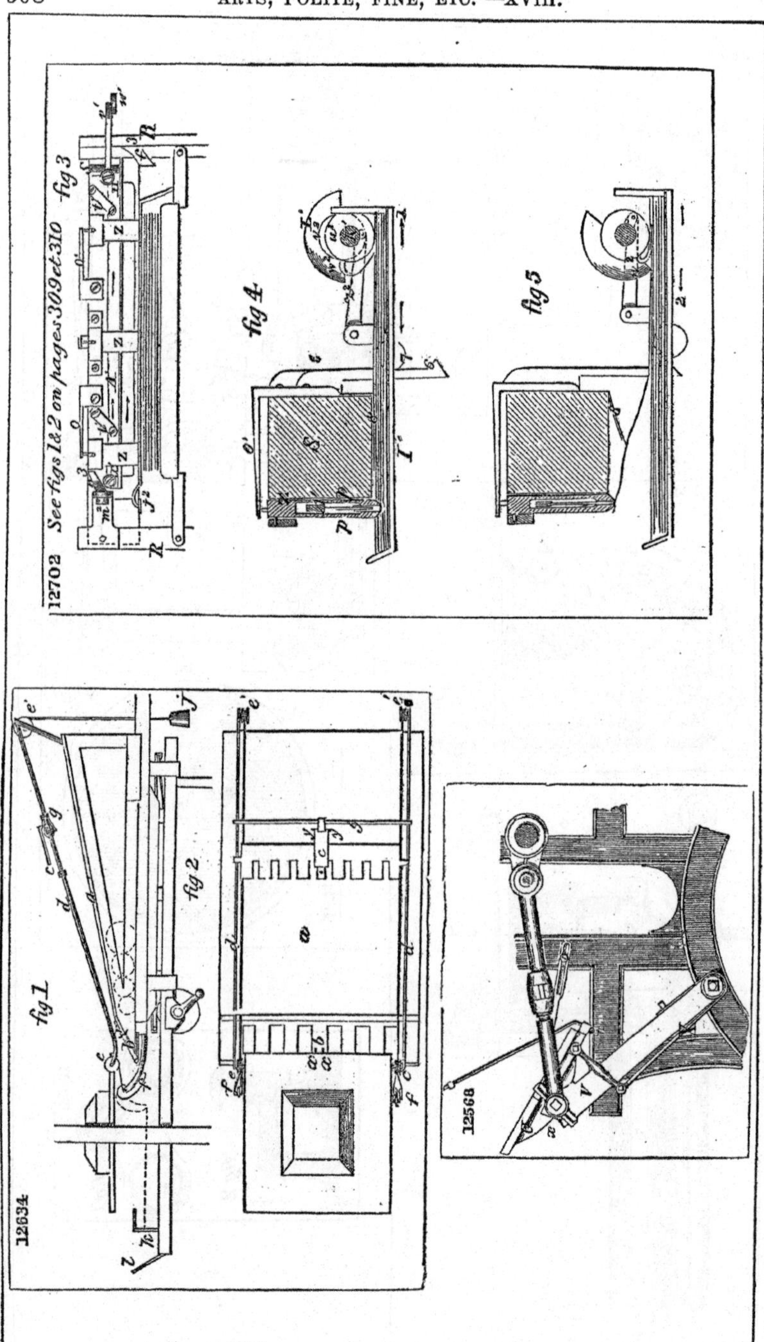
12702
See figs 1&2 on pages 309et310
fig 3
fig 4
fig 5
12634
fig 1
fig 2
12568

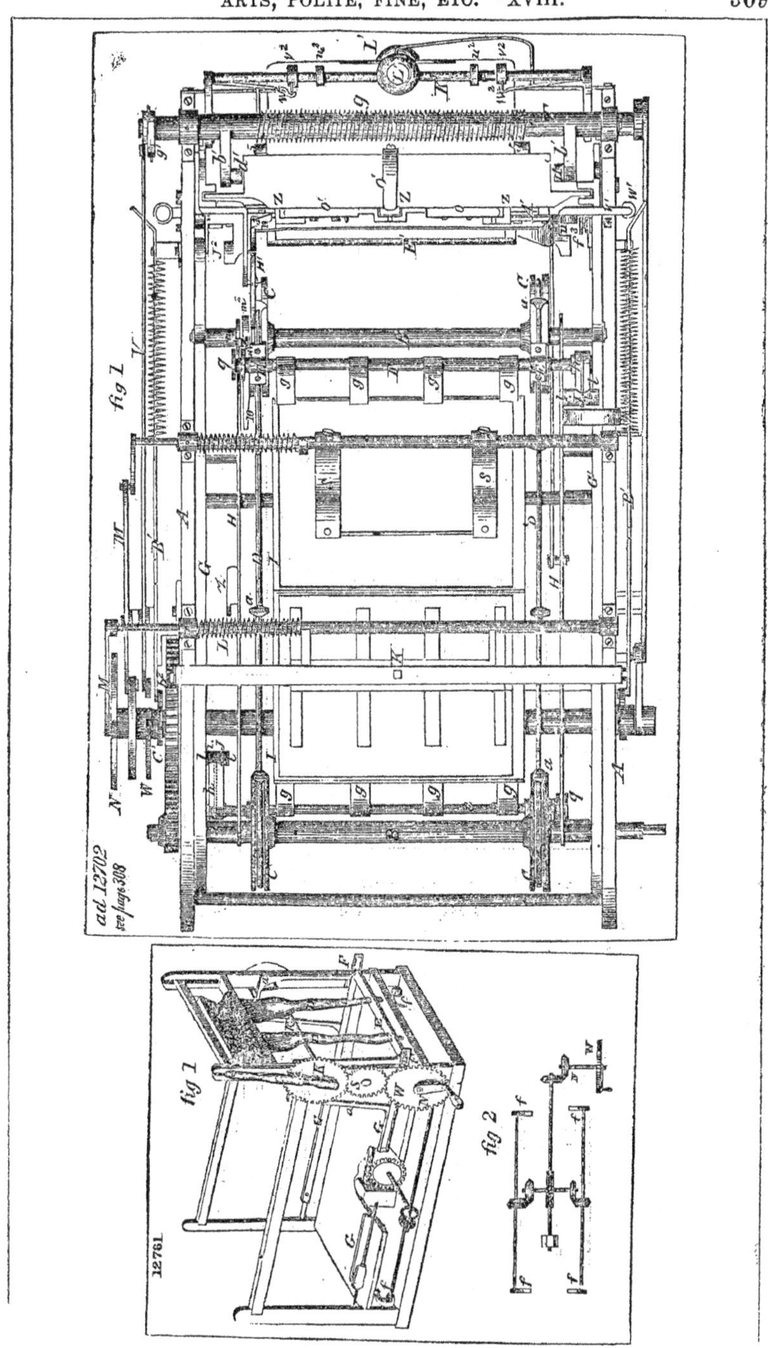
ad 12702
see page 308
fig 1
12761
fig 1
fig 2

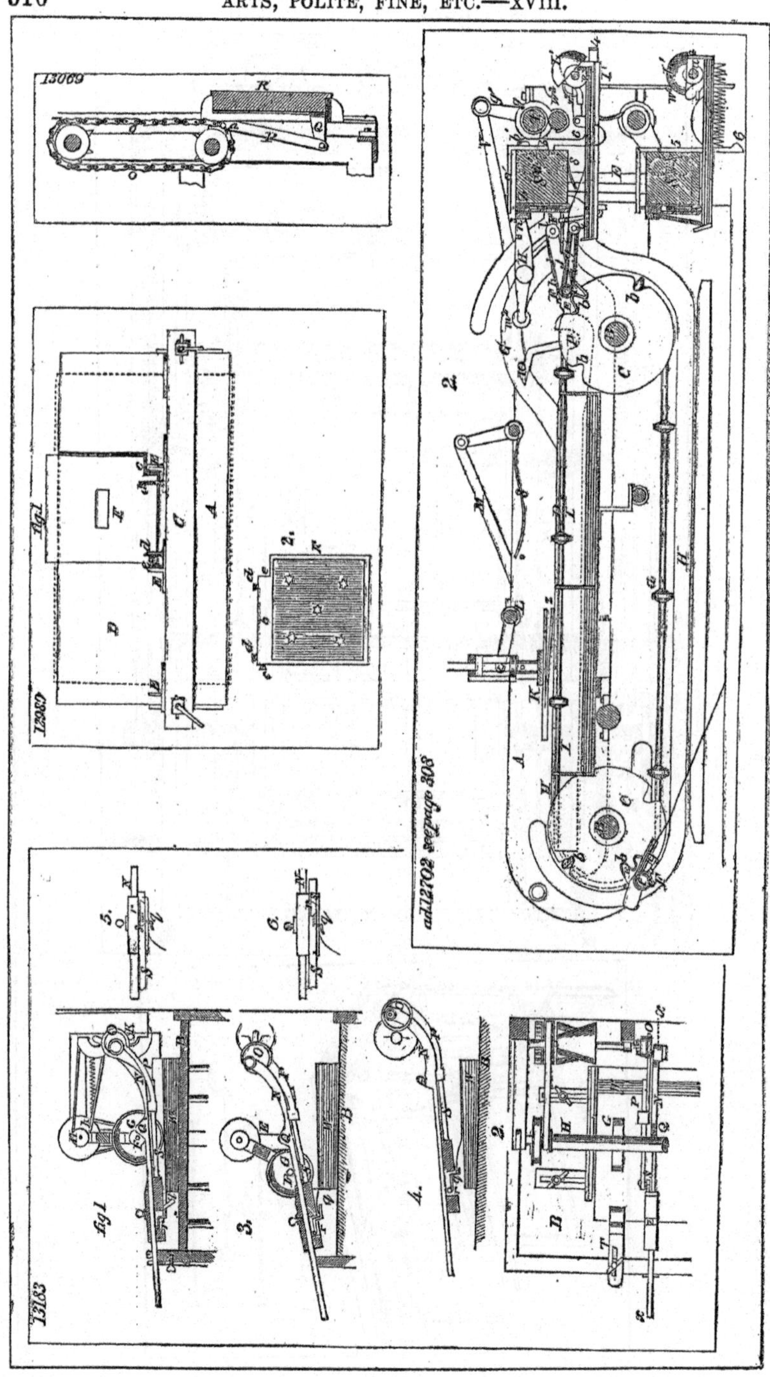
13069
13183
ad.12702 see page 308

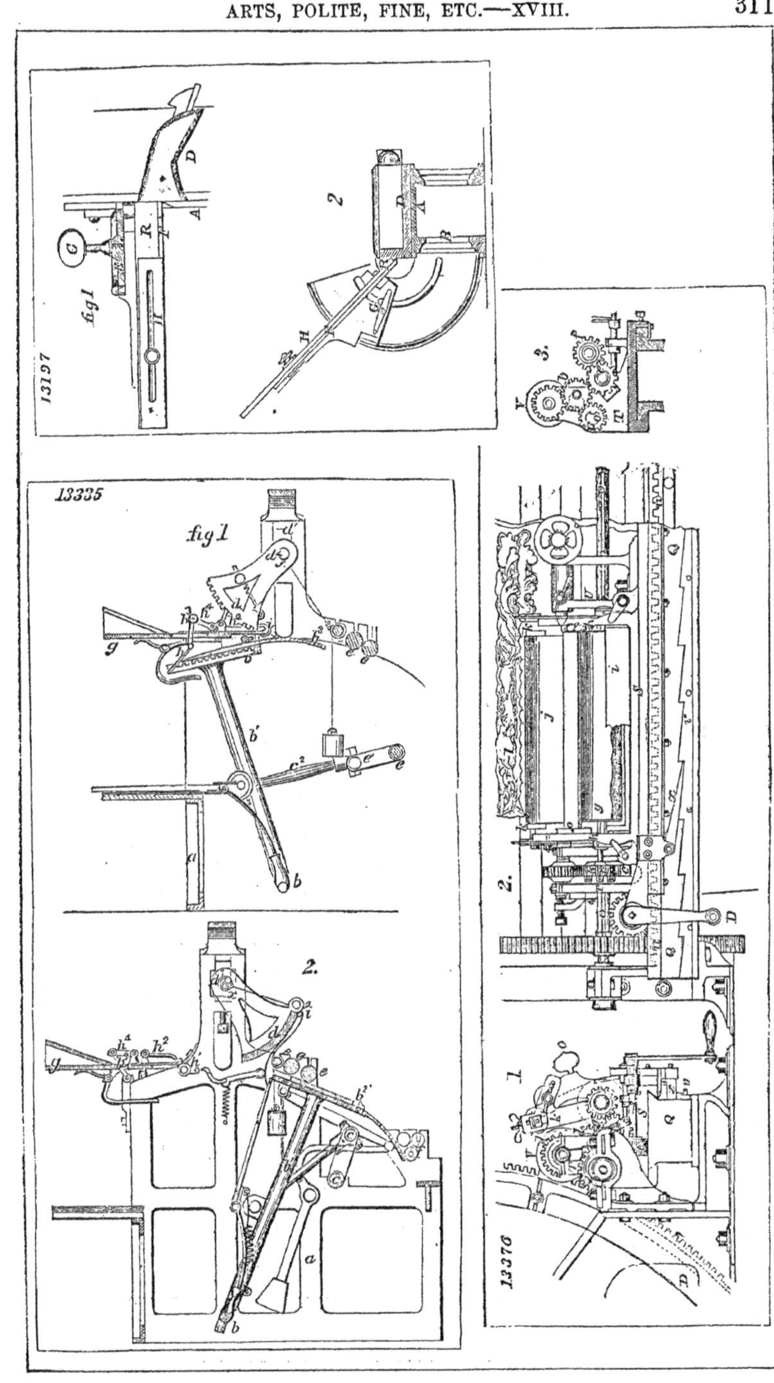

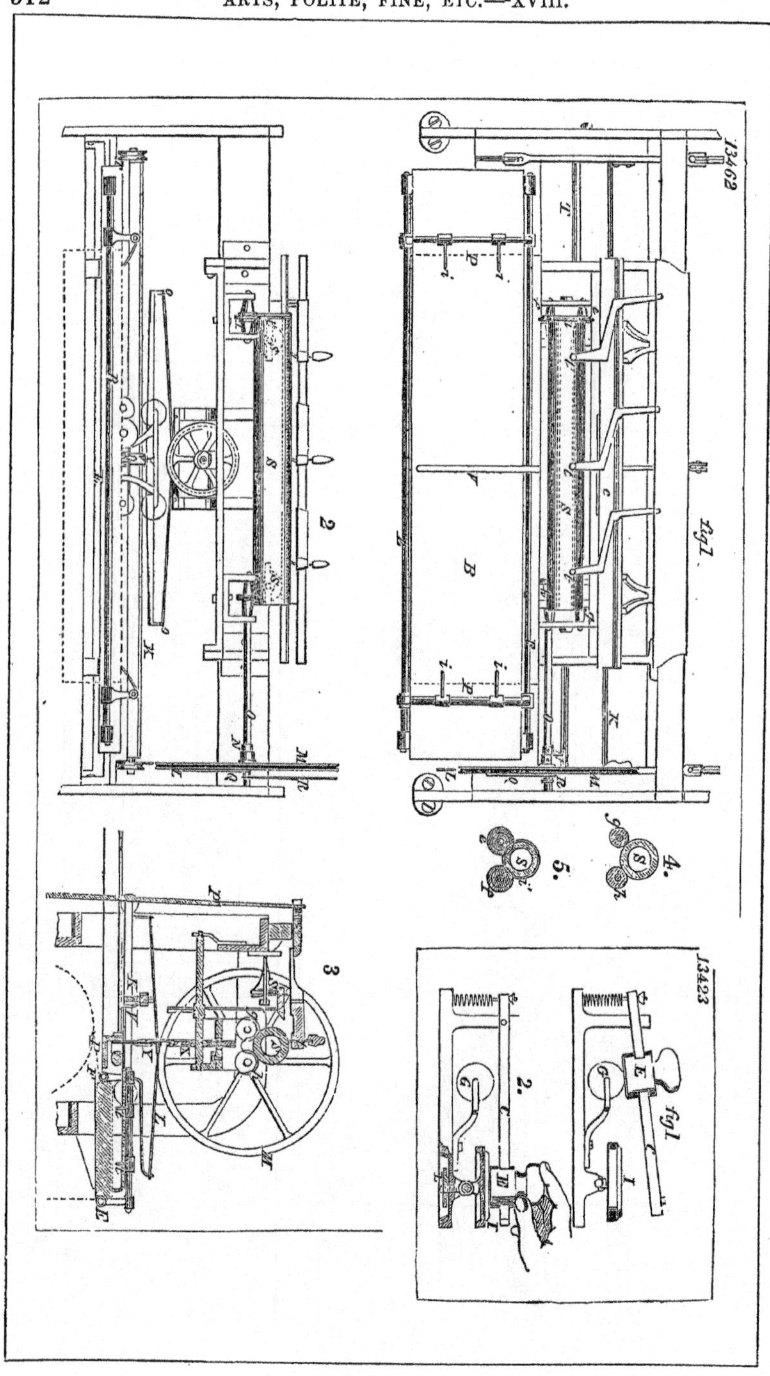
13462
fig 1
2
3
4.
5.
13423
fig 1
2.

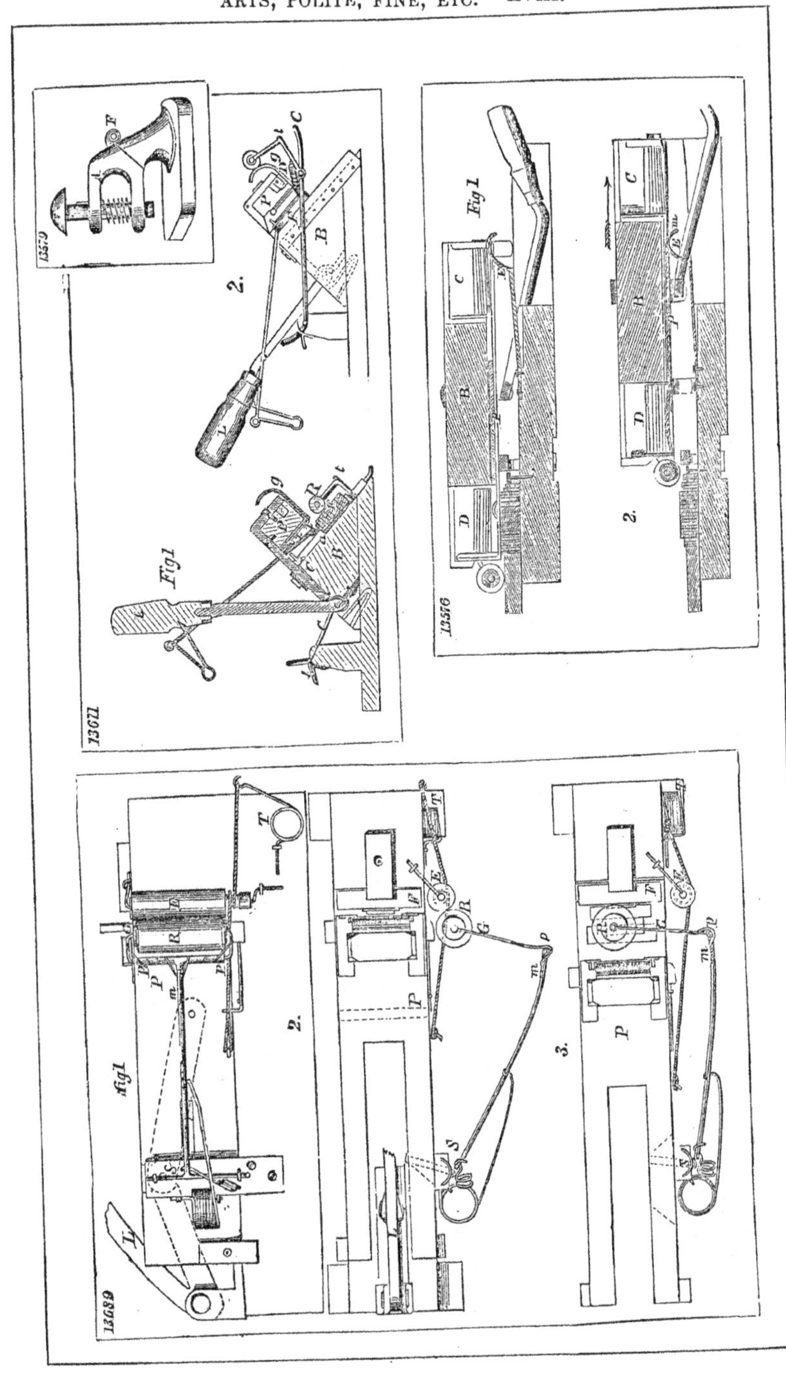

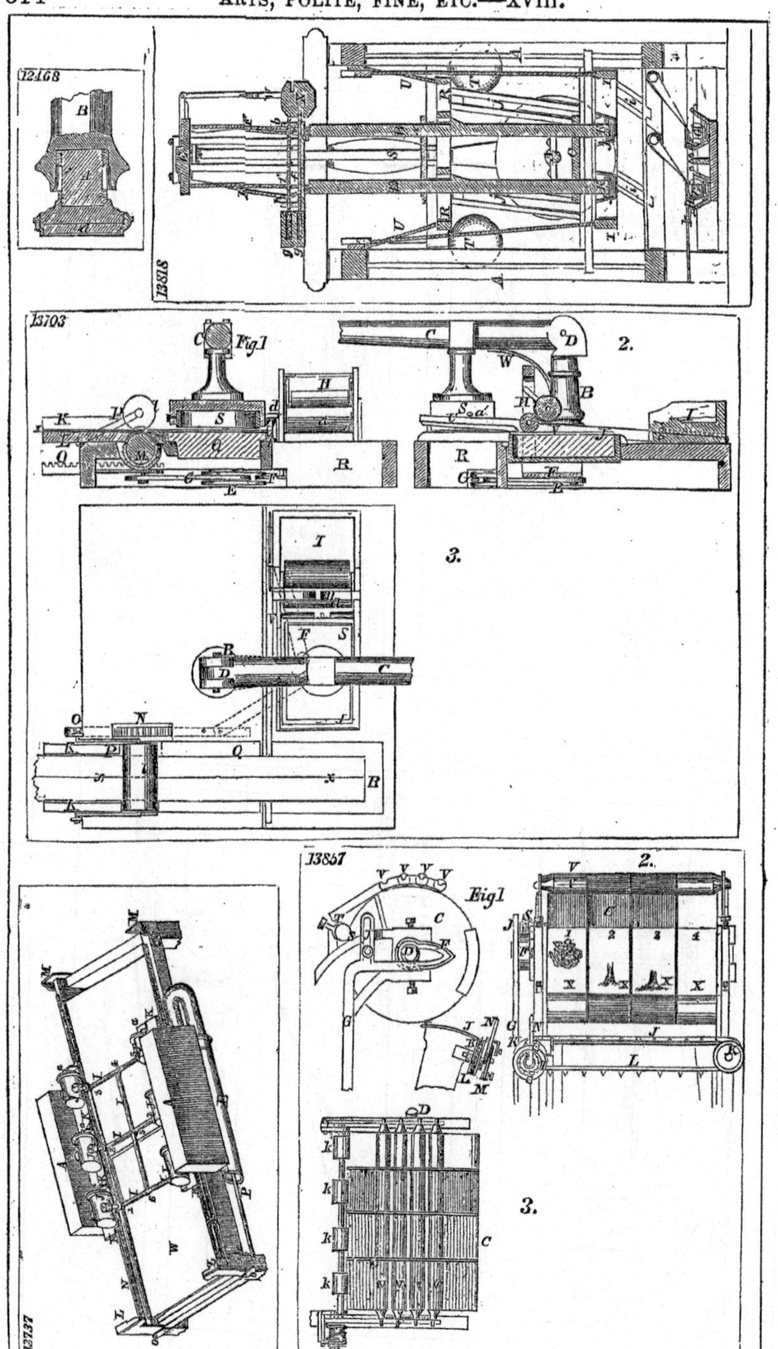
12168
13818
13103
Fig 1
2.
3.
13857
Fig 1
2.
3.
13737

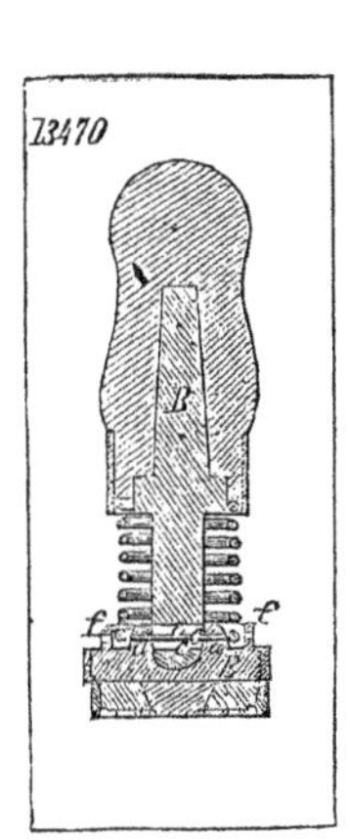
13470
B
f
f

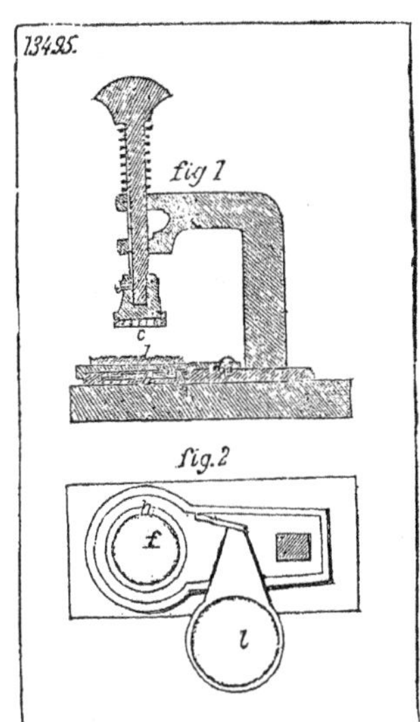
13425.
fig 1
c
fig.2
f
l

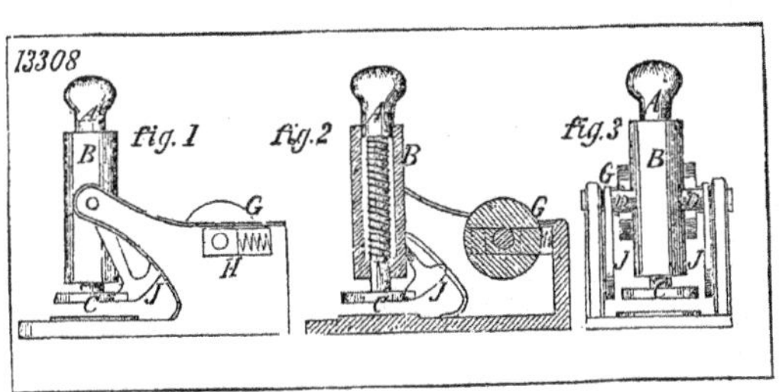
13308
fig.1
fig.2
fig.3
A
B
G
H
C
J

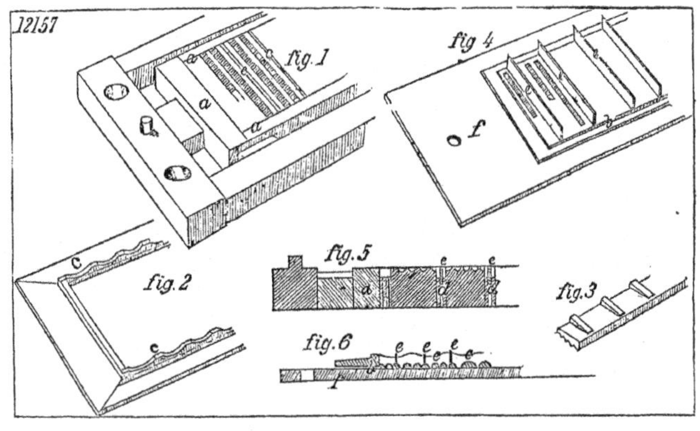
12157
fig.1
fig 4
f
fig.2
fig.5
fig.3
fig.6

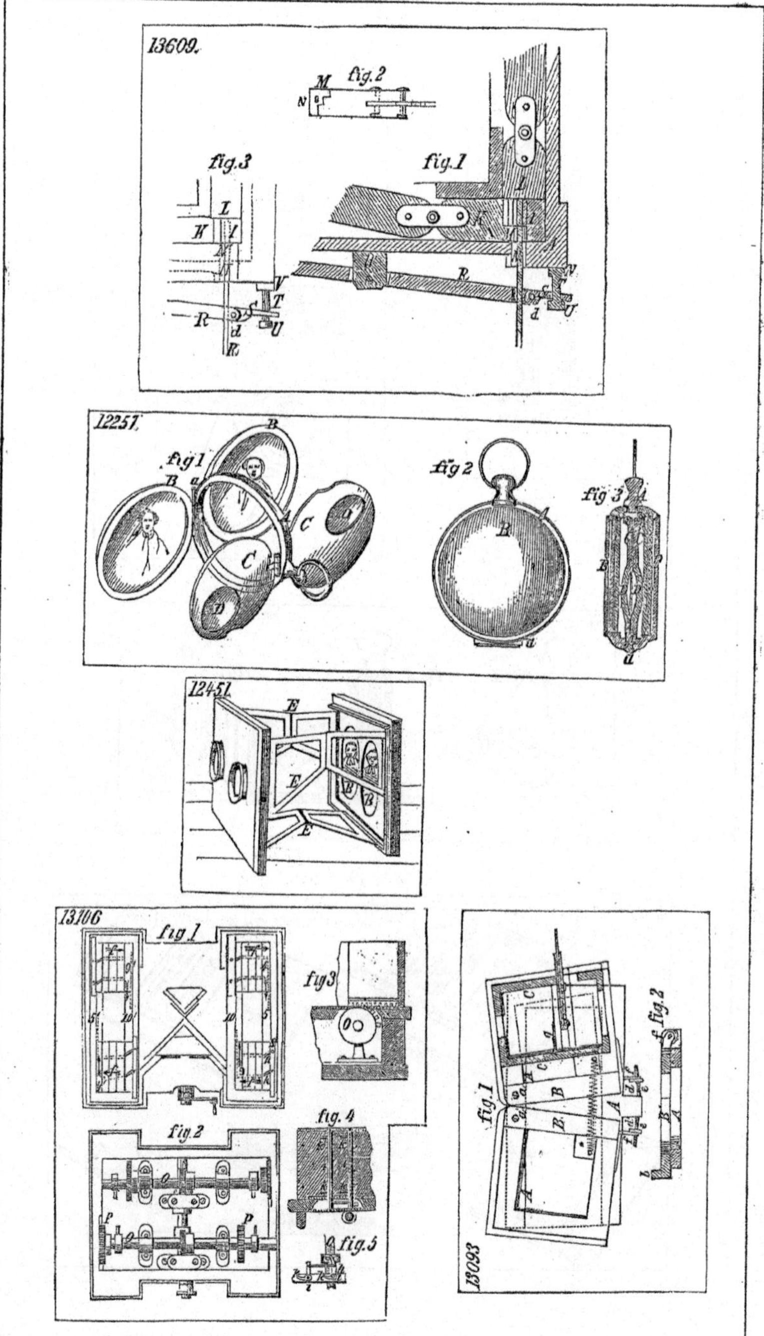
13609.
fig. 2
fig. 3
fig. 1
12257.
fig 1
fig 2
fig 3
12451.
13106
fig. 1
fig 3
fig. 2
fig. 4
fig. 5
13093
fig. 1
fig. 2

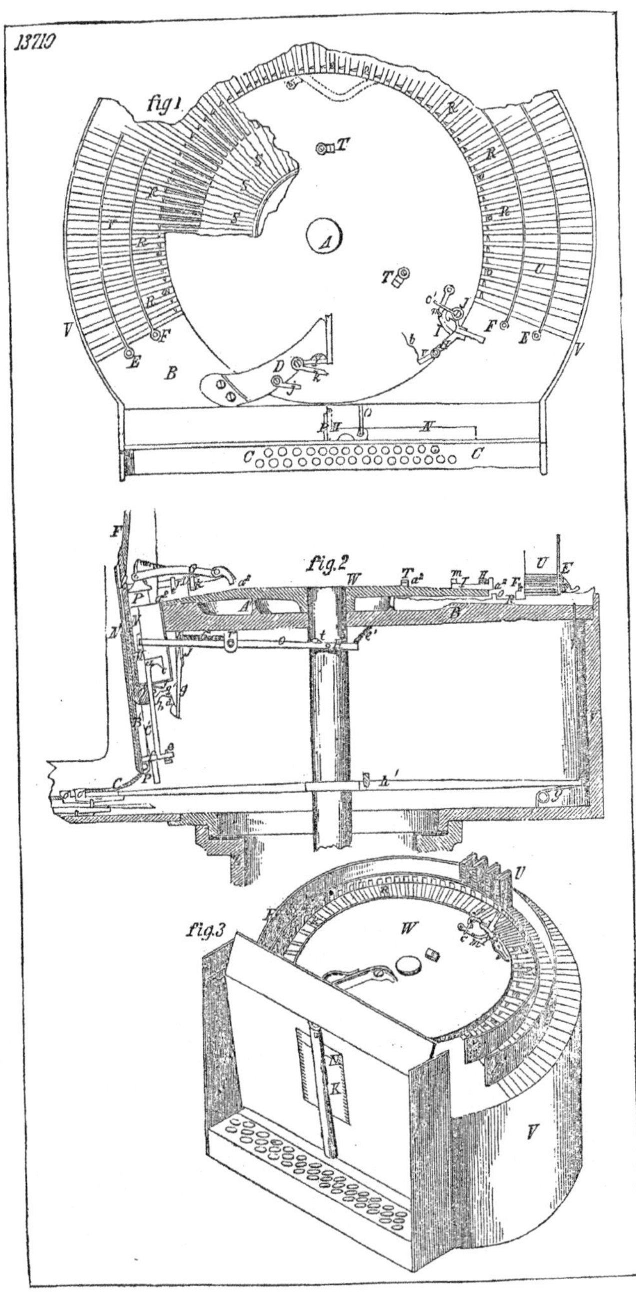
13710
fig 1.
fig.2
fig.3

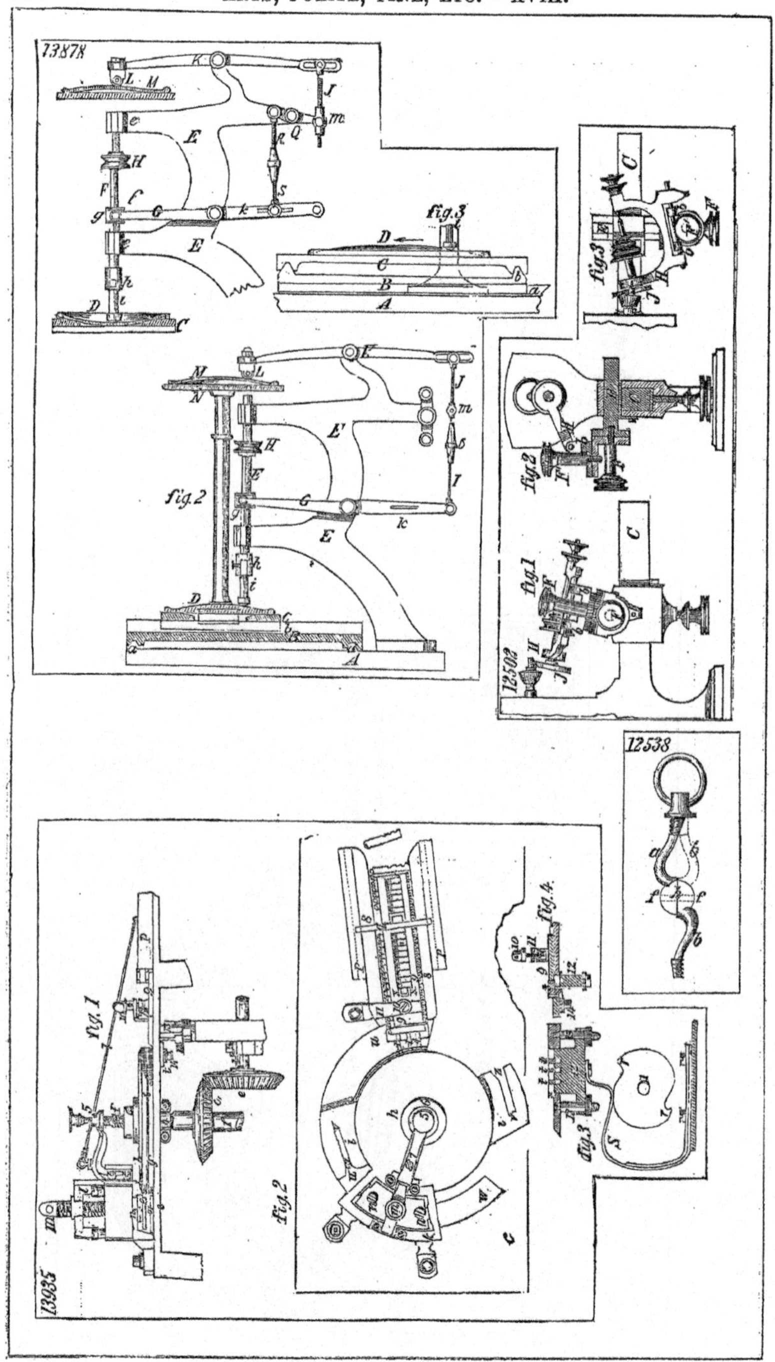

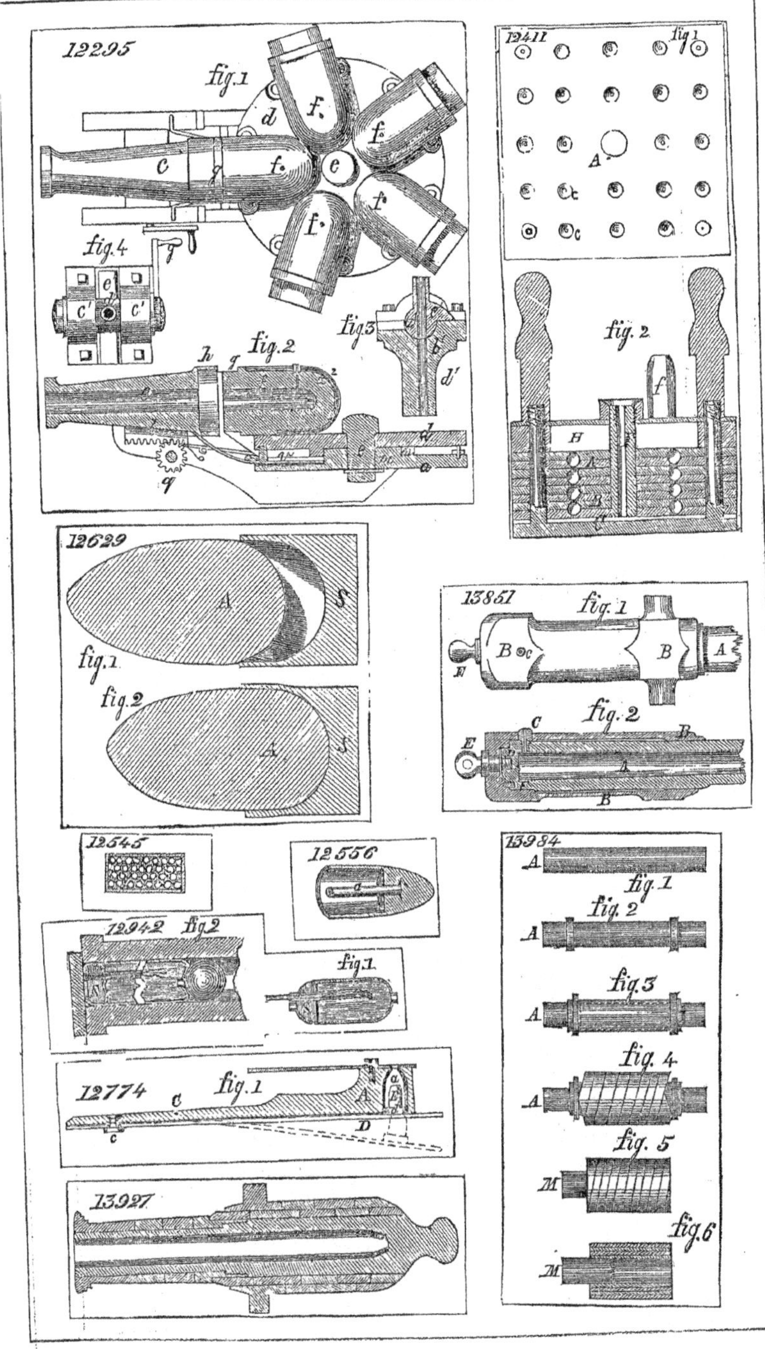
12295
fig.1
fig.4
fig.2
fig.3
12411
fig.1
fig.2
12629
fig.1
fig.2
13851
fig.1
fig.2
12545
12556
12942
fig.2
fig.1
12774
fig.1
13927
13984
fig.1
fig.2
fig.3
fig.4
fig.5
fig.6

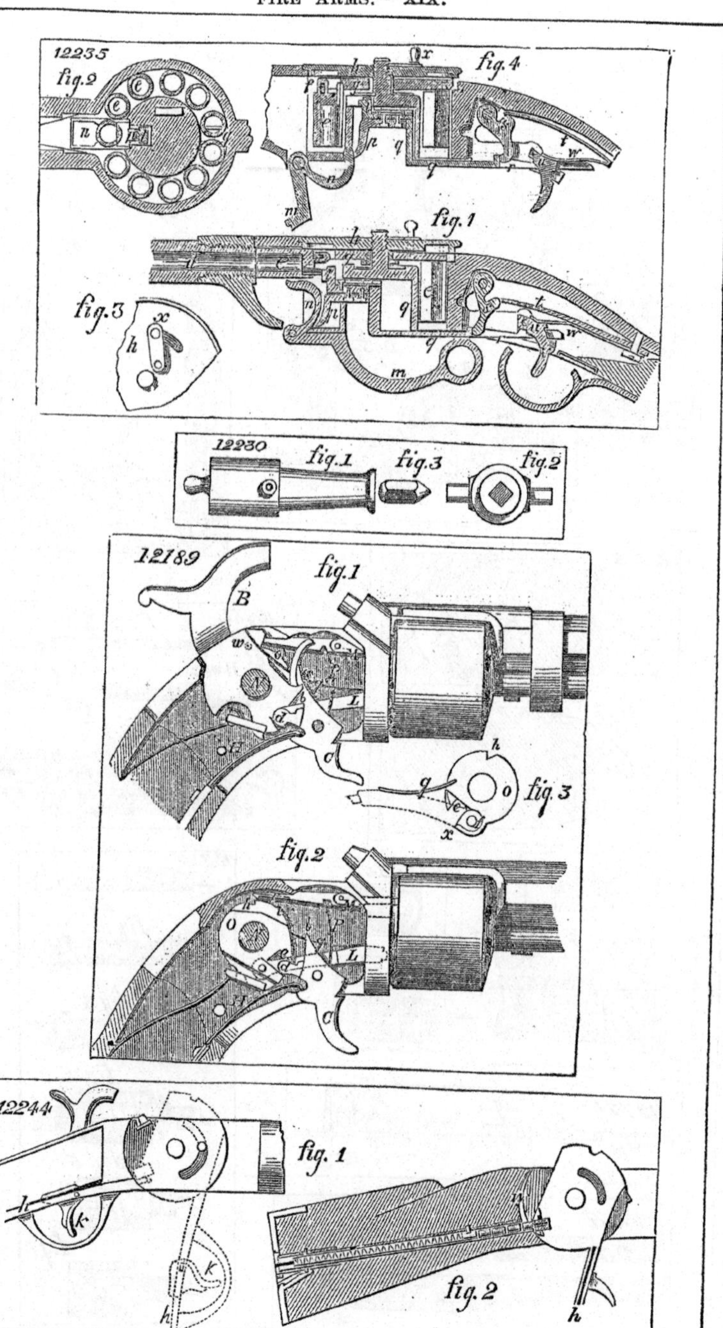
12235
fig. 2
fig. 4
fig. 1
fig. 3
12230
fig. 1
fig. 3
fig. 2
12189
fig. 1
fig. 3
fig. 2
12244
fig. 1
fig. 2

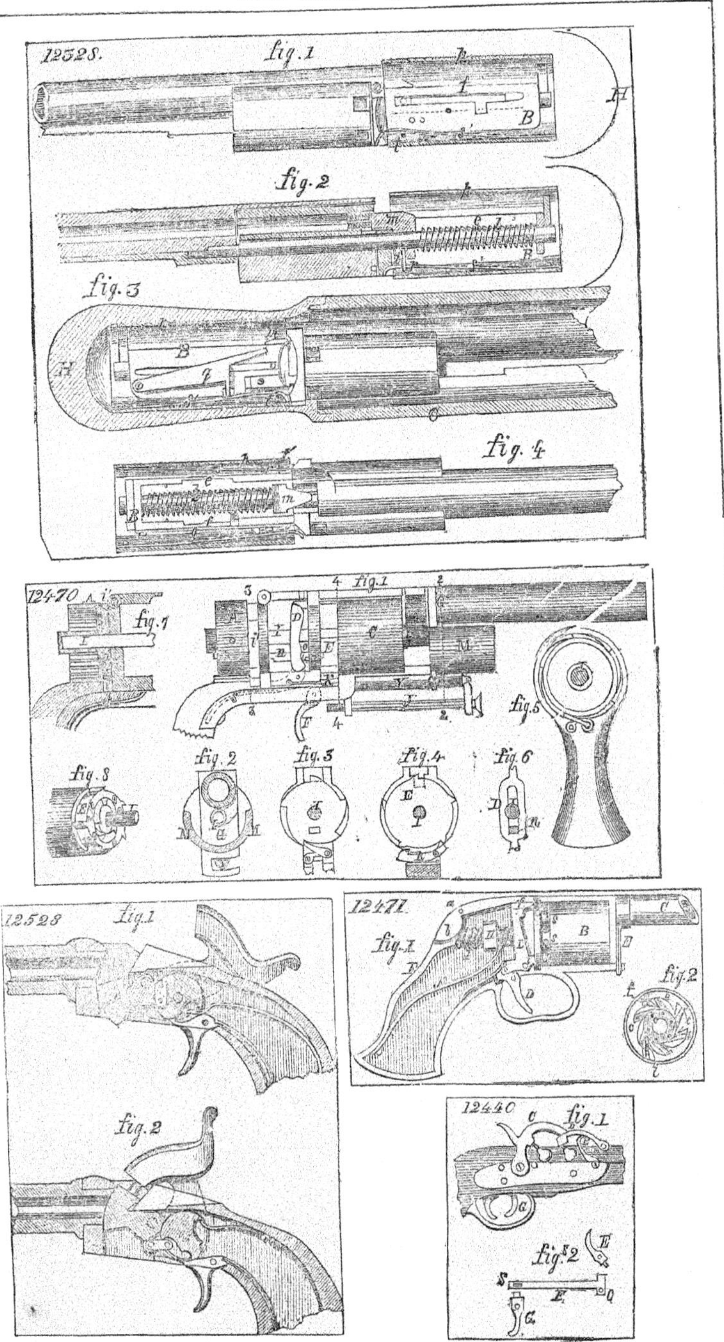
12328.
fig. 1
fig. 2
fig. 3
fig. 4
12470
fig. 1
fig. 2
fig. 3
fig. 4
fig. 5
fig. 6
fig. 7
fig. 8
12528
fig. 1
fig. 2
12471
fig. 1
fig. 2
12440
fig. 1
figs 2

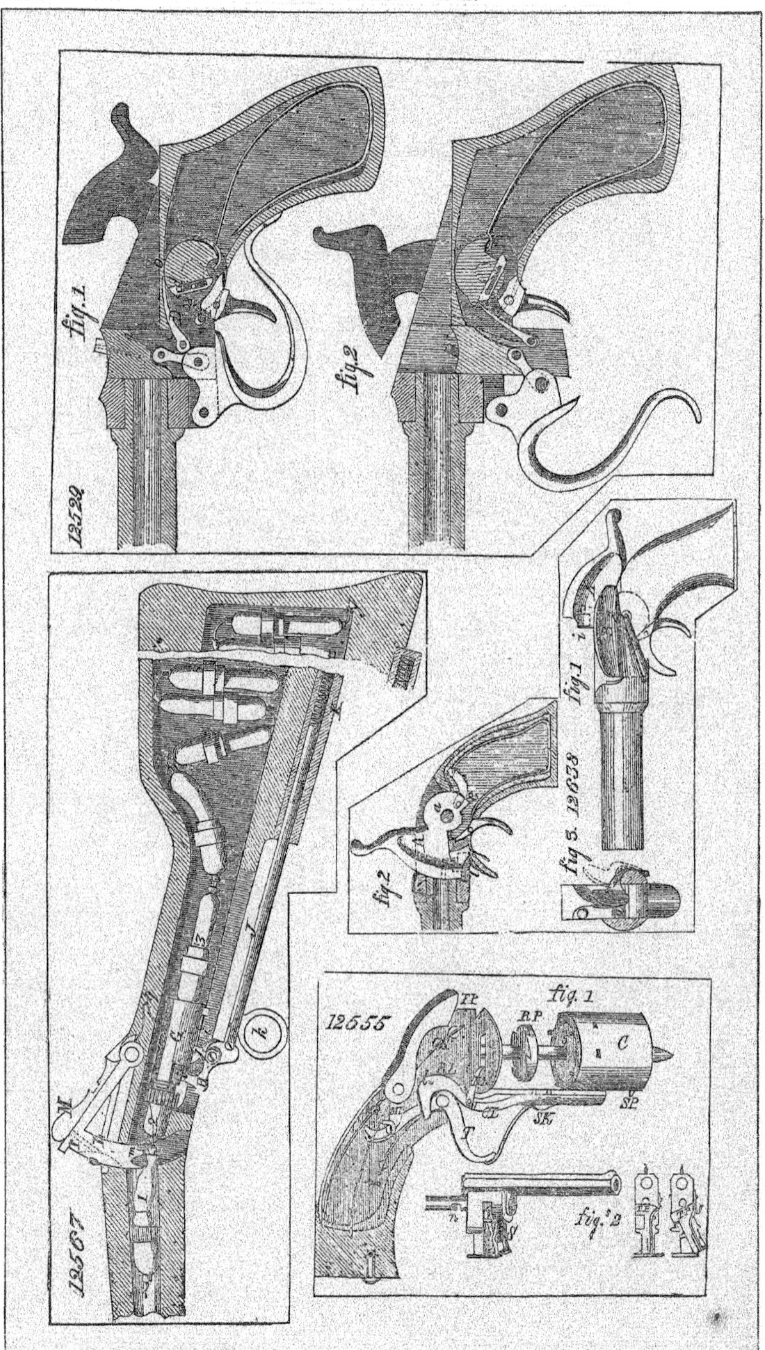

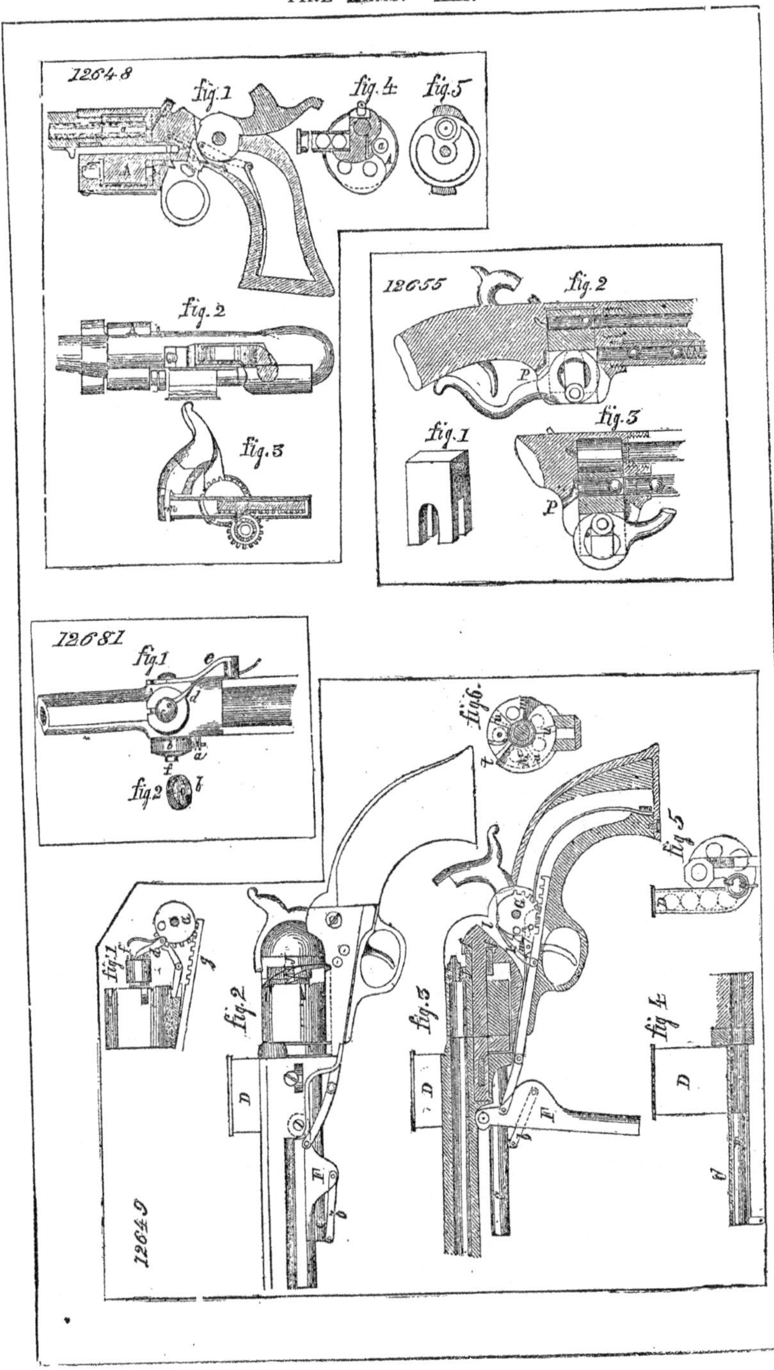
12648
fig. 1
fig. 4
fig. 5
fig. 2
fig. 3
12655
fig. 2
fig. 1
fig. 3
12681
fig. 1
fig. 2
fig. 6
fig. 5
fig. 1
fig. 2
fig. 3
fig. 4
12649

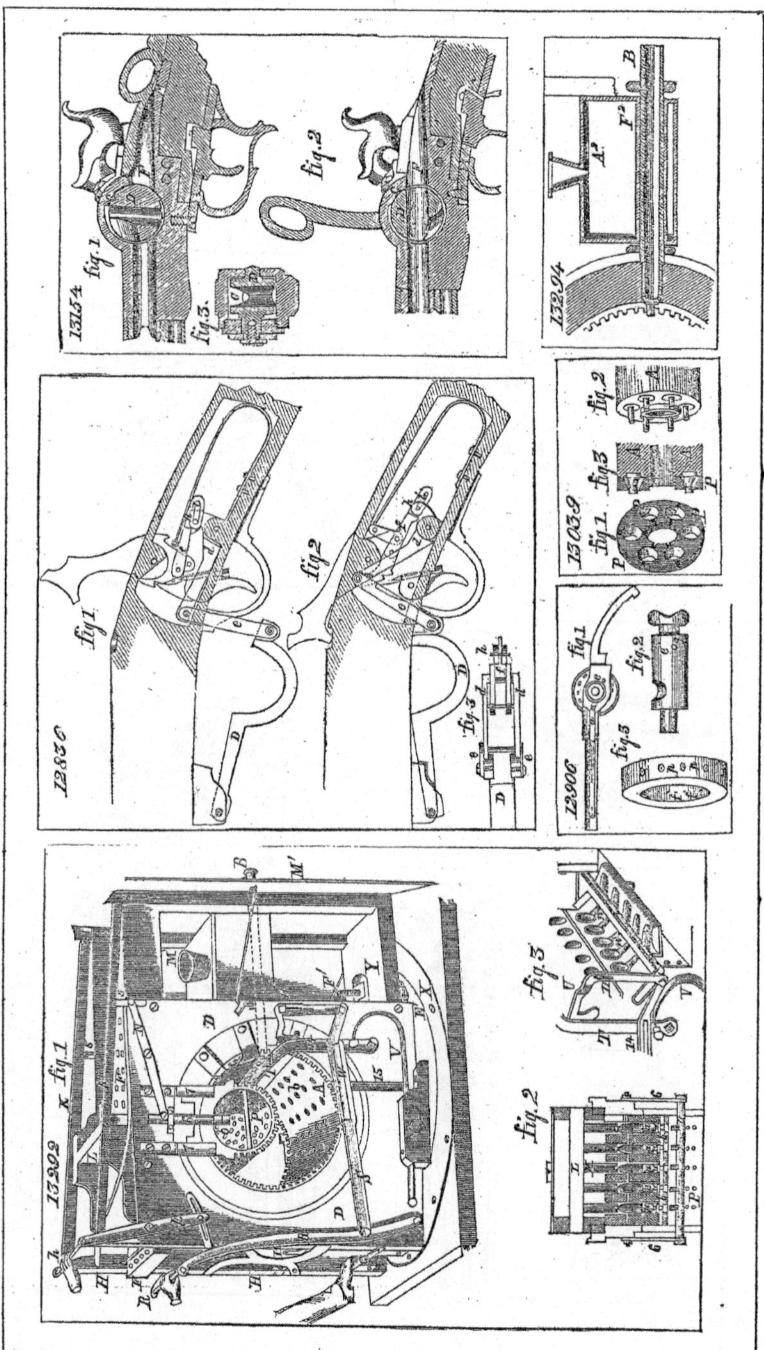
13154
fig.1
fig.2
fig.3.
13294
12836
fig.1
fig.2
fig.3
13039
fig.1
fig.3
fig.2
12906
fig.1
fig.2
fig.3
13292
fig.1
fig.2
fig.3

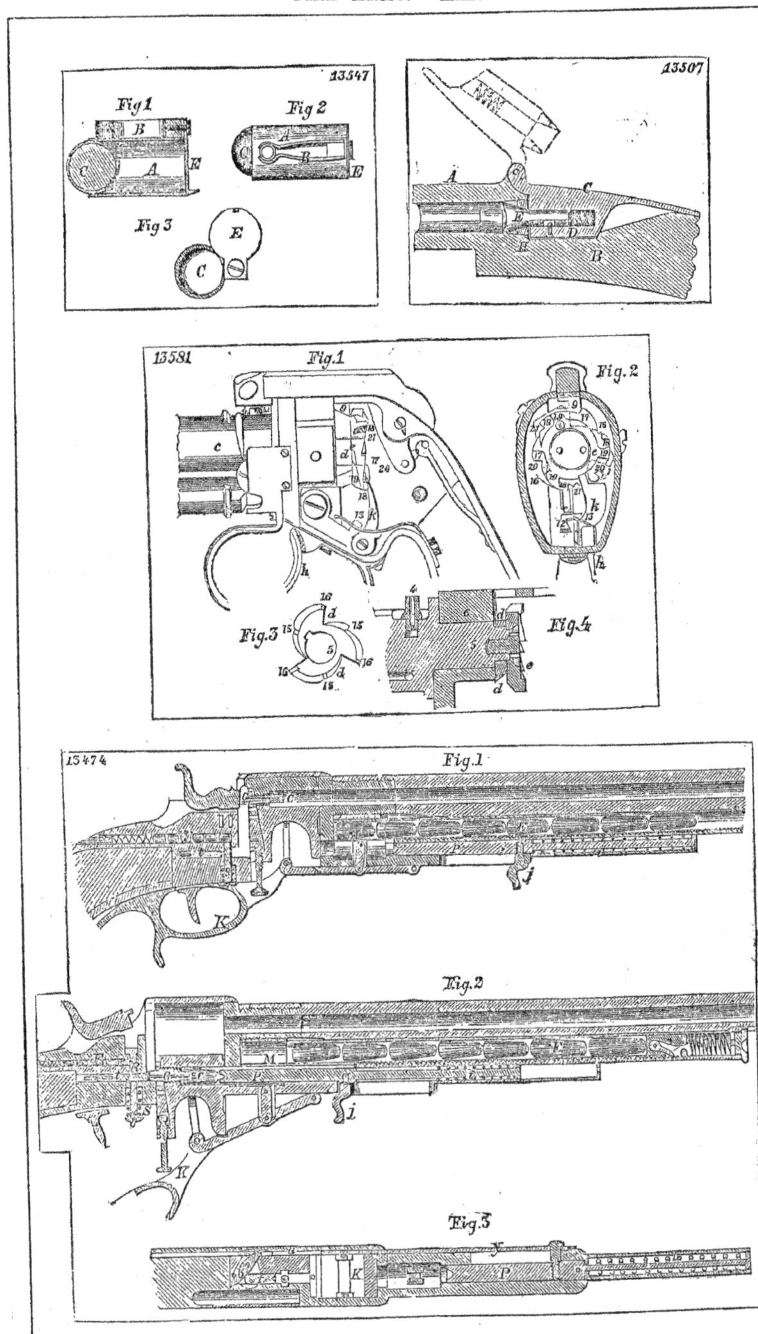
13547
Fig 1
Fig 2
Fig 3
13507
13581
Fig.1
Fig.2
Fig.3
Fig.4
13474
Fig.1
Fig.2
Fig.3

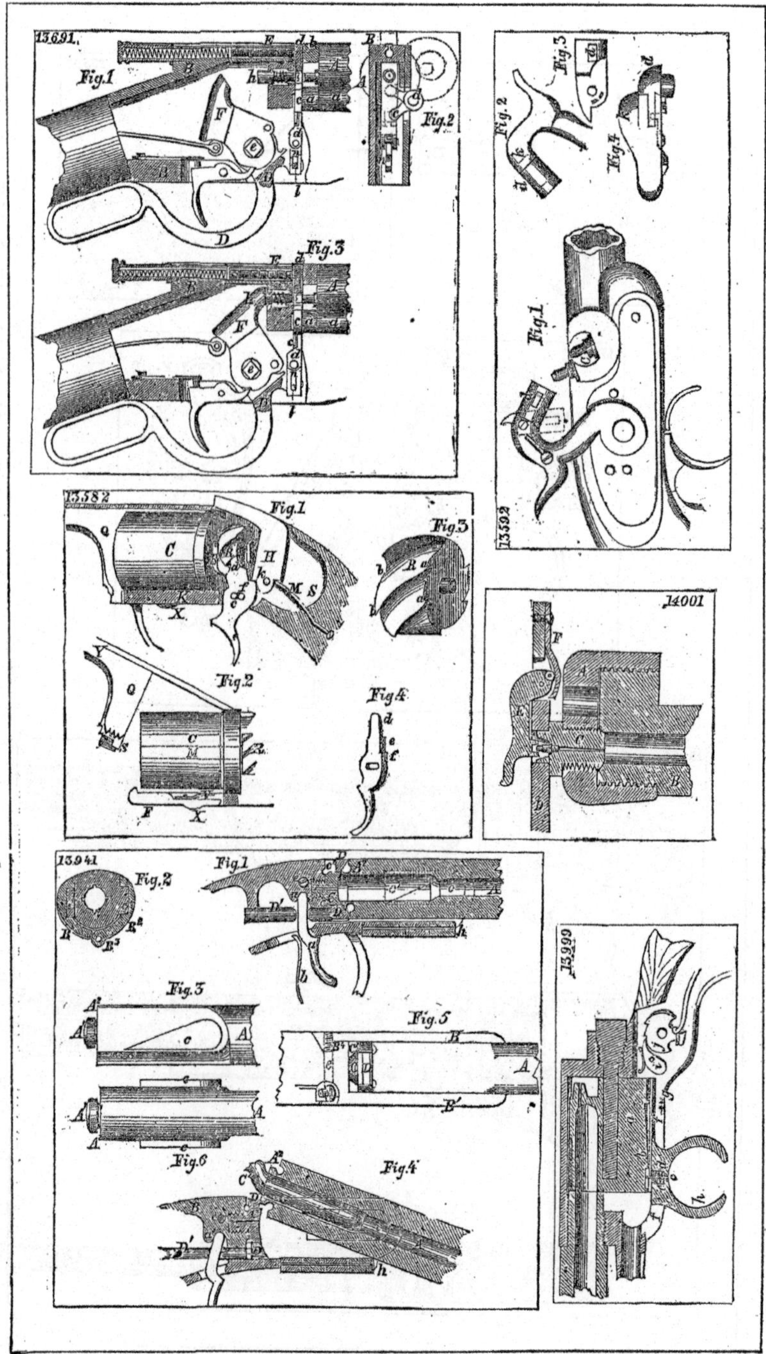
13691.
Fig.1
Fig.2
Fig.3
13592
Fig.1
Fig.2
Fig.3
Fig.4
13582
Fig.1
Fig.2
Fig.3
Fig.4
14001
13941
Fig.1
Fig.2
Fig.3
Fig.4
Fig.5
Fig.6
13999

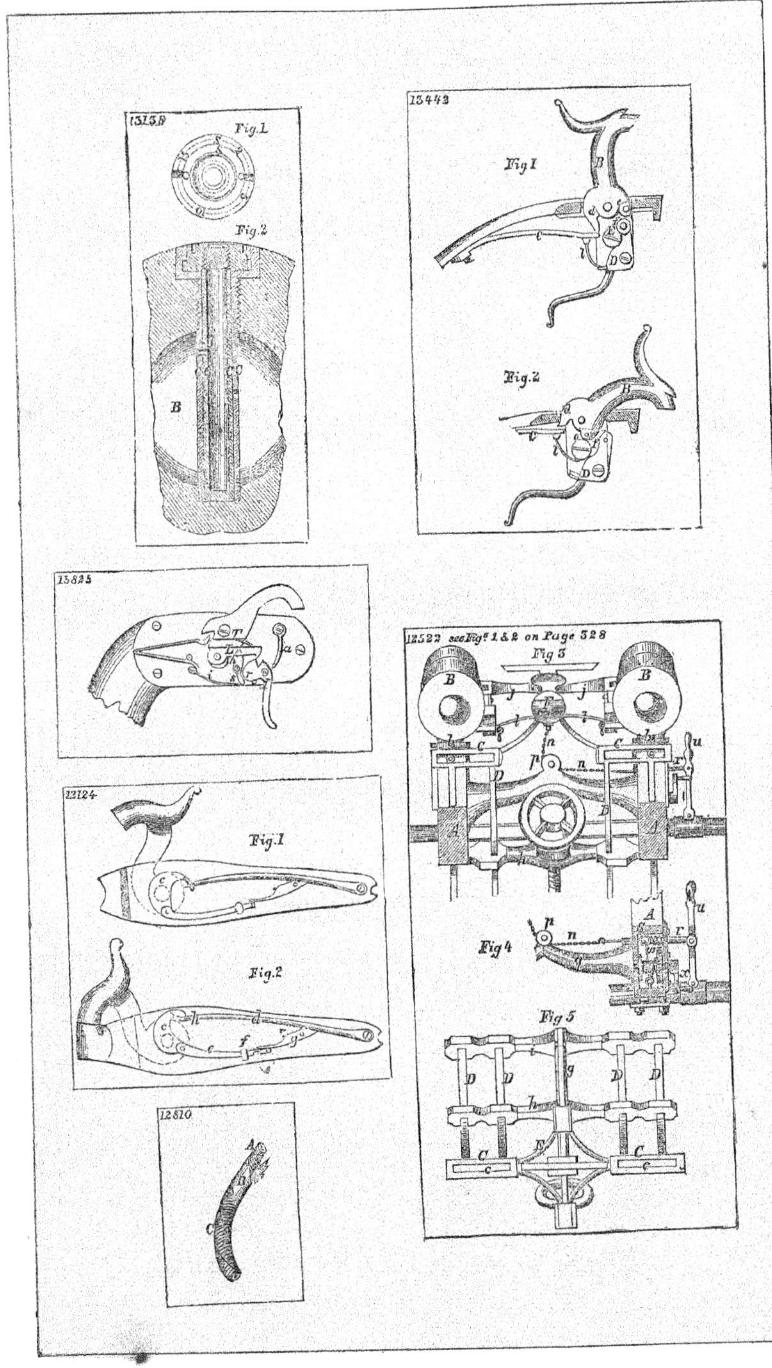
13138
Fig.1
Fig.2
B
13442
Fig 1
Fig.2
13825
12124
Fig.1
Fig.2
12810
12522 see Figs 1 & 2 on Page 328
Fig 3
Fig 4
Fig 5

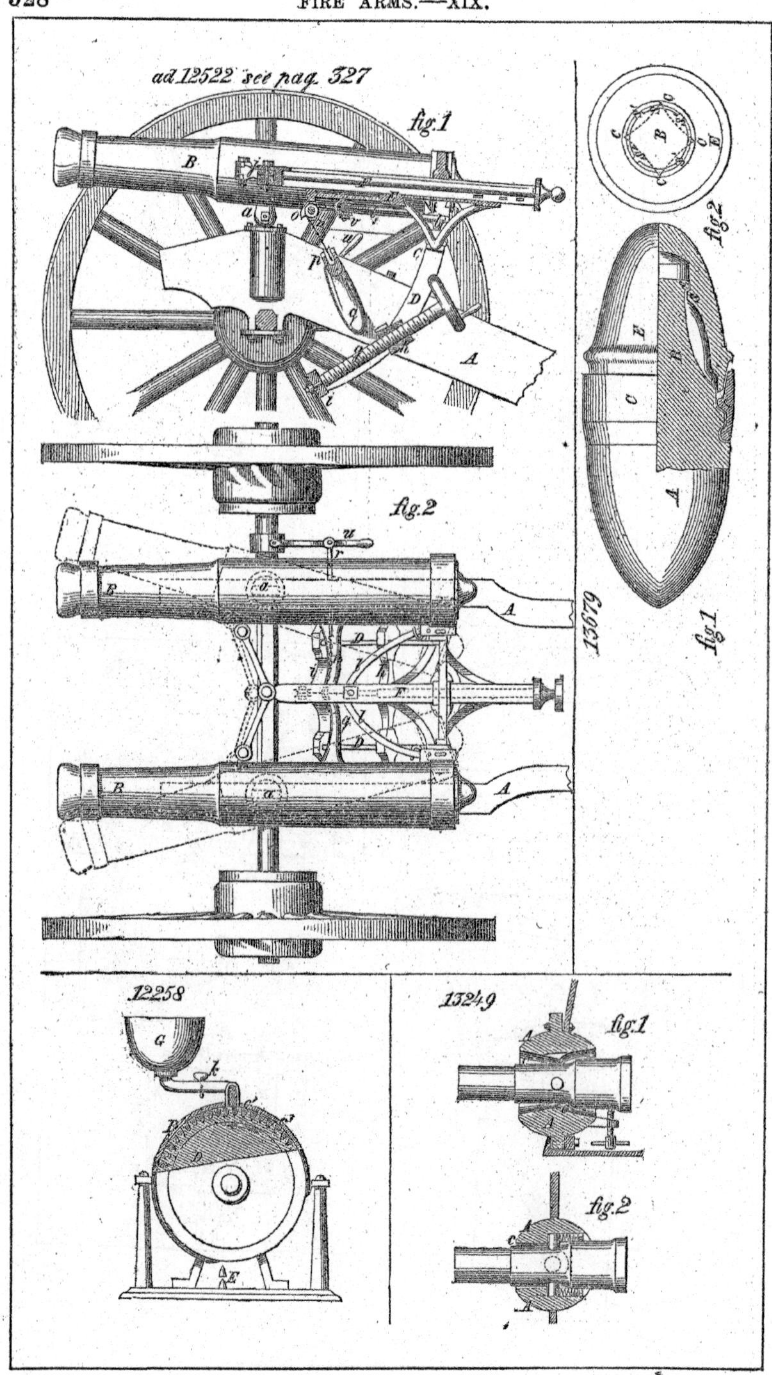
ad.12522 see pag. 327
fig.1
fig.2
13679
fig.1
fig.2
12258
13249
fig.1
fig.2

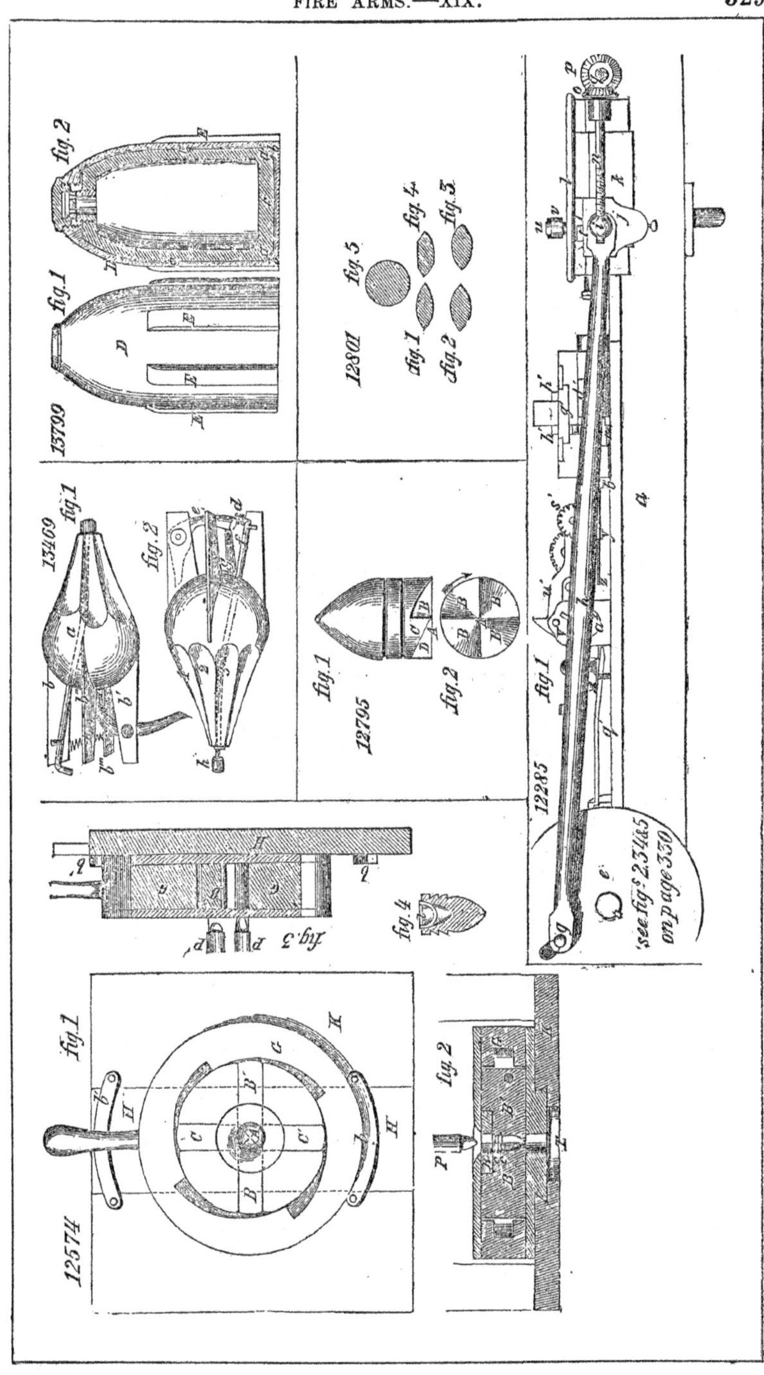
13799
fig. 1
fig. 2
13469
fig. 1
fig. 2
12801
fig. 5
fig. 1
fig. 2
fig. 4
fig. 3
12795
fig. 1
fig. 2
12285
fig. 1
see fig's 2,3,4&5 on page 330
fig. 3
fig. 4
12574
fig. 1
fig. 2

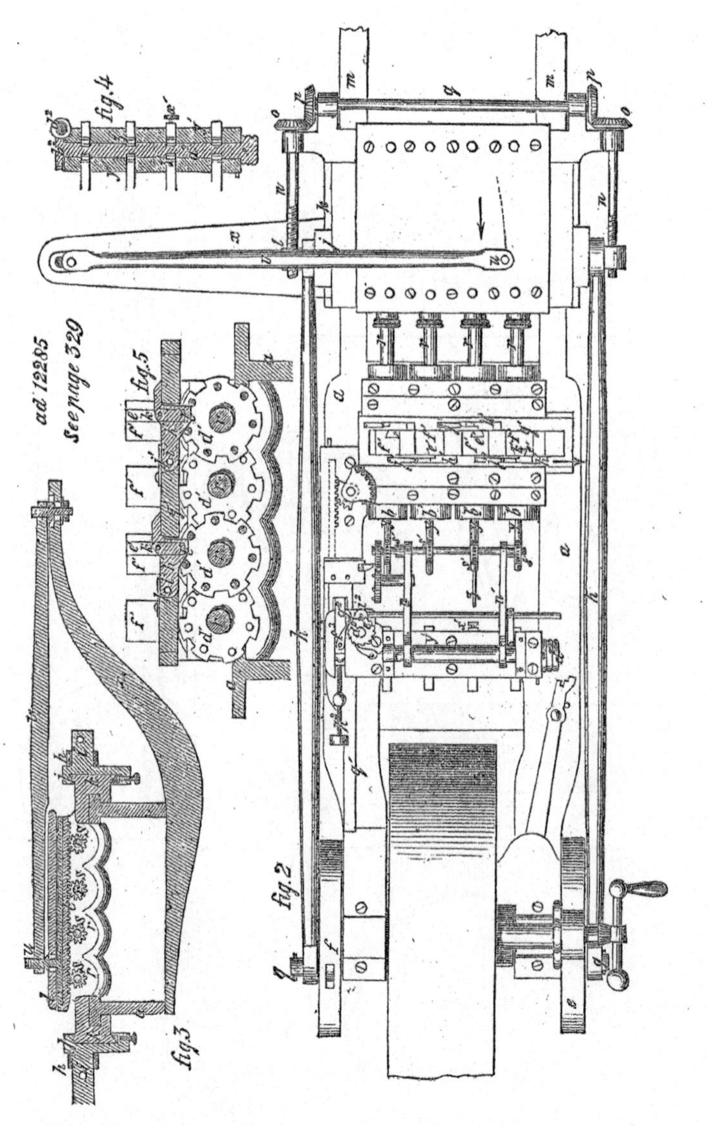

ad 12285
See page 329
fig. 2
fig. 3
fig. 4
fig. 5

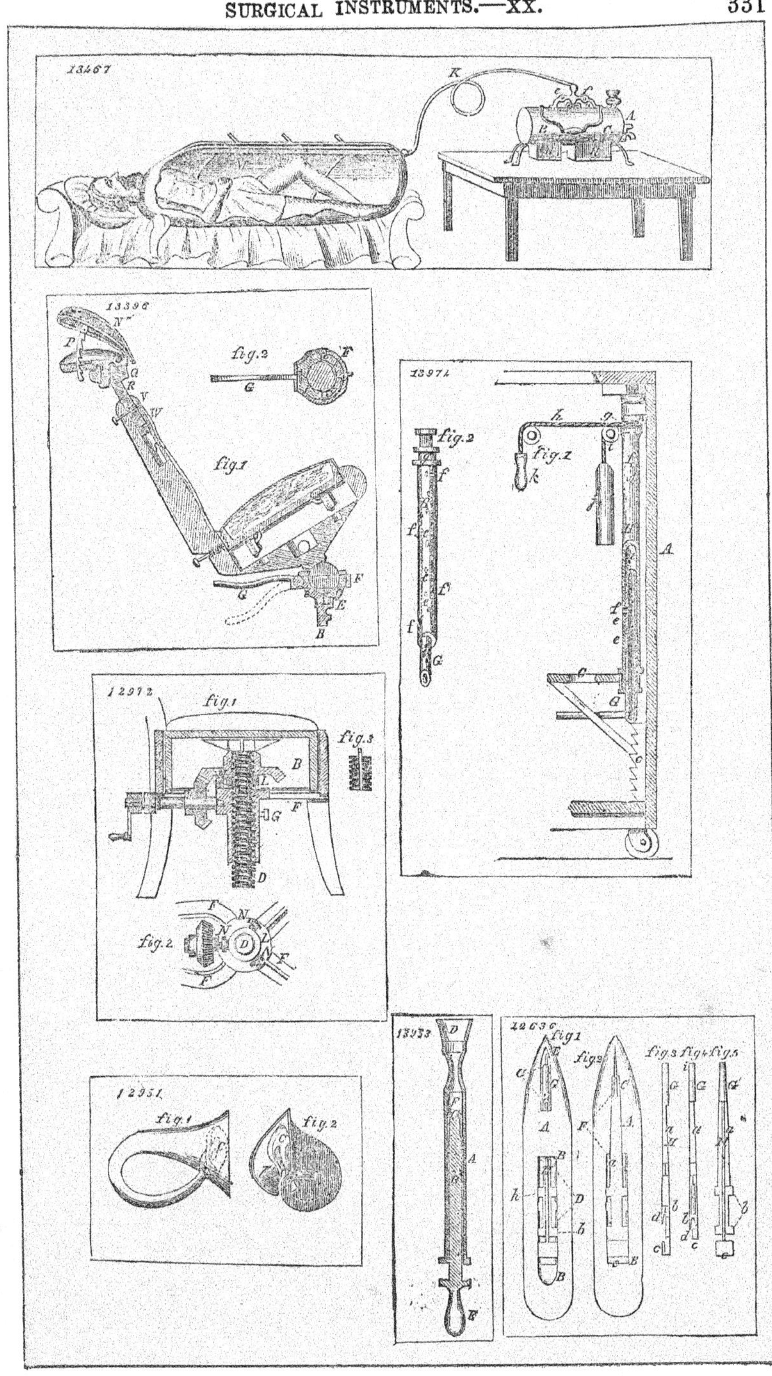

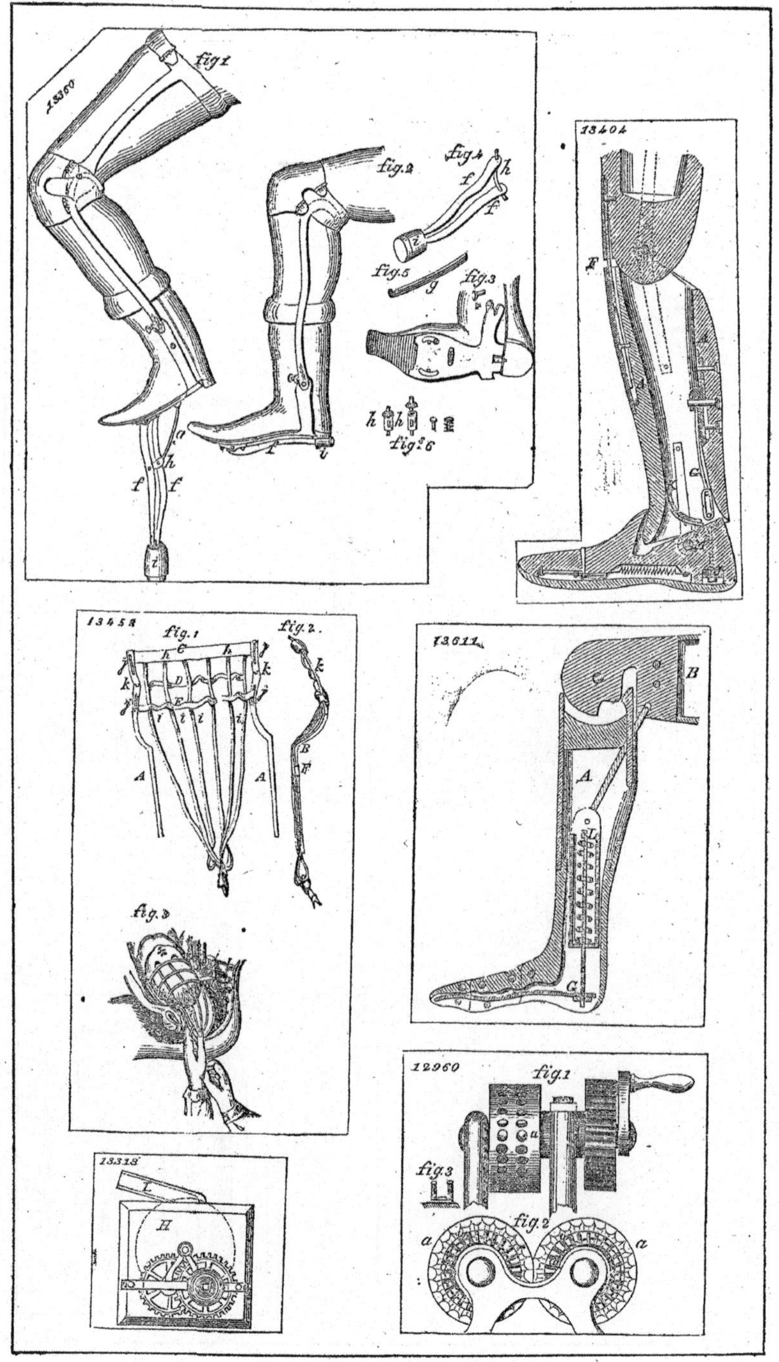
13360
fig.1
fig.2
fig.4
fig.5
fig.3
fig.6
13404
13452
fig.1
fig.2
fig.3
13611
12960
fig.1
fig.2
fig.3
13318

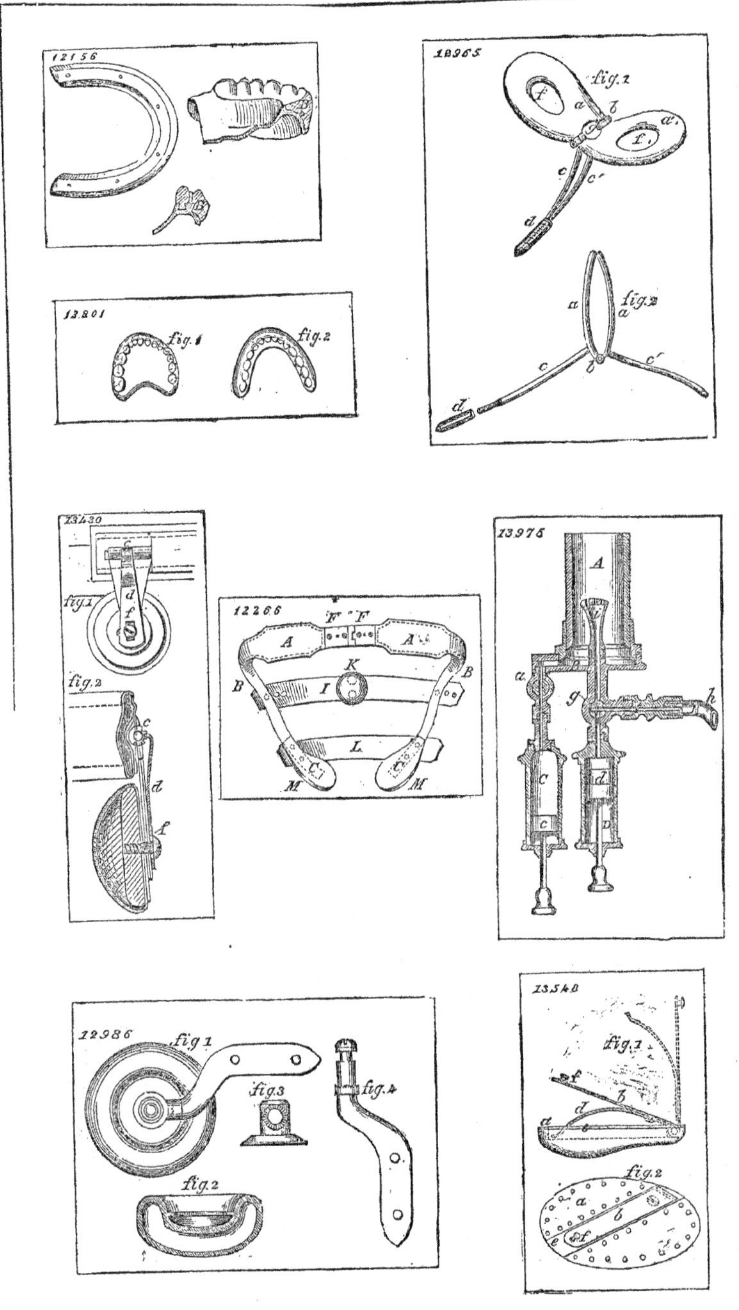
12156
12965
fig. 1
fig. 2
12801
fig. 1
fig. 2
13430
fig. 1
fig. 2
12266
13976
12986
fig 1
fig. 2
fig. 3
fig. 4
13548
fig. 1
fig. 2

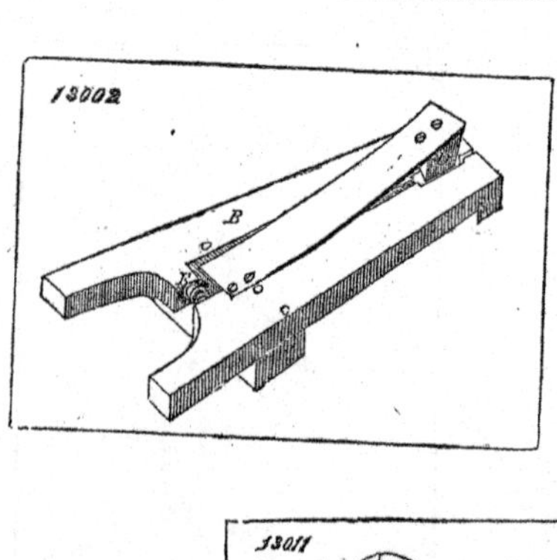
13002
B

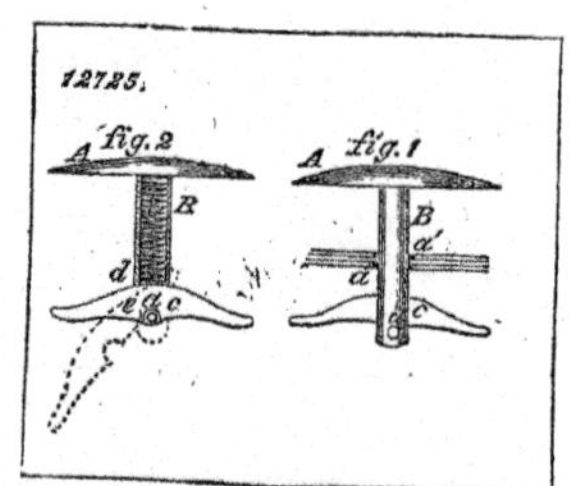
12725.
fig. 2
fig. 1
A
B

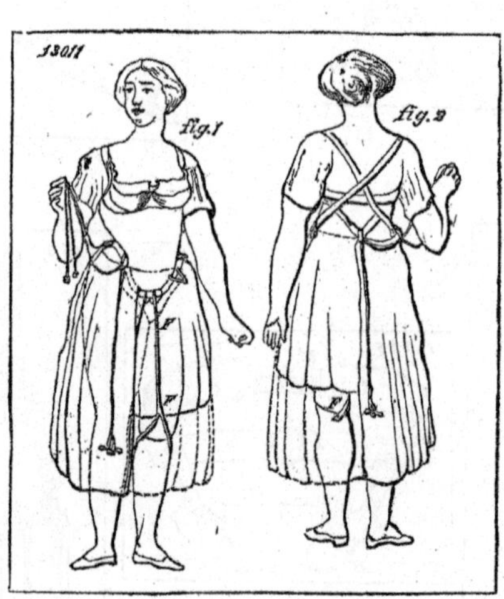
13011
fig. 1
fig. 2

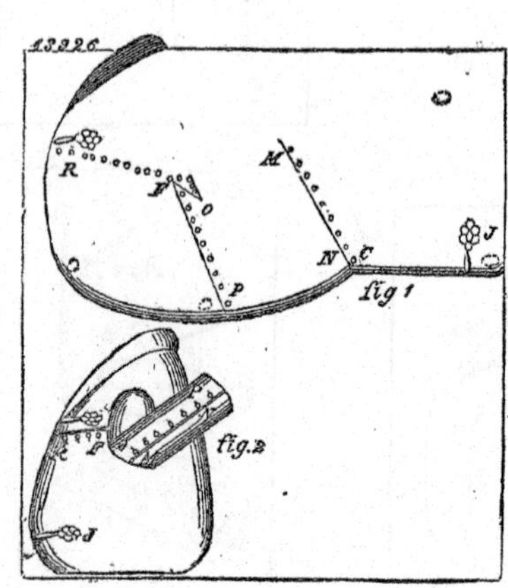
R
M
O
N
C
J
P
fig 1
fig. 2

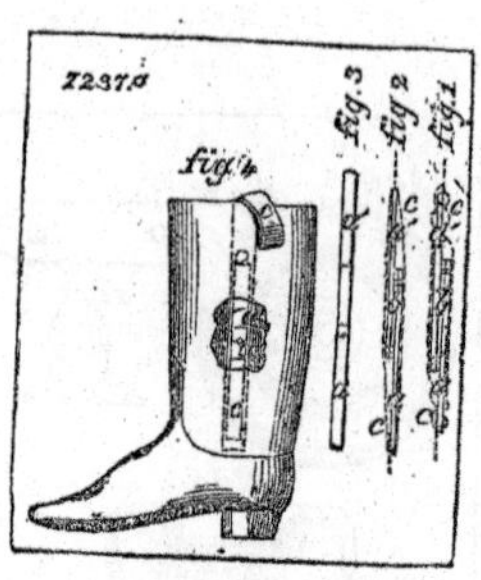
fig. 4
fig. 3
fig. 2
fig. 1

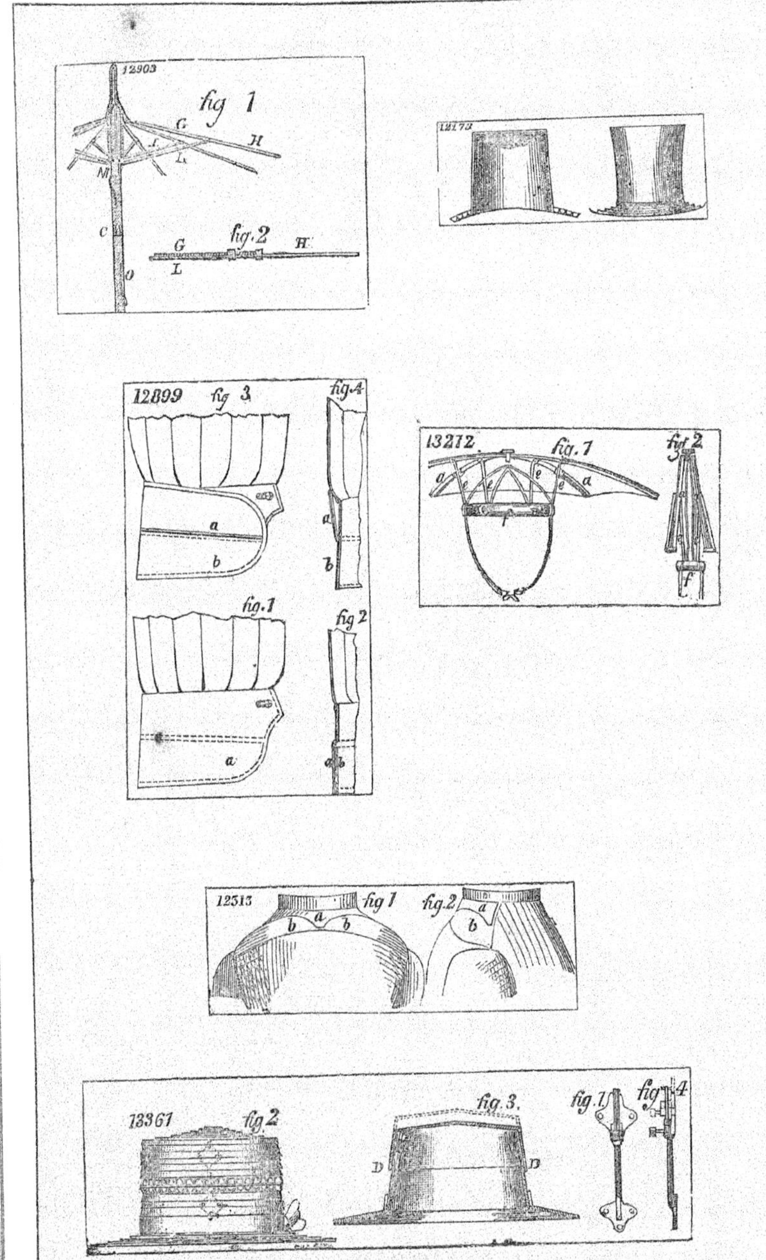
12903
fig 1
G
H
L
M
C
O
fig.2
12173
12899
fig 3
fig.4
a
b
fig.1
fig 2
13212
fig.1
fig 2
e
f
12313
fig 1
fig.2
13361
fig 2
fig.3.
D
fig.1
fig 4

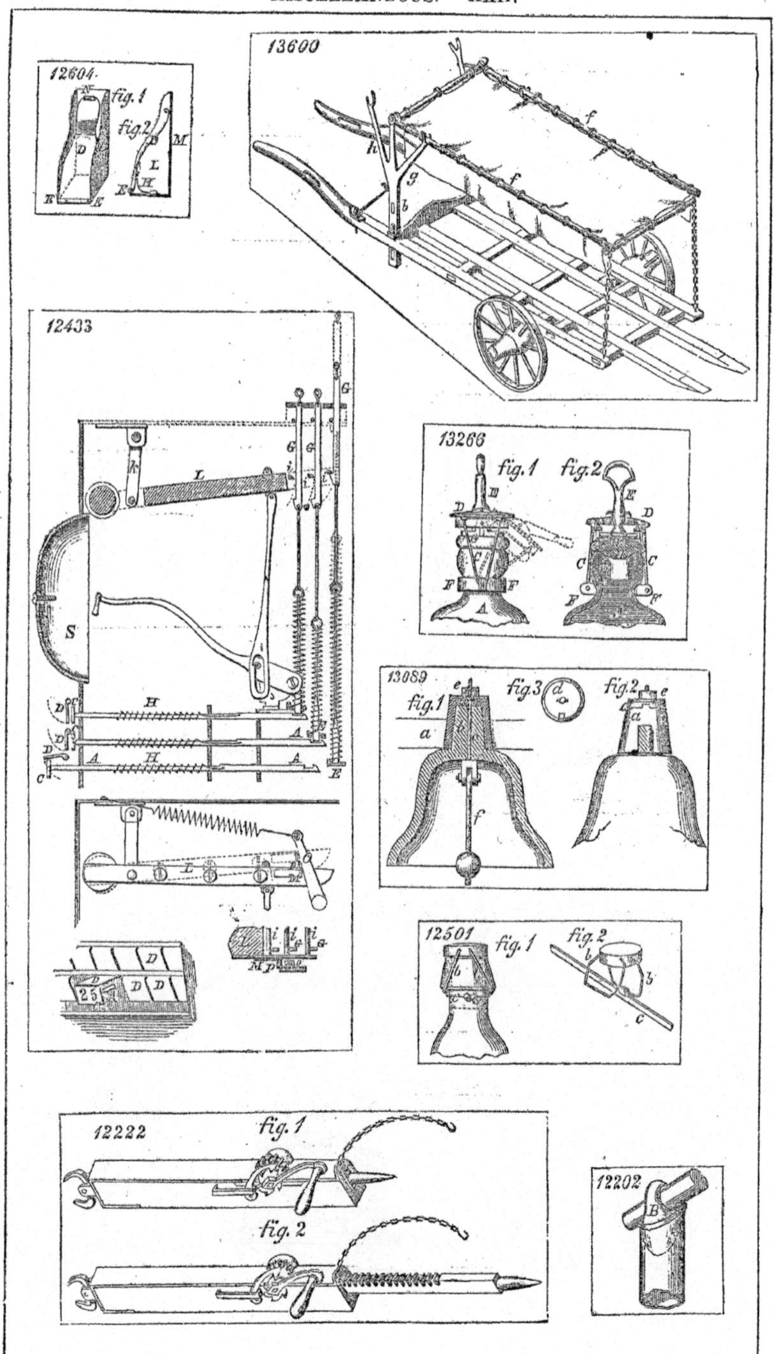
12604
fig. 1
fig. 2
13600
12433
13266
fig. 1
fig. 2
13089
fig. 1
fig. 2
fig. 3
12501
fig. 1
fig. 2
12222
fig. 1
fig. 2
12202

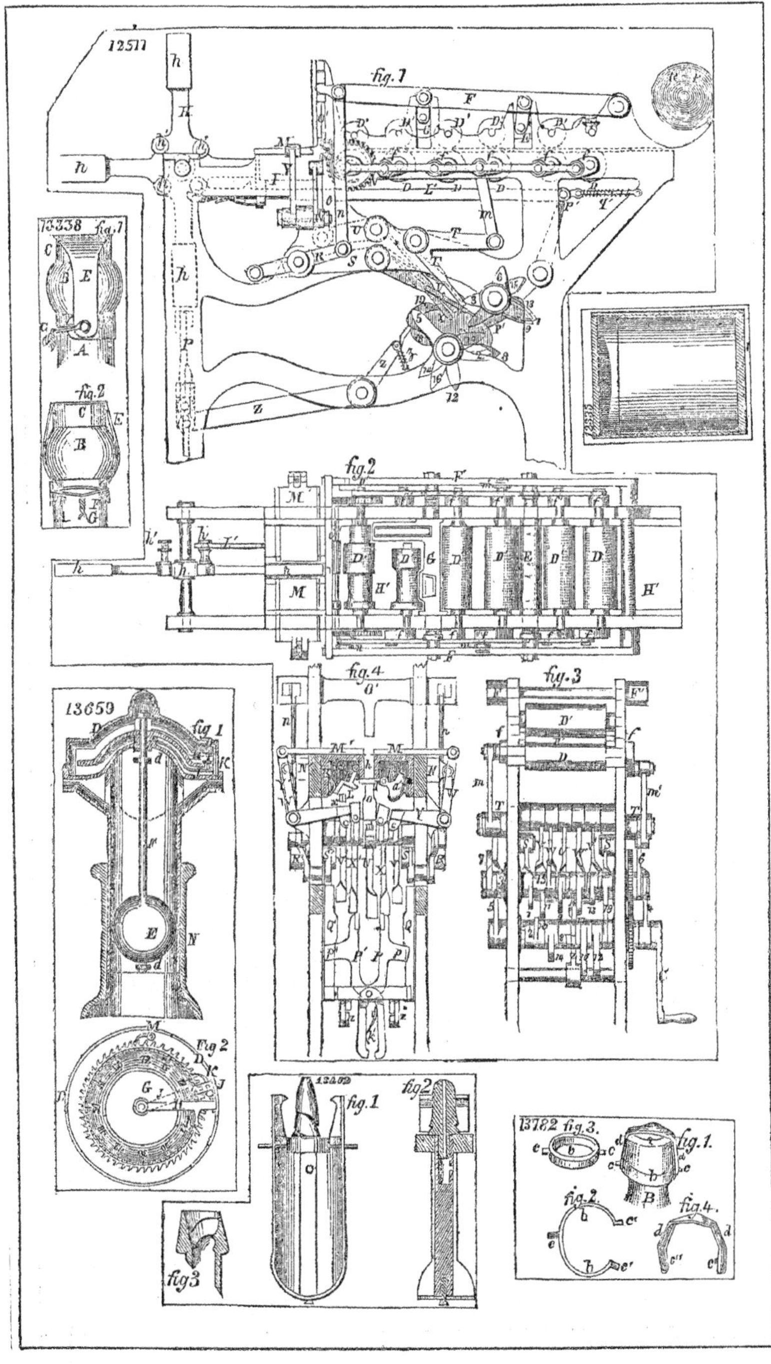

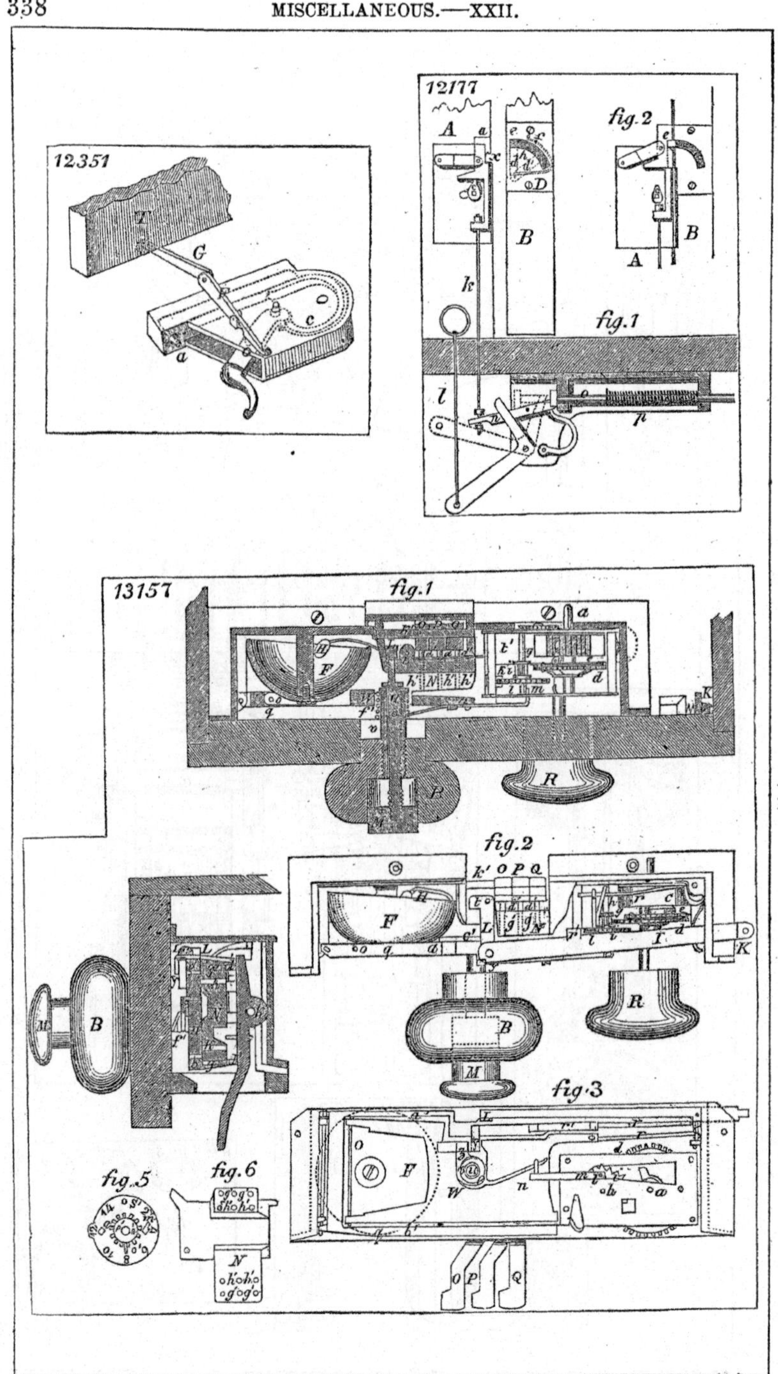
12351
12177
fig.2
fig.1
13157
fig.1
fig.2
fig.3
fig.5
fig.6

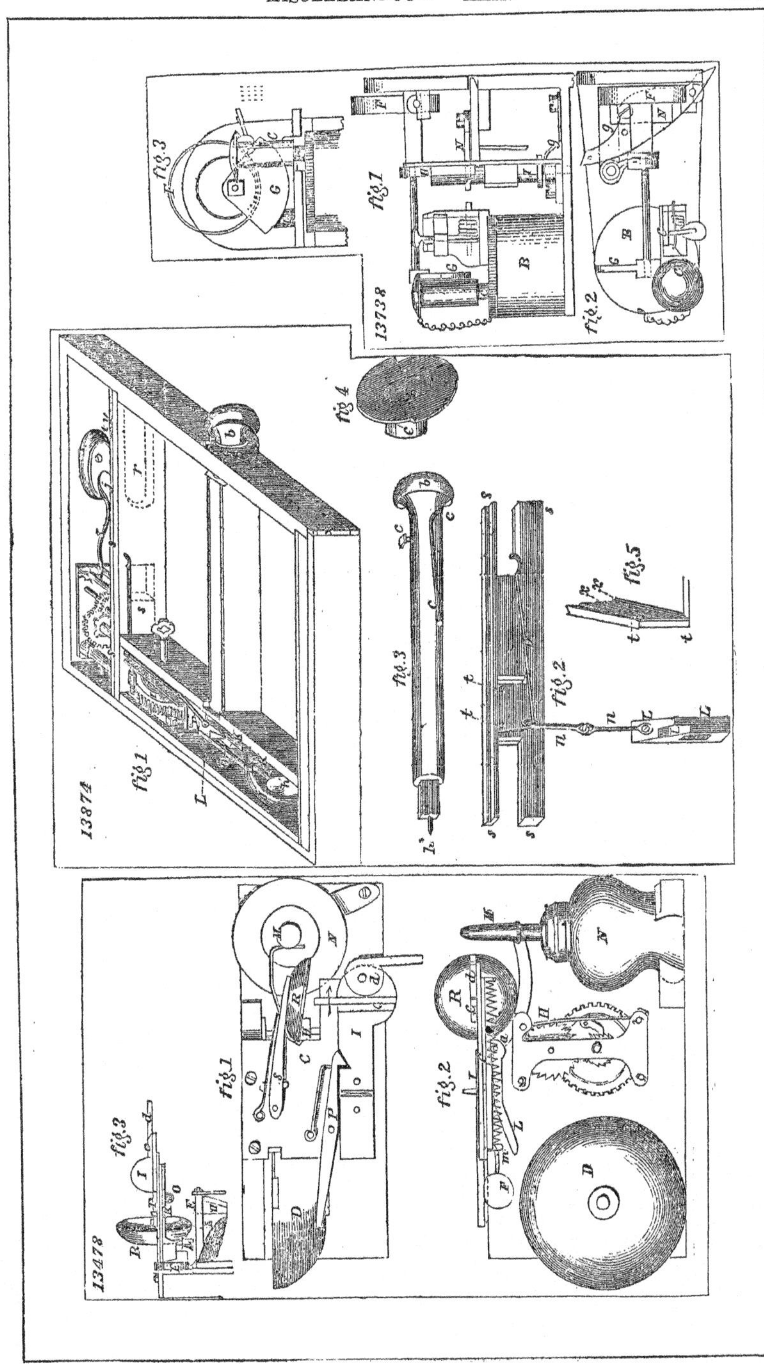

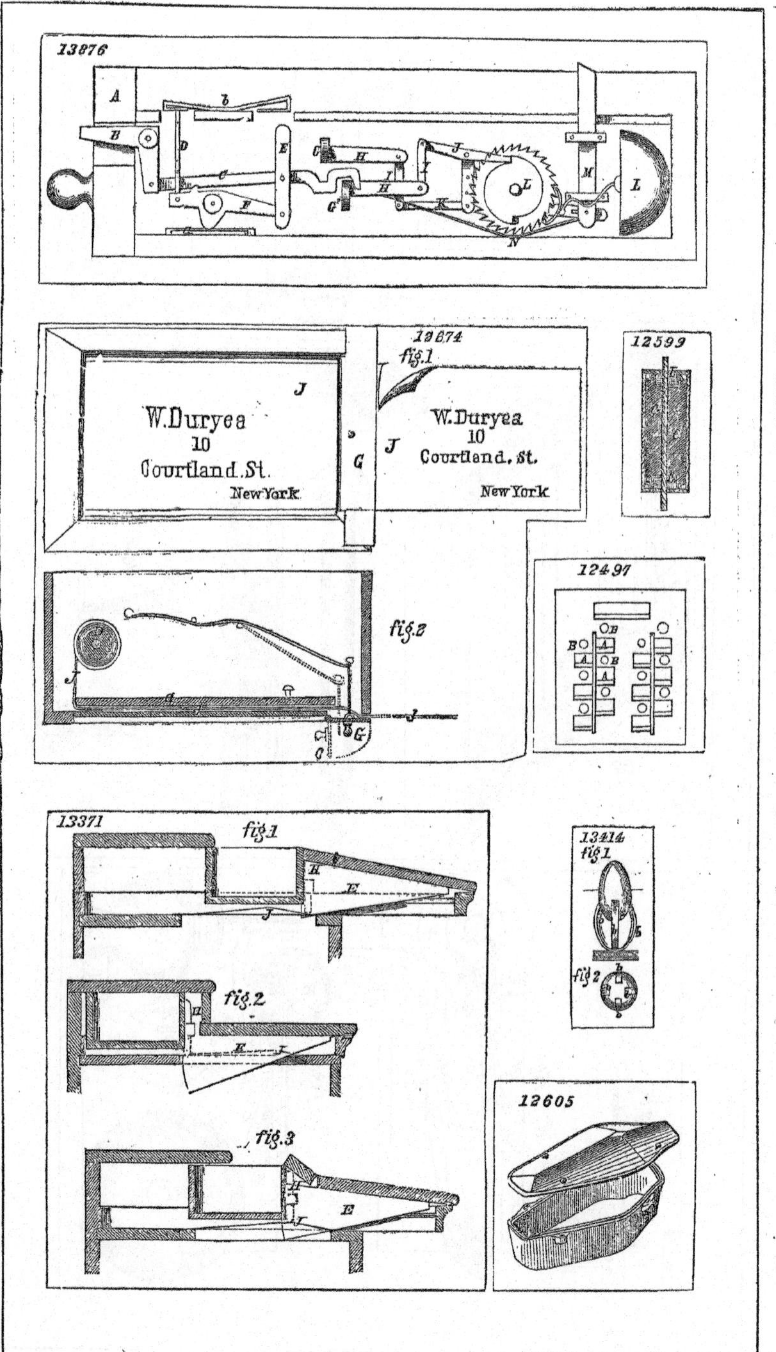
13876
12674
fig.1
W.Duryea
10
Courtland.St.
New York
W.Duryea
10
Courtland.St.
New York
fig.2
12599
12497
13371
fig.1
fig.2
fig.3
13414
fig.1
fig.2
12605

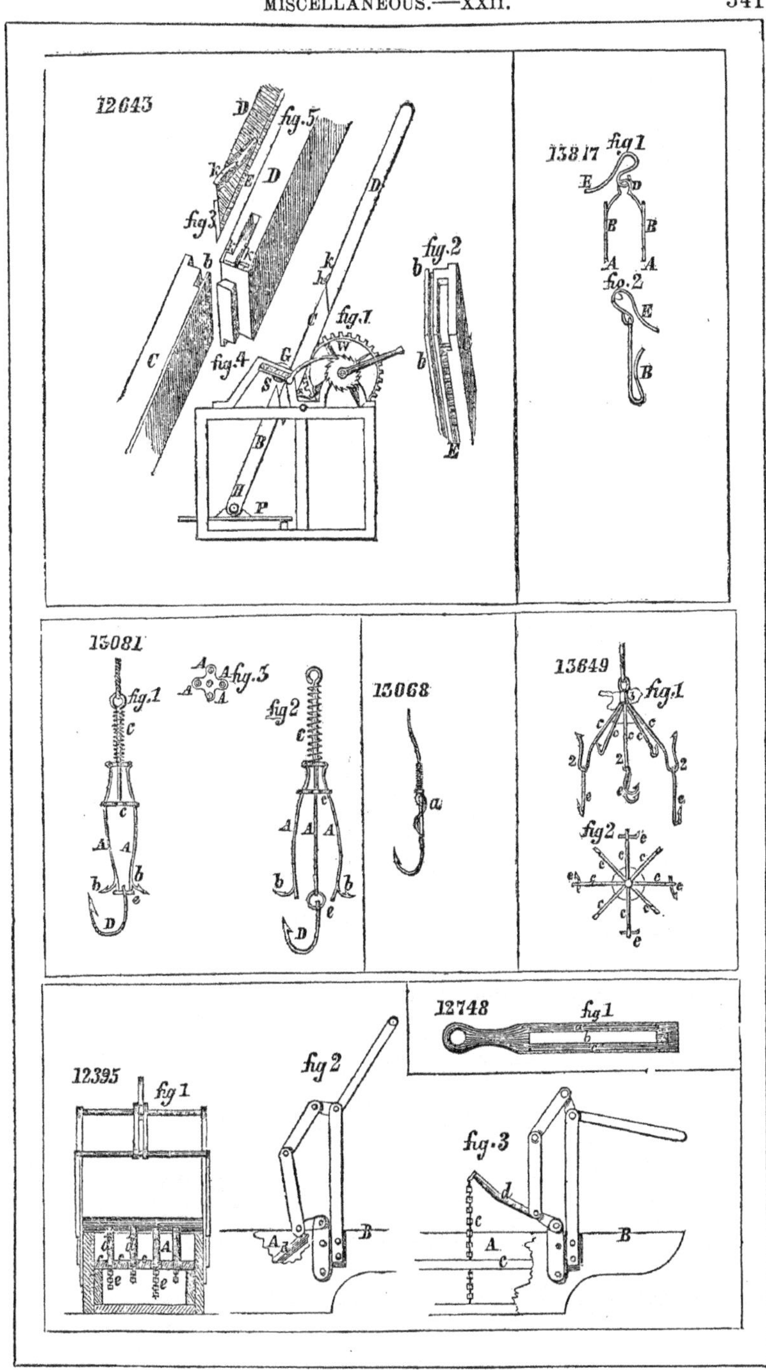
12643
13817
13081
13068
13649
12748
12395

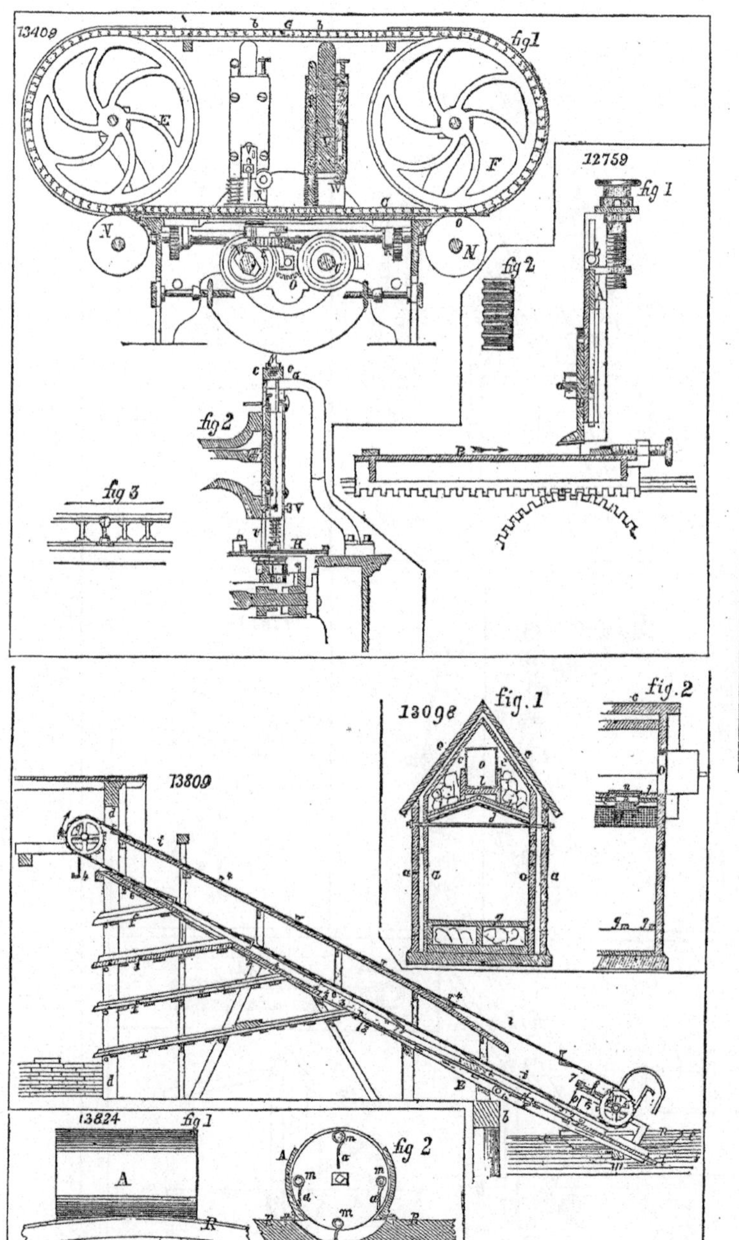
13409
fig 1
12759
fig 1
fig 2
fig 2
fig 3
13098
fig. 1
fig. 2
13809
13824
fig 1
fig 2
A
RICHMOND

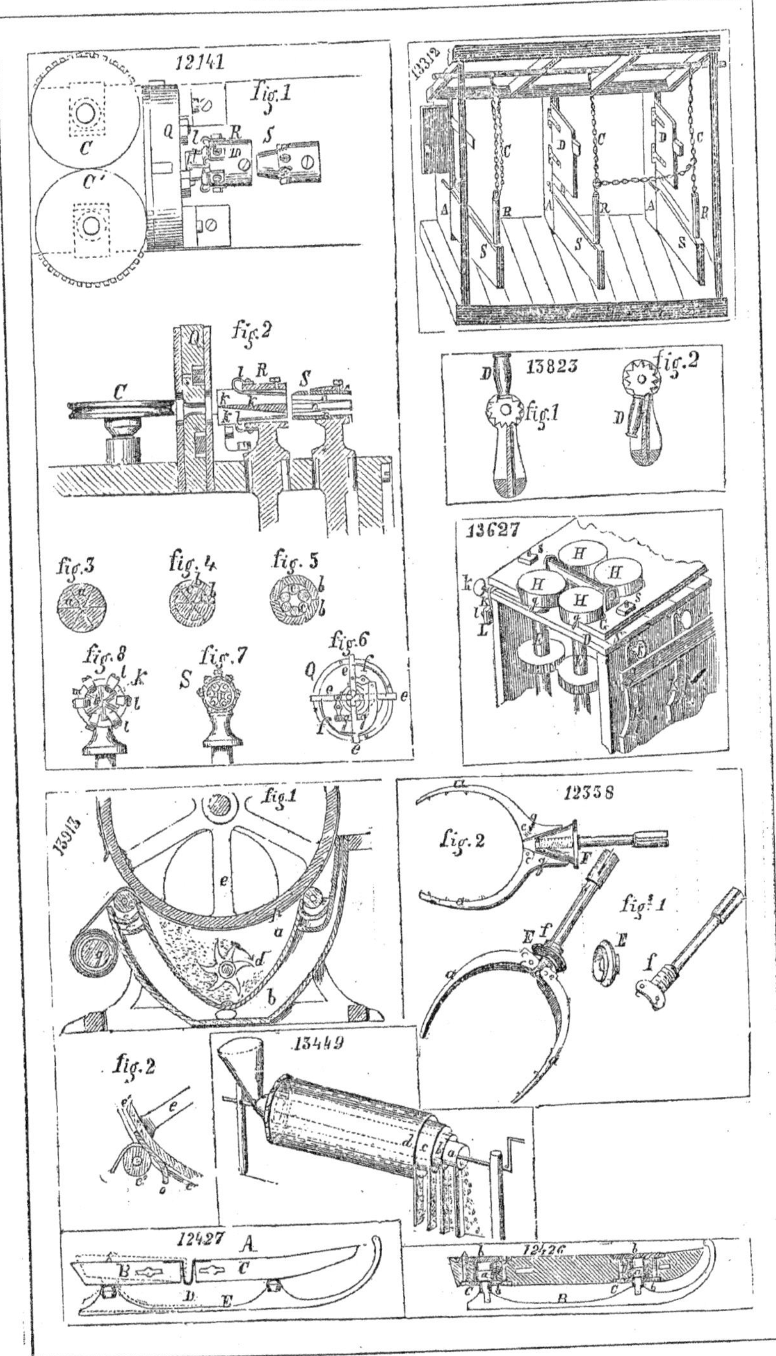
12141
fig.1
fig.2
fig.3
fig.4
fig.5
fig.6
fig.7
fig.8
13312
13823
13627
13913
12358
13449
12427
12426

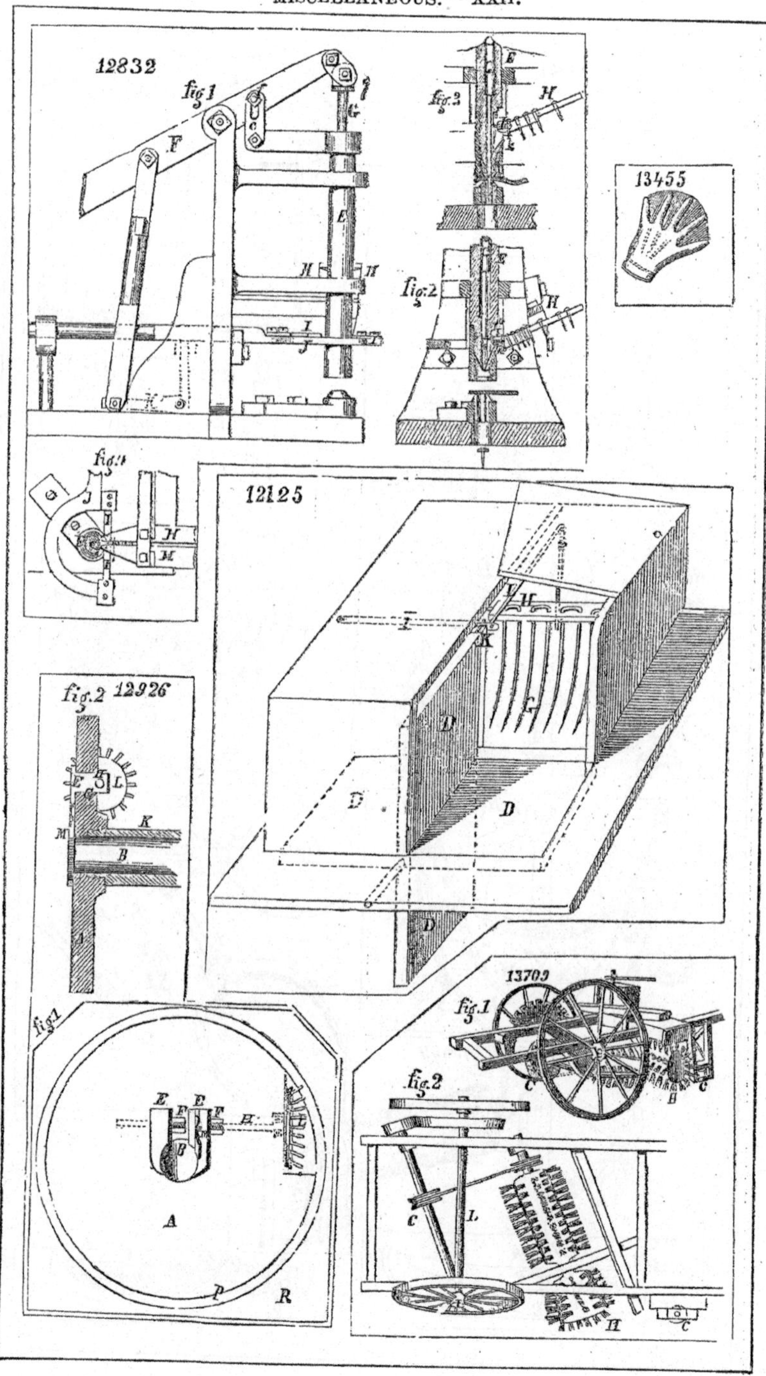
12832
fig 1
F
E
H
I
J
K
fig 3
fig 2
fig 4
M
13455
12125
D
G
L
12926
fig 2
B
fig 1
A
13709
fig 1
fig 2
C

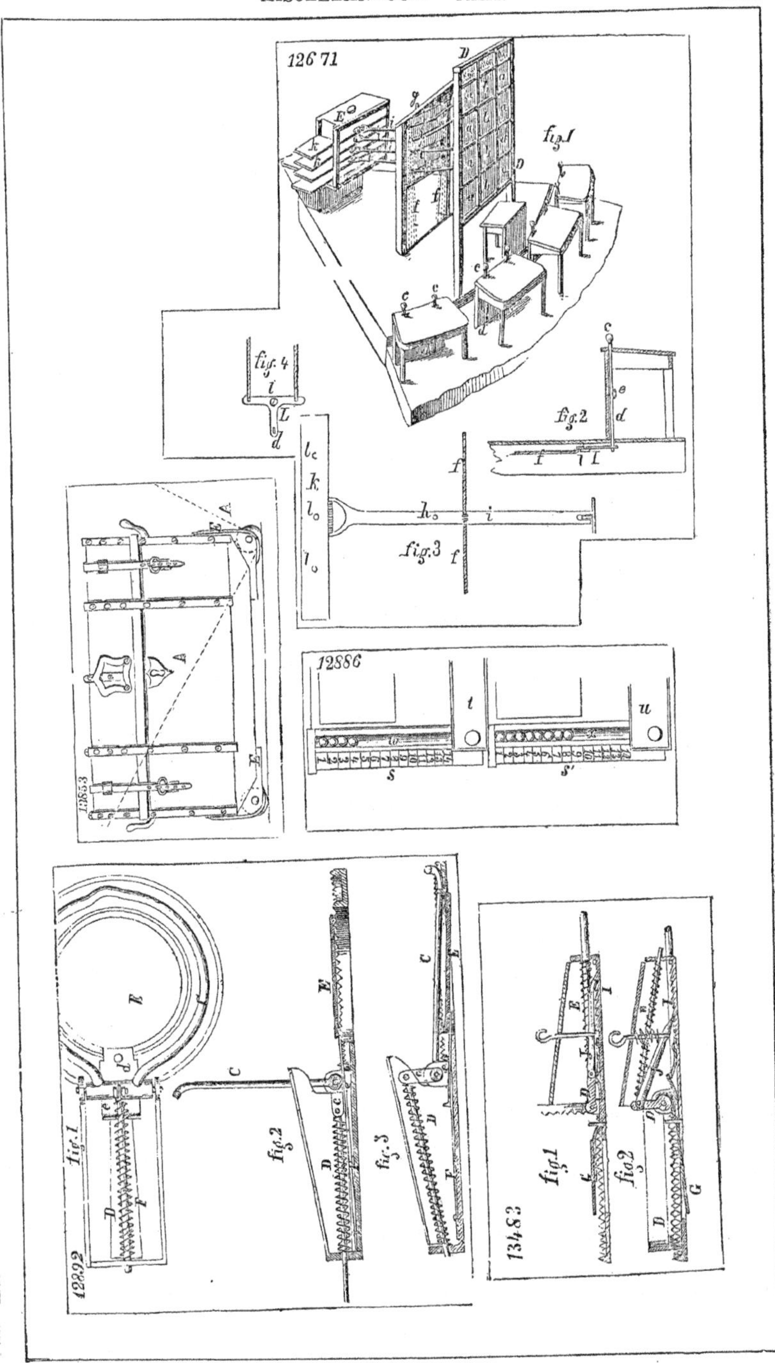
12671
fig.1
fig.2
fig.3
fig.4
12886
13853
12892
13483

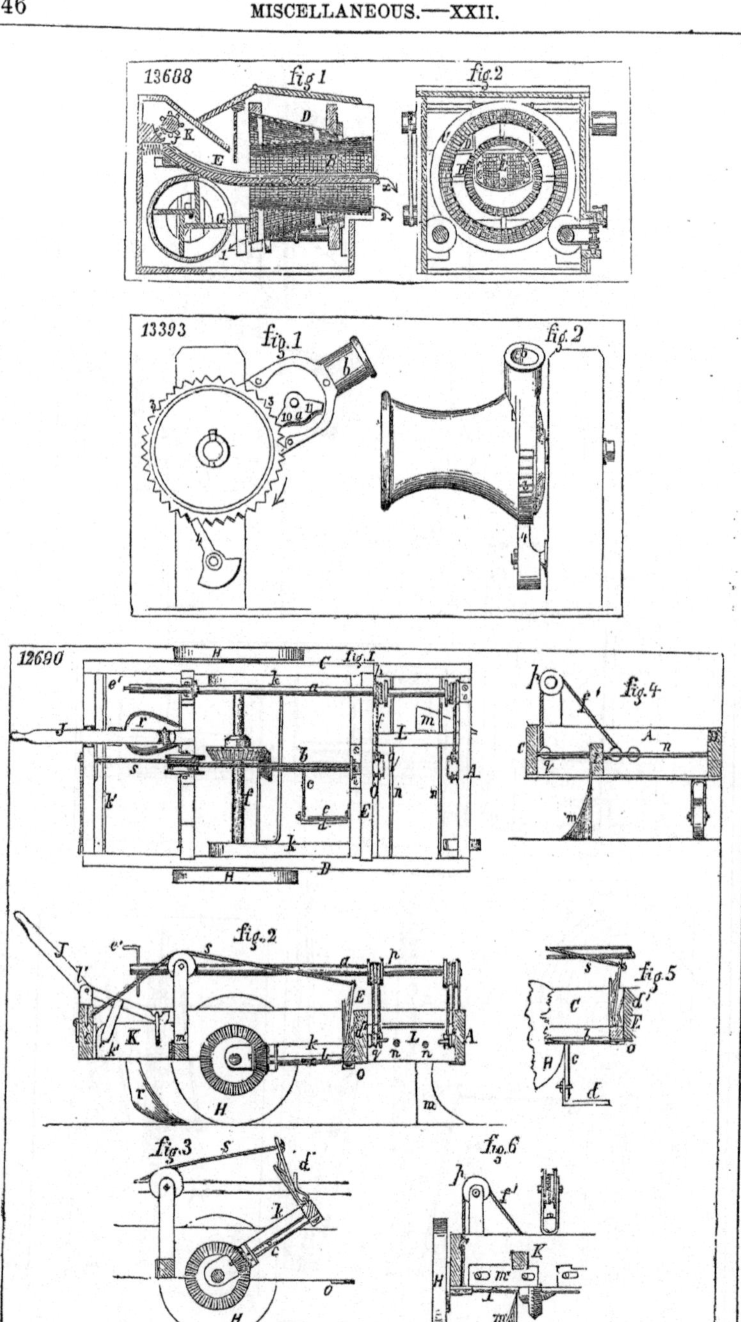
13688
fig 1
fig.2
13393
fig.1
fig.2
12690
fig.1
fig.4
fig.2
fig.5
fig.3
fig.6

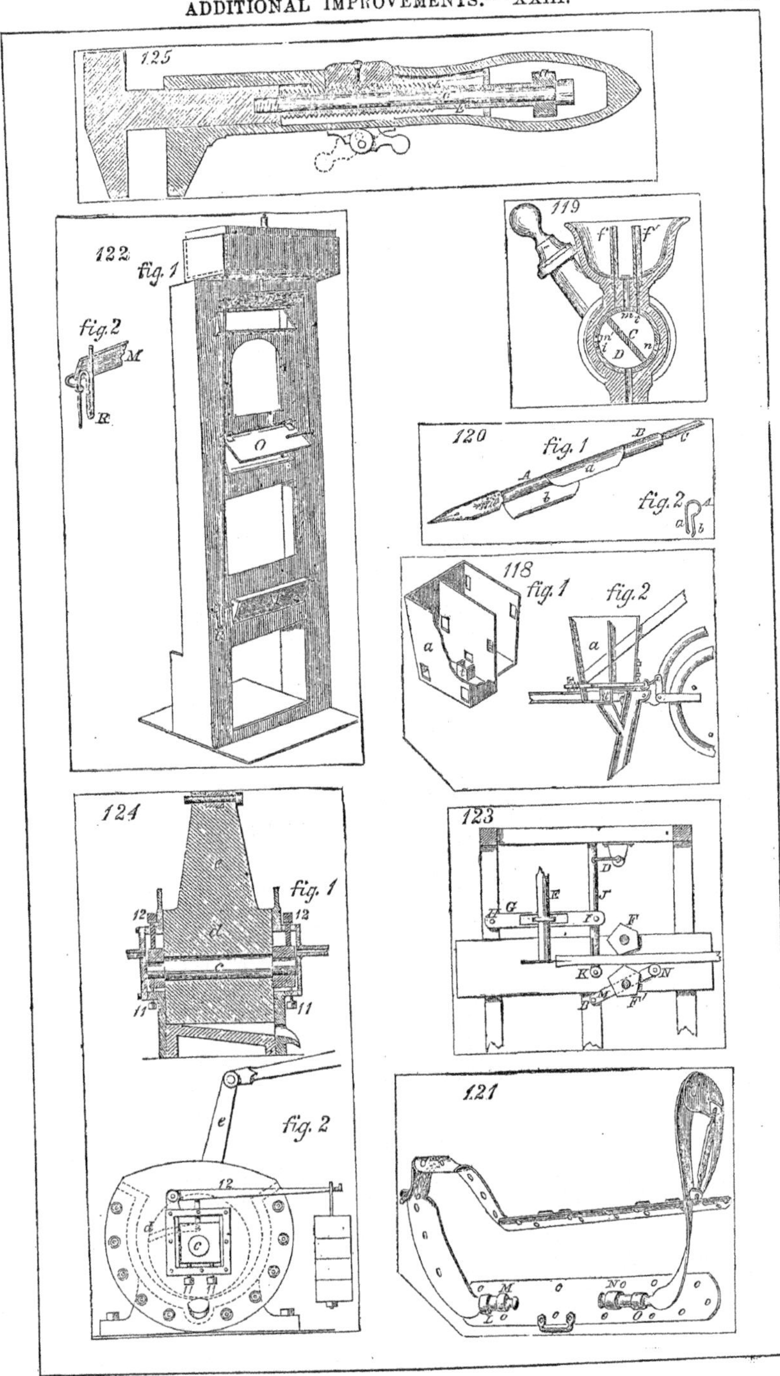
125
122
fig. 1
fig. 2
M
R
O
119
f
f'
C
D
n
120
fig. 1
A
a
b
D
C
fig. 2
118
fig. 1
fig. 2
a
124
fig. 1
12
d
c
11
fig. 2
e
123
G
E
F
K
N
M
F'
121
M
N
L
O

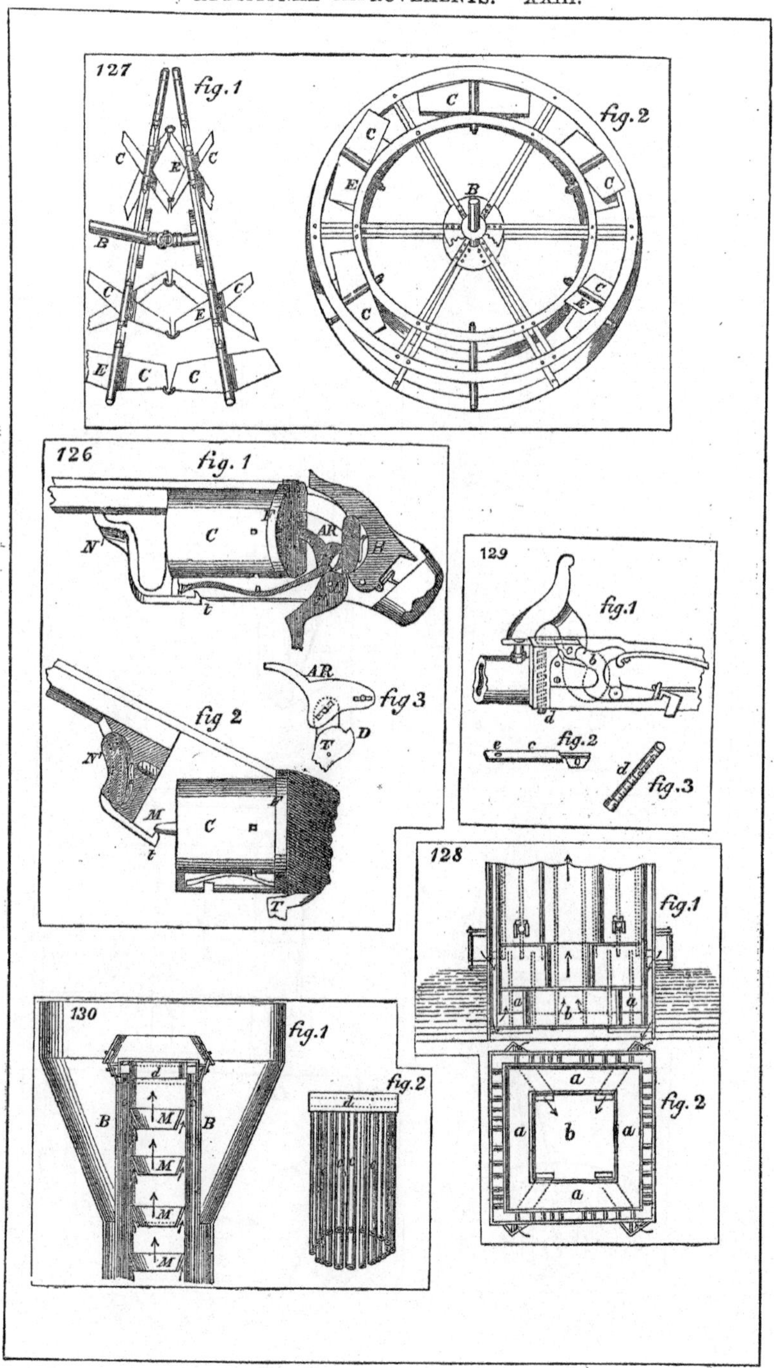
127
fig.1
C
E
C
B
C
C
E
E
C
C
fig.2
C
C
E
B
C
C
E
C
126
fig. 1
C
F
AR
H
N
t
fig 2
N
M
C
F
t
T
AR
fig 3
T
D
129
fig.1
c
b
d
e
c
fig.2
d
fig.3
128
fig.1
a
b
a
fig. 2
a
a
b
a
a
130
fig.1
d
B
M
B
M
M
M
fig.2
d

INDEX

TO THE

DESCRIPTIONS AND ILLUSTRATIONS.

[The descriptions contained in the second volume are indicated by the number II.]

CPSIA information can be obtained
at www.ICGtesting.com
Printed in the USA
BVOW08s1948241117
501190BV00024B/1159/P